For Anne

UNDERSTANDING CRANIOFACIAL ANOMALIES

UNDERSTANDING CRANIOFACIAL ANOMALIES

The Etiopathogenesis of Craniosynostoses and Facial Clefting

Edited by

MARK P. MOONEY, Ph.D.
Associate Professor
Departments of Oral Medicine and Pathology, Anthropology,
Surgery-Division of Plastic and Reconstructive Surgery, and Orthodontics
Cleft Palate-Craniofacial Center
University of Pittsburgh

MICHAEL I. SIEGEL, Ph.D.
Professor
Departments of Anthropology and Orthodontics,
Associate Director of Research at Cleft Plate-Craniofacial Center
University of Pittsburgh

⊗WILEY-LISS

A JOHN WILEY & SONS, INC., PUBLICATION

For ordering and cutomer service information please call 1-800-CALL-WILEY.

Library of Congress Cataloging-in-Publication Data:
Understanding craniofacial anomalies : etiopathogenesis of craniosynostoses and facial clefting / edited by Mark P. Mooney, Michael I. Siegel.
     p. cm.
    Includes bibliographical references and index.
    ISBN 0-471-38724-X (cloth : alk. paper)
    1. Face—Abnormalities. 2. Skull—Abnormalities. 3. Cleft palate. 4. Craniosynostoses.
  5. Face—Abnormalities—Genetic aspects. I. Mooney, Mark P. II. Siegel, Michael I.

RD763 .U53 2002
617.5′2—dc21

2001046949

Printed in the United States of America.

10 9 8 7 6 5 4 3 2 1

We would like to dedicate this book
to our families,
to our mentors,
to our students

CONTENTS

12 Cranial Base Dysmorphology and Growth in the Craniosynostoses   295

Alphonse R. Burdi, Ph.D.

13 Cranial Base Dysmorphology and Growth in Facial Clefting   307

*Timothy D. Smith, Ph.D., Mark P. Mooney, Ph.D.,
Annie M. Burrows, Ph.D., and Michael I. Siegel, Ph.D.*

14 Cranial Vault Dysmorphology and Growth in Craniosynostosis   321

Joan T. Richtsmeier, Ph.D.

FOREWORD

Understanding Craniofacial Anomalies: Etiopathogenesis of Craniosynostoses and Facial Clefting, edited by Mark P. Mooney and Michael I. Siegel, is a tour de force. It is a well-organized, well-referenced, up-to-date volume covering a wide range of topics written by numerous highly respected authorities in the field.

This book is divided into seven sections: (1) classification and terminology (three chapters); (2) embryogenesis and etiology (five chapters); (3) animal modeling (two chapters); (4) cranial vault and cranial base dysmorphology and growth disturbances (four chapters); (5) midfacial and mandibular dysmorphology and growth disturbances (three chapters); (6) regional dysmorphology and growth disturbances (two chapters); and (7) molecular biology studies and future directions (three chapters).

Understanding Craniofacial Anomalies was written to provide an understanding of two major entities: craniosynostosis and facial clefting. This book contains several perspectives: evolutionary, molecular, genetic, embryonic, growth and developmental, and dysmorphic. Analyses of the molecular, genetic, cellular, tissue, organismic, and population levels are included.

Various specialties are integrated around the theme of developmental biology and pathology. This book is very broadly conceived and will make an excellent textbook for undergraduate students, graduate students, preprofessional students, and professional students. Clinicians may use this book to improve their understanding of etiology and pathogenesis. *Understanding Craniofacial Anomalies* should be of great interest to molecular biologists, geneticists, craniofacial biologists, pathologists, specialists in growth and development, craniofacial surgeons, plastic surgeons, oral and maxillofacial surgeons, neurosurgeons, otolaryngologists, many dental specialists, and speech and hearing specialists.

This wonderful volume was created through the tireless efforts of Mark P. Mooney and Michael I. Siegel. They are to be congratulated on their magnificent achievement.

M. MICHAEL COHEN, Jr., D.M.D., Ph.D.

Professor of Oral and Maxillofacial Sciences, Pediatrics, Community Health and Epidemiology, Health Service Administration, Sociology and Social Anthropology Dalhousie University Halifax, Nova Scotia, Canada

PREFACE

The impetus and rationale for this book grew out of our past experiences in teaching introductory craniofacial anomalies courses at the undergraduate, graduate, and graduate-professional levels. It is obviously a very diverse topic with multilayered educational and analytical approaches, and it was difficult to find one or two textbooks that would adequately satisfy the student population. Many of the current texts, although excellent in content and presentation, are either quite specific (i.e., concentrating on a single anomaly or early time frame) (e.g., Ross and Johnston, 1978; Melnick and Jorgenson, 1979; Melnick et al., 1980; David et al., 1982; Vig and Burdi, 1988; Persing et al., 1989, Sperber, 1989, 2001; Morris and Bardach, 1990; Cohen and MacLean, 2000; Malek, 2001; Wyszynski, 2001) or oriented to the clinical end of the distribution (Marsh and Vannier, 1985; Jones, 1988; Gorlin et al., 1990; Montoya, 1992; Turvey et al., 1996; Cohen, 1997; Posnick, 2000). In lieu of requiring all these books for a single course (and risk zero enrollment), we opted for the next-best scenario: that of soliciting chapters from many of these authors to develop a broad-based text that would fill the needs of an introductory craniofacial anomalies course.

This book is designed to provide an understanding of two major abnormalities, craniosynostosis and facial clefting, which fall under the general heading of human craniofacial anomalies. The book reviews and presents current nomenclature and classification schema (chapters 2 and 3) and up-to-date knowledge in the field. The text covers, in a parallel fashion, the major research facets of craniosynostosis and facial clefting at the molecular (Chapters 20, 21, and 22), genetic (Chapters 6, 7, and 8), cellular (Chapters 4 and 5), tissue (Chapters 4, 5, 20, and 21), organismic (Chapters 9 and 10), and population (Chapters 11 and 15) levels, as well as presenting thinking on animal modeling (Chapters 9 and 10). The description and evolution of normal growth patterns and regional growth disturbances and dysmorphogenesis in the cranial base (Chapters 11, 12, and 13), cranial vault (Chapters 11 and 14), midface and mandible (Chapters 15, 16, and 17), dentition (Chapters 15 and 18), and vocal tract (Chapter 19) are covered with data from the human clinical setting.

The design and chapter organization of this book was also based, in part, on the recent philosophical shifts in craniofacial biology (see Chapter 22) from the standard dichotomous basic science and clinical research paradigm to a more unified and applied patient-centered approach. The move away from the treatment modality approach toward a more didactic approach integrates the various specialities around the theme of developmental biology/pathology as we consider the impact of craniosynostoses and facial clefts on other aspects of the craniofacial complex. While many of the emerging health care professionals have excellent clinical skills, they typically have had very little exposure to the literature of the etiopathogenesis of craniofacial anomalies. Integrating this literature could improve their research perspective, clinical management of patients, and cross-discipline interactions.

We envision that this volume will be used either as a textbook for a craniofacial anomalies course or as a general reference book. The audience for such a book would consist of young pre- and postprofessionals in a variety of related fields who need a good basic science overview of craniofacial anomalies (e.g., plastic and craniofacial surgery, oral and maxillofacial surgery, neurosurgery, orthodontics, pedodontics, otolaryngology, speech and hearing science, biological anthropology, human genetics, developmental biology, and so forth). This text, in combination with other clinical courses and experiences, should produce well-rounded applied morphologists.

MARK P. MOONEY, Ph.D.

MICHAEL I. SIEGEL, Ph.D.

Pittsburgh, PA

REFERENCES

Cohen, M. M., Jr. (1997). *The Child with Multiple Birth Defects*, 2nd ed. (New York: Oxford University Press).

Cohen, M. M., Jr., and MacLean, R. E. (eds.) (2000). *Craniosynostosis: Diagnosis, Evaluation, and Management*, 2nd ed. (New York: Oxford University Press).

David, J. D., Poswillo, D., and Simpson, D. (1982). *The Craniosynostoses: Causes, Natural History, and Management* (Berlin: Springer-Verlag).

Gorlin, R. J., Cohen, M. M., Jr., and Levin, L. S. (1990). *Syndromes of the Head and Neck* (New York: Oxford University Press).

Jones, K. L. (1988). *Smith's Recognizable Patterns of Human Malformations* (Philadelphia: W. B. Saunders).

Malek, R. (2001). *Cleft Lip and Palate: Lesions, Pathophysiology, and Primary Treatment* (London: Martin Dunitz, Ltd.).

Marsh, J. L., and Vannier, M. W. (1985). *Comprehensive Care for Craniofacial Deformities* (St. Louis: C. V. Mosby, Co.).

Melnick, M., and Jorgenson, R. (eds.) (1979). *Developmental Aspects of Craniofacial Dysmorphology* (New York: Alan R. Liss).

Melnick, M., Bixler, D., and Shields, E. D. (eds.) (1980). *Etiology of Cleft Lip and Cleft Palate* (New York: Alan R. Liss).

Montoya, A. (ed.) (1992). *Craniofacial Surgery* (Bologna, Italy: Monduzzi Editore).

Morris, H. L., and Bardach, J. (eds.) (1990). *The Multi Disciplinary Management of Cleft Lip and Palate* (Philadelphia: W. B. Saunders).

Persing, J. A., Edgerton, M. T., and Jane, J. A. (eds.) (1989). *Scientific Foundations and Surgical Treatment of Craniosynostosis* (Baltimore: Williams and Wilkins).

Posnick, J. C. (2000). *Craniofacial and Maxillofacial Surgery in Children and Young Adults*, vols. 1 and 2 (Philadelphia: W. B. Saunders).

Ross, R. B., and Johnston, M. C. (1978). *Cleft Lip and Palate* (Huntington, NY: Robert E. Krieger Publishing Co.).

Sperber, G. H. (1989). *Craniofacial Embryology*, 4th ed. (London: Wright Publishing Co.).

Sperber, G. H. (2001). *Craniofacial Development* (Hamilton: B. C. Decker).

Turvey, T. A., Vig, K. W. L., and Fonseca, R. L. (1996). *Facial Clefts and Craniosynostosis: Principles and Management* (Philadelphia: W. B. Saunders).

Vig, K. W. L., and Burdi, A. R. (eds.) (1988). *Craniofacial Morphogenesis and Dysmorphogenesis* Craniofacial Growth Series, Monograph 21 (Ann Arbor, MI: Center of Human Growth and Development).

Wyszynski, D. F. (ed.) (2001). *Cleft Lip and Palate: From Origin to Treatment* (Oxford: Oxford University Press).

CONTRIBUTORS

PETER T. BRONSKY, D.M.D., M.S., Research Associate Dental Research Center, Department of Orthodontics, School of Dentistry, University of North Carolina, Chapel Hill, NC 27599

ALPHONSE R. BURDI, Ph.D., Professor, Department of Cell and Developmental Biology, Center for Human Growth and Development, University of Michigan Medical School, Ann Arbor, MI 48109-0616

ANNE M. BURROWS, Ph.D., D.PT., Assistant Professor, Department of Physical Therapy, Duquesne University, Pittsburgh, PA 15282, Adjunct Research Associate Professor, Department of Anthropology, University of Pittsburgh, Pittsburgh, PA 15260,

M. MICHAEL COHEN, Jr., D.M.D., Ph.D., Professor, Departments of Oral and Maxillofacial Pathology, Pediatrics, Community Health and Epidemiology, Health Services Administration, Sociology and Social Anthropology, Dalhousie University, Halifax, Nova Scotia-B3H 3J5

VIRGINIA M. DIEWERT, Ph.D., Professor and Head, Department of Oral Health Sciences, Faculty of Dentistry, University of British Columbia, Vancouver, B.C., Canada V6T 1Z3

EDWARD F. HARRIS, Ph.D., Professor, Department of Orthodontics, College of Dentistry, The Health Science Center, University of Tennessee, Memphis, TN 38163

THOMAS C. HART, D.D.S., Ph.D., Associate Professor, Department of Oral Medicine, School of Dental Medicine, Department of Human Genetics, Graduate School of Public Health, University of Pittsburgh, Pittsburgh, PA 15261

ETHYLIN WANG JABS, M.D., Dr. Frank V. Sutland Professor of Pediatric Genetics, Professor of Pediatrics, Medicine, and Plastic Surgery, Director of the Center for Craniofacial Development and Disorders, McKusick-Nathans Institute of Genetic Medicine, The Johns Hopkins University School of Medicine, Baltimore, MD 21287-3914

TINA JASKOLL, Ph.D., Professor, Department of Developmental Genetics, University of Southern California, Los Angeles, CA 90089-0641

MALCOM C. JOHNSTON D.D.S., Ph.D., Professor Emeritus, Dental Research Center, Department of Orthodontics, School of Dentistry, Department of Cell Biology and Anatomy, School of Medicine, University of North Carolina, Chapel Hill, NC 27599

MARILYN C. JONES, M.D., Adjunct Professor, Department of Pediatrics, University of California–San Diego, Director, Dysmorphology and Genetics/Cleft Palate and Craniofacial Treatment Programs, Childre's Hospital of San Diego

SCOTT LOZANOFF, Ph.D., Professor and Chair, Department of Anatomy and Reproductive Biology, John A. Burns School of Medicine, University of Hawaii at Manoa, Honolulu, Hawaii 96822

MARY L. MARAZITA, Ph.D., F.A.C.M.G., Assistant Dean for Research, Head, Division of Oral Biology, Professor, Oral and Maxillofacial Surgery, School of Dental Medicine, Professor, Human Genetics, Graduate School of Public Health, University of Pittsburgh, Pittsburgh, PA 15261

MICHAEL MELNICK, D.D.S., Ph.D., Professor, Department of Developmental Genetics, University of Southern California, Los Angeles, CA 90089-0641

MARK P. MOONEY, Ph.D., Associate Professor, Departments of Oral Medicine and Pathology, Anthropology, Surgery, Division of Plastic and Reconstructive Surgery, Orthodontics, Cleft and Palate-Craniofacial Center, University of Pittsburgh, Pittsburgh, PA 15261

PATRICIA K. MONOSON, Ph.D., Associate Dean for Academic and Student Affairs, College of Health Related Professions, Professor, Department of Communicative Disorders, University of Arkansas for Medical Sciences, Little Rock, AR 72205-7199

ROY C. OGLE, Ph.D., Associate Professor, Departments of Neurological Surgery, Cell Biology, and Plastic Surgery, Director of Craniofacial Research, University of Virginia, Charlottesville, VA 22908

LYNNE A. OPPERMAN, Ph.D., Assistant Professor, Department of Biomedical Sciences, Baylor College of Dentistry, Texas A&M University Health Sciences Center, Dallas, TX 75246

SALLY J. PETERSON-FALZONE, Ph.D., Professor, Center for Craniofacial Anomalies, University of California San Francisco, San Francisco, CA 94143

JOAN T. RICHTSMEIER, Ph.D., Professor, Department of Anthropology, The Pennsylvania State University, University Park, PA 16802

R. BRUCE ROSS, D.D.S., Msc., Professor Emeritus, Deptartment of Dentistry, The Hospital for Sick Children, Professor Emeritus, University of Toronto, Toronto, ON, Canada M56 1X8

VANDANA SHASHI, M.D., Assistant Professor of Pediatrics, Section on Medical Genetics, Wake Forest University School of Medicine, Winston-Salem, NC 27157

ROBERT J. SHPRINTZEN, Ph.D., Professor, Department of Otolaryngology, Communication Disorder Unit, and Pediatrics, Upstate Medical University, Syracuse, NY 13210

JOSEPH R. SIEBERT, Ph.D., Research Associate Professor, Departments of Laboratories and Pathology, Children's Hospital and Regional Medical Center, University of Washington School of Medicine, Seattle, WA 98105

MICHAEL I. SIEGEL, Ph.D., Professor, Departments of Anthropology and Orthodontics, Associate Director of Research at the Cleft Palate-Craniofacial Center, University of Pittsburgh, Pittsburgh, PA 15260

HAROLD SLAVKIN, D.D.S., Ph.D., Dean, The G. Donald and Marian James Montgomery Professor of Dentistry, School of Dentistry, University of Southern California, Los Angeles, CA 90089-0641

TIMOTHY D. SMITH, Ph.D., Associate Professor, School of Physical Therapy, Slippery Rock University, Slippery Rock, PA 16057, Adjunct Research Associate Professor, Department of Anthropology, University of Pittsburgh, Pittsburgh, PA 15260

Geoffrey H. Sperber, Ph.D., Professor Emeritus, Department of Dentistry, Faculty of Medical and Oral Health Sciences, University of Alberta, Edmonton, AB T6G 2N8

Daris R. Swindler, Ph.D., Emeritus Professor, Department of Anthropology, University of Washington, Seattle, WA 98020

Katherine W. L. Vig, D.D.S., M.S., Ph.D., Professor and Chair, Department of Orthodontics, College of Dentistry, The Ohio State University, Columbus, OH 43218-2357

PART I

CLASSIFICATION AND TERMINOLOGY

CHAPTER 1

OVERVIEW AND INTRODUCTION

MARK P. MOONEY, Ph.D., and MICHAEL I. SIEGEL, Ph.D.

1.1 INTRODUCTION

As your knowledge and experience in morphology (i.e., shape) and dysmorphology (i.e., abnormal shape) increase, you will find yourself scrutinizing the craniofacial and somatic morphology of the people around you, noticing subtle and not-so-subtle asymmetries, deviations, and abnormalities in everyone. Are these morphologies the result of normal genetic and developmental processes or are they due to pathological conditions that pro duce morphologies outside the range of normal phenotypic variability? To answer these questions, you need to appreciate the normal variability of human facial form.

1.2 NORMAL CRANIOFACIAL MORPHOLOGY

Biological variation is a natural consequence of sexually reproducing organisms. Such variation has allowed populations to genetically adapt to changing environments. Enlow and Gans (1996) suggest that the human face probably has more basic, divergent kinds of facial patterns than most other species, save domesticated dogs

(Stockard, 1941). This may be related to a number of factors including the unique evolutionary changes in the primate brain and dentition (see Chapters 11 and 15), human biological adaptations and developmental acclimatizations to extreme environments (Steegman, 1970; Harrison et al., 1988; Bogin, 1988), and the fact that humans are essentially a "domesticated" species (Brace and Montagu, 1977).

Humans possess culture or the ability to alter their environment through technology and behavior. This ability somewhat reduces natural selection pressures and morphological canalization and homogeneity, which may subsequently increase phenotypic craniofacial and dental variability (Brace and Montagu, 1977; Chapter 18). One salient example of this process is the problem of cephalopelvic disproportion (CPD) during parturition. Historically, the mortality rate of mothers and/or fetuses during parturition was extremely high, and even today in some emerging countries infant mortality rates approach 30% (Kusiako et al., 2000). Although not all cases of infant morality can be attributed to CPD, it is still thought to play a significant role in birth complications (Kusiako et al., 2000). In countries where populations have

Understanding Craniofacial Anomalies: The Etiopathogenesis of Craniosynostoses and Facial Clefting, Edited by Mark P. Mooney and Michael I. Siegel, ISBN 0-471-38724-x Copyright © 2002 by Wiley-Liss, Inc.

increased access to medical technology (e.g., cesarean sections, prenatal ultrasound scans, and so on), infant mortality rates are approximately 11% (Meirowitz et al., 2001). These findings suggest that, in part, more individuals with large heads (and, conversely, mothers with small pelvises) are surviving. Thus, a greater frequency of large-headed phenotypes, which might normally have died during the birthing process, remain in the gene pool. This increases the range of normal phenotypic variability and heterogeneity in facial form in these populations. It is also an example of domestication or artificial selection, which is a process that increases phenotypical variability in a population (Brace and Montagu, 1977).

Normal human craniofacial shape extremes range from (1) dolichocephaly (long, narrow head form, cranial width-to-length index less than 0.75) with a leptoprosopic (long, narrow, protrusive) face shape; to (2) mesocephaly and a mesoprosopic face shape (proportional width-to-length head and facial forms with a cranial width-to-length index between 0.75 and 0.80); to (3) brachycephaly (wide, short globular head forms, cranial width-to-length index greater than 0.80) with a euryprosopic (broad, flat, less protrusive facial form) face shape (Krogman, 1978; Enlow, 1990; Enlow and Gans, 1996) (Fig. 1.1). Traditionally, these morphologies were associated with different human populations and geographic regions: Subsaharan African–derived populations were on average more dolichocephalic, European- and Middle Eastern–derived populations were on average more mesocephalic, and Asian-derived populations were on average more brachycephalic. However, the greatly increased genetic admixture among these populations has increased phenotypic heterogeneity and created distribution ranges of these head forms within populations (Enlow and Gans, 1996). Other genetic factors that may also influence normal craniofacial morphology include gender- and age-dependent growth processes (Bogin, 1988; Enlow, 1990; Enlow and Gans, 1996; Sarnat, 2001).

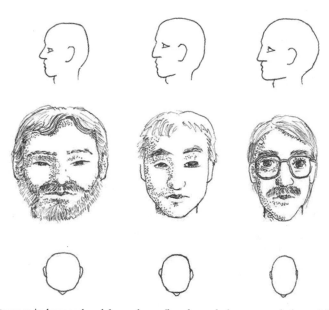

Figure 1.1   Extremes in human head form shape (brachycephaly, mesocephaly, and dolichocephaly).

As can be seen, there is a wide spectrum of human craniofacial morphologies that are all within the range of normal human variation. This diversity is produced by an interaction of normal genetic and epigenetic factors such as developmental acclimatizations to extreme environments (Enlow, 1990; Steegman, 1970; Harrison et al., 1988; Bogin, 1988). It has also been suggested that populations with certain morphologies may be predisposed or at risk for craniofacial anomalies based, in part, on facial, palatal, or cranial vault growth rates and morphologies (Burdi et al., 1972; Juriloff and Trasler, 1976; Trasler and Machado, 1979; Ross and Johnston, 1978; Siegel and Mooney, 1986; Johnston and Bronsky, 1995; Vergato et al., 1997). This underlying phenotypic variability makes the study and treatment of human craniofacial pathologies and dysmorphologies difficult and problematic but very exciting and challenging nonetheless.

1.3 CRANIOFACIAL DYSMORPHOLOGY

An understanding of craniofacial anomalies involves an appreciation of both the underlying, wide spectrum of normal craniofacial morphology and the overlying or interfering dysmorphology. Jones (1988) and Cohen (1997) suggest that craniofacial anomalies should be interpreted from the viewpoint of developmental anatomy and pathology (see also Chapters 4 and 5). It is important to determine which of the

multiple anomalies represent the earliest or primary defect in morphogenesis and if all of the anomalies can be traced to a single problem in morphogenesis. Knowledge of these relationships is instrumental in determining the etiology of craniofacial anomalies, understanding the pathogenesis of these conditions, assessing recurrence risks, and designing therapies and managements for the prevention and treatment of these cases (Ross and Johnston, 1978; Melnick and Jorgenson, 1979; Melnick et al., 1980; David et al., 1982; Marsh and Vannier, 1985; Vig and Burdi, 1988; Jones, 1988; Persing et al., 1989; Sperber, 1989, 2001; Morris and Bardach, 1990; Gorlin et al., 1990; Montoya, 1992; Turvey et al., 1996; Cohen, 1997; Cohen and MacLean, 2000; Posnick, 2000; Malek, 2001; Wyszynski, 2001).

Craniofacial anomalies can be divided into three categories: malformations, deformations, and disruptions (Spranger et al., 1982; Jones, 1988; Cohen, 1997) (Table 1.1; Fig. 1.2). Malformations are morphological defects of an organ, part of an organ, or a larger region of the body resulting from an intrinsically abnormal developmental process. Deformations are abnormal formations or positioning of a part of the body caused by nondisruptive mechanical forces. Disruptions are morphological defects of an organ, part of an organ, or a larger region of the body resulting from a breakdown of, or an interference with, an originally normal developmental process (Table 1.1).

Depending on the developmental timing and severity of the primary craniofacial

TABLE 1.1 Three Types of Dysmorphogenesis

Type of Anomaly	Developmental Process	Craniofacial Examples
Malformation	Abnormal development of tissue	Cleft lip and palate, microcephaly
Deformation	Unusual forces on normal tissue	Positional plagiocephaly, Robin sequence
Disruption	Breakdown of normal tissue	Hemifacial microsomia, rare facial clefts

Source: Adapted from Cohen, 1997.

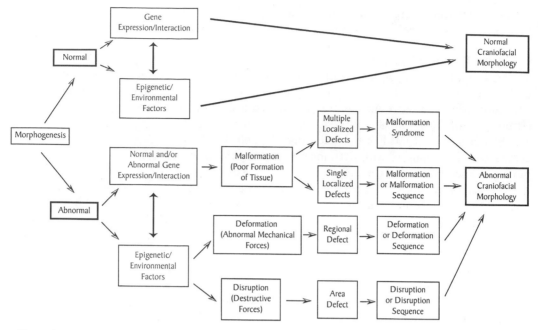

Figure 1.2 Different morphogenetic pathways producing normal and abnormal craniofacial morphologies.

anomaly, consistent patterns of multiple anomalies may be observed. These are referred to as syndromes or sequences (for an excellent discussion of these two concepts, see Cohen, 1997). A syndrome is a pattern of multiple anomalies that are pathogenetically related and not known to represent a single sequence (e.g., Down, Crouzon, or Apert syndrome) (Spranger et al., 1982; Jones, 1988; Cohen, 1997). In contrast, a sequence is a pattern of multiple anomalies derived from a single known or presumed prior anomaly or mechanical factor (e.g., Robin or amniotic band sequence) (Spranger et al., 1982; Jones, 1988; Cohen, 1997).

While syndromes are usually related to multiple localized malformations, sequences can be associated with all three types of structural defects (Table 1.2; Fig. 1.2). Malformation sequences may occur following a single localized poor formation of tissue that initiates a chain of subsequent defects. Typically these defects can affect

various growth centers of the craniofacial complex (i.e., brain, synchondroses, globes, cranial nerves, and craniofacial musculature and vasculature) (Enlow, 1990) and usually result in the most severe craniofacial anomalies (Table 1.2; see Chapters 4 and 5).

Deformation sequences may occur following unusual mechanical forces (such as intrauterine constraint from malpositioning or amniotic bands) on previously normal tissue, which may result in altered morphogenesis, usually of the molding type (Jones, 1988). These defects usually happen during the fetal period, only affect growth sites (sutures, cortical surfaces, and craniofacial musculature and vasculature) (Enlow, 1990), and result in less severe anomalies than malformations (Table 1.2). Disruption sequences may be related to a disruption of normal tissue (from mechanical, vascular, or infectious origin) and subsequent dysmorphogenesis or degeneration of tissue. Disruptions may affect

TABLE 1.2 Comparative Features of the Three Types of Dysmorphogenesis

Features	Malformations	Deformations	Disruptions
Time of occurrence	Embryonic	Fetal	Embryonic
Level of disturbance	Organ	Region	Area
Perinatal mortality	+	−	+
Phenotypic variability	Moderate	Mild	Extreme
Multiple etiologies	Very frequent	Less common	Less common
Growth center disruption	+	−	+
Growth site disruption	+	+	+
Spontaneous correction	−	+	−
Postural correction	−	+	−
Surgical correction	+	±	+
Relative recurrence rate	High	Low	Extremely low
Approximate frequency in newborns	2–3%	1–2%	1–2%

Source: Adapted from Cohen, 1997.

either growth centers or growth sites. The resultant craniofacial dysmorphogenesis and the ability to surgically correct it will vary depending on location. A comparison of the various features of these structural defects is presented in Table 1.2 and Figure 1.2.

1.4 CONCLUSIONS

Normal human craniofacial morphology develops as a complex consequence of genetic and environmental interactions. In contrast, craniofacial anomalies may occur as a result of early embryonic problems in tissue formation or later biomechanical or disruptive problems with normally differentiated fetal tissue. Subsequent, secondary craniofacial anomalies are typically a consequence as well. These dysmorphologies occur over a normal developmental substrate.

This chapter presents a short introduction into normal and abnormal craniofacial morphology. It is not our intent to present an exhaustive discussion of these topics, since there are many excellent generalized (Melnick and Jorgenson, 1979; Melnick et al., 1980; Enlow, 1990; Enlow and Gans, 1996; Jones, 1988; Montoya, 1992; Cohen, 1997) and specialized (Ross and Johnston, 1978; Melnick et al., 1980; David et al., 1982; Marsh and Vannier, 1985; Vig and Burdi, 1988; Sperber, 1989, 2001; Persing et al., 1989; Gorlin et al., 1990; Morris and Bardach, 1990; Turvey et al., 1996; Cohen and MacLean, 2000; Malek, 2001; Posnick, 2000; Wyszynski, 2001) textbooks available. Instead, our intent is to provide an appreciation of the interaction of these processes as an adjunct to understanding the etiopathogenesis and research paradigms of craniofacial anomalies as presented in the following chapters.

1.5 ACKNOWLEDGMENTS

This work was supported in part by a Comprehensive Oral Health Research Center of Discovery (COHRCD) program/project grant from NIH/NIDCR (P60 DE13078) to the Center for Craniofacial Development and Disorders, Johns Hopkins University, Baltimore, and a publication assistance award from the Richard D. and Mary Jane Edwards Endowed Publication Fund, Faculty of Arts and Sciences, University of Pittsburgh.

REFERENCES

Bogin, B. (1988). *Patterns of Human Growth* (Cambridge: Cambridge University Press).

Brace, C. L., and Montagu, A. (1977). *Human Evolution*, 2nd ed. (New York: Macmillan).

Burdi, A., Feingold, M., Larsson, K. S., Leck, I., Zimmerman, E. F., and Fraser, F. C. (1972). Etiology and pathogenesis of congenital cleft lip and cleft palate, an NIDR state of the art report. *Teratology 6*, 255–270.

Cohen, M. M. Jr. (1997). *The Child with Multiple Birth Defects*, 2nd ed. (New York: Oxford University Press).

Cohen, M. M., Jr., and MacLean, R. E. (eds.) (2000). *Craniosynostosis: Diagnosis, Evaluation, and Management*, 2nd ed. (New York: Oxford University Press).

David, J. D., Poswillo, D., and Simpson, D. (1982). *The Craniosynostoses: Causes, Natural History, and Management* (Berlin: Springer-Verlag).

Enlow, D. H. (1990). *Handbook of Facial Growth* (Philadelphia: W. B. Saunders).

Enlow, D. H., and Gans, M. G. (1996). *Essentials of Facial Growth* (Philadelphia: W. B. Saunders).

Gorlin, R. J., Cohen, M. M., Jr., and Levin, L. S. (1990). *Syndromes of the Head and Neck* (New York: Oxford University Press).

Harrison, G. A., Tanner, J. M., Pilbeam, D. R., and Baker, P. T. (1988). *Human Biology* (New York: Oxford University Press).

Johnston, M. C., and Bronsky, P. T. (1995). Prenatal craniofacial development: new insights on normal and abnormal mechanisms. *Crit. Rev. Oral Biol. Med. 6*, 25–79.

Jones, K. L. (1988). *Smith's Recognizable Patterns of Human Malformations* (Philadelphia: W. B. Saunders).

Juriloff, D. M., and Trasler, D. G. (1976). Test of the hypothesis that embryonic face shape is a causal factor in genetic predisposition to cleft lip in mice. *Teratology 14*, 35–41.

Krogman, W. M. (1978). *The Human Skeleton in Forensic Medicine* (Springfield: Charles C. Thomas).

Kusiako, T., Ronsmans, C., and Van der Paal, L. (2000). Perinatal mortality attributable to complications of childbirth in Matlab, Bangladesh. *Bull. World Health Organ. 78*, 621–627.

Malek, R. (2001). *Cleft Lip and Palate: Lesions, Pathophysiology, and Primary Treatment* (London: Martin Dunitz, Ltd.).

Marsh, J. L., and Vannier, M. W. (1985). *Comprehensive Care for Craniofacial Deformities* (St. Louis: C. V. Mosby).

Meirowitz, N. B., Ananth, C. V., Smulian, J. C., and Vintzileos, A. M. (2001). Effect of labor on infant morbidity and mortality with preterm premature rupture of membranes: United States population-based study. *Obstet. Gynecol. 97*, 494–498.

Melnick, M., and Jorgenson, R. (eds.) (1979). *Developmental Aspects of Craniofacial Dysmorphology* (New York: Alan. R. Liss).

Melnick, M., Bixler, D., and Shields, E. D. (eds.) (1980). *Etiology of Cleft Lip and Cleft Palate* (New York: Alan R. Liss).

Montoya, A. (ed.) (1992). *Craniofacial Surgery* (Bologna, Italy: Monduzzi Editore).

Morris, H. L., and Bardach, J. (eds.) (1990). *The Multidisciplinary Management of Cleft Lip and Palate* (Philadelphia: W. B. Saunders).

Persing, J. A., Edgerton, M. T., and Jane, J. A. (eds.) (1989). *Scientific Foundations and Surgical Treatment of Craniosynostosis* (Baltimore: Williams and Wilkins).

Posnick, J. C. (2000). *Craniofacial and Maxillofacial Surgery in Children and Young Adults*, vols. 1 and 2 (Philadelphia: W. B. Saunders).

Ross, R. B., and Johnston, M. C. (1978). *Cleft Lip and Palate* (Huntington, NY: Robert E. Krieger Publishing Co.).

Sarnat, B. G. (2001). Effects and noneffects of personal environmental experimentation on postnatal craniofacial growth. *J. Craniofac. Surg. 12*, 205–217.

Siegel, M. I., and Mooney, M. P. (1986). Palatal width growth rates as the genetic determinant of cleft palate induced by vitamin A. *J. Craniofac. Genet. Devel. Biol. Supplement 2*, 187–191.

Sperber, G. H. (1989). *Craniofacial Embryology*, 4th ed. (London: Wright Publishing Co.).

Sperber, G. H. (2001). *Craniofacial Development* (Hamilton: B. C. Decker).

Spranger, J. W., Benirschke, K., Hall, J. G., Lenz, W., Lowry, R. B., Opitz, J. M., Pinsky, L., Schwarzacher, H. G., and Smith, D. W. (1982). Errors of morphogenesis: concepts and terms. *J. Pediatr. 100*, 160–165.

Steegman, A. T., Jr. (1970). Cold adaptation and the human face. *Am. J. Phys. Anthrop. 32*, 243–250.

Stockard, C. R. (1941). *The Genetic and Endocrine Basis for Differences in Form and Behavior. American Anatomical Memoirs 19* (Philadelphia: Wistar Institute of Anatomy and Biology Press).

Trasler, D. G., and Machado, M. (1979). Newborn and adult face shapes related to mouse cleft lip predisposition. *Teratology 19*, 197–206.

Turvey, T. A., Vig, K. W. L., and Fonseca, R. L. (1996). *Facial Clefts and Craniosynostosis: Principles and Management* (Philadelphia: W. B. Saunders).

Vergato, L. A., Doerfler, R. J., Mooney, M. P., and Siegel, M. I. (1997). Mouse palatal width growth rates as an "at risk" factor in the development of cleft palate induced by hypervitaminosis A. *J. Craniofac. Genet. Devel. Biol. 17*, 204–210.

Vig, K. W. L., and Burdi, A. R. (eds.) (1988). *Craniofacial Morphogenesis and Dysmorphogenesis*, Craniofacial Growth Series Monograph 21 (Ann Arbor, MI: Center of Human Growth and Development).

Wyszynski, D. F. (ed.) (2001). *Cleft Lip and Palate: From Origin to Treatment* (Oxford: Oxford University Press).

CHAPTER 2

TERMINOLOGY AND CLASSIFICATION OF CRANIOSYNOSTOSIS

MARILYN C. JONES, M.D.

2.1 INTRODUCTION

This chapter addresses the current systems for classifying craniosynostosis. The confusing language that is applied to this group of conditions is more easily understood if you have some sense of the evolution of the systems and some understanding of how they complement one another. In large measure, the systems reflect the needs of those that developed and used them. Thus, practitioners interested in treatment use a classification based upon shape (a *morphologic system*). Those addressing natural history and recurrence risk focus on the etiology/cause (a *clinical genetic system*). The more recent advances in molecular biology elucidating the genes and pathways that impact sutural development have spawned yet a third system that classifies conditions based on the specific mutation in the specific gene accounting for the developmental abnormality (a *molecular genetic system*). All systems have value and communicate important, but often incomplete, information about a specific individual or clinical problem. Classification systems are evolving to reflect a greater

depth of understanding of the problem of craniosynostosis.

The purpose of this chapter is to provide a framework for understanding current terminology and to anticipate changes in future terminology. The earliest reports of craniosynostosis were descriptive and form the basis for the morphologic and clinical genetic systems for classification. It is perhaps difficult to understand that at the turn of the 20th century physicians were consulted more for their knowledge of the natural history of disease and disease processes than for their ability to treat a condition (Bliss, 1999). Interventions were limited. Classification schemes based on understanding of morbid anatomy provided the first useful framework for understanding the pathology and natural history of a particular set of findings or conditions. It is the persistent clinical utility of some of these systems that accounts for their survival in the present.

2.2 HISTORICAL PERSPECTIVE

In their recent monograph on craniosynostosis, Cohen and MacLean provide an

Understanding Craniofacial Anomalies: The Etiopathogenesis of Craniosynostoses and Facial Clefting, Edited by Mark P. Mooney and Michael I. Siegel, ISBN 0-471-38724-x Copyright © 2002 by Wiley-Liss, Inc.

extensive and encyclopedic review of terminology and the systems for classification that have evolved since the beginning of written history (Cohen and MacLean, 2000). It is not the intent of this chapter to duplicate their effort; however, this chapter draws heavily from various sections in the book. The first system for classification was published by Virchow in 1851 (Cohen and MacLean, 2000). Drawing on previous work of Otto, Virchow's system was based on head shape and consisted of four major types of heads: large heads (macrocephaly), small heads (microcephaly), long heads (dolichocephaly), and short heads (brachycephaly). The latter two relate to craniosynostosis and stem from the observation that skull growth continues in the dimensions perpendicular to a fused suture as a consequence of compensatory growth along adjacent sutures.

Virchow correlated skull shape with fusion in specific sutures and codified several descriptive terms that are still in wide general usage including *dolichocephaly* for sagittal synostosis, *trigonocephaly* for metopic synostosis, and *plagiocephaly* for unilateral coronal synostosis. The latter is currently also used to designate deformational changes in the skull resulting in asymmetry. Table 2.1 sets forth the most common terms in clinical usage. Surgeons still rely heavily on morphology in planning operative interventions.

Shortly after the turn of the 20th century, the French pediatrician, Apert, reported a patient with craniosynostosis and syndactyly, coining the term *acrocephalosyndactyly* (ACS) to describe the condition (Apert, 1906). Crouzon followed with a report on hereditary craniofacial dysostosis (Crouzon, 1912). Although the cause of most craniosynostosis remained largely unknown for most of the 20th century, the genetic etiology of many of the complex or syndromic synostoses became apparent because of observations of recurrences in families. In addition, by 1980, limb defects were repetitively observed in association with synostosis such that 80% of syndromes associated with craniosynostosis involved abnormalities of the limbs (Cohen, 1980).

In attempting to reconcile the descriptive terminology with emerging genetic understanding, a classification of acroce-

TABLE 2.1 Morphologic Nomenclature in Common Usage

Term	Meaning	Suture Involved
Dolichocephaly	Long head	Sagittal suture
Scaphocephaly	Keel-shaped head	Sagittal suture
Acrocephaly	Pointed head	Coronal suture
		Coronal/lambdoid or all sutures
Brachycephaly	Short head	Coronal suture
Oxycephaly	Tower-shaped head	Coronal/lambdoid or all sutures
Turricephaly	Tower-shaped head	Coronal suture
Trigonocephaly	Triangular-shaped head	Metopic suture
Plagiocephaly	Asymmetric head	Unilateral coronal
		Unilateral lambdoid or positional
Kleeblattschädel	Clover-leaf skull	Multiple but not all sutures
Craniofacial dysostosis[a]	Midface deficiency	Craniosynostosis with involvement of cranial base sutures

[a] Although craniofacial dysostosis refers specifically to Crouzon syndrome, the term is often used to describe any condition with midface deficiency and craniosynostosis.

phalosyndactylies developed, designating Apert syndrome ACS type I, Pfeiffer syndrome ACS type II, and Saethre-Chotzen syndrome ACS type III (McKusick, 1975). Carpenter syndrome was acrocephalo*poly*syndactyly. As the nomenclature became increasingly cumbersome and the phenotypic distinctions among disorders became less clear, clinicians reverted to an eponymic system to avoid imprecise Latin and Greek descriptors and to acknowledge the contribution of the physician who either described or was somehow associated with the condition in question. These eponyms are in broad clinical usage and remain the most efficient system for describing a clinical phenotype (Table 2.2).

In the early 1990s, the genes for some of the most common or well-defined craniosynostosis syndromes began to be identified. Soon afterward, Muenke and colleagues (1997) described the first disorder in which identification of the responsi-

ble mutation preceded description of the phenotype. Although there was some initial enthusiasm for changing nomenclature to reflect molecular pathogenesis, it became clear that both clinical and molecular heterogeneity existed for many disorders. For example, Pfeiffer syndrome can be produced by mutations in two different genes (FGFR1 and FGFR2) while identical mutations in the same gene, such as Cys278Phe in FGFR2, produce Crouzon syndrome in some patients and Pfeiffer syndrome in others. Table 2.3 sets forth the current status of phenotypes produced by mutations in various genes.

Wilkie (1997) has reviewed how a molecular focus can lead to better understanding of the developmental pathogenesis of sutural fusion. Moreover, molecular diagnosis is increasingly utilized in clinical practice both for clinical diagnosis of affected individuals and prenatal diagnosis in at-risk families.

Several recent attempts have been made

TABLE 2.2 Commonly Used Clinical Genetic Classification

Diagnostic Category	Name of Disorder	Etiology
Isolated craniosynostosis	Morphologically described	Unknown, uterine constraint, or FGFR3 mutation
Syndromic craniosynostosis	Antley-Bixler syndrome	Unknown
	Apert syndrome	Usually one of two common mutations in FGFR2
	Baere-Stevenson syndrome	Mutation in FGFR2 or FGFR3
	Baller-Gerold syndrome	Mutation in TWIST heterogeneous
	Carpenter syndrome	Unknown
	Craniofrontonasal dysplasia	Unknown gene at Xp22
	Crouzon syndrome	Numerous different mutations in FGFR2
	Crouzonomesodermoskeletal syndrome	Mutation in FGFR3
	Jackson-Weiss syndrome	Mutation in FGFR2
	Muenke syndrome	Mutation in FGFR3
	Pfeiffer syndrome	Mutation in FGFR 1 or numerous mutations in FGFR2
	Saethre-Chotzen syndrome	Mutations in TWIST
	Shprintzen-Goldberg syndrome	Mutation in FBN1

TABLE 2.3   Molecular Genetic Classification

Gene	Mutation	Phenotype
FGFR1	755C→G	Pfeiffer syndrome (milder phenotype)
FGFR2	Multiple	Apert syndrome, Baere-Stevenson syndrome, Crouzon syndrome, Jackson-Weiss syndrome, Pfeiffer syndrome (severe phenotype)
FGFR3	Multiple	Baere-Stevenson syndrome, Crouzonodermoskeletal syndrome, Muenke syndrome
MSX2	Pro148His	Boston-type synostosis
TWIST	Multiple	Baller-Gerold syndrome, Saethre-Chotzen syndrome

to fuse molecular and clinical genetic terminology (Biesecker, 1998; Cohen and MacLean, 1999). Cohen and MacLean list the common designation for the clinical phenotype, followed by the McKusick number (http://www.ncbi.nlm.nih.gov/Omim/), followed by the gene, followed by the specific mutation. For example, Crouzon syndrome (123500), FGFR2, Cys278Phe would distinguish Crouzon syndrome patients from those with Pfeiffer syndrome (101600), FGFR2, Cys278Phe who have the same mutation. Whereas this nomenclature might well be appropriate for a written text, it is cumbersome for common parlance.

2.3   CONCLUSIONS

Each of the systems in current usage has limitations. A child with plagiocephaly might have either a positional abnormality with an excellent prognosis for restoration of form without surgical intervention or unicoronal craniosynostosis requiring operative management. Isolated coronal craniosynostosis might be the result of prolonged constraint in utero with a negligible recurrence risk (Graham et al., 1980; Graham, 1988) or mutation in FGFR3 with a risk for development of sensorineural hearing loss and possibly an unaffected carrier parent (Gripp et al., 1998). Mutations in FGFR3 produce phenotypes as disparate as Baere-Stevens cutis gyrata

syndrome and thanatophoric dysplasia (a lethal dwarfing condition). Until a system develops to communicate not only the molecular pathogenesis but also information regarding phenotype, prognosis, and recurrence risk specific to the patient/family in question, it is unlikely that clinicians will use a uniform terminology. Practitioners in the area of craniosynostosis are well advised to be familiar with all three systems.

REFERENCES

Apert, E. (1906). De l'acrocéphalosyndactylie. *Bull. Soç. Méd. Hôp. Paris 36*, 1310.

Biesecker, L. G. (1998). Lumping and splitting: molecular biology in the genetics clinic. *Clin. Genet. 53*, 3–7.

Bliss, M. (1999). *William Osler: A Life in Medicine* (New York: Oxford University Press).

Cohen, M. M., Jr. (1980). Perspectives on craniosynostosis. *West. J. Med. 132*, 507–513.

Cohen, M. M., Jr., and MacLean, R. E. (1999). Should syndromes be defined phenotypically or molecularly? Resolution of the dilemma. *Am. J. Med. Genet. 86*, 203–204.

Cohen, M. M., Jr., and MacLean, R. E. (2000). *Craniosynostosis: Diagnosis, Evaluation, and Management* (New York: Oxford University Press).

Crouzon, O. (1912). Dysostose crâniofaciale héréditaire. *Bull. Soç. Méd. Hôp. Paris 33*, 545–555.

Graham, J. M., Jr. (1988). *Smith's Recognizable Patterns of Human Deformation* (Philadelphia: W. B. Saunders).

Graham, J. M., Jr., Badura, R., and Smith, D. W. (1980). Coronal craniostenosis: fetal head constraint as one possible cause. *Pediatrics 65*, 995.

Gripp, K. W., McDonald-McGinn, D. M., Gaudenz, K., Whitaker, L. A., Bartlett, S. P., Glat, P. M., Cassileth, L. B., Mayro, R., Zackai, E. H., and Muenke, M. (1998). Identification of the first genetic cause of isolated unilateral coronal synostosis: a unique mutation in the fibroblast growth factor receptor 3 (FGFR3). *J. Pediatr. 132*, 714–716.

McKusick, V. A. (1975). *Mendelian Inheritance in Man* (Baltimore: The Johns Hopkins University Press).

Muenke, M., Gripp, K. W., McDonald-McGinn, D. M., Gaudenz, K., Whitaker, L. A., Bartlett, S. P., Markowitz, R. I., Robin, N. H., Nwokoro, N., Mulvihill, J. J., Losken, H. W., Mulliken, J. B., Guttmacher, A. E., Wilroy, R. S., Clarke, L. A., Hollway, G., Adès, L. C., Haan, E. A., Mulley, J. C., Cohen, M. M., Jr., Bellus, G. A., Francomano, C. A., Moloney, D. M., Wall, S. A., Wilkie, A. O. M., and Zackai, E. H. (1997). A unique point mutation in the fibroblast growth receptor 3 (FGFR3) gene defines a new craniosynostosis syndrome. *Am. J. Hum. Genet. 60*, 555–564.

Wilkie, A. O. M. (1997). Craniosynostosis: genes and mechanisms. *Hum. Molec. Genet. 6*, 1647–1656.

CHAPTER 3

TERMINOLOGY AND CLASSIFICATION OF FACIAL CLEFTING

ROBERT J. SHPRINTZEN, Ph.D.

3.1 INTRODUCTION

Semantics are often a problem, even in science, which has been particularly true in the study of orofacial clefts. Once the word *cleft* has become attached to an anomaly, it becomes equated to any fusion failure in the orofacial complex. Clefts of the palate posterior to the incisive foramen often occur together with clefts of the lip and maxillary alveolus anterior to the incisive foramen. However, these two different types of clefts are not the same diseases, although it has been the tendency for clinicians and researchers to equate them. Systems for classifying clefts have always included both the anterior and posterior variety of fusion failures, but it is possible that these types of categorizations are not warranted. The purpose of this chapter is to discuss the classification of cleft lip, cleft palate, or both, in terms of what is currently known about the etiology of orofacial clefting and the timing of embryonic development.

3.2 HISTORICAL CLASSIFICATIONS OF FACIAL CLEFTS

In order to determine the relative value of classification systems, it must first be determined why such categorical approaches are necessary. Do they serve the same purpose for all clinicians, or are they discipline specific? Do they simplify descriptions from one professional to another, or do they oversimplify variable expression? Do they fulfill a research purpose, or do they confound good research? Are classifications actually used in the day-to-day management of patients, and if so, how?

The early attempts at classifying cleft anomalies coincided with the "modern" era of surgery following World War I when surgeons began applying new facial reconstruction procedures for battle trauma to children with congenital anomalies. At the time, little was known about the etiology of clefting, and categorizations were primarily a tool for descriptive purposes. Davis and Ritchie (1922) devised a classification system with three groups of cleft types. Group 1 clefts were isolated clefts of the lip (without inclusion of the maxillary alveolus). Group 2 clefts were inclusive of all

Understanding Craniofacial Anomalies: The Etiopathogenesis of Craniosynostoses and Facial Clefting, Edited by Mark P. Mooney and Michael I. Siegel, ISBN 0-471-38724-x Copyright © 2002 by Wiley-Liss, Inc.

defects of the palate from the incisive foramen posterior, although submucous clefts were not mentioned. Group 3 clefts were all complete clefts of the lip and palate (unilateral or bilateral).

Perhaps the earliest widely accepted system for categorizing orofacial clefts was that of Veau (1931). Veau divided cleft anomalies into four subgroups, but they were different than those offered by Davis and Ritchie (1922). By describing Veau cleft types, clinicians were presumably able to communicate the anatomic nature of the defect in their patient. This type of system might also have some research application by allowing scientists to classify clefts according to an operational definition that permitted the grouping together of classes of patients for clinical studies. The Veau classification divided clefts into four subtypes (Fig. 3.1):

Type I: Clefts of the soft palate posterior to the hard palate

Type II: Complete clefts of the palate from the incisive foramen posteriorly through the soft palate

Type III: Complete unilateral cleft lip and cleft palate

Type IV: Complete bilateral cleft lip and cleft palate

The Veau classification is purely anatomical in its perspective, designed by a surgeon in an attempt to specify particular cleft types, but it fails to do this well, and it also fails to be etiologically relevant.

Other anatomical classifications have also been developed more recently. Kernahan (1971) suggested a pictorial representation using a striped Y schematic diagram that represents the lip, palate, and maxillary alveolus. Kriens (1989) suggested a labeling system called LAHSHAL (a palindromic acronym for Lip, Alveolus, Hard palate, Soft palate, Hard palate, Alveolus, Lip) that assigned a letter to each portion of orofacial anatomy con-

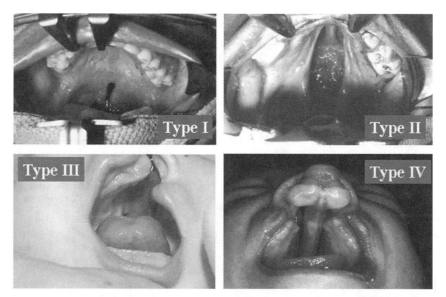

Figure 3.1 The Veau classification system demonstrated by four anatomically different clefts, including a cleft of the soft palate only (type I), a cleft of the hard and soft palate (type II), a complete unilateral cleft lip and palate (type III), and a complete bilateral cleft of the lip and palate (type IV).

noting the location of the cleft. The use of upper- and lowercase letters would connote if the cleft were complete or incomplete (uppercase for complete, lowercase for incomplete). As an example, LAS would connote a complete cleft of the left lip, complete cleft of the left alveolus, and complete cleft of the soft palate. SHAL would connote a complete right-sided unilateral cleft lip and palate. The LAHSHAL system was devised for easy retrieval of data regarding cleft type for the proper grouping of patients.

3.3 DO THESE CLASSIFICATIONS REALLY SERVE A PURPOSE?

No. It is this author's opinion that other than some limited descriptive purpose, classification systems such as those of Veau, Kernahan, and Kriens have very little value. As evidence of this position, I would point out that very few clinicians utilize these systems. Such systems also have very limited presence in the scientific literature. More importantly, the anatomically based systems do not recognize a number of important factors related to clefts that are of critical importance in their diagnosis and management—for example, the etiologic heterogeneity of clefting and the high degree of anatomic variability of cleft that are relevant to surgical management.

3.3.1 Etiologic Heterogeneity

As has been reported in a number of studies, cleft palate and cleft lip and palate are etiologically heterogeneous conditions (Jones, 1988; Rollnick and Pruzansky, 1981; Shprintzen et al., 1985b). Although the large majority of cases were thought to be multifactorial in etiology (Fraser, 1970), more recent studies have shown that many clefts are actually parts of genetic,

chromosomal, or teratogenic syndromes (Jones, 1988; Rollnick and Pruzansky, 1981; Shprintzen et al., 1985b). Multifactorial inheritance has never been convincingly demonstrated by any hard evidence, whereas direct genetic links to clefting are known to occur in over 400 separate genetic and chromosomally based conditions, as well as a number of teratogenic disorders, including fetal alcohol syndrome and fetal hydantoin syndrome. Fraser (1970) reported that approximately 3% of newborns had clefts as part of a multiple anomaly syndrome. However, Shprintzen and associates (1985b) reported that approximately 8% of all children with overt or submucous clefts of the palate had a single multiple anomaly syndrome— velocardiofacial syndrome (VCFS)—and that several other genetic syndromes were also common.

Examination of the cleft itself, or of its type (category) as might be described by any of the previously described systems, will not typically yield clues that would assist in the delineation of the primary diagnosis. Clefts represent failures for the palate or lip and palate to fuse normally, but it is not possible to ascertain from visual examination what caused the fusion failure. It has been reported that the etiology of the cleft has a major effect on treatment outcome and even the choice of specific surgical procedures (Shprintzen et al., 1985b; Shprintzen, 1994). For example, cleft repair in VCFS rarely results in a successful speech outcome (Shprintzen, 1982). Nearly all clefts in VCFS are of the secondary palate, but there is nothing anatomically obvious that would distinguish the palatal cleft in VCFS from those on children who do not have VCFS. The same cleft type in a child with Stickler syndrome or isolated cleft palate is likely to result in a good surgical outcome (Shprintzen, 1982). Therefore, simply categorizing the cleft as a Veau, type I, has no relevance to surgical outcome and no relevance to primary diagnosis or the biological mechanisms that made the cleft occur.

3.3.2 Anatomic Variability

Clinicians who see many individuals with clefts understand that the degree and location of clefts are highly variable from case to case, and therefore the categories applied by most systems tend to reduce the descriptive power of the clinician. For example, clefts of the lip may be complete or incomplete, and may involve the maxillary alveolus, which may also be completely or incompletely cleft (Fig. 3.2). Overt clefts of the secondary palate may involve only the tip of the uvula, may extend to the incisive foramen, or to any length of the palate in between these two points (Fig. 3.3). In cases where both the primary and secondary palates are involved, the variability of cleft type is essentially limitless. The primary palate may have a unilateral cleft

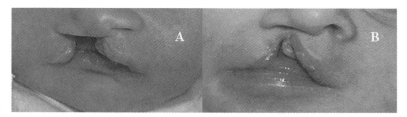

Figure 3.2 Complete versus incomplete unilateral clefts of the lip showing the anatomic variability found within the same type of cleft.

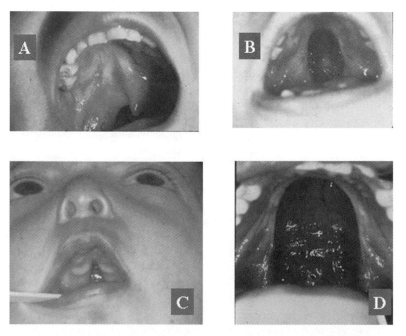

Figure 3.3 Anatomic variability within clefts of the secondary palate only, including a submucous cleft with overt clefting of the very posterior end of the soft palate (A), a wide U-shaped cleft of the soft palate and posterior hard palate (B), a narrow V-shaped cleft of the same portion of the palate as in B (C), and a very wide U-shaped cleft of the entire secondary palate (D).

while the secondary palate is cleft bilaterally, or the primary palate may have an incomplete cleft and the secondary palate a complete cleft, or vice versa. Applying categories to this type of variation oversimplifies a complex system.

3.4 BACK TO SEMANTICS

Perhaps what is really needed in the way of a classification system is to find some consistency in language, and then to apply that consistent language to a system that will be useful for clinical management and research.

3.4.1 The Language

The language used to describe clefts should be anatomically relevant, but most important, it must be consistent to allow data recall. Fortunately, with today's computer technology and database software, there is no need for codes (letters, numbers, or symbols), but there is a need for consistency. The basic language to be used should be recognized by all clinicians, regardless of their professional field of study.

Because the orofacial structures form embryologically in two distinct stages, it makes sense to group the structures anterior to the incisive foramen and the structures posterior to the incisive foramen as separate units. Most scientists who study clefts prefer to refer to the anterior structures (the lip and maxillary alveolus) as the primary palate and the posterior structures as the secondary palate. Within these groups, it makes the most sense to refer to each individual structure by name. Within the primary palate are the lip and the maxillary alveolus, and the posterior palate contains the hard palate and the soft palate (or velum). Within each of these subgroupings, the following types of clefts may occur: complete, incomplete, and submucous. In addition, the primary palate and

the hard palate may be cleft unilaterally or bilaterally. There may also be atypical clefts, which should simply be described according to the type of anomaly (lateral cleft, oblique cleft, etc.). There is no longer the need to use cryptic codes. The power of modern computers allows the retrieval of whole or part words from databases, word-processed reports, or spreadsheets.

3.4.2 The Most Important Purpose and Technique for Categorization

The importance of categorizing orofacial anomalies is related to the fact that clefting is etiologically heterogeneous. Both clinical management and research data are dependent on an accurate classification of individual cases. When collecting clinical data, it would be important to draw conclusions about treatment outcomes based on the primary diagnosis, not the cleft type. However, within each individual diagnosis, it would be possible to draw conclusions about the effects of cleft type, especially for those syndromes that have both clefts of the secondary palate alone and clefts of both the primary and secondary palates.

3.4.3 Clinical Management

The frequency of associated anomalies in children with clefts of the palate or lip and palate is high (Rollnick and Pruzansky, 1981; Shprintzen et al., 1985b; Jones, 1988). When a cleft occurs as part of a larger symptom complex, it is often true that the other anomalies will affect the management of the cleft anomaly. Therefore, in making decisions regarding the effectiveness of treatment (whether at the clinical or research level), it is important to determine what factors are responsible for the outcome. For example, maxillary deficiency has been attributed to early surgical repair of the palate, but it has been pointed out that many syndromes associated with cleft palate have primary maxillary hypoplasia as a clinical feature (Shprintzen, 1982, 1997). If all treatment outcomes are attributed to the cleft and not to other intrinsic factors, then both research and clinical decisions will often be incorrect. For clinical research, it is critical to make sure that outcomes following treatment are related only to the treatment. It is therefore important to verify that a study population is as homogeneous as possible. Because the population of inidividuals with clefts is so heterogeneous, a necessary technique for categorization becomes the differentiation between primary etiologies.

The first level of distinction should be to determine if the child has associated anomalies or no associated anomalies (an isolated cleft). Although the absence of associated anomalies does not imply that all isolated clefts are of the same etiology and therefore homogeneous, at the time of this writing, there has not been a specific single-gene link to a large number of cases of nonsyndromic clefting in humans. Therefore, further categorization of nonsyndromic clefts is not possible at this point even though it is likely that there are a number of single-gene causes of clefting of the palate or lip and palate. However, additional categorization of syndromic clefts is essential.

A number of multiple anomaly syndromes express the full range of cleft anomalies—that is, mixing of cleft types (secondary palate alone versus primary palate and secondary palate) occurs in a relatively small number of disorders, but is not typically observed in nonsyndromic familial clefting. Table 3.1 lists syndromes that have mixing of cleft type. Clinicians should be aware that first-degree relatives (i.e., parent–child or full siblings), one of whom has a cleft of the secondary palate only while another has a cleft of the primary palate (with or without a cleft of the secondary palate), almost certainly have a syndrome.

3.4.4 Categorizing Syndromic Clefts

There are more than 400 known syndromes that have cleft palate and/or cleft lip as a clinical finding. For each syndrome, the implication of the presence of the cleft is different. For example, class III malocclusion may be caused by maxillary deficiency in the presence of a normal or large mandible, but is less likely to occur in cases where the mandible is hypoplastic or retruded because of an abnormally open skull base angle (platybasia). Patients with a number of multiple anomaly syndromes are almost certain to have a class III malocclusion following lip and palate repair because of the intrinsic maxillary hypoplasia and normal mandibular size, including del(18p), AEC (Hay-Wells) syndrome, and Kallmann syndrome. In these multiple anomaly disorders, maxillary deficiency after palate repair will certainly occur because maxillary deficiency would happen even without palate repair. However, in Robinow syndrome, there is marked platybasia, which results in retrognathia in many cases. In cases of Robinow syndrome with clefting, class III malocclusion is much less likely (Fig. 3.4). In syndromic clefts, it may be that the type of cleft is less important than the overall pattern of craniofacial growth or malformation.

3.4.5 Additional Anatomic Differentiation

Another problem with categorization is how to classify submucous cleft palate or occult submucous cleft palate. Cleft lip is an obvious malformation, and cleft palate is generally an easy anomaly to identify by oral examination. It is likely that the majority of submucous cleft palates go undetected, primarily because the anomaly rarely results in detectable symptoms. Several studies have shown that the large majority of individuals with submucous cleft palate have normal speech (Shprint-

TABLE 3.1   Syndromes That May Have Either Clefts of the Primary Palate and Secondary Palate or Clefts of the Secondary Palate Only

Syndrome	Cleft Lip (Cleft Alveolus)	Cleft Lip and Cleft Palate	Cleft Palate Only
Aarskog syndrome	Uncommon	Uncommon	Uncommon
AEC syndrome	Least common	Occasional	Common
CHARGE assocation	Uncommon	Uncommon	Common
Craniofrontonasal syndrome	Common	Common	Common
Cri du chat syndrome	Occasional	Occasional	Common
Cryptophthalmos syndrome	Occasional	Occasional	Occasional
del(18p)	Common	Common	Common
Down syndrome	Uncommon	Occasional	Occasional
EEC syndrome	Occasional	Common	Common
Escobar syndrome	Uncommon	Occasional	Common
Fetal alcohol syndrome	Uncommon	Occasional	Common
Filiform adhesions with clefting syndrome	Occasional	Common	Common
Holoprosencephaly	Common	Common	Common
Kallmann syndrome	Common	Common	Common
Lenz syndrome	Uncommon	Uncommon	Occasional
Mohr syndrome	Common	Uncommon	Occasional
Niikawa-Kuroki syndrome	Occasional	Common	Common
Oculoauriculovertebral spectrum	Occasional	Common	Common
Oculodentodigital syndrome	Uncommon	Occasional	Occasional
Opitz syndrome	Occasional	Common	Common
Popliteal pterygium syndrome	Occasional	Common	Common
Rapp-Hodgkin syndrome	Uncommon	Occasional	Common
Robinow syndrome	Uncommon	Occasional	Occasional
Treacher Collins syndrome	Uncommon	Occasional	Common
Trisomy 13	Occasional	Common	Common
Trisomy 18	Uncommon	Uncommon	Occasional
van der Woude syndrome	Common	Common	Common
Velocardiofacial syndrome	Uncommon	Uncommon	Common
Wolf-Hirschhorn syndrome	Occasional	Occasional	Common

zen et al., 1985a; Weatherly-White et al., 1972) and that there is only a small increase in the frequency of middle ear disease (Schwartz et al., 1985). Although many clinicians do not regard bifid uvula as synonymous with submucous cleft palate, research has shown that the large majority of people with bifid uvula do have submucous cleft palate (Shprintzen et al., 1985a). These cases are largely omitted from incidence and prevalence statisics. It is not clear what the true frequency of clefting is in the general population; nor is it clear what percentage of cleft cases includes submucous clefts. The prevalence of submucous clefts is of major importance when it is considered that associated anomalies and syndromes occur more often with submucous cleft palate than any other form of clefting.

The occult submucous cleft palate is the least recognized form of clefting. The term

Figure 3.4 Normal occlusion in an adult with Robinow syndrome and complete cleft lip and palate. He has not had orthodontic management.

occult submucous cleft was first coined by Kaplan in a 1975 paper. Kaplan reported on four cases with velopharyngeal insufficiency and hypernasal speech without overt cleft palate; nor did these patients have the classic landmarks of submucous cleft palate (bifid uvula, notching of the hard palate, or midline muscular diastasis). The individuals in the study were referred for pharyngeal flap surgery. During the operation, Kaplan carefully dissected the nasal mucosa of the palate and examined the musculature. He found that all four of the patients had malalignment of the palatal musculature similar to that seen in obvious submucous clefts even though they had normal oral examinations. Kaplan therefore referred to the palatal anomaly as the occult (i.e., mysterious) submucous cleft and suggested that the only way this anomaly could be detected was by surgical dissection. Several years later, it was found that individuals with occult submucous cleft palate had a characteristic appearance of the nasal surface of the soft palate that was detectable by nasopharyngoscopy (Croft et al., 1978; Shprintzen, 1979; Lewin et al.,

1980). The only detectable cases of occult submucous cleft palate are those that result in hypernasal speech. It is not known what percentage of the population has occult submucous cleft palate and normal speech. There is no current classification system that accounts for occult clefts.

It is of interest to note that in the original article published by Kaplan (1975), three of the four patients clearly had velocardiofacial syndrome. VCFS was not delineated until 1978 (Shprintzen et al., 1978), but it has been shown that a high percentage of children with VCFS have occult submucous cleft palate (Shprintzen, 2001). Occult submucous cleft may also be found in many other multiple anomaly syndromes and in isolation. In some circumstances, the anomaly is isolated (nonsyndromic) and is unmasked by adenotonsillectomy (Croft et al., 1981). If there are cases that would not have been symptomatic but for adenoidectomy, this is a clear indication that occult submucous cleft palate goes undetected in many cases, if not the majority of cases.

3.5   IS THERE A USEFUL SYSTEM?

The etiologic and anatomic variability demonstrated above points out an interesting anathema to categorization. The purpose of categorizing clefts is to lump them together by some common denominator in order to make the category meaningful to a clinician or researcher. But the grouping together of two patients with clefts because of one common denominator (broadly anatomic cleft type) eradicates the many differences between people who would be in the same category, such as etiology or specific anatomic anomaly. Consider the following hypothetical case. Two brothers share the mutant gene for van der Woude syndrome. The mutant gene, located on the long arm of chromosome 1 at the q32 band, causes one child to have a complete bilateral cleft lip and cleft palate. His brother has a submucous cleft palate and no cleft lip. Although these siblings share at least half of their genes and have the same mutant gene that is causing their cleft condition, they would be categorized separately according to all currently existing systems.

Conversely, two children with clefts of the secondary palate are put into the same category. However, one has Beckwith-Wiedemann syndrome and the other has Stickler syndrome. Children with Beckwith-Wiedemann syndrome have somatic overgrowth, mandibular prognathism, possible abdominal hernias (including omphalocele), a risk of Wilms' tumor, hypotonia, and often mild cognitive and developmental impairment. Growth, both somatic and craniofacial, is different than normal, but class III malocclusions in Beckwith-Wiedemann syndrome are related to mandibular excess, not maxillary deficiency. The child with Stickler syndrome has no developmental or cognitive impairment, and rather than mandibular prognathism, micrognathia with class II malocclusion is common. In this case, although the cleft category is the same, the prognosis for facial growth after palate repair is completely different and not dependent on the type of anatomical defect.

Researchers who might study somatic growth (height and weight) in children with cleft palate and selected subjects according to cleft category (a common strategy in research for over 30 years) would get very different results depending on the mixture of specific genetic disorders in their clinic population. Because many studies of children with clefts have tended to involve relatively small samples sizes, the presence of only several cases of a particular genetic syndrome might alter the outcome. If there were a number of patients with Beckwith-Wiedemann syndrome, then heights and weights would be higher than normal.

More likely, however, is that heights and weights would be lower than normal. In children with cleft palate (Veau types I or II clefts), it has been demonstrated that the combination of velocardiofacial syndrome and Stickler syndrome constitute over 13% of all such cleft cases (Shprintzen et al., 1985b). Individuals with VCFS and Stickler syndrome are often of relatively short stature as children, although they are not often of severely short stature. Neither of these syndromes is characterized by significant facial dysmorphology. In the case of Stickler syndrome, there are few other obvious anomalies that can be identified on routine physical examination. The same is true in VCFS, especially if there is no history of congenital heart disease. Therefore, many patients with VCFS and Stickler syndrome may be undetected, especially at cleft palate centers where genetics services are not provided for every patient. A large variation in population homogeneity will clearly skew measurement statistics away from normal values, and these two syndromes, although the most common associated with cleft palate, are not the only ones that result in shorter than normal stature. Similar effects will be seen in other research measures, such as intellect, psychological issues, or head circumference.

3.5.1 Population Heterogeneity: The Instigator or Nemesis of Classification Systems?

Perhaps the primary message of this chapter is that the population of individuals who have clefts of the palate or clefts of the lip and palate is extremely heterogeneous. The population with clefts can be comprised of hundreds of genetic, chromosomal, or teratogenic diseases, or even by extrinsic factors that cause disruptions (as in amniotic disruption sequence). Perhaps it was this recognition that individuals with clefts are not all the same that led early clinicians to see the need for some means of simpler description. However, as discussed earlier, the variation in clefting anomalies comes from multiple sources: anatomic variation and etiologic variation. Clefting is merely a symptom of an interruption of an embryologic process, and as such, it is possible to draw an analogy regarding the need for categorization systems that may seem absurd, but are in fact perfectly valid.

It is possible to focus on any symptom of human illness and build a categorization system for it. Why not a categorization system for sneezing? Sneezing can be caused by many disorders, including upper respiratory infections, allergies, a foreign body in the nose, and abnormal nerve stimulation. Sneezes (like clefts) are easily identified as being sneezes. But even sneezes vary. Some are strong and violent; some are weak and stifled. There may be single sneezes, or series of repeated sneezes. Clinicians decide that there is a reason to categorize sneezes according to their strength, duration, and frequency. This is done, although there are hundreds of pathogens that result in sneezing, ranging from corona virus to pepper. In some cases, there is no treatment (as for viruses), in some cases there is treatment with medications (as for bacteria or allergies), and in some cases, there is common sense (not inhaling pepper).

The reality is that the degree of variability among the population of individuals with clefts is as great, if not greater, than for sneezing. It is impossible to simplify something that is very complex. Although many clefts are treated the same regardless of the etiology, this may be because we have not yet become sufficiently sophisticated to treat each case individually in the most efficient manner related to cause or structure. In some cases, we have begun to understand some of these variations that demand different modes of management. Operations that may be applied in some cases of clefts cannot be applied in syndromic cases.

Pharyngeal flaps are highly effective in treating the velopharyngeal insufficiency that often occurs after repair of cleft palate. However, pharyngeal flap should not be applied to patients with Treacher Collins syndrome because of the high potential for upper airway obstruction (Shprintzen et al., 1992). It has already been mentioned that primary repair of the palate in velocardiofacial syndrome rarely results in a successful speech outcome. Should we then continue to repair the palate in VCFS at 1 year of age as we do for most other children with clefts?

3.5.2   The Use of Modern Databases

If a classification system is to serve a useful purpose, it must account for the splitting of the population of individuals with clefts into the most finite categories possible. Although the specific cleft type should be recorded, more important is the primary diagnosis. There are many commercial databases currently available that can be constructed to retrieve information in a manner that will permit several layers of information to be included. Databases and spreadsheets can be indexed on any field of information, so it is not important to store the information in any particular order. Therefore, information about any clinical population can be entered into a database at any time and in any order without affecting its subsequent method of retrieval. Cleft type can be described in any degree of detail required to fully identify the anomaly. The language used to describe the cleft anomaly should be consistent and unambiguous. Using databases, it is typically best to divide data into the smallest possible units to avoid ambiguity. Databases can be designed many different ways in order to extract the same information as shown in Table 3.2, but in general, smaller and tightly confined cells are easier to fill in and more practical for retrieving information.

TABLE 3.2   Different Forms of Databases Describing the Same Patients

Case Number	Cleft Type
1	Incomplete soft palate
2	Complete left sided cleft lip and alveolus
3	Incomplete left sided cleft lip and alveolus, complete bilateral cleft of secondary palate
4	Incomplete cleft hard palate, complete cleft soft palate

Case Number	Lip		Alveolus		Hard Palate		Soft Palate
	Left	Right	Left	Right	Left	Right	
1	Intact	Intact	Intact	Intact	Intact	Intact	Incomplete cleft
2	Complete	Intact	Complete	Intact	Intact	Intact	Intact
3	Incomplete	Intact	Incomplete	Intact	Complete	Complete	Complete
4	Intact	Intact	Intact	Intact	Incomplete	Intact	Complete cleft

Case Number	Lip	Alveolus	Hard Palate	Soft Palate
1	Normal	Normal	Normal	Incomplete
2	Complete, left	Complete, left	Normal	Normal
3	Incomplete, left	Incomplete, left	Complete, bilateral	Complete
4	Normal	Normal	Incomplete, left	Complete

Case Number	Lip			Alveolus			Hard Palate			Soft Palate		
	N	I	C	N	I	C	N	I	C	N	I	C
1	+			+			+				+	
2			L			L	+			+		
3		L			L				Bilateral			+
4	+			+				L				+

Note: N, normal; I, incomplete; C, complete; L, left.

Is it important that the data stored be consistent from one center to another? It might seem obvious that if research is to be compared that the method of data storage and retrieval be the same. Attempts have been made through professional organizations to seek this uniformity, but as demonstrated in Table 3.2, the same information can be retrieved using more than one method. It is therefore only important that clinicians can communicate the exact nature of the anomaly to one another, not that the data be in the exact same form, much like 0.25 is exactly the same as $\frac{1}{4}$. The use of a standardized database or retrieval system is likely to be unrealistic, since many centers have already been using a large variety of systems to describe tens of thousands of patients.

3.6   CONCLUSIONS

Categorization of cleft type, a long-sought-after goal, has been tacitly accepted as important for both clinical care and research. In this chapter, I have attempted to show that such categories are not as important as previously assumed and that the emphasis has been in the wrong place. Issues such as the cause of the cleft and the presence of associated anomalies or syndromes have far more influence on treatment and research outcomes than the anatomic nature of the cleft. This can be easily demonstrated by looking at cases with the same syndrome but different cleft types in relation to their speech, facial growth, psychological adjustment, and cognitive skills. By focusing our attention on the cleft as a disease rather than the cleft as a symptom, we may not be seeing the forest for the trees.

REFERENCES

Croft, C. B., Shprintzen, R. J., Daniller, A. I., and Lewin, M. L. (1978). The occult submucous cleft palate and the musculus uvuli. *Cleft Palate J. 15*, 150–154.

Croft, C. B., Shprintzen, R. J., and Ruben, R. J. (1981). Hypernasal speech following adenotonsillectomy. *Otolaryngol. Head Neck Surg. 89*, 179–188.

Davis, J. S., and Ritchie, H. P. (1922). Classification of congenital clefts of the lip and palate. *J. Am. Med. Assoc. 79*, 1323.

Fraser, F. C. (1970). The genetics of cleft lip and palate. *Am. J. Hum. Genet. 22*, 336–352.

Jones, M. C. (1988). Etiology of facial clefts: prospective evaluation of 428 patients. *Cleft Palate J. 25*, 16–20.

Kaplan, E. N. (1975). The occult submucous cleft palate. *Cleft Palate J. 12*, 356–368.

Kernahan, D. A. (1971). The striped Y-A symbolic classification for cleft lips and palates. *Plast. Reconstr. Surg. 47*, 469–470.

Kriens, O. (1989). LAHSHAL. A concise documentation system for cleft lip, alveolus, and palate diagnoses. In O. Kriens (ed.), *What Is a Cleft Lip and Palate? A Multidisciplinary Update* (Stuttgart: Thieme), pp. 30–34.

Lewin, M. L., Croft, C. B., and Shprintzen, R. J. (1980). Velopharyngeal insufficiency due to hypoplasia of the musculus uvulae and occult submucous cleft palate. *Plast. Reconstr. Surg. 65*, 585–591.

Rollnick, B. R., and Pruzansky, S. (1981). Genetic services at a center for craniofacial anomalies. *Cleft Palate J. 18*, 304–313.

Schwartz, R. H., Hayden, G. F., Rodriquez, W. J., Shprintzen, R. J., and Cassidy, J. W. (1985).

The bifid uvula: Is it a marker for an otitis-prone child? *Laryngoscope 95*, 1100–1102.

Shprintzen, R. J. (1979). Hypernasal speech in the absence of overt or submucous cleft palate: the mystery solved. In R. Ellis and R. Flack (eds.), *Diagnosis and Treatment of Palato Glossal Malfunction* (London: College of Speech Therapists), pp. 37–44.

Shprintzen, R. J. (1982). Palatal and pharyngeal anomalies in craniofacial syndromes. *Birth Defects 18*, 53–78.

Shprintzen, R. J. (1994). Instrumental assessment of velopharyngeal valving. In R. J. Shprintzen, and J. Bardach (eds.), *Cleft Palate Speech Management: A utidisciplinary Approach* (St. Louis: Mosby), pp. 221–256.

Shprintzen, R. J. (1997). *Genetics, Syndromes, and Communication Disorders* (San Diego: Singular Publishing).

Shprintzen, R. J. (2001). Velo-cardio-facial syndrome. In S. B. Cassidy and J. Allanson (eds.), *Clinical Management of Common Genetic Syndromes* (New York: Wiley), pp. 495–516.

Shprintzen, R. J., Goldberg, R. B., Lewin, M. L., Sidoti, E. J., Berkman, M. D., Argamaso, R. V., and Young, D. (1978). A new syndrome involving cleft palate, cardiac anomalies, typical facies, and learning disabilities: velo-cardio-facial syndrome. *Cleft Palate J. 15*, 56–62.

Shprintzen, R. J., Schwartz, R., Daniller, A., and Hoch, L. (1985a). The morphologic significance of bifid uvula. *Pediatrics 75*, 553–561.

Shprintzen, R. J., Siegel-Sadewitz, V. L., Amato, J., and Goldberg, R. B. (1985b). Anomalies associated with cleft lip, cleft palate, or both. *Am. J. Med. Genet. 20*, 585–596.

Shprintzen, R. J., Singer, L., Sidoti, E. J., and Argamaso, R. V. (1992). Pharyngeal flap surgery: postoperative complications. *Int. Anesthesiol. Clin. 30*, 115–124.

Veau, V. (1931). *Division Palatine* (Paris: Masson).

Weatherley-White, R. C., Sakura, C. Y., Jr., Brenner, L. D., Stewart, J. M., and Ott, J. E. (1972). Submucous cleft palate. Its incidence, natural history, and indications for treatment. *Plast. Reconstr. Surg. 49*, 297–304.

PART II

EMBRYOGENESIS AND ETIOLOGY

CHAPTER 4

CRANIOFACIAL EMBRYOGENESIS: NORMAL DEVELOPMENTAL MECHANISMS

GEOFFREY H. SPERBER, Ph.D.

4.1 INTRODUCTION

The advent of advanced diagnostic imaging techniques, of molecular biology and the sequencing of the human genome has revolutionized our understanding of the developmental mechanisms involved in embryogenesis. Derangement of any of these complex mechanisms results in dysmorphogenesis and consequent anomalies that are detailed in the previous chapters. The diagnosis, prognosis, treatment, and prevention of developmental disorders are increasingly based on the revelations of genetics and molecular biology as the etiologic basis of dysmorphic syndromes.

Fundamental to understanding abnormal development is a need to comprehend the complexities of normal development. Consideration of the very early stages of embryogenesis, subcellular molecular biological mechanisms, differentiation, cytogenesis, histogenesis, and morphogenesis, which constitute enormous and burgeoning fields of study, must necessarily be greatly condensed in this overview. Appreciation of these underlying developmental phenomena should be constantly borne in mind in understanding the expression of the template of coded instructions (the gene) contained in the chromosomal deoxyribonucleic acid (DNA) of the initial zygote formed by the uniting of the parental gametes. The transcription of the coded instructions by signal transduction into morphologically identifiable structures is still an enigma to be elucidated. Complex processes controlling cellular adhesion, motility, proliferation, differentiation, mechanical support, and morphogenesis are implicated in converting the initial unicellular zygote into a multicellular and multiorganismal embryo and fetus.

The identification of specific genes in a number of different diseases and the rapidly accelerating ability to identify defects underlying dysmorphic syndromes allow a whole new level of understanding of their genesis (Thesleff, 1998; Sperber, 1999; Gorlin et al., 2001). Fundamental insights into pathophysiology, diseases, and dysmorphology are being revealed by molecular cell biology (Hart et al, 2000; Young et al., 2000).

The prenatal imaging capabilities of embryoscopy, fetoscopy, and ultrasonography have vaulted gestational developmental phenomena into the field of concern of

Understanding Craniofacial Anomalies: The Etiopathogenesis of Craniosynostoses and Facial Clefting, Edited by Mark P. Mooney and Michael I. Siegel, ISBN 0-471-38724-x Copyright © 2002 by Wiley-Liss, Inc.

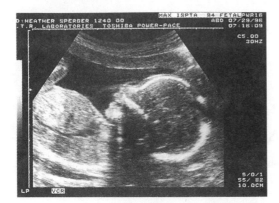

Figure 4.1   Ultrasonograph of head and thorax of a normal 18-week-old female fetus in utero.

the clinician (Banacerraf and Mulliken, 1993; Pretorius and Nelson, 1994; Thieme et al., 2000; Wong et al., 2001) (Fig. 4.1). Fetology is rapidly becoming a clinical discipline, and the fetus has become a potential patient. Chorionic villus sampling and amniocentesis provide advance information on the chromosomal and genetic status of the developing fetus. Reparative fetal surgery for congenital anomalies is now feasible.

4.2   EARLY DEVELOPMENT

Basic to an understanding of adult anatomy is an insight into the morphogenesis occurring in the embryo and fetus prior to birth and into the continuing postnatal development of the craniofacial complex in the infant and child. Comprehension of the manifold changes undergone by organ systems during their development will lead to an understanding of the dysmorphology exhibited in many syndromes of congenital anomalies that the clinician is called on to diagnose and treat.

The 3 billion base pairs of nucleotides forming DNA molecules in the human genome, consisting of a haploid number of 23 chromosomes, contain between 30,000 and 40,000 protein-coding alleles of genes (International Human Genome Sequencing Consortium, 2001). Every gene may produce one or more transcription products that are translated into structural, regulatory, or enzymatic proteins. The genes provide the blueprint for approximately a million proteins that form the estimated 10^{14} cells of approximately 220 different types in the human body (Nature Genome Information Centre, 2000: see http://www.nature.com/genomics/ and http://genetics.nature.com). The tissue types combine to create the nearly 4000 named structures in human anatomy. Genes encode proteins, which then engage in intricate interactions through signal transductions to differentiate cells into tissues. The selective expression of subsets of genes through transcription into proteins, which create structural elements of cells and polypeptides that form growth factors and morphogens, accounts for the mechanisms of differentiation of the early pluripotential cells into the myriad of cell types found in later development (Fig. 4.2). Genetic control of development is effected through signaling information (Francis-West et al., 1998; Massugué and Chen, 2000). Identification of these products of gene expression is revealing the complexity of control of cell differentiation, survival, migration, and death (apoptosis). Cell migrations and cell fusions (fusomorphogenesis) are fundamental to organogenesis (Shemer and Podbilewicz, 2000). A representative sample of the increasing number of growth factors is listed in Table 4.1.

Critical to the process of development and morphogenesis are concentration gradients and diffusion patterns of growth factors, selective permeability of cell membranes, and intercellular communication. The density of cells, a factor of their mitotic rate and adhesiveness, governed by receptor glycoproteins on cell membranes, determines the aggregation of cell types into

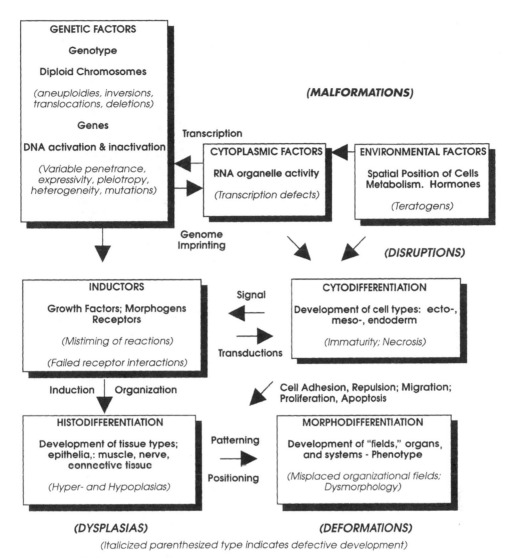

Figure 4.2   Schema of embryogenesis. (*Source*: From Sperber, 2001.)

tissues and their subsequent development into organs.

Cell-cell and cell-extracellular matrix interactions are essential elements in the control of development. Growth factors stimulate cell proliferation and differentiation by acting through specific receptors on responsive cells. Most of these growth factors are present and active throughout life, assuming different roles at different times, displaying remarkable conservation of mechanism and function. Growth factors play analogous roles in embryogenesis, in the immune system, and during inflammation and healing, giving rise to the concept of ontogenic inflammation, by which normal embryonic development may act as a prototype model for homeostasis and healing in the adult (Ferguson et al., 1999).

Pattern formation in the embryo,

TABLE 4.1 Signaling and Growth Factors

Factor	Abbreviation	Derivation	Action
Bone morphogenetic proteins	BMPs (1–7)	Pharyngeal arches; frontonasal mass	Mesoderm induction; dorsoventral organizer; skeletogenesis; neurogenesis
Brain-derived neurotrophic factor	BDNF	Neural tube	Stimulates dorsal root ganglia anlagen
Epidermal growth factor	EGF	Various organs; salivary glands	Stimulates proliferation and differentiation of many cell types
Fibroblastic growth factors (1–19)	FGFs	Various organs and organizing centers	Neural and mesoderm induction; stimulates proliferation of fibroblasts, endothelium, myoblasts, and osteoblasts
Hepatocyte growth factor	HGF	Pharyngeal arches	Cranial motor axon growth; angiogenesis
Homeodomain proteins	Hox-a, Hox-b, Pax	Genome	Craniocaudal and dorsoventral patterning
Insulin-like growth factors 1 and 2	IGF-1, IGF-2	Sympathetic chain ganglia	Stimulates proliferation of fat and connective tissues and metabolism
Interleukin-2 Interleukin-3 Interleukin-4	IL-2 IL-3 IL-4	White blood cells	Stimulates proliferation of T lymphocytes; hematopoietic growth factor; B-cell growth factor
Lymphoid enhancer factor 1	Lef1	Neural crest; mesencephalon	Regulates epithelial-mesenchymal interactions
Nerve growth factor	NGF	Various organs	Promotes axon growth and neuron survival
Platelet-derived growth factor	PDGF	Platelets	Stimulates proliferation of fibroblasts, neurons, smooth muscle cells, and neuroglia
Sonic hedgehog	Shh	Various organs	Neural plate and craniocaudal patterning, chondrogenesis
Transcriptional factors	TFs	Intermediate gene in mesoderm induction casade	Stimulates transcription of actin gene
Transforming growth factor-α	TGF-α	Various organs	Promotes differentiation of certain cells
Transforming growth factor-β (activin A, activin B)	TGF-β	Various organs	Mesoderm induction; potentiates or inhibits responses to other growth factors
Vascular endothelial growth factor	VEGF	Smooth muscle cells	Stimulates angiogenesis
Wingless	Wnt	Genome	Pattern formation; organizer

whereby specific cell types are generated at appropriate locations and shape specification, is determined by homeobox (Hox) genes that regulate other genes involved more directly in cell proliferation, migration, and differentiation (Sharpe, 1995). The position-specifying Hox genes account for segmental development of the craniofacial complex, and by their identification, anomalous development may be traced to specific deficiencies of these genes. The three primary germ layers, ectoderm, mesoderm, and endoderm, which form the early embryonic germ disc, are the bases of all subsequent tissue and organ formation (Table 4.2; Figs. 4.3 and 4.4). The intermediate mesodermal germ layer is deficient in two locations in the trilaminar germ disc: at the prechordal plate cranially and at the cloacal plate caudally, demarcating the sites of the future entrance and exit of the gut (Fig. 4.5). The persistent contact of ectoderm and endoderm at these two locations determines the future dissolution of these two plates, which are denoted as the oropharyngeal and cloacal membranes, respectively, and perforate at the mouth and anus. The oropharyngeal membrane sinks into a central depression, the stomodeum, around which the future face is built.

The ectoderm of the germ disc forms a neural plate that elevates along the axis of the elongating disc to create bilateral neural folds. These folds conjoin at multiple sites in the median plane to form the neural tube—a process termed *neurulation* (Streit et al., 2000; Nakatsu et al., 2000; Colas and Schoenwolf, 2001). The homeobox gene Pax 6 functions as a master regulator gene for initiating eye formation. A single optic primordium forms at the cranial end of the neural plate. The prechordal plate suppresses, by Sonic hedgehog (Shh) signaling, the median part of the optic primordium so that bilateral primordia form. Lack of suppression by interruption of Shh signals leads to cyclopia (Chiang et al., 1996).

4.3 NEURAL CREST TISSUE

Mesenchyme forming the connective tissues is mesodermal in origin in the postcranial region, but peculiarly in the head, most mesenchyme arises from neural crest cells—ectodermal cells located at the margins of the neural folds in the transition zone between the neuroectoderm and the epidermis (Garcia-Castro and Bronner-Fraser, 1999; La Bonne and Bronner-Fraser, 1999). Epitheliomesenchymal transformation is a key factor in embryogenesis, and its control accounts for precision of facial molding, or lack thereof in malformations (Boshart et al., 2000). During neural tube closure, these crest cells migrate into the underlying tissues as mesenchyme (hence ectomesenchyme), forming a lineage of pluripotential stem cells that give rise to diverse tissues (Chai et al., 2000; Perris and Perissinotto, 2000) (Fig. 4.3C). The embryonic prominences of the face and neck are formed by the migration and proliferation of neural crest tissue. The segmental pattern of migrating neural crest is foreshadowed by hindbrain neuromere segments, termed *rhombomeres*, which are determined by the Hox genes (Krumlauf, 1993; Birgbauer et al., 1995; Müller and O'Rahilly, 1980, 1997) (Fig. 4.6). It is the brain underlying the future face that is a key component of cephalogenesis.

Neural crest tissue from the first two rhombomeres migrates ventrally into the first pharyngeal arch; crest tissue from rhombomere 4 migrates into the second arch, and from rhombomeres 6 and 7, into the third, fourth, and sixth arches. The neural crest tissue of rhombomeres 3 and 5 suffer an apoptopic fate before migrating and does not contribute to the arches. The cephalic neural crest provides the precursors of cartilage, bone, muscles, and connective tissues of the head. The vasculature of the head is derived from mesoderm-derived endothelial precursors, while neural crest provides the pericytes and

TABLE 4.2 Chronology of major features of craniofacial development

Carnegie Stage	Postconception Age (days)	Craniofacial Features
6	14	Primitive streak appears; oropharyngeal membrane forms
8	17	Neural plate forms
9	20	Cranial neural folds elevate; otic placode appears
10	21	Neural crest migration commences; fusion of neural folds; otic pit forms
11	24	Frontonasal prominence swells; first arch forms; wide stomodeum; optic vesicles form; anterior neuropore closes; olfactory placodes appear
12	26	Second arch forms; maxillary prominences appear; lens placodes commence; posterior neuropore closes; adenohypophysial pouch appears
13	28	Third arch forms; dental lamina appears; fourth arch forms; oropharyngeal membrane ruptures
14	32	Otic and lens vesicles present; lateral nasal prominences appear
15	33	Medial nasal prominences appear; nasal pits form—widely separated, face laterally
16	37	Nasal pits face ventrally; upper lip forms on lateral aspect of stomodeum; lower lip fuses in midline; retinal pigment forms; nasolacrimal groove appears, demarcating nose; neurohypophysial evagination
17	41	Contact between medial nasal and maxillary prominences, separating nasal pit from stomodeum; upper lip continuity first established; vomeronasal organ
18	44	Primary palate anlagen project posteriorly into stomodeum; distinct tip of nose develops; eyelid folds form; retinal pigment; nasal pits move medially; nasal alae and septum present; mylohyoid, geniohyoid, and genioglossus muscles form
19	47–48	Nasal fin disintegrates (failure of disintegration predisposes to cleft lip); the rima oris of the mouth diminishes in width; mandibular ossification commences
20	50–51	Lidless eyes migrate medially; nasal pits approach each other; ear hillocks fuse
22	54	Eyelids thicken and encroach upon the eyes; the auricle forms and projects; the nostrils are in definitive position
23	56–57	Eyes are still wide apart but eyelid closure commences; nose tip elevates; face assumes a human fetal appearance; head elevates off the thorax; mouth opens; palatal shelves elevate; maxillary ossification commences
Fetus	60	Palatal shelves fuse; deciduous tooth buds form; embryo now termed a fetus

smooth muscle cells of the face and forebrain (Etchevers et al., 2001).

The olfactory, lens, and otic ectodermal placodes are the precursors of the sensory components of the nose, eyes, and inner ears, respectively, around which neural crest tissue builds the superstructures of nasal capsule, sclera, eyelids, and otic capsule. Neural crest gives rise to the main proximal portions of the ganglia of the

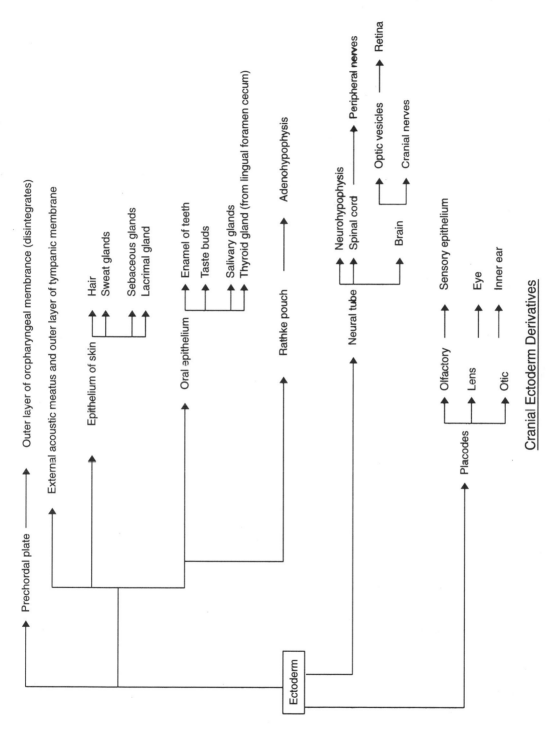

Figure 4.3 Flow charts of derivatives of germ layers. (*Source:* From Sperber, 2001.)

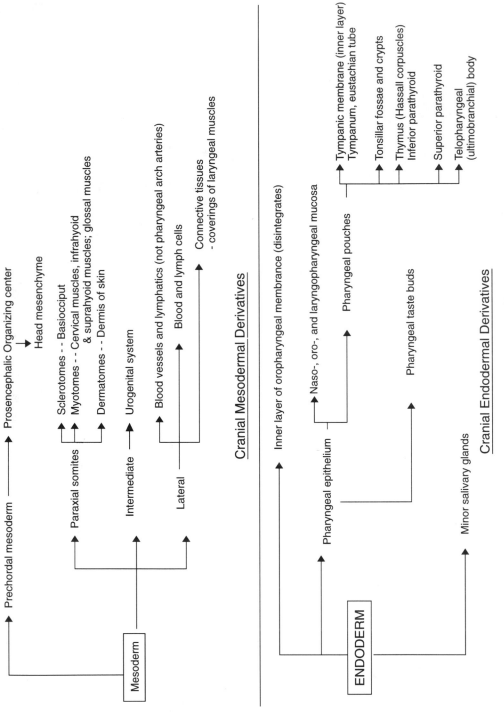

Cranial Mesodermal Derivatives

Cranial Endodermal Derivatives

Figure 4.3 *Continued*

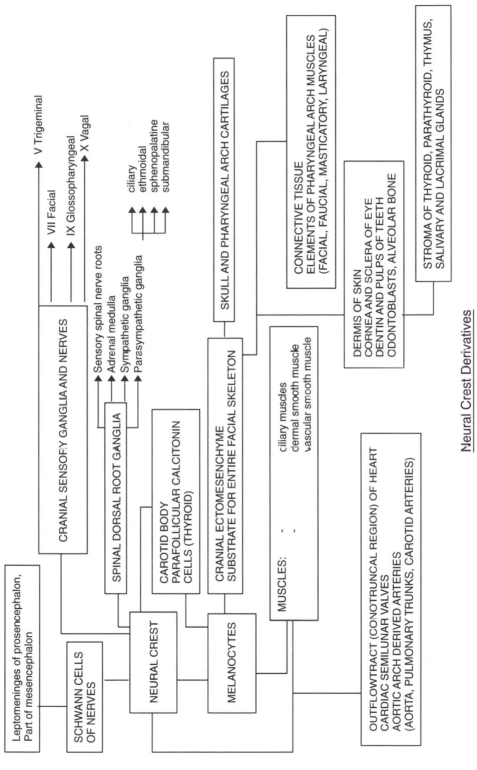

Neural Crest Derivatives

Figure 4.3 *Continued*

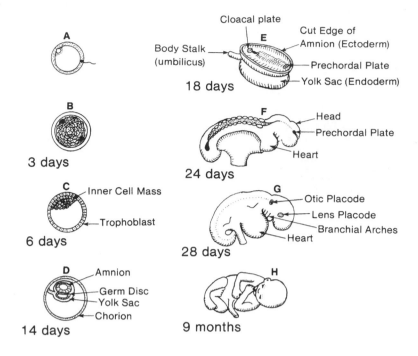

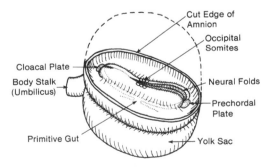

A Spermatozoon penetrating ovum to form zygote	E Primary germ layers forming in germ disc
B Morula stage of blastula	F Somite-stage embryo
C Blastocyst with inner cell mass	G Postsomite-stage embryo
D Fetal membranes in chorion	H Full-term fetus

Figure 4.4 Diagrammatic synopsis of embryogenesis.

Figure 4.5 Diagram of 21-day-old embryonic disc elevating neural folds, encompassed by fetal membranes.

trigeminal (V), facial (VII), glossopharyngeal (IX), and vagus (X) nerves. Epipharyngeal placodes, incorporating ectoderm, contribute to the distal portions of these ganglia (Le Douarin et al., 1986; Baker and Bronner-Fraser, 2001).

Deficiencies in neural crest tissue migration or proliferation account for a wide range of craniofacial malformations, the neurocristopathies, manifested in a variety of syndromes. Abnormalities in form, function, or apoptosis of neural crest cells range from von Recklinghausen's neurofibromatosis through Treacher Collins to DiGeorge and Klein-Waardenburg syndromes (Wong et al., 1999; Wallis and Muenke, 1999; Dixon et al., 2000).

4.4 CRANIOFACIAL DEVELOPMENT

The prior presence of the brain determines the subsequent development of the craniofacies. The rostral parts of the brain—prosencephalon and mesencephalon—are specified by the orthodenticle homologues Otx-1 and Otx-2, while the Hox genes specify the rhombencephalon.

Most of the skeletal and connective tissues of the craniofacial complex and pharyngeal arch apparatus are dependent on mesencephalic and rhombencephalic neural crest tissue migrating as ectomesenchyme into ventral regions of the future skull, face, and neck (Kanzler et al., 2000) (Fig. 4.6). Any defect in the quantity and quality of migrating ectomesenchyme manifests itself in nearly every congenital anomaly observed clinically, ranging from major holoprosencephaly to the most minor clefts of the lips and dimples in the cheeks (Schutte and Murray, 1999). Premature involution of embryonic arteries is another cause of arrested development (Noden, 1991) (Fig. 4.7), leading to dysplasias and dystopias, the abnormal formation and location of structures.

During the critical 21st to 31st days of

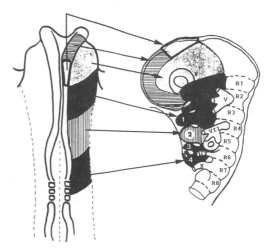

Figure 4.6 Schematic depiction of the fate map of neural crest tissue segments (left) and their facial and pharyngeal arch destinations (right). R1–R8 = rhombomers; 1–4 = pharyngeal arches; V, VII, IX, X = cranial nerves.

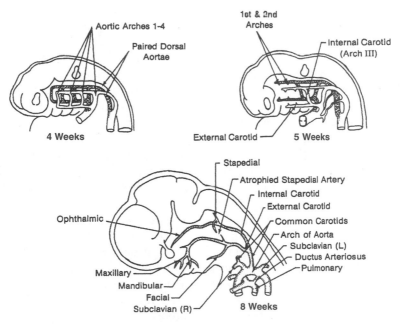

Figure 4.7 Development of cranial arterial system from aortic arches at 4, 5, and 8 weeks.

development, the central stomodeum is surrounded by five prominences formed by underlying neural crest migration and proliferation displaying selective regional characteristics. Ectomesenchyme from the mesencephalic region contributes to the leptomeninges and the frontal and nasal bones. Neural crest cells around the midbrain interact with ectoderm to form part of the trigeminal, facial, glossopharyngeal, vestibulocochlear, and vagal nerve ganglia, joined by neuroblasts from the maxillomandibular placodes (Stark et al., 1997). Lateral evaginations of the diencephalon, the optic vesicles, induce lens placodes in the surface ectoderm, around which neural crest cells migrate to form the scleral and choroid optic coats and the facial frontonasal prominence innervated by the ophthalmic division of the trigeminal nerve. More caudally located migrating neural crest cells encounter pharyngeal endoderm in their ventral migration around the aortic arterial arches, contributing to their walls and forming the mandibular and maxillary prominences and the pharyngeal arches. Neural crest cells deep to the ectodermal otic placode (precursor of the inner ear) join the underlying paraxial mesoderm to form the cartilaginous otic capsule. The otic placodes induce the underlying vestibulocochlear nerves, and if deficient, will result in defects of the inner ear (Groves and Bronner-Fraser, 2000).

Springing from the first pharyngeal arch on each side are the maxillary and mandibular prominences (innervated, respectively, by the maxillary and mandibular divisions of the trigeminal nerve); these prominences form inferior and lateral boundaries to the wide stomodeum, whose superior boundary is the frontonasal prominence (Fig. 4.8). On the inferolateral corners of the frontonasal prominence, bilateral nasal placodes develop, precursors of olfactory epithelium that forms neurons extending into the olfactory bulbs. Defective or absent nasal placodal development will produce anosmia but curiously may have no effect on nasal and central facial development (Braddock et al., 1995).

The nasal placodes appear to sink into the face to form nasal pits that become the anterior nares consequent to the rising elevation of the horseshoe-shaped median and lateral nasal prominences (Fig. 4.8). The posterior aspect of each nasal pit is initially separated from the oral cavity by the oronasal (bucconasal) membrane, which disintegrates by the end of the fifth week to form the posterior choanae. Failure of membrane disintegration leads to choanal atresia, one of the most common congenital nasal anomalies (approximately 1:8000 births), which in 90% of cases is caused by bone rather than membranous blockage (English, 1990). Membranous atresia tends to occur more posteriorly than does bony atresia. As the neonate is an obligatory nose-breather, bilateral choanal atresia frequently leads to fatal asphyxiation at birth.

Elevation of the lateral nasal prominences creates the alae of the nose. Defects of development may be in the midline, producing arrhinia, or a bifid nose, varying from a simple depression to complete separation of both nostrils. Other nasal malformations include degrees of aplasia of the alae, as well as atresia of the nasal fossa(e) (Nishimura, 1993).

The medial tip of the maxillary prominence is initially separated from the inferolateral aspect of the median nasal prominence by an epithelial nasal fin that degenerates, allowing maxillary mesenchyme to merge with median nasal mesenchyme. Persistence of the nasal fin may contribute to clefting of the upper lip and anterior palate.

The initially widely separated median nasal prominences merge in the midline with an intervening premaxillary prominence, from which are derived the tip of the nose, the columella, the philtrum, the labial tuberculum of the upper lip, the frenulum, and the entire primary palate (Sperber,

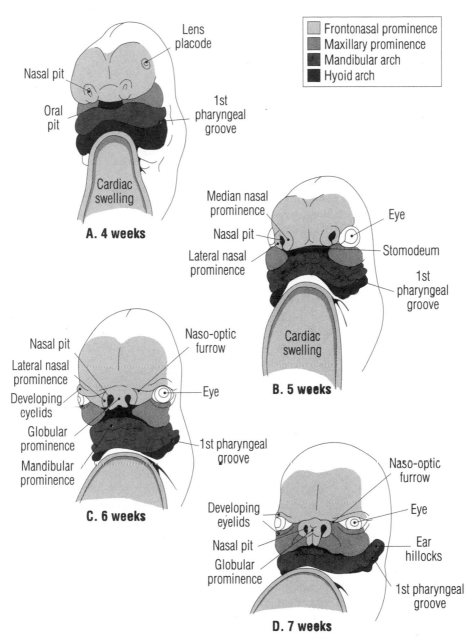

Figure 4.8 Stages of facial formation at (A) 4 weeks, (B) 5 weeks, (C) 6 weeks, and (D) 7 weeks. (*Source*: From Sperber, 2001.)

2001). The central premaxillary prominence is overgrown by the two lateral maxillary prominences to provide continuity to the upper lip, accounting for its maxillary nerve innervation (Viers, 1973). The philtrum and Cupid's bow shape of the upper lip form between the third and fourth intrauterine months (i.e., much later than the melding of the maxillary prominences) as a result of collagen condensation in the midline to produce the philtral groove. The philtrum may be congenitally absent when the upper lip lacks a Cupid's bow outline, as in fetal alcohol syndrome (Autti-Ramo et al., 1992; Astley et al., 1999).

The presence of congenital pits or fistulae, usually bilateral, in the lower lip is indicative of van der Woude syndrome, manifesting clefting of the upper lip and/or palate (Onofre et al., 1997). This autosomal-dominant syndrome has been ascribed to deletion on chromosome 1q32-41 (Bocain and Walker, 1987). Congenital pits of the upper lip are extremely rare (Ozgur and Tuncbilek, 2000).

Clefting of the upper lip (cheiloschisis) is one of the most frequent of all congenital anomalies; its unilateral incidence (usually on the left) varies among different racial groupings, indicating its inherited character: It is highest in frequency among Mongoloid peoples, intermediate in incidence among whites, and least frequent in blacks (varying from 1:500 to 1:2000 births). The anomaly appears more commonly in males and has been ascribed to inadequate neural crest tissue migration to the lip area. The degree of clefting varies enormously; the anomaly is rarely median, a characteristic of a major holoprosencephaly syndrome. Lip clefts may coincidentally be associated with cleft palate, which is inherited separately (Wyszynski, 2002).

The primitive wide stomodeal aperture is reduced by migrating mesenchyme fusing the maxillary and mandibular prominences to form the corners of the definitive mouth. Inadequate ectomesenchyme results in macrostomia (unilateral or bilateral), a form of facial clefting, while excessive neural crest tissue produces microstomia or astomia (Fig. 4.9), which is usually associated with other congenital anomalies (e.g., agnathia and synotia). The lower lip is rarely defective, but if so, it is clefted in its midline (Oostrom et al., 1996).

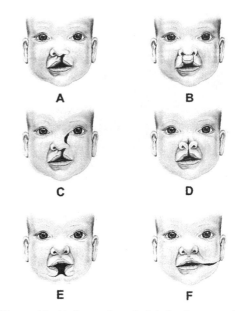

Figure 4.9   Defects of orofacial development: (A) unilateral cleft lip; (B) bilateral cleft lip; (C) oblique facial cleft; (D) median cleft lip and nasal defect; (E) median mandibular cleft; (F) unilateral macrosomia. (*Source*: From Sperber, 2001.)

The occurrence of double lips is often more apparent than real, usually being a redundancy of mucosal lining.

In the facial depression between the lateral nasal prominence and the maxillary prominence on each side, a solid rod of epithelial cells develops that sinks into the subjacent mesenchyme. The ends of the rod establish connections with the nasal pit and the developing conjunctival sac, and subsequently canalize to form the nasolacrimal sac and duct. The line of this duct may persist as an oblique facial cleft of varying degrees of severity on one or both sides of the face (Fig. 4.9).

The migration of the eyes from their initial lateral location toward the midline occurs as a result of a relative narrowing of the intervening frontonasal prominence and expansion of the lateral aspects of the face. Inadequate transition (morphokinesis) of the eyes produces hypertelorism, while their overmigration creates hypotelorism, which in its ultimate expression fuses the eyes into a median cyclopia (O'Rahilly and Müller, 1989) (Fig. 4.10). Rarely, defective optic vesicle formation results in microphthalmos, anophthalmos, or the extremely rare absent lens (aphakia).

The medial epicanthal folds of the

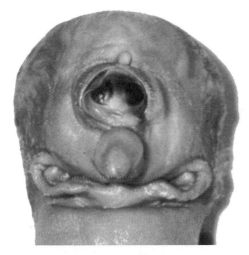

Figure 4.10 Malformed face of aborted human fetus featuring cyclopic eyes, median proboscis, astomia, agnathia, and synotia, in which the external ears are in their embryonic cervical position. The brain is holoprosencephalic. (*Source*: From Sperber, 1989.)

eyelids develop as a variably expressed genetic trait of high frequency in Oriental races, but are strongly expressed in only 3% of Caucasians (Farkas, 1994). Strong expression of the medial epicanthal trait is a feature, among other pathognomic characteristics, of trisomy 21 (Down syndrome), which is represented in 75% of Down syndrome patients (Farkas et al., 1991). The developmental mechanisms of the presence of an extra chromosome 21 in manifesting this and other traits that characterize this syndrome are still to be elucidated. This holds true for other genetically determined and aneuploidy syndromes that produce craniofacial abnormalities.

4.4.1 Adenohypophysial (Rathke) and Seessel Pouches

In the roof of the primitive stomodeum, a deep diverticulum of ectoderm develops that gives rise to adenohypophysis (Dattani and Robinson, 2000). This pouch, extending from the posterior edge of the nasal septum toward the diencephalon of the forebrain, meets its outpouching, the neurohypophysis, to form the fused pituitary gland (Fig. 4.11). The adenohypophysial pouch normally loses its contact with the oral

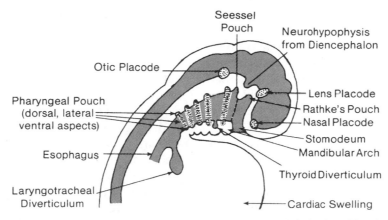

Figure 4.11 Median sagittal section revealing neural tube and pharyngeal derivatives. The site of the disintegrated oropharyngeal membrane is depicted as a dotted line.

ectoderm, but may persist as the cranio-pharyngeal canal, in which cysts or tumors develop in the body of the sphenoid bone. A further small outpouching of the foregut, Seessel's pouch, composed of endoderm, projects toward the brain and is represented in the adult by the pharyngeal bursa, a depression in the nasopharyngeal tonsil.

4.4.2 The Mouth

The initially shallow depression of the stomodeum becomes the familiarly deep oral cavity by the forward growth of the facial prominences surrounding the stomodeum. The stomodeum becomes the common oronasopharyngeal chamber and entrance to the gut by disintegration of the dividing oropharyngeal membrane on the 28th day, providing continuity between the mouth and pharynx. Persistence of the oropharyngeal membrane leads to an extremely rare form of pharyngeal atresia (Chandra et al., 1974).

While the oropharyngeal membrane demarcates the junction of ectoderm and endoderm in the embryo, this line of division is difficult to trace subsequently, owing to extensive changes occurring in oropharyngeal development. Since the hypophysis originates from the adenohypophysial pouch anterior to the oropharyngeal membrane, a measure of the depth of oral growth is provided by its adult location. The presumed site of the original oropharyngeal membrane in the adult is an imaginary oblique plane from the posterior border of the body of the sphenoid bone through the tonsillar region to the sulcus terminalis of the tongue.

The division of the stomodeal chamber into separate oral and nasal cavities occurs with development of the palate and nasal septum. The oral cavity and entire intestinal tract are sterile at birth. As soon as oral feeding commences, a bacterial flora is present that is characteristic of the person throughout life and to which an early resistance to commensal infection is established.

4.4.3 Pharyngeal Arches

Pharyngeal arches are a critical and prominent feature of the developing face. They consist of a series of bulges in the neck region within which skeletal and muscular elements form. These tissues derive from different embryonic cell types resulting from integration of both neural crest dependent and independent patterning mechanisms (Veitch et al., 1999; Graham and Smith, 2001).

Ventrally migrating neural crest cells interact with lateral extensions of pharyngeal endoderm, surround the six aortic arch arteries, and form six pharyngeal arches (Krumlauf, 1993; Wendling et al., 2000). Only five pairs of arches are discernible, however, since the fifth is extremely transitory. Contiguous arches are separated externally by ectodermal grooves and internally by endodermal pharyngeal pouches (Fig. 4.12). Components of these sequentially decreasing arches, forming stacked ring-like collars around the neck, confer an initially repeating pattern that is grossly distorted in subsequent development. This contortion of derivatives of the arches and their internal lining of pharyngeal pouches accounts for the confusing complexity of anomalies of the cervicopharyngeal region that is presented in many syndromes.

Common to the initial arches are a central cartilage rod surrounded by a pharyngeal mesodermal component, accompanied by a nerve element along the aortic arch artery (Caton et al., 2000). From these archetypal structures representing the primitive gill (branchial) system of chordates are derived a bewildering array of adult structures reflecting their evolutionary adaptation to new functions (Fig. 4.13).

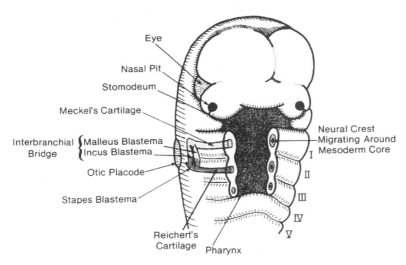

Figure 4.12 Diagrammatic sectioned view of facial and pharyngeal arch development.

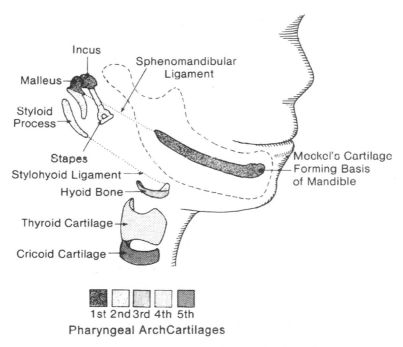

Figure 4.13 Derivatives of the pharyngeal arch cartilages.

The fates of the archetypal elements in each of the pharyngeal arches are outlined in Table 4.3. The consequences of maldevelopment of the derivatives of the pharyngeal system are listed in Table 4.4.

4.4.4 Pharyngeal Pouches and Cervical Grooves

A series of indentations (cervical grooves on the external cervical surface and

TABLE 4.3 Derivatives of the Pharyngeal System

Pharyngeal Arch	Ectodermal Groove	Endodermal Pouch	Skeleton	Viscera	Artery	Muscles	Motor Nerve	Sensory Nerve
1st (Mandibular)	External acoustic meatus: ear hillocks; pinna	Auditory tube; middle ear tympanum	Meckel's cartilage: malleus; incus (mandible template)	Body of tongue	External carotid artery Maxillary artery	Masticatory M Tensor tympani, mylohyoid, anterior digastric	Mandibular Division V Trigeminal	V Lingual nerve
2nd (Hyoid)	Disappears	Tonsillar fossa	Stapes, styloid process; superior hyoid body	Midtongue: thyroid gland anlage; tonsil	Stapedial artery (disappears)	Facial, stapedius, hyoid, posterior digastric	VII Facial	VII Chorda tympani
3rd Arch	Disappears	Inferior parathyroid 3; thymus	Inferior hyoid body, greater cornu hyoid	Root of tongue: fauces; epiglottis; thymus Inferior parathyroid 3	Interior carotid artery	Stylopharyngeus	IX Glossopharyngeal	IX Glossopharyngeal
4th Arch	Disappears	Superior parathyroid 4	Thyroid and laryngeal cartilages	Pharynx; epiglottis Superior parathyroid 4	Aorta (L) Subclavian (R)	Pharyngeal constrictors Levator palatini Palatoglossus Palatopharyngeus Cricothyroid	X Superior laryngeal nerve Vagus	X Auricular nerve to external acoustic meatus
6th Arch	Disappears	Telopharyngeal (ultimopharyngeal) body (cyst) Calcitonin C cells	Cricoid, arytenoid, corniculate cartilages	Larynx	Pulmonary arteries Ductus arteriosus	Laryngeal muscles Pharyngeal constrictors	X inferior laryngeal nerve Vagus	X Vagus
Postpharyngeal region Somites 4 Occipital somites			Tracheal cartilages SCLEROTOMES Basioccipital bone			Trapezius; sternomastoid MYOTOMIC MUSCLE Intrinsic tongue muscles Styloglossus } Extrinsic Hyoglossus } tongue Genioglossus } muscles	XI Spinal accessory XII Hypoglossal	
Prechordal somites Upper cervical somites			Nasal capsule Nasal septum Cervical vertebrae			Extrinsic ocular muscles Geniohyoid; infrahyoid muscles	{ III. Oculomotor { IV Trochlear { VI Abducens Spinal nerves C1, C2	

TABLE 4.4 Possible Anomalies of the Pharyngeal System

Arch	Ectodermal Groove	Endodermal Pouch	Skeleton	Artery	Muscles	Nerve
1st (Mandibular)	Aplasia, atresia, stenosis, duplication of external acoustic meatus	Diverticulum of auditory tube; aplasia, atresia, stenosis of tube	Aplasia/dysplasia of malleus, incus, mandible	Hypoplastic/absent external carotid and maxillary arteries	Deficient masticatory/facial muscles	Absent mandible nerve
2nd (Hyoid)	Cervical (pharyngeal) cleft, sinus, cyst, fistula	Tonsillar sinus, pharyngeal fistula	Aplasia/dysplasia of stapes, styloid process	Persistent stapedial artery	Deficient facial, stapedial muscles	Deficient facial, chorda tympani nerve
3rd Arch	Cervical cleft, sinus, cyst, fistula	Cervical thymus: thymic cyst; aplasia parathyroid 3; aplasia thymus (DiGeorge anomaly)	Defective hyoid bone	Hypoplastic/absent internal carotid artery		Deficient glossopharyngeal nerve
4th Arch	Cervical cleft, sinus, cyst, fistula	Fistula/sinus from pyriform sinus; aplasia parathyroid 4	Congenital laryngeal stenosis, cleft, atresia (Fraser syndrome)	Double aortae: aortic interruption; right aorta	Deficient faucial muscles	Deficient vagus nerve
6th Arch		Aplasia of calcitonin C cells (DiGeorge anomaly)		Aorticopulmonary septation anomalies (DiGeorge anomaly)		

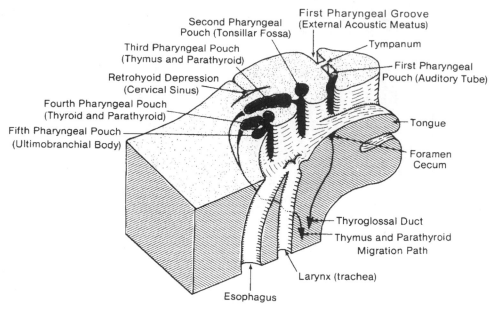

Figure 4.14 Schematic depiction of pharyngeal pouch derivatives and their migration paths.

pharyngeal pouches on the internal pharyngeal surface) delineate the pharyngeal arches (Fig. 4.14). While the cervical grooves are lined by ectoderm, the pharyngeal pouches are lined by endoderm, with an intervening mesodermal layer forming closing membranes between the arches.

At the dorsal ends of the first, second, and fourth cervical grooves, thickening of the ectoderm develops the epipharyngeal placodes, which give rise to the distal portions of the ganglia of the facial, glossopharyngeal, and vagus nerves. Their proximal portions are neural crest derivatives. The first cervical groove deepens to form the external acoustic meatus, meeting the outward extension of the first pharyngeal pouch (auditory tube) at the site of the intervening closing membrane that persists as the tympanic membrane. This membrane is accordingly composed of all three primary germ layers: ectoderm, mesoderm, and endoderm.

The remaining cervical grooves (second through fourth) are obliterated by caudal growth of the second arch, the hyoid operculum, over the collectively sunken grooves that form a retrohyoid depression. Failure to obliterate these cervical grooves completely results in cervical (branchial) fistulae leading from the pharynx to the outside, or a branchial (cervical) sinus or cyst, forming a closed sac (Albers, 1963) (Fig. 4.15). The second pharyngeal pouch is attenuated to form the palatine tonsillar fossa, which is invaded by lymphatic tissue to form the palatine tonsil (Slipka, 1992).

4.4.5 The Skull

Development of the skull reflects its extraordinarily complex origins in accommodating its numerous functions of protecting the brain (neurocranium); housing the special sense organs of sight, sound, smell, and taste; and providing for respiration and mastication (the viscerocranium).

The neurocranium is developmentally divided into the vault or calvaria, formed from membranous bone (the desmocra-

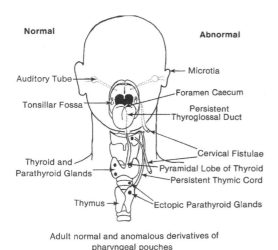

Adult normal and anomalous derivatives of pharyngeal pouches

Figure 4.15 Adult normal and anomalous derivatives of pharyngeal pouches.

Embryonic origins of the skull.

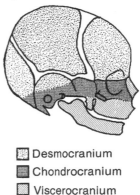

▨ Desmocranium
▨ Chondrocranium
▨ Viscerocranium

Main developmental divisions of the skull

Figure 4.16 Embryonic origins of the skull.

nium), and the basicranium, formed in cartilage (the chondrocranium). The facial skeleton (viscerocranium) ossifies intramembranously (Fig. 4.16). The molecular biology of ossification is extremely complex, involving a host of skeletal growth factors and their receptors, which are currently being elucidated (Canalis, 2000; Goode, 2000).

Initial neurocranial development is dependent on the formation of a membrane surrounding the brain, whose prior existence is essential for normal development. Brain absence (anencephaly) results in acalvaria (Sperber et al., 1986). The surrounding membrane, derived from a combination of mesoderm and neural crest ectomesenchyme, subdivides into an outer ectomeninx and an inner endomeninx. The ectomeninx provides an outer osteogenic layer, in which bones form, and an inner dura mater. The endomeninx subdivides into the outer arachnoid and the inner pia mater.

The ossification centers that develop in the membrane form the frontal, parietal, squamous temporal, and squamous occipital bones. The intervening areas form fibrous sutures and fontanelles, designated anterior, posterior, anterolateral, and posterolateral. The primary chondrogenic centers are the basioccipital, orbitotemporal, otic, and ethmoid regions. Endochondral ossification ensues to form the basicranial bones of the sphenoid, petrous temporal, and basioccipital (Fig. 4.17). Defects of calvarial intramembranous ossification are recognized as cranium bifidum and foramina parietala parmagna (Cargile et al., 2000; Wuyts et al., 2000). Premature closure of the calvarial sutures comprise a constellation of craniosynostoses that are due to genetic mutations of the fibroblast growth factor receptors (FGFR1, 2, 3) and the TWIST and MSX2 genes (Cohen and MacLean, 2000; Warren et al., 2001). They are dealt with in Chapters 6, 9, and 21.

The facial skeleton is subdivided into an upper third, predominantly of neurocranial composition and incorporating the orbits; a middle third, incorporating the nasal complex, maxillae, zygomata, temporal bones, and ears; and a lower third, composed of the mandible. The masticatory apparatus, formed by the jaws, temporomandibular joint, and teeth, is interposed

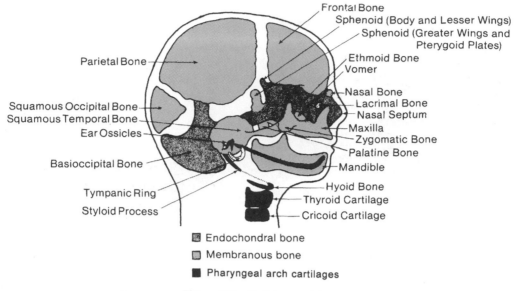

Figure 4.17 Skull bone origins.

between the middle and lower thirds of the face.

The total skull is thus a mosaic of individual components, each having different characteristics of growth, development, maturation, and function. Despite their individuality, each of the units is so integrated with the other that coordination of their growth is required for normal development (Shapiro and Robinson, 1980; Hanken and Hall, 1993). Failure of this correlation, or aberration of inception or faulty growth of an individual component, results in distorted craniofacial relationships reflected in numerous dysmorphic syndromes (Gorski et al., 2000).

Developmental defects of the face and jaws are relatively common, whereas congenital defects of the skull base and sense capsules (nasal and otic) are relatively rare. During postnatal growth of congenital craniofacial defects, three general patterns of development may occur: maintenance of the defective growth pattern; catch-up growth, minimizing the defect; or marked worsening of the derangement with increasing age.

Of great significance to facial development is the normal closing of the foramen cecum in the anterior cranial fossa at the ethmofrontal junction. Abnormal patency of this foramen allows a pathway for neural tissue to herniate into the nasal region, providing the basis of encephaloceles, gliomas, and dermoid cysts that cause gross disfigurement of facial features (Fig. 4.18) (David, 1993; Hoving, 1993). Conversely, congenital invaginations of the face constitute dermal sinuses and fistulas that often exist owing to cranial bone deficiencies (Sessions, 1982).

Afflictions of cartilage growth producing a reduced cranial base result in a dished deformity of the middle third of the facial skeleton and a rounding or brachycephalization of the neurocranium. Such diverse conditions as achondroplasia, cretinism, and Down syndrome (trisomy 21) all produce a similar characteristic facial malformation by virtue of their inhibiting effect on chondrocranial growth (Kjaer et al., 1994, 1999). Of great significance in determining normal facial growth is the nasal septal cartilage (Fig. 4.19), which,

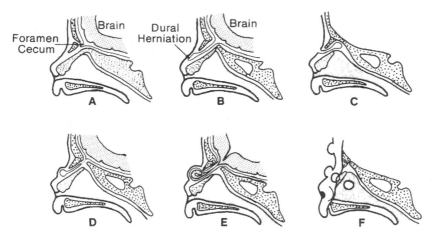

Figure 4.18 Patterns of meningoencephalocele herniation through the foramen cecum creating facial anomalies: (A) embryonic stage, dura within the foramen cecum; (B) dural herniation through the foramen cecum; (C) dermoid sinus through foramen cecum; (D) dermoid cyst with stalk through foramen cecum; (E) encephalocele through foramen cecum; (F) cyst sites in frontonasal region.

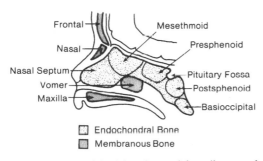

Figure 4.19 Fetal basichondrocranial cartilages and adjoining membrane bones.

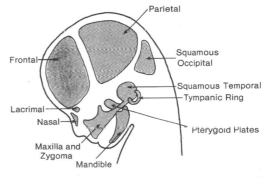

Figure 4.20 Ossification sites of membranous skull bones.

if deficient, adversely influences nasal and midfacial morphology.

Other skeletal facial defects are the result of neural crest tissue deficiencies (clefts, mandibulofacial dysostosis), morphokinetic arrest or overmigration (hypertelorism and retrognathism), amniotic bands (furrows, clefts), or hemorrhaging from fetal blood vessels (microtia, hemifacial microsomia) (Melnick and Jorgenson, 1979).

Facial bones develop intramembranously from ossification centers in the neural crest mesenchyme of the facial

prominences. During the third intrauterine month, centers appear for the frontal, nasal, lacrimal, palatine, zygomatic, and maxillary bones (Fig. 4.20) (Kjaer, 1989; Kjaer et al., 1999; Sandikcioglu et al., 1994). Separate membranous centers appear for the medial and lateral pterygoid plates, the pterygoid hamulus, and the greater wing of the sphenoid bone, apart from its endochondral centers in the basicranial cartilage. The squamous portion of the temporal bone ossifies intramembranously from a single center, while the tympanic ring arises from four centers, and the styloid process develops endochondrally from the second arch cartilage. The mandible ossifies intramembranously from a single center on each side (Merida-Velasco et al., 1993).

4.4.6   The Palate

Three elements make up the secondary definitive palate: the two lateral palatal processes projecting into the stomodeum from the maxillary prominences and the primary palate derived from the premaxillary (frontal) prominence. These elements are initially widely separated owing to the advancing edges of the lateral palatal processes being deflected down on either side of the tongue, which occupies most of the stomodeal chamber (Kjaer, 1997). Concomitantly, the cartilaginous nasal septum is descending from the roof of the stomodeum as a feature of nasal capsular development (Fig. 4.21).

As a result of growth of the stomodeum at the beginning of the fetal period (eighth week), the tongue drops below the level of the bases of the lateral palatal processes and the primary palate. Mouth-opening reflexes and extrinsic tongue muscle activity are implicated in the withdrawal of the tongue from between the vertical shelves. Absence of functioning of the hyoglossus muscle that depresses the tongue, resulting from Hoxa gene mutations, prevents palatal shelf lifting and consequent clefting (Barrow and Capecchi, 1999). In dramatic fashion, in a very short period the unencumbered vertical palatal processes flow like a wave into the horizontal plane, enabling them to establish contact with each other in the midline, with the primary palate anteriorly, and with the lower edge of the nasal septum. Fusion of the palatal

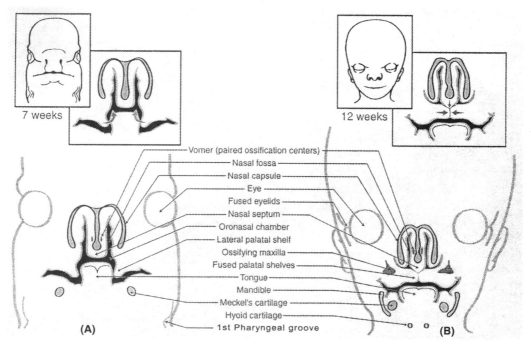

Figure 4.21   Coronal sections of head at 7 weeks (A); 12 weeks (B) revealing nasal capsule and palate development.

shelves is dependent on epithelial-mesenchymal transformation (Lavrin and Hay, 2000). A number of genes (Hoxa1, Hoxa2) and signaling factors (TGFB3, EGF, matrix metalloproteinases) are involved in the molecular and mechanical processes of palatal fusion (Kaartinen et al., 1997; Cui et al., 1998; Miettinen et al., 1999; Morris-Wiman et al., 2000; see also Chapters 5, 7, and 21).

Ossification commences anteriorly very soon after palatal fusion to form the hard palate, comprised of the palatine processes of the maxillae, disputed premaxillary bones, and the horizontal palates of the palatine bones (Kjaer, 1992; Sildu, 1994). The unossified posterior soft palate and uvula are invaded by mesenchyme from the first and fourth pharyngeal arches (Cohen et al., 1993) to provide the tensor palatini, levator palatini, and uvular muscles supplied, respectively, by the nerves of these arches, the mandibular division of the trigeminal and vagus nerves. Congenital palsy of these muscles and cleft palate abnormalities are thus of complex embryologic origin (Cohen et al., 1994).

Clefting of the palate is one of the most frequent congenital malformations, the basis of which can be traced directly to the complex embryology of the palate. All degrees of clefting may occur, ranging from the nondysfunctional submucous cleft to the major incapacitating forms of combined cheilouranoschisis (Fig. 4.22). Palatal clefting is multifactorial in its etiology despite its hereditary familial associations (Hibbert and Field, 1996).

Bifid uvula, the least severe form of clefting, occurs relatively frequently but is usually clinically insignificant. Increasingly severe soft and hard palate clefts advance anteriorly to involve the nasal fossa(e), incurring speech and swallowing defects. Hard palate clefts may be unilateral or bilateral with respect to the nasal fossae. Nasal septal attachment or detachment from the palate has clinical consequences not only for feeding and breathing but for growth of the palate and middle third of the face.

Severe palatal clefting that is deflected to the left or right of the premaxilla (or both) may be coincidentally associated with clefts of the lip, compounding the clinical problems. Palatal clefting has a pattern of familial inheritance, has a greater frequency in females, and is a feature of a great number of congenital defect syndromes, among which are mandibu-

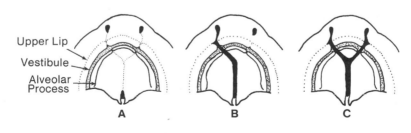

Cleft Palate Variations

A Bifid uvula
B Unilateral cleft palate and lip
C Bilateral cleft palate and lip

Figure 4.22 Cleft palate variations: (A) bifid uvula; (B) unilateral cleft palate and lip; (C) bilateral cleft palate and lip.

lofacial dysostosis (Treacher Collins syndrome), micrognathia (Pierre Robin sequence), orodigitofacial dysostosis, and holoprosencephaly.

REFERENCES

Albers, G. D. (1963). Branchial anomalies. *J. Am. Med. Assoc. 183*, 399.

Astley, S. J., Magnuson, S. I., Omnell, L. M., and Clarren, S. K. (1999). Fetal alcohol syndrome changes in craniofacial form with age, cognition and timing of ethanol exposure in the macaque. *Teratology 59*, 163–172.

Autti-Ramo, I., Gaily, E., and Granstrom, M. L. (1992). Dysmorphic features in offspring of alcoholic mothers. *Arch. Dis. Child 67*, 712–716.

Baker, C. V. H., and Bronner-Fraser, M. (2001). Vertebrate cranial placodes. I. Embryonic induction. *Devel. Biol. 232*, 1–61.

Banacerraf, B. R., and Mulliken, J. B. (1993). Fetal cleft lip and palate: sonographic diagnosis and postnatal outcome. *Plast. Reconstr. Surg. 92*, 1045–1051.

Barrow, J. R., and Capecchi, M. (1999). Compensatory defects associated with mutations in Hoxa1 restore normal palatogenesis in Hoxa2 mutants. *Development 126*, 5011–5026.

Birgbauer, E., Sechrist, J., Bronner-Fraser, M., and Fraser, S. (1995). Rhombomeric origin and rostrocaudal reassortment of neural crest cells revealed by intravital microscopy. *Development 121*, 935–945.

Bocain, M., and Walker, A. P. (1987). Lip pits and deletion 1q32–41. *Am. J. Med. Genet. 26*, 437–443.

Boshart, L., Vlot, C., and Vermeij-Keers, C. (2000). Epithelio-mesenchymal transformation in the embryonic face. *Eur. J. Plast. Surg. 23*, 217–223.

Braddock, S. R., Grafe, M. R., and Jones, K. L. (1995). Development of the olfactory nerve: its relationship to the craniofacies. *Teratology 51*, 252–256.

Canalis, E. (ed.) (2000). *Skeletal Growth Factors* (Philadelphia: Lippincott Williams & Wilkins).

Cargile, C. B., McIntosh, I., Clough, M. V., Rutberg, J., Yaghmai, R., Goodman, B. K., Chen, X. N., Korenberg, J. R., Thomas, G. H., and Geraghty, M. T. (2000). Delayed membranous ossification of the cranium associated with familial translocation (2;3) (p15;q12). *Am. J. Med. Genet. 92*, 328–335.

Caton, A., Hacker, A., Naeem, A., Livet, J., Maina, F., Bladt, F., Klein, R., Birchmeir, C., and Guthrie, S. (2000). The branchial arches and HGF are growth-promoting and chemo-attractant for cranial motor axons. *Development 127*, 1751–1766.

Chai, Y., Jiang, X., Ito, Y., Bringas, P., Jr., Han, J., Rowitch, D. H., Soriano, P., McMahon, A. P., and Sucov, H. M. (2000). Fate of the mammalian cranial neural crest during tooth and mandibular morphogenesis. *Development 127*, 1671–1679.

Chandra, R., Yadava, V. N., and Sharma, R. N. (1974). Persistent buccopharyngeal membrane. *Plast. Reconstr. Surg. 54*, 678.

Chiang, C., Litingtung, Y., Lee, E., Young, K. E., Corden, J. L., Westphal, H., and Beachy, P. A. (1996). Cyclopia and defective axial patterning in mice lacking Sonic hedgehog gene function. *Nature 383*, 407–413.

Cohen, M. M., Jr., and MacLean, R. E. (eds.) (2000). *Craniosynostosis: Diagnosis, Evaluation and Management*, 2nd ed. (Oxford: Oxford University Press).

Cohen, S. R., Chen, L., Trotman, C. A., and Burdi, A. R. (1993). Soft palate myogenesis: a developmental field paradigm. *Cleft Palate Craniofac. J. 30*, 441–446.

Cohen, S. R., Chen, L. L., Burdi, A. R., and Trotman, C. A. (1994). Patterns of abnormal myogenesis in human cleft palate. *Cleft Palate Craniofac. J. 31*, 345–350.

Colas, J.-F., and Schoenwolf, G. C. (2001). Towards a cellular and molecular understanding of neurulation. *Devel. Dyn. 221*, 117–145.

Cui, X.-M., Warburton, D., Zhao, J., Crowe, D. L., and Shuler, C. F. (1998). Immunohistochemical localization of TGF—type II cell receptor and TGF3 during palatogenesis in vivo and in vitro. *Int. J. Devel. Biol. 42*, 817–820.

Dattani, M. T., and Robinson, I. C. (2000). The molecular basis for developmental disorders of the pituitary gland in man. *Clin. Genet. 57,* 337–346.

David, D. J. (1993). Cephaloceles: classification, pathology and management—a review. *J. Craniofac. Surg. 4,* 192.

Dixon, J., Brakebusch, C., Fässler, R., and Dixon, M. J. (2000). Increased levels of apoptosis in the prefusion neural folds underlie the craniofacial disorder, Treacher-Collins syndrome. *Hum. Molec. Genet. 9,* 1473–1480.

English, G. M. (1990). Congenital anomalies of the nose, nasopharynx and paranasal sinuses. In G. M. English (ed.), *Otolaryngology* (Philadelphia: J. B. Lippincott).

Etchevers, H. C., Vincent, C., Le Douarin, N. M., and Couly, G. F. (2001). The cephalic neural crest provides pericytes and smooth muscle cells to all blood vessels of the face and forebrain. *Development 128,* 1059–1068.

Farkas, L. G. (ed.) (1994). *Anthropometry of the Head and Face,* 2nd ed. (New York: Raven Press).

Farkas, L. G., Posnick, J. C., and Hreczko, T. (1991). Anthropometry of the head and face in 95 Down syndrome patients. In C. J. Epstein (ed.), *The Morphogenesis of Down Syndrome* (New York: Wiley-Liss), pp. 53–97.

Ferguson, C., Alpern, E., Miclau, T., and Helms, J. A. (1999). Does adult fracture repair recapitulate embryonic skeletal formation? *Mech. Devel. 87,* 57–66.

Francis-West, P. H., Ladher, R., Barlow, A., and Graveson, A. (1998). Signalling interactions during facial development. *Mech. Devel. 75,* 3–28.

Garcia-Castro, M., and Bronner-Fraser, M. (1999). Induction and differentiation of the neural crest. *Curr. Opin. Cell Biol. 11,* 695–698.

Goode, J. (ed.) (2000). *The Molecular Basis of Skeletogenesis,* Novartis Foundation, vol. 232 (Chichester, UK: Wiley).

Gorlin, R. J., Cohen, M. M., Jr., and Hennekam, R. C. M. (2001). *Syndromes of the Head and Neck,* 4th ed. (Oxford: Oxford University Press).

Gorski, J. L., Estrada, L., Hu, C., and Liu, Z. (2000). Skeletal-specific expression of Fgd1 during bone formation and skeletal defects in faciogenital dysplasia (FGDY: Aarskog syndrome). *Devel. Dyn. 218,* 573–586.

Graham, A., and Smith, A. (2001). Patterning the pharyngeal arches. *Bioessays 23,* 54–61.

Groves, A. K., and Bronner-Fraser, M. (2000). Competence, specification and commitment in otic placode induction. *Development 127,* 3489–3499.

Hanken, J., and Hall, B. K. (eds.) (1993). *The Skull. Vol. 1. Development* (Chicago: University of Chicago Press).

Hart, T. C., Marazita, M. Z., and Wright, J. T. (2000). The impact of molecular genetics on oral health paradigms. *Crit. Rev. Oral Biol. Med. 11,* 26–56.

Hibbert, S. A., and Field, J. K. (1996). Molecular basis of familial cleft lip and palate. *Oral Dis. 2,* 238–241.

Hoving, E. W. (1993). Frontoethmoidal encephaloceles. A study of their pathogenesis. Doctoral dissertation, Rijksuniversiteit, Groningen, Netherlands.

International Human Genome Sequencing Consortium (2001). Initial sequencing and analysis of the human genome. *Nature 409,* 860–921.

Kaartinen, V., Cui, X. M., Heisterkamp, N., Groffen, J., and Shuler, C. F. (1997). TGF3 regulates transdifferentiation of medial edge epithelium during palatal fusion and associated degradation of the basement membrane. *Devel. Dyn. 209,* 255–260.

Kanzler, B., Foreman, R. K., Labosky, P. A., and Mallo, M. (2000). BMP signaling is essential for development of skeletogenic and neurogenic cranial neural crest. *Development 127,* 1095–1104.

Kjaer, I. (1989). Prenatal skeletal maturation of the human maxilla. *J. Craniofac. Genet. Devel. Biol. 9,* 257.

Kjaer, I. (1992). Human prenatal palatal shelf elevation related to craniofacial skeletal maturation. *Eur. J. Orthod. 14,* 26.

Kjaer, I. (1997). Mandibular movements during elevation and fusion of palatal shelves evaluated from the course of Meckel's cartilage. *J. Craniofac. Genet. Devel. Biol. 17,* 80–85.

Kjaer, I., Keeling, J. W., and Graem, N. (1994). Cranial base and vertebral column in human anencephalic fetuses. *J. Craniofac. Genet. Devel. Biol. 14*, 235–244.

Kjaer, I., Keeling, J. W., and Fischer-Hansen, B. (1999). *The Prenatal Human Cranium— Normal and Pathologic Development* (Copenhagen: Munksgaard).

Krumlauf, R. (1993). Hox genes and pattern formation in the branchial region of the vertebrate head. *Trends Genet. 9*, 106–112.

La Bonne, C., and Bronner-Fraser, M. (1999). Molecular mechanisms of neural crest formation. *Ann. Rev. Cell Devel. Biol. 15*, 81–112.

Lavrin, I. G., and Hay, E. D. (2000). Epithelial-mesenchymal transformation, palatogenesis and cleft palate. *Angle Orthod. 70*, 181–182.

Le Douarin, N. M., Fontaineperus, J., and Couly, G. (1986). Cephalic ectodermal placodes and neurogenesis. *Trends Neurosci. 9*, 175–180.

Massagué, J., and Chen, Y.-G. (2000). Controlling TGF-β signaling. *Genes Devel. 14*, 627–644.

Melnick, M., and Jorgenson, R. (eds.) (1979). *Developmental Aspects of Craniofacial Dysmorphology* (New York: Alan R. Liss).

Merida-Velasco, J. A., Sanchez-Montesinos, I., Espin-Ferra, J., Garcia-Garcia, J. D., and Roldan-Schilling, V. (1993). Developmental differences in the ossification process of the human corpus and ramus mandibulae. *Anat. Rec. 235*, 319–324.

Miettinen, P.J., Chin, J. R., Shum, L., Slavkin, H. C., Shuler, C. F., Derynck, R., and Werb, Z. (1999). Epidermal growth factor receptor function is necessary for normal craniofacial development and palate closure. *Nat. Genet. 22*, 69–73.

Morris-Wiman, J., Burch, H., and Basco, E. (2000). Temporospatial distribution of matrix metalloproteinase and tissue inhibitors of matrix metalloproteinases during murine secondary palate morphogenesis. *Anat. Embryol. 202*, 129–141.

Müller, F., and O'Rahilly, R. (1980). The human chondrocranium at the end of the embryonic period proper, with particular reference to the nervous system. *Am. J. Anat. 159*, 33–58.

Müller, F., and O'Rahilly, R. (1997). The timing and sequence of appearance of the neu-romeres and their derivatives in staged human embryos. *Acta Anat. 158*, 83–99.

Nakatsu, T., Uwabe, C., and Shiota, K. (2000). Neural tube closure in humans initiates at multiple sites: evidence from human embryos and implications for the pathogenesis of neural tube defects. *Anat. Embryol. 201*, 455–466.

Nishimura, Y. (1993). Embryological study of nasal cavity development in human embryos with reference to congenital nostril atresia. *Acta Anat. 147*, 140–144.

Noden, D. M. (1991). Development of craniofacial blood vessels. In R. N. Feinberg, G. K. Sherer, and R. Auerbach (eds.), *The Development of the Vascular System. Vol. 14, Issues Biomed.* (Basel: Karger), pp. 1–24.

Onofre, M. A., Brosce, H. B., and Taga, R. (1997). Relationship between lower lip fistulae and cleft lip and/or palate in Van der Woude syndrome. *Cleft Palate Craniofac. J. 34*, 261–265.

Oostrom, C. A. M., Vermij-Keers, C., Gilbert, P. M., and Van der Meulen, J. C. (1996). Median cleft of the lower lip and mandible: case reports, a new embryologic hypothesis and subdivision. *Plast. Reconstr. Surg. 97*, 313–319.

O'Rahilly, R., and Müller, F. (1989). Interpretation of some median anomalies as illustrated by cyclopia and symmelia. *Teratology 40*, 409–421.

Ozgur, F., and Tuncbilek, G. (2000). Bilateral congenital pits of the upper lip. *Ann. Plast. Surg. 45*, 658–661.

Perris, R., and Perissinotto, D. (2000). Role of the extracellular matrix during neural crest migration. *Mech. Devel. 95*, 3–21.

Pretorius, D. H., and Nelson, T. R. (1994). Prenatal visualization of cranial sutures and fontanelles with three-dimensional ultrasonography. *J. Ultrasound Med. 13*, 871–876.

Sandikcioglu, M., Molsted, K., and Kjaer, I. (1994). The prenatal development of the human nasal and vomeral bones. *J. Craniofac. Genet. Devel. Biol. 14*, 124–134.

Schutte, B. C., and Murray, J. C. (1999). The many faces and factors of orofacial clefts. *Hum. Molec. Genet. 8*, 1853–1859.

Sessions, R. B. (1982). Nasal dermal sinuses: new concepts and explanations. Laryngoscope *92*, 1.

Shapiro, R., and Robinson, F. (1980). *The Embryogenesis of the Human Skull* (Cambridge, MA: Harvard University Press).

Sharpe, P. T. (1995). Homeobox genes and orofacial development. *Connec. Tiss. Res. 32*, 17.

Shemer, G., and Podbilewicz, B. (2000). Fusomorphogenesis: cell fusion in organ formation. *Devel. Dyn. 218*, 30–51.

Sildu, A. M. (1994). Prenatal sagittal growth of the osseous components of the human palate. *J. Craniofac. Genet. Devel. Biol. 14*, 252–256.

Slipka, J. (1992). The development and involution of tonsils. *Adv. Otorhinolaryngol. 47*, 1–4.

Sperber, G. H. (1989). *Craniofacial Embryology* (London: Wright).

Sperber, G. H. (1999). Pathogenesis and morphogenesis of craniofacial anomalies. *Ann. Acad. Med. Singapore 28*, 703–713.

Sperber, G. H. (2001). *Craniofacial Development* (Hamilton: B. C. Decker Inc.).

Sperber, G. H., Honore, L. H., and Johnson, E. S. (1986). Acalvaria, holoprosencephaly and facial dysmorphia syndrome. In M. Melnick (ed.), *Current Concepts in Craniofacial Anomalies* (New York: Alan R. Liss), pp. 318–329.

Stark, M. R., Sechrist, J., Bronner-Fraser, M., and Marcelle, C. (1997). Neural tube/ectoderm interactions are required for trigeminal placode formation. *Development 124*, 4287–4295.

Streit, A., Berliner, A. J., Papanayotou, C., Sirulnik, A., and Stern, C. D. (2000). Initiation of neural induction by FGF signaling before gastrulation. *Nature 406*, 74–78.

Thesleff, I. (1998). The genetic basis of normal and abnormal craniofacial development. *Acta Odont. Scand. 56*, 321–325.

Thieme, G., Manco-Johnson, M. L., and Cioffi-Ragan, D. (2000). In obstetrics, 3-D imaging solves clinical problems. *Diag. Imaging Suppl. 3-D Ultrasound*, 8–19.

Veitch, E., Begbie, J., Schilling, T. F., Smith, M. M., and Graham, A. (1999). Pharyngeal arch pattterning in the absence of neural crest. *Curr. Biol. 9*, 1481–1484.

Viers, W. (1973). Transmedian innervation of the upper lip: an embryologic study. *Laryngoscope 83*, 1.

Wallis, D. E., and Muenke, M. (1999). Molecular mechanisms of holoprosencephaly. *Molec. Genet. Metab. 68*, 126–138.

Warren, S. M., Greenwald, J. A., Spector, J. A., Bouletreau, P., Mehrara, B. J., and Longaker, M. T. (2001). New developments in cranial suture research. *Plast. Reconstr. Surg. 107*, 523–540.

Wendling, O., Dennefeld, C., Chambon, P., and Mark, M. (2000). Retinoid signaling is essential for patterning the endoderm of the third and fourth pharyngeal arches. *Development 127*, 1553–1562.

Wong, F. K., Karsten, A., Larson, O., Huggare, J., Hagberg, C., Larsson, C., The, B. T., and Linder-Aronson, S. (1999). Clinical and genetic studies of Van der Woude syndrome in Sweden. *Acta Odont. Scand. 57*, 72–76.

Wong, G. B., Mulliken, J. B., Benacerraf, B. R. (2001). Prenatal sonographic diagnosis of major craniofacial anomalies. *Plast. Reconstr. Surg. 108*, 1316–1333.

Wuyts, W., Reardon, W., Preis, S., Homfray, T., Rasore-Quartino, A., Christians, H., Willems, P. J., and Van Hul, W. (2000). Identification of mutations in the MSX2 homeobox gene in families affected with foramina parietalia permagna. *Hum. Molec. Genet. 9*, 1251–1255.

Wyszynski, D. F. (ed.) (2002). *Cleft Lip and Palate: From Origin to Treatment* (Oxford: Oxford University Press).

Young, D. L., Schneider, R. A., Hu, D., and Helms, J. A. (2000). Genetic and teratogenic approaches to craniofacial development. *Crit. Rev. Oral Biol. Med. 11*, 304–317.

WEB SITES

Web Sites: Hugo Gene nomenclature committee:
http://www.gene.ucl.ac.uk/nomenclature
Human Genome Resources; Genomes Guide:
http://www.ncbi.nlm.nih.gov/genome/guide
Embryo Images:
http://www.med.unc.edu/embryo_images

CHAPTER 5

CRANIOFACIAL EMBRYOGENESIS: ABNORMAL DEVELOPMENTAL MECHANISMS

MALCOLM C. JOHNSTON, D.D.S., Ph.D., and PETER T. BRONSKY, D.M.D., M.S.

5.1 INTRODUCTION

This chapter is primarily concerned with alterations in developmental mechanisms that lead to various craniofacial malformations. Particular emphasis is placed on the human relevance of the studies reviewed.

There have been very rapid advances in our understanding of normal developmental mechanisms, especially at the molecular level. This new information is providing numerous insights into the relationships between development and evolution, which are clarifying abnormal developmental mechanisms. While such new information is widely disseminated throughout this text, certain aspects relevant to abnormal mechanisms are discussed throughout this chapter.

Very early developmental alterations in germ layer formation lead to closely related brain and facial malformations that are sometimes referred to as holoprosencephalies. These include cyclopia, arhinencephaly, certain types of cleft and palate and, apparently, fetal alcohal syndrome (FAS). The apparent role of excessive or premature mesodermal cell death in the

pathogenesis of the otocephalic ("ear head") malformations, including mandibular loss (agnathia), is discussed in this chapter.

Defects in the somewhat complex mechanisms involved in head segmentation, including close integration of brain, branchial arch and neural crest development, may be largely responsible for certain malformations including retinoic acid syndrome (RAS) and hemifacial microsomia. Treacher Collins syndrome appears to be rather different, and evidence for the role of premature cell death in the cranial sensory ganglia in the pathogenesis of the syndrome is presented in this chapter.

Cell death has been noted as being involved in a number of the above malformations. The embryo utilizes cell death as a normal mechanism at numerous points in development. This type of cell death is usually termed *programed cell death* and is sometimes referred to as *apoptosis*. It is involved in segmentation, such as in the head segmentation noted above, and in the formation of digits. It is also involved in the elimination of cells that are no longer needed, as in pharyngeal arch mesodermal

Understanding Craniofacial Anomalies: The Etiopathogenesis of Craniosynostoses and Facial Clefting, Edited by Mark P. Mooney and Michael I. Siegel, ISBN 0-471-38724-x Copyright © 2002 by Wiley-Liss, Inc.

core cells after the surrounding mesenchyme has been vascularized or in sensory ganglion neuroblasts after peripheral sites have been innervated, as well as in many other aspects of development. It is when cell death is excessive or the timing of cell death is abnormal that developmental abnormalities occur, such as in RAS and Treacher Collins animal models.

The possible alterations in growth, morphogenetic movements, and merging and fusion found in the pathogenesis of facial clefting are discussed in this chapter, and limited comparisons are made to similar problems in neural tube defects (NTDs). The proposed mechanisms are explored by which manipulation of environmental factors such as folic acid supplementation may prevent NTDs and facial clefts. Cigarette smoking related to clefts is also discussed. The etiology of many of the more common malformations, such as cleft lip and palate, appear to conform to the multifactorial threshold concept, and this is covered in terms of developmental alterations. Studies of monozygotic (MZ) twins discordant for facial clefts may shed some light on the multifactorial threshold nature of facial clefting.

A number of chapters in this book discuss the involvement, or possible involvement, of various molecular defects including those related to facial clefts and craniosynostoses. The developmental changes by which the altered function of regulatory molecules may lead to these abnormalities will be discussed in this chapter.

5.2 SOME RELEVANT ASPECTS OF CRANIOFACIAL EVOLUTION AND DEVELOPMENT

Since the mid-1980s, the methods of molecular biology have led to rapid progress in our understanding of normal developmental mechanisms that are particularly relevant to the origins of a number of craniofacial malformations. Although many of them are covered in earlier publications of the authors (Johnston and Bronsky, 1995; Johnston, 1997) and others (e.g., Graham, et al., 1996; Kontges and Lumsden, 1996; Hunt et al., 1999; Francis-West et al., 1998; Young et al., 2000), as well as in the current text, a brief review of the molecular regulation of development is undertaken here. Emphasis is placed on those aspects most important to abnormal craniofacial development.

As described in Chapter 4 and later in this chapter, normal development is accomplished by the progressive differentiation of embryonic cells, which is greatly aided by embryonic cell movements and shape changes (morphogenetic movements) that bring groups of cells from one part of the embryo into close proximity to those in other parts. Inductive interactions between these cell groups are of such a nature that the responding cells are capable of some differentiation on their own, even without the continued presence of the inducing tissue. The molecular regulation of such interactions, as well as the mechanisms by which pattern development occurs within a population of cells to give rise to individual structures such as bones, muscles, and teeth, is now beginning to be understood. A large number of signaling molecules have been identified, as well as many of the target genes that mediate their effects (Fig. 5.1). Figure 5.1 is an oversimplified picture of these molecules; while there is much redundancy (e.g., one signaling molecule may substitute for another if one of the group is nonfunctional), the individual members of each group may differ widely in location, function, and time of expression, so combining them may be somewhat misleading.

There are two major types of signaling molecules: (1) those that enter the nucleus and combine with receptors, with the complex acting directly on genes to alter their function, and (2) other molecules that alter gene function indirectly through cell surface receptors. Both types of molecules are illustrated in Figure 5.2. Steroids, thy-

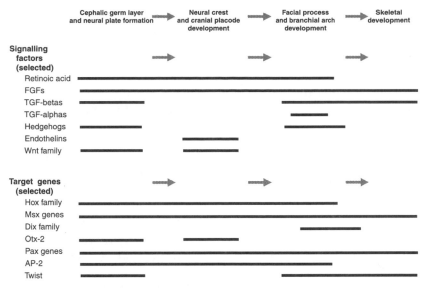

Figure 5.1   Signaling factors and target genes at different stages of development. Both were chosen because of their importance in craniofacial development and human craniofacial malformations. While derived from the same ancestral genes, the individual genes of a family may have different functions, and they act at different locations and times. The chart, however, gives some idea of the location and times that they function.

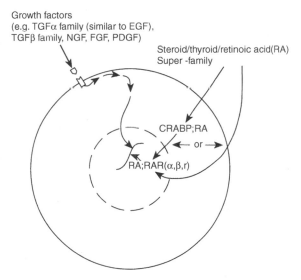

Figure 5.2   Schematic diagram illustrating two main groups of signaling molecules involved in interactions between embryonic cell populations (Source: Johnston, 1993).

roxin, and retinoic acid gain direct access to the nucleus, and, after binding to receptor molecules, the complex binds with DNA to alter gene function, while the growth factors bind to cell surface receptors whose altered configuration initiates a sometimes complex series of molecular interactions that eventually alters gene function. The factors illustrated were among the earliest discovered and most intensively studied. It should be noted that in this chapter we adhere to the convention of capitalizing human signaling molecules and genes and lowercasing those for experimental animals.

The eventual targets of these signaling molecules are frequently homeotic genes that produce transcription factors containing a segment ("box") that binds the factors to target genes in the same cell, thereby permitting them to alter the activity of that gene. As noted in the caption for Figure 5.1, there are families of such genes that are derived from the same ancestral genes. Transcription factors apparently remain within the cell's nucleus.

Duplication and retention of chromosomal material has led to a great deal of redundancy, and only a small amount of genetic material is ever activated. This redundancy has, however, permitted the modification of duplicated genes without threatening the survival of the organism. An example of this duplication and modification is the hox gene family, which produces transcription factors and is located on a single chromosome in the fruit fly, Drosophila, in a contiguous head-to-tail sequence. The same family of genes has been duplicated on four different chromosomes in mammals. The duplicated genes on the different chromosomes, while still involved in head and trunk segmentation, have different functions appropriate to the mammalian embryo. These are considered in some detail later in this chapter.

Similar evolutionary changes in genes regulating the formation of growth factors and their receptors have also occurred, so again there may be closely related molecules with different functions. The names of these growth factors relate primarily to their discovery and usually have very little to do with their embryonic functions. For this reason, only the abbreviated names are used here. Their involvement with cell surface receptors is schematically presented in Figure 5.2. The transforming growth factor A (TGF-A) family is closely related to epidermal growth factor (EGF), and presumably these factors are derivatives of

the same ancestral gene. The TGF-β family, nerve growth factor (NGF), fibroblast growth factor (FGF), and platelet-derived growth factor (PDGF) are similarly related to each other. There are also duplications and modifications to individual factors such as FGF to form families (FGF1, FGF2, etc.).

Another layer of complexity is added by duplication and modification of receptors for signaling molecules. The discovery that specific molecular modifications are related to specific malformations has been an enormous breakthrough. Examples are the modified FGFR2 receptors in Crouzon and Apert syndromes and the modified FGFR3 receptor found in achondroplasia (see Chapters 6 and 21). In addition to providing accurate diagnostic information, these mutations will undoubtedly provide information about both normal and abnormal suture (Crouzon and Apert syndromes) and cartilage formation (achondroplasia), as is discussed later. Msx-1 is involved in another type of craniosynostosis. Abnormal functioning of many of the other signaling factors and homeobox genes illustrated in Figure 5.1 are involved in abnormal development of craniofacial structures (see below).

Many signaling molecules must act through cell surface receptors, such as the growth factors in Figure 5.2. These molecules typically act through very small distances and are often called paracrine factors. They include the bone morphogenetic protein (BMP) factors, which are members of the TGF-B family. One of the most important of the nongrowth factor molecules is the protein sonic hedgehog (shh), which is involved in the patterning of developing structures, including separation of the eye fields through the suppression of Pax-6 in the anterior neural plate, as described below, and in pattern formation in the limb bud and facial processes.

Of the second group of regulatory molecules (Figure 5.2), the corticosteroids and retinoic acid (and other retinoids) have long been known to be potent inducers of craniofacial malformations, and, while less well documented, abnormal levels of thyroxin have also been implicated. Some aspects of their roles in normal development are now beginning to be understood.

Retinoic acid syndrome (RAS) is of very considerable interest in that it permits analysis of the developmental effects of large doses of 13-cis-retinoic acid (13-cis-RA, Accutane), a commonly used drug effective in the treatment of severe cystic acne that has biological effects virtually identical to retinoic acid. Detailed comparisons of the effects of the compound on humans and experimental animals have been possible, permitting insights not only into the pathogenesis of the malformations but also into the role of the compound in normal development. These insights include the role of retinoic acid in the regulation of programed cell death, the alteration of which appears to be a major primary defect in the pathogenesis of RAS and other malformations.

5.3 THE HOLOPROSENCEPHALIES AND GERM LAYER FORMATION

The term *holoprosencephaly* (literally, single-cavity forebrain) describes a grouping of malformations in which there is at least partial failure of formation of the lateral cerebral ventricles (Siebert et al., 1990). The primary defect is underdevelopment of the anterior neural plate and its derived forebrain. Further examination of the developmental origins holoprosencephalies is warranted. This necessitates a brief review of some relevant aspects of germ layer formation.

Germ layer formation is schematically presented in Figure 5.3. It now appears that the entire embryo is derived from the upper layer of the two-layer germ disc (Rosenquist, 1966; Nicolet, 1970; Sulik et al., 1994). Cells in the midline form the primitive streak from which cells intercalate themselves into the lower layer and push it aside to form the endoderm and notochord. Other cells from the primitive streak migrate as individuals into the space between the upper layer and the forming endoderm to form the middle germ layer or mesoderm, while cells remaining on the surface form the ectoderm (Figure 5.3A). The notochord and its anterior extension, the prechordal plate, separate from the endoderm and come to lie in the same plane as the mesoderm (Figure 5.3B).

The notochord (and the prechordal plate) together with the adjacent (paraxial) mesoderm are collectively termed the *chordamesoderm* and this structure is responsible for altering the differentiation of the overlying ectoderm so that it forms the neural plate. As noted in Chapter 4, this induction is a common embryonic phenomenon in which, after the stimulus has passed, the responding tissue is at least capable of some further differentiation on its own. Much of the fundamental work was done many years ago by embryologists who worked primarily with amphibian embryos. A major difference between amphibian and other vertebrate embryos is that the amphibian embryonic cell derives its energy exclusively from its own yolk platelets rather than from an external source such as the yolk sac of fish, reptiles, and birds and the placenta of mammalian embryos. This provided the embryologist with very large and easily manipulated cells and tissues capable of survival in simple salt solutions. Unfortunately, this also gives the amphibian embryo its unusual plump, relatively featureless appearance, which made embryologists working on higher vertebrate slow to accept the results.

The most extensively studied of the holoprosencephaly malformations is cyclopia (single midline eye, see Fig. 5.4D). While cyclopia is rare in humans, its spectacular appearance and wide distribu-

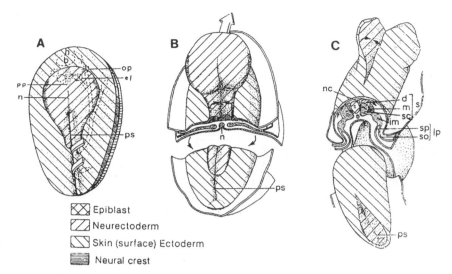

Figure 5.3 Sketches summarizing development from germ layer through neural tube formation. It now appears that only the upper layer (epiblast) of the two-layer embryonic disc will form the embryo, with the lower layer (hypoblast) contributing only to support tissues (e.g., the placenta) of the embryo. In (A), the notochord (n) and its rostral (anterior) extension, the prechordal plate (pp), as well as associated pharyngeal endoderm form as a single layer. Prospective mesodermal cells migrate (arrows in A) through the primitive streak (ps) and insert themselves between the epiblast and endoderm. Epiblast cells remaining on the surface become the ectoderm. Cells of the notochord, prechordal plate and adjacent mesoderm (the chordamesoderm) induce overlying cells to form the neural plate (neuroectoderm). Endoderm anterior to the prechordal plate is involved in the induction of the anterior neural plate. Only later does the notochord separate from the neural plate (B) while folding movements and differential growth (arrows in B and C) continue to shape the embryo. Other abbreviations are as follows: h, heart primordium; b, buccal plate; op, olfactory placode; ef, eye field; nc, neural crest; s, somite and lp, lateral plate; d, dermatome; m, myotome; s, sclerotome. (Source: Modified from Johnston and Sulik, 1979.)

tion in the animal kingdom has drawn considerable attention, and its pathogenesis is relatively well understood. The holoprosencephalies apparently include a much broader spectrum of malformations (Figure 5.4A–D), such as the very common fetal alcohol syndrome (FAS) (Figure 5.4A) at the mild end of the spectrum.

A key observation of the amphibian embryologists was that the notochord and prechordal plate were involved in a bilateralization process that resulted in the neural plate almost completely dividing into two halves, with a thinner floor plate in the midline (see Holtfreter and Hamburger, 1955, for a review). Adelmann (1936) also showed that experimental removal of the prechordal plate resulted in failure of the midline structures of the anterior neural plate, leading to cyclopia similar to the human cyclopia illustrated in Figure 5.4D. The eye fields are in contact in the midline and form a symmetrical midline eye. Recently, it has been shown (Hatta et al., 1991) that a mutation in zebrafish results in failure of the prospective floor plate to respond to the prechordal plate, which leads to cyclopia. Molecular studies have provided some insight into the role of signaling molecules in this bilateralization phenomenon. For example, Chiang and colleagues (1996) have shown that mice lacking the sonic hedgehog gene (shh) demonstrate cyclopia and abnormal axial patterning.

As will be seen below, shh is involved in many aspects of development, including primary palate development. Normally shh is produced by the notochord and suppresses Pax-6 in the medial portion of the anterior neural plate, leading to floor plate

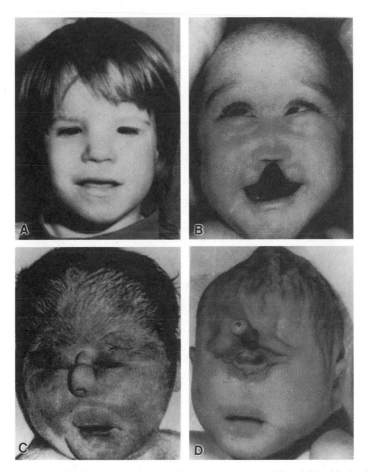

Figure 5.4 A–D Examples of holoprosencephalies of increasing severity. The child with fetal alcohol syndrome (A) apparently is a mild form; arrhinencephaly (B) is of intermediate severity; ethmocephaly (C) is the most severe expression. Eye size apparently decreases with increasing severity in A through C. The single large eye seen in cyclopia perfecta (D) probably arises by a somewhat different mechanism. (Sources: A, courtesy of Sterling Clarren; B, from Ross and Johnston (1972); C, from Taysi and Tinaztepe (1972); and D, from Gorlin et al. (2001).

formation and separation of the eye fields. These malformations are caused by Veratrum and related alkaloids (Keeler and Binns, 1968). Cholesterol is crucial for the synthesis of shh, and alkaloids that resemble cholesterol interfere with its synthesis. However, Cooper and associates (1998) showed that sufficient amounts of shh were synthesized for the normal induction of neural plate in alkaloid (jervine)-treated chick embryos and that the cyclopia resulted from inadequate response of the neural plate, as was the case in the zebrafish mutant already noted. They speculated

that the patched protein involved in the response to shh contains a sterol-sensing domain that may be affected.

While midline derivatives of the neural plate are affected in all holoprosencephaly cases (Fig. 5.4), the effects on eye development can be rather different. In most cases (Fig. 5.4A–C), the eyes are progressively reduced in size with increasing severity of the malformations and remain "unfused." In humans, such malformations may be caused by single gene defects, trisomies 13 and 19, and, apparently, ethanol. Studies of a C57B1 mouse model for the FAS (Sulik et al., 1981; Sulik and Johnston, 1982) shed some light on the pathogenesis. C57B1 mice have a very low spontaneous incidence of FAS-like malformations, indicating genetic predisposition (see Johnston and Bronsky, 1995, and Section 5.4 in this chapter).

Exposure to large maternal doses of alcohol in the mouse at the time of germ layer formation (approximately day 17 in the human embryo) results in malformations similar to those in humans with fetal alcohol syndrome. These consist of under-development of midline structures such as the philtrum of the lip. There is a consider-able amount of cell death in the anterior neural plate associated with a deficiency of underlying mesoderm, particularly in the midline region. With increasing deficiency of the neural plate, the olfactory placodes (which appear to be at least partly derived from lateral portions of the anterior neural

folds) (Holtfreter and Hamburger, 1955; Couly and Le Douarin, 1985) come to lie progressively closer to the midline, even-tually coming into contact (Fig. 5.5). Decreased space between the placodes apparently leads to deficiencies in struc-tures between them, first affecting those closest to the midline, such as philtrum of the lip in humans, which is deficient or absent in FAS (Fig. 5.4A).

The most frequent facial malformations in trisomy-13 holoprosencephalies are clefts of the lip and palate that are characteri-zed by small premaxillary derivatives (e.g., the premaxillary segment in bilateral cleft lip and palate). In 1977, one of the authors (MCJ) had the opportunity of examining a series of human arrhinence-phalic embryos in the Japanese Kyoto col-lection in which the normally formed lateral portions of the olfactory placodes were

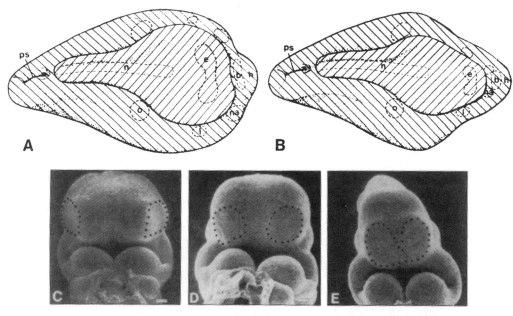

Figure 5.5 Maternal ethanol treatment narrows the anterior neural plate (compare B to control embryo, A), which results in the olfactory (nasal) placodes (na) (dotted outlines in C–E) being located abnormally close to the midline (compare B, D and E to control embryos, A and C). Also, the eye fields (e) are reduced. Contact of the placodes, as in E, leads to arrhinencephaly-like malformations (Fig. 5.4B). Other structures (or their precursors) indicated are the buccopharyngeal membrane (b), heart (h), lens placode (l), otic placode (o), notochord (n) and primitive streak (ps). (Sources: A, B modified from Johnston and Sulik, 1979; C,D, and E from Sulik, 1984.

fused together in the midline, such as the lateral halves of eye vesicles in the cyclopic embryos. Actually, in both cases it is failure of the primordia to separate—rather than fusion—that causes these malformations. In familial cases (presumably determined by single genes), the midline defect in the mildest cases can be so precise that single symmetrical midline incisors may be formed, again from essentially normal lateral structures. The mechanisms involved in the reduction of the medial nasal processes in the less severe holoprosencephaly cases are discussed further in Section 5.6.1.

If the etiology of the Japanese holoprosencephalies were similar to that of Caucasian holoprosencephalies, most of them would have resulted from trisomies 13 and 19. Other than the observations on the Kyoto embryos, almost the only studies on pathogenesis have been the ethanol studies. Ethanol diffuses rapidly throughout the embryo and apparently has many diverse effects. It is known to destabilize cell membranes (Chin et al., 1979), which could interfere with many cellular functions such as cell division and cell migration, both of which involve major membrane turnover. The cell death noted in the neural plate could be a primary effect or secondary to the underlying mesodermal deficiency, since the differentiation and growth of the neural plate is known to depend on the underlying mesoderm (Kallen, 1956). The deficient mesodermal cells do not originate directly from the adjacent prechordal plate but rather from the primitive streak, and they must reach this location by first migrating laterally forward and then medially.

Ethanol affects migrating neural crest cells at a later time (see below) as well as migrating neuroblasts in the central nervous system (Clarren, 1996). Of particular interest in relation to ethanol effects on cell migration is the cell adhesion molecule, L1. This molecule is very sensitive to low levels of ethanol in vitro (Ramanathan et al., 1996) and neurite outgrowth in culture is severely

affected. Neural crest cells have high levels of at least one form of this molecule (Chen et al., 2001). Some of the effects of ethanol in very late development may result from excessive cell death at the time of synapse formation (Ikonomidou et al., 2000). These authors attribute the cell death to effects of ethanol on neurotransmitters such as GABA, but the reductions they observe could be secondary to failed synapse formation resulting from primary effects on L1.

At the opposite end of the spectrum is hypertelorism (excessive distance between the eyes). The most common human hypertelorisms are the frontonasal dysplasias or medial cleft face syndrome (Gorlin et al., 2001). It appears that there is a midline defect that prevents the two halves of the face from coming together, which is consistent with the expanded corpus callosum and other features (DeMeyer, 1967). In methotextrate-induced medial facial clefts, the primary developmental defect appears to be midline hemorrhage (Darab et al., 1987). Hu and Helms (1999) used induced overexpression of shh to increase the amount of frontonasal tissue (frequently in the form of duplications), which appears to be quite unlike the common human hypertelorisms. A rare form of hypertelorism (Griegs) is associated with hypersyndactyly (extra fused digits). The defective gene, GLI3 (e.g., Mo et al., 1997; Shin et al., 1999), is related to the hedgehog family, but to the knowledge of the authors a transgenic mouse model has not yet been produced. It is possible that quantitative differences in SHH may, at least in part, account for racial and other variations in interocular width. Other factors that may play a role in these relationships are discussed further in Section 5.6.1.

5.4 OTOCEPHALIES

Before considering the somewhat related otocephalic malformations, it is necessary

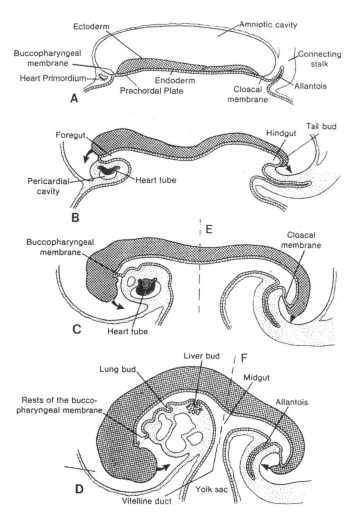

Figure 5.6 Pharyngeal arch formation. Growth of the anterior neural plate (compare to Fig. 5.3B) carries the developing forebrain up and over the primordia of the mouth and heart (buccopharyngeal membrane and heart primordium in A). As a consequence, portions of the embryo lateral to these structures become the walls of the pharynx while the developing heart becomes the floor of the pharynx. (Source: Modified from Sadler, 2000.)

to briefly review relevant aspects of pharyngeal arch formation. After completion of germ layer formation in the head region, the anterior neural plate grows rapidly forward, upward and over the developing oral region and heart primordia (Figs. 5.3B and 5.6B–D). This converts the more lateral tissues into the walls of a tube-like structure, the primitive pharynx, while the heart primordium forms the floor. The mesoderm in the pharyngeal walls begins to segment, and the outer and inner endoderms form grooves between the segments and thereby the initial rudiments of the branchial (visceral, pharyngeal) arches. As will be seen, the mesoderm is primarily involved in the formation of the endothelial cells that line the aortic arch blood vessels as they carry blood from the heart to the head and the rest of the developing embryo. The mesenchyme, which will form the rest of the walls of the aortic arch blood

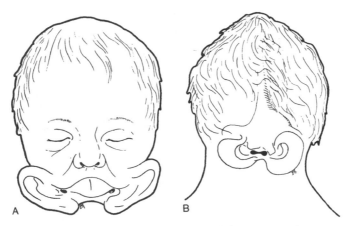

Figure 5.7 Otocephaly ("ear head"). In the more severe from (B), little more than external ears are apparent. In a milder form (A), derivatives of the skeletal and dental portions of the first arch (mandible, etc.) are absent (agnathia). In specimens such as that illustrated in (A), the morphology of the upper lip has not been accurately described. (Source: From Duhamel, 1966.)

vessels, is derived from the neural crest cells that migrate into the arches, surrounding the mesodermal cores (Johnston and Listgarten, 1972; Noden, 1983a; Etchevers et al., 2001).

The otocephalies (Fig. 5.7) are a series of malformations, whose name (literally, "ear head") describes the most severe form (Fig. 5.7B). Although rare, their pathogenesis is of considerable interest. Wright and Wagner (1934) studied otocephaly in the guinea pig and concluded that it was probably a neural crest problem. A more extensive study has been conducted by Juriloff and colleagues (1985) in a substrain of C57B1 mice, which has a balanced chromosomal translocation that otherwise has little effect on development. They found that in the less severely affected embryos the first evidence of cell death was in the mesodermal cores of the first branchial arch. Although the crest cells appear initially normal, their continued survival depends on their being vascularized by capillary endothelial buds, whose formation is dependent on the mesodermal cores. In normal development, when the crest mesenchyme has been vascularized, the remaining core cells undergo programed cell death (Johnston and Listgarten, 1972), suggesting the possibility that balanced translocation hastens the cell death. Eventual sloughing of the whole arch would lead to the malformation depicted in Figure 5.7A.

In more severe cases in the study by Juriloff and associates, cell death is also found in the mesoderm underlying the neural tube. This could account for the more severely affected cases (e.g., Fig. 5.7B). The midfaces in less severely affected cases look very similar to those of the FAS C57B1 mouse model (K. Sulik, personal communication), which, as noted earlier, are sometimes observed in untreated controls in mice of this strain. It appears possible that an inherent weakness in the mesoderm of C57B1 embryos may be exacerbated by the balanced translocation or ethanol.

Knockouts for a number of genes affecting anterior head formation cause malformations not unlike otocephaly (see Kuratani et al., 1997, for a review). These genes include Otx2, Pax6, Lim-1, and Hesx-1. Some of them have broad effects and can affect many tissues such as mesoderm, epithelia, and neural crest cells. The anterior portions of the head are under a different type of developmental control than more posterior (caudal) regions. This was the conclusion of embryologists working on germ layer formation in amphibian embryos who wrote about "head organizers" and "trunk-tail organizers" (see Holtfreter and Hamburger, 1955, for a review). In mammals at least, this developmental control appears to involve portions of the hypoblast (anterior visceral endoderm, AVE) that are extra-embryonic; that is displaced rostrally by the gastrulating "definitive" endoderm and prechordal plate. This would be the first tissue to gastrulate in amphibia and is equivalent to the head organizer used in Holtfreter's experiments. There is some type of interaction between the AVE and the adjacent portion of the prospective anterior neural plate, both of which produce a homeobox gene, Lim-1 (Shawlot et al., 1999). Knockouts for Lim-1 develop heads with no structures anterior to the ear (Shawlot and Behringer, 1995). Other genes expressed in the AVE are Hesx-1 and Otx-2. Otx-2 knockout malformations are often less severe (Kuritani et al.). Pax-6 is expressed in the anterior edge of the anterior neural plate, the eye cup, and the olfactory placode. Pax (paired homeobox) genes are widely distributed and affect many aspects of development. Matsuo and colleagues (1993), and Osumi-Yamashita and colleagues (1997) extensively studied the developmental changes in a pax-6 mutant (sey) in the rat. They observed alterations in crest cell migration among other changes. The Wnt genes are involved in midbrain and hindbrain specifications and many other aspects of development.

5.5 SYNDROMES WITH MAJOR CREST CELL INVOLVEMENT

Retinoic acid syndrome malformations first appeared shortly after the introduction of the drug Accutane (13-cis-retinoic acid) used for the treatment of severe cystic acne (Lammer et al., 1985). These malformations,

in addition to crest cell involvement in head segmentation and other findings, have led to an explosion of interest in neural crest cells and a reevaluation of the developmental alterations leading to a number of craniofacial syndromes. A number of malformations have major crest cell involvement, and these are described below. They are summarized in Table 5.1.

5.5.1 Aspects of Normal Neural Crest Cell Development Relevant to Craniofacial Malformations

The neural crest is an unusual, highly versatile group of cells arising from the neural folds during or after the formation of the neural tube (Fig. 5.3C). These cells migrate extensively and differentiate into a wide range of derivatives. In addition to forming pigment cells and most of the peripheral nervous system, as they do elsewhere in the body, head neural crest cells form most of the skeletal and connective tissues in the face and pharyngeal regions. In the head and neck, they migrate primarily under the surface ectoderm, and when they reach the pharyngeal arches, they surround the mesodermal cores already present (Fig. 5.8A–D). In the upper face, they form almost all the mesenchyme between the surface ectoderm and the underlying forebrain and eye.

Gans and Northcutt (1983) were the first investigators to suggest that neural crest and placodal cells were key to the transition from chordates to vertebrates. They were particularly concerned with placodally derived peripheral receptors, some of which, they reasoned, required some sort of skeletal protection. There is now a growing consensus (Donaghue et al., 2000; Zimmer, 2000; Shimeld and Holland, 2000; Neidert et al., 2001) that the first mineralized skeletal elements in vertebrates were teeth, which contained dentin formed by neural crest cells. The prevertebrate chordates (e.g., Amphioxus) are filter feeders and do not need teeth. However, they have a group

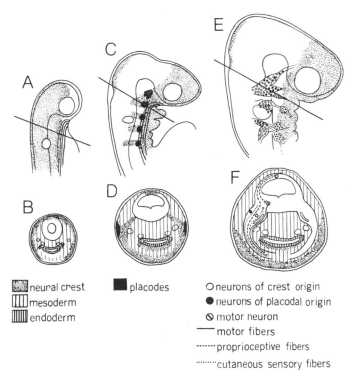

Figure 5.8 Diagrams illustrating the distribution of crest cells (stipple) during (A, B) and at the end (C, D) of crest cell migration. The planes of sections are illustrated in the lateral views. The distribution of crest and placodal cells in the sensory ganglia of different cranial nerves is illustrated in (E) and (F). (Source: Modified from Johnston and Hazelton, 1972.)

of cells at the junction between the neural plate and surface ectoderm in the anterior head region that express many of the same regulatory genes (Holland et al., 1994; Ferrier end Holland, 2001), and these authors have speculated that they may be the precursors of vertebrate neural crest cells. They also have an epithelial structure at the anterior end of the head with neurons that send their axons to the forebrain, which appears to be the forerunner of the olfactory placode (Ferrier and Holland, 2001).

The fossil conodonts (cone-shaped teeth) are thought to be representative of the earliest vertebrates, more primitive in most ways than the cyclostomes (e.g., the lamprey), although they had some dermal bone. As predators, they had, in addition to teeth, at least rudimentary eyes and some

sense of smell. Since the chordates are filter feeders, they have no need for teeth, and, while they have a notochord and segmented muscle, they do not have crest cells or other skeletal structures. Thus, the innovations provided by crest cells are considered to be pivotal for the evolution of vertebrates. In this scheme, the ability of mesoderm to form mineralized skeletal elements would have come later. The cyclostomes are considered to be in some ways regressive. The role of neural crest cells in these most primitive living vertebrates has been well documented (Newth, 1956; Langille and Hall, 1988), and it has recently been shown (Neidert et al., 2001) that they have at least some of the molecular characteristics of crest cells in higher vertebrates.

It is important to note that although the chordates, such as the amphioxus, have branchial arches and gill slits that they use for filter feeding, they do not have crest cells, although some of the genes (e.g., Pax) involved in vertebrate crest cell development are already present in the epithelia of the arches (Ferrier and Holland, 2001; Kozmik et al., 1999). In fact, the arches are already present at the time of crest cell migration in the chick, complete with lateral mesoderm and cell-free spaces into which crest cells migrate (Johnston, 1966; Pratt et al., 1975). Veitch and associates (1999) have studied the development of branchial arches in the chick after extirpation of neural crest cells and have found that the pattern of regulatory molecules in the arch epithelia and mesodermal cores is essentially normal.

The onset of crest cell migration is signaled by a switch from N-cadherin intercellular binding sites to the more labile H-cadherin binding sites. The timing varies considerably in different species, both with respect to developmental age and the timing of neural tube closure. The onset is relatively early in the rodent, during the early stages of neural tube closure, and much later in the chick, after the neural folds have made contact. The timing in the human embryo is intermediate. The large cell-free space, noted above for the chick embryo, is not seen in the rodent embryo, perhaps because the earlier migrating crest cells keep the opening space filled up. The space along which the crest cells migrate has an abundance of highly hydrated hyaluronic acid (Pratt et al., 1975), which may provide room for the cells to move, and a network of collagen and fibronectin along which the crest cells pull themselves as they migrate. Recently, it has been shown (Chen et al., 2001) that the migrating crest cells have high levels of one form of the L1 adhesion molecule noted above. This molecule is necessary for neurite outgrowth in culture (Wilkemeyer and Charness, 1998) and may

be the primary means by which crest cells attach to extracellular fibers in their migration path.

As more has been learned about the characteristics of migrating neural crest, several markers have become available for their identification in mammalian embryos. Among these markers are the cellular retinoic acid binding protein (CRAPBI), (Dencker et al., 1990; Graham et al., 1994), several retinoic acid receptors (e.g., RAR-β, Osumi-Yamashita et al., 1992; RAR-A, Lohnes et al., 1993), and expression of the regulatory genes AP-2 (e.g., Neidert et al., 2000; Richman et al., 1997) and Hoxa-2, crest cells caudal to the first arch only. While the functions of these molecules is not well understood, they provide good markers and are relevant to abnormal crest development as is discussed below. Some progress has been made on the development of markers that can be used to identify crest cell derivatives. In a study by Chiani and colleagues (1997), a reporter gene that could be turned on by cre-mediation was used. While impressive in many respects, there were obvious false positives and false negatives. At least the study appears to be a step in the right direction.

Some recent studies clarify the types of signaling interactions to which crest cells respond in their new environments. Of particular interest are studies showing that lateral plate mesoderm and the epithelia of the branchial arches with which crest cells come into contact produce a signaling molecule, endothelin-1, and that crest cells contain the receptor (endothelin-A receptor) for this molecule (Kurihara et al., 1994; Thomas et al., 1998; Clouthier et al., 1998, 2000). Endothelin-A receptor–deficient mice develop malformations similar to those described below as having "major crest cell involvement." These malformations include cardiovascular outflow tract defects (Kurihara et al., 1995a; Clouthier et al., 1998), deficiencies of the thyroid and thymus glands (Kurihara et al., 1995b), as well as craniofacial malformations involving the middle and external ear and other structures (Clouthier et al., 2000). Further, the mechanisms by which they alter crest cell development are beginning to be understood (Thomas et al., 1998; Clouthier et al., 2000), and these are discussed further below.

Of considerable potential relevance to understanding the mechanisms of a large number of craniofacial malformations are recent findings on the effects of the hindbrain and derived neural crest on head segmentation (Hunt et al., 1991; Noden, 1991a,b; Graham et al., 1996). Segmentation in the hindbrain is probably, in turn, determined by the underlying chordamesoderm (see above; and Smith and Schoenwolf, 1998). The segmental organization of the head is depicted in the diagram in Figure 5.9. The duplication of the drosophila Hox gene segment on four mammalian chromosomes has already been noted and is illustrated in Figure 5.10.

In both Drosophila and mammals, it appears the increasing addition of hox gene activity with the progression caudally in the hindbrain and derived neural crest determines the fate of the individual pharyngeal arches. Perhaps the most dramatic evidence is provided by two independent knockout studies for the gene Hox-a2 in which the embryos develop second mandibular skeletal apparatus in place of the hyoid skeletal apparatus, indicating that the addition of Hox-a2 activity to the second arch neural crest cells changes their response to the inductive effects of the pharyngeal epithelia. These results explain earlier findings showing that when mandibular arch crest (Noden, 1983a) is used to replace hyoid crest, a second mandibular skeletal apparatus is formed. Also, it was found that when trunk crest was used to replace cranial crest, it migrated normally into the visceral arches and formed normal neural elements, but it was unable to respond to the pharyngeal arch epithelia. Crest cells migrating into the upper face appear to be basically the same as mandibular arch crest cells in that they form mandibular arch skeletal elements when used to replace mandibular crest, but respond differently to the upper face inductive influences.

Crest cells demonstrate plasticity in other ways. For example, if the hindbrain is rotated, the crest cells migrate into the nearest arches. While they initially maintain the hox gene expression of the neuromeres from which they (with the exception of the hyoid arch crest), eventually adopt hox expression appropriate to their new arch (Hunt et al., 1998). Crest cells show enormous capacity for repair. Other examples are shown by experimental removal of crest, in which very large deficits are repaired by increased proliferation and the recruitment of crest cells from other areas. It appears that reductions of crest cell populations by 80% or more may be necessary in order to attain malformations (Johnston, 1964; McKee and Ferguson, 1984). While some of the repair may result from redifferentiation of adjacent neural plate/tube cells, many of the cells arrive from more distant crest populations and they differentiate according to their new environment.

5.5.2 Retinoic Acid Syndrome

Since the retinoids (the normal biologically active retinoic acid and its related compounds, including vitamin A, the dietary precursor of retinoic acid) had long been known to be potent teratogens, the drug Accutane (13-cis-retinoic acid) was marketed for the treatment of severe cystic acne with the provision that it was not to be taken during pregnancy. Although 13-cis-RA has the same biological actions as its parent compound, it was less toxic. However, many accidental exposures occurred, resulting in a surprisingly high incidence of very severe malformations involving craniofacial structures. Some of these defects are illustrated in Figure

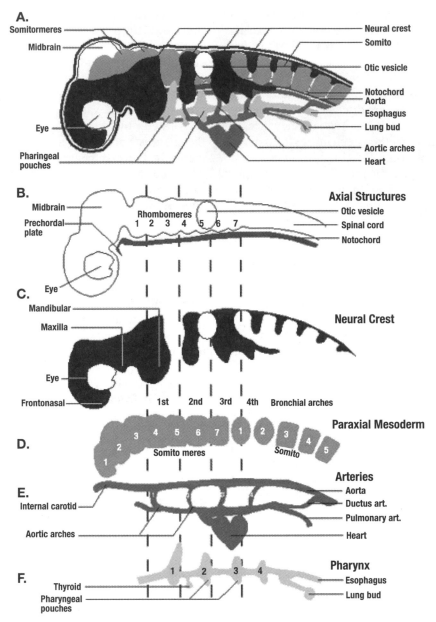

Figure 5.9 Schematic presentation of the segmental relationships of different components in the embryonic head and neck. In the upper diagram the surface ectoderm has been removed, illustrating the relationships of the different components, while the remaining diagrams illustrate the components individually. (Source: Modified from Noden, 1991b.)

5.11A. They consist of abnormalities of the external and middle ear, sometimes mandibular under-development and cleft palate, cerebellar defects, and defects in the cardiovascular system (outflow tract defects) and the pharyngeal glands, including the thymus and parathyroid glands. Such malformations are usually fatal within the first few years of life.

The reason for the unexpectedly severe nature of RAS malformations apparently relates to the very poor ability of humans

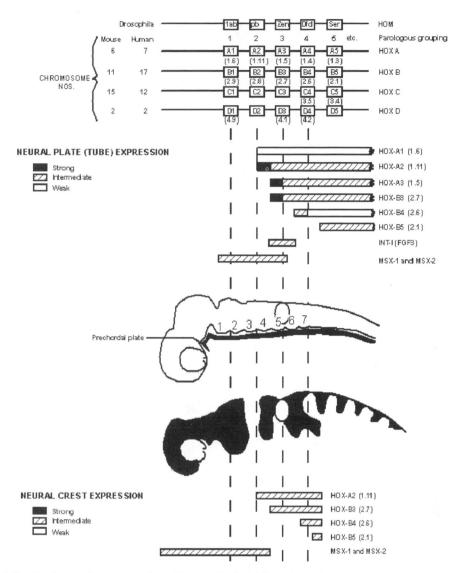

Figure 5.10 The homeotic gene complex of Drosophila (HOM) has been duplicated more or less intact in four different complexes on different chromosomes of the mouse and human. The same head-to-tail (rostrocaudal) sequence in the chromosomes has been preserved and corresponds roughly to rostrocaudal gene expression in the neural plate (tube) and neural crest, which is derived from the neural plate (tube). The newer terminology is utilized for individual genes, with the older terminology used for mice in parentheses. Depending on chromosomal positioning, the genes are arranged in paralogous groups 1 through 4 and, in general, these genes are expressed in a sequential overlapping cascade with one or more active genes being added every two neuromeres. (Source: Portions of illustrations modified from Noden, 1991b.)

to clear retinoic acid metabolites (Webster et al., 1986). The primary animal used to test for teratogenicity was the rat, which has no difficulty clearing the metabolites. In humans, the metabolite, 4-oxo-13-cis-RA, reaches blood levels three to five times higher than the parent compound (Brazzell et al., 1983). With each successive dose the levels "stair-step" higher, since the metabolite from the earlier dose(s) has not yet

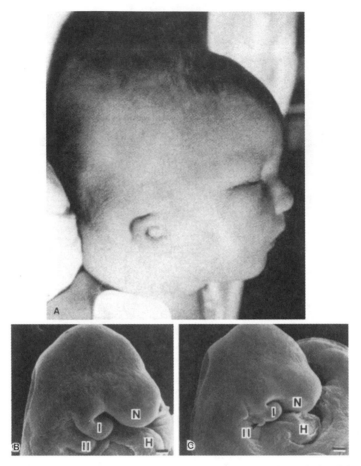

Figure 5.11 (A). Child with retinoic acid syndrome. In addition to severe hypoplasia of the external ear, the mandible is underdeveloped and the anterior neck structures are hypoplastic. (B and C). Scanning EMs of embryos from control (B) and retinoic acid-treated (C) mice. (Source: A, Courtesy of Dr. Paul Fernhoff; B and C, from Webster et al., 1986.)

cleared. When tested in whole embryo culture, the parent compound and the metabolite are equally teratogenic (Webster, 1985; Goulding and Pratt; 1986; Webster et al., 1986). At the concentration found in humans, 13-cis-RA had only mild growth effects, but with the addition of an equal amount of 4-oxo-13-RA, the resulting malformations were virtually identical to those of the in vivo model for RAS (Webster et al., 1986) (Fig. 5.11B,C).

The timing of exposure for the most severe facial malformations in the studies by Webster and associates (1986), Goulding and Pratt (1986), and others coincided with the onset and period of migration of first and second arch crest cells (about day 21 in human embryos) while the sensitive period for cardiovascular malformations coincided with the migration of third and forth arch crest cells (about human embryo day 23). Before discussing the pathogenesis of these malformations, some of the more recent studies on neural crest and head segmentation are appropriate.

As seen in Figure 5.11B,C, by far the most severely affected structure is the hyoid arch (Fig. 5.11C), which apparently results primarily from the massive death of crest cells in hyoid arch crest cells prior to their migration (Sulik et al., 1988). There is normally an area of programed cell death in the neural folds between the first and second arches at a slightly later stage. Graham and associates (1993) have provided evidence that the gene Msx-2 is involved in programed cell death and have concluded that crest cell death in the neural folds is related to the elimination of crest cells where they are not needed. Others have shown that RA not only induces Msx-2 but also upregulates RAR-β (Fig. 5.2), which binds RA to the genes that it activates.

RA also has effects on migrating neural crest cells (Webster et al., 1986; Pratt et al., 1987) and in cell culture (Thorogood et al., 1982; Smith-Thomas et al., 1987) with the appearance of blebs on the cell surfaces and other changes. This is consistent with the fact that migrating crest cells have high CRABPI levels (Dencker et al., 1990; Maden et al., 1991, 1992)—a compound thought to protect cells from high levels of RA through binding. Migrating crest cells demonstrate high levels of C14-labeled RA following maternal administration (Dencker, 1987). Some effects on neural crest cells occur when RA is administered only after the onset of crest cell migration, as when it is used to cause defects involving third and fourth crest cell derivatives (Webster et al., 1986). These include defects of the outflow tract (conotruncus) of the cardiovascular system and the pharyngeal glands, which closely mimic those of RAS. The significance of this association, which also occurs after early administration of thalidomide and the administration of ethanol at the time second and third arch crest cells are migrating (Daft et al., 1986), was greatly clarified by the findings of Kirby and colleagues (1983) that cranial neural crest cells form much of the mesenchyme of the conotruncal septum and that extirpation of these cells prior to migration leads to cardiac malformations of the

conotruncal (outflow tract) type (Bockman et al., 1989).

5.5.3 Other Malformations with Major Crest Cell Involvement

A number of the malformations considered below are summarized in Table 5.1. Their features are compared to those of RAS. Together, this group could be termed *otofacial malformations* (see, for example, Gorlin et al., 1990). Their similarities and differences are also summarized at the end of this section and an attempt is made to reconcile the many divergent interpretations of these malformations.

5.5.3.1 DiGeorge Syndrome DiGeorge syndrome (Fig. 5.12) bears many similarities to RAS. It has been demonstrated that the gene deletions for DiGeorge syndrome and the similar velocardiofacial syndrome (essentially, DiGeorge syndrome without the pharyngeal gland problems) are at the same locus on chromosome 22 (Driscoll et al., 1992a). DiGeorge syndrome is now sometimes referred to as the 22q11.2 deletion syndrome (Emanuel et al., 2001) or "catch 22". DiGeorge syndrome is also associated with maternal alcoholism (Ammann et al., 1982). Whether these cases had the deletion is not known.

The effect of ethanol on neural crest cells is noted above. The malformations produced are very similar to those of DiGeorge syndrome (Sulik et al., 1986; Daft et al., 1986). Treatment at the time that first and second arch crest cells are migrating or preparing to migrate results in massive cell death (Sulik et al., 1986). It is not clear how much of this cell death is because migrating crest cells are vulnerable or the degree to which other factors play a role (see discussion of FAS). In the study by Sulik and colleagues, many of the dead cells were far from the neural folds, indicating that they died during or at the end of their migrations. In this respect, the

TABLE 5.1 Three Syndromes with Apparent Major Primary Cranial Neural Crest Involvement, Compared with a Syndrome (Treacher Collins) with Apparent Primary Ganglionic Placodal Cell Death and Secondary Localized Crest Cell Involvement

Syndrome	Developmental Abnormalities						Etiology	Possible Pathogenesis
	Ear	Facial Bones	Cardiovascular Malformations	Pharyngeal Glands	Facial Clefts (variable incidence)	Other Associated Malformations (variable incidence)		
Retinoic acid syndrome (Fig. 5.11)	Primary middle and external ear	Mandibular and other deficiencies	Conotruncal defects	Thymus, etc., deficient or absent	Cleft palate (8%)	Brain (particularly cerebellum)	13-cis-retinoic acid (isotretinoin, Accutane)	Cranial neural crest —kills second arch crest before migration along with other neural plate cells; retards migration? Other cell populations in some cases
DiGeorge syndrome (Fig. 5.12)	External and middle ear	Variable maxillary and mandibular deficiencies	Conotruncal defects	Thymus, etc., deficient or absent	Cleft lip and/or palate (10%)	Brain	Heterogeneous Frequently associated with chromosome deletion (22q11.2), sometimes with ethanol	Cranial neural crest —ethanol kills before and during migration
Hemifacial microsomia (Fig. 5.13)	Primarily middle and external ear	Usually asymmetrical deficiencies of mandible, squamous temporal, and other bones	Conotruncal defects	No reported abnormalities	Cleft lip and/or palate (7–22%)	Eye, brain (in severe cases), vertebrae in oculo-auriculo-vertebral variant	Heterogeneous, multi-factorial	Cranial neural crest —selective death of second arch crest retards migration; sometimes other cell populations affected
Treacher Collins syndrome (Fig. 5.14)	External, middle, ear	Symmetrical deficiencies or absence of zygoma, underdevelopment of posterior maxilla and mandible	No increase in cardiovascular malformations	No reported abnormalities	Cleft palate (35%)	Rarely with limb defects, e.g., Nager syndrome	Dominant gene (TCOF-1)	Ganglionic placodal cell death with secondary crest cell involvement

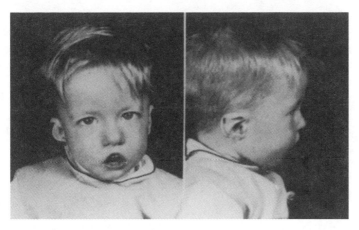

Figure 5.12 Child with DiGeorge syndrome. In addition to the external ear malformation, the mandible is somewhat underdeveloped, and, in contrast to retinoic acid syndrome, the upper lip is short, particularly in its central portion. (Source: From Kretchmer et al., 1969.)

studies of the L1 adhesion molecule noted above are particularly relevant. The L1 adhesion molecule is particularly sensitive to ethanol (Ramanahan et al., 1996), very low concentrations of which interfere with neurite outgrowth (Wilkemeyer et al., 1998), and one form of the molecule is found at high levels in migrating neural crest cells (Chen et al., 2001).

There has recently been a great deal of interest in the possibility that a number of gene knockouts produce DiGeorge syndrome. Certainly, these malformations include defects of conotruncal septation, thymus gland, and so forth. Hox-A2 and Hox-A3 are normally expressed in the third and fourth arch crest as well as in the pharyngeal ectoderm and endoderm of these arches (Couly et al., 1998). Chisaka and Cappechi (1991) found skeletal and pharyngeal gland abnormalities in a Hox-A3 knockout that were somewhat similar to those of DiGeorge syndrome. More recently, Allen and associates (2001) reduced the migration of crest cells into the heart by interfering with BMP-2/4 formation, producing defects similar to those of DiGeorge syndrome. These malformations have focused a great deal of attention on the cardiac crest and indicate how basic research findings are stimulated by such malformations and vice versa.

Very recently (Lindsay et al., 1999; Lindsay and Baldini, 2001), a mouse model heterozygous for the 22q11.2 deletion was developed. While neural crest migration into the pharyngeal (branchial) arches appeared to be normal, there was severe underdevelopment of the fourth arch aortic vessels in all of the embryos. Some of the embryos showed variable recovery. Using a molecular marker of smooth muscle cells, researchers were able to show that the recovery depended on whether smooth muscle cells formed the outer walls of the vessels. These cells are derived from neural crest (Johnston and Listgarten, 1972; Noden, 1988; Etchevers et al., 2001) and appear to become closely associated with aortic arch vessels at a very early stage (late day 9 in the mouse). If there were insufficient crest cells to perform this function, cells arriving later might be too late to provide it. The numbers of neural crest cells were examined at a fairly late stage (10.5 days), possibly giving a small number of crest cells ample time to make up for earlier deficiencies (see above), or for other crest cells to migrate into the affected arches. It is difficult to see why the fourth aortic arch vessels should be more affected than more caudal aortic arches. The pharyngeal glands were apparently not affected. While a goosecoid-like gene from the deleted region turned out to be negative in a transgenic mouse (Wakamiya et al., 1998), a transcription factor, Tbx, produced the similar cardiovascular developmental alterations (Lindsay et al., 2001). There are approximately 15 genes in the deletion.

5.5.3.2 Thalidomide Malformations
The preponderance of limb defects following thalidomide exposure is related to timing. In Europe, however, there were enough early exposures (for a timetable, see Nowack, 1965) to document a large number of malformations very similar to those of RAS. Johnston and Bronsky (1995) have extensively reviewed evidence for crest cell involvement.

Recently, studies addressing this question have been reported (Jacobsson, 1997; Jacobsson and Granstrom, 1997). Thalidomide was administered using protocols similar to the retinoic acid studies and very similar malformations were produced. Other studies of the effects of thalidomide on craniofacial development are considered in Section 5.5.3.3 and are further considered in Section 5.5.3.4.

5.5.3.3 Hemifacial Microsomia
Also extensively reviewed by Johnston and Bronsky (1995) was the possible pathogenesis of hemifacial microsomia (Fig. 5.13), a malformation with apparent multifactorial etiology. It has a number of key similarities to RAS as well as important differences. Both have major ear involvement of a similar nature, although it is usually bilateral in RAS and unilateral in hemifacial microsomia. Hemifacial microsomia has a high incidence of conotruncal defects (perhaps 50–60%), (Gorlin et al., 1990; M. Cohen, personal communication) but few, if any, pharyngeal gland abnormalities. Also, the cerebellar malformations found

in RAS are not seen in hemifacial microsomia. Numerous other proposals (Poswillo, 1973) for the pathogenesis have been put forward, and these have been thoroughly discussed by Johnston and Bronsky (1995).

5.5.3.4 Treacher Collins Syndrome
A single dominant gene (TCOF-1) causes Treacher Collins syndrome. Although penetrance is somewhat variable, malformations are usually very consistent (Ross and Johnston, 1972; Shprintzen, 1982; Marres et al., 1995; Gorlin et al., 2001). Figure 5.14 shows some of the characteristic malformations, including notching of the outer eyelid, surface depression in the zygomatic region, underdeveloped mandible, and minor abnormalities of the external ear. Radiographic and surgical studies reveal that the mandibular reduction is largely confined to its proximal portions and characteristic reductions or absence of the zygomatic bone (Fig. 5.14C). Clefts of the soft palate only are very common, and there is a fairly high incidence of inner ear deafness. About the only malformations that are not regional are minor abnormalities of the other pharyngeal arch derivatives (Shprintzen, 1982).

Studies of the pathogenesis of RAS defects in the mouse model have also provided information relevant to Treacher Collins syndrome (Fig. 5.14). In the portion of their study dealing with the later period of treatment necessary for conotruncal defects, Webster and colleagues (1986) found other malformations very similar to those of Treacher Collins syndrome. There is severe deficiency of bone in the zygomatic region of some infants following exposure to isotretinoin (E. Lammer, personal communication). Further studies on the pathogenesis of the defects in the mouse model (Sulik et al., 1987, 1989) showed that the initial abnormalities were largely limited to the distal portion of the trigeminal ganglion (Fig. 5.14C, D), which

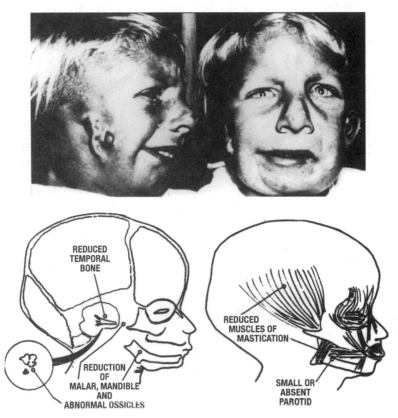

Figure 5.13 Hemifacial microsomia. In addition to the malformation of the external ear, as seen in this patient, many regional structures are usually deficient (patient and diagram). These include the middle ear ossicles, the squamous portion of the temporal bone, the mandible, the muscles of mastication, and the parotid gland. The malformations are often largely limited to one side, as in the patient illustrated. (Source: From Poswillo, 1973.)

corresponds to that portion of the ganglion populated by neuroblasts derived from the ganglionic placodes (Fig. 5.8C, D). It appears that the cell death is so massive that it interferes with the development of the surrounding crest cell mesenchyme, which gives rise to the zygoma and posterior portions of the maxillary prominence and mandibular arch, thereby producing defects similar to those in Treacher Collins syndrome. Deficiencies in the zygomatic region are sometimes observed after RA exposure (E. Lammer, personal communication).

Other authors (Osumi-Yamachita et al., 1992; Evrard et al., 2000) have detected cell death in the same region following similar retinoid treatment. However, they concluded that they were dealing only with neural crest cells. There are, of course, neural crest cells in the area, such as those that initiate the maxillary process, but we believe that death in these cells is probably secondary. Using thalidomide as the teratogen in monkeys at the same stage of development, Poswillo (1975) produced virtually identical embryonic changes, but concluded that the resulting malformations were consistent with hemifacial microsomia. After repeating the Poswillo study, the conclusion of Newman and Hendrickx (1981) that the resulting malformations more closely resembled Treacher Collins

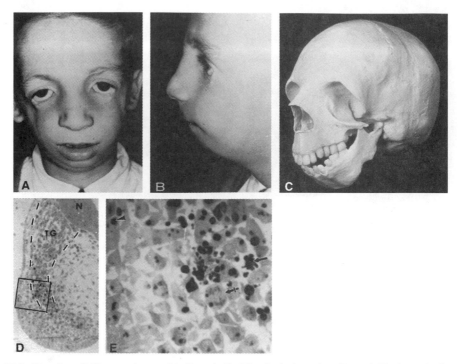

Figure 5.14 Treacher Collins syndrome (A, B and C). Histological section (D and E) through the trigeminal ganglion (TG) of a mouse embryo 6 hours after maternal 13-cis-RA administration illustrates pyknotic debris (arrows in E) which shows a distribution similar to that of the placodal cells in Figure 5.8E and 5.8F. (Sources: A and B from Ross and Johnston, 1972; C. Courtesy of S. Pruzansky and P. Tessier; D and E from Johnston, 1990.)

syndrome is more in line with the studies by Sulik and associates (and the more recent Jacobsen studies, noted above).

In humans, Dixon (1996) reported that Treacher Collins mutation is in the TCOF-1 gene, a finding confirmed by Wise and associates (1997). TCOF-1 normally produces treacle, a phosphoprotein involved in nucleolar organization, and possibly in the transport of materials from the nucleus to the cytoplasm (Isaac et al., 2000). The TCOF-1 gene is widely expressed in the embryo, but it was difficult to localize in the whole mount preparations presented in another article (Dixon et al., 1997). It appears to reach high levels in the neural plate and the epithelia of the branchial arches. In a biochemical analysis of branchial arch tissue, Jones and colleagues (1999) found high levels of treacle, the gene product.

Dixon and associates (1997) produced an animal model that was hemizygous for a nonfunctional TCOF-1 gene. Massive cell death was found in the neural folds that normally contain the high levels of TCOF-1 expression, as well as in many other areas. Although they did not comment on it, one area where there appears to have been very severe cell death was the prospective trigeminal placode, as shown in the one section (their Fig. 15.5D) passing through that structure. Also, there were severe reductions in the cranial ganglia, especially the trigeminal ganglion, which is consistent with this cell death and our observations on placodal ganglionic cells. However, the researchers also noted reductions in the spinal ganglia that were purely of neural crest origin with no contributions from ganglia. The features of the malformations least consistent with Treacher Collins syndrome were the very severe craniofacial malformations that included exencephaly, anophthalmia, and abnormal nasal and maxillary structures.

There were abnormalities of the structures involved in Treacher Collins syndrome. While there were some characteristic abnormalities, it is difficult to determine whether they were merely secondary to the other malformations. For example, the portion of the maxilla comparable to the human zygomatic bone was reduced but not absent, as it is in many Treacher Collins patients. Also, the mandible was only slightly affected. Fibroblasts from Treacher Collins patients appear to be similar to those of controls (Isaac et al., 2000), despite their heterozygosity for the abnormal TCOF-1 gene. Apparently, the output of a single allele is, however, not sufficient for normal embryonic development—a condition termed *haploinsufficiency* and considered to be a form of dominant inheritance. Also, it should be pointed out that almost all genes known to produce malformations function only as regulatory genes.

5.5.4 Summary of the Otofacial Malformations

These concepts and other information about four craniofacial syndromes are summarized in Table 5.1. Much new information has been developed since Poswillo's pioneering studies of animal models for hemifacial microsomia and Treacher Collins syndrome, prompting a reevaluation of the pathogenesis of these and related malformations. Much of the new information results from two major areas of endeavor. First is the demonstration of a major role for crest cells in normal cardiac conotruncal septum formation and experimental crest cell reductions leading to cardiovascular malformations of the conotruncal variety very similar to those found in RAS, hemifacial microsomia, and other craniofacial malformations. Second is the intense activity related to the roles of RA in normal devel-

opment and the pathogenesis of malformations in animal models for human RAS.

The first three malformations listed in Table 5.1 show a number of common features (e.g., conotruncal cardiovascular malformations) that suggest major crest cell involvement. Research on animal models for RAS (Fig. 5.11) indicates that there are major effects on crest cells, particularly hyoid (second) arch crest, just before they migrate from rhombomere 4. There is also massive cell death in adjacent portions of the rhombomere, apparently accounting for RAS cerebellar defects. Also, the area of programed cell death between rhombomere 4 that is normally involved in the separation of the second arch crest from the fourth and fifth arch crest may be expanded. Crest cells are affected during their migration phase. In all these and other sensitive cells, there is a very high correlation between the cellular retinoic acid binding protein (CRABPI) and RA vulnerability. CRABPI is found in very high levels in rhombomere 4 and in derived hyoid arch crest cells during their migrations. Later effects on ganglionic placodal cells may lead to malformations characteristic of Treacher Collins syndrome, some of which are occasionally seen in RAS.

Human DiGeorge syndrome (Fig. 5.12; Table 5.1) has a heterogeneous etiology including deletion of 22q11.2. A large number of these children are born to alcoholic mothers, and the syndrome can be produced in animal models with ethanol. Ethanol appears to have major effects on many migrating cells including CNS neuroblasts and, apparently, crest cells and gastrulating mesodermal cells. In the DiGeorge model it appears to kill cranial crest cells before and during their migrations, and these effects can account for virtually all of the characteristics of DiGeorge syndrome.

There has recently been considerable interest in the pathogenesis of DiGeorge syndrome, generated by studies on transgenic mice, although the findings are somewhat controversial. Chisaka and Capecchi (1991) reported conotruncal and pharyngeal gland deficiencies in a transgenic mouse with a knockout for Hox-A3 (see Fig. 5.10) that they felt closely resembled DiGeorge syndrome. Perhaps the most interesting transgenic mouse was that developed recently by Lindsay and colleagues (2001). They studied the development of mouse embryos with a deletion found in DiGeorge patients (see above). They used genetic markers to identify cells of neural crest origin compared to those destined to form smooth muscle. However, as pointed out, this may be a bit misleading, in that the smooth muscle cells are derived from neural crest and appear to interact with blood vessels very early. A delay in their arrival could mean that they were too late to normally interact with the vessels. Neural crest cells can easily make up for severe deficiencies by rapid proliferation. Also, the pharyngeal glands appear not to be affected. The very recent identification of the gene responsible for the changes still has the same problems as the deletion.

Hemifacial microsomia (Fig. 5.13; Table 5.1) is a malformation that appears to have a multifactorial etiology. Unlike RAS and DiGeorge syndrome, there are no reports of pharyngeal gland (e.g., thymus) deficiencies. The hemifacial malformation could be quantitative in that conotruncal crest cells have much longer migration paths than those participating in pharyngeal gland development. This same quantitative difference is presumably the reason for the lack of pharyngeal gland abnormalities in the (apparently) milder form of DiGeorge syndrome, the velocardiofacial syndrome, which has the conotruncal malformations but not the pharyngeal gland deficiencies.

Finally, a late effect of RA on mouse embryos, apparently limited to ganglionic placodal cells (where there is massive cell death) but with secondary effects on regional crest cells, mimics Treacher Collins syndrome (Fig. 5.14; Table 5.1), the features of which are sometimes seen in humans following 13-cis-RA exposure. Given the major difference in etiology, there is a surprisingly close resemblance between the RA animal model and human Treacher Collins syndrome, which results from a single dominant gene. Associated malformations are limited and almost all can be accounted for by ganglionic placodal cell death, which raises the problem of interpreting the mouse model hemizygous for the defective TCOF-1 gene. Some of the reasons for the difference in phenotype could be a problem in that the trigeminal placodal cells may be the most affected, but the specific malformations are largely masked by the more widespread effects.

Given the etiological diversity between (and even within) those syndromes with apparent primary crest involvement, the differences between them should not be surprising. For example, if the basic defect in hemifacial microsomia was only expressed in migratory cells with high CRABPI levels, it could account for both the crest-related (facial and conotruncal) and gastrulating mesoderm-related (vertebral) malformations. On the other hand, such a defect would not cause the cerebellar defects seen in RAS, where CRAPBI levels are high but little or no cell migration is involved.

5.6 DEVELOPMENT ALTERATIONS IN FACIAL CLEFTING WITH EMPHASIS ON THEIR MULTIFACTORIAL ETIOLOGY

As with the discussion of other malformations, some brief comments on normal primary and secondary palate development relevant to the following considerations on clefting are necessary.

5.6.1 Primary Palate Formation and Cleft Lip with or without Cleft Palate

The term *primary palate* (Fig. 5.15) is used in different ways by different authors. The *embryonic primary palate*, a term used by teratologists and many embryologists to describe the initial separation between oral and nasal cavities, includes those portions of the medial nasal, lateral nasal, and maxillary processes that contribute to the separation of the cavities. Only the more posterior portion of this embryonic primary palate actually contributes to the definitive palate (that portion anterior to the incisive foramen), with the rest forming portions of the upper lip and alveolus. Unlike the secondary palate, the exact limits of the primary palate are difficult to define (Johnston and Sulik, 1979). Primary and secondary palate formation share many common features, and we feel that this less formal definition is more useful with respect to both normal and abnormal (e.g., cleft lip) development. The narrowest definition (Sadler, 2000) limits the primary palate to the portion of the hard palate derived from the frontonasal process, which is derived from the mammalian medial nasal processes and the material in between (sometimes called the intermaxillary segment). It lies anterior to that portion of the palate derived from the embryonic secondary palate.

5.6.1.1 Normal Primary Palate Development

Formation of the primary palate appears to be virtually identical in all higher vertebrates (Fig. 5.15). It begins with the formation of the olfactory placode, which is a thickening of the surface ectoderm. As already noted, the placode is probably derived, at least in part, from the neural folds (Couly et al., 1993). It eventually separates from the neural folds, and neural crest mesenchyme cells further separate it from the underlying forebrain and eye. Almost as soon as the placode begins to thicken, growth at its margins begins to form the medial and lateral nasal processes (Fig. 5.15A, D). Part of this growth appears to be related to a morphogenetic movement ("curling up") at the edges of the placode (Smuts, 1981) and part is related to rapid proliferation of the underlying mesenchyme and more lateral epithelium. At the same time, the maxillary process develops from the proximal end of the first visceral arch (Fig. 5.15A, D).

Of considerable interest have been studies examining the interactions between the facial process epithelium and the underlying mesenchyme in chick embryos. Minkoff and Kuntz (1977, 1978) and Minkoff (1991) have shown that the proliferation rates in the facial processes are maintained at higher levels than in adjacent structures. Also, it has been shown that this mesenchyme is very dependent on overlying epithelium for its survival (Minkoff, 1991). Richman (1992) has shown that the addition of FGF will support facial process mesenchyme in culture. Crossley and Martin (1995) have shown that FGF8 m-RNA is expressed at high levels in the lateral and medial edges of the nasal placode, which overlies the rapidly proliferating mesenchyme in the MNP and LNP. Similar high levels were found in apical ectodermal ridges (AERs) of limb buds, and Crossley and Martin noted that the appropriate FGF receptor (FGFR2) is widely expressed in the interacting mesenchymes. While FGF8 is also expressed in the epithelia of the maxillary prominence and mandibular arch, principally at the junction of these structures (Crossley and Martin, 1995), FGF8 was also found in epithelia at the points where the pharyngeal endoderm makes contact with the surface ectoderm, including the prospective ganglionic placodes.

Wall and Hogan (1995) also examined the distribution of FGF8 and correlated it

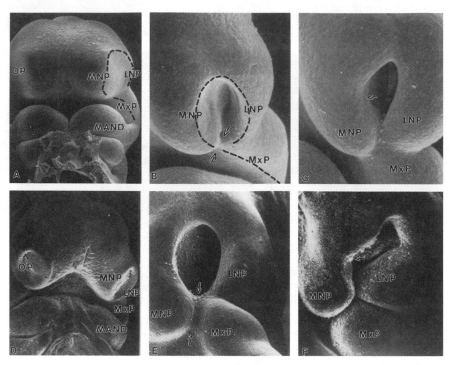

Figure 5.15 Scanning electron micrographs comparing the development of the primary palate in mouse (A, B, and C) and human (D, E, and F) embryos at different stages. In A and D the lateral edge of the olfactory placode (OP) is beginning to curl forward to initiate development of the lateral nasal process (LNP). In B and E, the medial nasal process (MNP) first makes contact (between arrows) with the maxillary process (MxP). Later (C and F), contact between the MNP and LNP is established. The distal portion of the MNP appears to bend laterally (arrows in C and F), which presumably increases the probability for the contacts. The broken lines added for this publication represent the approximate locations of FGF8 and Shh. (Sources: A, B, and C from Sulik and Schoenwolf, 1985; D, E, and F, from Jirasek, 1983.)

with the expression of BMP4 and 7, as well as with sonic hedgehog (shh). Shh was found in the facial ectoderm, particularly in the area of the presumptive olfactory placode, and was also found in the caudal (posterior) ectoderm of the hyoid arch, where it was coextensive with FGF8 and BMP7. This is the area where there will be an extensive outgrowth to form the opercular fold. These data (Crossley and Martin, 1995; Wall and Hogan, 1995) were confirmed and extended by Helms and colleagues (1997) and Hu and Helms (1999). They further characterized the location of FGF8 in the facial processes of chicks, and a later publication from the same group (Young et al., 2000) extended the findings

to mouse embryos. The distributions of FGF8 and shh in the facial epithelia is virtually identical, and they are approximated by dashed lines in Figure 5.15B and C. The relative roles of FGF8 and shh are unclear.

It now appears more obvious that research on limb development can tell us a great deal about how the face develops. It was shown some time ago by Richman and Tickle (1989) that the epithelia of facial processes and limb buds are interchangeable with the mesenchyme cells directing the outgrowth. Both limb and face development involves retinoic acid signaling (Wedden, 1991). From numerous studies on limb development (for a review, see Johnson and Tabin, 1997), it now appears clear that FGF8 in the apical ectodermal ridge is involved in maintaining the underlying mesenchyme cells in an undifferentiated, rapidly proliferating state; that it induces shh in the immediately caudal mesenchyme (as compared to its epithelial distribution in the face and branchial arches), which in turn determines that the smaller digits (e.g., the index finger) will be on the caudal side and the larger digits (e.g., the thumb) will be on the medial side; and that a number of "paralogous" hox genes are involved in determining the locations of different skeletal elements.

A number of investigators (Tamarin et al., 1984; Wedden, 1991; Richman and Delgado, 1995) have shown that it is possible to produce primary palate clefts in the chick with retinoic acid. Helms and associates (1997) found that treatment with retinoic acid severely reduces the expression of sonic hedgehog (shh), which is in the same signaling pathway, but leaves FGF8 expression intact (actually, it appears to be considerably reduced). The growth of the facial processes is reduced, resulting in clefts of the primary palate. Some of the studies described by Long and colleagues (2000) are difficult to interpret, such as the studies in which extirpation of epithelia results in both clefts and loss of shh. While extirpation of the epithelium sounds like a simple experiment, the vascularization of the facial processes would undoubtedly cause severe hemorrhage (even with efforts to decrease it), making it a major insult, which would change many aspects of development. Also, the use of beads soaked in RA in this publication resulted in a depression of growth in the region of the beads but not a facial cleft.

Morphogenetic movements also appear to be involved in primary palate formation (Fig. 5.16) (Bronsky et al., 1986; Johnston and Bronsky, 1991, 1995). The enhancement of the "curling up" of the lateral portion of the olfactory placode (Smuts, 1981) was already noted. In Smut's study,

the changes in the placode occurred immediately after administration of ATP to the medium in which the primary palates were being cultured. Also, the movements (Fig. 5.16) of the lateral nasal process, as well as lateral bending of the lower half of the medial nasal process, appear to be too rapid to be accounted for only by growth, suggesting morphogenetic movements are involved. All of these morphogenetic changes are inhibited by maternal hypoxia, leading to cleft lip in mice (see below). The distribution of terminal web in the subsurface cytoplasm is illustrated in Figure 5.16, and it largely coincides with points of flexure. The possible role of the terminal web in the generation of morphogenetic movements has been discussed in detail for the primary palate (Johnston and Bronsky, 1995) and for neural tube closure (Sadler, 1979). It appears that much of the folding of the neural tube is related to differential growth of the adjacent epithelia (Smith and Schoenwolf, 1998), and the same types of differential growth may be involved in primary palate formation.

In the mouse embryo, contact between the medial nasal processes on one side and the combined lateral nasal and maxillary processes on the other is preceded by major alterations in the fusion epithelia (Fig. 5.17) similar to that found in the more extensively studied secondary palate and in the neural folds (see below). These changes consist of cell death and sloughing of the surface layer of peridermal cells followed by alterations in the remaining basal cells, which apparently promote contact and adhesion of the processes. Eventually, mesenchyme cells replace these basal cells, and, in fact, the basal cells may contribute to the mesenchyme by transformation (Sun et al., 1998). The mesenchyme permits consolidation of the primary palate through the formation of collagen and other connective tissue components. Abnormal development of this epithelium also appears to be involved in clefts of the primary palate.

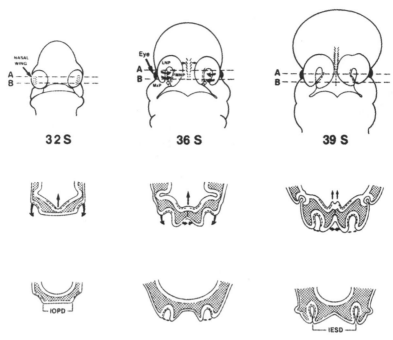

Figure 5.16 Schematic diagram of morphogenetic movements apparently involved in primary palate development. Embryos of 32, 36, and 39 somites are depicted with sections through the upper (A) and lower (B) portions of the primary palate. Closure is dependent on the lateral nasal and maxillary processes making contact with the medial nasal process. In addition to high rates of cell proliferation, this contact appears to be enhanced by a major morphogenetic movement of the LNP initiated as a curling forward of the lateral portion of the olfactory placode to form the LNP or nasal wing. Midline invagination of forebrain may be involved in bringing the MNPs toward the midline. Terminal web positions indicated by broken lines may provide much of the driving force for the morphogenetic movements. (Source: Bronsky and Johnston, in preparation.)

Figure 5.18A shows the locations of the facial processes at the completion of primary palate formation in a human embryo. Note that both the lateral nasal and maxillary processes contribute to the fusion area. In Figure 5.18B, the approximate contributions of the derivatives of the facial processes are illustrated. Note that the lateral nasal process contributes to the floor of the nose and upper part of the lip. Not shown is the apparent "overflowing" of maxillary processes over the median nasal process to form outer portions of the lip, which is responsible for the facial hair that forms the "mustache." Consequently, individuals with bilateral cleft lip have no facial hair on the portion of the lip derived from the medial nasal process.

At this point, it is worth making some additional comments about the size of the midface (Fig. 5.18). As noted in the section dealing with the holoprosencephalies, the amount of tissue between the olfactory placodes (i.e., the median nasal processes and the tissue between them) is determined to a large extent by placode positioning, which in turn is determined by the width of the anterior neural plate. It was also noted that SHH and other members of the hedgehog family play a role in the separation of the two halves of the anterior neural plate and also partially regulate the amount of tissue forming between the placodes. By the time the facial processes are forming, the picture is further complicated by the presence of the cerebral hemispheres to

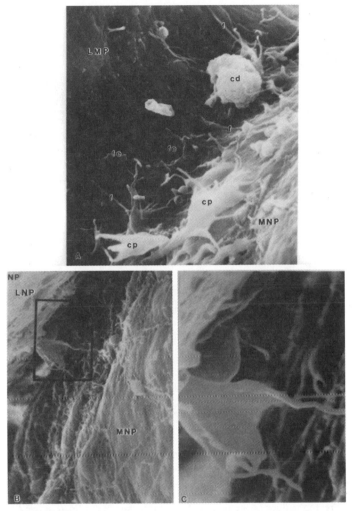

Figure 5.17 (A) In normal primary palate development surface peridermal cells are lost from the area of contact (comparable to area between arrows in Fig. 5.15B and 5.15E) and the underlying basal epithelial cells becoming very active with the formation of flattened edges (fe) and cell processes (cp). This activity may promote contact and adhesion of the facial prominences. It is absent or greatly reduced in mouse embryos of two (CL/Fr and A/WySn) high CL(P) incidence strains and following administration of some teratogens. (B and C) Severely reduced epithelial activity in a Cl/Fr embryo. The area is the same as that of the C57 embryo illustrated in A but the CL/Fr embryo is slightly older. Most peridermal cells persist, and the microvilli-rich margin (pcm) of such a cell is indicated in (B). (A) cell process reaching from the lateral nasal process (LNP) to the medial nasal process (MNP) is illustrated at higher magnification in (C). (Source: (A) From Millicovsky and Johnston, 1981a; (B and C) are from Millicovsky et al., 1982.)

which the developing nasal pits are attached. They appear to be drawn together by a sort of merging phenomenon, variations of which may alter the distance between the pits and even the medial nasal processes, as seen in frontonasal dysplasia, which sometimes exhibits midfacial clefting. Midline defects result from insufficient space between the nasal pits. It appears that the derivatives of medial nasal processes are laid down in a lateral to medial direction, and the last elements to be formed are

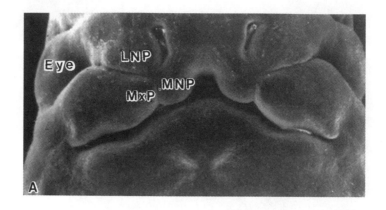

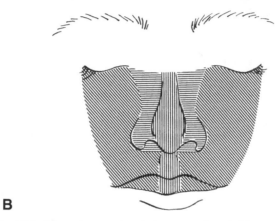

B

Figure 5.18 Scanning EM of the primary palate in human embryo is illustrated in (A), and the distribution of the derivatives of the embryonic facial processes in the diagram of the adult face in (B). The medial nasal processes are indicated by horizontal hatching, the lateral nasal processes by oblique hatching, and the maxillary processes by stipple. Note that the lateral nasal process contributes to the epithelial seam and, therefore, to the floor of the nose and the upper part of the lip. (Source: Modified from Sulik and Johnston, 1982.)

those adjacent to the midline. This would explain the missing philtrum and single central incisor of the least severe holoprosencephalies. These distances may be important in the pathogenesis of cleft lip.

5.6.1.2 *Cleft Lip with or without Cleft Palate* By far the most common of the severe craniofacial malformations are facial clefts (Fig. 5.19) and neural tube closure defects (e.g., anencephaly). While many of these malformations are components of syndromes such as the facial clefts

in holoprosencephalies already mentioned, the majority have only a low incidence of minor defects, often similar to those seen in the general population. Most of the malformations associated with anencephaly are secondary to the primary defect, which occurs very early. In contrast, facial clefting occurs relatively late in development (human postfertilization age 30–35 days for the lip and 50–60 days for the palate) and is often secondary to the prior developmental alterations that lead to the associated malformations. Obviously, neural tube

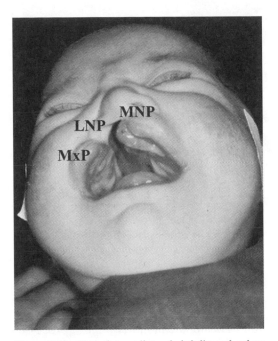

Figure 5.19 Complete unilateral cleft lip and palate infant. The approximate locations of the medial (MNP), lateral (LNP), and maxillary process (MxP) derivatives are indicated. Fusion between these processes occurred normally on the non-cleft side. (Source: Courtesy, The Hospital for Sick Children, Toronto.)

closure and lip and palate formation are developmental weak points.

The concept of multifactorial threshold etiology, developed many years ago by Falconer (1965), Carter (1976, 1977), and others, is helpful in understanding how multiple genetic and environmental factors may sometimes work together in the pathogenesis of these very common malformations (see Fraser, 1976, regarding facial clefts). The concept is presented diagrammatically in Figure 5.20. The application of this concept to the pathogenesis of facial clefting is the focus here.

Many factors play a role in getting the facial processes into contact. One of the earliest studies indicating the nature of a genetically predisposing factor was conducted by Trasler (1968). She found that growth direction of the facial processes that is less favorable for contact of the processes in genetically susceptible A/J mice may be seen as a genetic factor (Trasler, 1968) in contrast to the genetically resistant C57B1 mice (see Fig. 5.21). Preliminary studies by Johnston and Hunter (1989) on twins

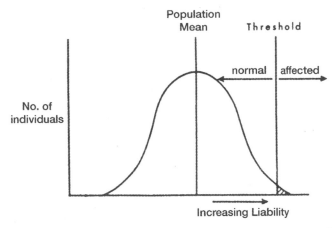

Figure 5.20 Distribution of individuals in a population (Caucasians) with different liabilities for cleft lip with or without cleft palate (CL (P)) according to Carter, 1976. If the combined effects of all genetic or genetic and/or environmental factors (liability) affecting the trait exceeds a certain threshold, the facial processes are no longer capable of complete contact and fusion and the individual falls beyond the threshold for CL(P). The area under the curve beyond the threshold represents the percent affected (here it is about 0.1%). This liability multifactorial threshold model has a number of attractive features. (Source: From Ross and Johnston, 1972.)

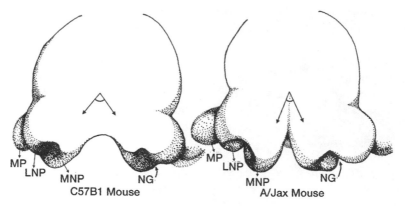

Figure 5.21   Growth directions of the medial and lateral nasal processes in A/J and C57Bl embryos. The direction of growth of the medial nasal process in the A/J embryo more closely parallels the midline, presumably putting it at greater risk of failing to contact the lateral nasal and maxillary processes during primary palate formation. (Source: From Trasler, 1968.)

discordant for clefts implicated small medial nasal processes somewhat similar to the trisomy-13 holoprosencephalies noted earlier. Small medial nasal processes would also predispose to failure of their contact with the lateral nasal and maxillary processes.

In their study of the pathogenesis of cleft lip induced by phenytoin in A/J mice, Sulik and associates (1979) found that suppression of the subepithelial process meshwork (see above) was associated with retardation in lateral nasal process growth. This would predispose to cleft lip as would the more generalized RA-induced retardation of facial process growth noted above.

Numerous factors contribute to the shaping of the face including the position of the olfactory placodes relative to the midline. The relative width of the prechordal plate and the floor plate of the anterior neural plate and tube, probably along with portions of the forebrain between the floor plate and the developing placode, appears to determine the initial position of the placodes. In a rare form of hypertelorism (Grieg syndrome), the abnormal gene is related to the hedgehog family. Since no animal model has yet been developed, the mechanism by which the altered gene leads

to the hypertelorism is unknown. By the time the medial and lateral nasal processes have formed, the cerebral hemispheres are forming, and they appear to be drawn to the midline by a morphogenetic movement (Fig. 5.16; 36S to 39S), secondarily affecting placode positioning (Fig. 5.16). It is not clear whether failure of this "drawing together" is responsible for the more common forms of hypertelorism such as fronto-nasal dysplasia (or midfacial clefting).

After contact of the facial processes, there may be problems related their fusion. Spontaneous suppression of activity in fusion epithelia was also observed in two genetically susceptible mouse strains (Millicovsky et al., 1982; Forbes et al., 1989), and, while it is a bit of a stretch, this could be the type of alteration resulting from the TGF-A variants observed in humans with cleft lip and/or palate. TGF-A (Fig. 5.2) is considered to be an embryonic form of EGF. It is known to be involved in the regulation of epithelial differentiation. For example, it is produced by basal cells of the epidermis, which then use it to regulate their own proliferation (a rare example of an autocrine growth factor). In patients with psoriasis, a scaly skin condition, the *excess* production of epidermal cells is linked to *increased* production of TGF-A caused by an underlying inflammation. Also, if TGF-A is linked to a keratin promoter in transgenic mouse embryos, elevated levels in the epidermis are associated with excessive thickening (Vasser and Fuchs, 1991). Apparently, these mice do not have cleft lip. The regulation of this activity and its possible role in facial clefting are considered in Section 5.6.2. TGF-A and its receptor, EGFR, have been demonstrated immunohistochemically in the tips of the growing medial and lateral nasal processes as well as in the maxillary process (Iamaroon et al., 1996). A knockout for EGFR, the receptor for TGF-A, resulted in a number of malformations including cleft palate, a small mandible, and a narrow face, but not cleft lip (Miettinen et al., 1999).

There have been a very large number of genetic studies on families with cleft lip and/or palate (Fogh-Andersen, 1942; Chung et al., 1986, 1987; Marazita et al., 1992; Mitchell and Risch, 1992) (see Chapter 7). Most researchers agree that many cases are multifactorial. An exciting new approach was developed by Ardinger et al. (1989) in which they examined DNA fragments of families with cleft lip and palate for restriction fragment-length polymorphisms (RFLPs) indicating rare variants of genes that might interfere with normal development. Ardinger et al. (1989) found unusual variants for the signaling factor TGF-A. There has been considerable support for the involvement of TGF-A in secondary palate formation (Abbott and Pratt, 1987; Abbott et al., 1998), and Shiang and colleagues (1993) found an association in human families with cleft palate. TGF-A has also been demonstrated to be present in the mouse primary palate (Iamaroon et al., 1996). The association between TGF-A and cleft lip has been confirmed in some populations (Chenevix-Trench et al., 1992; Holder et al., 1992) but not in others (Vintiner et al., 1992; Wyszinski et al., 1996).

In a more extensive study that included the same subjects used by Ardinger et al. (1989), Lidral and associates (1998) found that the relationship between TGF-A and cleft lip, only held for subjects with a family history of cleft lip. However, they did find an association with MSX1 and TGF-β3. Only the association with MSX1 was found for cleft palate subjects. The candidate genes were selected for their known involvement or earlier evidence of association. Sometimes, mutations (van der Boogaard, 2000) or knockouts (Satokata and Maas, 1994; van der Boogaard et al., 2000) for the genes were known to produce facial clefts.

This approach is ripe for further exploitation. With the rapid expansion of knowledge related to normal development, many more candidate genes will be selected. Some may not be found in all populations. One promising area for research is the examination of interaction between these genes and environmental factors.

It is unlikely that environmental factors play much of a role in most cases and it is possible to have facial clefts secondary to the combined effects of predisposing genes without the contribution of environmental factors. However, these small effects may be enough to move a genetically predisposed embryo over the threshold or pull the embryo back to the normal side (Fig. 5.20). Examples are the use of maternal hyperoxia or vitamin supplementation to decrease the incidence of spontaneous clefting in mice (as also, vitamin supplementation in humans).

There are few well-documented environmental factors affecting clefting incidence in either humans or experimental animals (for a review, see Wyszinski et al., 1996) (Chapter 8). While there are some conflicting studies on the effects of cigarette smoking, the most careful studies indicate that it is at least a mild predisposing factor. Most retrospective studies are plagued by maternal memory bias, and the numbers available for prospective studies are small. Prospective studies have been conducted in Sweden, where habits such as cigarette smoking are routinely recorded in prenatal clinics. These studies (Erikson et al., 1979; Keels, 1991; Keels et al., 1991, 1992; Kallen, 1997) indicate there is a rise in clefting associated with cigarette smoking that is loosely correlated with the number of cigarettes smoked per day. Recently, a massive retrospective study (Chung et al., 2000) using birth records from the National Center for Health Statistics (New York, NY) in which both cigarette smoking and clefts were recorded (2207 cleft lip and/or palate and 4414 controls) found approximately the same increase as well as a progressive increase with numbers of cigarettes smoked per day.

Other large positive studies have been published by Khoury and colleagues (1987), and Wyszynski and colleagues (1997). Several studies (Hwang et al., 1995; Shaw et al., 1996; Romitti et al., 1999) have examined the interactions between the presence of unusual TGF-A variants in the child and cigarette smoking in the mother, both showing even higher risks. Werler and associates (1990) found no correlation between facial clefting and cigarette smoking. Lieff and colleagues (1999) reanalyzed data from the same sample with the same result. In light of all the other studies noted, this result is surprising, but sample selection or inadequate records are possible problems. At the other end of the spectrum is a study (Savitz et al., 1991) showing a positive correlation between *paternal* cigarette smoking and clefts.

The effects of cigarette smoking on prenatal growth have been well documented (Longo, 1980). Smoking appears to exert its effects on prenatal development by interfering with the electron transport chain (Fig. 5.22), primarily by reducing the amount of oxygen available for accepting electrons at the end of the chain (Longo, 1980). When the antimetabolite 6-aminonicotinamide (6-AN) is incorporated into the NADH dehydrogenase at the head of the chain, it

interferes with its acceptance of electrons from high-energy intermediates. In both cases, the reduced electron flow results in a reduction of ATP formation, the main source of energy for numerous cellular functions including morphogenetic movements. It should be noted that there are many toxic chemicals in cigarette smoke, but a screen for abnormal variants in metabolizing enzymes for one of them was negative (Hartsfield et al., 2000).

Although they were conducted many years ago, the studies of Landauer (1954) and Landauer and Sopher (1970) are probably still relevant to the problem of smoking and cleft lip. They used a number of agents to interfere with the function of the electron transport chain (Fig. 5.22) in chick embryos. They used 6-aminonicotinamide, 3-acetyl pyridine, and boric acid (a rat poison) to interfere with the function of NADH dehydrogenase, the first enzyme in the chain, which is responsible for accepting electrons from high-energy intermediates. As the electrons pass down the chain, they produce ATP through oxidative phosphorylation. At the end of the chain, the electrons are absorbed in the formation of water from hydrogen and oxygen. The researchers were able to produce a wide range of malformations including limb defects and clefts of the primary palate. Through the use of high-energy intermediates such as ascorbate (vitamin C) to introduce electrons at points beyond the block (Fig. 5.22), they were able to reduce the incidence of malformations, presumably by increasing ATP production.

Cigarette smoking could interfere with the functioning of the electron transport chain in at least two ways. As already noted, smoking drastically reduces the amount of oxygen available to the embryo, which would remove the electron acceptor oxygen at the end of the chain. Also, the carbon monoxide in the cigarette smoke binds tightly to hemoglobin, thereby making less available for the transport of oxygen. Carbon monoxide also has nega-

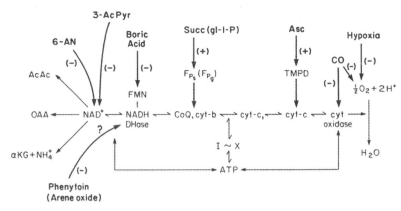

Figure 5.22   Schematic presentation of the site of action of teratogens and energy sources as antiteratogens in electron transport. Electrons are donated from high-energy intermediates to NAD+ (oxidized nicotinamide adenine dinucleotide) at the head of chain. The electrons move down the chain through a series of enzymes including NADH dehydogenase. As they move down the chain, ATP is formed. The electrons are removed at the end of the chain by combination with oxygen. The nicotinic acid analogues 6-An and 3-acetyl pyridine interfere with the function of NAD, while boric acid and, perhaps, phenytoin (produces cleft lip in mice) block electron transport by binding to NADH dehydrogenase. These alterations result in various malformations such as clefts of the primary palate in chick embryos and mice. Increased carbon monoxide from cigarette smoke interferes with the function of cytochrome oxidase and the delivery of oxygen by binding to hemoglobin. Perhaps the most serious consequence of smoking is hypoxia, which severely reduces the supply of oxygen to embryonic tissues. The electron flow is therefore dammed up with the reduced capacity to remove electrons from the end of the chain. Maternal hypoxia produces cleft lip in mice, as does exposure to CO. Experimentally, it is possible to ameliorate the effects of 6-AN and boric acid in chick embryos through the use of high-energy intermediates such as succinate which donates electrons to coenzyme Q (CoQ), and ascorbate (vitamin C), which donates electrons to cytochrome C (cyt-c), thereby by passing the block. (Source: From Bronsky, 1985.)

tive effects on cytochrome oxidase, the last enzyme in the chain.

Trasler and colleagues (1978) have used 6-AN to induce clefting in C57B1 mice, and Millicovsky and Johnston (1981b) and Bronsky and associates (1986) have used maternal hypoxia achieved by lowering oxygen levels in environmental chambers for inducing cleft lip in A/WySn and CL/Fr mice. Bronsky and colleagues (1986) showed that interference with lateral nasal process development was associated with epithelial cell death, and they discussed the possibility that this cell death might be related to the high-energy requirements of the epithelium and its possible role in morphogenetic movements in a very rapid growth spurt seen in this process just before contact with the medial nasal promi-

nence. Proposed morphogenetic movements are summarized in Figure 5.16. Carbon monoxide also induces cleft lip in A/J mice (Bailey et al., 1995). Not only does carbon monoxide interfere with oxygen delivery by forming carboxyhemoglobin, it also interferes with the function of cytochrome oxidase at the end of the chain (Fig. 5.22). Finally, Millicovsky and Johnston (1981a, 1981b) reduced the incidence of clefting in A/J and A/WySn mice through maternal hyperoxia, as did Bronsky (1985) for CL/Fr mice.

Two additional comments are inorder. The first deals with the failure to implicate cigarette smoking in the etiology of NTDs. This probably relates to the fact that the early embryo utilizes anaerobic glycolysis to generate ATP (Hunter and Sadler,

1988). The second comment relates to the studies of Petter et al. (1977) on the role of hypoxia in limb defects in rabbits. They concluded that hypoxia stimulated the early release of large erythroblasts, which plugged small blood vessels in the limb bud leading to hemorrhage. While we did not detect hemorrhage in our hypoxia studies, Hetzl and Brown (1973) found it in phenytoin-induced cleft lip in CL/Fr mice and hemorrhages were also found by Diewert and Shiota (1990) in human embryos with spontaneous cleft lip.

Phenytoin is a potent inducer of cleft lip in A/J mice (Massey, 1966; Sulik et al., 1979; Hansen and Hodes, 1983). However, evidence relating phenytoin to human cleft lip is conflicting, and it is unlikely to be responsible for many cases. Khoury and colleagues (1993) conducted the most extensive study, in which they found that the incidence for cleft lip was approximately double that of controls. Many years ago Hanson and associates (1976) linked phenytoin to mild developmental changes and coined the term *fetal hydantoin syndrome* to describe them.

One study (King and Irgens, 1996) used the Medical Birth Registry of Norway to examine the relation between anticonvulsant drugs and cleft lip and spina bifida. King & Irgens (1996) reported that the incidence of cleft lip went from three times the incidence in controls before 1981, down to control levels from 1981 through 1992 (odds ratio 1:3.0 to odds ratio 1:1.1), while the odds ratio for spina bifida went from 1:1.5 to 1:4.4 (i.e., an increase from 1.5 times controls to 4.4 times controls). These changes coincide with the switch from phenytoin to carbamazepine as the primary drug. Win one, lose one! These findings are in agreement with a study conducted by (Wray et al., 1982) on A/J mice. In contrast to phenytoin, which is a potent teratogen of cleft lip in this strain, only a very modest increase was found with carbamazepine. The timing of administration in their study

would be too late for spina bifida. Phenytoin is still widely used in the United States.

There is one study (Shaw and Lammer, 1999) linking alcohol consumption and cleft lip with or without cleft palate. While low consumption levels showed no increase in risk, higher levels of consumption (five or more drinks at one sitting) showed an increased risk.

The most obvious possible explanation for the difficulty in identifying individual genetic and environmental factors is that many different factors may be operating in individual cases, each with a relatively small effect.

One approach is to examine possible interactions, such as the study conducted by Hwang and colleagues (1995), where it was found that the risk of having a cleft was much higher if the embryo had the abnormal TGF-A variant and the mother smoked cigarettes. These results have been confirmed by Shaw and associates (1996).

Romitti and colleagues (1999) conducted a more extensive study. They found interactions between TGF-A, MSX1, cigarette smoking, and alcohol consumption and clefting. Risk estimates were elevated for cleft palate with variants at the TGF-A and MSX1 sites. Risk estimates for cleft lip were most elevated for the combination of alcohol and MSX1.

Many coordinated developmental mechanisms are involved in the attainment of contact and fusion, and there are many possible errors leading to clefting. An examination of the liability:threshold model (Fig. 5.20) reveals that there may, in fact, be opportunities for prevention. The area under the curve beyond the threshold is very small, indicating that most clefts are not far over the threshold. This is also supported by studies of monozygotic twins where half the pair are discordant for the defect, presumably with the affected twin just over the threshold and the unaffected twin just on the normal side. Small differences in environment are presumably the deciding factors in these and some other cases. For example, maternal smoking could push a genetically predisposed embryo over the threshold, while not smoking could pull another embryo back. A carefully conducted study by the British Medical Research Council (M. R. C. Vitamin Study Research Group, 1991) has shown conclusively that prenatal dietary supplements containing folic acid reduce the recurrence rates of neural tube defects, and there is some evidence (Briggs, 1976: Tolarova, 1982, 1990; Shaw et al., 1995a) that folate supplements may also be effective for cleft lip prevention.

5.6.2 Secondary Palate Formation and Cleft Palate

Even before primary palate formation is completed, the secondary palate is begin-

ning to form. As noted above, FGF8 and sonic hedgehog expression are found along the medial edge of the maxillary prominence (Fig. 5.14). Presumably, they are also involved in growth of what will be the palatal shelves.

In mammals, the palatal shelves first grow downward beside the tongue (Fig. 5.23A, C, and E). At about 50 to 60 days in the human, the tongue is withdrawn from between the shelves and they elevate, contact in the midline, and fuse. There is a very large volume of literature on mechanisms involved in secondary palate development and the alterations leading to cleft palate (see, e.g., Zimmerman, 1984; Greene and Weston, 1994).

It used to be thought that all the cells in the epithelial seam underwent cell death. Waterman and Meller (1974) showed clearly that in the human embryo the

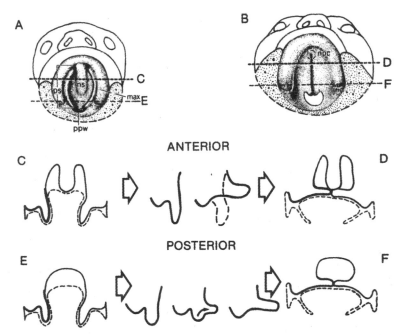

Figure 5.23   Normal secondary palate development before (A, C, E) and after (B, D, F) palate shelf (ps) elevation. Other abbreviations are as follows: ns, nasal septum; max, maxillary prominence; ppw, posterior pharyngeal wall; npc, nasopalatine canal. Broken lines indicate sections of the tongue. Many developmental abnormalities may lead to cleft palate (CP). A large number of human CP cases appear to result from problems related to getting the shelves elevated above the tongue. Problems of epithelial differentiation (Fig. 5.16) may contribute to clefting in some cases. (Source: From Johnston et al., 1975.)

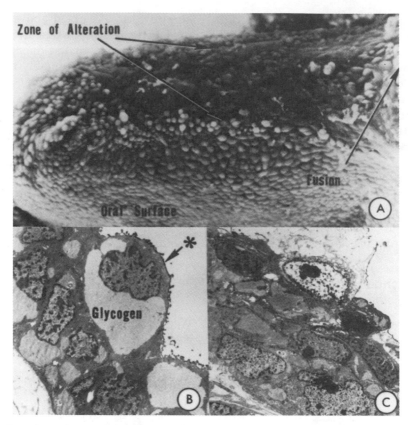

Figure 5.24 Scanning and transmission electron micrographs of the same human palatal shelf. An asterisk in B, a section through the oral surface epithelium, marks the metal coating necessary for scanning EM. The surface epithelial peridermal cells are being lost from the zone of alteration (epithelium sectioned in C), and the underlying epithelia of the opposing shelves will contact and adhere to each other. It now has been shown that many of these epithelial cells later transform into mesenchymal cells, possibly making palatal fusion a relatively error-free mechanism. (Source: From Waterman and Meller, 1974.)

peridermal cells of fusion epithelium died and apparently sloughed off, while while the underlying epithelial cells appeared to become very active in contact and adhesion of the palatal shelves (Fig. 5.24). Waterman (1976) also showed that very similar changes were involved in the contact and adhesion of the neural folds in neural tube closure. More recently, Fitchett and Hay (1989) provided ultrastructural and histochemical evidence that these basal cells underwent transformation into mesenchymal cells, and this was further substantiated by cell marking studies (Shuler et al., 1992).

Molecular studies on the epithelial change indicate that TGF-β3 is involved in the transformation. Kaartainen and associates (1995) found that in TGF-β3 null mutants, there is partial failure of the breakdown of the epithelial seam and that it is closely associated with persistence of the epithelial basement membrane. Also, they found that addition of TGF-β3 to cultures containing the null mutant shelves resulted in complete breakdown of the basement membrane of the medial epithelial seam and normal fusion of the cultured shelves (see also Taya et al., 1998). Finally, Sun and colleagues (1998) showed that TGF-β3 promotes transformation of the medial edge epithelial seam of chick embryo palatal shelves (which normally remain unfused).

Secondary palate formation appears to be such a high-risk developmental event that it seems unusual that cleft palate is not even more common. The shelves grow down beside the tongue, which must be removed before they can elevate and make contact (Fig. 5.23). Since changes in the programing of the epithelium is on a fixed schedule, any delay in elevation for fusion and breakdown of the epithelial seam could be critical. Also, the shelves have to be sufficiently large and correctly positioned in order to make contact at all. Further compounding the problem is the fact that secondary palate formation is one of the last embryonic changes to occur, so many previous developmental alterations could predispose to cleft palate, as evidenced by the enormous number of syndromes of which cleft palate is a component (Gorlin et al., 1990).

Not surprisingly, cleft palate is very easy to induce in experimental animals (summarized in Zimmerman, 1989; Greene and Weston, 1994). One of the earliest studies was conducted by Fraser and Fainstat (1951). They produced a very high incidence of cleft palate with cortisone, a corticosteroid (Fig. 5.2). Again, the retinoids were potent producers of cleft palate. While there is little evidence linking any of the myriads of teratogens to human cleft palate, evidence for genetic predisposition exists. As seen in cleft lip and palate, cleft palate by itself is probably a multifactorial trait.

There is some information about the possible nature of genetic predisposition on facial clefting. These include: 1) the identification of unusual TGF-A isoforms in individuals with cleft palate (Shiang et al., 1993); 2) unusually wide faces which have been in a Finnish subpopulation with a very high incidence of cleft palate (Saxen, 1976) and in different strains of rodents (Siegel and Mooney, 1986; Vergato et al., 1997), where rapid transverse facial growth could move the palatal shelves further

apart and prevent fusion during palatogenesis (Siegel and Mooney, 1986; Vergato et al., 1997); 3) tongue-tie in a familial form of cleft palate in Iceland, which could inhibit protrusion of the tongue during shelf elevation (Moore et al., 1991); 4) evidence of a large tongue in MZ twins discordant for cleft palate (Johnston and Hunter, 1989; Johnston and Bronsky, 1995), and 5) mandibular retrognathia in Pierre Robin syndrome, which interferes with tongue descent during palatal shelf elevation (Poswillo, 1996). In an animal model for Pierre Robin syndrome, Trasler and associates (1956) and Poswillo (1966) showed that amniotic sac puncture held the face down against the chest wall, inhibiting mouth opening and mandibular growth and causing wide palatal clefts to form. There is some evidence for reduced mandibular size in many cleft cases (Ross and Johnston, 1972; Johnston and Bronsky, 1995). Cleft palate often results from gene knockouts (Long et al., 2000), but so far these are usually unlike the common human forms. Miettinen and colleagues (1999) have found that a knockout for EGFR, the receptor for TGF-A, produces cleft palate, a small mandible, a narrow face, and numerous other malformations, but not cleft lip. Hopefully, improvements in our understanding of the normal mechanisms of secondary palate formation will lead to more rapid progress in our understanding of cleft palate formation.

There is very little evidence linking human cleft palate with environmental factors. As is the case with cleft lip, this may be largely due to minor individual effects in a multifactorial etiology. Investigations of cleft palate are more difficult than those of cleft lip. Their less obvious nature leads to their being omitted on birth certificates.

Of considerable interest are recent studies on mice linking the neurotransmitter gamma-amino-butyric acid (GABA) to secondary palate development. Some years ago, studies conducted by Zimmerman and

Wee (1984) suggested that GABA and other neurotransmitters might be involved in palate shelf elevation. In more recent studies, Culiat and associates (1993, 1994, 1995) noted that disruption of the cp1 locus resulted in cleft palate. This locus includes genes coding for GABA receptors, and the researchers (Culiat et al., 1995) showed that the addition of a transgene coding for the β3 subunit of the type A GABA receptor resulted in normal development. Further, Condie and colleagues (1997) demonstrated that a knockout of the GABA-producing enzyme GAD67 results in a very similar cleft of the palate. Apparently, both sets of cleft palate mice have no other skeletal malformations. Diazepam (Valium), a benzodiazepine that alters GABA signaling, induces cleft palate in mice (Miller and Becker, 1975; Wee and Zimmerman, 1983). There is also some evidence linking diazepam to facial clefts in humans (Saxen and Saxen, 1975; Safra and Oakley, 1975).

5.7 ABNORMAL DEVELOPMENTAL MECHANISMS IN THE FORMATION OF TISSUES AND ORGANS

The number of abnormalities involving nonskeletal tissues is relatively low. Most of this discussion concentrates on skeletal tissues.

5.7.1 Abnormalities of Tooth Formation

Tooth formation involves numerous interactions of epithelia and mesenchyme (see also Chapters 4 and 18). The mesenchyme is of neural crest origin, which is already programed for appropriate responses to the inductive influences of the various epithelia it encounters as it moves into the facial region. Also noted are the limb bud-like growth centers that are set up after the crest cells have arrived. Similar to the apical ectodermal ridge, much of the subsequent programing is taken over by the growth centers of the facial processes, such as those at the ends of the mandibular arch processes, which grow forward essentially "free-ended" most obvious in avian embryos. The situation is obviously more complicated than that of limb buds because of the merging and fusion involved, and, in the cases of the medial and lateral nasal processes, the olfactory placode.

The initial tooth rudiment is the dental lamina, which is a linear thickening from which the individual tooth buds are derived. Formation of the dental lamina begins before the merging and fusion in the primary palate occurs. Errors in merging and fusion lead to disruptions in the dental lamina that are most obvious in facial clefts (see also Chapter 18 for a more complete discussion), including clefts of the mandible. Even when fusion appears to have been normal, the high incidence of tooth abnormalities (e.g., lateral incisors) forming in the region of fusion suggests that they are areas prone to error. There has been remarkable progress in research related to our understanding of the molecular controls involved in tooth formation (Maas and Bei, 1997; Thesleff, 1998; Peters and Balling, 1999; Cohen, 2000).

As noted previously, many gene knockouts result in missing teeth. It should be emphasized, however, that commonly missing teeth in humans, such as lateral incisors, second bicuspids and third molars, apparently have multifactorial etiology (see Chapter 18).

5.7.2 Development of Cartilage and Bone

This section reviews some of the newer information on the development of cartilage and bone, particularly as it relates to such malformations as achondroplasia, and Crouzon and Apert syndromes. This section also presents new findings on roles

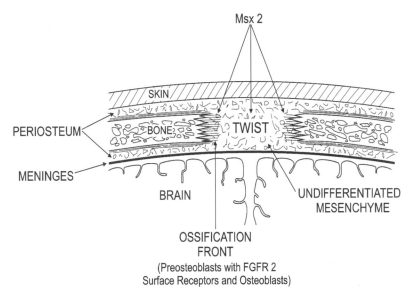

Figure 5.25 Distribution of some of the regulatory molecules involved in suture development. Fibroblast growth factor receptors (FGFRs) on mesenchyme cells enhance their differentiation into osteoblasts when they bind FGFs, and are repressed by Twist. Msx also enhances the differentiation into osteoblasts and is involved in cell death, which may be significant in early suture closure. The mutated FGFR2s involved in the synostoses found in Crouzon and Apert syndromes show increased activity (gain of function) when bound to FGF, and MSX2 mutations involved in Boston-type synostosis are also dominant gain-of-function mutations. The TWIST mutation in Opitz syndrome is nonfunctional, and the one functional gene is unable to produce enough Twist to prevent the synostosis, an example of dominant haploinsufficiency.

of fibroblast growth factor receptors (FGFRs) and other molecules in normal and abnormal cartilage and suture development (see also Chapters 2, 6, and 20).

FGFs have been shown to stimulate early differentiation of mesenchyme cells into postmitotic chondroblasts, thereby slowing cartilage growth chondroblasts (Deng et al., 1996; Webster and Donoghue, 1996). A mutation in FGFR3 is responsible for almost all cases of achondroplasia (Shiang et al., 1994) apparently due to increased activity (gain of function) of the FGF/FGFR3 complex.

The distribution of signaling factors, receptors, and homeotic genes involved in suture development is illustrated schematically in Figures 5.1 and 5.25. Gain-of-function mutations of FGFR2 appear to result in craniosynostosis by stimulating undifferentiated mesenchyme cells to dif-

ferentiate into osteoblasts, thereby enhancing bone formation and premature suture closure. Different mutations give rise to different syndromes, and sometimes variations of the same syndrome (for a review, see Chapter 6; Cohen, 1997). Crouzon syndrome results from a wide variety of mutations (Reardon et al., 1994; Jabs et al., 1994), whereas the vast majority of Apert cases have the same mutation (Park et al., 1995). In Boston-type craniosynostosis, it is an abnormality in MSX2 that results in synostosis (Jabs et al., 1994). A less common form of synostosis is that found in Opitz syndrome, where the abnormality is in the TWIST gene (El Ghouzzi et al., 1996; Howard et al., 1996). The developmental abnormalities in the Boston-type craniosynostosis and Crouzon syndrome are largely confined to the sutures of the cranium and the face and cranium, respectively. In Apert and Opitz syndromes, many other structures are abnormal.

A series of studies (Kim et al., 1998; Rice et al., 1999, 2000) has been conducted on mouse calvarial sutures in vivo and in vitro in an attempt to elucidate the roles of the above mutations in suture closure. The distribution of some of the molecules involved is presented diagrammatically in Figures 5.1 and 5.25. The Twist protein, FGF2, and small amounts of Msx2 are found in the median portion of the suture. Larger amounts of Msx2 are found in the mesenchyme closer to the ossification front. The FGFR2 receptor is found close to the ossification front. FGF4, which is widely distributed, stimulates both proliferation in the suture as well as osteoblast differentiation. The FGFR2 mutation is thought to cause increased activity (gain of function) of the FGF4/FGFR4 complex, resulting in enhanced osteoblast differentiation and premature fusion. The Twist protein suppresses the function of the FGFR, thereby stimulating proliferation and inhibiting differentiation. The Twist mutation inactivates the gene, and the remaining normal gene does not produce enough protein (haploinsufficiency), leading to premature suture closure. The role of Msx2 is not as clear. It is known to be involved in the regulation of cell death and is upregulated by Bmp4, which is also present in the suture. Both Msx and Bmp stimulate mesenchymal cells to differentiate into osteoblasts. The MSX2 mutation in Boston-type craniosynotosis is a gain-of-function mutation (i.e., when complexed with its ligand, it shows increased activity).

5.8 CONCLUDING REMARKS

Since the authors wrote their 1995 review of normal and abnormal mechanisms involved in craniofacial development, there have been enormous advances in our understanding of the molecular basis of normal developmental mechanisms such as embryonic induction, growth regulation, and pattern formation including the regulation of programed cell death (apoptosis). This information is being applied to the problems of abnormal craniofacial development with considerable success.

As might be expected, progress on the single-gene defects has been the most successful. This includes our understanding of the developmental alterations in cyclopia, DiGeorge syndrome, suture closure defects, and abnormalities of cartilage development.

Although rare, the development of cyclopia has attracted a disproportionate amount of attention. As shown many years ago, the primary defect is in the interaction between the prechordal plate (an anterior extension of the notochord) and the overlying neural plate. Now it is known that a signaling molecule, sonic hedgehog (Shh), is produced by the prechordal plate, and it suppresses a homeotic gene, Pax-6, in the overlying neural plate. This portion of the neural plate then takes a different path of development, leading to the formation of the floor plate and separation of the eye fields. The genetic defect can either be in the inducing or responding tissue, but both lead to the same end result: the floor plate does not form and the eye fields fail to separate, leading to the formation of a large, single, midline eye. A rare but well-known teratogen, *Veratrum californicum*, found in a range plant, apparently causes cyclopia in sheep by interfering with cholesterol synthesis, which in turn is required for the synthesis of Shh.

Cyclopia is included in the holoprosencephalies, a group of malformations that derives its name from the tendency to form single-cavity forebrains. However, in the other holoprosencephalies the eyes remain separate. Most human cases result from trisomy 13 (an extra chromosome for the 13th largest pair) and trisomy 19 in which there is underdevelopment of the anterior neural plate, but the eye fields are separated. With increasing severity, the eyes become progressively smaller, and the olfactory placodes, which are derived from the anterior neural plate, come to lie increasingly close to the midline. The decreasing space reduces the size of the medial nasal processes, resulting in cleft lip and eventually arrhinencephaly ("no nose") when the placodes come into contact, thereby preventing medial nasal process development. In these cases, the mechanism may be quite different from that found in cyclopia. In an animal model for the very common fetal alcohol syndrome, which appears to be at the mild end of the holoprosencephaly spectrum, there is cell death in the anterior neural plate associated with a severe reduction in the underlying mesoderm.

Recent studies on an adhesion molecule, L1, may shed some light on the problem. Cell migration depends on "make-and-break" adhesions that permit the cells to pull themselves along a substratum. This L1 molecule is exquisitely sensitive to very low levels of ethanol, which when added to cell culture media prevents cell movement and neurite outgrowth (axon formation, which is fundamentally similar to cell migration). In vivo, ethanol has major effects on migrating neural crest cells, as well as on migrating neuroblasts and synapse formation in the developing central nervous system. Crest cells have recently been shown to have very high levels of one form of L1. This raises the interesting possibility that at least part of the primary problem in noncyclopia holoprosencephalies is related to retarded migration of gastrulating mesodermal cells.

The continued application of the findings of developmental biologists studying Drosophila is providing enormous advances in our understanding of normal and abnormal vertebrate development. Through the duplication and subsequent alteration genes involved in Drosophila development, families of regulatory molecules have evolved. Members of these families, which bear a close resemblance to their "ancestral" Drosophila parent gene, may have overlapping or quite different functions as shown mostly by transgenic mice in which one or more regulatory genes has been knocked out. Such information is being used to identify candidate genes for studies of single-gene human malformations such as craniosynostosis, or for variations in the candidate genes in malformations of multifactoial etiology.

An interesting example of the administration of such technology has been the search for the developmental origins of DiGeorge syndrome. First came the observations that cranial neural crest cells contribute to the septal mesenchyme of the outflow portion of the developing heart and that extirpation of these cells leads to outflow tract cardiovascular defects such as the tetrology of Fallot. Environmental factors such as ethanol and retinoic acid that interfere with crest cell development also produced these defects, along with the pharyngeal gland and the middle and external ear defects seen in DiGeorge syndrome. Then came studies of knockouts for various regulatory molecules involved in neural crest development such as Hoxb 3 and endothelin and its receptor. These studies produced outflow tract defects as well as other defects (e.g., in pharyngeal glands, and the external and middle ear) somewhat similar to those found in DiGeorge syndrome. Interestingly, the primary defects in the endothelin (or its receptor) had their most severe effects on the third and fourth branchial arches.

Similar developmental changes were observed in transgenic mice with the chromosomal deletion found in DiGeorge syndrome. In this study, the developmental alterations seemed to involve only a subpopulation of crest cells those contributing to the development of the aortic arch vessels. Finally, a knockout for one of the genes from the deleted region, Tbx (a gene known to be involved in craniofacial development), was found to produce the same developmental defects. Actually, the defects observed in the latter studies produced only cardiovascular malformations and not pharyngeal gland abnormalities, and therefore are more similar to velocardiofacial (Sphrintzen) syndrome, another syndrome resulting from the 22q deletion.

The results of studies related to Treacher Collins syndrome are unclear. Here the abnormal gene (TCOF-1) does not appear to be a regulatory gene, as are most of the others involved in abnormal development, but the normal gene product, treacle, appears to be involved in the transport of materials between the nucleolus and the cytoplasm. Study of transgenic mice with the TCOF-1 gene revealed extensive cell death in the neural folds as well as in other parts of the neural plate and apparently in the ganglionic placodes. The authors concluded that the malformations, which were extensive and often dissimilar to those of Treacher Collins syndrome, resulted from abnormalities in the neural crest. In a retinoic acid–induced model, the authors concluded that the primary defect was in the portion of the trigeminal ganglion derived from its ganglionic placode. The trigeminal placodes in the TCOF-1 mouse also appeared to have undergone extensive cell death.

Animal models for retinoic acid syndrome indicate severe cell death in the hyoid arch neural crest prior to migration, which appears to be an expansion of the cell death that normally occurs between the first and second arch crest cells. Migrating crest cells appear also to be affected. The middle and external ear are severely malformed. Much more serious, in terms of survival, are defects in the cardiovascular outflow tract and in the pharyngeal glands, including severe deficiencies or absence of the thymus and parathyroid glands. There are also defects in cerebellar development.

Despite their shortcomings, studies of transgenic mice with specific gene alterations are necessary for studying the precise developmental alterations involved in single-gene defects. One apparent shortcoming of molecular studies involving both normal and abnormal development is an overemphasis on the role of neural crest cells. There are now many excellent markers for crest and other types of cells involved in craniofacial development, so identifying the different cell populations in mammalian embryos is relatively easy. Also, many aspects of normal craniofacial development that may be important in the development of a number of malformations may not be receiving as much attention as they deserve. Examples are the roles of intercellular matrices and morphogenetic movements.

Hemifacial microsomia, easily the third most common severe facial malformation after clefts of the lip and/or palate, bears a number of similarities to retinoic acid and DiGeorge syndromes, as well as important differences. They have similar ear and cardiovascular defects but no pharyngeal gland problems. In spite of the high incidence in humans, no animal models with spontaneous hemifacial microsomia have been identified, in contrast to cleft lip and/or palate. Neither have any environmental factors been identified, although the etiology appears to be multifactorial.

Much greater progress have been made in studies of cleft lip and/or palate, which are also multifactorial problems. Here, numerous animal models with spontaneous clefting are available, including several mouse models that have been extensively studied. Studies on normal development have provided much information regarding the mechanisms involved in normal primary and secondary palate formation. These studies have provided information about how the initial plan for midfacial development is dependent on the forebrain, how growth is regulated by molecular signaling, the apparent role of morphogenetic movements, as well as information about the mechanisms involved in fusion of the facial processes after contact is made. Such studies have suggested numerous candidate genes. Since they are presumed to have individual small effects, a major focus has been identification of variants that might alter development in a manner predisposing to clefts. Examples include TGF-B3, which is normally involved in the breakdown of the epithelial seam during fusion, and MSX1, a mutation of which has been associated with cleft palate in humans; and an Msx1 knockout in mice that results in similar malformations.

Variants of TGF-β3 have been associated with cleft lip and variants of MSX1 with both cleft lip and cleft palate. There are numerous other possibilities, one example being GABA signaling. Interference with GABA signaling has been shown to produce cleft palate in transgenic mice, both by knocking out one of the GABA receptors or interference with an enzyme involved in GABA synthesis. These knockouts seem to be remarkably specific in that no associated malformations were detected.

The fact that environmental factors may play a substantial role in cleft lip and palate, as well as the fact that most embryos about to develop cleft are probably not far over the threshold, provides significant possibilities for prevention. Already, there appears to be a degree of prevention provided by folic acid supplementation and the clear identification of negative environmental factors such as cigarette smoking (with its interaction with predisposing genetic factors) provides further opportunities for prevention. The fact that it is at least theoretically possible that some cases of clefting may involve only genetic factors does not rule not possibilities for prevention, especially when the genetic factors can be overcome by environmental manipulation. Experimentally, it has been shown that maternal hyperoxia reduces the incidence of spontaneous cleft lip in CL/Fr mice, and, although this manipulation is probably impractical in humans, the use of high-energy intermediates to increase ATP production could achieve similar results.

REFERENCES

Abbott, B. D., and Pratt, R. M. (1987). Retinoids and EGF alter embryonic mouse palatal epithelial and mesenchymal cell differentiation in organ culture. *J. Craniofac. Genet. Devel. Biol. 7*, 219–240.

Abbott, B. D., Probst, M. R., Perdew, G. H., and Buckalew, A. R. (1998). AH receptor, ARNT, glucocorticoid receptor, EGF receptor, EGF, TGF alpha, TGF beta 1, TGF beta 2, and TGF beta 3 expression in human embryonic palate, and effects of 2,3,7,8-tetrachlorodibenzo-p-dioxin (TCDD). *Teratology 58*, 30–43.

Acampora, D., Merlo, G. R., Paleari, L., Zerega, B., Postiglione, M. P., Mantero, S., Bober, E., Barbieri, O., Simeone, A., and Levi, G. (1999). Craniofacial, vestibular and bone defects in mice lacking the distal-less-related gene Dlx5. *Development 126*, 3795–3809.

Adelmann, H. B. (1936). The problem of cyclopia. *Quart. Rev. Biophys. 1*, 161–182.

Allen, S. P., Bogardi, J. P., Barlow, A. J., Mir, S A , Qayyum, S R., Verbeek, F. J., Anderson, R. H., Francis-West, P. H., Brown, N. A., and Richardson, M. K. (2001). Misexpression of noggin leads to septal defects in the outflow tract of the chick heart. *Devel. Biol. 235*, 98–109.

Ammann, A. J., Wara, D. W., Cowan, A. J., Barrett, D. J., and Stechm, E. R. (1982). The DiGeorge syndrome and the fetal alcohol syndrome. *Am. J. Dis Child 136*, 906–908.

Ardinger, H. H., Buetow, K. H., Bell, G. I., Bardach, J., Vandemark, D. R., and Murray, J. C. (1989). Association of genetic variation of the transforming growth factor-alpha gene with cleft lip and palate. *Am. J. Hum. Genet. 45*, 348–353.

Bailey, L. J., Johnston, M. C., and Billet, J. (1995). Effects of carbon monoxide and hypoxia on cleft lip in A/J mice. *Cleft Palate Craniofac. J. 32*, 14–19.

Beaty, T. H., Maestri, N. E., Hetmanski, J. B., Wyszynski, D. F., Vanderkolk, C. A., Simpson,

J. C., McIntosh, I., Smith, E. A., Zeiger, J. S., Rahmond, G. V., Panny, S. R., Tifft, C. J., Lewanda, A. F., Crstion, C. A., and Wulfsberg, E. A. (1997). Testing for interaction between maternal smoking and TGFA genotype among oral cleft cases born in Maryland 1992–1996. *Cleft Palate Craniofac. J. 34*, 447–454.

Bockman, D. E., Redmond, M. E., and Kirby, M. L. (1989). Alteration of early vascular development after ablation of cranial neural crest. *Anat. Rec. 225*, 209–217.

Brazzell, R. K., Vane, F. M., Ehmann, C. W., and Colburn, W. A. (1983). Pharmacokinetics of isotretinoin during repetitive dosing to patients. *Eur. J. Pharmacol. 24*, 695–702.

Briggs, R. M. (1976). Vitamin supplementation as a possible factor in the incidence of cleft lip/palate deformities in humans. *Clin. Plast. Surg. 3*, 647–652.

Bronsky, P. T. (1985). Effects of hypoxia on facial prominence development in CL/Fr mice. M.S. thesis, University of North Carolina.

Bronsky, P. T., Johnston, M. C., and Sulik, K. K. (1986). Morphogenesis of hypoxia-induced cleft lip in CL/Fr mice. *J. Craniofac. Genet. Devel. Biol. 2*, 113–128.

Bronsky, P. T., and Johnston, M. C. (in preparation). Craniofacial morphogenetic movements: Their role in normal and abnormal prenatal development.

Brunet, C. L., Sharpe, P. M., and Ferguson, M. W. J. (1995). Inhibition of TGF-B3 (but not TGF-B1 or TGF-B2) activity prevents normal mouse embryonic palate fusion. *Int. J. Devel. Biol. 39*, 345–355.

Carter, C. O. (1976). Genetics of common single malformations. *Br. Med. Bull. 32*, 21–26.

Carter, C. O. (1977). Monogenic disorders. *J. Med. Genet. 14*, 316–320.

Chen, R., Bouck, J. B., Weinstock, G. M., and Gibbs, R. A. (2001). Comparing vertebrate whole-genome shotgun reads to the human genome. *Genome Res. 11*, 1807–1816.

Chenevix-Trench, G., Jones, K., Green, A. C., Duffy, D. L., and Martin, N. G. (1992). Cleft lip with or without cleft palate: associations with transforming growth factor and retinoic acid receptor loci. *Am. J. Hum. Genet. 51*, 1337–1385.

Chiang, C., Litingtun, Y., Lee, E., and Young, K. E. (1996). Cyclopia and defective axial patterning in mice lacking Sonic hedgehog gene function. *Nature 383*, 407–413.

Chin, J. H., Goldstein, D. A., and Parsons, L. M. (1979). Fluidity and lipid composition of mouse biomembranes during adaption to ethanol. *Alcohol. Clin. Exp. Res. 3*, 47–49.

Chisaka, O., and Capecchi, M. R. (1991). Regionally restricted developmental defects resulting from targeted disruption of the mouse homeobox gene hox-1.5. *Nature 350*, 473–479.

Chung, C. S., Bixler, D., Watanabe, T., Koguchi, H., and Fogh-Andersen, P. (1986). Segregation analysis of cleft lip with or without cleft palate: a comparison of Danish and Japanese data. *Am. J. Hum. Genet. 39*, 603–611.

Chung, C. S., Mi, M. P., and Beechert, A. M. (1987). Genetic epidemiology of cleft lip with or without cleft palate in the population of Hawaii. *Genet. Epidemiol. 4*, 415–423.

Chung, K. C., Kowalski, C. P., Kim, H. M., and Buchman, S. R. (2000). Maternal cigarette smoking during pregnancy and the risk of having a child with cleft lip/palate. *Plast. Reconstr. Surg. 105*, 485–491.

Clarren, S. (1996). Neuropathology of the fetal alcohol syndrome. In J. R. West (ed.), *Alcohol and Brain Development* (New York: Oxford University Press).

Clouthier, D. E., Hosoda, K., Richardson, J. A., Williams, S. C., Yanagisawa, H., Kuwaki, T., Kumada, M., Hammer, R. E., and Yanagisawa, M. (1998). Cranial and cardiac neural crest defects in endothelin-A receptor-deficient mice. *Development 125*, 813–824.

Clouthier, D. E., Williams, S. C., Yanagisawa, H., Wieduwilt, M., Richardson, J. A., and Yanagisawa, M. (2000). Signaling pathways crucial for craniofacial development revealed by endothelin-A receptor-deficient mice. *Devel. Biol. 217*, 10–24.

Cohen, M. M., Jr. (1997). Short-limb skeletal dysplasias and craniosynostosis: What do they have in common? *Pediatr. Radiol. 27*, 442–446.

Cohen, M. M., Jr. (2000). Craniofacial disorders caused by mutations in homeobox genes MSX-1 and MSX-2. *J. Craniofac. Genet. Devel. Biol. 20*, 19–25.

Condie, B. G., Bain, G., Gottlieb, D. J., and Cappechi, M. R. (1997). Cleft palate in mice with a targeted mutation in the gamma-aminobutyric acid-producing enzyme glutamic acid decarboxylase 67. *Proc. Natl. Acad. Sci. USA 94*, 11451–11455.

Conley, M. E., Beckwith, J. B., Mancer, J. F., and Tenchkoff, L. (1979). The spectrum of the DiGeorge syndrome. *J. Pediatr. 94*, 883–890.

Conlon, R. A., and Rossant, J. (1992). Exogenous retinoic acid rapidly induces anterior ectopic expression of murine hox-2 genes in vivo. *Development 116*, 357–368.

Cooper, M. K., Porter, J. A., Young, K. E., and Beachy, P. A. (1998). Teratogen-mediated inhibition of target tissue response to shh signaling. *Science 280*, 1603–1607.

Couly, G. F., and Le Douarin, M. N. (1985). Mapping of the early neural primordium in quail-chick chimeras I. Developmental relationships between placodes, facial ectoderm, and prosencephalon. *Devel. Biol. 110*, 423–439.

Couly, G. F., Coltey, P. M., and Le Dourain, M. N. (1993). The triple origin of skull in higher verebrates: a study in quail-chick chimeras. *Development 117*, 409–429.

Couly, G., Grapin-Botton, A., Coltey, P., Ruhin, B., and Le Douarin, M. N. (1998). Determination of the identity of the derivatives of the cephalic neural crest: incompatibility between Hox gene expression and lower jaw development. *Development 125*, 3445-3459.

Crossley, P. H., and Martin, G. R. (1995). The mouse Fef8 encodes a family of polypeptides and is expressed in direct outgrowth and patterning in the developing embryo. *Development 121*, 439–451.

Culiat, C. T., Stubbs, L, Nicholls, R. D., Montgomery, C. S., Russell, L. B., Johnson, D. K., and Rinchik, E. M. (1993). Concordance between isolated cleft palate in mice and alterations within a region including the gene encoding the beta 3 subunit of the type A gamma-aminobutyric acid receptor. *Proc. Natl. Acad. Sci. USA 90*, 5105–5109.

Culiat, C. T., Stubbs, L. J., Montgomery, C. S., Russell, L. B., and Rinchik, E. M. (1994). Phenotypic consequences of deletion of the gamma 3, alpha 5, or beta 3 subunit of the type A gamma-aminobutyric acid receptor in mice. *Proc. Natl. Acad. Sci. USA 91*, 2815–2818.

Culiat, C. T., Stubbs, L. J., Woychik, R. P., Russell, L. B., Johnson, D. K., and Rinchik, E. M. (1995). Deficiency of the beta 3 subunit of the type A gamma-aminobutyric acid receptor causes cleft palate in mice. *Nat. Genet. 11*, 344–346.

Daft, P. A., Johnston, M. C., and Sulik, K. K. (1986). Abnormal heart and great vessel development following acute ethanol exposure in mice. *Teratology 33*, 93–104.

D'Amico-Martel, A., and Noden, D. M. (1983). Contributions of placodal and neural crest cells to avian cranial peripheral ganglia. *Am. J. Anat. 66*, 445–468.

Darab, D. J., Sciote, J., Minkoff, R., and Sulik, K. K. (1987). Pathogenesis of the median facial clefts in mice treated with methotrexate. *Teratology 36*, 77–86.

Davideau, J. L., Demri, P., Gu, T. T., Simmons, D., Nessman, C., Forest, N, MacDougall, M., and Berdal, A. (1999). Expression of DLX5 during human embryonic craniofacial development. *Mech. Devel. 81*, 183–186.

DeMeyer, W. (1967). Median cleft face syndrome: differential diagnosis of cranium bifidum ocultum, hypertelorism, and median cleft nose. *Neurology 17*, 961–971.

Dencker, L. (1979). Embryonic-fetal localization of drugs and nutrients. In T. V. N. Persaud (ed.), *Advances in the Study of Birth Defects, Vol. 1, Teratogenic Mechanisms* (Lancaster, Great Britain: MTP Press Ltd.), pp. 1–18.

Dencker, L., Annerwall, E., Bush, C., and Eriksson, E. (1990). Localization of specific retinoid-binding sites and expression of cellular retinoic-acid-binding protein (CRABP) in the early mouse embryo. *Development 110*, 343–352.

Dencker, L., d'Argy, R., Danielsson, B. R., Ghantous, H., and Sperber, G. O. (1987). Saturable accumulation of retinoic acid in neural and neural crest derived cells in early embryonic development. *Dev. Pharmacol. Ther. 10*, 212–223.

Deng, C., Winshaw-Boris, A., Zhou, F., Kuo, A., and Leder, P. (1996). Fibroblast growth factor receptor 3 is a negative regulator of bone growth. *Cell 84*, 911–921.

Depew, M. J., Liu, J. K., Long, J. E., Presley, R., Meneses, J. J., Pedersen, R. A., and Rubenstein, J. L. (1999). Dlx5 regulates regional development of the branchial arches and sensory capsules. *Development 126*, 3831–3846.

Diewert, V. M., and Shiota, K. (1990). Morphological observations in normal primary palate and cleft lip embryos in the Kyoto collection. *Teratology 41*, 663–677.

Diewert, V. M., and Wang, K. Y. (1992). Recent advances in primary palate and midface morphogenesis research. *Crit. Rev. Oral Biol. Med. 4*, 111–130.

Dixon, M. J. (1996). Treacher Collins syndrome. *Hum. Molec. Genet. 5*, 1391–1396.

Dixon, J., Hovannes, K., Shiang, R., and Dixon, M. J. (1997). Sequence analysis, identification of evolutionary conserved motifs and expression analysis of murine tcof1 provide further evidence for a potential function for the gene and its homologue, TCFO1. *Hum. Molec. Genet. 6*, 727–737.

Dixon, J., Brakebush, C., Fassler, R., and Dixon, M. J. (2000). Increased levels of apoptosis in the prefusion neural folds underlie the craniofacial disorder, Treacher Collins syndrome. *Hum. Molec. Genet. 9*, 1473–1480.

Donaghue, P. C., Forey, P. L., and Aldrirge, R. J. (2000). Conodont affinity and chordate phylogeny, *Biol. Rev. 75*, 191–251.

Drillien, C. M., Ingram, T. T. S., and Wilkinson, E. M. (eds.) (1966). *The Causes and Natural History of Cleft Lip and Palate* (Edinburgh and London: E & S Livingstone).

Driscoll, D. A., Burdoff, M. L., and Emanuel, B. S. (1992a). A genetic etiology for the DiGeorge syndrome: consistent deletions and microdeletions of 22q11. *Am. J. Hum. Genet. 50*, 924–933.

Driscoll, D. A., Spinner, N. B., and Burdoff, M. N. (1992b). Deletions and microdeletions of 22q11.2 in velo-cardiofacial syndrome. *Am. J. Med. Genet. 44*, 261–268.

Duhamel, B. (1966). *Morphogenèse Pathologique* (Paris: Masson et Cie).

El Ghouzzi, V., LeMerrer, M., Perrin-Schmitt, F., Lajeunie, E., Benit, P., Renier, D., Bourgeois, P., Bolcato-Beilemin, A. L., Munnich, A., and Bonaventure, A. (1996). Mutations of the TWIST gene in the Saethre-Chotzen syndrome. *Nature Genet. 15*, 42–46.

Emanuel, B. S., McDonald-McGinn, D., Saitta, S. C., and Zackai, E. H. (2001). The 22q11.2 deletion syndrome. *Adv. Pediatr. 48*, 39–73.

Erikson, A., Kallen, B., and Westerholm, P. (1979). Cigarette smoking as an etiologic factor in cleft lip and palate. *Am. J. Obstet. Gynecol. 135*, 348–351.

Etchevers, H. C., Vincent, C., LeDouarin, N. M., and Couly, G. F. (2001). The cephalic neural crest provides pericytes and smooth muscle cells to all blood vessels in the face and forebrain. *Development 128*, 1059–1068.

Evrard, L., Vanmuylder, N., Dourov, N., Hermans, C., Biermans, J., Werry-Huet, A., Rooze, M., and Louryan, S. (2000). Correlation of HSP110 expression with all-trans retinoic acid-induced apoptosis. *J. Craniofac. Genet. Devel. Biol. 20*, 183–192.

Falconer, D. S. (1965). The inheritance of liability to certain diseases, estimates from incidence among relatives. *Ann. Hum. Genet. 29*, 51–57.

Farrall, M., and Holder, S. (1992). Familial recurrence-pattern analysis of cleft lip with or without cleft palate. *Am. J. Hum. Genet. 50*, 270–277.

Farrall, M., Waldo, K., Li, Y. X., and Kirby, M. L. (1999). A novel role for cardiac neural crest in heart development. *Trends Cardiovasc. Med. 9*, 214–220.

Ferrier, D. E., and Holland, P. W. (2001). Sipunculan ParaHox genes. *Evol. Dev. 3*, 263–270.

Fitchett, J. E., and Hay, E. D. (1989). Medial edge epithelium transforms to mesenchyme after embryonic palatal shelves fuse. *Devel. Biol. 131*, 455–474.

Fogh-Andersen, P. (1942). *Inheritance of Harelip and Cleft Palate* (Copenhagen: Nyt Nordisk Forlag Arnold Busck).

Forbes, D. P., Steffek, A. J., and Klepacki, M. (1989). Reduced epithelial surface activity is related to a higher incidence of facial clefting in A/WynSn mice. *J. Craniofac. Genet. Devel. Biol. 9*, 271–283.

Foerst-Potts, L., and Sadler, T. W. (1997). Disruption of Msx-1 and Msx-2 reveals roles for these genes in craniofacial, eye, and axial development. *Devel. Dyn. 209*, 70–84.

Francis-West, P. H., Tatla, T., and Brickell, P. M. (1994). Expression patterns of the bone morphogenetic protein genes Bmp-4 and Bmp-2 in the developing chick face suggest a role in outgrowth of the primordia. *Devel. Dyn. 201*, 168–178.

Francis-West, P. H., Ladher, R., Barlow, A., and Graveson, A. (1998). Signaling interactions during facial development. *Mech. Devel. 75*, 3–28.

Fraser, F. C. (1976). The multifactorial/threshold concept—uses and misuses. *Teratology 14*, 267–280.

Fraser, F. C. (1980). The genetics of cleft lip and palate. In R. M. Pratt and T. W. Christiansen (eds.), *Current Research Trends in Prenatal Craniofacial Development* (North Holland: Elsevier).

Fraser, F. C., and Fainstat, T. D. (1951). The production of congenital defects in the offspring of pregnant mice treated with cortisone: a program report. *Pediatrics 8*, 527–531.

Fukiishi, Y., and Morriss-Kay, G. M. (1992). Migration of cranial neural crest cells to the pharyngeal arches and heart in the rat embryo. *Cell Tiss. Res. 268*, 1–8.

Furtwangler, J. A., Hall, S. H., and Koskinen-Moffett, J. K. (1985). Sutural morphogenesis in the mouse calvaria: the role of apoptosis. *Acta Anat. 124*, 74–80.

Gans, G., and Northcutt, R. G. (1983). Neural crest and the origin of vertebrates: a new head. *Science 220*, 268–274.

Gaunt, S. J., Blum, M., and De Roberts, E. M. (1993). Expression of the mouse goosecoid gene during mid-embryogenesis may mark mesenchymal cell lineages in the developing head, limbs and body wall. *Development 117*, 769–778.

Gendron-Maguire, M., Mallo, M., Zhang, M., and Gridley, T. (1993). Hoxa-2 mutant mice exhibit homeotic transformation of skeletal elements derived from cranial neural crest. *Cell 75*, 1317–1331.

Golding, J. P., Trainor, P., Krumlauf, R., and Gassmann, M. (2000). Defects in pathfinding by cranial neural crest cells in mice lacking the neuregulin receptor ErbB4. *Nat. Cell. Biol. 2*, 103–109.

Gorlin, R. J., Cohen, M. M., Jr., and Levin, K. S. (1990). *Syndromes of the Head and Neck*, 3rd ed. (Oxford, England: Oxford University Press).

Gorlin, R. J., Cohen, M. M., Jr., and Hennekam, R. C. M. (2001). *Syndromes of the Head and Neck*, 4th ed. (Oxford, England: Oxford University Press).

Goulding, E. H., and Pratt, R. M. (1986). Isotretinoin teratogenicity in mouse whole embryo culture. *J. Craniofac. Genet. Devel. Biol. 6*, 99–112.

Graham, A., Francis-West, P., Brickell, P., and Lumsden, A. (1994). The signalling molecule BMP4 mediates apoptosis in the rhombencephalic neural crest. *Nature 372*, 684–686.

Graham, A., Heyman, I., and Lumsden, A. (1993). Even-numbered rhombomeres control the apoptotic elimination of neural crest cells from odd-numbered rhombomeres in the chick hindbrain. *Development 119*, 233–245.

Graham, A., Koentegnes, G., and Lumsden, A. (1996). Neural crest apoptosis and establishment of craniofacial pattern: an honorable cell death. *Molec. Cell Neurosci. 8*, 76–83.

Greene, R. M., and Weston, W. M. (1994). Craniofaical embryology. In M. Cohen (ed.), *Mastery of Surgery* (New York: Little, Brown and Co.), pp. 459–470.

Grigoriou, M., Tucker, A. S., Sharpe, P. T., and Pachnis, V (1998). Expression and regulation of Lhx6 and Lhx7, a novel subfamily of LIM homeodomain encoding genes, suggests a role in mammalian head development. *Development 125*, 2063–2074.

Gruss, P., and Walther, C. (1992). Pax in development. *Cell 69*, 719–722.

Halford, M. M., Armes, J., Buchert, M., Meskenaite, V., Grail, D., Hibbs, M. L., Wilks, A. F., Farlie, P. G., Newgreen, D. F., Hovens, C. M., and Stacker, S. A. (2000). Ryk-deficient mice exhibit craniofacial defects associated with perturbed Eph receptor crosstalk. *Nat. Genet. 24*, 414–418.

Hall, B. K., and Hörstadius, S. (1988). *The Neural Crest* (London: Oxford University Press).

Hansen, D. K., and Hodes, M. E. (1983). Comparative teratogenicity of phenytoin among several inbred strains of mice. *Teratology 28*, 175–179.

Hanson, J. W., Myrianthopoulos, N. C., Sedgwick, H., and Smith, D. W. (1976). Risks to the offspring of women treated with the hydantoin anticonvulsant, with emphasis on the fetal hydantoin syndrome. *J. Pediatr. 89*, 662–668.

Hartsfield, J. K., Jr., Hickman, T. A., Everett, E. T., Shaw, G. M., Lammer, E. J., and Finnell, R. A. (2000). Analysis of the EPHX1 113 polymorphism and GSTM1 homozygous null polymorphism and oral clefting associated with maternal smoking. *Am. J. Med. Genet. 102*, 21–24.

Hassell, J. R., Greenberg, J. H., and Johnston, M. C. (1977). Inhibition of cranial neural crest cell development by vitamin A in the cultured chick embryo. *J. Embryol. Exp. Morphol. 39*, 267–271.

Hatta, K., Kimel, C. B., and Ho, R. K. (1991). The cyclops mutation blocks specification on the floor plate of the zebrafish central nervous system. *Nature 350*, 399–341.

Hay, S., and Wehrung, D. A. (1970). Congenital malformations in twins. *Am. J. Hum. Genet. 22*, 662–678.

Helms, J. A., Kim, Z. H., Hu, D., Minkoff, R. M., Thaller, C., and Eichle, G. (1997). Sonic hedgehog participates in craniofacial morphogenesis and is down-regulated by teratogenic doses of retinoic acid. *Devel. Biol. 187*, 25–35.

Hetzel, S., and Brown, K. S. (1975). Facial clefts and lip hematomas in mouse fetuses given diphenylhydantoin. *J. Dent. Res. 54*, 83.

Hogan, B. L., Blessing, M., Winnier, G. E., Suzuki, N., and Jones, C. M. (1994). Growth factors in development: the role of TGF-β related polypeptide signaling molecules in embryogenesis. *Development 103* (suppl.), 53–60.

Holder, S. E., Vintiner, G. M., Farren, G., Malcolm, S., and Winter, R. M. (1992). Confirmation of an association between RFLPs at the transforming growth factor-alpha locus and nonsyndromic cleft lip and palate. *J. Med. Genet. 29*, 390–392.

Holland, P. W. H., Garcia-Fernandez, J., Williams, N. A., and Sidow, A. (1994). Gene duplications and the origins of vertebrate development. *Development 103* (suppl.), 125–133.

Holtfreter, J., and Hamburger, V. (1955). Amphibians. In D. H. Willier, P. A. Weiss, and V. Hamburger (eds.), *Analysis of Development* (Philadelphia–London: W. B. Saunders Co.), pp. 230–296.

Höstadius, S. (1950). *The Neural Crest* (Oxford, England: Oxford University Press).

Howard, T. D., Paznekas, W. A., Green, E. D., Chiang, L. C., Ma, N., Oritiz de Luna, R. I., Delgado, G., Gonzales-Ramos, M., Kline, A. D., and Jabs, E. W. (1996). Mutations in TWIST, a basic helix-loop-helix transcription factor in Saethre-Chotzen syndrome. *Nat. Genet. 15*, 36–41.

Hu, D., and Helms, J. A. (1999). The role of sonic hedgehog in normal and abnormal craniofacial morphogenesis. *Development 126*, 4873–4884.

Hunt, P., Wilkinson, D., and Krumlauf, R. (1991). Patterning the vertebrate head: murine hox 2 genes mark distinct subpopulations of premigratory and migrating cranial neural crest. *Development 112*, 43–50.

Hunt, P., Clarke, J. D., Buxton, P., Ferretti, P., and Thorogood, P. (1998). Stability and plasticity of neural crest patterning and branchial arch Hox code after extensive cephalic crest rotation. *Devel. Biol. 198*, 82–104.

Hunter, E. S., III, and Sadler, T. W. (1988). Embryonic metabolism of fetal fuels in whole embryo culture. *Toxicol. In Vitro 2*, 163–167.

Hwang, S. J., Beaty, T. H., Panny, S. R., Street, N. A., Joseph, J. M., Gordon, S., McIntosh, I., and Francomano, C. A. (1995). Association study of transforming growth factor alpha (TGFA) TaqI polymorphism and oral clefts: indication of gene–environment interaction in a population based sample of infants with birth defects. *Am. J. Epidemiol. 141*, 629–636.

Iamaroon, A., Yait, B., and Diewert, V. M. (1996). Cell proliferation and expression of EGF, TGF alpha, and EGF receptor in the developing primary palate. *J. Dent. Res. 75*, 1534–1539.

Ikonomidou, C., Stefovska, V., and Turski, L. (2000). Neuronal death enhanced by N-methyl-D-aspartate antagonists. *Proc. Natl. Acad. Sci. U. S. A. 97*, 12885–12890.

Isaac, C., Marsh, K. L., Paznekas, W. A., Dixon, J., Dixon, M. J., Jabs, E. W., and Meier, U. T. (2000). Characterization of the nucleolar gene product, treacle, in Treacher Collins syndrome. *Mol. Biol. Cell. 11*, 3061–3071.

Izpisúa-Belmonte, J. C., De Robertis, M., Storey, K. G., and Stern, C. D. (1993). The homeobox gene goosecoid and the origin of organizer cells in the early chick blastoderm. *Cell 74*, 645–659.

Jabs, E. W., Muller, U., Li, X., Ma, L., Luo, W., Haworth, I. S., Klisak, I., Sparkes, R., Warman, M. L., and Mulliken, J. B. (1993). A mutation in the homeodomain of the human MSX2 gene in a family affected with dominant autosomal craniosynostosis. *Cell 75*, 443–450.

Jabs, E. W., Li, X., Scott, A. F., Meyers, G., Chen, W., Eccles, M., Mao., J. I., Charnas, L. R., Jackson, C. E., and Jaye, M. (1994). Jackson-Weiss and Crouzon syndromes are allelic with mutations in fibroblast growth factor receptor 2. *Nat. Genet. 8*, 275–279.

Jacobsson, C. (1997). Teratological studies on craniofacial malformations. *Swed. Dent. J. 21* (suppl.), 3–84.

Jacobsson, C., and Granstrom, G. (1997). Clinical appearance of spontaneous and induced first and second branchial arch syndromes. *Scand. J. Plast. Reconstr. Surg. Hand. Surg. 31*, 123–125.

Jarvis, B. L., Sulik, K. K., and Johnston, M. C. (1990). Congenital malformations of the external, middle and inner ear produced by isotretinoin exposure in fetal mice. *Otol. Head Neck Surg. 102*, 391–401.

Jirasek, J. E. (1983). *Atlas of Human Prenatal Morphogenesis* (Hingham, MA.: Kluwer Martinus).

Johnson, R. L., and Tabin, C. J. (1997). Molecular models for vertebrate limb development. *Cell 90*, 979–990.

Johnston, C. (1981). Cigarette smoking and the outcome of human pregnancies: a status report on the consequences. *Clin. Toxicol. 18*, 189–209.

Johnston, M. C. (1964). Fetal malformations in chick embryos resulting from removal of neural crest. *J. Dent. Res. 43* (spec. issue), 822.

Johnston, M. C. (1966). A radioautographic study of the migration and fate of cranial neural crest cells in the chick embryo. *Anat. Rec. 156*, 143–156.

Johnston, M. C. (1990). Embryology of the head and neck. In J. McCarthy (ed.), *Plastic Surgery* (Philadelphia: W. B. Saunders), pp. 31–46.

Johnston, M. C. (1993). Understanding human development. In R. E. Stevenson, J. G. Hall, and R. M. Goodman (eds.), *Human Malformations and Related Anomalies* (New York and Oxford: Oxford University Press), pp. 30–64.

Johnston, M. C. (1997). Developmental biology of the mouth, palate, and pharynx. In T. L. Tewfik and V. M. Der Kaloustian (eds.), *Congenital Anomalies of the Ear, Nose, and Throat* (New York: Oxford University Press), pp. 17–38.

Johnston, M. C., Vig, K., and Ambrose, L. (1981). Neurocristopathy as a unifying concept: clinical correlations. *Adv. Neurol. 29*, 97–104.

Johnston, M. C., and Bronsky, P. T. (1991). Animal models for craniofacial malformations. *J. Craniofac. Genet. Devel. Biol. 11*, 227–229.

Johnston, M. C., and Bronsky, P. T. (1995). Prenatal craniofacial development: new insights on normal and abnormal mechanisms. *Crit. Rev. Oral Biol. Med. 6*, 368–422.

Johnston, M. C., and Hazelton, R. B. (1972). Embryonic origins of facial structures related to oral sensory and motor function. In J. F. Bosma (ed.), *Oral Sensory Perception, The Mouth of the Infant* (Springfield, IL: Charles C. Thomas), pp. 6–27.

Johnston, M. C., and Hunter, W. S. (1989). Cleft lip and/or palate in twins: evidence for two major cleft lip groups. *Teratology 39*, 461.

Johnston, M. C., and Listgarten, M. A. (1972). The migration interaction and early differentiation of oro-facial tissues. In H. S. Slavkin and L. A. Bavetta (eds.), *Development Aspects of Oral Biology* (New York: Academic Press), pp. 55–80.

Johnston, M. C., and Sulik, K. K. (1979). Some abnormal patterns of development in the craniofacial region. *Birth Defects Orig Artic Ser. 15*, 23–42.

Johnston, M. C., Bhakdinaronk, A., and Reid, Y. C. (1974). An expanded role for the neural crest in oral and pharyngeal development. In J. F. Bosma (ed.), *Oral Sensation and Percep-*

tion—*Development in the Fetus and Infant* (Washington, DC: U.S. Government Printing Office), pp. 37–52.

Johnston, M. C., Hassell, J. R., and Brown, K. S. (1975). The embryology of cleft lip and cleft palate. *Clin. Plast. Surg. 2*, 195–203.

Johnston, M. C., Noden, D. M., Hazelton, R. D., Coulombre, J. L., and Coulombre, A. J. (1979a). Origins of avian ocular and periocular tissues. *Exp. Eye Res. 29*, 27–43.

Johnston, M. C., Sulik, K. K., and Dudley, K. H. (1979b). Genetic and metabolic studies of the differential sensitivity of A/J and C57Bl/6J mice to phenytoin ("Dilantin")-induced cleft lip. *Teratology 19*, 33A.

Johnston, M. C., Bronsky, P. T., and Millicovsky, G. (1990). Embryogenesis of cleft lip and palate. In J. McCarthy (ed.), *Plastic Surgery* (Philadelphia: W. B. Saunders), pp. 114–126.

Jones, M. C. (1990). The neurocristopathies: reinterpretation based upon the mechanism of abnormal morphogenesis. *Cleft Palate J. 27*, 136–140.

Juriloff, D. M. (1986). Major genes that cause cleft lip in mice: progress in the construction of a congenic strain and in linkage mapping. *J. Craniofac. Genet. Devel. Biol. 2* (suppl.), 55–66.

Juriloff, D. M., Sulik, K. K., Roderick, T. H., and Hogan, B. K. (1985). Genetic and developmental studies of a new mouse mutation that produces otocephaly. *J. Craniofac. Genet. Devel. Biol. 5*, 121–145.

Kaartainen, V., Vonckin, J. W., Shuler, C., Warburton, D., Bu, D., and Heisterkamp, N. (1995). Abnormal lung development and cleft palate in mice lacking TGFbeta3 indicates defects of epithelial-mesenchymal interaction. *Nat. Genet. 11*, 415–521.

Kallen, B. (1956). Contribution to the knowledge of the regulation of the proliferation processes in the vertebrate brain during ontogenesis. *Acta Anat. (Basel) 27*, 351.

Kallen, K. (1997). Maternal smoking and orofacial clefts. *Cleft Palate Craniofac. J. 34*, 11–16.

Keeler, R. F., and Binns, W. (1968). Teratogenic compounds of *Veratrim californicum* (Durand) V. Comparison of cylopian effects of steroidal alkaloids from the plant and structurally related compounds from other sources. *Teratology 1*, 5–10.

Keels, M. A. (1991). The Role of Maternal Cigarette Smoking in the Etiology of Cleft Lip With or Without Cleft Palate. Ph.D. dissertation, University of North Carolina.

Keels, M. A., Savitz, D. A., and Stamm, J. W. (1991). An etiologic study of cleft palate in humans. *J. Dent. Res. 68* (special issue), 431.

Khoury, M. J., Botto, L., Waters, G. D., Mastroiacovo, P., Castilla, E., and Erickson, J. D. (1993). Monitoring for new multiple congenital anomalies in the search for human teratogens. *Am. J. Med. Genet. 46*, 460–466.

Khoury, M. K., Weinstein, A., Parry, S., Holtzman, N. A., Lindsay, P. K., Farrel, K., and Eisenberg, M. (1987). Maternal cigarette smoking and oral clefts: a population based study. *Am. J. Public Health 77*, 623–625.

Kim, H. J., Rice, D. P., Kettunen, P. J., and Thesleff, I. (1998). FGF-, BMP- and Shh-mediated signalling pathways in the regulation of cranial suture morphogenesis and calavrial bone development. *Development 125*, 1241–1251.

King, P. B., Lie, R. T., and Irgens, L. M. (1996). Spina bifida and cleft lip among newborns of Norwegian women with epilepsy: changes related to the use of anticonvulsants. *Am. J. Public Health 86*, 1454–1456.

Kirby, M. L., Gale, T. F., and Stewart, D. E. (1983). Neural crest cells contribute to normal aorticopulmonary septation. *Science 220*, 1059–1061.

Kontges, G., and Lumsden, A. (1996). Rhombencephalic neural crest segmentation is preserved throughout craniofacial ontogeny. *Development 122*, 3220–3242.

Kozmik, Z., Holland, N. D., Kalousova, A., Paces, J., Schubert, M., and Holland, L. Z. (1999). Characterization of an amphioxus paired box gene, AmphiPax2/5/8: developmental expression patterns in optic support cells, nephridium, thyroid-like structures and pharyngeal gill slits, but not in the midbrain-hindbrain boundary region. *Development 126*, 1295–1304.

Kretchmer, R., Say, B., Brown, D., and Rosen, S. F. (1969). Congenital aplasia of the thymus gland (DiGeorge's syndrome). *N. Engl. J. Med. 279*, 1295.

Kuratani, S. G., and Wall, N. A. (1992). Expression of Hox 2.1 protein in restricted populations of neural crest cells and pharyngeal ectoderm. *Devel. Dyn. 195*, 15–28.

Kuratani, S., Matsuo, I., and Aizawa, S. (1997). Developmental pattern and evolution of the mammalian viscercranium: genetic insights into comparative morphology. *Devel. Dyn. 209*, 139–155.

Kurihara, Y., Kurihara, H., Suzuki, H., Kodama, T., Maemura, K., Nagai, R., Oda, H., Kuwaki, T., Cao, W. H., and Kamada, N. (1994). Elevated blood pressure and craniofacial abnormalities in mice deficient in endothelin-1. *Nature. 368*, 703–710.

Kurihara, Y., Kurihara, H., Oda, H., Maemura, K., Nagai, R., Ishikawa, T., and Yazaki, Y. (1995a). Aortic arch malformations and ventricular septal defect in mice deficient in endothelin-1. *J. Clin. Invest. 96*, 293–300.

Kurihara, Y., Kurihara, H., Maemura, K., Kuwaki, T., Kumada, M., and Yazaki, Y. (1995b). Impaired development of the thyroid and thymus in endothelin-1 knockout mice. *J Cardiovasc Pharmacol. 26 Suppl 3*, S13–S16.

Lammer, E. J., Chen, D. T., and Hoar, R. M. (1985). Retinoic acid embryopathy. *N. Engl. J. Med. 313*, 837.

Lanctôt, C., Lamolet, B., and Drouin, J. (1997). The bicoid-related homeoprotein Ptx1 defines the most anterior domain of the embryo and differentiates posterior from anterior lateral mesoderm. *Development 124*, 2807–2817.

Lanctôt, C., Moreau, A., Chamberland, M., Tremblay, M. L., and Drouin, J. (1999). Hindlimb patterning and mandible development require the Ptx1 gene. *Development 126*, 1805–1810.

Landauer, W. J. (1954). The chemical production of developmenal abnormalites and phenocopies in chicken embryos. *J. Cell Comp. Physiol. 43*, 26.

Landauer, W. J., and Sopher, D. (1970). Succinate, glycerophosphate and ascorbate as sources of cellular energy as antiteratogens. *J Embryol. Exp. Morphol. 24*, 282.

Langille, R. M., and Hall, B. K. (1988). Role of the neural crest in development of the trabeculae and branchial arches in embryonic sea lamprey, Petromyzon marinus (L). *Development 102*, 301–309.

Le Douarin, N. M. (2001). Early neurogenesis in Amniote vertebrates. *Int. J. Devel. Biol. 45* (spec. issue), 373–378.

Le Douarin, N. M., and Kalcheim, C. (1992). *The Neural Crest* (Cambridge: Cambridge University Press).

Lieff, S., Olshan, A. F., Werler, M., Strauss, R. P., Smith, J., and Mitchell, A. (1999). Maternal cigarette smoking during pregnancy and risk of oral clefts in newborns. *Am. J. Epidemiol. 150*, 683–694.

Lidral, A. C., and Reising, B. (1999). The role of MSX1 in the etiology of congenital hypodontia. *J. Dent. Res. 78*, A3529.

Lidral, A. C., Murray, J. C., Buetow, K. H., Basart, A. B., Schearer, H., Shiang, R., Naval, A., Layda, E., and Magee, K. (1997). Studies of the candidate genes TGFB2, MSX1, TGFA, and TGFB3 in the etiology of cleft lip and palate in the Philippines. *Cleft Palate Craniofac. J. 34*, 1–6.

Lidral, A. C., Romitti, P. A., Basart, A. M., Doetschman, T., Leysens, N. J., Daack-Hirsch, R. T., Semina, E. V., Johnson, L. R., Machida, J., Burds, A., Parnell, T. J., Rubenstein, J. L. R., and Murray, J. C. (1998). Association of MSX1 and TGFB3 with nonsyndromic clefting in humans. *Am. J. Hum. Genet. 63*, 557–568.

Lindsay, F. A., and Baldini, A. (2001). Recovery from arterial growth delay reduces penetrance of cardiovascular defects in mice delected for the DiGeorge syndrome region. *Hum. Molec. Genet. 10*, 997–1002.

Lindsay, E. A., Botta, A., Jurecic, V., Carattini-Rivera, S., Cheah, Y. C., Rosenblatt, H. M., Bradley, A., and Baldini, A. (1999). Congenital heart disease in mice deficient for the DiGeorge syndrome region. *Nature. 401*, 379–383.

Lindsay, E. A., Vitelli, F., Su, H., Morishima, M., Huyuh, T., Pramparo, T., Jurcic, V., Ogunrinu, G., Sutherland, H. F., Scrambler, P. J., Bradley, A., and Baldini, A. (2001). Tbx1 haploinsufficiency in the DiGeorge syndrome region causes aortic arch defects in mice. *Nature 410*, 97–101.

Lohnes, D., Kastner, P., Dierich, A., Mark, M., LeMeur, M., and Chambon, P. (1993). Function of retinoic acid receptor gamma in the mouse. *Cell. 73*, 643–658.

Longo, L. D. (1980). Environmental pollution and pregnancy: risks and uncertainties for the fetus and infant. *Am. J. Obstet. Gynecol. 137*, 162.

Lu, M. F., Pressman, C., Dyer, R., Johnson, R. L., and Martin, J. F. (1999a). Function of Rieger syndrome gene in left-right asymmetry and craniofacial development. *Nature 401*, 276–278.

Lu, M. F., Cheng, H. T., Kern, M. J., Potter, S. S., Tran, B., Diekwisch, T. G., and Martin, J. F. (1999b). Prx-1 functions cooperatively with another paired-related homeobox gene, prx-2, to maintain cell fates within the craniofacial mesenchyme. *Development 126*, 495–504.

Lu, H., Jin, Y., and Tipoe, G. L. (2000). Alteration in the expression of bone morphogenetic protein-2,3,4,5 mRNA during pathogenesis of cleft palate in BALB/c mice. *Arch. Oral Biol. 45*, 133–140.

Maas, R., and Bei, M. (1997). The genetic control of early tooth development. *Crit. Rev. Oral Biol. Med. 8*, 4–39.

Maden, M., Hunt, P., Eriksson, U., Kuroiwa, A., Krumlauf, R., and Summerbell, D. (1991). Retinoic acid-binding protein, rhombomeres and the neural crest. *Development 111*, 35–44.

Maden, M., Horton, C., Graham, A., Leonard, L., Pizzey, J., and Siegenthaler, G. (1992). Domains of cellular retinoic acid binding protein I (CRABP I) expression in the hindbrain and neural crest of the mouse embryo. *Mech. Devel. 37*, 12–23.

Marazita, M. L., Hu, D.-N., Spence, M. A., Liu, Y.-E., and Melnick, M. (1992). Cleft lip with or without cleft palate in Shanghai, China: evidence for an autosomal major locus. *Am. J. Hum. Genet. 5*, 648–653.

Marres, H. A. M., Cremers, C. W. R. J., Dixon, M. J., Huygens, P. L. M., and Joosten, F. B. M. (1995). The Treacher Collins syndrome: a clinical, radiological and genetic linkage study of two pedigrees. *Arch. Otoloaryngol. 121*, 509–514.

Massey, K. M. (1966). Teratogenic effect of diphenylhydantoin sodium. *J. Oral Ther. Pharmacol. 2*, 380–385.

Matsuo, T., Osumi-Yamashita, N., Noji, S., Ohuchi, H., Koyama, E., Myokai, F., Matsuo, N., Taniguchi, S., Doi, H., and Iseki, S. (1993). A mutation in the Pax-6 gene in rat small eye mutation is associated with impaired migration of midbrain crest cells. *Nat. Genet. 3*, 299–304.

McGonnell, I. M., Clarke, J. D. W., and Tickle, C. (1998). Fate map of the developing chick face: analysis of expansion of facial primordia and establishment of the primary palate. *Devel. Dyn. 212*, 102–118.

McKee, G. J., and Ferguson, M. W. (1984). The effects of mesencephalic neural crest cell extirpation on the development of chicken embryos. *J. Anat. 139*, 491–512.

Miettinen, P. J., Chin, J. R., Shum, L., Slavkin, H. C., Shuler, C. F., Derynch, R., and Werb, Z. (1999). Epidermal growth factor receptor function is necessary for normal craniofacial development and palate closure. *Nat. Genet. 22*, 69–73.

Miller, R. P., and Becker, H. A. (1975). Teratogenicity of oral diazepam and diphenylhydantoin in mice. *Toxicol. Appl. Pharmacol. 32*, 53–61.

Millicovsky, G., and Johnston, M. C. (1981a). Active role of embryonic facial epithelium: new evidence of cellular events in morphogenesis. *J. Embryol. Exp. Morphol. 63*, 53–66.

Millicovsky, G., and Johnston, M. C. (1981b). Hyperoxia greatly reduces the incidence of cleft lip and palate in A/J mice. *Science 212*, 671–672.

Millicovsky, G., and Johnston, M. C. (1981c). Hyperoxia and hypoxia in pregnancy: simple experimental manipulation alters the incidence of cleft lip and palate in CL/Fr mice. *Proc. Natl. Acad. Sci. USA 78*, 5722–5723.

Millicovsky, G., Ambrose, L. J. H., and Johnston, M. C. (1982). Developmental alterations associated with spontaneous cleft lip and palate in CL/FR mice. *Am. J. Anat. 164*, 29–44.

Minkoff, R. (1991). Cell proliferation during formation of the embryonic facial primordia. *J. Craniofac. Genet. Devel. Biol. 11*, 251–261.

Minkoff, R., and Kuntz, A. J. (1977). Cell proliferation during morphogenetic change: analyses of frontonasal morphogenesis in the chick embryo explaining DNA labeling indices. *J. Embryol. Exp. Morphol. 40*, 101–113.

Minkoff, R., and Kuntz, A. J. (1978). Cell prolif-eration and cell density of mesenchyme in the maxillary process and adjacent regions during facial development in the chick embryo. *J. Embryol. Exp. Morphol. 46*, 65–74.

Mitchell, L. E., and Risch, N. (1992). Mode of inheritance of nonsyndromic cleft lip with or without cleft palate: a reanalysis. *Am. J. Hum. Genet. 51*, 323–332.

Mo, R., Freer, A. M., Zinyk, D. L., Crackower, M. A., Michaud, J., Heng, H. H., Chik, K. W., Shi, X. M., Tsui, L. C., Cheng, S. H., Joyner, A. L., and Hui, C. (1997). Specific and redun-dant functions of Gli2 and Gli3 zinc finger genes in skeletal patterning and develop-ment. *Development 124*, 113–123.

Moore, G., Williamson, R., Jensson, O., Chambers, J., Takakubo, F., and Newton, R. (1991). Localization of a mutant gene for cleft palate and ankyloglossia in an x-linked Icelandic family. *J. Craniofac. Genet. Devel. Biol. 11*, 372–376.

Morriss, G. (1972). Morphogenesis of the mal-formation induced in rat embryo by mater-nal hypervitaminosis A. *J. Anat. 113*, 241.

Morriss, G. M., and Thorogood, P. V. (1978). An approach to cranial neural crest cell migration and differentiation in mammalian embryos. In M. H. Johnson (ed.), *Develop-ment in Mammals*, vol. 3 (Amsterdam: North Holland), pp. 363–412.

Morriss-Kay, G. (1993). Retinoic acid and cran-iofacial development: molecules and mor-phogenesis. *Bio Essays 15*, 9–15.

Morriss-Kay, G. M., Murphy, P., and Davidson, D. R. (1991). Effects of retinoic acid excess on expression of hox-2.9 and krox-20 and on morphological segmentation in the hind-brain of mouse embryos. *EMBO J. 10*, 2985–2995.

M. R. C. Vitamin Study Research Group (1991). Prevention of neural tube defects: results of the medical research council vitamin study. *Lancet 338*, 131.

Mulinare, J., Cordero, H. F., Erickson, J., and Berry, R. (1988). Periconceptional use of multivitamins and the occurrence of neural tube defects. *J. Am. Med. Assoc. 260*, 3141–3145.

Murray, J. C. (1995). Face facts: genes, environ-ment, and clefts. *Am. J. Hum. Genet. 57*, 227–232.

Murray, J. C., Daack-Hirsch, S., Buetow, K. H., Munger, R., Espina, L., Paglinawan, N., Villanueva, E., Rary, J., Magee, K., and Magee, W. (1997). Clinical and epidemiologic studies of cleft lip and palate in the Philip-pines. *Cleft Palate Craniofac. J. 34*, 7–10.

Nataf, V., Lecoin, L., Eichmann, A., and Le Douarin, N. M. (1996). Endothelin-B recep-tor is expressed by neural crest cells in the avian embryo. *Proc. Natl. Acad. Sci. USA 93*, 9645–9650.

Neidert, A. H., Panopoulou, G., and Langeland, J. A. (2000). Amphioxus goosecoid and the evolution of the head organizer and pre-chordal plate. *Evol. Dev. 2*, 303–310.

Neidert, A. H., Virupannavar, V., Hooker, G. W., and Langeland, J. A. (2001). Lamprey Dlx genes and early vertebrate evolution. *Proc. Natl. Acad. Sci. USA 98*, 1665–1670.

Newman, L. M., and Hendrickx, A. G. (1981). Fetal ear malformations induced by maternal ingestion of thalidomide in the bonnet monkey *(Macaca radiata). Teratology 23*, 351–364.

Newth, D. R. (1956). On the neural crest of lamprey embryos. *J. Embryol. Exp. Morphol. 4*, 358–375.

Nichols, D. H. (1986). Formation and distribu-tion of neural crest mesenchyme to the first pharyngeal arch region of the mouse embryo. *Am. J. Anat. 176*, 221–231.

Nicolet, G. (1970). Analyse autoradiographique de la localisation des différentes ébauches présomptives dans la ligne primitive de l'embryon de poulet. *J. Embryol. Exp. Morphol. 23*, 79–108.

Noden, D. M. (1975). An analysis of the migra-tory behaviour of avian cephalic neural crest cells. *Devel. Biol. 42*, 106–130.

Noden, D. M. (1983a). The role of the neural crest in patterning of avian cranial skeletal, connective, and muscle tissues. *Devel. Biol. 96*, 144–165.

Noden, D. M. (1983b). The embryonic origins of avian craniofacial muscles and associated connective tissues. *Am. J. Anat. 186*, 257–276.

Noden, D. M. (1988). Interactions and fates of avian craniofacial mesenchyme. *Development 103* (suppl.), 121–140.

Noden, D. M. (1991a). Vertebrate craniofacial development: the relation between ontogenetic process and morphological outcome. *Brain Behav. Evol. 38*, 190–225.

Noden, D. M. (1991b). Cell movements and control of patterned tissue assembly during craniofacial development. *J. Craniofac. Genet. Devel. Biol. 11*, 192–213.

Noden, D. M., Poelmann, R. E., and Gittenberger-de Groot, A. C. (1995). Cell origins and tissue boundaries during outflow tract development. *Trends Cardiovasc. Med. 5*, 69–73.

Nowack, E. (1965). Die sensible phase bei der thalidomide embryopathie. *Humangenetik 1*, 516–536.

Nuckolls, G. H., Shum, L., and Slavkin, H. C. (1999). Progress toward understanding craniofacial malformations. *Cleft Palate Craniofac. J. 36*, 12–26.

Osumi-Yamashita, N., Iseki, S., Noji, S., Nohno, T., Koyama, E., Taniguchi, S., Doi, H., and Eto, K. (1992). Retinoic acid treatment induces the ectopic expression of retinoic acid receptor β gene and excessive cell death in the embryonic mouse face. *Devel. Growth and Differ. 34*, 199–209.

Osumi-Yamashita, N., Ninomiya, Y., Doi, H., and Eto, K. (1994). The contribution of both forebrain and midbrain crest cells to the mesenchyme in the frontonasal mass of mouse embryos. *Devel. Biol. 164*, 409–419.

Osumi-Yamashita, N., Ninomiya, Y., and Eto, K. (1996). Rhombomere formation and hindbrain crest cell migration from prorhomberic origins in mouse embryos. *Devel. Growth and Differ. 38*, 107–118.

Osumi-Yamashita, N., Kuritani, S., Ninomya, Y., Aoki, K., Iseki, S., Chareonvit, S., Doi, H., Fujiwara, M., Watanabe, T., and Eto, K. (1997). Cranial anomaly of homzygous rSey rat is associated with a defect in the migration pathway of midbrain crest cells. *Devel. Growth and Differ. 39*, 53–67.

Park, W. J., Theda, C., Maestri, N. E., Meyers, G. A., Fryburg, J. S., Dufresne, C., Cohen, M. M. Jr., and Jabs, E. W. (1995). Analysis of phenotypic features and FGFR2 mutations in Apert syndrome. *Amer. J. Hum. Genet. 57*, 321–328.

Patterson, S. B., Johnston, M. C., and Minkoff, R. (1984). An implant labeling technique employing sable hair probes as carriers for ³H-thymidine: applications to the study of facial morphogenesis. *Anat. Rec. 210*, 515–536.

Pauli, R. M., Pettersen, J. C., Arya, S., and Gilbert, E. F. (1983). Familial agnathia-holoprosencephaly. *Am. J. Med. Genet. 4*, 677–698.

Peters, H., and Balling, R. (1999). Teeth: where and how do we make them. *Trends Genet. 15*, 59–65.

Peters, H., Neubuser, A., Kratochwil, K., and Balling, R. (1998). Pax9-deficient mice lack pharyngeal pouch derivatives and teeth and exhibit craniofacial and limb abnormalities. *Genes Devel. 12*, 2735–2747.

Petter, C., Bourban, J., Maltier, J. D., and Jost, A. (1977). Simultaneous prevention of blood abnormalities and hereditary congenital amputations in brachydactylous rabbit stock. *Teratology 151*, 49–58.

Plant, M. R., MacDonald, M. E., Grad, L. I., Ritchie, S. J., and Richman, J. M. (2000). Locally released retinoic acid repatterns the first branchial arch cartilages in vivo. *Devel. Biol. 222*, 12–26.

Poswillo, D. (1966). Observations of fetal posture and causal mechanisms of congenital deformity of the palate, mandible, and limbs. *J. Dent. Res. 45*, 584–589.

Poswillo, D. (1973). The pathogenesis of the first and second branchial arch syndrome. *Oral Surg. 35*, 302–328.

Poswillo, D. (1975). The pathogenesis of the Treacher-Collins syndrome (manibulofacial dysostosis). *Br. J. Oral Surg. 13*, 1–26.

Pratt, R. M., Larson, M. H., and Johnston, M. C. (1975). Migration of cranial neural crest cells in a cell-free, hyaluronate-rich space. *Devel. Biol. 44*, 298–305.

Pratt, R. M., Goulding, E. H., and Abbott, B. D. (1987). Retinoic acid inhibits migration of cranial neural crest cells in the cultured mouse mebryos. *J. Craniofac. Genet. Devel. Biol. 7*, 205–217.

Ramanathan, R., Wilkemeyer, M. F., Mittal, B., Perides, G., and Charness, M. E. (1996).

Alcohol inhibits cell-cell adhesion mediated by human L1. *J. Cell Biol. 133*, 381–390.

Reardon, W., Winter, R. M., Rutland, P., Pulleyn, L. J., Jones, B. M., and Malcolm, S. (1994). Mutations in the fibroblast growth factor receptor 2 gene cause Crouzon syndrome. *Nat. Genet. 8*, 98.

Rice, D. P. C., Kim, H. J., and Thesleff, I. (1999). Apoptosis in calvarial bone and suture development. *Eur. J. Oral Sci. 107*, 265–275.

Rice, D. P. C., Aberg, T., Chan, Y., Tang, Z., Kettunen, P. J., Pakarinen, L., Maxson, R. E., and Thesleff, I. (2000). Integration of FGF and TWIST in calvarial bone and suture development. *Development 127*, 1845–1855.

Richman, J. M. (1992). The role of retinoids in normal and abnormal embryonic craniofacial morphogenesis. *Crit. Rev. Oral Biol. Med. 4*, 93–109.

Richman, J. M., and Delgado, J. (1995). Locally released retinoic acid leads to facial clefts in the chick embryo but does not alter the expression of receptors for fibroblast growth factor. *J. Craniofac. Genet. Devel. Biol. 15*, 190–204.

Richman, J. M., and Tickle, C. (1989). Epithelia are interchangeable between facial primordia of chick embryos and morphogenesis is controlled by the mesenchyme. *Devel. Biol. 136*, 201–210.

Richman, J. M., and Tickle, C. (1992). Epithelial-mesenchymal interactions in the outgrowth of limb buds and facial primordia in chick embryos. *Devel. Biol. 154*, 299–308.

Richman, J. M., Herbert, M., Matovinovic, E., and Walin, J. (1997). Effect of fibroblast growth factors on outgrowth of facial mesenchyme. *Devel. Biol. 189*, 135–147.

Rijli, F. M., Mark, M., Lakkaraju, S., Dierich, A., Dollé, P., and Chambon, P. (1993). A homeotic transformation is generated in the rostral branchial region of the head by disruption of hoxa-2, which acts as a selector gene. *Cell 75*, 1333–1349.

Romitti, P., Lidral, A. C., Munger, R., Daack-Hirsch, S., Basart, A. M., Millholin, L., Hanson, L., Burns, T., and Murray, J. C. (1999). Maternal pregnancy exposures and candidate genes for nonsyndromic cleft lip and palate: evaluation of gene–environment interactions from a population-based case-control study. *Teratology 59*, 39–50.

Rosenquist, G. C. (1966). A radioautographic study of labelled grafts in the chick blastoderm: development from primitive streak stages to state 12. *Contrib. Embryol. 38*, 71–100.

Ross, R. B., and Johnston, M. C. (1972). *Cleft Lip and Palate* (Baltimore: Williams & Wilkins).

Rowe, A., Richman, J. M., and Brickell, P. M. (1992). Development of the spatial pattern of retinoic acid receptor-β transcripts in embryonic chick facial primordia. *Development 114*, 805–813.

Sadler, T. W. (1979). Culture of early somite mouse embryos during organogenesis. *J. Embryol. Exp. Morphol. 49*, 17–25.

Sadler, T. W. (2000). Susceptible periods during embryogenesis of the heart and endocrine glands. *Environ. Health Perspect. 108* (suppl. 3), 555–561.

Safra, M. J., and Oakley, G. P., Jr. (1975). Association between cleft lip with or without cleft palate and prenatal exposure to diazepam. *Lancet 2*, 478–480.

Satokata, I., and Maas, R. (1994). Msx1 deficient mice exhibit cleft palate and abnormalities of craniofacial and tooth development. *Nat. Genet. 6*, 348–356.

Savitz, D. A., Schwingl, P. J., and Keels, M. A. (1991). Influence of paternal age, smoking, and alcohol consumption on congenital anomalies. *Teratology 44*, 429–440.

Saxen, I. (1976). Etiological factors in cleft lip and cleft palate. *Duodecim 92*, 224–226.

Saxen, I., and Saxen, L. (1975). Letter: association between maternal intake of diazepam and oral clefts. *Lancet 2*, 498.

Sela-Donenfeld, D., and Kalcheim, C. (1999). Regulation of the onset of neural crest migration by coordinated activity of BMP4 and Noggin in the dorsal neural tube. *Development 126*, 4749–4762.

Semina, E. V., Reiter, R., Leysens, N. J., Alward, W. L., Small, K. W., Datson, N. A., Siegel-Bartelt, J., Bierke-Nelson, D., Bitoun, P., Zabel, B. U., Carey, J. C., and Murray, J. C. (1996). Cloning and characterization of a novel bicoid-related homeobox transcription factor gene, RIEG, involved in Rieger syndrome. *Nat. Genet. 14*, 392–399.

Shaw, G. M., and Lammer, E. F. (1999). Maternal periconceptional alcohol consumption and risk for orofacial clefts. *J. Pediatr. 134*, 298–303.

Shaw, D., Ray, A., Marazita, M., and Field, L. (1993). Further evidence of a relationship between the retinoic acid receptor alpha locus and nonsyndromic cleft lip with or without cleft palate. *Am. J. Hum. Genet. 53*, 1156–1157.

Shaw, G. M., Lammer, E. F., Wasserman, C. R., O'Malley, C. D., and Tolarova, M. M. (1995a). Risks of orofacial clefts in children born to women using multivitamins containing folic acid periconceptionally. *Lancet 346*, 393–396.

Shaw, G. M., Wasserman, C. R., O'Malley, C. D., Lammer, E. J., and Finnell, R. H. (1995b). Orofacial clefts and maternal anticonvulsant use. *Reprod. Toxicol. 9*, 97–98.

Shaw, G. M., Wasserman, C. R., Lammer, E. J., O'Malley, C. D., Murray, J. C., Basart, A. M., and Tolarova, M. M. (1996). Orofacial clefts, parental cigarette smoking, and transforming growth factor-alpha gene variants. *Am. J. Hum. Genet. 58*, 551–561.

Shawlot, W., and Behringer, R. R. (1995). Requirement for Lim1 in head-organizer function. *Nature 374*, 425–430.

Shawlot, W., Wakamiya, M., Kwan, K. M., Kania, A., Jessell, T. M., and Behringer, R. R. (1999). Lim1 is required in both primitive streak-derived tissues and visceral endoderm for head formation in the mouse. *Development 126*, 4925–4932.

Shiang, R., Lidral, A. C., Ardinger, H. H., Murray, J. C., and Buetow, K. H. (1992). Association of TGFA DNA variants with cleft lip and palate (OFC2). *Cytogenet. Cell Genet. 58*, 1872.

Shiang, R., Lidral, A. C., Ardinger, H. H., Buetow, K. H., Romitti, P. A., and Munger, R. G. (1993). Association of transforming growth-factor alpha gene polymorphisms with nonsyndromic cleft palate only (CPO). *Am. J. Hum. Genet. 53*, 836–843.

Shiang, R., Thompson, L. M., Zhu, Y. Z., Church, D. M., Fielder, T. J., and Bocian, M. (1994). Mutations in the transmembrane domain of FGFR3 cause the most common genetic form of dwarfism, achondroplasia. *Cell 78*, 335–342.

Shields, E. D., Bixler, D., and Fogh-Andersen, P. (1979). Facial clefting in Danish twins. *Cleft Palate J. 16*, 1–16.

Shilling, T. F. (1994). Genetic analysis of craniofacial development in the vertebrate embryo. *BioEssays 19*, 459–468.

Shimeld, S. M., and Holland, P. W. H. (2000). Vertebrate innovations. *Proc. Natl. Acad. Sci. USA 97*, 4449–4452.

Shin, S. H., Kogerman, P., Lindstrom, E., Toftgard, R., and Biesecker, L. G. (1999). GLI3 mutations in human disorders mimic Drosophila cubitus interruptus protein functions and localization. *Proc. Natl. Acad. Sci. USA 96*, 288–304.

Shiota, K., Nakatsu, T., and Irie, H. (1993). Computerized three-dimensional reconstruction of the brain of normal and holoprosencephalic human embryos. *Birth Defects 29*, 261–271.

Shokeir, M. H. K. (1977). The Goldenhar syndrome: a natural history. *Birth Defects Original Article Senes 13*, 67.

Shprintzen, R. J. (1982). Palatal and pharyngeal anomalies in craniofacial syndromes. *Birth Defects Original Article Senes 18*, 53.

Shprintzen, R. J., Goldberg, R. B., Lewin, H. L., Sidoti, E. J., Berkman, M. D., and Argamaso, R. V. (1978). A new syndrome involving cleft palate, cardiac anomalies, typical facies, and learning disabilities: velo-cardio-facial syndrome. *Cleft Palate J. 15*, 56–62.

Shprintzen, R. J., Siegel-Sadewitz, V. L., Amato, J., and Goldberg, R. B. (1985). Anomalies associated with cleft lip, cleft palate, or both. *Am. J. Med. Genet. 20*, 585–596.

Shuler, C. F., Halpern, D. E., Guo, Y., and Sank, A. C. (1992). Medial edge epithelium fate traced by cell lineage analysis during epithelial-mesenchymal transformation in vivo. *Devel. Biol. 154*, 318–330.

Siebert, J. R., Cohen, M. M., Jr., and Sulik, K. K. (1990). *Holoprosencephaly: An Overview and Atlas of Cases* (New York: Wiley-Liss).

Siegel, M. I., and Mooney, M. P. (1986). Palatal width growth rates as the genetic determinant of cleft palate induced by vitamin A. *J. Craniofac. Genet. Devel. Biol. Suppl. 2*, 187–191.

Simeone, A., Acampora, D., Gulisano, M., Stornaluolo, A., and Boncinelli, E. (1992).

behavior of neural crest cells in vitro. *Devel. Biol. 123*, 276–281.

Smuts, M. K. (1981). Rapid nasal pit formation in mouse stimulated by ATP-containing medium. *J. Exp. Zool. 216*, 409–414.

Sperber, G. H. (2000). *Craniofacial Embryology*, 5th ed. (London: Wright), pp. 31–57, 132–143.

Stark, M. R., Biggs, J. J., Schoenwolf, G. C., and Rao, M. S. (2000). Characterization of avian frizzled genes in cranial placode development. *Mech. Devel. 93*, 195–200.

Steffek, A. J., King, C. T., and Derr, J. E. (1967). The comparative pathogenesis of experimentally induced cleft palate. *J. Oral. Ther. 3*, 9–16.

Sulik, K. K. (1984). *Critical Periods for Alcohol Teratogenesis in Mice, with Special Reference to the Gastrulation Stage of Embryogenesis.* Ciba Foundation Symposium no. 105 (London: Pitman Books, Ltd.).

Sulik, K. K., and Schoenwolf, G. C. (1985). Highlights of craniofacial morphogenesis in mammalian embryos, as revealed by scanning electron microscopy. *Scan. Electron Microsc. 4*, 1735–1752.

Sulik, K. K., and Johnston, M. C. (1982). Embryonic origin of holoprosencephaly: interrelationship of the eveloping brain and face. *Scan. Electron Microsc. 1*, 309–322.

Sulik, K. K., and Johnston, M. C. (1983). Sequence of developmental changes following ethanol exposure in mice: craniofacial features of the fetal alcohol syndrome (FAS). *Am. J. Anat. 166*, 275–289.

Sulik, K. K., Johnston, M. C., Ambrose, L. J. H., and Dorgan, D. (1979). Phenytoin (Dilantin)-induced cleft lip: a scanning and transmission electron microscopic study. *Anat. Rec. 195*, 243–256.

Sulik, K. K., Johnston, M. C., and Webb, M. A. (1981). Fetal alcohol syndrome: embryogenesis in a mouse model. *Science 214*, 936–938.

Sulik, K. K., Johnston, M. C., Daft, P. A., and Russell, W. E. (1986). Fetal alcohol syndrome and DiGeorge anomaly: critical ethanol exposure periods for craniofacial malformations as illustrated in an animal model. *Am. J. Med. Genet. 2*, 191–194.

Sulik, K. K., Johnston, M. C., Smiley, S. J., Spieght, H. S., and Jarvis, B. E. (1987).

Mandibulofacial dysostosis (Treacher Collins syndrome): a new proposal for its pathogenesis. *Am. J. Med. Genet. 27*, 359–372.

Sulik, K. K., Cook, C. S., and Webster, W. S. (1988). Teratogens and craniofacial malformations: relationships to cell death. *Development 103* (suppl.), 213–232.

Sulik, K. K., Smiley, S. J., Turvey, T. A., Speight, H. S., and Johnston, M. C. (1989). Pathogenesis involving the secondary palate and mandibulofacial dystotosis and related syndromes. *Cleft Palate J. 26*, 209–216.

Sulik, K. K., Dehart, D. B., Inagaki, T., Carson, J. L., Vrablic, T., Gesteland, K., and Schoenwolf, G. C. (1994). Morphogenesis of the murine node and notochordal plate. *Devel. Dyn. 201*, 260–278.

Sulik, K. K., Dehart, D. B., Rogers, J. M., and Chernoff, N. (1995). Teratogenicity of low doses of all-trans retinoic acid in presomite mouse embryos. *Teratology 51*, 398–403.

Sun, D., Vanderburg, C. R., Odierna, G. S., and Hay, E. D. (1998). TGFbeta3 promotes transformation of chicken palate medial edge epithelium to mesenchyme in vitro. *Development 125*, 95–105.

Sun, D., Baur, S., and Hay, E. D. (2000). Epithelial-mesenchymal transformation is the mechanism for fusion of the craniofacial primordia involved in morphogenesis of the chicken lip. *Devel. Biol. 228*, 337–349.

Takahashi, Y., and LeDouarin, N. (1900). cDNA cloning of a quail homeobox gene and its expression in neural crest-derived mesenchyme and lateral plate mesoderm. *Proc. Natl. Acad. Sci. USA 87*, 7482–7486.

Tamarin, A., Crawley, A., Lee, J., and Tickle, C. (1984). Analysis of upper beak defects in chicken embryos following treatment with retinoic acid. *J. Embryol. Exp. Morphol. 84*, 105–123.

Tan, S. S., and Morriss-Kay, G. M. (1986). Analysis of cranial neural crest cell migration and early fates in postimplantation rat. *J. Embryol. Exp. Morphol. 98*, 21–58.

Tassabehji, M., Read, A. P., Newton, V. E., Patton, M., Gruss, P., and Harris, R. (1993). Mutations in the PAX3 gene causing Waardenburg syndrome type 1 and type 2. *Nat. Genet. 3*, 26–30.

Taya, Y., O'Kane, S., and Ferguson, M. (1998). Pathogenesis of cleft palate in TGFbeta3 knockout mice. *Development 126*, 3869–3879.

Taysi, K., and Tinaztepe, K. (1972). Trisomy D and the cyclops malformation. *Am. J. Dis. Child 124*, 170.

ten Berge, D., Brouwer, A., el Bahi, S., Guenet, J. L., Robert, B., and Meijlink, F. (1998). Mouse Alx3: an aristaless-like homeobox gene expressed during embryogenesis in ectomesenchyme and lateral plate mesoderm. *Devel. Biol. 199*, 11–25.

Thesleff, I. (1998). Two genes for missing teeth. *Nature 13*, 379–380.

Thorogood, P., Smith, L., Nicol, A., McGinty, R., and Garrod, D. (1982). Effects of vitamin A on the behavior of migratory neural crest cells in vitro. *J. Cell Sci. 57*, 331–350.

Tolarova, M. (1982). Periconceptional supplementation with vitamins and folic acid to prevent recurrence of cleft lip. *Lancet 2*, 217.

Tolarova, M. (1987). Orofacial clefts in Czechoslovakia. *Scand. J. Plast. Reconstr. Surg. 21*, 19–25.

Tolarova, M. (1990). Genetics, gene carriers, and environment. In J. Bader (ed.), *Proceedings of Conference of Risk Assessment in Dentistry* (Chapel Hill, NC: University of North Carolina Department of Dental Ecology), pp. 116–129.

Torczynski, E., Jacobiec, F. A., Johnston, M. C., Font, R. L., and Madewell, J. A. (1977). Synophthalmia and cyclopia: a histopathologic, radiographic, and organogenetic analysis. *Documenta Ophthalmologica 44*, 311–378.

Tosney, K. W. (1982). The segregation and early migration of cranial neural crest cells in the avian embryo. *Devel. Biol. 89*, 13–24.

Townsley, J. T., and Johnston, M. C. (eds.) (1991). Research advances in prenatal craniofacial development. *J. Craniofac. Genet. Devel. Biol. 11*, 378–492.

Trasler, D. G., (1968). Pathogenesis of cleft lip and its relation to embryonic face shape in A/J and C57B1 mice. *Teratology 1*, 33–50.

Trasler, D. G. Walker, B. E., and Fraser, F. C. (1956). Congenital malformations produced by amniotic-sac puncture. *Science 124*, 439–440.

Trasler, D. G., Reardon, C. A., and Rajchgot, H. (1978). A selection experiment for distinct types of 6-aminonicotinamide-induced cleft lip in mice. *Teratology 18*, 49–54.

Tucker, A. S., Yamada, G., Grigoriou, M., Pachnis, V., and Sharpe, P. T. (1999). Fgf-8 determines rostral-caudal polarity in the first branchial arch. *Development 126*, 51–61.

Tyan, M. L. (1982). Differences in the reported frequencies of cleft lip plus cleft lip and palate in Asians born in Hawaii and the continental United States. *Proc. Soc. Exp. Biol. Med. 171*, 41–48.

Tyler, M. S., and Hall, B. K. (1977). Epithelial influences on skeletogenesis in the mandible of the embryonic chick. *Anat. Rec. 118*, 229–240.

Uyemura, K., Asou, H., Yazaki, T., and Takeda, Y. (1996). Cell-adhesion proteins of the immunoglobulin superfamily in the nervous system. *Essays Biochem. 31*, 37–48.

van den Boogaard, M. J., Dorland, M., Beemer, F. A., and van Amstel, H. K. (2000). MSX1 mutation is associated with orofacial clefting and tooth agenesis in humans. *Nat. Genet. 24*, 342–343.

Vasser, R., and Fuchs, E. (1991). Transgenic mice provide new insights into the role of TGF-alpha during epidermal development. *Genes Devel. 5*, 714–727.

Veitch, E., Begbie, J., Schilling, T. F., Smith, M. M., and Graham, A. (1999). Pharyngeal arch patterning in the absence of neural crest. *Curr. Biol. 9*, 1481–1484.

Vergato, L. A., Doerfler, R. J., Mooney, M. P., and Siegel, M. I. (1997). Mouse palatal width growth rates as an "at risk" factor in the development of cleft palate induced by hypervitaminosis A. *J. Craniofac. Gen. Devel. Biol. 17*, 204–210.

Vermeij-Keers, C. (1990). Craniofacial embryology and morphogenesis: normal and abnormal. In M. Stricker, J. C. van der Meulin, B. Raphael, R. Mazzola, and D. E. Tolhurst (eds.), *Craniofacial Malformations* (Edinburgh: Churchill Livingston).

Vintiner, G. M., Holder, S. E., Winter, R. M., and Malcolm, S. (1992). No evidence of linkage between the transforming growth factor-alpha gene in families with apparently autosomal dominant inheritance of cleft lip and palate. *J. Med. Genet. 29*, 393–397.

Wakamiya, M., Lindsay, E. A., Rivera-Perez, J. A., Baldini, A., and Behringer, R. R. (1998). Functional analysis of Gscl in the pathogenesis of the DiGeorge and velocardiofacial syndromes. *Hum. Mol. Genet. 12*, 1835–1840.

Wall, N. A., and Hogan, B. L. (1995). Expression of bone morphogenetic protein-4 (BMP-4), bone morphogenetic Protein-7 (BMP-7), fibroblast growth factor-8 (FGF-8) and sonic hedgehog (SHH) during branchial arch development in the chick. *Mech. Devel. 53*, 383–392.

Wang, K. Y., Juriloff, D. M., and Diewert, V. M. (1995). Deficient and delayed primary palatal fusion and mesenchymal bridge formation in cleft lip-susceptible strains of mice. *J. Craniofac. Genet. Devel. Biol. 15*, 99–116.

Waterman, R. E. (1976). Topographic changes along the neural fold associated with neurulation in the hamster and mouse. *Am. J. Anat. 146*, 151–156.

Waterman, R. E., and Meller, S. M. (1974). Alterations in the epithelial surface of human palatal shelves prior to and during fusion: a scanning electron microscopic study. *Anat. Rec. 180*, 111–135.

Webb, J. F., and Noden, D. M. (1993). Ectodermal placodes: contributions to the development of the vertebrate head. *Am. Zool. 33*, 434–447.

Webster, M. K., and Donoghue, D. J. (1996). Constitutive activation of fibroblast growth factor receptor 3 by the transmembrane domain point mutation found in achondroplasia *EMBO J. 15*, 520–527.

Webster, W. S. (1985). Isotretinoin embryopathy: the effects of isotretinoin and 4-oxo-isotretioin post-implantation rat embryos in vitro. *Teratology 31*, 211A.

Webster, W. S., Johnston, M. C., Lammer, E. J., and Sulik, K. K. (1986). Isotretinoin embryopathy and the cranial neural crest: an in vivo and in vitro study. *J. Craniofac. Genet. Devel. Biol. 6*, 211–222.

Wedden, S. E. (1987). Epithelial-mesenchymal interactions in the development of chick facial primordia and the target of retinoid action. *Development 99*, 341–351.

Wedden, S. E. (1991). Effects of retinoids on chick face development. *J. Craniofac. Genet. Devel. Biol. 11*, 326–337.

Wee, E. L., and Zimmerman, E. F. (1983). Involvement of GABA in palate morphogenesis and its relation to diazepam in two mouse strains. *Teratology 28*, 15–22.

Werler, M. M., Lammer, E. J., Rosenberg, L., and Mitchell, A. A. (1990). Maternal cigarette smoking during pregnancy in relation to oral clefts. *Am. J. Epidemiol. 132*, 926–928.

Wilke, T. A., Gubbels, S., Schwartz, J., and Richman, J. M. (1997). Expression of fibroblast growth factor receptors (FGFR1, FGFR2, FGFR3) in the developing head and face. *Devel. Dyn. 210*, 41–52.

Wilkemeyer, M. F., and Charness, M. E. (1998). Characterization of ethanol-sensitive and insensitive fibroblast cell lines expressing human L1. *J. Neurochem. 71*, 2382–2391.

Wilkinson, D. G. (1993). Molecular mechanisms of segmental patterning in the vertebrate hindbrain and neural crest. *BioEssays 15*, 499–505.

Willhite, C., Hill, R. M., and Irving, D. W. (1986). Isotretinoin-induced craniofacial malformations in humans and hamsters. *J. Craniofac. Genet. Devel. Biol. 2* (suppl.), 193–209.

Williams, G. T., and Smith, C. A. (1993). Molecular regulation of apoptosis: genetic controls on cell death. *Cell 74*, 777–779.

Wise, C. A., Chiang, L. C., Paznekas, W. A., Sharma, M., Musy, M. M., Ashley, J. A., Lovett, M., and Jabs, E. W. (1997). TCOF1 encodes a putative nucleolar phosphate that exhibits mutations in Treacher Collins syndrome throughout its coding region. *Proc. Natl. Acad. Sci. USA 94*, 310–315.

Wray, S. D., Hassell, T. M., Phillips, C., and Johnston, M. C. (1982). Preliminary study of the effects of carbamazepine on congenital orofacial defects in offspring of A/J mice. *Epilepsia 23*, 101–110.

Wright, S., and Wagner, K. (1934). Types of subnormal development of the head from inbred strains of guinea pigs and their bearing on the classification and interpretation of vertebrate monsters. *Am. J. Anat. 54*, 383.

Wyszynski, D. F., and Beaty, T. H. (1996). Review of the role of potential teratogens in the origin of human nonsyndromic oral clefts. *Teratology 53*, 309–317.

Wyszynski, D. F., Beaty, T. H., and Maestri, N. E. (1996). Genetics of nonsyndromic oral clefts revisited. *Cleft Palate Craniofac. J. 33*, 406–417.

Wyszinski, D. F., DeMaestri, N., Lewanda, A. F., McIintosh, I., Smith, E. A., Garcia-Delgado, C., Vinageras-Guarmeros, E., Wolfsberg, E., and Beaty, T. H. (1997a). No evidence for linkage to cleft lip with or without cleft palate to a marker near the transforming growth factor alpha locus in two populations. *Hum. Hered. 47*, 101–109.

Wyszynski, D. F., Duffy, D. L., and Beaty, T. H. (1997b). Maternal cigarette smoking and oral clefts: a meta-analysis. *Cleft Palate Craniofac. J. 34*, 206–210.

Yamagishi, H., Garg, V., Matsuoka, R., Thomas, T., and Srivastava, D., (1999). A molecular pathway revealing a genetic basis for human cardiac and craniofacial defects. *Science 283*, 1158–1161.

Yamaguchi, T. P., Bradley, A., McMahon, A. P., and Jones, S. (1999). A Wnt5a pathway underlies outgrowth of multiple structures in the vertebrate embryo. *Development 126*, 1211–1223.

Yanagisawa, H., Yanagisawa, M., Kapur, R. P., Richardson, J. A., Williams, S. C., Clouthier, D. E., de Wit, D., Emoto, N., and Hammer, R. E. (1998). Dual genetic pathways of endothelin-mediated intercellular signaling revealed by targeted disruption of endothelin converting enzyme-1 gene. *Development 125*, 825–836.

Young, D. L., Schneider, R. A., Hu, D., and Helms, J. A. (2000). Genetic and teratogenic approaches to craniofcial development. *Crit. Rev. Oral Biol. Med. 11*, 304–317.

Zhang, J., Hagopian-Donaldson, S., Serbedzija, G., Elsemore, J., Plehn-Dujowich, D., McMahon, A. P., Flavell, R. A., and Williams, T. (1996). Neural tube, skeletal and body wall defects in mice lacking transcription factor AP-2. *Nature 381*, 238–241.

Zhao, Y., Guo, Y. J., Tomac, A. C., Taylor, N. R., Grinberg, A., Lee, E. J., Huang, S., and Westphal, H. (1999). Isolated cleft palate in mice with a targeted mutation of the LIM homeobox gene lhx8. *Proc. Natl. Acad. Sci. USA 96*, 15002–15006.

Zheng, Z. S., Polakowska, R., Johnson, A., and Goldsmith, L. A. (1992). Transcriptional control of epidermal growth factor receptor by retinoic acid. *Cell Growth Differ. 3*, 225–232.

Zimmer, C. (2000). In search of vertebrate origins: beyond brain and bone. *Science 287*, 1576–1579.

Zimmerman, E. L. (ed.). (1984). Palate development: normal and abnormal, cellular and molecular aspects. *Curr. Topics Devel. Biol. 19*, 1–224.

Zimmerman, E. F., and Wee, E. L. (1984). Role of neurotransmitters in palate development. *Curr. Topics Devel. Biol. 19*, 37–63.

CHAPTER 6

GENETIC ETIOLOGIES OF CRANIOSYNOSTOSIS

ETHYLIN WANG JABS, M.D.

6.1 INTRODUCTION

In this chapter, current knowledge of the genetic etiologies of the craniosynostoses is presented. In the past, the discipline of genetics described the epidemiology, clinical aspects, and patterns of inheritance of the craniosynostoses and their associated syndromes. In the 1990s, with the availability of molecular tools through the Human Genome Project, understanding of the genetic etiologies of the craniosynostoses increased exponentially. For most of the common syndromes and for some cases of nonsyndromic craniosynostosis, the disease genes and their functions now are known. The genetic approach used to elucidate the molecular pathogenesis of the craniosynostoses can be applied to other human craniofacial malformations.

6.2 GENETIC EPIDEMIOLOGY OF CRANIOSYNOSTOSES

Craniosynostosis is a major malformation, with a birth prevalence of approximately 1 in 2100 to 3000 newborns, reported from a variety of racial and ethnic groups (Chung and Myrianthopoulos, 1975; Hunter and Rudd, 1976, 1977; Lajeunie et al., 1995, 1996, 1998; Lammer et al., 1987a, 1987b). Nonsyndromic craniosynostosis (craniosynostosis as an isolated feature) is more frequent than syndromic craniosynostosis (craniosynostosis as a feature associated with other malformations). Sagittal synostosis is the most common of the nonsyndromic synostoses, with a birth prevalence of 1 in 5000. Approximately 72% of sagittal cases are sporadic, with a male preponderance of 3.5:1. No paternal or maternal age effects have been reported. Six percent of cases are familial and usually inherited as an autosomal dominant condition with 38% penetrance.

Isolated coronal synostosis occurs half as frequently as isolated sagittal synostosis (94 versus 190 per million newborns, respectively). Sporadic cases account for approximately 61% of all coronal cases, with a female preponderance of 2:1 and an advanced paternal age effect. More coronal than sagittal synostosis cases are familial. Ten to 14% of nonsyndromic coronal cases are inherited as an autosomal dominant condition, with 60% penetrance. Most autosomal dominant craniosynostosis syndromes involve the coronal suture. When craniosynostosis is a feature of a syndrome,

Understanding Craniofacial Anomalies: The Etiopathogenesis of Craniosynostoses and Facial Clefting, Edited by Mark P. Mooney and Michael I. Siegel, ISBN 0-471-38724-x Copyright © 2002 by Wiley-Liss, Inc.

it is usually associated with facial, limb, ear, and/or heart malformations.

6.3 PHENOTYPES AND GENOTYPES OF CRANIOSYNOSTOSES

The most common craniosynostosis syndromes include the autosomal dominant Crouzon, Apert, and Saethre-Chotzen syndromes. The classical clinical descriptions of these three conditions are distinctive, and Apert syndrome is the most recognizable because of the severity of the hand and feet syndactyly. Other diagnoses such as Pfeiffer, Jackson-Weiss, and Beare-Stevenson cutis gyrata syndromes have overlapping craniofacial features between each other and with Crouzon and Apert syndromes. The phenotypic variability of these conditions represents the phenotypic spectrum associated with fibroblast growth receptor 2 (FGFR2) mutations. Pfeiffer, Jackson-Weiss, and Beare-Stevenson syndromes and nonsyndromic coronal craniosynostosis can be caused by mutations in other members of the same receptor family, demonstrating genetic heterogeneity.

Saethre-Chotzen and Robinow-Soaurf syndromes have significant overlap in their physical features and represent phenotypic variability within the same condition. These syndromes are allelic, both with mutations in the TWIST gene that codes for a transcription factor with a DNA binding and helix-loop-helix domains. Other less common craniosynostosis syndromes include craniosynostosis, Boston type, with a mutation in the MSX2 gene that codes for a transcription factor with a homeobox domain.

The distinctions for some of these clinical diagnoses have become less obvious as more families and larger families with phenotypic variability have been analyzed for their molecular pathogenesis. The careful delineation of the clinical features for each condition has given the molecular biologist clues when to test for phenotypic variability due to different alleles and/or genetic heterogeneity. Information on the clinical description and molecular definition of these conditions can be accessed through Online Mendelian Inheritance in Man database (http://www.ncbi.nlm.nih.gov/Omim).

6.4 FIBROBLAST GROWTH FACTOR RECEPTORS

Fibroblast growth factor receptors are transmembrane receptor tyrosine kinase proteins. They are composed of an extracellular domain with three immunoglobulin-like domains (IgI, IgII, and IgIII), a transmembrane domain, and a split tyrosine kinase domain (Fig. 6.1) (Jaye et al., 1992; Johnson and Williams, 1993). The conformation of the Ig-like domains is maintained in part by disulfide bonds created between cysteine residues (Plotnikov et al., 1999). There are four tyrosine kinase receptors of this protein family. Three (FGFR1, FGFR2, and FGFR3) are known to be mutated in 15 to 20% of all craniosynostosis cases. The FGFRs are coded by genes with approximately 20 exons. The genomic structure of FGFR2 is the largest of this receptor gene family (Ingersoll et al., 2002). Most of the mutations occur in exons IIIa and IIIc that code for the IgIII domain and the linker region between the IgII and IgIII domains.

These receptors also have multiple alternative spliceoforms or isoforms. FGFR2 is normally present in alternative isoforms with the same amino acid sequence for the first half of the IgIII domain coded by exon IIIa and alternative amino acid sequences for the second half of the IgIII domain coded by exon IIIb or exon IIIc. The resulting isoforms are designated keratinocyte growth factor receptor (KGFR) and bacterially expressed kinase (BEK), respectively. KGFR is expressed primarily in epithelial

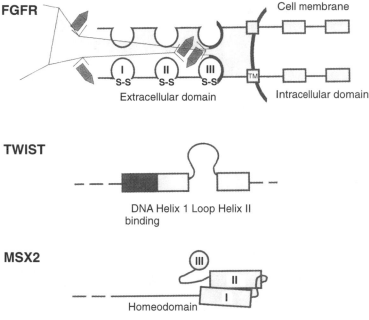

Figure 6.1 Diagrams of proteins coded by craniosynostosis disease genes. Fibroblast growth factor receptor (FGFR) is composed of an extracellular domain with three immunoglobulin-like domains (IgI, IgII, IgIII) with disulfide bonds (S-S), transmembrane domain (TM), and intracellular split tyrosine kinase domain (rectangular boxes). The second half of IgIII (thickened portion of IgIII) is coded by alternative exons IIIb and IIIc. The complex of heparan sulfate–containing proteoglycans (branching structure) and the fibroblast growth factor (gray arrows) ligands activate the dimerized FGFRs. TWIST consists of a DNA binding domain and helix-loop-helix region. The MSX2 homeodomain is comprised of an N-terminus followed by helices I, II, and III. (*Source*: Adapted from Lewanda and Jabs, 2001.)

cell types, whereas BEK is expressed in osteoblasts and other cell types. Each isoform has different ligand specificities (Dionne et al., 1990; Dell and Williams, 1992; Miki et al., 1992).

The crucial ligand binding domain is created by the IgII and IgIII domains and their linker region. The ligands for the receptors are fibroblast growth factors (FGFs), of which there are at least 23, each of which bind with varying specificity and affinity (Naski and Ornitz, 1998; Shimada et al., 2001). The FGFs are monomeric and with heparan sulfate–containing proteoglycans (HSPGs) induce the receptors to homodimerize and heterodimerize (Ornitz et al., 1992; Ornitz and Leder, 1992). Formation of this FGF-FGFRs-HSPG complex results in autophosphorylation of intracellular tyrosine residues, which then serve to stimulate the catalytic activity of the receptors and recruit sites for downstream signaling proteins including FRS2, Grb2, Gab1, phosphatidylinositol 3-kinase, and Ras/mitogen-activated protein kinase or phospholipase C-γ1 (Olivier et al., 1993; Blaikie et al., 1994; Ong et al., 2001).

The expression patterns of FGFR1–3 in the developing calvarium predict that if these receptors were mutated, craniosynostosis would result. Each of these receptors, as well as their ligands, has its own pattern of expression. Prior to ossification of the human skull, FGFR1 is expressed in the head mesenchyme, and FGFR2 is expressed in the epidermis and mesenchymal condensations. The FGFR3 IIIc isoform is present at a lower level in the epidermis and basal head mesenchyme (Delezoide et al., 1998). During ossification of the membranous skull vault, FGFR1 and FGFR2 and to a lower level FGFR3 are expressed in the mesenchyme, before osteoblast secretion of osteoid, around the mineralized bone. At 18 weeks, sutures appear in the periphery of the fused ossification centers or skull bones. There are high levels of FGFR1 and FGFR2 and lower levels of FGFR3 IIIc present at the osteogenic front of sutures.

6.5 CROUZON SYNDROME AND FGFR2 EXTRACELLULAR DOMAIN MUTATIONS

Crouzon syndrome (OMIM 123500) is characterized by craniosynostosis, ocular proptosis, shallow orbits, and maxillary hypoplasia (Fig. 6.2A) (Crouzon, 1912; Kreiborg, 1981; Cohen and MacLean, 2000). This condition occurs in approximately 1 per 25,000 to 65,000 newborns and accounts for 4.8% of all cases of cran-

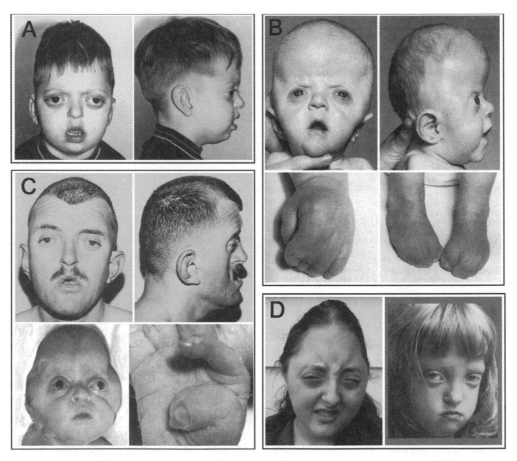

Figure 6.2 Craniosynostosis syndromes with FGFR extracellular mutations. (A) Crouzon syndrome with brachycephaly, ocular proptosis, and maxillary hypoplasia (Cohen, 1975). (B) Apert syndrome with turribrachycephaly, ocular proptosis, downslanting palpebral fissures, maxillary hypoplasia, and syndactyly (Muenke and Wilkie, 2001). (C) Pfeiffer syndrome type 1 with brachycephaly, ocular hypertelorism, maxillary hypoplasia, and low-set ears (top; Cohen, 1975). Cloverleaf skull deformity in type 2 and broad thumb (Cohen and MacLean, 2000). (D) Jackson-Weiss syndrome with acrocephaly, ocular proptosis, and maxillary hypoplasia (Jackson et al., 1976).

iosynostosis. Crouzon syndrome is inherited as a highly variable autosomal dominant condition, and approximately 67% of cases are familial. In sporadic cases increased paternal age at the time of conception and mutations exclusively in the paternally derived gene have been demonstrated (Glaser et al., 2000). Germinal mosaicism has been reported (Rollnick, 1988). Phenotypic variability in this condition can range from ocular proptosis and midface hypoplasia with no craniosynostosis to a cloverleaf skull deformity in members of the same family. Brachycephaly is most common, although scaphocephaly and trigonocephaly have also been observed. Craniofacial findings that occur in the majority of cases include exotropia, lateral palatal swellings, calcification of stylohyoid ligament, and cervical spine anomalies. Patients often have exposure conjunctivitis, mild to moderate hearing loss, obligatory mouth breathing, and central nervous system abnormalities including progressive hydrocephalus associated with chronic tonsillar herniation and jugular foramen stenosis with venous obstruction. Intelligence is usually normal. Limb anomalies include minimal metacarpophalangeal profile pattern defects, carpal fusions, and stiffness of the elbows with some cases of radial head subluxation. Crouzon syndrome is often distinguished from the other craniosynostosis conditions by the absence of limb defects.

More than 26 different mutations primarily in exons IIIa and IIIc, which code for the ligand binding, extracellular domain of FGFR2, have been identified in Crouzon syndrome patients (Table 6.1) (Reardon et al., 1994; Passos-Bueno et al., 1999). The most common mutations are recurrent missense mutations and include the loss and gain of a cysteine residue such as Cys278Phe, Cys342Arg, Cys342Ser, Cys342Trp, Cys342Tyr, or Ser354Cys, and a splice site mutation Ala344Ala (1032G > A). Presumably these mutations perturb the formation of the ligand binding domain of FGFR2. The unpaired cysteine residues result in destabilization of IgII and IgIII domains. The silent Ala344Ala mutation creates a new donor splice site and thereby alters the receptor with a deletion in the IgIIIc domain (Li et al., 1995). There are also a few small in-frame insertion and deletion mutations. There are rare cases of mutations outside IgII and IgIII domains in IgI and tyrosine kinase domains (Kan et al., 2002).

6.6 APERT SYNDROME AND FGFR2 Ser252Trp AND Pro253Arg MUTATIONS

Apert syndrome (OMIM 101200) is characterized by craniosynostosis, midface hypoplasia, and symmetric syndactyly of the hands and feet (Fig. 6.2B) (Apert, 1906; Blank, 1960; Cohen and MacLean, 2000). This condition has been estimated to occur in 1 per 65,000 to 150,000 newborns and in 4.5% of all craniosynostosis cases. Approximately 98% of cases are new mutations because the severe malformations associated with this condition result in reduced likelihood of mating. Advanced paternal age and germinal mosaicism have been reported. The origin of new mutations is exclusively paternal as in Crouzon syndrome. Apert syndrome newborns have coronal synostosis, a widely patent midline calvarial defect extending from the glabella to the posterior fontanelle, and megalencephaly. Adult measurements of the cranial volume are increased in Apert syndrome and reduced in Crouzon syndrome.

Apert syndrome is associated with central nervous syndrome abnormalities including nonprogressive ventriculomegaly, hypoplasia of the corpus callosum, agenesis of the septum pellucidum, and cavum septum pellucidum. Hydrocephalus is less often associated with Apert syndrome than with a cloverleaf skull deformity or Crouzon syndrome. Intelligence ranges from normal to severe mental retardation, with the majority of Apert

TABLE 6.1 Craniosynostosis Conditions and Mutations[a]

Condition Diagnosis OMIM Number	Phenotype	Gene Symbol and Location OMIM Number	Nucleotide Change	Type Region	Protein Amino Acid Change	Mechanism of Action
Crouzon syndrome 123500	Shallow orbit, proptosis, hypertelorism, maxillary hypoplasia Limb normal	FGFR2 exon IIIa (exon 8 or U) exon IIIc (exon 10 or D) 176943	Missense, splicing, in-frame deletion, in-frame insertion	Tyrosine kinase receptor Extracellular IgII–IgIII domain	C342Y, C342R, C278F, A344A	Constitutive activation
Apert syndrome 101200	Acrocephaly, wide open fontanelle at birth, hypertelorism, shallow orbit, maxillary hypoplasia, cleft palate Hand and foot syndactyly	FGFR2 exons IIIa and IIIc 176943	Missense, splicing, Alu repeat insertion	Tyrosine kinase receptor Extracellular IgII–IgIII domain	S252W, P253R	Ectopic ligand-dependent activation
Pfeiffer syndrome 136350.0001	Brachycephaly, shallow orbit, proptosis, maxillary hypoplasia, cloverleaf skull Broad thumb and great toe, medial deviation, cutaneous syndactyly	FGFR1 exon IIIa 136350	Missense	Tyrosine kinase receptor Extracelluar IgII–IgIII linker	P252R	
Pfeiffer syndrome 101600	Brachycephaly, shallow orbit, proptosis, maxillary hypoplasia, cloverleaf skull Broad thumb and great toe, medial deviation, cutaneous syndactyly	FGFR2 exons IIIa and IIIc 176943	Missense, splicing, in-frame deletion	Tyrosine kinase receptor Extracelluar IgII–IgIII domain	C342Y, C342R, C278F	Constitutive activation

Disorder	Gene/exon	Mutation type	Protein domain	Mutation	Effect
Antley-Bixler-like syndrome 207410	FGFR2 exons IIIa and IIIc 176943	Missense	Tyrosine kinase receptor Extracellular IgII–IgIII domain	S351C, W290C	Constitutive activation
Jackson-Weiss syndrome 123150	FGFR2 exons IIIa and IIIc 176943	Missense	Tyrosine kinase receptor Extracelluar IgII–IgIII domain	Q289P, C342R, A344G	Constitutive activation
Jackson-Weiss syndrome 123150	FGFR1 exons IIIa 136350.0)01	Missense	Tyrosine kinase receptor Extracelluar IgII–IgIII linker	P252R	
Beare-Stevenson cutis gyrata syndrome 123790	FGFR2 exon 11 176943	Missense	Tyrosine kinase receptor Juxta- or transmembrane domain	S372C, Y375C	
Beare-Stevenson cutis gyrata syndrome 123790	FGFR3 exon IIIa 176943	Missense	Tyrosine kinase receptor Extracellular IgII–IgIII linker	P252R	
Crouzon syndrome and acanthosis nigricans 134934.001	FGFR3 exon TM 134934	Missense	Tyrosine kinase receptor Transmembrane domain	A391E	

Phenotype details:
- Antley-Bixler-like syndrome 207410: Brachycephaly, maxillary hypoplasia Humeroradial synostosis, femoral bowing
- Jackson-Weiss syndrome 123150: Variable craniosynostosis, maxillary hypoplasia Broad great toe and medial deviation, syndactyly, bony fusion
- Jackson-Weiss syndrome 123150: Variable craniosynostosis, maxillary hypoplasia Broad great toe and medial deviation, syndactyly, bony fusion
- Beare-Stevenson cutis gyrata syndrome 123790: Variable craniosynostosis Cutis gyrata, acanthosis nigricans
- Beare-Stevenson cutis gyrata syndrome 123790: Mild phenotype
- Crouzon syndrome and acanthosis nigricans 134934.001: Crouzon syndrome, acanthosis nigricans, hydrocephalus, choanal stenosis, jaw cementoma

TABLE 6.1 *Continued*

Condition	Phenotype	Gene		Protein		
Diagnosis OMIM Number		Symbol and Location OMIM Number	Nucleotide Change	Type Region	Amino Acid Change	Mechanism of Action
Nonsyndromic coronal craniosynostosis; Crouzon-like, Pfeiffer-like, Jackson-Weiss-like syndromes (Muenke syndrome; craniosynostosis, Adelaide type) 602849, 600593	Variable, uni- or bicoronal synostosis Brachydactyly	FGFR3 exon IIIa 134934.0014	Missense	Tyrosine kinase receptor Extracellular IgII–IgIII linker	P250R	
Saethre-Chotzen syndrome 101400	Variable brachycephaly, facial asymmetry, ptosis, maxillary hypoplasia, prominent ear crus Brachydactyly, cutaneous syndactyly	TWIST 3' UTR and exon 1 601622	Nonsense, in-frame duplication, missense, microdeletion, change in 3' UTR	Helix-loop-helix transcription factor 3' UTR, entire protein	Y103X, P139ins7	Haploinsufficiency
Saethre-Chotzen with bifid great toe (Robinow-Sorauf syndrome) 180750	Variable brachycephaly, facial asymmetry, ptosis, maxillary hypoplasia, prominent ear crus Bifid great toe, brachydactyly, cutaneous syndactyly	TWIST exon 1 601622.009	Insertion frame shift	Helix-loop-helix transcription factor Helix II	R154K + 88 amino acids	
Craniosynostosis, Boston type 12310.001	Variable craniosynostosis, supraorbital recession Short first metatarsals	MSX2 exon 2 604757	Missense	Homeobox transcription factor Homeodomain	P148H	Enhanced DNA binding affinity

[a] The most common conditions, phenotypic features, and mutations are listed.

cases with normal IQs to mild retardation. Craniofacial features include more frequent or severe ocular hypertelorism, retrusion of the supraorbital rim, downslanting palpebral fissures, low-set ears, short bulbous nose, flat nasal bridge, trapezoidal-shaped mouth, cleft of the soft palate, and dental malocclusion in Apert syndrome compared to Crouzon syndrome. Cervical fusion occurs, especially of C5 to C6 in Apert syndrome and C2 to C3 in Crouzon syndrome. Hearing deficits including frequent bouts of otitis media are common. The fact that the craniofacial features are similar although more severe in Apert syndrome than Crouzon syndrome is consistent with the two conditions being allelic, both with mutations in FGFR2.

The clinical features most characteristic of Apert syndrome are the syndactylous anomalies of the hands and feet. Usually digits 2, 3, and 4 are fused (type 1), although all 5 digits can be fused (type 3). The pattern of syndactyly is symmetric, but may not be congruent between the hands and feet, with the hands usually more affected. Other skeletal findings include those of the shoulders, humeri, elbows, hips, knees, rib cage, and spine. Cardiovascular and genitourinary abnormalities occur in 10% and 9.6% respectively, and respiratory and gastrointestinal problems may also occur. Cutaneous manifestations of dimpling, sweating, or acne can present in 70% of cases. These extraskeletal features that are not observed in Crouzon syndrome suggest that although these are allelic conditions, Apert and Crouzon syndromes are due to mutations that have variable functional effects.

More than 200 or 99% of cases with Apert syndrome are associated with either of two FGFR2 mutations, Ser252Trp or Pro253Arg (Table 6.1) (Park et al., 1995; Wilkie et al., 1995). These mutations are located in the IgII and IgIII linker region, which is coded by exon IIIa and is the same functional domain where some mutations

occur for Crouzon syndrome. The Ser252Trp mutation is more common, accounting for 70% of cases, and has been suggested to be more often associated with more severe syndactyly, while the Pro253Arg mutation may be more associated with cleft of the palate (Slaney et al., 1996). These two mutations are distinct from the mutations found in Crouzon and all other craniosynostoses, except for one Pfeiffer syndrome case reported with the Ser252Trp mutation (Passos-Bueno et al., 1997). There are two Apert syndrome cases due to Alu insertions (Oldridge et al., 1999). One causes a splice site mutation with an insertion in intron IIIb and results in substitutions at -3-4 nucleotides of the exon IIIc acceptor splice site. Splicing of exon IIIc was examined in RNA extracted from the patient's fibroblasts and keratinocytes. Ectopic expression of KGFR in the fibroblast lines was observed. The increased KGFR expression correlated with the increased severity of limb abnormalities.

These missense and insertion mutations increase the affinity of FGFR2 for FGFs (Anderson et al., 1998). The Ser252Arg mutation enhances the binding of a subset of FGFs including FGF2, and the Pro253Arg mutation increases affinity for any FGF (Ibrahimi et al., 2001). At the cellular level, these mutations lead to an increase in calvarial cell differentiation, subperiosteal bone matrix, and premature calvarial ossification (Lomri et al., 1998).

6.7 PFEIFFER SYNDROME AND FGFR2 EXTRACELLULAR MUTATIONS AND PARALOGOUS MUTATIONS IN FGFR1 AND FGFR3

Pfeiffer syndrome (OMIM 101600) is characterized by craniosynostosis, midface hypoplasia, broad thumbs and great toes, and variable syndactyly (Fig. 6.2C) (Pfeiffer, 1964, 1969; Cohen, 1993). The prevalence in newborns has not been determined, but is estimated to be more frequent than Apert syndrome. It is inherited as an autosomal dominant condition with sporadic cases and families with variable expressivity. The origin of new mutations has been shown to be paternal. Clinical subtypes have been suggested (Cohen, 1993). Type 1, representing the classic description of this syndrome, is associated with brachycephaly involving the coronal sutures and normal intelligence and is compatible with life. Type 2 is more severe than the other types and is described with a cloverleaf skull, severe central nervous system involvement, severe ocular proptosis, elbow ankylosis/synostosis, broad thumbs, and great toes. A midline calvarial defect may be present during infancy as in Apert syndrome. All cases reported of this type have been sporadic. Type 3 is severe and similar to type 2, but the cloverleaf skull is absent. A very short anterior cranial base is present.

Elbow involvement is common in types 2 and 3, and occasionally in type 1. Brachydactyly is a common finding and can be associated with mild tissue syndactyly, especially of the second and third digits. The thumbs and toes can be broad and deviated with brachymesophalangy. Other features reported such as cervical and lumbar fusions and short humeri in types 2 and 3 are milder in severity than Apert syndrome, but progressive hydrocephalus is more frequent in Pfeiffer and Crouzon syndromes. The overlap between these types and Crouzon and Apert syndromes is consistent with these three syndromes representing the phenotypic variability of the same condition or allelic conditions.

Pfeiffer syndrome is due to more than 32 different mutations also in the ligand binding, extracellular domain of FGFR2 (Table 6.1) (Passos-Bueno et al., 1999). The most common mutations are the same missense mutations as those associated with Crouzon syndrome. For more than nine of these mutations including Ser267Pro,

Phe276Val, Cys278Phe, Cys342Arg, Cys342Ser, Cys342Trp, Cys342Tyr, Ser351Cys, and Ser354Cys, the identical nucleotide change can be found in Crouzon syndrome patients (Rutland et al., 1995). The splice site mutation common in Pfeiffer syndrome is intron IIIb −2A > G instead of the Ala344Ala (1032G > A) commonly found in Crouzon syndrome. This splice acceptor site mutation was shown to generate ectopic KGFR expression in fibroblasts of Pfeiffer patients at a level less than that of Alu insertion splice site mutation in Apert syndrome (see Section 6.6), thus correlating with severity of limb involvement and delineating a phenotype-genotype correlation (Oldridge et al., 1999).

Two mutations, Ser351Cys and Trp290Cys, are found in severe cases of Pfeiffer syndrome that resemble the clinical diagnosis of Antley-Bixler syndrome (OMIM 207410), except they lack the distinctive features of femoral bowing, fractures, and autosomal recessive inheritance that are present in the latter condition. Pfeiffer syndrome can also be due to a Pro252Arg or Pro250Arg mutation in other members of the same gene family, FGFR1 or FGFR3, respectively (Muenke et al., 1994; Bellus et al., 1996). The phenotype in cases with the FGFR1 Pro252Arg mutation is thought to be more mild than that of cases with the FGFR2 mutations. The paralogous FGFR2 Pro253Arg mutation was noted above as a common mutation found in Apert syndrome.

In Pfeiffer syndrome, there are a few cases with mutations in IgII and tyrosine kinase domains (Kan et al., 2002).

6.8 JACKSON-WEISS SYNDROME AND FGFR2 EXTRACELLULAR DOMAIN MUTATIONS

Jackson-Weiss syndrome (OMIM 123150) is characterized by craniosynostosis, midface hypoplasia, and abnormalities of the feet (Fig. 6.2D) (Jackson et al., 1976). The condition was described in a large Amish kindred with 138 affected individuals. Phenotypic variability was notable in this family, with the entire range of the cephalosyndactyly syndromes represented except Apert syndrome. Skull and facial abnormalities could be absent. The only consistent finding was the presence of bony abnormalities of the feet detected clinically or radiologically. Mild syndactyly of the second and third toes and medial deviations of broad great toes were noted, and minimal manifestations included short, broad first metatarsal, abnormally shaped tarsal bone, and calcaneocuboid fusion. The phenotypic characterization of the extended pedigree demonstrates that Crouzon, Jackson-Weiss, and Pfeiffer syndromes represent a graded continuum of severity and are allelic conditions. The mutation associated with the original Amish kindred is identical to an FGFR2 Ala344Gly mutation found in Crouzon syndrome patients (Table 6.1) (Jabs et al., 1994). Other Jackson-Weiss syndrome cases have Cys278Phe, Gln289Pro, Cys342Arg, and Cys342Ser mutations (Passos-Bueno et al., 1999). Recently, one case was reported to have the FGFR1 Pro252Arg mutation, which is also reported in Pfeiffer syndrome (Roscioli et al., 2000).

6.9 BEARE-STEVENSON CUTIS GYRATA SYNDROME AND FGFR2 JUXTA- OR TRANSMEMBRANE DOMAIN MUTATIONS

Beare-Stevenson cutis gyrata syndrome (OMIM 123790) is characterized by craniosynostosis, corrugated furrowed skin, acanthosis nigricans, ear defects, digital anomalies, umbilical stump, anogenital anomalies, and early death (Fig. 6.3A). Less than 10 cases of this rare autosomal dominant condition with associated advanced paternal age have been reported (Przylepa et al., 1996). Cloverleaf skull deformity may be present. The craniofacial features of ocular proptosis and hypertelorism, midface hypoplasia, choanal atresia, and palatal anomalies are common to this and the other allelic craniosynostosis syndromes described above. The disease gene for Beare-Stevenson syndrome is FGFR2, but the associated Ser372Cys and Tyr375Cys mutations occur in the juxta- or transmembrane domain rather than the extracellular domain where mutations are found for all the previously described craniosynostosis syndromes (Table 6.1). One mildly affected case has been reported with the FGFR3 extracellular Pro250Arg mutation (Roscioli et al., 2001).

6.10 CROUZON SYNDROME WITH ACANTHOSIS NIGRICANS OR CROUZONODERMOSKELETAL SYNDROME AND FGFR3 Ala391Glu TRANSMEMBRANE DOMAIN MUTATION

Crouzonodermoskeletal syndrome is a term used to describe patients with the craniofacial features of Crouzon syndrome and acanthosis nigricans, the dermatologic condition of verrucous hyperplasia (Fig. 6.3B) (OMIM 134934.0011). The craniosynostosis and skin disorder are constant features, and additional findings include jaw cementomas and severe choanal atresia and hydrocephalus (Meyers et al., 1995). Skeletal findings such as brachydactyly and widening of the interpediculate distances of the spine associated with achondroplasia, the most common form of dwarfism in humans, were also detected. These subtle features of achondroplasia were sought and discovered in this condition after its associated Ala391Glu mutation in the transmembrane domain of FGFR3 was discovered to be in the same gene and the same functional domain as the Gly380Arg mutation found in virtually

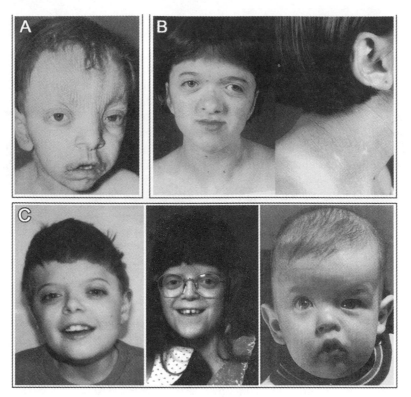

Figure 6.3   Craniosynostosis syndromes with FGFR2 transmembrane or FGFR3 mutations. (A) Beare-Stevenson syndrome with cutis gyrata of the facial skin (Beare et al., 1969). (B) Crouzon syndrome with acanthosis nigricans (Muenke et al., 1998). (C) Variable phenotype of bi- and unicoronal synostosis in patients with FGFR3 Pro250Arg mutation (Paznekas et al., 1998; Muenke et al., 1997).

all achondroplasia patients (Table 6.1) (Schweitzer et al., 2001).

6.11   FGFR3 Pro250Arg-ASSOCIATED CORONAL SYNOSTOSIS SYNDROME OR MUENKE SYNDROME OR CRANIOSYNOSTOSIS ADELAIDE TYPE

Unlike the other syndromes that are defined by phenotype, this condition is defined by a FGFR3 Pro250Arg mutation in the extracellular, ligand binding domain (Fig. 6.3C; Table 6.1) (OMIM 134934.0014, 600593). The associated phenotypic spectrum is highly variable with some patients with Crouzon, Pfeiffer, Jackson-Weiss, or Saethre-Chotzen–like features (Muenke et al., 1997). The most characteristic features include unilateral or bilateral coronal syn-ostosis, downslanting palpebral fissures, ptosis, midface hypoplasia, brachdactyly, and carpal/tarsal coalition (Graham et al., 1998). Other patients have nonsyndromic craniosynostosis, macrocephaly, or normal craniums. Studies have shown the presence of this FGFR3 mutation in 31% (8/26) of cases with nonsyndromic coronal synosto-sis and in 11% (4/37) of cases with non-syndromic unicoronal synostosis. Some of these cases were previously thought to be due to intrauterine constraint (Moloney et al., 1997; Gripp et al., 1998).

6.12   PATHOPHYSIOLOGY OF FGFR MUTATIONS IN CRANIOSYNOSTOSIS

More than 60 different FGFR1–3 muta-tions have been found in more than 560 reported craniosynostosis cases (Table 6.1)

(Passos-Bueno et al., 1999). Most of the mutations occur in exons IIIa and IIIc, which code for the IgIII domain. Many are recurrent missense mutations resulting in the loss or gain of cysteine residues. A few are splice site mutations and in-frame insertions and deletions. Biochemical cellular assays and animal model studies have shown that FGFR mutations result in craniosynostosis through the mechanism of gain of function (Neilson and Friesel, 1996).

In vivo experiments using FGFR2/Neu chimeric receptors in transformed NIH 3T3 cells suggest that loss or gain of cysteine residues lead to intermolecular disulfide bridges between mutant receptors and ligand-independent constitutive activation (Galvin et al., 1996; Robertson et al., 1998) (see Chapters 9 and 20). Mouse models for Crouzon syndrome, Bulgy-eye (Bey) heterozygous mice, are viable and fertile but show facial shortening with increased interorbital distance and precocious closure of several cranial sutures (Carlton et al., 1998). These mice were created with a retroviral insertion in the intergenic region between the tandem Fgf3 and Fgf4 genes that resulted in increased transcript levels of these genes in the cranial sutures. These data suggest that increased expression of the ligands would lead to increased activation of their receptors.

Studies of the Apert FGFR2 Ser252Pro and Pro253Arg mutations demonstrate enhanced affinity for FGF ligands and a ligand-dependent mechanism, especially under conditions of limiting ligand (see Section 6.6) (Anderson et al., 1998; Ibrahimi et al., 2001). Heterozygotic abrogation of Fgfr2 exon IIIc in mice causes a splicing switch, resulting in a gain-of-function mutation (Hajihosseini et al., 2001). The consequence is ectopic Kgfr expression, and the phenotype of the mutant mouse has overlapping features with Apert and Pfeiffer syndromes. The mice are affected with neonatal growth retardation and death, coronal synostosis, ocular proptosis, precocious sternal fusion, and abnormalities in secondary branching in several organs that undergo branching morphogenesis.

6.13 SAETHRE CHOTZEN SYNDROME AND ROBINOW-SORAUF SYNDROME AND TWIST MUTATIONS

Saethre-Chotzen syndrome (OMIM 101400) is characterized by cranial asymmetry, craniosynostosis, low frontal hairline, ptosis, deviated nasal septum, midface hypoplasia, and small ears with posterior rotation and prominent crura (Fig. 6.4) (Saethre, 1931; Chotzen, 1932). Hand and feet anomalies include brachydactyly with partial cutaneous syndactyly usually of the second and third fingers. These phenotypic features are usually subtle for this common autosomal dominant condition. The clinical features of ptosis, deviated nasal septum, and prominent crura in Saethre-Chotzen syndrome that are distinct from those of other craniosynostosis syndromes validate the disease gene is different from the other craniosynostosis genes.

More than 51 different mutations have been found in the transcription factor TWIST (Table 6.1) (El Ghouzzi et al., 1997; Howard et al., 1997; Gripp et al., 2000). Some mutations are deletions and nonsense mutations and others are insertions, duplications, and missense. Patients with large deletions are more likely to have mild to moderate mental retardation. Bifid halluces and duplication of the great toe in association with the craniofacial features of Saethre-Chotzen syndrome had been considered a separate clinical entity, Robinow-Sorauf syndrome (OMIM 180750). Detection of a frame-shift mutation in a family with Robinow-Sorauf syndrome demonstrated that this condition is allelic. The subtle overlapping features of coronal synostosis, midface hypoplasia, and digital brachydactyly and syndactyly among Saethre-Chotzen and the other craniosyn-

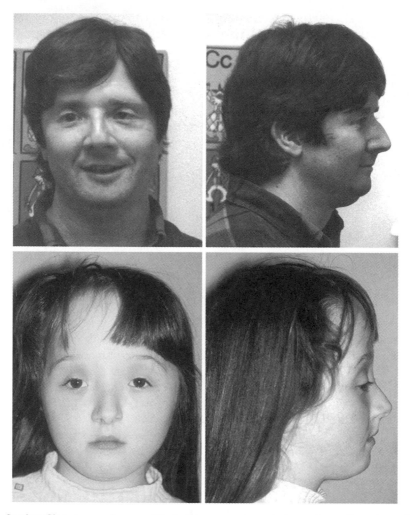

Figure 6.4 Saethre-Chotzen syndrome with facial asymmetry, ptosis, and maxillary hypoplasia (Paznekas et al., 1998).

ostosis syndromes suggest that TWIST and the fibroblast growth factor receptors interact in the same molecular pathway involved in the development of the skull (Paznekas et al., 1998; Jabs, 2001).

TWIST codes for a transcription factor with a DNA binding and helix-loop-helix domain (Fig. 6.1). TWIST is activated to bind to its DNA target(s) when its helix-loop-helix region dimerizes. It plays an important role in the development of the mesodermal layer and mesodermally derived tissues and organs and is expressed in the primary osteoblasts of the newborn mouse calvaria (Murray et al., 1992). The pathogenesis of Saethre-Chotzen syndrome is mediated through the mechanism of loss of function as evidenced by deletion of TWIST in some patients. Nonsense mutations can lead to a premature stop codon and unstable mutant transcript. Point mutations may also cause loss of TWIST function by altering the ability of the protein to dimerize or bind to its DNA target(s) (El Ghouzzi et al., 2000, 2001). Missense mutations involving the helical domains lead to loss of TWIST heterodimerization with the E12 basic helix-loop-helix protein and altered ability of the TWIST protein to localize in the nucleus. The TWIST Y103X mutation is a true null allele, and mutant osteoblasts with this mutation show increased cell growth and ability to form bone-like nodular structures (Yousfi et al., 2001). Thus, the premature ossification of cranial sutures in Saethre-Chotzen syndrome is due to increased osteoblastic cell proliferation and differentiation that differs from pathogenesis of Apert syndrome, which is associated with only increased osteoblastic differentiation and not proliferation.

Further evidence of loss of function of the TWIST gene in Saethre-Chotzen syndrome patients is from mice heterozygous for the loss of Twist. These mice have abnormalities of the skull and limbs, which are the same tissues and organs involved in Saethre-Chotzen syndrome (Bourgeois et al., 1998). Furthermore, in these mice, Fgfr2 protein expression is altered and appears ectopically at the cranial sutures, suggesting that FGFRs may mediate a signal downstream of TWIST (Rice et al., 2000). Based on both clinical and molecular results, loss of function of the TWIST gene may lead to gain of function or activation of FGFRs in the development of craniosynostosis.

6.14 CRANIOSYNOSTOSIS, BOSTON TYPE, AND MSX2 Pro148His MUTATION

Craniosynostosis, Boston type (OMIM 604757), is characterized by recession of the supraorbital region in relation to the anterior surface of the cornea, myopia, or hyperopia, seizures, and short first metatarsals (Fig. 6.5). This autosomal dominant condition was described in a single New England family (Warman et al., 1993). Cranial morphology was variable with forehead retrusion, frontal bossing, turribrachycephaly, or cloverleaf skull. A cleft of the soft palate or a triphalangeal thumb was noted in one member. Intelligence was normal.

MSX2 is a transcription factor containing a homeobox domain that is expressed in the neural crest–derived mesenchyme of the branchial arches and at birth in the osteogenic fronts and mesenchymal cells of the sutures (Fig. 6.1). A Pro148His mutation in the homeobox domain was found in each affected family member (Table 6.1) (Jabs et al., 1993). This mutation causes increased DNA binding affinity or a gain of function (Ma et al., 1996). Mice with more than 13 copies of an MSX2 transgene died at birth from severe craniofacial malformations but no craniosynostosis (Winograd et al., 1997). Mice with no more than two copies showed premature fusion and ectopic bone formation (Liu et al., 1995). In contrast, mice heterozygous for the loss of Msx2 developed defects in ossification of the skull (Satokata et al., 2000). Mutations of MSX2 haploinsufficiency in humans result in parietal foramina (Fig. 6.5) (Wilkie et al., 2000; OMIM 123101). These results demonstrate that skull development is finely calibrated by MSX2 dosage.

6.15 INTEGRATION OF DEVELOPMENTAL, CLINICAL, AND MOLECULAR GENETICS OF CRANIOSYNOSTOSES

The progress in elucidating the genetic etiologies of common craniosynostosis syndromes has been impressive in the last decade, yet many challenges still remain. Further work will be needed from not only clinical and molecular geneticists but also developmental biologists, biochemists, and cell biologists. The integration of their skills and knowledge will define the exact genes

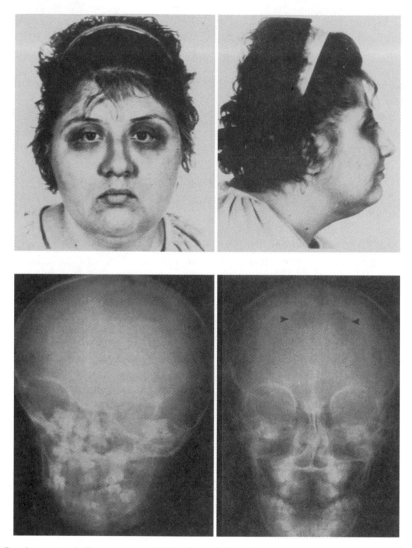

Figure 6.5   Craniosynostosis, Boston type, with brachycephaly and frontal orbital recession (top; Muller et al., 1993) and parietal foramina (bottom; Wilkie et al., 2000).

and proteins and their functions and interactions in the complex and intricate developmental pathway(s) involved in skull development. There are already clues for other potential candidate genes that may interact directly or modify indirectly the actions of FGFRs, TWIST, and MSX2 (Jabs, 2001). However, even more challenging will be the elucidation of factors involved in nonsyndromic craniosynostosis. Thus, further integration will be required to identify not only the genetic but also the environmental factors. It surely will be exciting to watch the further elucidation of the pathogenesis of craniosynostosis in the next decade.

ACKNOWLEDGMENTS

This work was supported in part by NIH grants P60 DE13078 and RO1 DE11441.

REFERENCES

Anderson, J., Burns, H. D., Enriquez-Harris, P., Wilkie, A. O., and Heath, J. K. (1998). Apert syndrome mutations in fibroblast growth factor receptor 2 exhibit increased affinity for FGF ligand. *Hum. Molec. Genet. 7*, 1475–1483.

Apert, E. (1906). De l'acrocephalosyndactylie. *Bull. Soc. Med. Paris 23*, 1310–1330.

Beare, J. M., Dodge, J. A., and Nevin, N. C. (1969). Cutis gyratum, acanthosis nigricans and other congenital anomalies. A new syndrome. *Br. J. Dermatol. 81*, 241–247.

Bellus, G. A., Gaudenz, K., Zackai, E. H., Clarke, L. A., Szabo, J., Francomano, C. A., and Muenke, M. (1996). Identical mutations in three different fibroblast growth factor receptor genes in autosomal dominant craniosynostosis syndromes. *Nat. Genet. 14*, 174–176.

Blaikie, P., Immanuel, D., Wu, J., Li, N., Yajnik, V., and Margolis, B. (1994). A region in Shc distinct from the SH2 domain can bind tyrosine-phosphorylated growth factor receptors. *J. Biol. Chem. 269*, 32031–32034.

Blank, C. E. (1960). Apert's syndrome (a type of acrocephalosyndactyly)—observations on a British series of thirty-nine cases. *Ann. Hum. Genet. 24*, 151–164.

Bourgeois, P., Bolcato-Bellemin, A. L., Danse, J. M., Bloch-Zupan, A., Yoshiba, K., Stoetzel, C., and Perrin-Schmitt, F. (1998). The variable expressivity and incomplete penetrance of the twist-null heterozygous mouse phenotype resemble those of human Saethre-Chotzen syndrome. *Hum. Molec. Genet. 7*, 945–957.

Carlton, M. B., Colledge, W. H., and Evans, M. J. (1998). Crouzon-like craniofacial dysmorphology in the mouse is caused by an insertional mutation at the Fgf3/Fgf4 locus. *Devel. Dyn. 212*, 242–249.

Chotzen, F. (1932). Eine eigenartige familiaere entwicklungsstoerung (Akrocephalosyndaktylie, Dysostosis craniofacialis und Hypertelorismus). *Mschr. Kinderheilk 55*, 97–122.

Cohen, M. M., Jr. (1975). An etiologic and nosologic overview of craniosynostosis syndromes. *Birth Defects 11*, 137–189.

Cohen, M. M., Jr. (1993). Pfeiffer syndrome update, clinical subtypes, and guidelines for differential diagnosis. *Am. J. Med. Genet. 45*, 300–307.

Cohen, M. M. Jr., and MacLean, R. E. (eds.) (2000). *Craniosynostosis: Diagnosis, Evaluation, and Management*, 2nd ed. (Oxford: Oxford University Press).

Chung, C. S., and Myrianthopoulos, N. C. (1975). Factors affecting risks of congenital malformations. I. Analysis of epidemiologic factors in congenital malformations. Report from the Collaborative Perinatal Project. *Birth Defects Orig. Article Ser. 11*, 1–22.

Crouzon, O. (1912). Dysostose cranio-faciale hereditaire. *Bull. Soc. Med. Hop. Paris 33*, 545–555.

Delezoide, A. L., Benoist-Lasselin, C., Legeai-Mallet, L., Le Merrer, M., Munnich, A., Vekemans, M., and Bonaventure, J. (1998). Spatio-temporal expression of FGFR 1, 2 and 3 genes during human embryo-fetal ossification. *Mech. Devel. 77*, 19–30.

Dell, K. R., and Williams, L. T. (1992). A novel form of fibroblast growth factor receptor 2. Alternative splicing of the third immunoglobulin-like domain confers ligand binding specificity. *J. Biol. Chem. 267*, 21225–21229.

Dionne, C. A., Crumley, G., Bellot, F., Kaplow, J. M., Searfoss, G., Ruta, M., Burgess, W. H., Jaye, M., and Schlessinger, J. (1990). Cloning and expression of two distinct high-affinity receptors cross-reacting with acidic and basic fibroblast growth factors. *EMBO J. 9*, 2685–2692.

El Ghouzzi, V., Le Merrer, M., Perrin-Schmitt, F., Lajeunie, E., Benit, P., Renier, D., Bourgeois, P., Bolcato-Bellemin, A. L., Munnich, A., and Bonaventure, J. (1997). Mutations of the TWIST gene in the Saethre-Chotzen syndrome. *Nat. Genet. 15*, 42–46.

El Ghouzzi, V., Legeai-Mallet, L., Aresta, S., Benoist, C., Munnich, A., de Gunzburg, J., Bonaventure, J. (2000). Saethre-Chotzen mutations cause TWIST protein degradation or impaired nuclear location. *Hum. Mol. Genet. 22*, 813–819.

El Ghouzzi, V., Legeai-Mallet, L., Benoist-Lasselin, C., Lajeunie, E., Renier, D.,

Munnich, A., and Bonaventure, J. (2001). Mutations in the basic domain and the loop-helix II junction of TWIST abolish DNA binding in Saethre-Chotzen syndrome. *FEBS Lett. 492*, 112–118.

Galvin, B. D., Hart, K. C., Meyer, A. N., Webster, M. K., and Donoghue, D. J. (1996). Constitutive receptor activation by Crouzon syndrome mutations in fibroblast growth factor receptor (FGFR)2 and FGFR2/Neu chimeras. *Proc. Natl. Acad. Sci. USA 93*, 7894–7899.

Glaser, R. L., Jiang, W., Boyadjiev, S. A., Tran, A. K., Zachary, A. A., Van Maldergem, L., Johnson, D., Walsh, S., Oldridge, M., Wall, S. A., Wilkie, A. O., and Jabs, E. W. (2000). Paternal origin of FGFR2 mutations in sporadic cases of Crouzon syndrome and Pfeiffer syndrome. *Am. J. Hum. Genet. 66*, 768–777.

Graham, J. M., Jr., Braddock, S. R., Mortier, G. R., Lachman, R., Van Dop, C., and Jabs, E. W. (1998). Syndrome of coronal craniosynostosis with brachdactyly and carpal/tarsal coalition due to Pro250Arg mutation in FGFR3 gene. *Am. J. Med. Genet. 77*, 322–329.

Gripp, K. W., McDonald-McGinn, D. M., Gaudenz, K., Whitaker, L. A., Bartlett, S. P., Glat, P. M., Cassileth, L. B., Mayro, R., Zackai, E. H., and Muenke, M. A. (1998). Identification of a genetic cause for isolated unilateral coronal synostosis: a unique mutation in the fibroblast growth factor receptor 3. *J. Pediatr. 132*, 714–716.

Gripp, K. W., Zackai, E. H., and Stolle, C. A. (2000). Mutations in the human TWIST gene. *Hum. Mut. 15* (2), 150–155.

Hajihosseini, M. K., Wilson, S., De Moerlooze, L., and Dickson, C. (2001). A splicing switch and gain-of-function mutation in FgfR2-IIIc hemizygotes causes Apert/Pfeiffer- syndrome-like phenotypes. *Proc. Natl. Acad. Sci. USA 98*, 3855–3860.

Howard, T. D., Paznekas, W. A., Green, E. D., Chiang, L. C., Ma, N., Ortiz de Luna, R. I., Garcia Delgado, C., Gonzalez-Ramos, M., Kline, A. D., and Jabs, E. W. (1997). Mutations in TWIST, a basic helix-loop-helix transcription factor, in Saethre-Chotzen syndrome. *Nat. Genet. 15*, 36–41.

Hunter, A. G., and Rudd, N. L. (1976). Craniosynostosis. I. Sagittal synostosis: its genetics and associated clinical findings in 214 patients who lacked involvement of the coronal suture(s). *Teratology 14*, 185–193.

Hunter, A. G., and Rudd, N. L. (1977). Craniosynostosis. II. Coronal synostosis: its familial characteristics and associated clinical findings in 109 patients lacking bilateral polysyndactyly or syndactyly. *Teratology 15*, 301–309.

Ibrahimi, O. A., Eliseenkova, A. V., Plotnikov, A. N., Yu, K., Ornitz, D. M., and Mohammadi, M. (2001). Structural basis for fibroblast growth factor receptor 2 activation in Apert syndrome. *Proc. Natl. Acad. Sci. USA 98*, 7182–7187.

Ingersoll, R., Paznekas, W. A., Tran, A. K. N., Scott, A. F., Jiang, G., and Jabs, E. W. (2002). Fibroblast growth factor receptor 2 (FGFR2) genomic sequence and variations. *Cytogenet. Cell Genet. 94*, 121–126.

Jabs, E. W. (2001). A TWIST in the fate of human osteoblasts identifies signaling molecules involved in skull development. *J. Clin. Invest. 107*, 1075–1077.

Jabs, E. W., Muller, U., Li, X., Ma, L., Luo, W., Haworth, I. S., Klisak, I., Sparkes, R., Warman, M. L., Mulliken, J. B., Snead, M. L., and Maxson, R. (1993). A mutation in the homeodomain of the human MSX2 gene in a family affected with autosomal dominant craniosynostosis. *Cell 75*, 443–450.

Jabs, E. W., Li, X., Scott, A. F., Meyers, G., Chen, W., Eccles, M., Mao, J. I., Charnas, L. R., Jackson, C. E., and Jaye, M. (1994). Jackson-Weiss and Crouzon syndromes are allelic with mutations in fibroblast growth factor receptor 2. *Nat. Genet. 8*, 275–279.

Jackson, C. E., Weiss, L., Reynolds, W. A., Forman, T. F., and Peterson, J. A. (1976). Craniosynostosis, midfacial hypoplasia and foot abnormalities: an autosomal dominant phenotype in a large Amish kindred. *J. Pediatr. 88*, 963–968.

Jaye, M., Schlessinger, J., and Dionne, C. A. (1992). Fibroblast growth factor receptor tyrosine kinases: molecular analysis and signal transduction. *Biochim. Biophys. Acta. 1135*, 185–199.

Johnson, D. E., and Williams, L. T. (1993). Structural and functional diversity in the FGF receptor multigene family. *Adv. Cancer Res.* *60*, 1–41.

Kan, S.-H., Elanko, N., Jahnson, D., Correjo-Roldan, L., Cook, J., Reich, E. W., Tomkins, S., Verloes, A., Twigg, S. R. F., Rannan-Eliya, S., McDonald-McGinn, D. M., Zackai, E. H., Wall, S. A., Muenbe, M., and Wiltie, A. O. M. (2002). Genomic screening of fibroblast growth-factor receptor 2 reveals a wide spectrum of mutations in patients with syndromic crawosynostosis. *Am. J. Hum. Genat.* *70*, 472–486.

Kreiborg, S. (1981). Crouzon syndrome. A clinical and roentgencephalometric study. *Scand. J. Plast. Reconstr. Surg.* *18* (Suppl.), 1–198.

Lajeunie, E., Le Merrer, M., Bonaiti-Pellie, C., Marchac, D., and Renier, D. (1995). Genetic study of nonsyndromic coronal craniosynostosis. *Am. J. Med. Genet.* *55*, 500–504.

Lajeunie, E., Le Merrer, M., Bonaiti-Pellie, C., Marchac, D., and Renier, D. (1996). Genetic study of scaphocephaly. *Am. J. Med. Genet.* *62*, 282–285.

Lajeunie, E., Le Merrer, M., Marchac, D., and Renier, D. (1998). Syndromal and nonsyndromal primary trigonocephaly: analysis of a series of 237 patients. *Am. J. Med. Genet.* *75*, 211–215.

Lammer, E. J., Cordero, J. F., Wilson, M. J., Oimette, D., and Ferguson, S. (1987a). Investigation of a suspected increased prevalence of craniosynostosis—Colorado, 1978–1982. *Proc. Greenwood Genet. Ctr. 6*, 126–127.

Lammer, E. J., Cordero, J. F., Wilson, M. J., Oimette, D., and Ferguson, S. (1987b). Document EPI-83-56-2. Public Health Service–CDC–Atlanta, April 8, 1987.

Lewanda, A. F., and Jabs, E. W. (2001). Craniosynostosis. In D. L. Rimoin, J. M. Connor, R. E. Pyeritz, and B. R. Korf (eds.), *Emery and Rimoin's Principles and Practice of Medical Genetics, 4th ed.* (London: Harcourt Publishers Limited).

Li, X., Park, W. J., Pyeritz, R. E., and Jabs, E. W. (1995). Effect on splicing of a silent FGFR2 mutation in Crouzon syndrome. *Nat. Genet.* *9*, 232–233.

Liu, Y. H., Kundu, R., Wu, L., Luo, W., Ignelzi, M. A. Jr., Snead, M. L., and Maxson, R. E., Jr. (1995). Premature suture closure and ectopic cranial bone in mice expressing Msx2 transgenes in the developing skull. *Proc. Natl. Acad. Sci. USA 92*, 6137–6141.

Lomri, A., Lemonnier, J., Hott, M., de Parseval, N., Lajeunie, E., Munnich, A., Renier, D., and Marie, P. J. (1998). Increased calvaria cell differentiation and bone matrix formation induced by fibroblast growth factor receptor 2 mutations in Apert syndrome. *J. Clin. Invest. 101*, 1310–1317.

Ma, L., Golden, S., Wu, L., and Maxson, R. (1996). The molecular basis of Boston-type craniosynostosis: the Pro148→His mutation in the N-terminal arm of the MSX2 homeodomain stabilizes DNA binding without altering nucleotide sequence preferences. *Hum. Molec. Genet. 5*, 1915–1920.

Meyers, G. A., Orlow, S. J., Munro, I. R., Przylepa, K. A., and Jabs, E. W. (1995). Fibroblast growth factor receptor 3 (FGFR3) transmembrane mutation in Crouzon syndrome with acanthosis nigricans. *Nat. Genet. 11*, 462–464.

Miki, T., Bottaro, D. P., Fleming, T. P., Smith, C. L., Burgess, W. H., Chan, A. M., and Aaronson, S. A. (1992). Determination of ligand-binding specificity by alternative splicing: two distinct growth factor receptors encoded by a single gene. *Proc. Natl. Acad. Sci. USA 89*, 246–250.

Moloney, D. M., Wall, S. A., Ashworth, G. J., Oldridge, M., Glass, I. A., Francomano, C. A., Muenke, M., and Wilkie, A. O. (1997). Prevalence of Pro250Arg mutation of fibroblast growth factor receptor 3 in coronal craniosynostosis. *Lancet 349*, 1059–1062.

Muenke, M., and Wilkie, A. O. M. (2001). Craniosynostosis syndromes. In C. R. Scriver, A. L. Beaudet, W. S. Sly, and D. Valle (eds.), *The Metabolic and Molecular Bases of Inherited Disease, 8th ed.* (New York: McGraw-Hill).

Muenke, M., Schell, U., Hehr, A., Robin, N. H., Losken, H. W., Schinzel, A., Pulleyn, L. J., Rutland, P., Reardon, W., Malcolm, S., and Winter, R. M. (1994). A common mutation in the fibroblast growth factor receptor 1 gene in Pfeiffer syndrome. *Nat. Genet. 8*, 269–274.

Muenke, M., Gripp, K. W., McDonald-McGinn, D. M., Gaudenz, K., Whitaker, L. A., Bartlett, S. P., Markowitz, R. I., Robin, N. H., Nwokoro, N., Mulvihill, J. J., Losken, H. W., Mulliken, J. B., Guttmacher, A. E., Wilroy, R. S., Clarke, L. A., Hollway, G., Ades, L. C., Haan, E. A., Mulley, J. C., Cohen, M. M. Jr., Bellus, G. A., Francomano, C. A., Moloney, D. M., Wall, S. A., Wilkie, A. O., and Zackai, E. H. (1997). A unique point mutation in the fibroblast growth factor receptor 3 gene (FGFR3) defines a new craniosynostosis syndrome. *Am. J. Hum. Genet. 60*, 555–564.

Muenke, M., Francomano, C. A., Cohen, M. M., Jr., and Jabs, E. W. (1998). Fibroblast growth factor receptor related skeletal disorders: Craniosynostosis and dwarfism syndromes. In Section on Genetic Basis of Congenital Malformations (ed. Jabs, E. W.) of *Principles of Molecular Medicine* (ed. Jameson, L.), Humana Press, pp. 1029–1038.

Muller, U., Warman, M. L., Mulliken, J. B., and Weber, J. L. (1993). Assignment of a gene locus involved in craniosynostosis to chromosome 5qter. *Hum. Mol. Genet. 2*, 119–122.

Murray, S. S., Glackin, C. A., Winters, K. A., Gazit, D., Kahn, A. J., and Murray, E. J. (1992). Expression of helix-loop-helix regulatory genes during differentiation of mouse osteoblastic cells. *J. Bone Miner. Res. 7*, 1131–1138.

Naski, M. C., and Ornitz, D. M. (1998). FGF signaling in skeletal development. *Front. Biosci. 3*, D781–794.

Neilson, K. M., and Friesel, R. (1996). Ligand-independent activation of fibroblast growth factor receptors by point mutations in the extracellular, transmembrane, and kinase domains. *J. Biol. Chem. 271*, 25049–25057.

Oldridge, M., Zackai, E. H., McDonald-McGinn, D. M., Iseki, S., Morriss-Kay, G. M., Twigg, S. R. F., Johnson, D., Wall, S. A., Jiang, W., Theda, C., Jabs, E. W., and Wilkie, A. O. M. (1999). De novo Alu element insertions in FGFR2 identify a distinct pathological basis for Apert syndrome. *Am. J. Hum. Genet. 76*, 446–461.

Olivier, J. P., Raabe, T., Henkemeyer, M., Dickson, B., Mbamalu, G., Margolis, B., Schlessinger, J., Hafen, E., and Pawson, T. (1993). A Drosophila SH2–SH3 adaptor protein implicated in coupling the sevenless tyrosine kinase to an activator of Ras guanine nucleotide exchange, Sos. *Cell 73*, 179–191.

Ong, S. H., Hadari, Y. R., Gotoh, N., Guy, G. R., Schlessinger, J., and Lax, I. (2001). Stimulation of phosphatidylinositol 3-kinase by fibroblast growth factor receptors is mediated by coordinated recruitment of multiple docking proteins. *Proc. Natl. Acad. Sci. USA 98*, 6074–6079.

Ornitz, D. M., and Leder, P. (1992). Ligand specificity and heparin dependence of fibroblast growth factor receptors 1 and 3. *J. Biol. Chem. 267*, 16305–16311.

Ornitz, D. M., Yayon, A., Flanagan, J. G., Svahn, C. M., Levi, E., and Leder, P. (1992). Heparin is required for cell-free binding of basic fibroblast growth factor to a soluble receptor and for mitogenesis in whole cells. *Molec. Cell. Biol. 12*, 240–247.

Park, W. J., Theda, C., Maestri, N. E., Meyers, G. A., Fryburg, J. S., Dufresne, C., Cohen, M. M. Jr., and Jabs, E. W. (1995). Analysis of phenotypic features and FGFR2 mutations in Apert syndrome. *Am. J. Hum. Genet. 57*, 321–328.

Passos-Bueno, M. R., Sertie, A. L., Zatz, M., and Richieri-Costa, A. (1997). Pfeiffer mutation in an Apert patient: How wide is the spectrum of variability due to mutations in the FGFR2 gene? *Am. J. Med. Genet. 71*, 243–245.

Passos-Bueno, M. R., Wilcox, W. R., Jabs, E. W., Sertie, A. L., Alonso, L. G., and Kitoh, H. (1999). Clinical spectrum of fibroblast growth factor receptor mutations. *Hum. Mut. 14*, 115–125.

Paznekas, W. A., Cunningham, M. L., Howard, T. D., Korf, B. R., Lipson, M. H., Grix, A. W., Feingold, M., Goldberg, R., Borochowitz, Z., Aleck, K., Mulliken, J., Yin, M., and Jabs, E. W. (1998). Genetic heterogeneity of Saethre-Chotzen syndrome, due to TWIST and FGFR mutations. *Am. J. Hum. Genet. 62*, 1370–1380.

Pfeiffer, R. A. (1964). Dominant erbliche Akrocephalosyndaktylie. *Z. Kinderheilk 90*, 301–302.

Pfeiffer, R. A. (1969). Associated deformities of the head and hands. *Birth Defects 5*, 18–34.

Plotnikov, A. N., Schlessinger, J., Hubbard, S. R., and Mohammadi, M. (1999). Structural basis for FGF receptor dimerization and activation. *Cell 98*, 641–650.

Przylepa, K. A., Paznekas, W., Zhang, M., Golabi, M., Bias, W., Bamshad, M. J., Carey, J. C., Hall, B. D., Stevenson, R., Orlow, S., Cohen, M. M., Jr., and Jabs, E. W. (1996). Fibroblast growth factor receptor 2 mutations in Beare-Stevenson cutis gyrata syndrome. *Nat. Genet. 13*, 492–494.

Reardon, W., Winter, R. M., Rutland, P., Pulleyn, L. J., Jones, B. M., and Malcolm, S. (1994). Mutations in the fibroblast growth factor receptor 2 gene cause Crouzon syndrome. *Nat. Genet. 8*, 98–103.

Rice, D. P., Aberg, T., Chan, Y., Tang, Z., Kettunen, P. J., Pakarinen, L., Maxson, R. E., and Thesleff, I. (2000). Integration of FGF and TWIST in calvarial bone and suture development. *Development 127*, 1845–1855.

Robertson, S. C., Meyer, A. N., Hart, K. C., Galvin, B. D., Webster, M. K., and Donoghue, D. J. (1998). Activating mutations in the extracellular domain of the fibroblast growth factor receptor 2 function by disruption of the disulfide bond in the third immunoglobulin-like domain. *Proc. Natl. Acad. Sci. USA 95*, 4567–4572.

Rollnick, B. R. (1988). Germinal mosaicism in Crouzon syndrome. *Clin. Genet. 33*, 145–150.

Roscioli, T., Flanagan, S., Kumar, P., Masel, J., Gattas, M., Hyland, V. J., and Glass, I. A. (2000). Clinical findings in a patient with FGFR1 P252R mutation and comparison with the literature. *Am. J. Med. Genet. 93*, 22–28.

Roscioli, T., Flanagan, S., Mortimore, R. J., Kumar, P., Weedon, D., Masel, J., Lewandowski, R., Hyland, V., and Glass, I. A. (2001). Premature calvarial synostosis and epidermal hyperplasia (Beare-Stevenson syndrome-like anomalies) resulting from a P250R missense mutation in the gene encoding fibroblast growth factor receptor 3. *Am. J. Med. Genet. 101*, 187–194.

Rutland, P., Pulleyn, L. J., Reardon, W., Baraitser, M., Hayward, R., Jones, B.,

Malcolm, S., Winter, R. M., Oldridge, M., Slaney, S. F., Poole, M. D., and Wilkie, A. O. M. (1995). Identical mutations in the FGFR2 gene cause both Pfeiffer and Crouzon syndrome phenotypes. *Nat. Genet. 9*, 173–176.

Saethre, M. (1931). Ein Beitrag zum Turmschaedelproblem (Pathogenese, Erblichkeit und Symptomatologie). *Dtsch. Z. Nervenheilk 119*, 533–555.

Satokata, I., Ma, L., Ohshima, H., Bei, M., Woo, I., Nishizawa, K., Maeda, T., Takano, Y., Uchiyama, M., Heaney, S., Peters, H., Tang, Z., Maxson, R., and Maas, R. (2000). Msx2 deficiency in mice causes pleiotropic defects in bone growth and ectodermal organ formation. *Nat. Genet. 24*, 391–395.

Schweitzer, D. N., Graham, J. M., Jr., Lachman, R. S., Jabs, E. W., Okajima, K., Przylepa, K. A., Shanske, A., Chen, K., Neidich, J. A., and Wilcox, W. R. (2001). Subtle radiographic findings of achondroplasia in patients with Crouzon syndrome with acanthosis nigricans due to an Ala391Glu substitution in FGFR3. *Am. J. Med. Genet. 98*, 75–91.

Shimada, T., Mizutani, S., Muto, T., Yoneya, T., Hino, R., Takeda, S., Takeuchi, Y., Fujita, T., Fukumoto, S., and Yamashita, T. (2001). Cloning and characterization of FGF23 as a causative factor of tumor-induced osteomalacia. *Proc. Natl. Acad. Sci. USA 98*, 6500–6505.

Slaney, S. F., Oldridge, M., Hurst, J. A., Morriss-Kay, G. M., Hall, C. M., Poole, M. D., and Wilkie, A. O. M. (1996). Differential effects of FGFR2 mutations on syndactyly and cleft palate in Apert syndrome. *Am. J. Hum. Genet. 58*, 923–932.

Warman, M. L., Mulliken, J. B., Hayward, P. G., and Muller, U. (1993). Newly recognized autosomal dominant disorder with craniosynostosis. *Am. J. Med. Genet. 46*, 444–449.

Wilkie, A. O., Slaney, S. F., Oldridge, M., Poole, M. D., Ashworth, G. J., Hockley, A. D., Hayward, R. D., David, D. J., Pulleyn, L. J., Rutland, P., Malcolm, S., Winter, R. M., and Reardon, W. (1995). Apert syndrome results from localized mutations of FGFR2 and is allelic with Crouzon syndrome. *Nat. Genet. 9*, 165–172.

Wilkie, A. O., Tang, Z., Elanko, N., Walsh, S., Twigg, S. R., Hurst, J. A., Wall, S. A., Chrzanowska, K. H., and Maxson, R. E., Jr. (2000). Functional haploinsufficiency of the human homeobox gene MSX2 causes defects in skull ossification. *Nat. Genet. 24*, 387–390.

Winograd, J., Reilly, M. P., Roe, R., Lutz, J., Laughner, E., Xu, X., Hu, L., Asakura, T., Vander Kolk, C., Strandberg, J. D., and Semenza, G. L. (1997). Perinatal lethality and multiple craniofacial malformations in MSX2 transgenic mice. *Hum. Molec. Genet. 6*, 369–379.

Yousfi, M., Lasmoles, F., Lomri, A., Delannoy, P., and Marie, P. J. (2001). Increased bone formation and decreased osteocalcin expression induced by reduced Twist dosage in Saethre-Chotzen syndrome. *J. Clin. Invest. 107*, 1153–1161.

CHAPTER 7

GENETIC ETIOLOGIES OF FACIAL CLEFTING

MARY L. MARAZITA, Ph.D., F.A.C.M.G.

7.1 INTRODUCTION

This chapter summarizes current evidence regarding genetic etiologies for syndromic and nonsyndromic orofacial clefts. Chapter 8 summarizes environmental etiologies, Chapter 10 summarizes evidence from animal models, and Chapter 21 provides a framework for understanding the molecular genetic underpinnings of orofacial growth and development.

Orofacial clefts, particularly cleft lip with or without cleft palate (CL/P) and isolated cleft palate (CP), are major public health problems, affecting 1 in every 500 to 1000 births worldwide. Other types of facial clefts are much more rare (see Chapter 3), and aside from a few case reports, there are no detailed etiologic investigations. Therefore, the focus of this chapter is on CL/P and CP. Many research groups have studied the etiology of CL/P and CP with considerable success for syndromic forms and limited success for nonsyndromic forms. It is clear that CL/P and CP can occur as part of Mendelian syndromes, that certain chromosomal anomalies include CL/P or CP in the phenotype, and that certain teratogens can increase the risk of having an offspring with an orofacial cleft (Romitti et al., 1999) (see Chapter 8). However, note that phenotypes of known etiology comprise only a small portion of all individuals with orofacial clefts (Jones, 1988; Gorlin et al., 1990; Murray, 1995; Schutte and Murray, 1999).

An understanding of the etiologies of orofacial clefting is essential for effective genetic counseling of families with affected members. Identification of a marker locus, or a locus or loci directly involved in clefting, will someday allow prereproductive counseling for affected individuals and their relatives, and prenatal diagnosis will be possible. Although clefts can be surgically repaired, it would be very desirable for parents to prepare for the facial appearance and early care of an infant with a cleft. In addition, with advances in prenatal surgery, it may eventually be possible to repair clefts in utero, bypassing the disadvantages of current treatment modalities (scarring, speech problems, lengthy dental treatment, etc.). Even further in the future, gene therapy may be possible.

7.2 GENETIC ETIOLOGIES OF SYNDROMIC OROFACIAL CLEFTS

7.2.1 Terms Used in Syndromology

In the attempts to make useful clinical classifications of individuals with congenital

Understanding Craniofacial Anomalies: The Etiopathogenesis of Craniosynostoses and Facial Clefting, Edited by Mark P. Mooney and Michael I. Siegel, ISBN 0-471-38724-x Copyright © 2002 by Wiley-Liss, Inc.

anomalies, there are a number of arbitrary and overlapping terminologies in use that are not necessarily used consistently. For example, the distinction between a syndromic versus a nonsyndromic form of orofacial clefting is not always clear. *Syndrome* is used in clinical genetics to refer to a constellation of anomalies that co-occur, presumably due to a common underlying etiology that may or may not be genetic. For example, there are a number of named syndromes that are due to teratogenic exposures. However, in the orofacial cleft literature, sometimes *syndrome* is used in the clinical genetics sense, but sometimes it is used to describe only familial or genetic forms of orofacial clefts.

To further complicate matters, the terms *isolated* and *nonisolated* are often used to distinguish between individuals with only one physical finding and individuals with multiple anomalies. This can be appropriate, particularly if it is not clear whether the multiple anomalies are due to the same etiology. However, there is no consistent agreement as to what constitutes multiple anomalies, with some authors including only major anomalies and others including minor anomalies as well. Furthermore, there are a number of syndromes known to be genetic (e.g., van der Woude and Stickler syndromes) in which there is variable expressivity of the genes: Some individuals carrying the affected gene have isolated clefts (i.e., no other anomaly), and some individuals have clefts plus other physical findings.

For the purposes of this chapter, I follow the syndrome classifications in use in clinical genetics. It is important to note that the explosion of knowledge in human genetics is changing the face of syndromology and will make arguments about terminology moot. As the underlying genes are identified that lead to the clinically classified syndromes, less arbitrary classification of individuals with isolated or multiple anomalies will be possible, along with a clearer appreciation of the underlying etiologies.

The next two sections summarize genetic etiologies underlying some of the more common syndromes important for orofacial clefting (see also the recent review in Schutte and Murray, 1999). Chapter 8 summarizes environmental etiologies. Comprehensive coverage of all clefting syndromes is beyond the scope of this chapter, and excellent compendiums with further details on both common and rare clefting syndromes are to be found by Cohen (1978) and Gorlin and colleagues (1990). Specific genetic information is available online through the home page of the National Center for Biotechnology Information (NCBI, http://www.ncbi.nlm.nih.gov), which serves as an online site to access the many useful databases of human genes and sequence. For further information on syndromes inherited in a Mendelian fashion, refer to the Online Mendelian Inheritance in Man database (OMIM, accessible through NCBI).

7.2.2 Association of Orofacial Clefts with Other Anomalies

A number of researchers have investigated the occurrence of additional anomalies with orofacial clefting in order to determine the relative proportions of isolated and nonisolated cases. There is a wide range of estimates, from about 3% of cases having additional anomalies to as high as 65% (Shprintzen et al., 1985). Consistent among studies is that CP cases are more likely to have associated anomalies than are CL/P cases. Jones (1988) evaluated 428 consecutive cleft palate clinic patients (93% clefts and 7% velopharyngeal insufficiency or atypical clefts): 14.3% of CL/P cases and 54.7% of CP cases had multiple anomalies; about half of those had recognizable syndromes. Shprintzen and colleagues (1985) evaluated 1000 cases and found that 61% of CP cases had other major anomalies compared to 45% of CL/P and 41% of CL. Again, approximately half of the cases with other anomalies had recognized syndromes, sequences, or associations. Clearly, it is imperative that any study of genetic etiologies in orofacial clefts include careful delineation of associated anomalies and an attempt to identify syndromic cases.

7.2.3 Chromosomal Rearrangements

Orofacial clefting is seen as part of the phenotype in a wide variety of rearrangements of many chromosomes, including trisomies, duplications, deletions, microdeletions, or cryptic rearrangements (Brewer et al., 1998, 1999). Rearrangements that can include a CL/P are deletions of 4p (Wolf-Hirschorn syndrome), 4q or 5p (cri du chat syndrome); duplications of 3p, 10p, and 11p; and trisomy 13 or 18 (and trisomy 9 mosaic). CP is seen with deletions of 4q and 7p; duplications of 3p, 7p, 7q, 8q, 9q, 10p, 11p, 14q, 17q, 19q; and trisomy 9 or 13. See Gorlin and co-workers (1990) for details on associated anomalies and birth prevalences. Furthermore, recent reviews of congenital anomalies seen with chromosomal abnormalities identified five regions that have higher rates of orofacial cleft associations than would be expected by background rates: deletions of 1q, 4p16-14, 4q31-35; or duplications of 3p26-21, 10p15-11 (Brewer et al., 1998, 1999).

The role of microdeletions and other cryptic rearrangements in orofacial cleft etiology has recently been recognized. Such small rearrangements are notable in cleft etiology because they are often transmitted within families, unlike the larger rearrangements that are more likely to be de novo. Microdeletions of 22q11.2 are now known to be the common etiology for at least three clinically classified syndromes—DiGeorge

syndrome, velocardiofacial syndrome, and conotruncal anomaly face syndrome (Swillen et al., 2000)—with an incidence of about 1 in 4000 (Devriendt et al., 1998), and cleft palate is seen in as many as 80% of cases (Ryan et al., 1997), although this estimate is likely to be biased by the ascertainment of cases through cleft palate centers. A microdeletion in this region has also been noted in 7 of 23 patients with velopharyngeal insufficiency (Zori et al., 1998). In addition, cryptic chromosome deletions or duplications involving telomeric regions have recently been recognized as an etiology underlying several patterns of anomalies. The following such cryptic rearrangements are associated with orofacial clefting or high arched palate: deletion 1p36 (Slovotinek et al., 1999), 8p23 (Digilio et al., 1998), 18q23 (Strathdee et al., 1997), and 22q13 (Prasad et al., 2000). Interestingly, Schutte and colleagues (1999) also reported microdeletions at 1q32-q41 in a family with apparent van der Woude syndrome.

In summary, many types of chromosomal rearrangements, duplications, and deletions involving either large or small chromosomal segments can include an orofacial cleft in their phenotypic expression. Such observations have been used to target regions for genetic study in nonsyndromic clefting (see Section 7.3.2). Furthermore, cataloging and obtaining genotype/phenotype correlations for the various microdeletions and other cryptic rearrangements has led to fascinating discoveries, such as the realization that multiple clinical entities were all due to microdeletions in the same 22q11 region, and that the usually Mendelian clinical phenotype of van der Woude syndrome could also be due to a microdeletion in 1q31-42. Such discoveries were made possible by the enhanced degree of resolution allowed by the new techniques at the intersection of cytogenetics and molecular genetics.

7.2.4 Single-Gene Etiologies

Almost 300 syndromes have been described in which a cleft of the lip and/or palate is a feature (Gorlin et al., 1990; Lettieri, 1993). About half of those syndromes are due to Mendelian inheritance of alleles at a single genetic locus, and great strides have been made in recent years in mapping genes for such Mendelian disorders. Analogous to the diversity seen in chromosomal abnormalities leading to clefts, every possible Mendelian pattern is observed in the syndromes that include orofacial clefts in their phenotypes. About 50% follow autosomal-recessive inheritance, 40% autosomal-dominant, and 10% X-linked (either recessive or dominant). Complications commonly seen in other Mendelian disorders are also seen in clefting syndromes, such as reduced penetrance, variability expressivity, imprinting, allelic heterogeneity, locus heterogeneity, and so forth. Further complicating syndrome delineation is that some patterns of anomalies can be due to either cytogenetic rearrangements or Mendelian segregation.

As mentioned previously, a full description of the many syndromes that can include a cleft is beyond the scope of this chapter. Therefore, I focus on some of the recent successes in identifying genes for cleft syndromes, and refer the reader to the online data resources (NCBI) for complete details (see also Cohen, 1978; Gorlin et al., 1990; Schutte and Murray, 1999).

Of the total 150 Mendelian clefting syndromes, approximately 30 genes have been cloned (Schutte and Murray, 1999). These genes fall into various classes, including transcription factors (GLI3, 7p13; PAX3, 2q35—Waardenburg syndrome; SIX3, 2p21—holoprosencephaly 2; SOX9, 17q24.3-q25.1—Campomelic dysplasia), extracellular matrix proteins (COL2A1, 12q13.1-q13.2—Stickler syndrome type I; COL11A2, 1p21—Stickler syndrome type II; GPC3, Xp22—Simpson-Golabi-Behmel

syndrome), and cell signaling molecules (FGFR2, 10q26—Apert-Crouzon syndrome; PTCH, 9q22.3—basal cell nevus syndrome; SHH, 7q36—holoprosencephaly 3). Other syndromes have been mapped to a specific chromosomal location but not yet cloned, for example, X-linked cleft palate with ankyloglossia (CPX), mapped by linkage to Xq21-q22 (Forbes et al., 1996), and van der Woude syndrome (VDWS), mapped to 1q32 (Schutte et al., 1996).

One of the major reasons to map and clone genes for syndromic forms of clefting is to help develop strategies for delineating the etiology of nonsyndromic clefting. Van der Woude syndrome is a clearly Mendelian syndrome that has a phenotype only slightly more complicated than isolated clefting—that is, families segregating the VDWS gene exhibit orofacial clefts (CL/P or CP), paramedian lip pits of the lower lip, and sometimes hypodontia. VDWS follows an autosomal-dominant inheritance pattern, with both reduced penetrance (individuals carrying the gene who show no phenotypic features) and variable expressivity (individuals expressing the phenotype may have a cleft or lip pits or both, with varying degrees of severity). Furthermore, VDWS is rare among syndromic forms of clefting in that both CP and CL/P are seen in the same families. The VDWS gene has been localized but not yet cloned.

7.2.5 Summary

As is obvious from the preceding sections, diversity is the defining characteristic for the genetic etiologies underlying syndromic forms of orofacial clefts. Genes and/or chromosomal rearrangements on virtually every chromosome can lead to syndromes including clefts, as can each possible mode of single-gene inheritance. This diversity in mechanism highlights the fact that the processes leading to the development of the oral cavity and face are complex and sensitive to disturbances at multiple time points and/or within multiple genetic domains.

7.3 GENETIC ETIOLOGIES OF NONSYNDROMIC OROFACIAL CLEFTS

Early estimates of the genetic contribution to nonsyndromic orofacial clefts ranged from about 12 to 20%, with the remainder attributed to environmental factors or gene–environment interactions (Fogh-Andersen, 1968, 1971). Estimates from more recent studies suggest that about 20 to 50% may be more realistic (Chung et al., 1986; Marazita et al., 1984; Murray, 1995; Wyszynski et al., 1996). Two general approaches have been taken to investigate genetic factors involved in nonsyndromic facial clefting: large-scale family studies and linkage/association studies with specific genetic markers.

7.3.1 Family Studies

The first published description of a family with several affected members was in 1757 (Trew, 1757). Rischbieth (1910) summarized pre-1900 publications of familial cases of cleft lip ("hare lip") and cleft palate (abridged facsimile of Rischbieth, 1910, and commentary putting Rischbieth's conclusions into historical perspective are provided by Melnick, 1997). Rischbieth (1910) and the other members of the Galton Laboratory concluded that the inheritance of cleft lip and cleft palate was an expression of general physical and racial degeneracy that could be traced to poor protoplasm (Melnick, 1997), while Bateson (1909) and other proponents of Mendelism included "hare lip" as one of a group of "dominant hereditary diseases and malformations" (Bateson, 1909).

Fogh-Andersen (1942) was the first investigator to collect a systematic data set of cleft families and to evaluate the observed inheritance patterns. He concluded that the CL/P families were consistent with segregation of alleles at a single genetic locus with variable penetrance, and that CP families were consistent with autosomal-dominant inheritance with greatly reduced penetrance.

In the 1960s and 1970s, a specific statistical model termed the multifactorial threshold model (MF/T) was described and proposed (Carter, 1976; Fraser, 1976). Under the MF/T model, the occurrence of a cleft depends on a very large number of genes, each of equal, minor, and additive effect, plus environmental factors. An accumulation of these genes and environmental factors is tolerated by the developing fetus to a point, termed the *threshold*, beyond which there is the risk of malformation. This model has testable predictions, and, in theory, could explain many of the features observed for orofacial clefts in families. Several large series of cleft families were presented as evidence in favor of the MF/T (Woolf et al., 1963, 1964; Woolf and Gianas, 1971; Carter et al., 1982a, 1982b; Hu et al., 1982). However, none of these early studies attempted any statistical tests of the MF/T model. Studies that attempted to test the predictions of the MF/T followed (Bear, 1976; Melnick et al., 1980; Mendell et al., 1980; Marazita et al., 1984) but were inconclusive.

In the late 1970s and early 1980s, investigators began to apply segregation analysis methods that allow explicit statistical tests of the MF/T versus the major locus alternative (mixed models). In most published segregation analyses of CL/P under the mixed model, the MF/T could be rejected in favor of either a mixed model (major locus plus multifactorial background) (Marazita et al., 1984; Chung et al., 1986) or a major locus alone (Hecht et al., 1991b; Marazita et al., 1992; Nemana et al., 1992). Most of these studies were conducted in Caucasian populations, although there were a few in Asian populations (Marazita et al., 1992).

There are many fewer segregation analyses of nonsyndromic cleft palate alone than there are of CL/P. There are only three published studies: one in Hawaii (Chung et al., 1974) and two in Caucasian populations (Demenais et al., 1984; Clementi et al., 1997). Chung and colleagues (1974) and Demenais and associates (1984) could not distinguish between MF/T and major locus models; Clementi and co-workers (1997) concluded that a single recessive major locus with reduced penetrance was sufficient to explain their data. There is also a significant subset of nonsyndromic CP families in which CP is X-linked, as evidenced by descriptive studies such as those by Rushton (1979) and Rollnick and Kaye (1986, 1987). The X-linked form of CP includes ankyloglossia, and has been confirmed by linkage analysis and physical mapping (Forbes et al., 1996).

In summary, segregation analyses of orofacial clefts, both CL/P and CP, have consistently resulted in evidence for genes of major effect. Although such studies seem to imply a single major locus, hypotheses of multiple interacting loci or genetic heterogeneity cannot be ruled out, and indeed, were not explicitly tested in any of the published segregation analyses to date. Analyses of recurrence risk patterns (Farrall and Holder, 1992; Mitchell and Risch, 1992; Fitzpatrick and Farrall, 1993; Christensen and Mitchell, 1996) have been consistent with oligenic models with approximately four to seven interacting loci. With evidence that orofacial cleft family history patterns are consistent with one or a few loci, there are now many groups attempting to identify those genes using the positional cloning approach, beginning with linkage and association analyses.

7.3.2   Linkage/Association Studies

The procedures for mapping, cloning, and characterizing genes are now well established, with many successes for rare Mendelian traits. If nonsyndromic orofacial clefts can be shown to be linked to or associated with a marker of known genetic location, it would be powerful support for a Mendelian genetic contribution to the etiology. However, only in recent years have investigators attempted such studies because nonsyndromic clefting was considered to follow the MFT model and thus would not be amenable to a linkage approach. With emerging statistical evidence from human family studies and from knockout mouse experiments that one or a few genes can explain clefting etiology, linkage and association studies were launched in a variety of populations.

Linkage analyses assess the cosegregation of alleles at a genetic locus of known chromosomal location (marker) and a disease locus. Different marker alleles thus cosegregate with the disease allele in different families, and the overall frequencies of the marker alleles calculated from population-based samples need not vary between affected and control groups. In this situation, the two loci are said to be in linkage equilibrium—that is, linked but not associated. In contrast, if allele frequencies differ significantly between the affected and control groups, the specific allele at the marker or candidate locus is said to be associated with the disease at the population level, with the most common interpretation of an association being linkage disequilibrium. Association methods are used as an adjunct to linkage approaches for gene mapping, especially for complex traits. There is a variety of statistical methods, both parametric and model-free, that are used to assess linkage and association (see Ott, 1991, for review of methods).

Gene mapping studies of orofacial clefts have utilized both linkage and association methods. Candidate loci or regions on seven chromosomes (1, 2, 4, 6, 14, 17, and 19) have positive linkage or association results in CL/P, CP, or both (see recent reviews by Wyszynski et al., 1996, and Carinici et al., 2000). There are a few additional loci and chromosomal regions that have only negative results reported in the literature and are not presented in detail here. Also, there are many studies for some loci and few studies for others—this is not a reflection of the strength of the evidence for any particular locus but merely a reflection of the interest in particular loci.

7.3.2.1 *Chromosome 1* Given a possible (but controversial) increased risk of clefting with folate deficiencies (Tolarova and Harris, 1995; Hayes et al., 1996; Czeizel and Hirschberg, 1997; Mills, 1999; see also Chapter 8), the methylene tetrahydrofolate reductase (MTHFR) gene on chromosome 1p36 has been investigated. MTHFR is the major gene involved in the effect of folate leading to neural tube defects; however, neither Shaw and colleagues (1998) nor Gaspar and colleagues (1999) found any allelic associations with clefting and MTHFR. Interestingly, subtle terminal deletions in 1p36 have recently been recognized, with up to 40% of such patients having clefting or high arched palate (Slovotinek et al., 1999; Wu et al., 1999).

7.3.2.2 *Chromosome 2* The first positive association published with orofacial clefts was a population-based association between CL/P and a TaqI restriction site polymorphism in the transforming growth factor alpha locus on chromosome 2p13 (TGFA) (Ardinger et al., 1989). Interestingly, this locus was studied as a candidate because of its involvement in CP in the mouse. The TGFA association with CL/P has since been replicated in several studies (Chenevix-Trench et al., 1991, 1992; Holder et al., 1992; Stoll et al., 1992, 1993; Sassani et al., 1993; Feng et al., 1994; Field et al., 1994; Jara, 1995a, 1995b; Maestri et al., 1997; Lidral et al., 1998; Pezzetti et al., 1998), but several other studies have failed to confirm the association (Hecht et al., 1991a; Hwang et al., 1995; Beaty et al., 1997; Shaw et al., 1996; Lidral et al., 1997; Wyszynski et al., 1997a; Scapoli et al., 1998; Christensen et al., 1999). An association with TGFA has also been reported with CP (Shiang et al., 1993; Hwang et al., 1995), although most studies of TGFA and CP failed to find an association (Stoll et al., 1992, 1993; Shaw et al., 1996; Lidral et al., 1997; Beaty et al., 1997; Maestri et al., 1997; Lidral et al., 1998; Christensen et al., 1999). There is also evidence from some studies that alleles at TGFA and maternal smoking interact to increase risk to CL/P in the fetus (Hwang et al., 1995; Shaw et al., 1996).

There are many possible reasons for the conflicting TGFA association study results. Although the majority of the studies are in Caucasian populations, the data sets differ in the proportion of cases with positive family histories, in the proportions of CL versus CLP versus CP, and in the type of data collected (case control, versus nuclear triads, versus nuclear familes, versus extended kindreds). In general, studies with fewer familial cases were less likely to exhibit a positive association with TGFA. There were no positive results in Asians, although there were only two studies (Filipinos, Lidral et al., 1997; Chinese, Marazita et al., 2001)—too few to draw any conclusion about TGFA and ethnicity. A meta-analysis of pre-1996 studies (Mitchell, 1997) concluded that there was positive evidence of association between CL/P and TGFA in Caucasians (OR 1.43, 95% C.I. 1.12–1.80). The meta-analysis found significant heterogeneity between Caucasian studies in the allele frequencies of cases but not controls. Thus, heterogeneity between studies is unlikely to be due to ethnicity differences, and is, rather, more likely to be due to the differing proportions of familial and/or severe cases of clefting (Mitchell, 1997).

7.3.2.3 Chromosome 4 Loci on two regions of chromosome 4 have shown positive linkage or association results with clefting: MSX1 (homeobox 7; 4p16) and anonymous markers in region 4q31. MSX1 was first examined in humans because an MSX1-deficient knockout mouse can include cleft palate (Satokata and Maas, 1994). Lidral and co-workers (1998) found a positive association result between CL/P and CP with MSX1, although Marazita and colleagues (2001) found no positive linkage or association results with CL/P and the marker. Furthermore, a nonsense mutation in exon 1 of MSX1 (Ser104stop) was recently identified in a Dutch family with tooth agenesis plus orofacial clefting (Van den Boogaard et al., 2000).

Regarding the 4q31 region, although not meeting the strict criterion for significance (i.e., an LOD ≥ 3.0), Beiraghi and colleagues (1994) found suggestive LOD scores of about 2.0 for two closely linked markers (D4S192, D4S175) in one large CL/P family. Mitchell and co-workers (1995) found evidence for association between CL/P and one of the markers (D4S192). However, Blanton and associates (1996) found significant evidence against linkage of CL/P to these same markers, as did Marazita and colleagues (2001) for additional closely linked markers in the region (D4S194, D4S175).

7.3.2.4 Chromosome 6 The first published association studies for orofacial clefts evaluated association with alleles at loci within the major histocompatibility system (HLA in humans, H2 in mice). HLA (chromosome 6p36) was examined because susceptibility to cortisone-induced cleft palate in some mouse strains is associated primarily with genotypes at the H2 locus (Bonner and Slavkin, 1975). Although several studies were done in Caucasian and Asian populations (Bonner et al., 1978; Van Dyke et al., 1980, 1983; Watanabe et al., 1984), no overall positive associations between HLA and CL/P or CP were found, and thus are not presented in detail here.

Eiberg and colleagues (1987) did the first linkage study in CL/P, reporting a significant linkage with F13A (chromosome 6p23-25). Other studies have also found positive results with either F13A or closely linked anonymous markers (Carinici et al., 1995; Scapoli et al., 1997; Prescott et al., 1998; Sakata et al., 1999), while three studies could reject linkage (Hecht et al., 1993; Vintiner et al., 1992, 1993; Blanton et al., 1996). Translocations involving 6p24.1-25 have also been observed (Donnai et al., 1992; Davies et al., 1998) with orofacial clefts as part of the phenotype.

7.3.2.5 Chromosome 14 Transforming growth factor beta 3 (TGF-β3; 14q24) has been investigated in both CL/P and CP because the TGF-β3 knockout mouse has cleft palate (Kaartinen et al., 1995). There are positive results from association studies of CP (Maestri et al., 1997) and borderline positive results (p values between 0.05 and 0.10) with CL/P (Maestri et al., 1997; Lidral et al., 1998). However, other studies of TGF-β3 have been negative (CL/P: Lidral et al., 1997, 1998; CP: Lidral et al., 1998).

7.3.2.6 Chromosome 17 Retinoic acid receptor alpha (RARA, 17q21.1) showed a significant association with CL/P in an Australian population (Chenevix-Trench et al., 1992) but no association in a British population (Vintiner et al., 1993). A study in an Indian population (West Bengal: Shaw et al., 1993) did not find a significant association, but results were borderline and were consistent with RARA acting as a modifier of severity. Two linkage studies could reject linkage to the locus (Shaw et al., 1993; Vintiner et al., 1993).

7.3.2.7 Chromosome 19

Positive linkage or association results were found with CL/P and BCL3, a proto-oncogene located at 19q13.1, or closely linked anonymous markers (Stein et al., 1995; Amos et al., 1996; Wyszynski, 1997b; Maestri et al., 1997). Negative results have also been reported for CL/P (Lidral et al., 1998; Martinelli et al., 1998) and for CP (Maestri et al., 1997; Lidral et al., 1998). A balanced chromosomal translocation involving 19q13.3 has also been reported in a family with CL/P (Yoshiura et al., 1998).

7.3.2.8 Genome-Wide Scans for Cleft Loci

Analyses of recurrence patterns in CL/P (Farrall and Holder, 1992; Mitchell and Risch, 1992) suggest that there may be about four to seven genetic loci involved in nonsyndromic CL/P. Given the contradictory results from candidate locus approaches to date and given the availability of dense maps of markers, studies of CL/P are now turning to genome-wide scans to simultaneously search for multiple regions involved in clefting. The first genome-wide search was recently published using 92 United Kingdom CL/P sib pairs (Prescott et al., 1998, 2000). Initially, 11 loci with suggestive results were identified using a 10 cM mapping panel, 10 of which were confirmed using a 5 cm map (Prescott et al., 2000). Additional genome scans in other populations and in larger sample sizes are necessary to confirm these results. There have not yet been any genome-wide scans for isolated CP.

7.3.3 Summary

Family studies are consistent with one or a few loci exerting major effects on nonsyndromic orofacial cleft etiology. To summarize the linkage and association studies to date, no single locus has clearly emerged as the "necessary" locus for development of nonsyndromic clefts. On the contrary, the genetic etiology appears complex, with several loci showing significant results in at least some studies.

To date, most mapping studies of nonsyndromic clefting have employed a candidate locus approach—that is, identifying genetic loci that are either plausibly involved in the development of the craniofacies or that are implicated from studies of animal models of clefting. These studies have had some successes in identifying loci associated or linked to CL/P or CP, but the results are not consistent across study populations. The emerging consensus for the etiology of orofacial clefting is that of complexity (Murray, 1995; Wyszynski et al., 1996)—probably multiple genetic loci, some of which may be susceptibility loci, some of which may be modifying loci, some of which may be "necessary" loci (following Greenberg, 1993; Greenberg and Doneshka, 1996).

REFERENCES

Amos, C., Gasser, D., and Hecht, J. T. (1996). Nonsyndromic cleft lip with or without cleft palate: new BCL3 information. *Am. J. Hum. Genet. 59*, 743–744.

Ardinger, H. H., Buetow, K. H., Bell, G. I., Bardach, J., VanDemark, D. R., and Murray, J. C. (1989). Association of genetic variation of the transforming growth factor alpha gene with cleft lip and palate. *Am. J. Hum. Genet. 45*, 348–353.

Bateson, W. (1909). *Mendel's Principles of Hereditary* (Cambridge: Cambridge University Press).

Bear, J. C. (1976). A genetic study of facial clefting in northern England. *Clin. Genet. 9*, 277–284.

Beaty, T. H., Maestri, N. E., Hetmanski, J. B., and Wyszynski, D. F. (1997). Testing for interaction between maternal smoking and TGFA genotype among oral cleft cases born in Maryland 1992–1996. *Cleft Palate Craniofac. J. 34*, 447–454.

Beiraghi, S., Foroud, T., Diouhy, S., Bixler, D., Conneally, P. M., Delozier-Blanchet, D., and Hodes, M. E. (1994). Possible location of a major gene for cleft lip and palate to 4q. *Clin. Genet. 46*, 255–256.

Blanton, S. H., Crowder, E., Malcolm, S., Winter, R., Gasser, D. L., Stal, S., Mulliken, J., and Hecht, J. T. (1996). Exclusion of linkage between cleft lip with or without cleft palate and markers on chromosomes 4 and 6 (letter). *Am. J. Hum. Genet. 58*, 239–241.

Bonner, J. J., and Slavkin, H. C. (1975). Cleft palate susceptibility linked to histocompatibility-2 (H-2) in the mouse. *Immunogenetics 2*, 213–218.

Bonner, J. J., Terasaki, P. I., Thompson, P., Holve, L. M., Wilson, L., Ebbin, A. J., and Slavkin, H. C. (1978). HLA phenotype frequencies in individuals with cleft lip and/or cleft palate. *Tiss. Antigens 12*, 228–232.

Brewer, C., Holloway, S., Zawalnyski, P., Schinzel, A., and FitzPatrick, D. (1998). A chromosomal deletion map of human malformations. *Am. J. Hum. Genet. 63*, 1153–1159.

Brewer, C., Holloway, S., Zawalnyski, P., Schinzel, A., and FitzPatrick, D. (1999). A chromosomal duplication map of malformations: regions of suspected haplo- and triplolethality—and tolerance of segmental aneuploidy—in humans. *Am. J. Hum. Genet. 64*, 1702–1708.

Carinici, F., Pezzetti, F., Scapoli, L., Padula, E., Baciliero, U., Curioni, C., and Tognon, M. (1995). Nonsyndromic cleft lip and palate: evidence of linkage to a microsatellite marker on 6p23. *Am. J. Hum. Genet. 56*, 337–339.

Carinici, F., Pezzetti, F., Scapoli, L., Martinelli, M., Carinci, P., and Tognon, M. (2000). Genetics of nonsyndromic cleft lip and palate: a review of international studies and data regarding the Italian population. *Cleft Palate Craniofac. J. 37*, 33–40.

Carter, C. O. (1976). Genetics of common single malformations. *Br. Med. Bull. 32*, 21–26.

Carter, C. O., Evans, K., Coffey, R., Roberts, J. A., Buck, A., and Roberts, M. F. (1982a). A family study of isolated cleft palate. *J. Med. Genet. 19*, 329–331.

Carter, C. O., Evans, K., Coffey, R., Roberts, J. A., Buck, A., and Roberts, M. F. (1982b). A three generation family study of cleft lip with or without cleft palate. *J. Med. Genet. 19*, 246–261.

Chenevix-Trench, G., Jones, J., Green, A., and Martin, N. (1991). Further evidence for an association between genetic variation in transforming growth factor alpha and cleft lip and palate. *Am. J. Hum. Genet. 48*, 1012–1013.

Chenevix-Trench, G., Jones, J., Green, A., Duffy, D. L., and Martin, N. (1992). Cleft lip with or without cleft palate: associations with transforming growth factor-alpha and retinoic acid receptor loci. *Am. J. Hum. Genet. 51*, 1377–1385.

Christensen, K., and Mitchell, L. E. (1996). Familial recurrence-pattern analysis of nonsyndromic isolated cleft palate—a Danish registry study. *Am. J. Hum. Genet. 58*, 182–190.

Christensen, K., Olsen, J., Norgaard-Pedersen, B., Basso, O., Stovring, H., Milhollin-Johnson, L., and Murray, J. C. (1999). Oral clefts, transforming growth factor alpha gene variants, and maternal smoking: a population-based case-control study in Denmark, 1991–1994. *Am. J. Epidemiol. 149*, 248–255.

Chung, C. S., Ching, G. H. S., and Morton, N. E. (1974). A genetic study of cleft lip and palate in Hawaii. II. Complex segregation analysis and genetic risks. *Am. J. Hum. Genet. 26*, 177–188.

Chung, C. S., Bixler, D., Watanabe, T., Koguchi, H., and Fogh-Andersen, P. (1986). Segregation analysis of cleft lip with or without cleft palate: a comparison of Danish and Japanese data. *Am. J. Hum. Genet. 39*, 603–611.

Clementi, M., Tenconi, R., Forabosco, P., Calzolari, E., and Milan, M. (1997). Inheritance of cleft palate in Italy. Evidence for a major autosomal recessive locus. *Hum. Genet. 100*, 204–209.

Cohen, M. M., Jr. (1978). Syndromes with cleft lip and cleft palate. *Cleft Palate J. 15*, 306–328.

Czeizel, A. E., and Hirschberg, J. (1997). Orofacial clefts in Hungary. *Folia Phoniatr. Logop. 49*, 111–116.

Davies, A. F., Imaizumi, K., Mirza, G., Stephens, R. S., Kuroki, Y., Matsuno, M., and Ragoussis, J. (1998). Further evidence for the involvement of human chromosome 6p24 in the aetiology of orofacial clefting. *J. Med. Genet. 35*, 857–861.

Demenais, F., Bonaiti-Pelie, J., Briard, M. L., and Feingold, J. (1984). An epidemiological and genetic study of facial clefting in France. II. Segregation analysis. *J. Med. Genet. 21*, 436–440.

Devriendt, K., Fryns J.-P., Mortier, G., Van Thienen, M.-N., and Keymolen, K. (1998). The annual incidence of DiGeorge/velo-cardiofacial syndrome. *J. Med. Genet. 35*, 789–790.

Digilio, M. C., Marino, B., Guccione, P., Giannotti, A., Mingarelli, R., and Dallapiccola, B. (1998). Deletion 8p syndrome. *Am. J. Med. Genet. 75*, 534–536.

Donnai, D., Heather, L. J., Sinclair, P., Thakker, Y., Scambler, P. J., and Dixon, M. J. (1992). Association of autosomal dominant cleft lip and palate and translocation 6p23;9q22.3. *Clin. Dysmorphol. 1*, 89–97.

Eiberg, H., Bixler, D., Fogh-Andersen, P., Conneally, P. M., and Mohr, J. (1987). Suggestion of linkage of a major locus for non-syndromic orofacial cleft with F13A and tentative assignment to chromosome 6. *Clin. Genet. 32*, 129–132.

Farrall, M., and Holder, S. (1992). Familial recurrence-pattern analysis of cleft lip with or without cleft palate. *Am. J. Hum. Genet. 55*, 932–936.

Feng, H., Sassani, R., Bartlett, S. P., Lee, A., Hecht, J. T., Malcolm, S., and Winter, R. M. (1994). Evidence, from family studies for linkage disequilibrium between TGFA and a gene for nonsyndromic cleft lip with or without cleft palate. *Am. J. Hum. Genet. 55*, 932–936.

Field, L. L., Ray, A. K., and Marazita, M. L. (1994). Transforming growth factor alpha (TGFA): a modifying locus for nonsyndromic cleft lip with or without cleft palate? *Eur. J. Hum. Genet. 2*, 159–165.

Fitzpatrick, D., and Farrall, M. (1993). An estimation of the number of susceptibility loci for isolated cleft palate. *J. Craniofac. Devel. Biol. 13*, 230–235.

Fogh-Andersen, P. (1942). *Inheritance of Harelip and Cleft Palate* (Copenhagen: Nyt Nordisk Forlag, Arnold Busck).

Fogh-Andersen, P. (1968). Increasing incidence of facial clefts: genetically or nongenetically determined. In J. J. Longacre (ed.), *Craniofacial Anomalies Pathogenesis and Repair* (Philadelphia: Lippincott), pp. 27–29.

Fogh-Andersen, P. (1971). Epidemiology and etiology of clefts. In D. Bergsma (ed.), *Third Conference on the Clinical Delineation of Birth Defects* (Baltimore: Williams and Wilkins), pp. 50–53.

Forbes, S. A., Brennan, L., Richardson, M., Coffey, A., Cole, C. O., Gregory, S. G., Bentley, D. R., Mumm, S., Moore, G. E., and Stanier, P. (1996). Refined mapping and YAC contig construction of the X-linked cleft palate and ankyloglossia locus (CPX) including the proximal S-Y homology breakpoint within Xq21.3. *Genomics 31*, 36–43.

Fraser, F. C. (1976). The multifactorial threshold concept—uses and misuses. *Teratology 14*, 267–280.

Gaspar, D. A., Pavanello, R. C., Zatz, M., Passos-Bucno, M. R., Andre, M., Steman, S., Wyszynski, D. F., and Matiolli, S. R. (1999). Role of the C677T polymorphism at the MTHFR gene on risk to nonsyndromic cleft lip with/without cleft palate: results from a case-control study in Brazil. *Am. J. Med. Genet. 87*, 197–199.

Gorlin, R. J., Cohen, M. M., Jr., and Levin, L. S. (1990). *Syndromes of the Head and Neck*, 3rd ed. Oxford Monographs on Medical Genetics No. 19 (New York: Oxford University Press).

Greenberg, D. A. (1993). Linkage analysis of "necessary" disease loci versus "susceptibility" loci. *Am. J. Hum. Genet. 52*, 135–143.

Greenberg, D. A., and Doneshka, P. (1996). Partitioned association-linkage test: distinguishing "necessary" from "susceptibility" loci. *Genet. Epidemiol. 13*, 243–252.

Hayes, C., Werler, M. M., Willett, W. C., and Mitchell, A. A. (1996). Case-control study of periconceptional folic acid supplementation and oral clefts. *Am. J. Epidemiol. 143*, 1229–1234.

Hecht, J. T., Wang, Y., Blanton, S. H., Michels, V. V., and Daiger, S. P. (1991a). Cleft lip and palate: no evidence of linkage to transforming growth factor alpha. *Am. J. Hum. Genet.* *49*, 682–686.

Hecht, J. T., Yang, P., Michels, V. V., and Buetow, K. H. (1991b). Complex segregation analysis of nonsyndromic cleft lip and palate. *Am. J. Hum. Genet.* *49*, 674–681.

Hecht, J. T., Wang, Y., Connor, B., Blanton, S. H., and Daiger, S. P. (1993). Nonsyndromic cleft lip and palate: no evidence of linkage to HLA or factor 13A. *Am. J. Hum. Genet.* *52*, 1230–1233.

Holder, S. E., Vintiner, G. M., Farren, B., Malcolm, S., and Winter, R. M. (1992). Confirmation of an association between RFLPs at the transforming growth factor–alpha locus and nonsyndromic cleft lip and palate. *J. Med. Genet.* *29*, 390–392.

Hu, D. N., Li, J. H., Chen, H. Y., Chang, H. S., Wu, B. X., Lu, Z. K., Wang, D. Z., and Liu, X. G. (1982). Genetics of cleft lip and palate in China. *Am. J. Hum. Genet.* *34*, 999–1002.

Hwang, S.-J., Beaty, T. H., Panny, S. R., Street, N. A., Joseph, J. M., Gordon, S., McIntosh, I., and Francomano, C. A. (1995). Association study of transforming growth factor alpha TaqI polymorphism and oral clefts: indication of gene-environment interaction in a population-based sample of infants with birth defects. *Am. J. Epidemiol.* *141*, 629–636.

Jara, L., Blanco, R., Chiffelle, I., Palomino, H., and Carreño, H. (1995a). Evidence for an association between RFLPs at the transforming growth factor alpha locus and nonsyndromic cleft lip/palate in a South American population. *Am. J. Hum. Genet.* *56*, 339–341.

Jara, L., Blanco, R., Chiffelle, I., Palomino, H., and Carreño, H. (1995b). Association between alleles of the transforming growth factor alpha locus and cleft lip and palate in the Chilean population. *Am. J. Med. Genet.* *57*, 548–551.

Jones, M. C. (1988). Etiology of facial clefts: prospective evaluation of 428 patients. *Cleft Palate J.* *25*, 16–20.

Kaartinen, V., Voncken, J. W., Shuler, C., Warburton, D., Bu, D., Heisterkamp, N., and Groffen, J. (1995). Abnormal lung development and cleft palate in mice lacking TGF-beta-3 indicates defects of epithelial-mesenchymal interaction. *Nat. Genet.* *11*, 415–421.

Lettieri, J. (1993). Lip and oral cavity. In R. E. Stevenson, J. G. Hall, and R. M. Goodman (eds.), *Human Malformations and Related Anomalies*. Oxford Monographs on Medical Genetics No. 27 (New York: Oxford University Press).

Lidral, A. C., Murray, J. C., Buetow, K. H., Basart, A. M., Schearer, H., Shiang, R., Naval, A., Layda, E., Magee, K., and Magee, W. (1997). Studies of the candidate genes TGFB2, MSX1, TGFA and TGFB3 in the etiology of cleft lip and palate in the Philippines. *Cleft Palate J. 34*, 1–6.

Lidral, A. C., Romitti, P. A., Basart, A. M., Doetschman, T., Leysens, N. J., Daack-Hirsch, S., and Semina, E. V. (1998). Association of MSX1 and TGFB3 with nonsyndromic clefting in humans. *Am. J. Hum. Genet. 63*, 557–568.

Maestri, N. E., Beaty, T. H., Hetmanski, J., Smith, E. A., McIntosh, I., Wyszynski, D. F., Liang, K.-Y., Duffy, D. L., and VanderKolk, C. (1997). Application of transmission disequilibrium tests to nonsyndromic oral clefts: including candidate genes and environmental exposures in the models. *Am. J. Med. Genet.* *73*, 337–344.

Marazita, M. L., Spence, M. A., and Melnick, M. (1984). Genetic analysis of cleft lip with or without cleft palate in Denmark. *Am. J. Med. Genet. 19*, 9–18.

Marazita, M. L., Hu, D. N., Spence, M. A., Liu, Y. E., and Melnick, M. (1992). Cleft lip with or without cleft palate in Shanghai, China: evidence for an autosomal major locus. *Am. J. Hum. Genet. 51*, 648–653.

Marazita, M. L., Field, L. L., Cooper, M. E., Tobais, R., Maher, B. S., and Liu, Y. (in press). Non-syndromic cleft lip with or without cleft palate in China: assessment of candidate regions. *Cleft Plate-Craniofac. J.*

Martinelli, M., Scapoli, L., Pezzetti, F., Carinci, F., Carinci, P., Baciliero, U., and Padula, E. (1998). Suggestive linkage between markers on chromosome 19q13.2 and nonsyndromic orofacial cleft malformation. *Genomics 51*, 177–181.

Melnick, M. (1997). Cleft lip and palate etiology and its meaning in early 20th century England: Galton/Pearson vs. Bateson; polygenically poor protoplasm vs. Mendelism. *J. Craniofac. Genet. Devel. Biol. 17*, 65–79.

Melnick, M., Bixler, D., Fogh-Andersen, P., and Conneally, P. M. (1980). Cleft lip ± cleft palate: an overview of the literature and an analysis of Danish cases born between 1941 and 1968. *Am. J. Med. Genet. 6*, 83–97.

Mendell, N. R., Spence, M. A., Gladstien, K., Brunette, J., Stevens, A., Clifford, E., and Georgiade, N. (1980). Multifactorial/threshold models and their application to cleft lip and cleft palate. In M. Melnick, D. Bixler, and E. D. Shields (eds.), *Etiology of Cleft Lip and Cleft Palate* (New York: Alan R. Liss), pp. 387–406.

Mills, J. L. (1999). Folate and oral clefts: Where do we go from here? New directions in oral clefts research. *Teratology 60*, 251–252.

Mitchell, L. E. (1997). Transforming growth factor alpha locus and nonsyndromic cleft lip with or without cleft palate: a reappraisal. *Genet. Epidemiol. 14*, 231–240.

Mitchell, L. E., and Risch, N. (1992). Mode of inheritance of nonsyndromic cleft lip with or without cleft palate: a reanalysis. *Am. J. Hum. Genet. 51*, 323–332.

Mitchell, L. E., Healey, S. C., and Chenevix-Trench, G. (1995). Evidence for an association between nonsyndromic cleft lip with or without cleft palate and a gene located on the long arm of chromosome 4. *Am. J. Hum. Genet. 57*, 1130–1136.

Murray, J. C. (1995). Invited editorial: face facts: genes, environment, and clefts. *Am. J. Hum. Genet. 57*, 227–232.

Nemana, L. J. R., Marazita, M. L., and Melnick, M. (1992). A genetic analysis of cleft lip with or without cleft palate in Madras, India. *Am. J. Med. Genet. 42*, 5–10.

Ott, J. (1991). *Analysis of Human Genetic Linkage* (Baltimore: Johns Hopkins University Press).

Pezzetti, F., Scapoli, L., Martinelli, M., Carinci, F., Bodo, M., Carinci, P., and Tognon, M. (1998). A locus in 2p13-p14 (OFC2), in addition to that mapped in 6p23, is involved in nonsyndromic familial orofacial cleft malformation. *Genomics 50*, 299–305.

Prasad, C., Prasad, A. N., Chodirker, B. N., Lee, C., Dawson, A. K., Jocelyn, L. J., and Chudley, A. E. (2000). Genetic evaluation of pervasive developmental disorders: the terminal 22q13 deletion syndrome may represent a recognizable phenotype. *Clin. Genet. 57*, 103–109.

Prescott, N., Lees, M., Malcolm, S., and Winter, R. (1998). Non-syndromic cleft lip and palate: a genome-wide sib-pair analysis. *Am. J. Hum. Genet. 63* (suppl.), A1764.

Prescott, N. J., Lees, M. M., Winter, R. M., and Malcolm, S. (2000). Identification of susceptibility loci for nonsyndromic cleft lip with or without cleft palate in a two stage genome scan of affected sib-pairs. *Hum. Genet. 106*, 345–350.

Rischbieth, H. (1910). Hare-lip and cleft palate. In K. Pearson (ed.), *Treasury of Human Inheritance, Part IV* (London: Dulau), pp. 79–123.

Rollnick, B. R., and Kaye, C. I. (1986). Mendelian inheritance of isolated nonsyndromic cleft palate. *Am. J. Med. Genet. 24*, 465–473.

Rollnick, B. R., and Kaye, C. I. (1987). Letter to the editor: a response: a further X-linked isolated nonsyndromic cleft palate family with a nonexpressing obligate affected male. *Am. J. Med. Genet. 26*, 241.

Romitti, P. A., Lidral, A. C., Munger, R. G., Daack-Hirsch, S., Burns, T. L., and Murray, J. C. (1999). Candidate genes for nonsyndromic cleft lip and palate and maternal cigarette smoking and alcohol consumption: evaluation of genotype-environment interactions from a population-based case-control study of orofacial clefts. *Teratology 59*, 39–50.

Rushton, A. R. (1979). Sex-linked inheritance of cleft palate. *Hum. Genet. 48*, 179–181.

Ryan, A. K., Goodship, J. A., Wilson, D. I., Philip, N., Levy, A., Seidel, H., Schuffenhauer, S., Oechsler, H., Belohradsky, B., Prieur, M., Aurias, A., Raymond, F. L., Clayton-Smith, J., Hatchwell, E., McKeown, C., Beemer, F. A., Dallapiccola, B., Novelli, G., Hurst, J. A., Ignatius, J., Green, A. J., Winter, R. M., Brueton, L., Brondum-Nielsen, K., Stewart, F., VanEssen, T., Patton, M., Patterson, J., and Scambler, P. J. (1997). Spectrum of clinical features associated with interstitial chromosome 22q11 deletions: a European collaborative study. *J. Med. Genet. 34*, 798–904.

Sakata, Y., Tokunaga, K., Yonehara, Y., Bannai, M., Tsuchiya, N., Susami, T., and Takato, T. (1999). Significant association of HLA-B and HLA-DRB1 alleles with cleft lip with or without cleft palate. *Tiss. Ant. 53*, 147–152.

Sassani, R., Bartlett, S. P., Feng, H., Goldner-Sauve, A., Haq, A. K., Buetow, K. H., and Gasser, D. L. (1993). Association between alleles of the transforming growth factor alpha locus and the occurrence of cleft lip. *Am. J. Med. Genet. 45*, 565–569.

Satokata, I., and Maas, R. (1994). Msx1 deficient mice exhibit cleft palate and abnormalities of craniofacial and tooth development. *Nat. Genet. 6*, 348–355.

Scapoli, L., Pezzetti, F., Carinci, F., Martinelli, M., Carinci, P., and Tognon, M. (1997). Evidence of linkage to 6p23 and genetic heterogeneity in nonsyndromic cleft lip with or without cleft palate. *Genomics 43*, 216–220.

Scapoli, L., Pezzetti, F., Carinci, F., Martinelli, M., Carinci, P., and Tognon, M. (1998). Lack of linkage disequilibrium between transforming growth factor alpha taq I polymorphism and cleft lip with or without cleft palate in families from northeastern Italy. *Am. J. Med. Genet. 75*, 203–206.

Schutte, B. C., and Murray, J. C. (1999). The many faces and factors of orofacial clefts. *Hum. Molec. Genet. 8*, 1853–1859.

Schutte, B. C., Sander, A., Malik, M., and Murray, J. C. (1996). Refinement of the van der Woude gene location and construction of a 3.5 Mb YAC contig and STS map spanning the critical region in 1q32-q41. *Genomics 36*, 507–514.

Schutte, B. C., Basart, A. M., Watanabe, Y., Laffin, J. J., Coppage, K., Bjork, B. C., Daack-Hirsch, S., Patil, S., Dixon, M. J., and Murray, J. C. (1999). Microdeletions at chromosome bands 1q32-q41 as a cause of van der Woude syndrome. *Am. J. Med. Genet. 84*, 145–150.

Shaw, D., Ray, A., Marazita, M. L., and Field, L. L. (1993). Further evidence of a relationship between the retinoic acid receptor alpha locus and nonsyndromic cleft lip with or without cleft (letter). *Am. J. Hum. Genet. 53*, 1156–1157.

Shaw, G. M., Wasserman, C. R., Lammer, E. J., O'Malley, C. D., Murray, J. C., Basart, A. M., and Tolarova, M. M. (1996). Orofacial clefts, parental cigarette smoking, and transforming growth factor alpha gene variants. *Am. J. Hum. Genet. 58*, 551–561.

Shaw, G. M., Rozen, R., Finnell, R. H., Todoroff, K., and Lammer, E. J. (1998). Infant C677T mutation in MTHFR, maternal periconceptional vitamin use, and cleft lip. *Am. J. Med. Genet. 80*, 196–198.

Shiang, R., Lidral, A. C., Ardinger, H. H., Buetow, K. H., Romitti, P. A., Munger, R. G., and Murray, J. C. (1993). Association of transforming growth factor alpha gene polymorphisms with nonsyndromic cleft palate only. *Am. J. Hum. Genet. 53*, 836–843.

Shprintzen, R. J., Siegel-Sadewitz, V. L., Amato, J., and Goldberg, R. B. (1985). Anomalies associated with cleft lip, cleft palate, or both. *Am. J. Med. Genet. 20*, 585–595.

Slovotinek, A., Shaffer, F. L., and Shapria, S. K. (1999). Monosomy 1p36. *J. Med. Genet. 36*, 657–663.

Stein, J., Mulliken, J. B., Stal, S., Gasser, D. L., Malcolm, S., Winter, R., Blanton, S. H., Amos, C., Seemanova, E., and Hecht, J. T. (1995). Nonsyndromic cleft lip with or without cleft palate: evidence of linkage to BCL3 in 17 multigenerational familes. *Am. J. Hum. Genet. 57*, 257–272.

Stoll, C., Qian, J. F., Feingold, J., Sauvage, P., and May, E. (1992). Genetic variation in transforming growth factor alpha: possible association of Bam H1 polymorphism with bilateral sporadic cleft lip and palate (letter). *Am. J. Hum. Genet. 50*, 870–871.

Stoll, C., Qian, J. F., Feingold, J., Sauvage, P., and May, E. (1993). Genetic variation in transforming growth factor alpha: possible association of Bam H1 polymorphism with bilateral sporadic cleft lip and palate. *Hum. Genet. 9*, 81–82.

Strathdee, G., Sutherland, R., Jonsson, J. J., Sataloff, R., Kohonen-Corish, M., Grady, D., and Overhauser, J. (1997). Molecular characterization of patients with 18q23 deletions. *Am. J. Hum. Genet. 60*, 860–868.

Swillen, A., Vogels, A., Devriendt, K., and Fryns, J. P. (2000). Chromosome 22q11 deletion syndrome: update and review of the clinical features, cognitive-behavioral spectrum, and psychiatric complications. *Am. J. Med. Genet. 97*, 128–135.

Tolarova, M., and Harris, J. (1995). Reduced recurrence of orofacial clefts after periconceptional supplementation with high-dose folic acid and multivitamins. *Teratology 51*, 71–78.

Trew, C. J. (1757). Sistens plura exempla palati deficientis. *Nova Acta Physico-Medica Academi ae Caesarae Leopoldion-Carolinae 1*, 445–447.

Van den Boogaard, M. J., Dorland, M., Beemer, F. A., and van Amstel, H. K. (2000). MSX1 mutation is associated with orofacial clefting and tooth agenesis in humans. *Nat. Genet. 24*, 342–343.

Van Dyke, D., Goldman, A., Spielman, R., Zmijewski, C., and Oka, S. (1980). Segregation of HLA in sibs with cleft lip or cleft palate: evidence against genetic linkage. *Cleft Palate Craniofac. J. 17*, 189–193.

Van Dyke, D., Goldman, A., Spielman, R., and Zmijewski, C. (1983). Segregation of HLA in families with oral clefts: evidence against linkage between isolated cleft palate and HLA. *Am. J. Med. Genet. 15*, 85–88.

Vintiner, G. M., Holder, S. E., Winter, R. M., and Malcolm, S. (1992). No evidence of linkage between the transforming growth factor alpha gene in families with apparently autosomal dominant inheritance of cleft lip and palate. *J. Med. Genet. 29*, 393–397.

Vintiner, G. M., Lo, K. K., Holder, S. E., Winter, R. M., and Malcolm, S. (1993). Exclusion of candidate genes from a role in cleft lip with or without cleft palate: linkage and association studies. *J. Med. Genet. 30*, 773–778.

Watanabe, T., Ohishi, M., and Tashiro, H. (1984). Population and family studies of HLA in Japanese with cleft lip and cleft palate. *Cleft Palate J. 21*, 293–300.

Woolf, C. M., and Gianas, A. D. (1971). Congenital cleft lip: a genetic study of 496 propositi. *J. Med. Genet. 65*, 65–71.

Woolf, C. M., Woolf, R. M., and Broadbent, T. R. (1963). A genetic study of cleft lip and palate in Utah. *Am. J. Hum. Genet. 15*, 209–215.

Woolf, C. M., Woolf, R. M., and Broadbent, T. R. (1964). Cleft lip and heredity. *Plast. Reconstr. Surg. 34*, 11–14.

Wu, Y. Q., Heilstedt, H. A., Bedell, J. A., May, K. M., Starkey, D. E., McPherson, J. D., Shapira, S. K., and Shaffer, L. G. (1999). Molecular refinement of the 1p36 deletion syndrome reveals size diversity and a preponderance of maternally derived deletions. *Hum. Molec. Genet. 8*, 313–321.

Wyszynski, D. F., Beaty, T. H., and Maestri, N. E. (1996). Genetics of nonsyndromic oral clefts revisited. *Cleft Palate Craniofac. J. 33*, 406–417.

Wyszynski, D. E., Maestri, N. E., Lewanda, A., McIntosh, I., Smith, E., Garcia-Delgado, C., and Vinageras-Guarneros, E. (1997a). No evidence of linkage for cleft lip with or without cleft palate to a marker near the transforming growth factor alpha locus in two populations. *Hum. Hered. 47*, 101–109.

Wyszynski, D. E., Maestri, N. E., McIntosh, I., Smith, E., Lewanda, A., Garcia-Delgado, C., and Vinageras-Guarneros, E. (1997b). Evidence for an association between markers on chromosome 19q and non-syndromic cleft lip with or without cleft palate in two groups of multiplex families. *Hum. Genet. 99*, 22–26.

Yoshiura, K., Machida, J., Daack-Hirsch, S., Patil, S. R., Ashworth, L. K., Hecht, J. T., and Murray, J. C. (1998). Characterization of a novel gene disrupted by a balance chromosomal translocation t(2;19)(q11.2;q13.3) in a family with cleft lip and palate. *Genomics 54*, 231–240.

Zori, R. T., Boyar, F. Z., Williams, W. N., Gray, B. A., Bent-Williams, A., Stalker, H. J., Rimer, L. A., Nackashi, J. A., Driscoll, D. J., Rasmussen, S. A., Dixon-Wood, V., and Williams, C. A. (1998). Prevalence of 22q11 region deletions in patients with velopharyngeal insufficiency. *Am. J. Med. Genet. 77*, 8–11.

CHAPTER 8

ENVIRONMENTAL ETIOLOGIES OF OROFACIAL CLEFTING AND CRANIOSYNOSTOSIS

VANDANA SHASHI, M.D. and THOMAS C. HART, D.D.S., Ph.D.

8.1 INTRODUCTION

In this chapter, environmental factors that are etiologically related to orofacial clefts and craniosynostoses are reviewed. Complex conditions with important environmental and genetic factors are difficult to understand, but human and animal studies suggest that environmental factors are more likely to play a role in the causation of orofacial clefts than craniosynostoses, which are frequently Mendelian in inheritance. Historically, literature reports of isolated (nonsyndromic) cleft lip and palate supported a sporadic pattern of occurrence. Familial distributions raised the possibility of genetic factors, and with the increasing knowledge of genetics, there is no doubt a genetic component to orofacial clefting (reviewed in Chapter 7). However, simple Mendelian models cannot account for the genetic susceptibilities in the majority of orofacial clefts, and environmental agents are believed to be etiologically important in these instances. In reality, it is likely that the majority of environmental factors act in conjunction with genetic factors and other environmental exposures as stochastic events. Therefore, a discussion of environmental etiologies of orofacial clefts must include gene–environment interactions, which are currently incompletely understood in most cases.

The etiology of nonsyndromic orofacial clefting can be explained in terms of a continuous variable, *liability*, with a threshold value beyond which individuals will be affected. Both genetic and environmental factors determine liability, making the system multifactorial (Fraser, 1976). The recent model of a major susceptibility gene(s) that is modified by various teratogens (Murray, 1995; Wyszynski and Beaty, 1996) is compatible with the threshold effect proposed by Fraser (1976).

Contrary to orofacial clefts, many of the craniosynostoses have been found to be due to single-gene mutations. While there are known environmental agents such as teratogens that are causative of craniosynostosis, the list of these is short. There is very little information on gene–environment interaction in the causation of human craniosynostosis, and while there are genetic models of craniosynostosis, few

Understanding Craniofacial Anomalies: The Etiopathogenesis of Craniosynostoses and Facial Clefting, Edited by Mark P. Mooney and Michael I. Siegel, ISBN 0-471-38724-x Copyright © 2002 by Wiley-Liss, Inc.

appropriate animal models currently exist for environmentally induced craniosynostosis. For these reasons, the section on environmental etiologies of craniosynostosis is brief.

We preface this discussion with the following general statements about environmental etiologies of craniofacial abnormalities: Although epidemiological and case studies describe associations between various environmental factors and human craniofacial abnormalities, the role of the environment is complex and difficult to study because (1) large numbers of case-control studies are needed to establish a correlation and achieve statistical power (relative risk or odds ratio); (2) an environmental agent may be just one of several factors that ultimately leads to an abnormality (etiologic heterogeneity and the role of gene–environment interactions); and (3) it is difficult to assess exposures (especially in retrospective studies) due to bias and the difficulty in controlling for confounding factors (Wyszynski and Beaty, 1996).

Even in instances wherein causality is established, the effect of a particular teratogen is usually modified by several factors (Gorlin et al., 1990). They include differences in timing of exposure during embryonic development, differences in concentration or method of teratogen delivery, variations in susceptibility of individuals due to genetic factors, and synergistic interactions among various compounds. All these variables can lead to a continuum in the severity of congenital anomalies.

Animal studies have been helpful in corroborating human clinical observations and have shed light on pathogenetic mechanisms in the case of individual teratogens. While animal data offer valuable insight, extrapolation of findings to the human condition is usually not straightforward and translational applications to the human clinical setting are rare. We provide summaries of the data (human or animal or both) in instances where the evidence is conflicting or equivocal. The first section elaborates on the environmental etiologies of orofacial clefts.

8.2 NOMENCLATURE

Crucial to the understanding of the etiopathogenesis of craniofacial anomalies is the ability to distinguish between the terms *malformation*, *disruption*, and *deformation* (Spranger et al., 1982) (see also Chapters 1, 2, and 3). Malformations occur during the development of a particular organ and are caused by genetic or environmental factors, or a combination of the two. They result from abnormalities in organogenesis and thus tend to occur during the period of embryogenesis (first 8 weeks of pregnancy). Disruptions are the result of destructive processes that occur after the period of organogenesis. The underlying tissue/organ is normal morphogenetically. The ensuing abnormalities can be as significant as those caused by malformations. A classic example of a disruption is a cleft lip due to an amniotic band that interferes with lip closure. Most disruptions are thought to be environmental in etiology. A deformation occurs after normal organogenesis when mechanical forces distort a part of the body. As with disruptions, there is no intrinsic tissue abnormality. Deformations are due to direct local (uterine) environmental factors and usually resolve after birth. Most environmental factors that result in craniosynostoses and orofacial clefts cause malformations and disruptions, although deformations due to fetal head and mandibular constraint have been linked to craniosynostoses and oral clefts (Graham, 1983; Shahinian et al., 1998; Gorlin et al., 1990).

Craniofacial abnormalities, like other congenital anomalies, are classified into major and minor anomalies. A major anomaly has medical and social consequences. A minor anomaly has no significant medical or social burden. Orofacial clefts and most forms of craniosynostoses are major anomalies. They can occur as isolated defects (referred to frequently as nonsyndromic), or can be part of a *syndrome* (defined as a group of features seen together, with an underlying specific etiology), an *association* (nonrandom occurrence of features, with no specific cause), or a *sequence* (a series of anomalies that occur secondary to a primary event). Environmental agents associated with orofacial clefts and craniosynostoses can cause syndromes, associations, and sequences. Examples are fetal valproate syndrome caused by exposure to valproic acid in utero, the association of orofacial clefts and growth retardation with maternal smoking, and the well-known sequence of Pierre-Robin, wherein a primary event of fetal mandibular constraint leads to a cleft palate.

The term *multifactorial* is frequently applied to isolated (nonsyndromic) orofacial clefts. A multifactorial anomaly, as defined by Fraser (1976), is determined by the combined interaction of genetic and environmental factors (e.g., nonsyndromic orofacial clefts). Poylgenic inheritance refers to a large number of genes, each with a small effect, acting additively, as has been suggested in the genetic susceptibilities that lead to orofacial clefts.

The above categorization of craniofacial anomalies is important not only in the understanding of the etiopathogenesis but also in estimating prognosis and in counseling families regarding recurrence risks. We begin the discussion by elaborating the environmental etiologies of orofacial clefts.

8.3   ENVIRONMENTAL ETIOLOGIES OF OROFACIAL CLEFTS

For practical purposes, we have divided the better-known environmental factors linked to orofacial clefts into the following major groups:

1. *Teratogens:* All environmental agents that produce structural alteration after fertilization are termed teratogens. Maternal exposure to these agents during the period of craniofacial organogenesis could result in malformations or disruptions such as orofacial clefts. Teratogens include (a) prescription medications, associated metabolites, and dietary supplements; (b) recreational drugs; (c) toxins; and (d) hyperthermia.

2. *Maternal factors:* Lack of certain vitamins such as folic acid has been associated with a higher incidence of orofacial clefts. Alterations in maternal hormones are also thought to be correlated with orofacial clefting.

3. *Intrauterine factors:* An abnormality in the intrauterine environment, such as fetal mandibular constraint due to multiple pregnancy or oligohydramnios, can cause cleft palate. Similarly, the presence of amniotic bands around the developing fetus can result in orofacial clefts due to disruption.

Common examples of these environmental factors are summarized in Table 8.1. Detailed below are well-studied environmental etiological agents linked to orofacial clefts in humans. In many cases, animal studies have also been performed, and summaries of these findings follow human reports.

8.3.1   Medications

8.3.1.1   Anticonvulsants

Clinical Observations Anticonvulsants have been observed to be teratogenic since the 1960s. A two- to three-fold increase in anomalies (including clefts) was reported among babies prenatally exposed to anticonvulsants (Hanson and Smith, 1976; Janz, 1982), with the risk increasing with poly-

TABLE 8.1   Classification of Environmental Agents Linked to Orofacial Clefts and Craniosynostosis

Environmental Agents	Common Examples
Teratogens	Anticonvulsants, retinoic acid, vitamin A
prescription drugs, associated	Alcohol, cigarette smoke
metabolites and dietary supplements	Dioxin, organic solvents, plant toxins, food contaminants
recreational drugs	
toxins	
hyperthermia	
Maternal factors	Vitamin A, folic acid
nutritional deficiencies	Maternal hyperthyoidism
endocrine abnormalities	
Intrauterine factors	Fetal constraint
	Amniotic bands

therapy (Lindhout et al., 1984). This translates into a risk of 6 to 7% for a major congenital anomaly with anticonvulsant therapy, compared to a general population risk of 2 to 3% (Dieterich, 1979). Although some studies have suggested that there is a genetic predisposition to clefting in mothers who are epileptic (Starreveld-Zimmerman et al., 1974; Kelly et al., 1984; Durner et al., 1992), others have not found substantial evidence that epilepsy is by itself etiologically related to orofacial clefts (Janz, 1982; Friis, 1989; Hecht and Annegers, 1990; Holmes et al., 2001). The increased incidence of clefting in children whose mothers have seizures is believed to be due to exposure to antiepileptic medications prenatally (Friis, 1989; Samren et al., 1999; Hernandez-Diaz et al., 2000).

Among the anticonvulsants, phenytoin, valproic acid, trimethadione, and primidone have been associated with an increased risk of oral and facial clefts (Aase, 1974; Zackai et al., 1975; Hanson and Smith, 1976; Ardinger et al., 1988). They may cause a specific recognizable pattern of anomalies (with clefting as one of the features) that constitute a syndrome. Well-known examples are fetal valproate syndrome, fetal hydantoin syndrome, and fetal trimethadione syndrome (Zackai et al., 1975; Hanson and Smith, 1976; DiLiberti et al., 1984). A full description of

these syndromes is beyond the scope of this chapter. Comprehensive reviews of each of these can be found in Gorlin's textbook on *Syndromes of the Head and Neck* (1990). The risk of nonsyndromic cleft lip with or without cleft palate (CL/P), and to a lesser extent for cleft palate alone (CP), also appears to be increased in women exposed to these anticonvulsants (Abrishamchian et al., 1994; Hernandez-Diaz et al., 2000).

The pathogenesis of orofacial clefting with anticonvulsants is not completely clear. Since these medications affect folate metabolism, it is postulated that they may cause clefting by impairing folate absorption, by increasing the degradation of folate, or by interacting with enzymes involved in folate metabolism (Lambie and Johnson, 1985; Dansky et al., 1992; Dean et al., 1999). However, supplementation with folic acid does not appear to reduce the risk of congenital anomalies in these women, although it is routinely advised (Hernandez-Diaz et al., 2000). It has been suggested that anticonvulsants such as phenytoin and primidone may exert effects other than through depletion of folic acid, and other mechanisms including a direct toxic or vascular effect have been proposed (Finnell et al., 1997; Azarbayzani and Danielsson, 2001).

In summary, use of anticonvulsant medications during pregnancy is associated with an increased risk of congenital anomalies, including orofacial clefts in the off-spring. The exact mechanisms are still unclear; interference with folic acid metabolism and direct toxic and vascular effects have been proposed. Gene–environment interactions may modify the risk, making some women more vulnerable than others.

Animal Data Administration of anti-seizure medications (primidone, diazepam) to pregnant animals (mice, rats, hamsters) is associated with palatal defects with full-length or submucosal clefts (Miller and Becker, 1975). The metabolites of primidone, phenylethylmalonamide (PEMA), and phenobarbital are similar in many species including humans. In contrast to a dose-related teratogenicity seen for many other teratogens, in several animal models a dose-related teratogenic effect was not found for primidone or diazepam (McElhatton et al., 1977). Craniofacial anomalies including clefting can be induced in mice, to different degrees, by other anticonvulsant drugs including carbamazepine (CMZ), sodium valproate (NaV), and diphenylhydantoin (DPH). There is evidence of a drug-, dose-, and time-dependent increase in the observed frequencies of hydrocephalies, secondary palatal clefts, and submucous palatal clefts induced by these drugs. Additionally, dose–phenotype relationships are reported—for example, cleft lips were observed only in the highest dose level of DPH, while CMZ was the least teratogenic in mice (Eluma et al., 1984). Vigabatrin is a relatively recently introduced antiepileptic drug that enhances the brain levels of gamma-aminobutyric acid (GABA). Few data on the teratogenic effects of Vigabatrin appear, but prenatal exposure is associated with significant intrauterine growth retardation as well as mandibular and maxillary hypoplasia, arched palate, and cleft palate (Abdulrazzaq et al., 1997). In summary, animal data for many anticonvulsants support a causal role in orofacial clefting. These data suggest that the increased incidence of clefting in children of epileptic mothers taking anticonvulsant medications during pregnancy is due to prenatal exposure to anticonvulsants or their metabolites.

8.3.1.2 *Dihydrofolate Reductase Inhibitors*

Clinical Observations These medications displace folate from the enzyme dihydrofolate reductase and thereby block the conversion of folate to its more active metabolites. They have been used in the therapy of malignant diseases, psoriasis, and rheumatoid arthritis. Methotrexate and aminopterin are two dihydrofolate reductase inhibitors used commonly in clinical practice. Their teratogenic effects have been known since the early 1950s, and a variety of malformations have been reported including cranial and skeletal defects (Thiersch, 1952; Milunsky et al., 1968; Shaw and Steinbach, 1968). A characteristic facies also has been described, and several affected children were found to have CL/P (Thiersch, 1952; Milunsky et al., 1968; Shaw and Steinbach, 1968; Feldkamp and Carey, 1993). In a case-control study, Hernandez-Diaz and colleagues (2000) studied the effects of the dihydrofolate reductase inhibitors trimethoprim, sulfasalazine, and triamterene during the second or third month of pregnancy. The relative risk for oral clefts was 2.6 (95% CI = 1.1–6.1). The use of folic acid supplements diminished the adverse effects of the dihydrofolate reductase inhibitors. To summarize, dihydrofolate reductase inhibitors are associated with an increased risk of orofacial clefting, which may be part of a recognizable syndrome. The use of folic acid as part of a multivitamin supplement may reduce the risk of clefting.

Animal Data Prenatal exposure to methotrexate produces cleft palate, hydrocephalus, and limb defects in rats and

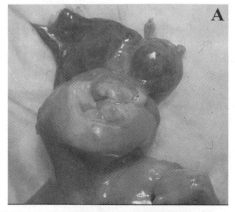

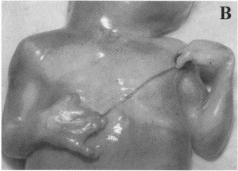

Figure 8.1A A 28 week gestation fetus with severe facial clefting due to disruption from amniotic bands. **Figure 8.1B** Amputation of digits and an amniotic band are visible. (Acknowledgement-Kindly provided by William G Wilson MD, University of Virginia, Charlottesville).

rabbits. While a spectrum of malformations has been reported, the specific malformation appears to be related to timing of exposure. In rabbits, methotrexate administration on days 8 and 9 led to universal litter resorption. Administration on days 10, 11, or 12 caused abnormalities, including hydrocephalus, microphthalmia, cleft lip and palate, while later administration caused mainly distal limb dysplasias (DeSesso and Jordan, 1977; Jordan et al., 1977; Sulik et al., 1988). Methotrexate-induced developmental toxicity in rabbits is reportedly ameliorated by 1-(p-tosyl)-3,4,4-trimethylimidazolidine, a functional analogue for tetrahydrofolate-mediated one-carbon transfer, suggesting that

impaired one-carbon transfer may be the fundamental cause of the fetal abnormalities seen with methotrexate (DeSesso and Goeringer, 1992).

8.3.1.3 Retinol and Retinoids

Clinical Observations Megadose vitamin A (retinol) use as a dietary supplement is widespread, but it is difficult to establish human teratogenicity as experience outcomes are poorly observed and epidemiological controls are lacking. High doses of vitamin A are associated with a differential propensity to induce cleft palate in rats, depending on the rate of palatal width development in the specific strain studied. This finding in rats may have implications for the differential rate of clefting in different human populations (Siegel and Mooney, 1986; Vergato et al., 1997). Based on animal studies and experience with vitamin A, avoiding long-term megadose vitamin A use in fertile women appears warranted (Geelen, 1979; Hayes et al., 1981; Rosa et al., 1986).

The vitamin A derivative retinoic acid (RA) is essential for normal embryonic development. Retinoic acid undergoes oxidative metabolism in the body to yield several retinoid metabolites. At high concentrations retinoids are teratogenic in humans, producing a syndrome of craniofacial malformations that includes cleft palate (retinoic acid syndrome) (Lammer et al., 1985; Abbott and Pratt, 1991). Despite knowledge of vitamin A congener teratogenicity in animal studies since 1954, striking human findings arose in the 1980s following isotretinoin (13-cis retinoic acid, Ro 4-3780, Accutane) marketing for oral treatment of severe acne (Rosa et al., 1986). As the clinical use of isotretinoin (ITR) and other retinoic acid (RA) derivatives has increased, so too have reports of adverse human pregnancy outcomes including cleft (Benke, 1984; Abbott and Pratt, 1987). The effects of retinoids on

Figure 8.2 Two year-old girl with fetal valproate syndrome. She has evidence of metopic suture prominence, due to premature synostosis.

craniofacial development are dependent on dose and the time of exposure (Collins et al., 1992, 1995; Hendrickx and Hummler, 1992). ITR half-life is less than a day, although a teratogenic metabolite, 4-oxo-isoretinoin, has a half-life of several days. The critical period for exposure appears to be 2 to 5 weeks postconception, although this is clinically inexact.

Several studies indicate that the human retinoic acid receptor (RAR-α) locus may be a CL/P susceptibility gene (Chenevix-Trench et al., 1993; Mitchell et al., 1995, 2001). Additionally, support for certain CL (P) susceptibility loci is increased when information on type of cleft, race, family history, or maternal smoking is incorporated as effect modifiers (Maestri et al., 1997). Nonetheless, not all studies support an association for RARα and craniofacial clefting (Stein et al., 1995; Chapter 7). However, a major CL (P)-causing gene has been localized to murine chromosome 11 in a region having linkage homology with human 17q21–24, possibly supporting reports of association of human CL (P) with the retinoic acid receptor alpha (RARα) locus (Juriloff and Mah, 1995).

Animal Data Because of the phenotypic similarities between human and animal models of RA-induced malformations, significant research in a variety of animal models has been performed to characterize RA teratogenesis. As a result, biological models of RA teratogenesis are emerging. Numerous studies have investigated hypervitamin A induction of craniofacial clefting in animal models (Geelen, 1979; Hayes et al., 1981; Siegel and Mooney, 1986; Vergato et al., 1997). These studies suggest that retinol (vitamin A) is either a proximate teratogen or a coteratogen with the vitamin A analogue all-trans retinoic acid. Retinoic acid (RA) is teratogenic in many species and is an effective inducer of cleft palate, although susceptibility, dose, and timing vary (Helms et al., 1997; Abbott and Pratt, 1991; Johnston and Bronsky, 1991; Sulik et al., 1988; Willhite et al., 1986; Makori et al., 1998). The in vivo oxidation of retinol appears to be an important factor in the teratogenic activity of high doses of vitamin A (Eckhoff et al., 1989), and many metabolites of retinol and retinoids appear to have teratogenic potential.

The diverse effects of retinoids on the development, growth, and homeostasis of vertebrate organisms are mediated in part by three distinct isoforms of retinoic acid receptors (RARs). These proteins, which are structurally and functionally closely related to thyroid hormone receptors, regulate patterns of gene expression in target tissues. By mutating these receptors in transgenic animals, RAR-α, RAR-β, and RAR-γ can be converted into potent negative transcriptional regulators that block wild-type RAR function. These mutant RARs, but not the wild-type receptors, actively repress the basal transcription level of target promoters. When expressed in transgenic mice, the most potent of these inhibitory receptor mutants is apparently able to disturb developmental processes by inducing a cleft palate in transgenic offspring (Damm et al., 1993). Craniofacial malformations similar to those induced by retinoids are observed in mice embryos lacking two retinoid receptor genes (RAR-γ and RAR-α), while loss of one receptor results in either normal development or mild abnormalities. These findings suggest that mutations leading to abnormalities in the structure or regulation of RA signaling pathway genes may be etiologically important in human congenital craniofacial abnormality, and the presence of mutations in these retinoid receptor genes can influence susceptibility to environmental teratogens. These receptors may also be targets of environmental teratogens. Such a pathogenesis model would be consistent with the multifactorial model as defined by Fraser (1976) and highlight the importance of gene–environment interactions in the etiology of craniofacial anomalies (Morriss-Kay and Sokolova, 1996).

The pathogenesis of cleft formation varies with the timing of teratogen exposure. Mice studies suggest alterations in mesenchyme development and the subsequent delay in palate shelve elevation are central to RA-induced cleft formation following exposure at the palate shelf outgrowth stage (Degitz et al., 1998). Mouse palatal shelves exposed in organ cultures to retinoids and epidermal growth factor (EGF) display altered medial epithelial cell morphology, blocking normal union of apposing shelves. Extensions of these studies in precontacting human palatal shelves maintained in organ culture and exposed to retinoic acid derivatives (13-cis RA and trans RA with or without EGF) suggest alterations in medial cell morphology could interfere with adhesion of the palatal shelves and may play a role in retinoid-induced cleft palate in the human embryo (Abbott and Pratt, 1991). Studies in Msx-1 knockout mice suggest a mechanistic interaction involving retinoic acid and Msx-1 in the etiology of retinoic acid–induced cleft palate (Nugent and Greene, 1998).

In mice, high doses of retinoic acid truncate the growth of the frontonasal and maxillary processes and thus produce bilateral clefting of the lip and palate. Inhibition of Sonic hedgehog and patched expression in the craniofacial primordia, and disruption of polarizing activity of these tissues, has been identified in RA-associated teratogenesis. These studies suggest that RA teratogenicity disrupts signaling centers in the epithelium that participate in craniofacial growth and patterning in the embryonic face. Additionally, the teratogenic effect of RA on craniofacial morphogenesis appears independent of its effects on Hox gene expression or neural crest cell migration (Helms et al., 1997). Animal model studies of retinoic acid syndrome (RAS) also indicate major involvement of neural crest cells. Later administration of retinoic acid prematurely and excessively kills ganglionic placodal cells and leads to a malformation complex similar to Treacher Collins syndrome (Johnston and Bronsky, 1995).

Transplacental pharmacokinetics are also an important variable in teratogenic-induced anomalies. Animal studies indicate all-trans RA and all-trans-4-oxoretinoic acid, both well-known teratogens, largely contribute to teratogenic outcomes.

In addition to aiding development of pathogenesis paradigms, animal models may permit development and testing of intervention strategies. In rodent models, ethanol and RA appear to affect similar cell populations, and both agents can induce comparable malformations. When administered during gastrulation, they cause a major insult to the anterior neural plate, which results in characteristic ocular, brain, and facial malformations comparable to those seen in fetal alcohol syndrome (Sulik et al., 1988). Decrease in RA-induced clefting was observed when mice had been pretreated with the alcohol dehydrogenase inhibitor 4-methylpyrazole (Collins et al., 1992). The teratogenic actions of retinoic acid on pregnant mice may also be reduced or even completely prevented by subsequent administration of methionine, providing a biological rationale for its consideration as a therapeutic agent (Lau and Li, 1995). The similarity observed in the malformation syndrome induced by both all-trans and 13-cis retinoic acids in the cynomolgus monkey and 13-cis retinoic acid embryopathy in humans suggests this macaque species may be a model for further developmental toxicity studies of vitamin A–related compounds (Hendrickx and Hummler, 1992). Determination of LOAEL (lowest observed adverse effect level) for developmental toxicity suggests a vitamin A safety recommendation of 30,000 IU/day as nonteratogenic for humans (Wiegand et al., 1998).

Retinol and retinoic acid teratogenesis provides a paradigm for studying gene–environment interactions. Dose, timing, genetic strain susceptibility, interactions with other agents, and the protective effects of some agents can be studied in animal models to understand how normal development is disrupted, permitting identification of specific molecular targets of teratogens. This understanding should provide the basis for quantitating risk, implementing effective avoidance strategies, and developing novel prevention strategies, particularly if individuals at increased risk (e.g., genetic predisposition) can be identified.

8.3.1.4 Glucocorticoids

Clinical Observations Corticosteroids are potent medications used in a vast variety of illnesses for their anti-inflammatory effects. Earlier studies indicated that the risk of cleft palate with maternal use of corticosteroids during pregnancy was small (<1%) (Stevenson et al., 1993). Three analytic studies in humans indicate that maternal corticosteroid use during early pregnancy might be associated with a three- to six-fold increase in the risk for orofacial clefts

(Rodriguez-Pinilla and Martinez-Frias, 1998; Czeizel and Rockenbauer, 1997; Arpino et al., 2000). In the study by Rodriguez-Pinilla and Martinez-Frias (1998), systemic corticosteroid therapy during the first trimester was associated with an increased risk of CL/P. Czeizel and Rockenbauer (1997) found an increased risk with both topical and systemic corticosteroid use during the first month of pregnancy.

Contrary to these studies, a review of case reports and case series found that treatment with corticosteroids in pregnancy posed little if any teratogenic risk to the fetus (Fraser and Sajoo, 1995). Another randomized trial recommended methylprednisone as a safe treatment for hyperemesis gravidarum during early pregnancy (Safari et al., 1998). In a recent population-based case-control study, the association between corticosteroid therapy during the periconceptual period and the occurrence of orofacial clefts, among other anomalies, was examined (Carmichael and Shaw, 1999). An increased risk for isolated CL/P (odds ratio 4.3, 95% CI = 1.1–17.2) and isolated CP (odds ratio 5.3, 95% CI = 1.1–26.5) was found. There was no increased risk for the other anomalies examined (neural tube defects, conotruncal heart defects, and limb reduction defects). These authors noted that previous case series and case reports that did not find a causal link between oral clefts and corticosteroid use were less reliable than case-control studies. They concluded that a possible causal association exists between the use of corticosteroids during early pregnancy and cleft lip and palate, and that these medications should be used with caution. To summarize, the data in humans regarding the association between orofacial clefting and maternal corticosteroid therapy are inconclusive. Further studies are needed to clarify this issue.

Animal Data In 1951, Fraser and Fainstat (1951) reported orofacial clefts could be induced in mice using corticosteroids. Over the next few years, additional studies in mice and hamsters confirmed that cleft palate can be induced by corticosteroid administration (Ballard et al., 1975; Shah, 1984; Bedrick and Ladda, 1978; Rowland and Hendrickx, 1983b; Kusanagi, 1984; Kopf-Maier, 1985; Marazita et al., 1988) (see also Chapters 7, 10, and 21). The teratogenicity of corticosteroids is perhaps influenced by the physiology of the glucocorticoid receptors (Brown and Hackman, 1985). These receptors are seen in large numbers in the palatal mesenchymal cells, and corticosterids may exert their teratogenic effects by inhibiting the growth of the palatal mesenchymal cells (Pratt, 1985). Other factors such as epidermal growth factor, TGF-α, TGF-β_1 and TGF-β_2 have also been shown to be involved in palatogenesis, proliferation, differentiation, and extracellular matrix production, and may act synergistically in the production of corticosteroid-induced cleft palate (Bedrick and Ladda, 1978; Abbott et al., 1992; Jaskoll et al., 1996). Abbott and associates (1999) reported that the glucocorticoid receptor and the dioxin receptor mediated the increased susceptibility when corticosterids were combined with dioxin in mouse embryos. The timing of the exposure, the different types of corticosteroids, and the dose have also been shown to influence risk (Biddle, 1978; Kusanagi, 1984; Sauerbier, 1986; Hansen et al., 1999). The role of genetics in the causation of clefting by corticosteroid therapy is supported by reports of increased susceptibility in certain strains of mice (Biddle and Fraser, 1979; Juriloff, 1995; Marazita et al., 1988; Montenegro et al., 1998).

To summarize, the animal data indicate a causal relationship between the use of corticosteroids and the occurrence of cleft palate. This effect is mediated by glucocorticoid receptors, and is influenced by epidermal and other growth factors, genetic susceptibility, as well as the timing and dose of the corticosteroid.

8.3.1.5 Miscellaneous Medications Linked to Orofacial Clefts in Animals

The teratogenic effects of a great number of medications have been evaluated in various animal models. In most cases, the teratogenic effects result in malformations of multiple systems in addition to craniofacial malformations. We have focused on results of craniofacial significance, primarily orofacial clefting. To maintain an appropriate perspective, it is important to appreciate that in addition to the drugs discussed, many others have been associated with orofacial clefting, particularly cleft palate. These include a variety of drugs used in cancer treatments, for example, 5-fluorouracil (Shuey et al., 1994), 1-beta-D-arabinofuranosylcytosine (Ortega et al., 1991), 5-Aza-2'-deoxycytidine (Branch et al., 1996), cyclophosphamide (Francis et al., 1990; Porter and Singh, 1988), bromodeoxyuridine (Bannigan et al., 1990), urethane (Nakane and Kameyama, 1986), procarbazine (Bienengraber et al., 1999), mercaptopurine (Platzek and Bochert, 1996), and hadacidin (Shah et al., 1991); anticonvulsants including dilantin, carbamazepine, and valproic acid (Buehler et al., 1994; Fritz, 1976; Miller and Becker, 1975; Wells, 1983); aspirin (Robertson et al., 1979; Shah et al., 1979); tranquilizers including benzodiazepine and phenobarbital (McElhatton et al., 1977; Walker and Patterson, 1974; Rodriguez Gonzalez and Friman Perez, 1985; Takeno et al., 1990); drugs with potassium channel blocking activity (almokalant, dofetilide, and d-sotalol) (Webster et al., 1996); organophosphorus anthelmintics (Yoshimura, 1987); cyproheptadine (Rodriguez Gonzalez et al., 1983); corticoids (Rowland and Hendrickx, 1983b); tetrachlorodibenzofuran (Weber et al., 1985); and trifluoperazine (Rodriguez Gonzalez and Friman Perez, 1985).

In addition to evaluating drugs as teratogens, some drugs and supplements have been evaluated to determine if they ameliorate the effect of teratogens. While many studies appear to have established a causative link between exposure to an agent and an undesirable clinical outcome, it is often difficult to compare results between studies and to extrapolate meaning for the human condition. Differences in dose, time of exposure, manner of exposure, and genetic differences between animals and between strains of the same animal complicate interpretation of findings. These limitations withstanding, animal models offer several important opportunities to design and test hypotheses in a scientifically rigorous manner. As a result, these studies have provided important insights into the role of teratogen-induced craniofacial malformations, particularly for cleft palate.

8.3.1.6 Animal Models of Intervention Studies

Several studies have investigated the utility of pre and/or posttreatment interventions to reduce or prevent the induction of craniofacial malformations (primarily clefting). For example, methylmercury is an effective inducer of cleft palate in ICR mice (Fuyuta et al., 1978; Yasuda et al., 1985). Subsequent treatment with tiopronin (2-mercaptopropionyl glycine) is effective in dramatically reducing the expected incidence of methylmercury-induced cleft palate (Fujimoto et al., 1979). More widely studied is the utility of B vitamins in reducing and/or ameliorating the adverse effect of certain teratogens. The propensity for beta-aminoproprionitrile (BAPN)–induced cleft palate in rat fetuses can be prevented both in number and severity by administration of vitamin B_6 before and simultaneously with BAPN (Jacobsson and Granstrom, 1997). Folinic acid and vitamin B6 plus vitamin B12 can effectively reduce valproic acid–induced malformations in mice and rats, but the protection is not complete, which may suggest the involvement of other factors.

Concern has also been noted that the dose levels utilized by preventive strategies should be carefully chosen, since high doses of the combined vitamins can actually increase the incidence of certain defects (Elmazar et al., 1992; Jacobsson, 1997). The observation that coadministration of methanol and low dietary folate to CD-1 mice was correlated with an increased cleft palate incidence in offspring suggests that folate supplementation may ameliorate the teratogenic effect of methanol. The mitigating effect on orofacial clefting of another vitamin, thiamine, was studied by Bienengraber and colleagues (1997), who found no beneficial treatment effect in Wistar rats. A prophylactic effect of thiamine initially tested as a highly dosed monotherapy was not verifiable. The investigators noted that the timing of treatment and dosage of thiamine must be considered not only in animal experiments but also when applying results to humans (Bienengraber et al., 1997).

In addition, prevention intervention models have also been studied. The spontaneous incidence of cleft lips and palates in CL/Fr mice was significantly reduced when the antipellagratic vitamin nicotinamide was administered at a low dose (0.5% in diet), but not at a higher dose (1.0%). When (carbonyl-14C) nicotinamide was given to pregnant mice, nicotinamide and small amounts of nicotinamide adenine dinucleotide (NAD^+), but not nicotinic acid, were detected chromatographically in the fetus and placenta, indicating that nicotinamide or NAD^+ acts directly on the fetus to suppress urethane-induced malformations (Gotoh et al., 1988).

8.3.2 Recreational Drugs

8.3.2.1 Cigarette Smoke

Clinical Observations Since the late 1950s, it has been recognized that tobacco smoke adversely affects pregnancy, with low birth weight being a frequent feature

(Simpson, 1957). In the Cardiff Births Survey of over 18,000 births, an excess of oral clefts was observed in the babies of mothers who smoked during pregnancy (Andrews and McGarry, 1972). Other studies since that time have shown an increased risk of orofacial clefts with tobacco smoke exposure during pregnancy (Saxen, 1974; Kelsey et al., 1978; Ericson et al., 1979; Khoury et al., 1989; Lorente et al., 2000a; Chung et al., 2000), although some found no such correlation (Malloy et al., 1989; Werler et al., 1990; Lieff et al., 1999). Two meta-analyses conclude that there is a small but significant increase in risk for both CL/P and CP with maternal smoking (Wyszynski et al., 1997; Maestri et al., 1997). These findings support the hypothesis that cigarette smoking, with the attendant hypoxia, may be a contributing environmental factor to genetic predisposition (Johnston and Bronsky, 1995). Recently, it has been reported that maternal smoking is associated with an increased risk of orofacial clefting more often in male fetuses than in females (Romitti et al., 1999), but this needs to be confirmed.

The dose-response effect of smoking and the interaction between tobacco smoke and specific genotypes for orofacial clefting have been studied in case-control studies (Hwang et al., 1995; Shaw et al., 1996). Mothers who smoked more than 10 cigarettes a day were found to have up to an eight-fold increase in risk of having offspring with nonsyndromic orofacial clefts when the infant had a rare allele of the transforming growth factor alpha gene (TGF-α) (Hwang et al., 1995). TGF-α has been reported to be associated with CL/P in other studies (Mitchell, 1997; reviewed in Schutte and Murray, 1999). In the study by Shaw and associates (1996), infants who were homozygous for a rare TGF-α allele and whose mothers smoked more than 20 cigarettes a day had an increased risk of clefting. A study by Romitti and co-workers (1999) did not confirm the asso-

ciation between the rare allele of TGF-α and clefts. Similarly, Beaty and colleagues (1997) and Christensen and colleagues (1999) were not able to find an association between TGF-α, maternal smoking, and orofacial clefts. Studies by Romitti and others (Maestri et al., 1997; Lidral et al., 1998) have reported an association between allelic variants of several genes (transforming growth factor beta 3 [TGFB3] and MSX1) and cleft lip/palate. Development of CL/P and CP may be influenced independently by maternal exposures and also by interaction of such exposures and specific genes such as allelic variants of MSX1 and TGFB3 (Lidral et al., 1998; Romitti et al., 1999).

In summary, maternal cigarette smoke exposure appears to be associated with a modest increase in risk of nonsyndromic orofacial clefting in the offspring. This effect may be mediated directly as well as by interaction with genes involved in palatal and lip closure.

Animal Data   Intermittent exposure of rats to sidestream cigarette smoke at concentrations several-fold greater than those encountered in smoky public indoor environments caused intrauterine growth retardation, but no detectable increase in craniofacial anomalies was found (Rajini et al., 1994). Subchronic administration of an aqueous extract of smokeless tobacco was associated with dose-related growth retardation and a significant decrease in ossification, but no craniofacial anomalies were reported (Paulson et al., 1991). In utero nicotine leads to growth retardation and impairs development of the nervous system, leading to gross and cellular dysmorphology, particularly of the forebrain, as well as the branchial arches, possibly leading to microcephaly and cleft palate, respectively, in term fetuses (Joschko et al., 1991).

Several mechanisms have been proposed to explain the pathogenesis of clefting in offspring exposed to cigarette smoke. In mouse models, smoking causes oxygen deficit by carbon monoxide interference with hemoglobin saturation (Bailey et al., 1995) and disruption of the electron transport chain. Such maternal hypoxia is believed to induce fetal malformations (Johnston and Bronsky, 1991). While the carcinogenic effects of tobacco products are well studied, the teratogenic potential of many tobacco metabolites are unknown. Cadmium, a component of cigarette smoke, produces cleft palate in hamsters (Ferm, 1971) and rats (Chernoff, 1973). Mutations in TGFB3 (which is thought to be protective against cadmium toxicity) may cause disruption of the mesenchymal shelf between the oral and nasal cavities, resulting in clefting (Romitti et al., 1999). Interestingly, in the murine model, TGF-α and TGFB3 are expressed in high levels in the medial edge of the palatal shelves at the time of fusion (Dixon and Feguson, 1992; Brunet et al., 1995). The finding that TGFB3 null mice have isolated cleft palates (Kaartinen et al., 1995; Proetzel et al., 1995) strengthens the hypothesis that this gene is causally related to orofacial clefts. MSX1 null mice have other craniofacial anomalies in addition to cleft palate, and it is thought that this gene is responsible for neural crest migration, which is essential for palatal fusion.

A number of carcinogens also have teratogenic effects, although the molecular mechanisms of these two processes may vary. This complicates studies trying to elucidate the etiology of teratogen-mediated effects. When administered to pregnant CD-1 mice during organogenesis, the carcinogen 4-(methylnitrosamino)-1-(3-pyridyl)-1-butanone (NNK) induces a low level of cleft palate not observed in controls. Pretreatment with the P450 inducer phenobarbital increases the number of fetuses with a cleft palate, although phenobarbital controls show no anomalies. The molecular mechanism of NNK teratogenesis and embryotoxicity appears to differ from that for its carcinogenicity (Winn et al., 1998).

8.3.2.2 Alcohol

Clinical Observations The adverse effects of alcohol on the developing fetus were first reported by Jones and colleagues in 1973 (see Jones, 1986). Fetal alcohol syndrome (FAS), seen in over half of babies whose mothers consume large amounts of alcohol (>3 ounces of absolute alcohol daily) during pregnancy, is now well recognized as consisting of growth retardation, cognitive deficits, and a characteristic facial appearance and is one of the most common causes of mental retardation in Western societies (Warner and Rosett, 1975; Jones, 1986; Streissguth, 1997). Orofacial clefts have been described in children with FAS (10%). It has been postulated that alcohol exerts its effects by interfering with cranial neural crest activities, perhaps causing cell death (Lammer et al., 1985; Sulik et al., 1988).

Apart from this syndromic association with FAS, alcohol consumption may be linked specifically with the occurrence of nonsyndromic orofacial clefts. Several studies have addressed this issue. Werler and associates (1991) observed a three-fold increase in cleft palate among mothers who consumed five or more alcoholic drinks a day. A three-fold increase in cleft lip and palate with a maternal daily drinking equivalent of 500 milliliters of wine was reported by Laumon and co-workers 1996). Shaw and Lammer (1999) found that there was no increased risk for orofacial clefts with low quantities of maternal alcohol consumption; however, the risk increased with consumption of five or more drinks per day. None of the infants had other features of FAS and thus the authors suggested that the risk of orofacial clefts might be increased in the absence of FAS. Lesser amounts of alcohol in other studies

(Khoury et al., 1989; Werler et al., 1991) have also not been found to be associated with an increased incidence of clefting, although one study did suggest an increased risk with smaller amounts of alcohol consumption (Munger et al., 1996). Recently, Lorente and colleagues (2000a) reported an increased risk of cleft palate, with no dose-effect relationship.

The role of the maternal genotype in modifying the susceptibility of the fetus to alcohol was addressed by Romitti and colleagues (1999), who found that among women who reported consuming more than four drinks per month during pregnancy, the risk for clefts was higher in infants who had a specific allelic variant in the MSX1 gene. This underscores the importance of the influence that the genotype has on the phenotypic effects of an environmental exposure.

In summary, it is well known that babies who are exposed to heavy alcohol consumption during pregnancy may have orofacial clefts as part of fetal alcohol syndrome. Although there is emerging evidence that large quantities of alcohol consumption may increase the risks of nonsyndromic clefts, there are several confounding factors that need to be clarified before a definite association can be made. Most of the studies are retrospective, and the frequency of self-reported alcohol consumption may be inaccurate. Many women who report alcohol consumption also report use of other substances such as tobacco, marijuana, and cocaine, which can also be teratogenic. Finally, infants with FAS may be missed and reported as having isolated clefting, thus erroneously establishing an association between alcohol exposure and nonsyndromic clefting.

Animal Data Animal studies indicate that alcohol significantly increases intrauterine growth retardation and induces craniofacial anomalies including arched palate and cleft palate (Sulik et al., 1986; Webster and Ritchie, 1991; Padmanabhan and Pallot, 1995). While human studies suggest a possible link between certain genotypes and alcohol-induced clefting propensity, animal studies clearly demonstrate strain-specific differences in susceptibility to the teratogenic effects of alcohol, indicating susceptibility is in part genetically determined (Gilliam et al., 1988). Furthermore, the spectrum of teratogenic effects depends on the dose, the timing (i.e., stage of embryonic development), and the number of exposures (Bolon et al., 1993). Additionally, the severity and frequency of several craniofacial fetal abnormalities are compounded when other environmental agents such as lithium, aspirin, retinyl acetate, or methylmercury are administered simultaneously with alcohol, underscoring the potential etiologic heterogeneity that may contribute to the threshold effect proposed by Fraser (Fraser, 1976; Lee, 1985; Sharma and Rawat, 1986; Padmanabhan and Pallot, 1995).

8.3.2.3 Miscellaneous Recreational Drugs in Animal Studies

Exposure to delta-9-tetrahydrocannabinol (THC) significantly reduces the body weight of surviving fetuses and is associated with an increase in cleft palate. THC-induced cleft palate is potentiated by SKF-525A (a general cytochrome P450 inhibitor) and phenobarbital treatment. While caffeine treatment has been associated with reduced fetal weight and a low incidence of cleft palate, studies in human populations do not support a link between caffeine and negative birth outcome (Elmazar et al., 1982). Caffeine has been shown to enhance the teratogenic effect for a variety of pharmaceutical agents including acetazolamide, mitomycin C, hydroxyurea, and 5-fluorouracil. Extrapolation of these findings in terms of human hazard is complicated by the general use of high-dose bolus exposures, which are not typical of human exposures, and the use of test systems that are not readily applicable to humans (Sivak, 1994; Elmazar et al., 1982).

8.3.3 Toxins

8.3.3.1 Organic Solvents/ Industrial Chemicals

Clinical Observations This chemically heterogeneous group of compounds is used in both domestic and occupational settings. These compounds are found in paints, varnishes, inks, cosmetics, cleaning agents (especially window cleaners), and products used by mechanics and metallurgists (Cicolella, 1992). Of the different types of solvents, those that are fat soluble may be more teratogenic, since they pass easily through biological membranes, such as the placenta (Wilson, 1977). In 1982, Holmberg and co-workers studied 388 mothers of children with oral clefts and 388 controls, and found that exposure to organic solvents in the first trimester was more common in the case mothers ($n = 14$) as compared to the control mothers ($n = 4$; $p < 0.05$). Lacquer petrol was the most common solvent, although xylene, acetone, and toluene were also recorded.

Cordier and associates (1992) interviewed the mothers of 325 infants with major malformations and compared them to 325 mothers of normal children, regarding exposure to solvents, detergents, disinfectants, work in a hospital, and work as a janitor. Mothers of children with oral clefts were seven to nine times more likely to have been exposed to organic solvents at the beginning of the pregnancy, after adjusting for socioeconomic status, age, and residential area. Other exposures, such as to detergents, were not statistically significant.

In a case-control study, Laumon and colleagues (1996) compared maternal exposure to an organic solvent in 200 mothers who had a baby with CL/P and 400 con-

trols, and found an odds ratio of 1.62 (95% CI = 1.14–2.52). In comparing subgroups of organic solvents, significance was found only with exposure to halogenated aliphatic solvents (odds ratio = 4.40, 95% CI = 1.41–16.51).

Cordier and co-workers (1997) studied the effect of glycol ether exposure on the occurrence of congenital anomalies in 984 cases and 1134 control subjects matched for place and date of birth. The malformations were classified into 22 subgroups. The association of cleft lip with glycol ether exposure was particularly strong (odds ratio 2.03, 95% CI = 1.11–3.73). The risk tended to increase with level of exposure.

In a recent study by Lorente (2000b), 100 mothers of infants with oral clefts were matched with controls who had healthy children ($n = 761$). Occupational exposures during the first trimester were recorded, and confounding variables such as maternal age and socioeconomic status were controlled for. Exposure to glycol ethers and aliphatic aldehydes was associated with an increased risk of CL/P (odds ratios 1.7 and 2.1, 95% CI = 0.9–3.3 and 0.8–5.9). Lorente cautioned that the study needed to be interpreted guardedly because of the small number of subjects.

Apart from orofacial clefts, organic solvents have been associated with miscarriages, neural tube defects, and hydrocephalus (McMartin et al., 1998). Khattak and colleagues (1999) prospectively followed 125 pregnancies exposed in the first trimester to various solvents and found a relative risk of 13.0 (95% CI = 1.8–99.5) for a major congenital defect, which occurred in women who were symptomatic from their exposure. However, in response to this study, the Public Affairs Committee of the Teratology Society felt that specific conclusions regarding the relationship between organic solvent exposure and reproductive risk could not be conclusively established until the specific exposure and maternal symptoms were correlated with developmental effects (Brent et al., 1999). Limiting factors in the study, such as a low rate of malformations in the control group, the grouping together of diverse anomalies (such as deformations and malformations), the inclusion of premature infants with full-term infants, and not controlling for parity and gender, were cited as factors that must be considered when interpreting the results of this important study. Nonetheless, an increase in major malformations in women exposed to organic solvents was observed.

In summary, exposure to organic solvents during pregnancy may be associated with an increased risk of CL/P. As with other environmental factors, timing and level of exposure, the "lumping" and "splitting" of individual solvents, and identification of anomalies at birth (reflecting prevalence rather than incidence) can all influence the findings in these epidemiological studies.

Animal Data Exposure to many industrial solvents in animal models is associated with a reduction in mean fetal body weight and craniofacial malformations including clefting. For example, delayed ossification and cleft palate have been demonstrated in CD-1 mice exposed to high levels of the solvent naphtha (McKee et al., 1990). The teratogenic effect of many solvents varies with the genotype of animal (strain) and with the dose, time of exposure, and xenobiotic metabolism of the agent (Marks et al., 1982; Hood and Ottley, 1985; Thomas et al., 1987; Couture et al., 1989; McKee et al., 1990; Schwetz et al., 1991; Ema et al., 1996b). Mice and rats exposed to the solvent mercaptan show differential responses for doses and times of exposure, with mice showing an increased incidence for total fetal abnormalities including cleft palate that was not observed in rats (Thomas et al., 1987).

Developmental toxicity of di-n-butyl phthalate (DMP) following administration during late pregnancy in rats includes reduced fetal weight and increased incidence of cleft palate (Ema et al., 1998). The DMP metabolites mono-n-butyl phthalate (MbeP) and MBP glucuronide are rapidly transferred to embryonic tissues, where their levels remain lower than those in maternal plasma. MBP, itself a potent teratogen, is likely to largely contribute to the embryotoxic effects of DBP. These studies also indicate that the susceptibility and spectrum of the developmental toxicity of MbeP vary with developmental stages at the time of administration (Ema et al., 1998).

Humans have exposure to several widely used industrial solvents such as methyl ethyl ketone (MEK) and methanol. Offspring of pregnant mice exposed to MEK show a dose-related reduction of mean fetal body weight, skeletal malformations, and cleft palate (Schwetz et al., 1991). At high levels, maternal inhalation of the alternative motor vehicle fuel methanol in a concentration-dependent manner induces developmental toxicity including clefting in pregnant CD-1 mice (Bolon et al., 1993; Rogers et al., 1993).

Perfluorodecanoic acid (PFDA) is a representative of the perfluorinated carboxylic acids used as commercial wetting agents and flame retardants. Signs of PFDA toxicity have been reported to resemble those seen after exposure to the dioxin TCDD; however, this may only occur at doses that are maternally toxic, and implications for human teratogenesis are unclear (Harris and Birnbaum, 1989).

8.3.3.2 Agricultural Chemicals

Clinical Observations Exposure to pesticides is common in many agricultural occupations. Common pesticides include herbicides such as 2,4,5-trichlorophenoxy-acetic acid (2,4,5-T), insecticides such as malathion, and fertilizers. Early studies reported an increased rate of congenital anomalies, including cleft palate, with the use of the herbicide 2,3,5-T (Field and Kerr, 1979; Nelson et al., 1979), but others do not support these findings (Thomas et al., 1987; Hanify et al., 1981). Some studies of agricultural pesticides found no evidence of an increase of orofacial clefts in exposed infants (McDonald et al., 1987; White et al., 1988). However, in a case-control study Gordon and Shy (1981) found a statistically significant association between oral clefts and exposure to all agricultural chemical agents combined (fertilizers, herbicides, and insecticides), as well as for selected chemicals (insecticides and herbicides). Similarly, an increase of all birth defects was observed in an Iowa population of 865 municipalities where the water level of the pesticide atrazine was elevated (Munger et al., 1992).

In another study, a cohort of over 7000 pregnancies that were exposed to the pesticide malathion was followed. Of these, 933 who had an adverse outcome and a control group of 1000 women were questioned further. Individual exposure indices for each week of pregnancy were created. For oral clefts after adjusting for confounding factors such as cigarette smoking, there was an increased risk (relative risk 3.35, 95% CI = 0.61–18.5) (Thomas et al., 1992).

In a study of maternal agricultural work, pesticide exposure, and the risk of birth defects in 1306 pairs of infants (581 oral clefts) Nurminen and associates (1995) found an increased risk (odds ratio 1.9, 95% CI = 1.1–3.5) for oral clefts with agricultural work. However, when agricultural work with pesticide exposure was compared with agricultural work without pesticide exposure, there was no increased risk of oral clefts. Garcia and co-workers (1999) analyzed occupational exposure to pesticides and the occurrence of congenital anomalies. Maternal exposure during the month before pregnancy and the first trimester correlated with an increased risk of congenital anomalies, including oral clefts (odds ratio 3.6, 95% CI = 1.11–9.01).

In conclusion, the published studies on agricultural chemical exposure and oral clefts have given some indication of a potential increased risk for orofacial clefts, but the epidemiological evidence does not allow for clear inferences to be made. These epidemiological studies are limited by bias, confounding factors, and the variable outcomes that can occur depending on the time and level of exposure.

Animal Data A number of chemicals, toxins, and pesticides commonly used in many agricultural occupations have been studied in animal models. The pesticide 2,3,7,8-tetrachlorodibenzo-p-dioxin (TCDD) is teratogenic in mice, inducing cleft palate and hydronephrosis at doses that are not overtly maternally toxic or embryotoxic. A number of TCDD congeners have been implicated as teratogenic agents capable of inducing cleft palate in animal models, with evidence for additive toxicity (e.g., with retinoic acid) and for teratogens exerting a direct effect on embryonic palatal tissue and on development of the secondary palate (Hassoun et al., 1984; Weber et al., 1985; Weber and Birnbaum, 1985; Birnbaum et al., 1989).

Several potential models of dioxin pathogenesis have been proposed. The aryl hydrocarbon receptor (AhR or dioxin receptor) is a ligand-activated transcription factor that is considered to mediate pleiotropic biological responses such as teratogenesis. Animal studies suggest AhR mediates TCDD-induced malformation of the palate and kidney in mouse embryos, although the mechanisms of its involvement differ for cleft palate and hydronephrosis. Mice lacking the Ah (dioxin) receptor show a loss of teratogenic response to TCDD (Mimura et al., 1997). Glucocorticoids and TCDD both readily cross the yolk sac and placenta, and appear in the developing secondary palate. Receptors for glucocorticoids and TCDD are present in the palate, and their levels in various mouse strains are highly correlated with their sensitivity to cleft palate induction. Whereas glucocorticoids may exert their teratogenic effect on the palate by inhibiting the growth of the palatal mesenchymal cells, TCDD may alter the terminal cell differentiation of the medial palatal epithelial cells (Pratt, 1985).

Hexabrominated naphthalenes (HBNs) are potent fetotoxic and teratogenic agents, producing similar effects to TCDD and other toxic polyhalogenated aromatic hydrocarbons such as polyhalogenated biphenyls, dibenzofurans, and dibenzo-p-dioxins. Polyhalogenated naphthalenes, polyhalogenated biphenyls, dibenzofurans, and dibenzo-p-dioxins have certain structural similarities, and studies in laboratory animals report similarities in their capacity to elicit certain toxicological responses (Miller and Birnbaum, 1986). Although specific animal studies can identify teratogenic potency, interpretation of findings and extrapolation to other models can be difficult. For example, while the teratogenic potency of 2,3,4,7,8-pentachlorodibenzofuran and of mixtures of polychlorinated dibenzo-p-dioxins and dibenzofurans can be demonstrated in mice, there are difficulties with risk assessment using TCDD toxic equivalency factors (Nagao et al., 1993). The teratogenic effects of metabolites of some known teratogens are actually greater than those of the parent compound. Differences in how a teratogen is metabolized and excreted have implications for differential in vivo toxicity. Animal studies indicate that differences between TCB and TCDD can be explained by a more rapid metabolism and excretion of TCB (d'Argy et al., 1987).

The propensity for rodenticides such as 6-aminonicotinamide to induce cleft lip and cleft palate in mice has been demonstrated but varies greatly depending on the genetic background of the target animal (Trasler and Ohannessian, 1983). Strain differences in susceptibility to teratogen-induced cleft palate by several different teratogens (e.g., to 6-aminonicotinamide and to cortisone) in the mouse support an important role for genetic modulation of susceptibility. Additionally, genetic susceptibility for certain teratogens appears to be under the influence of different genetic loci, suggesting that specific gene–environment associations underlie teratogen sensitivity (Biddle and Fraser, 1979).

8.3.3.3 Miscellaneous Toxin Animal Studies
In addition to agricultural and industrial toxins, a number of biocides, antifouling agents, antimicrobials, fungicides, and naturally occurring plant toxins have teratogenic properties. A number of naturally occurring plant and microbial agents have teratogenic potential. The plant toxin nigericin is associated with reduced fetal growth and increased prenatal death. Additional signs of toxicity to the conceptus include treatment-related extra ribs and delayed ossification. Exposures that result in gross and skeletal malformations, such as median facial cleft, are associated with observable maternal toxicity (Vedel-Macrander and Hood, 1986).

Many species of lupines contain quinolizidine or piperidine alkaloids known to be toxic or teratogenic to livestock, and a number of naturally occurring alkaloid-containing plants such as *Conium maculatum* (poison hemlock), *Nicotiana glauca* (tree tobacco), and *Lupinus formosus* (lunara lupine) can act as teratogens, causing craniofacial anomalies including clefting. Results of animal studies show a species-dependent differential susceptibility (Keeler, 1975; Omnell et al., 1990; Panter and Keeler, 1993; Shibata, 1993; Panter et al., 2000).

The molecular targets of some plant teratogens have been identified. Piperidine selectively induces skeletal malformations characterized by cleft palate in rat fetuses. Rat studies suggest piperidine is a strong inducer of the xenobiotic metabolizing enzyme cytochrome P-450, with characteristics resembling those of phenobarbital (Numazawa et al., 2000). Sonic hedgehog (Shh) plays an important role in morphogenesis of the craniofacial complex. Mutations in human and murine Shh cause midline patterning defects that are manifested in the head as holoprosencephaly and cyclopia. Plant teratogens such as jervine, which inhibit the response of tissues to Shh, also produce cyclopia (Hu and Helms, 1999).

In rat studies, fluoroquinolone antibacterials have been associated with a reduction in fetal weight, and cleft palate (Kim et al., 2000). Prenatal exposure to the fungicides (dinocap, thiabendazole) causes decreased fetal weight and cleft palate in mice, but the teratogenic effect for some fungicide compounds varies considerably between species (Gray et al., 1986). The dose needed to cause clefting is much lower than the dose needed to cause maternal toxicity in mice (Rogers et al., 1993). Chemicals that are teratogenic at exposures that are not maternally toxic are a particular risk because avoidance may be less likely. Benomyl, a benzimidazole fungicide, coupled with a protein-deficient diet, offers a teratogenic model with a spectrum of abnormalities similar to hypervitaminosis A but with a higher yield of specific craniocerebral anomalies (Ellis et al., 1987).

Citrinin, a fungal metabolite produced by several species of Penicillium and Aspergillus, has been found to contaminate foods used by animals and humans. When administered to rats during gestation, citrinin induces reduced fetal weight and an increased incidence of cleft palate (Reddy et al., 1982).

Teratogenicity studies of tri-n-butyltin acetate in rats show a dose-related induction of reduced fetal weight and craniofacial malformations including cleft palate (Noda et al., 1993). There is also a difference in the teratogenic potency and manifestation of malformations induced by different butyltin compounds, with the di-n-butyl group likely to be an important determinant of teratogeneity (Noda et al., 1993; Ema et al., 1998, 1996b).

Several forms of urea are teratogenic in animal models, but the prevalence of craniofacial malformations including clefting varies with the agent and the specific animal evaluated (Khera and Iverson, 1978; DePass and Weaver, 1982; Daston et al., 1987). Cleft palate is one of the most frequent malformations associated with induction of fetal malformations after treatment with methylnitrosourea or ethylnitrosourea, and a number of different mechanisms have been proposed (Daston et al., 1987; Ohnishi, 1989). While most animal studies of teratogen-induced craniofacial anomalies evaluate progeny of pregnant animals exposed to teratogens, evidence of male-mediated teratogenesis has been demonstrated in mice. The most common malformation to occur in male mice treated with ethylnitrosourea was cleft palate or cleft lip. Based on these results and other data, it has been proposed that a large fraction of spontaneously occurring malformations in fetuses may be due to mutagenized paternal germ cells (Nagao and Fujikawa, 1996).

8.3.4 Hyperthermia

8.3.4.1 Clinical Observations Elevation of maternal body temperature from sauna, infection, or hot tub bathing has been linked to fetal anomalies (Clarren et al., 1979; Pleet et al., 1981). The most vulnerable organ appears to be the central nervous system. Maternal hyperthermia was found in 11% of pregnancies with anencephaly and 7% of those with spina bifida in the above studies. An increased incidence of congenital anomalies, especially neural tube defects, has been reported in cases of maternal hyperthermia due to viral infections (Coffey and Jessup, 1963;

Kurppa et al., 1991). The facial changes seen when hyperthermia occurred prior to 7 weeks include cleft lip and palate. Zhang and Cai (1993) studied the association between common cold with or without fever in the first trimester and the occurrence of oral clefts. The odds ratios for cold with fever were elevated for both CL/P and CP. In a later study, mothers of children with unilateral cleft lip, bilateral cleft lip with or without cleft palate, and cleft palate only were asked about hyperthermia during early pregnancy. A high proportion reported febrile illnesses (24–33%) in all three groups (Peterka et al., 1994). Other studies have not found an association between hyperthermia due to maternal sauna exposure, electric bed heating, and congenital abnormalities (Saxen et al., 1982; Dlugosz et al., 1992).

8.3.4.2 Animal Data

Animal studies have found hyperthermia to be teratogenic, with oral clefts being one of the anomalies observed (Edwards, 1974; Germain et al., 1985). A rise in body temperature of 2.5°C for 1 hour was the threshold temperature that caused birth defects in both rats and guinea pigs (Germain et al., 1985). With higher temperatures, the length of time required to produce abnormalities was less. Increased body temperature was found to kill cells and inhibit mitosis in proliferating cells (Edwards, 1974). In summary, elevated maternal body temperature may pose a potential risk for fetal anomalies. The higher the body temperature and the longer the duration, the higher the risk appears to be.

8.4 MATERNAL FACTORS

8.4.1 Vitamin Deficiency

8.4.1.1 Clinical Observations

The role of nutrition in orofacial clefting is supported by the indirect evidence of increased clefting in geographic areas of poor socioeconomic status where nutrition is less than optimal (Cembrano et al., 1995; Murray et al., 1997) and lower incidence in areas of higher socioeconomic status (Chung et al., 1987; Cembrano et al., 1995; Croen et al., 1998).

Human studies have implicated vitamin A deficiency as one of the nutritional deficiencies that could be related to orofacial clefts. However, most studies have addressed the role of multivitamins (including vitamin A) in the causation of orofacial clefts (Conway, 1958; Tolarova and Harris, 1995), and there is a lack of information regarding the role of vitamin A deficiency by itself in orofacial clefts (Natsume et al., 1995). In 1958, it was reported that the administration of vitamins A, B, C, and D to pregnant women who had had a child with an oral cleft decreased the recurrence of clefting, compared to controls (Conway, 1958). Consumption of beta-carotene in the diet has been found to be associated with lower risk of CL/P (Natsume et al., 1995). Additionally, the occurrence of cleft lip and palate in humans with malabsorption syndrome, which is thought to be caused by hypovitaminosis A associated with folic acid and vitamin B_2 deficiencies, suggests dietary deficiencies may be an etiologic factor in some craniofacial clefting (Faron et al., 2001).

Subsequent studies reported that folic acid in particular was etiologically related to oral clefting, since women who took folate antagonists such as methotrexate, aminopterin, and anticonvulsants had offspring with orofacial clefts (Milunsky et al., 1968; Czeizel, 1976; Ardinger et al., 1988). In addition, mothers who consume supplemental multivitamins, including folic acid during pregnancy, have been found to have a reduced risk of congenital anomalies (including orofacial clefts) in their offspring (Tolarova and Harris, 1995; Shaw et al., 1995; Czeizel and Hirschberg, 1997; Itikala et al., 2001). These studies found a risk reduction of between 30 and 60% for CL/P and a much lower degree of risk reduction for CP, probably reflecting the fact that CP occurs more frequently with genetic abnormalities than CL/P. However, Werler and colleagues (1999) found a greater reduction in risk for CP (60%) as compared to CL/P (30%).

Contrary to these studies, others have shown no protective effect from clefting with periconceptual folic acid therapy (Bower and Stanley, 1992; Czeizel, 1993; Hayes et al., 1996). Attempts have been made to correlate the dose with the protective effects, but the results have been contradictory (Shaw et al., 1995; Czeizel et al., 1999).

Shaw and associates (1998a) studied the correlation between infant TGA-α phenotype, maternal multivitamin use (including folic acid), and risk of orofacial clefts. Infants who were homozygous for the variant allele and whose mothers did not take periconceptual vitamins had a higher risk of clefting (odds ratio ranged from 2.4 to 8.1, 95% CI = 0.69–27.7), with the lowest risk being for isolated CL/P and the highest for syndromic clefts. Homozygous infants whose mothers took folic acid had a smaller risk. Thus, they concluded that this was evidence of gene–nutrient (environment) interaction. Infants who are homozygous for the TGF-α variant may be more sensitive to the effects of cigarette smoke, which has been shown to decrease folate level in the blood (Shaw et al., 1996). Again, as with other association studies, there is evidence to the contrary. Christensen and colleagues (1999) found no increased risk for clefting in mothers who smoked and whose infants were homozygous for the variant TGF-α allele, although smoking by itself was associated with an increased risk. Another gene–environment interaction was examined by Munger and co-workers (1995) in a study in which infants who were homozygous for a mutation within the MSX1 gene and whose

mothers did not take vitamin supplementation were at increased risk for cleft palate.

A gene mutation within the methylenetetrahydofolate reductase gene (MTHFR) that causes a thermolabile variant of the gene (C677T) has also been studied as a risk factor for orofacial clefting. MTHFR plays a role in folate metabolism by catalyzing the reduction of 5,10-methylenetetrahydrofolate to 5-methylenetetrahydrofolate, which is the methyl donor for methionine synthesis from homocysteine. Homozygosity for the common C677T mutation is associated with reduced activity of the enzyme, resulting in mild to moderate elevation of homocysteine levels (Frosst et al., 1995).

Folate levels in red blood cells of mothers who had a child with an orofacial cleft were found by Munger and co-workers (1996) to be higher than in control mothers. This was thought to be due to abnormal folate metabolism in the mothers. Wong and colleagues (1999) found hyperhomocysteinemia in mothers who had children with nonsyndromic orofacial clefts, as compared to control women. A defect in methylation—the mechanism by which folate transfers methyl groups in more than 100 different reactions—was suggested. Mills and colleagues (1999) found an increased risk for cleft palate in infants homozygous for the MTHFR mutation in an Irish population and suggested that this could be an etiological link.

Contrary to these studies, Shaw and associates (1998b) did not find an increased incidence of orofacial clefting in infants who were homozygous for the C677T mutation. They also did not find evidence that there was an interaction between the MTHFR genotype and maternal use of multivitamins including folic acid. They concluded that the reduction in orofacial clefts with maternal folic acid use could be due to interactions with other folate-related genes. Wyszynski and Diehl (2000) reanalyzed the data from the study by Shaw and associates (1998b) and reported that maternal nonuse of vitamins was associated with an increased risk of CL/P (odds ratio = 4.0, 95% CI = 1.15–3.40), although the increased risk was not associated with the C677T genotype.

In summary, the current evidence indicates that the causative role of multivitamin (including folic acid) deficiencies in orofacial clefting is controversial. It is not clear if the protective effect shown in some studies is due to folic acid or other vitamins, if the dosage regimen would influence the protective effect, or if an underlying genetic factor(s) could modify the effects of vitamin supplementation.

8.4.1.2 Animal Data Several animal studies early in the 20th century showed that dietary deficiencies might be causal factors in orofacial clefting (Strauss, 1914; Hale, 1937). Mammalian embryos are protected against vitamin A (retinol) deficiency by maternal retinoid homeostasis until stored retinoids fall to very low levels. While severe congenital vitamin A deficiency is associated with several systemic malformations, it is only in situations of extreme congenital vitamin A deficiency that craniofacial clefting occurs, and the embryos are not viable (Wilson, 1977; Morriss-Kay and Sokolova, 1996).

Gestational folate deficiency has been associated with abnormal growth and development in both experimental animal and human studies, and has been postulated as a putative mechanism for the teratogenic effects of antiepileptic drugs (AEDs). Animal studies have shown that the administration of AEDs results in folate depletion and teratogenic effects (Dansky et al., 1992). Levels of maternal dietary folate intake have a direct effect on fetal folate stores in the mouse. Marginal dietary maternal folate intake is associated with a significant increase in cleft palate in offspring in CD-1 mice. Coadministration

of methanol with marginal dietary folate to dams dramatically increased cleft palate occurrence in offspring. Methanol is detoxified in part by a folate-dependent pathway, which may explain this synergistic effect, and has potential prevention implications (Sakanashi et al., 1996; Bianchi et al., 2000).

8.4.2 Maternal Hormone Influences

It has been observed that some congenital anomalies occur more frequently in one sex as compared to the other (Arena and Smith, 1978; Lubinsky, 1997). In mammalians (including humans), parental hormone levels may help determine offspring gender. Based on these two observations, it has been hypothesized that maternal hormone levels may be causatively linked to orofacial clefts (James, 2000). Mothers of children with CL/P were found to have a tendency to produce male offspring, and mothers of children with CP tended to have female children (James, 2000). The author suggested that these results be corroborated and then the hormonal effects of various teratogens involved in orofacial clefts (tobacco and alcohol) be examined. As of now, the evidence does not permit us to make any conclusions.

8.5 INTRAUTERINE FACTORS

8.5.1 Amniotic Bands

8.5.1.1 Clinical Observations Amniotic band sequence (ABS) encompasses a wide variety of congenital anomalies due to disruptions and deformations of the developing fetus by fibrous bands. About one-third have facial and oral clefts (Jones et al., 1974; Keller et al., 1978; Bagatin et al., 1997; Jabor and Cronin, 2000). ABS, also known as amniotic band syndrome, aberrant tissue band syndrome, or ADAM (amniotic deformity, adhesions, and mutilations) is a dramatic yet unusual condition, with a frequency of 1/1200 to 1/15,000 in live births and a much higher rate of occurrence in stillbirths (7–14%) (Seidman et al., 1989; Froster and Baird, 1993). Other associated anomalies include limb reduction defects, amputations, constriction bands, craniofacial disruptions such as encephalocele, anencephaly, and clefts, limb body wall deficiency, extrathoracic heart, and gastroschisis (Gorlin et al., 1990). In almost all instances, ABS is sporadic, with a low recurrence risk, although there are a few instances of familial recurrence (Lubinsky, 1983).

The prevailing hypothesis (exogenous or extrinsic theory) regarding the pathogenesis of ABS was put forward by Torpin in 1965. This theory is based on the premise that there is no inherent abnormality in the developing fetus. There is spontaneous rupture of the amnion (due to vascular insufficiency) around the fifth to sixth gestational week, resulting in the formation of numerous fibrous bands which then entangle the fetus. The bands interfere with normal development by constriction, disruption, and deformation. The developing nasal process may serve as an anchoring point for the bands, accounting for the high incidence of facial clefts and other craniofacial disruptions (Eppley et al., 1998). Swallowing of the bands by the fetus may result in interference with palatal closure. The extent of the defects is dependent on the timing of the amniotic band interference. When entanglement occurs prior to 45 days gestation, severe clefts of the face and defects of the brain and the calvarium are seen; an insult later on is hypothesized to result in more limb involvement and milder craniofacial involvement (Higginbottom et al., 1979).

Alternative hypotheses for the pathogenesis of ABS have been proposed. Streeter, in 1930, postulated that ABS in due to an inherent developmental defect occurring at the time of embryonic disk and amniotic cavity development (intrinsic or endogenous theory) (Bamforth, 1992). Incomplete obliteration of the extraembryonic coelem, leading to lack of support for the amniotic membranes and resulting in partial rupture, band formation, and oligohydramnios, was postulated by Levy (1998) It has also been suggested that the bands are not the cause of the abnormalities but are just a secondary effect due to scarring (Granick et al., 1987).

The pattern of craniofacial anomalies in ABS seems to support Torpin's exogenous theory. The facial clefts range from midfacial clefts to bizarre clefts that do not follow normal developmental planes. These oblique and asymmetrical clefts are compatible with a disruption of the migrating neural crest cells that form the early skeletal mesenchyme of the anterior calvarium, frontonasal and maxillary processes (Higginbottom et al., 1979). Frequently, these unusual clefts are accompanied by limb defects and other craniofacial disruptions. "Common" clefts that form along the normal fusion lines are uncommon in ABS (Coady et al., 1998), although they have been reported (Eppley et al., 1998; Jabor and Cronin, 2000).

In summary, amniotic bands are a well-known environmental and etiological factor for orofacial clefts. They are sporadic in occurrence and are not associated with a specific genetic condition or with a teratogenic agent. The pattern of clefting is unusual, with bizarre clefts that do not follow normal developmental lines of lip and palate closure. They are frequently accompanied by other craniofacial disruptions and limb reduction defects.

8.5.1.2 Animal Data Animal studies have produced orofacial clefts and limb constrictions by attempting to create conditions that simulate amniotic bands. Rowsell (1988) was able to produce limb constrictions of varying severity in rats using ligatures. Facial clefts have also been produced in fetal lambs by ligatures (Stelnicki et al., 1998). Creation of oligohydramnios resulted in cleft lip/palate in rats, due to decreased amniotic fluid (which is thought to occur following spontaneous amnion rupture in ABD). The oral clefting observed was thought to be due to deformation of the lower jaw against the chest, with the tongue being displaced between the unfused palatal processes (Poswillo, 1968; Kennedy and Persand, 1977).

8.5.2 Fetal Constraint

8.5.2.1 Clinical Observations
Pierre Robin sequence (cleft palate, micrognathia, and glossoptosis) is etiologically heterogeneous (Gorlin et al., 1990). Some instances of this sequence can be caused by oligohydramnios. It is postulated that reduced amniotic fluid volume results in compression of the chin against the chest, resulting in mandibular constraint (deformation). This leads to the tongue being impacted between the palatal shelves, causing a cleft. Amniotic bands can also cause mandibular constraint, leading to poor growth of the mandible and subsequently cleft palate.

8.5.2.2 Animal Data
In a rat model, Poswillo was able to produce Pierre Robin sequence by puncturing the amniotic sac and causing oligohydramnios. Along with a cleft palate, limb anomalies were also noted, which is a common association seen in children with this sequence (Poswillo, 1968).

8.6 ENVIRONMENTAL ETIOLOGIES OF CRANIOSYNOSTOSIS

Among the etiologies of craniosynostoses, the genetic causes are best known and well understood (see Chapters 6 and 20). A significant proportion are due to mutations in single genes (Cohen and MacLean, 2000).

Very few environmental factors have been etiologically linked to craniosynostoses. Examples include rickets, hyperthyroidism, and a few teratogens, which are discussed below (see also Chapter 9).

8.6.1 Rickets

8.6.1.1 Clinical Observations
Rickets is the failure of calcification of the growing bones in a child, due to deficiency of vitamin D. Vitamin D deficiency can occur from insufficient dietary intake, resistance to vitamin D, chronic renal failure, liver disease, and hypophosphatasia. All of these forms of rickets have been linked to craniosynostosis (Coleman and Foote, 1954; Fraser, 1957; Reilly et al., 1964; McCarthy and Reid, 1980). Although some of the above causes are genetic conditions, the pathogenesis of premature fusion of the sutures in rickets is due to abnormal bone metabolism, which is a feature seen in all types of rickets. Reilly and co-workers (1964) found that one-third of 59 children with rickets had premature fusion of the sutures. The severity of the craniosynostosis was directly proportional to the severity of the rickets. The length of the period of illness with rickets was also directly correlated: The longer the period, the more severe the synostosis.

It has been suggested that treatment of rickets with high doses of vitamin D might be a precipitating factor in producing synostosis (Carlsen et al., 1984), although this theory has been disputed (Clemens, 1984). In infantile hypophosphatasia, craniosynostosis is a common finding, with ocular proptosis occurring in severe cases (Brenner et al., 1969). Even in instances when the rickets is due to a genetic disorder, the craniosynostosis is secondary to the abnormality in calcium/phosphorous metabolism. Willis and Beaty (1997) described craniosynostosis in X-linked hypophosphatemic rickets and suggested that radiographic screening be offered to all patients with this condition. Thus, rickets is causatively linked to craniosynostoses.

8.6.1.2 *Animal Data* A mouse model exists for X-linked dominant hypophosphatasia, the most common form of vitamin D–resistant rickets (Roy et al., 1981). Such a model could be useful to study the role of vitamin D–resistant rickets in the development of craniosynostosis.

8.6.2 Hyperthyroidism

8.6.2.1 *Clinical Observations* Craniosynostosis has been reported in two infants of mothers with Graves' disease (thyrotoxicosis) (Leonard et al., 1987), at ages 4 months (bicoronal synostoses) and 15 months (pansynostoses). The latter infant had transient neonatal hyperthyroidism. It was thought that the maternal thyroid excess was causative of the premature fusion of sutures.

Several other reports have described craniostenosis in children who are hyperthyroid, either due to thyrotoxicosis or due to treatment with thyroid hormone (Riggs et al., 1972; Menking et al., 1972; Hollingsworth and Mabry, 1976; Daneman and Howard, 1980). Johnsonbaugh and colleagues (1978) reported craniosynostosis on radiologic evaluation in 10 out of 10 children who had thyrotoxicosis, compared to those who had congenital virilizing adrenal hyperplasia, precocious puberty, and normal children. The synostoses involved the sagittal, lambdoid, and upper third of the coronal suture. In the children with adrenal hyperplasia and precocious puberty, although skeletal maturation was advanced, premature suture fusion was not seen. Thus, the sutures may be very sensitive to excess thyroid hormone.

8.6.2.2 *Animal Data* There is generally a paucity of animal studies that have

addressed the role of hyperthyroidism in the causation of craniosynostosis. In a rat model, the interactions between hyperthyroidism and insulin-like growth factor (IGF-1) were examined. With the administration of thyroid hormone, there was accelerated closure of the sagittal suture. This was correlated with increased staining for IGF-1 along the sutural margins, indicating that IGF-1 may play a role in sagittal suture formation (Akita et al., 1996).

8.6.3 Teratogens

8.6.3.1 *Clinical Observations* Several teratogens have been implicated in the causation of craniosynostosis. These are case reports and series, with a lack of case-control studies. Phenytoin ingestion during pregnancy has been found to result in fusion of sagittal and coronal sutures (Char et al., 1978). Treatment of pregnant women with retinoids resulted in craniosynostosis as one of the features of retinoic acid embryopathy (three out of eight cases) (Lammer et al., 1985). Valproate, an anticonvulsant that is associated with cleft lip and palate, has also been associated with metopic ridging (Ardinger et al., 1988). Infants exposed to aminopterin/methotrexate prenatally had an increased incidence of craniosynostosis (Milunsky et al., 1968). In a report by Aleck and Bartley (1997), two out of four infants exposed to the antifungal fluconazole prenatally had craniosynostosis. All four infants had a specific phenotype that resembled an autosomal-recessive condition, Antley-Bixler syndrome, in which craniosynostosis is common. Cyclophosphamide, an alkylating agent used in cancer chemotherapy, was found to have coronal craniosynostosis in two out of seven exposed infants (Mutchinick et al., 1992; Enns et al., 1999). A recognizable phenotype for cyclophosphamide exposure in utero was delineated by Enns and associates (1999). Thus, teratogens are associated

with craniosynostosis, and often the craniosynostosis is one of multiple congenital anomalies.

8.6.3.2 Animal Data While several naturally occurring genetic forms of craniosynostosis exist, few teratogen-induced models of craniosynostosis exist (Roy et al., 1981; Cantrell et al., 1987; Mooney et al., 1993; Ignelzi et al., 1995) (see also Chapter 9). To permit studies of craniosynostosis, particularly regarding treatment, a number of animal models have been developed, but these are not classic models of teratogen-induced malformation. In the rabbit model, premature fusion of the sutures can be achieved by suture immobilization with topical application of methylcyanoacrylate adhesive and by osteoinduction using demineralized bone matrix (Persing et al., 1986; Duncan et al., 1992; Antikainen et al., 1992). Craniosynostosis can be induced in lambs by excising the coronal suture and promoting osteoinduction with demineralized bone augmented with bone morphogenic protein-2 and TGF-β (Stelnicki et al., 1998). Treatment of excised rat fetal calvaria with prooxidant produced suture fusion, while treatment with an iron chelator and antioxidant (deferoxamine) prevented suture fusion (Im et al., 1997).

8.6.4 Fetal Constraint

8.6.4.1 Clinical Observations Fetal head constraint due to prenatal factors such as abnormal positioning in utero, overcrowding due to multiple pregnancy, or oligohydramnios has been linked to craniosynosotosis (Graham, 1983; Higginbottom et al., 1980). Shahinian and colleagues (1998) found an association between increased duration of the first stage of labor and the development of lambdoid synostosis. The authors were of the opinion that the deformational forces imposed on the fetal head played a role in the pathogenesis of the lambdoid suture fusion. Bone deposition is reported to occur along the suture lines if the normal tensile forces that maintain patency of the suture are disturbed, as could happen with fetal head constraint (Graham, 1983). To summarize, fetal head constraint is a risk factor that predisposes to craniosynostoses later in development, perhaps by interfering with the normal process of bone deposition.

8.6.4.2 Animal Data In studies on pregnant mice, closing off of the uterine cervix close to the time of delivery, resulting in fetal overgrowth and constraint, resulted in plagiocephaly in the majority of offspring. Craniosynostosis was confirmed histologically in most cases (88%) (Koskinen-Moffett and Moffett, 1989).

8.7 SUMMARY

Craniofacial malformations are common birth defects that occur in humans, with facial clefting representing the majority of these defects. While many syndromes with cleft lip with or without cleft palate (CL/P) or with cleft palate (CP) are recognized, the majority of oral clefts fall into the category of nonsyndromic oral clefts, and the etiology of this group remains incompletely understood. Investigators agree that oral clefts are multifactorial in origin, with both genetic and environmental factors in their etiology. However, environmental agents as a group are not a major etiological factor in the causation of craniosynostoses. In both orofacial clefts and in craniosynostoses, these environmental etiologies are not as well defined or studied as the genetic causes. As the genetic basis of craniofacial growth and development is increasingly understood, it will be possible to identify specific molecular targets of environmental teratogens. This should permit development of accurate etiopathogenic models that will help in determining the physiological basis of these heterogeneous anomalies, permitting development of avoidance and intervention strategies for at-risk individuals.

Human studies in understanding the role of environmental agents in craniofacial abnormalities consist of two broad categories: epidemiological case-control studies and case reports/studies. The former group has the advantage of allowing large numbers to be studied and enabling comparison with control subjects. They have been useful in establishing associations that need to be explored further. However, a major limiting factor in these epidemiological studies is the difficulty in controlling for confounding variables—in fact, it would not be feasible to control for all possible confounding factors. Other limiting factors with epidemiological studies are described in Section 8.1. Case reports/studies (although they typically consist of smaller numbers of patients) have been more effective in indicating a causal relationship between specific teratogens and craniofacial anomalies (e.g., fetal alcohol syndrome). These are usually the consequence of clinical observations, and if replicated often enough, result in a causal relationship being established. Often, animal studies subsequently confirm the clinical observations. The lack of control subjects, small numbers of patients, and bias of ascertainment are some of the limiting factors in case studies. Keeping these limitations in mind, there is no doubt that both epidemiological and case studies can help to clarify the relationship between a specific environmental agent and congenital anomalies.

Animal models have been important resources in developing and formally testing specific hypotheses related to teratogen-induced clefting (Diewert, 1980; Johnston and Millicovsky, 1985) (see also Chapters 10 and 21). Animal models permit more rigorous and standardized testing designs and evaluate the roles of dose,

timing, and genetic background (Nelson and Holson, 1978; Jelinek, 1985; Freni and Zapisek, 1991; Bowman et al., 1995). While animal models have identified teratogens for oral clefts, in many instances their precise relevance for humans remains unclear (Wyszynski and Beaty, 1996). Nonetheless, studies of animal models have provided significant contributions that have advanced our understanding of the developmental basis for facial clefting through the analysis of the effects of treatments with teratogens, through the modulating effects of treatment with ameliorating agents, and more recently through gene disruption experiments. In addition to providing insight into the bases for abnormal craniofacial growth, genetic and teratogenic techniques are powerful tools for understanding the normal morphological, cellular, and biochemical aspects that underlie the developmental processes that generate and pattern the face (Diewert, 1980; Fraser, 1980; Shah, 1984; Young et al., 2000).

In conclusion, the environmental etiologies of orofacial clefts and craniosynostoses are complex. Environmental factors contribute more significantly to the etiology of orofacial clefts than craniosynostoses. Considerable progress has been made in understanding of some of the environmental factors involved, resulting in formulation of specific preventive and management strategies (e.g., alcohol, retinoic acid). In the future, elucidation of other variables such as gene–environment interactions, dose, and timing of exposure will be crucial in clarifying the pathogenesis and devising appropriate preventive and management interventions.

ACKNOWLEDGMENTS

We are grateful to Trish Metz for her editorial assistance in the final preparation of the manuscript.

REFERENCES

Aase, J. M. (1974). Anticonvulsant drugs and congenital abnormalities. *Am. J. Dis. Child* 127, 758.

Abbott, B. D., and Pratt, R. M. (1987). Human embryonic palatal epithelial differentiation is altered by retinoic acid and epidermal growth factor in organ culture. *J. Craniofac. Genet. Devel. Biol.* 7, 241–265.

Abbott, B. D., and Pratt, R. M. (1991). Retinoic acid alters epithelial differentiation during palatogenesis. *J. Craniofac. Genet. Devel. Biol.* 11, 315–325.

Abbott, B. D., Harris, M. W., and Birnbaum, L. S. (1992). Comparisons of the effects of TCDD and hydrocortisone on growth factor expression provide insight into their interaction in the embryonic mouse palate. *Teratology* 45, 35–53.

Abbott, B. D., Schmid, J. E., Brown, J. G., Wood, C. R., White, R. D., Buckalew, A. R., and Held, G. A. (1999). RT-PCR quantification of AHR, ARNT, GR, and CYP1A1 mRNA in craniofacial tissues of embryonic mice exposed to 2,3,7,8-tetrachlorodibenzo-p-dioxin and hydrocortisone. *Toxicol. Sci.* 47, 76–85.

Abdulrazzaq, Y. M., Bastaki, S. M., and Padmanabhan, R. (1997). Teratogenic effects of vigabatrin in TO mouse fetuses. *Teratology* 55, 165–176.

Abrishamchian, A. R., Khoury, M. J., and Calle, E. E. (1994). The contribution of maternal epilepsy and its treatment to the etiology of oral clefts: a population based case-control study. *Gen. Epidemiol.* 11, 343–351.

Akita, S., Hirano, A., and Fujii, T. (1996). Identification of IGF-I in the calvarial suture of young rats: histochemical analysis of the cranial sagittal sutures in a hyperthyroid rat model. *Plast. Reconstr. Surg.* 97, 1–12.

Aleck, K. A., and Bartley, D. L. (1997). Multiple malformation syndrome following fluconazole use in pregnancy: report of an additional patient. *Am. J. Med. Genet.* 72, 253–256.

Amicarelli, F., Tiboni, G. M., Colafarina, S., Bonfigli, A., Iammarrone, E., Miranda, M., and DiIlio, C. (2000). Antioxidant and GSH-related enzyme response to a single teratogenic exposure to the anticonvulsant phenytoin: temporospatial evaluation. *Teratology* 62, 100–107.

Andrews, J., and McGarry, J. M. (1972). A community study of smoking in pregnancy. *J. Obstet. Gynecol. Brit. Comm.* 79, 1057–1073.

Antikainen, T., Kallioinen, M., Warit, T., and Serlo, W. (1992). A new method for the creation and measurement of experimental craniosynostosis. *Childs Nerv. Sys.* (PMID 1288855) 8, 457–468.

Ardinger, H. H., Atkin, J. F., Blackston, R. D., Elsas, L. J., Clarren, S. K., Livingstone, S., Flannery, D. B., Pellock, J. M., Harrod, M. J., and Lammer, E. J. (1988). Verification of the fetal valproate syndrome phenotype. *Am. J. Med. Genet.* 29, 171–185.

Arena, J. F. P., and Smith, D. W. (1978). Sex liability to single structural defects. *Am. J. Dis. Child* 132, 970–972.

Arpino, C., Brescianini, S., Robert, E., Castilla, E. E., Cocchi, G., Cornel, M. C., de Vigan, C., Lancaster, P. A., Merlob, P., Sumiyoshi, Y., Zampino, G., Renzi, C., Rosano, A., and Mastroiacovo, P. (2000). Teratogenic effects of antiepileptic drugs: use of an International Database on Malformations and Drug Exposure (MADRE). *Epilepsia* 41, 1436–1443.

Azarbayzani, F., and Danielsson, B. R. (2001). Phenytoin-induced cleft palate: evidence for embryonic cardiac bradyarrhythmia due to inhibition of delayed rectifier K+ channels resulting in hypoxia-reoxygenation damage. *Teratology* 63, 152–160.

Bagatin, M., Der Sarkissian, R., and Larrabee, W. F., Jr. (1997). Craniofacial manifestations of the amniotic band syndrome. *Otolaryngol. Head Neck Surg.* 116, 525–528.

Bailey, L. J., Johnston, M. C., and Billet, J. (1995). Effects of carbon monoxide and hypoxia on cleft lip in A/J mice. *Cleft Palate Craniofac. J.* 32, 14–19.

Ballard, P. D., Hearney, E. E., and Smith, M. B. (1975). Induction of cleft palate in mice after ophthalmic administration of hydrocortisone. *Toxicol. Appl. Pharmacol.* 34, 358–361.

Bamforth, J. S. (1992). Amniotic band sequence: Streeter's hypothesis reexamined. *Am. J. Med. Genet.* 44, 280–287.

Bannigan, J. G., Cottell, D., and Morris, A. (1990). Study of the mechanisms of BUdRinduced cleft palate in the mouse. *Teratology 42*, 79–89.

Beaty, T. H., Maestri, N. E., Hetmanski, J. B., Wyszynski, D. F., Vanderkolk, C. A., Simpson, J. C., McIntosh, I., Smith, E. A., Zeiger, J. S., Raymand, G. V., Panny, S. R., Tifft, C. J., Lewanda, A. F., Cristion, C. A., and Wulfsberg, E. A. (1997). Testing for interaction between maternal smoking and TGFA genotype among oral cleft cases born in Maryland 1992–1996. *Cleft Palate Craniofac. J. 34*, 447–454.

Bedrick, A. D., and Ladda, R. L. (1978). Epidermal growth factor potentiates cortisone-induced cleft palate in the mouse. *Teratology 17*, 13–18.

Benke, P. J. (1984). The isotretinoin teratogen syndrome. *J. Am. Med. Assoc. 251*, 3267–3269.

Bianchi, F., Calzolari, E., Ciulli, L., Cordier, S., Gualandi, F., Pierini, A., and Mossey, P. (2000). Environment and genetics in the etiology of cleft lip and cleft palate with reference to the role of folic acid. *Epidemiologia e Prevenzione 24*, 21–27.

Biddle, F. G., and Fraser, F. C. (1976). Genetics of cortisone-induced cleft palate in the mouse-embryonic and maternal effects. *Genetics 84*, 743–754.

Biddle, F. G. (1978). Use of d-response relationships to discriminate between the mechanisims of cleft-palate induction by different teratogens: an argument for discussion. *Teratology 18*, 247–252.

Biddle, F. G., and Fraser, F. C. (1979). Genetic independence of the embryonic reactivity difference to cortisone- and 6-aminonicotinamide-induced cleft palate in the mouse. *Teratology 19*, 207–211.

Bienengraber, V., Fanghanel, J., Malek, F. A., and Kundt, G. (1997). Application of thiamine in preventing malformations, specifically cleft alveolus and palate, during the intrauterine development of rats. *Cleft Palate Craniofac. J. (PMID 10081572) 34*, 318–324.

Bienengraber, V., Malek, F., Fanghanel, J., and Kundt, G. (1999). Disturbances of palatogenesis and their prophylaxis in animal experiments. *Anat. Anz. 181*, 111–115.

Birnbaum, L. S., Harris, M. W., Stocking, L. M., Clark, A. M., and Morrissey, R. E. (1989). Retinoic acid and 2,3,7,8-tetrachlorodibenzo-p-dioxin selectively enhance teratogenesis in C57BL/6N mice. *Toxicol. Appl. Pharmacol. 98*, 487–500.

Bolon, B., Dorman, D. C., Janszen, D., Morgan, K. T., and Welsch, F. (1993). Phase-specific developmental toxicity in mice following maternal methanol inhalation. *Fund. Appl. Toxicol. 21*, 508–516.

Bower, C., and Stanley, F. J. (1992). Dietary folate and nonneural midline birth defects: no evidence of an association from a case-control study in western Australia. *Am. J. Med. Genet. 44*, 647–650.

Bowman, D., Chen, J. J., and George, E. O. (1995). Estimating variance functions in developmental toxicity studies. *Biometrics 51*, 1523–1528.

Branch, S., Francis, B. M., Brownie, C. F., and Chernoff, N. (1996). Teratogenic effects of the demethylating agent 5-aza-2'-deoxycytidine in the Swiss Webster mouse. *Toxicology 112*, 37–43.

Brenner, R. L., Smith, J. L., Cleveland, W. W., Bejar, R. L., and Lockhart, W. S. (1969). Eye signs of hypophosphatasia. *Arch. Ophthalmol. 81*, 614–617.

Brent, R. L., Chambers, C. D., Chernoff, G. F., Jones, K. L., and Miller, R. K. (1999). Pregnancy outcome following gestational exposure to organic solvents: a response. *Teratology 60*, 328–329.

Brown, K. S., and Hackman, R. M. (1985). The significance of receptor physiology for corticosterone-induced cleft palate in A/J mice. *J. Craniofac. Genet. Devel. Biol. 1* (suppl.), 299–304.

Brunet, C. L., Sharpe, P. M., and Ferguson, M. W. (1995). Inhibition of TGF-beta 3 (but not TGF-beta 1 or TGF-beta 2) activity prevents normal mouse embryonic palate fusion. *Int. J. Devel. Biol. 39*, 345–355.

Buehler, B. A., Rao, V., and Finnell, R. H. (1994). Biochemical and molecular teratology of fetal hydantoin syndrome. *Neurol. Clin. 12*, 741–748.

Cantrell, C., Baskin, G., and Blanchard, J. (1987). Craniosynostosis in two African green monkeys. *Lab. Anim. Sci. 37*, 631–634.

Carlsen, N. L. T., Krasilnikoff, P. A., and Eiken, M. (1984). Premature cranial synostosis in X-linked hypophosphatemic rickets: possible precipitation by 1-alpha-OH-cholecalciferol intoxication. *Acta. Paediatr. Scand. 73*, 149–154.

Carmichael, S. L., and Shaw, G. M. (1999). Maternal corticosteroid use and risk of selected congenital anomalies. *Am. J. Med. Genet. 86*, 242–244.

Cembrano, J. R. J., Vera, J. S. D. D., Joaquino, J. B., Ng, E. F., Tongson, T. L., Manalo, P. D., Fernandez, G. C., and Encarnacion, R. C. (1995). Familial risk of recurrence of clefts of the lip and palate. *Phillipp. J. Surg. Surg. Spec. 50*, 37–40.

Char, F., Herty, J. B., Wilson, R. S., and Dungen, W. T. (1978). Patterns of malformations in infants exposed to gestational anticonvulsants. Paper presented at the Birth Defects Meeting, San Francisco, June.

Chenevix-Trench, G., Jones, K., Green, A. C., Duffy, D. L., and Martin, N. G. (1993). Cleft lip with or without cleft palate: associations with transforming growth factor alpha and retinoic acid receptor loci. *Comment. Am. J. Hum. Genet. 53*, 1156–1157.

Chernoff, N. (1973). Teratogenic effects of cadmium in rats. *Teratology 8*, 29–32.

Christensen, K., Oslen, J., Norgaard-Pedersen, B., Basso, O., Stovring, H., Milhollin-Johnson, L., and Murray, J. C. (1999). Oral cleft, transforming growth factor alpha gene variants, and maternal smoking: a population-based case-control study in Denmark, 1991–1994. *Am. J. Epidemiol. 149*, 248–255.

Chung, C. S., Mi, M. P., and Beechert, A. M. (1987). Genetic epidemiology of cleft lip or without cleft palate in the population of Hawaii. *Genet. Epidemiol. 4*, 415–423.

Chung, K. C., Kowalski, C. P., Kim, H. M., and Buchman, S. R. (2000). Maternal cigarette smoking during pregnancy and the risk of having a child with cleft lip/palate. *Plast. Reconstr. Surg. 105*, 485–491.

Cicolella, A. (1992). *Les Ethers de Glycol: Etat Acruel des Connaissances: Perspectives de Recherche.* Cahiers de Notes Documentaires No. 148 (Nancy, France: Institut National de Recherche et de Sécurité).

Clarren, S. K., Smith, D. W., and Harvey, M. A. S. (1979). Hyperthermia—a prospective evaluation of a possible teratogenic agent in man. *J. Pediatr. 95*, 81.

Clemens, P. (1984). Premature cranial synosotsis and hypophosphatemic rickets. *Acta Paediatr. Scand. 73*, 857–862.

Coady, M. S., Moore, M. H., and Wallis, K. (1998). Amniotic band syndrome: the association between rare facial clefts and limb ring constrictions. *Plast. Reconstr. Surg. 101*, 640–649.

Coffey, V. P., and Jessup, W. J. E. (1963). Maternal influenza and congenital deformities. A follow-up study. *Lancet 1*, 748–751.

Cohen, M. M., and MacLean, R. E. (eds.) (2000). *Craniosynostosis: Diagnosis, Evaluation and Management* (New York: Oxford University Press).

Coleman, E. N., and Foote, J. B. (1954). Craniostenosis with familial rickets. *Br. Med. J. 1*, 561–562.

Collins, M. D., Eckhoff, C., Chahoud, I., Bochert, G., and Nau, H. (1992). 4-Methylpyrazole partially ameliorated the teratogenicity of retinol and reduced the metabolic formation of all-trans-retinoic acid in the mouse. *Arch. Toxicol. 66*, 652–659.

Collins, M. D., Tzimas, G., Burgin, H., Hummler, H., and Nau, H. (1995). Single versus multiple dose administration of all-trans-retinoic acid during organogenesis: differential metabolism and transplacental kinetics in rat and rabbit. *Toxicol. Appl. Pharmacol. 130*, 9–18.

Conway, H. (1958). Effect of supplemental vitamin therapy on the limitation of incidence of cleft lip and cleft palate in humans. *Plast. Reconstr. Surg. 22*, 450–453.

Cordier, S., Ha, M. C., and Ayme, S. (1992). Maternal occupational exposure and congenital malformations. *Scand. J. Work Environ. Health 18*, 11–17.

Cordier, S., Bergeret, A., Goujard, J., Ha, M. C., Ayme, S., Bianchi, F., Calzolari, E., De Walle, H. E., Knill-Jones, R., Candela, S., Dale, I., Dananche, B., de Vigan, C., Fevotte, J., Kiel, G., and Mandereau, L. (1997). Congenital malformation and maternal occupational exposure to glycol eithers. *Occupational*

Exposure and Congenital Malformations Working Group. *Epidemiology 8*, 355–363.

Couture, L. A., Harris, M. W., and Birnbaum, L. S. (1989). Developmental toxicity of 2,3,4,7,8-pentachlorodibenzofuran in the Fischer 344 rat. *Fund. Appl. Toxicol. 12*, 358–366.

Croen, L. A., Shaw, G. M., Wasserman, C. R., and Tolarova, M. M. (1998). Racial and ethnic variations in the prevalence of orofacial clefts in California, 1983–1992. *Am. J. Med. Genet. 79*, 42–47.

Czeizel, A. E. (1976). Diazepam, phenytoin and etiology of cleft lip and/or palate. *Lancet 1*, 810.

Czeizel, A. E. (1993). Prevention of congenital abnormalities by periconceptional multivitamin supplementation. *Br. Med. J. 306*, 1645–1648.

Czeizel, A. E., and Hirschberg, J. (1997). Orofacial clefts in Hungary. Epidemiological and genetic data, primary prevention. *Folia Phoniatr. Logop. 49*, 111–116.

Czeizel, A. E., and Rockenbauer, M. (1997). Population-based case-control study of teratogenic potential of corticosteroids. *Teratology 56*, 335–340.

Czeizel, A. E., Timar, L., and Sarkozi, A. (1999). Dose-dependent effect of folic acid on the prevention of orofacial clefts. *Pediatrics 104*, e66.

d'Argy, R., Dencker, L., Klasson-Wehler, E., Bergman, A., Darnerud, P. O., and Brandt, I. (1987). 3,3′4,4′-Tetrachlorobiphenyl in pregnant mice: embryotoxicity, teratogenicity, and toxic effects on the cultured embryonic thymus. *Pharmacol. Toxicol. 61*, 53–57.

Damm, K., Heyman, R. A., Umesono, K., and Evans, R. M. (1993). Functional inhibition of retinoic acid response by dominant negative retinoic acid receptor mutants. *Proc. Natl. Acad. Sci. USA 90*, 2989–2993.

Daneman, D., and Howard, N. J. (1980). Neonatal thyrotoxicosis: intellectual impairment and craniosynostosis in later years. *J. Pediatr. 97*, 257–259.

Dansky, L. V., Rosenblatt, D. S., and Andermann, E. (1992). Mechanisms of teratogenesis: folic acid and antiepileptic therapy. *Neurology 42*, 32–42.

Daston, G. P., Ebron, M. T., Carver, B., and Stefanadis, J. G. (1987). In vitro teratogenicity of ethylenethiourea in the rat. *Teratology 35*, 239–245.

Dean, J. C., Moore, S. J., Osborne, A., Howe, J., and Turnpenny, P. D. (1999). Fetal anticonvulsant syndrome and mutation in the maternal MTHFR gene. *Clin. Genet. 56*, 216–220.

Degitz, S. J., Francis, B. M., and Foley, G. L. (1998). Mesenchymal changes associated with retinoic acid induced cleft palate in CD-1 mice. *J. Craniofac. Genet. Devel. Biol. 18*, 88–99.

DePass, L. R., and Weaver, E. V. (1982). Comparison of teratogenic effects of aspirin and hydroxyurea in the Fischer 344 and Wistar strains. *J. Toxicol. Environ. Health 10*, 297–305.

DeSesso, J. M., and Jordan, R. L. (1977). Drug-induced limb dysplasias in fetal rabbits. *Teratology 15*, 199–211.

DeSesso, J. M., and Goeringer, G. C. (1992). Methotrexate-induced developmental toxicity in rabbits is ameliorated by 1-(p-tosyl)-3,4,4-trimethylimidazolidine, a functional analogue for tetrahydrofolate-mediated one-carbon transfer. *Teratology 45*, 271–283.

Dieterich, E. (1979). Antiepileptica—embryopathien. *Ergeb Inn Med. Kinderheilkd 43*, 93–107.

Diewert, V. M. (1980). Differential changes in cartilage cell proliferation and cell density in the rat craniofacial complex during secondary palate development. *Anat. Rec. 198*, 219–228.

DiLiberti, J. H., Farndon, P. A., Dennis, N. R., and Curry, C. J. (1984). The fetal valproate syndrome. *Am. J. Med. Genet. 19*, 473–481.

Dixon, M. J., and Ferguson, M. W. (1992). The effects of epidermal growth factor, transforming growth factors alpha and beta and platelet-derived growth factor on murine palatal shelves in organ culture. *Arch. Oral Biol. 37*, 395–410.

Dlugosz, L., Vena, J., Byers, T., Sever, L., Bracken, M., and Marshall, E. (1992). Congenital defects and electric bed heating in New York State: a register-based case-control study. *Am. J. Epidemiol. 135*, 1000–1011.

Duncan, B. W., Adzick, N. S., and Moelleken, B. R. (1992). An in utero model of craniosynostosis. *J. Craniofac. Surg. 3*, 70–76.

Durner, M., Greenberg, D. A., and Delgado-Escueta, A. V. (1992). Is there a genetic relationship between epilepsy and birth defects? *Neurology 42*, 63–67.

Eckhoff, C., Lofberg, B., Chahoud, I., Bochert, G., and Nau, H. (1989). Transplacental pharmacokinetics and teratogenicity of a single dose of retinol (vitamin A) during organogenesis in the mouse. *Toxicol. Lett. 48*, 171–184.

Edwards, M. J. (1974). The effects of hyperthermia on pregnancy and prenatal development. In D. H. M. Woollam and G. M. Morriss (eds.), *Experimental Embryology and Teratology* vol. 1 (London: Elek Science), pp. 90–133.

Ellis, W. G., Semple, J. L., Hoogenboom, E. R., Kavlock, R. J., and Zeman, F. J. (1987). Benomyl-induced craniocerebral anomalies in fetuses of adequately nourished and protein-deprived rats. *Teratogen. Carcinogen. Mutagen. 7*, 357–375.

Elmazar, M. M., McElhatton, P. R., and Sullivan, F. M. (1982). Studies on the teratogenic effects of different oral preparations of caffeine in mice. *Toxicology 23*, 57–71.

Elmazar, M. M., Thiel, R., and Nau, H. (1992). Effect of supplementation with folinic acid, vitamin B6, and vitamin B12 on valproic acid–induced teratogenesis in mice. *Fund. Appl. Toxicol. 18*, 389–394.

Eluma, F. O., Sucheston, M. E., Hayes, T. G., and Paulson, R. B. (1984). Teratogenic effects of dosage levels and time of administration of carbamazepine, sodium valproate, and diphenylhydantoin on craniofacial development in the CD-1 mouse fetus. *J. Craniofac. Genet. Devel. Biol. 4*, 191–210.

Ema, M., Kurosaka, R., Amano, H., and Ogawa, Y. (1996b). Comparative developmental toxicity of di-tri- and tetrabutyltin compounds after administration during late organogenesis in rats. *J. Appl. Toxicol. 16*, 71–76.

Ema, M., Miyawaki, E., and Kawashima, K. (1998). Further evaluation of developmental toxicity of di-n-butyl phthalate following administration during late pregnancy in rats. *Toxicol. Lett. 98*, 87–93.

Enns, G. M., Roeder, E., Chan, R. T., Ali-Khan Catts, Z., Cox, V. A., and Golabi, M. (1999). Apparent cyclophosphamide (cytoxan) embryopathy: a distinct phenotype? *Am. J. Med. Genet. 86*, 237–241.

Eppley, B. L., David, L., Li, M., Moore, C. A., and Sadove, A. M. (1998). Amniotic band facies. *J. Craniofac. Surg. 9*, 360–365.

Ericson, A., Kallen, B., and Westerholm, P. (1979). Cigarette smoking as an etiologic factor in cleft lip and palate. *Am. J. Obstet. Gynecol. 135*, 348–351.

Faron, G., Drouin, R., Pedneault, L., Poulin, L. D., Laframboise, R., Garrido-Russo, M., and Fraser, W. D. (2001). Recurrent cleft lip and palate in siblings of a patient with malabsorption syndrome, probably caused by hypovitaminosis associated with folic acid and vitamin B(2) deficiencies. *Teratology 63*, 161–163.

Feldkamp, M., and Carey, J. C. (1993). Clinical teratology counseling and consultation case report: low dose methotrexate exposure in the early weeks of pregnancy. *Teratology 47*, 533–539.

Ferm, V. H. (1971). Developmental malformations induced by cadmium. A study of timed injections during embryogenesis. *Biol. Neo. (PMID 5003301) 19*, 101–107.

Field, B., and Kerr, C. (1979). Herbicide use and incidence of neural-tube defects. *Lancet 1*, 1341–1342.

Finnell, R. H., Wlodarczyk, B. C., Craig, J. C., Piedrahita, J. A., and Bennett, G. D. (1997). Strain-dependent alterations in the expression of folate pathway genes following teratogenic exposure to valproic acid in a mouse model. *Am. J. Med. Genet. 70*, 303–311.

Francis, B. M., Rogers, J. M., Sulik, K. K., Alles, A. J., Elstein, K. H., Zucker, R. M., Massaro, E. J., Rosen, M. B., and Chernoff, N. (1990). Cyclophosphamide teratogenesis: evidence for compensatory responses to induced cellular toxicity. *Teratology 42*, 473–482.

Fraser, D. (1957). Hypophosphatasia. *Am. J. Med. 22*, 730–746.

Fraser, F. C. (1976). The multifactorial/threshold concept—uses and misuses. *Teratology 14*, 267–280.

Fraser, F. C. (1980). Animal models for craniofacial disorders. *Prog. Clin. Biol. Res. 46*, 1–23.

Fraser, F. C., and Fainstat, T. D. (1951). The production of congenital defects in the offspring of pregnant mice treated with cortisone: a progress report. *Pediatrics 8*, 527–533.

Fraser, F. C., and Sajoo, A. (1995). Teratogenic potential of corticosteroids in humans. *Teratology 51*, 45–46.

Freni, S. C., and Zapisek, W. F. (1991). Biologic basis for a risk assessment model for cleft palate. *Cleft Palate Craniofac. J. 28*, 338–346.

Friis, M. L. (1989). Facial clefts and congenital heart defects in children of parents with epilepsy: genetic and environmental etiologic factors. *Acta Neurol. Scand. 79*, 433–459.

Fritz, H. (1976). The effect of cortisone on the teratogenic action of acetylsalicylic acid and diphenylhydantoin in the mouse. *Experientia 32*, 721–722.

Frosst, P., Blom, H. J., Milos, R., Goyette, P., Sheppard, O. A., Mathews, R. G., Boers, G. J. H., den Heijer, M., Kluijtmans, L. A. J., van der Heuvel, L. P., and Rozen, R. (1995). A candidate genetic risk factor for vascular disease: a common mutation in methylenetetrahydrofolate reductase. *Nat. Genet. 10*, 111–113.

Froster, U. G., and Baird, P. A. (1993). Amniotic band sequence and limb defects data from a population-based study. *Am. J. Med. Genet. 46*, 497–500.

Fujimoto, T., Fuyuta, M., Kiyofuji, E., and Hirata, S. (1979). Prevention by tiopronin (2-mercaptopropionyl glycine) of methylmercuric chloride-induced teratogenic and fetotoxic effects in mice. *Teratology 20*, 297–301.

Fuyuta, M., Fujimoto, T., and Hirata, S. (1978). Embryotoxic effects of methylmercuric chloride administered to mice and rats during organogenesis. *Teratology 18*, 353–366.

Garcia, A. M., Fletcher, T., Benavides, F. G., and Orts, E. (1999). Parental agricultural work and selected congenital malformations. *Am. J. Epidemiol. 149*, 64–74.

Geelen, J. A. (1979). Hypervitaminosis A induced teratogenesis. *CRC Crit. Rev. Toxicol. 6*, 351–375.

Germain, M. A., Webster, W. S., and Edwards, M. J. (1985). Hyperthermia as a teratogen: parameters determining hyperthermia-induced head defects in the rat. *Teratology 31*, 265–272.

Gilliam, D. M., Kotch, L. E., Dudek, B. C., and Riley, E. P. (1988). Ethanol teratogenesis in mice selected for differences in alcohol sensitivity. *Alcohol 5*, 513–519.

Gordon, J. E., and Shy, C. M. (1981). Agricultural chemical use and congenital cleft lip and/or palate. *Arch. Environ. Health 36*, 213–221.

Gorlin, R. J., Cohen, M. M., and Levin, L. S. (1990). *Syndromes of the Head and Neck* (New York: Oxford University Press).

Gotoh, H., Nomura, T., Nakajima, H., Hasegawa, C., and Sakamoto, Y. (1988). Inhibiting effects of nicotinamide on urethane-induced malformations and tumors in mice. *Mut. Res. 199*, 55–63.

Graham, J. M., Jr. (1983). Alterations in head shape as a consequence of fetal head constraint. *Semin. Perinatol. 7*, 257–269.

Granick, M. S., Ramasastry, S., Vries, J., and Cohen, M. M., Jr. (1987). Severe amniotic band syndrome occurring with unrelated syndactyly. *Plast. Reconstr. Surg. 80*, 829–832.

Gray, L. E., Jr., Rogers, J. M., Kavlock, R. J., Ostby, J. S., Ferrell, J. M., and Gray, K. L. (1986). Prenatal exposure to the fungicide dinocap causes behavioral torticollis, ballooning and cleft palate in mice, but not rats or hamsters. *Teratogen. Carcinogen. Mutagen. 6*, 33–43.

Hale, F. (1937). Relation of maternal vitamin A deficiency to microphthalmia in pigs. *Tex. State J. Med. 33*, 228–232.

Hanify, J. A., Metcalf, P., Nobbs, C. L., and Worsley, K. J. (1981) Aerial spraying of 2, 4, 5-T and human birth malformations: an epidemiological investigation. *Science 212*, 349–351.

Hansen, D. K., LaBorde, J. B., Wall, K. S., Holson, R. R., and Young, J. F. (1999). Pharmacokinetic considerations of dexamethasone-induced developmental toxicity in rats. *Toxicol. Sci. 48*, 230–239.

Hanson, J. W., and Smith, D. W. (1976). The fetal hydantoin syndrome. *J. Pediatr. 87*, 285.

Harris, M. W., and Birnbaum, L. S. (1989). Developmental toxicity of perfluorodecanoic acid in C57BL/6N mice. *Fund. Appl. Toxicol. 12*, 442–448.

Hassoun, E., d'Argy, R., Dencker, L., and Sundstrom, G. (1984). Teratological studies on the TCDD congener 3,3′,4,4′-tetrachloroazoxybenzene in sensitive and nonsensitive mouse strains: evidence for direct effect on embryonic tissues. *Arch. Toxicol. 55*, 20–26.

Hayes, C., Werler, M. M., Willett, W. C., and Mitchell, A. A. (1996). Case-control study of periconceptional folic acid supplementation and oral clefts. *Am. J. Epidemiol. 143*, 1229–1234.

Hayes, W. C., Cobel-Geard, S. R., Hanley, T. R., Jr., Murray, J. S., Freshour, N. L., Rao, K. S., and John, J. A. (1981). Teratogenic effects of vitamin A palmitate in Fischer 344 rats. *Drug Chem. Toxicol. 4*, 283–295.

Hecht, J. T., and Annegers, J. F. (1990). Familial aggregation of epilepsy and clefting disorders: a review of the literature. *Epilepsia 31*, 574–577.

Helms, J. A., Kim, C. H., Hu, D., Minkoff, R., Thaller, C., and Eichele, G. (1997). Sonic hedgehog participates in craniofacial morphogenesis and is down-regulated by teratogenic doses of retinoic acid. *Devel. Biol. 187*, 25–35.

Hendrickx, A. G., and Hummler, H. (1992). Teratogenicity of all-trans-retinoic acid during early embryonic development in the cynomolgus monkey *(Macaca fascicularis)*. *Teratology 45*, 65–74.

Hernandez-Diaz, S., Werler, M. M., Walker, A. M., and Mitchell, A. A. (2000). Folic acid antagonists during pregnancy and the risk of birth defects. *New Engl. J. Med. 343*, 1608–1614.

Higginbottom, M. C., Jones, K. L., Hall, B. D., and Smith, D. W. (1979). The amniotic band disruption complex timing of amniotic rupture and variable spectra of consequent defects. *J. Pediatr. 95*, 544–549.

Higginbottom, M. C., Jones, K. L., and James, H. E. (1980). Intrauterine constraint and craniosynostosis. *Neurosurgery 6*, 39–44.

Hollingsworth, D. R., and Mabry, C. C. (1976). Congenital Graves disease: four familial cases with long term follow-up and perspective. *Am. J. Dis. Child 130*, 148–155.

Holmberg, P. C., Hernberg, S., Kurppa, K., Rantala, K., and Riala, R. (1982). Oral clefts and organic solvent exposure during pregnancy. *Int. Arch. Occup. Environ. Health 50*, 371–376.

Holmes, L. B., Harvey, E. A., Coull, B. A., Huntington, K. B., Khosbin, S., Hayes, A. M., and Ryan, L. M. (2001). The teratogenicity of anticonvulsant drugs. *New Engl. J. Med. 344*, 1132–1138.

Hood, R. D., and Ottley, M. S. (1985). Developmental effects associated with exposure to xylene: a review. *Drug Chem. Toxicol. 8*, 281–297.

Hu, D., and Helms, J. A. (1999). The role of sonic hedgehog in normal and abnormal craniofacial morphogenesis. *Development 126*, 4873–4884.

Hwang, S. J., Beaty, T. H., Panny, S. R., Street, N. A., Joseph, J. M., Gordon, S., McIntosh, I., and Francomano, C. A. (1995). Association study of transforming growth factor alpha (TGF alpha) TaqI polymorphism and oral clefts: indication of gene-environment interaction in a population-based sample of infants with birth defects. *Am. J. Epidemiol. 141*, 629–636.

Ignelzi, M. A., Liu, Y. H., Maxson, R. E., Jr., and Snead, M. L. (1995). Genetically engineered mice: a tool to understand craniofacial development. *Crit. Rev. Oral Biol. 6*, 181–201.

Iitikala, P. R., Watkins, M. L., Mulinare, J., Moore, C. A., and Liu, Y. (2001). Maternal multivitamin use and orofacial clefts in offspring. *Teratology 63*, 79–86.

Im, M. J., Winograd, J. M., Manson, P. N., and Vander Kolk, C. A. (1997). Iron-induced rat coronal suture fusion in vitro: the role of redox regulation. *J. Craniofac. Surg. 8*, 262–269.

Jabor, M. A., and Cronin E. D. (2000). Bilateral cleft lip and palate and limb deformities: a presentation of amniotic band sequence. *J. Craniofac. Surg. 11*, 388–393.

Jacobsson, C. (1997). Teratological studies on craniofacial malformations. *Swed. Dent. J. 121* (suppl.), 3–84.

Jacobsson, C., and Granstrom, G. (1997). Effects of vitamin B6 on beta-aminoproprionitrile-induced palatal cleft formation in the rat. *Cleft Palate Craniofac. J. 34*, 95–100.

James, L. F. (1999). Teratological research at the USDA-ARS poisonous plant research laboratory. *J. Nat. Tox (PMID 10091129) 8*, 63–80.

James, W. H. (2000). Are oral clefts a consequence of maternal hormone imbalance? Evidence from the sex ratios of sibs of probands. *Teratology 62*, 342–345.

Janz, D. (1982). On major malformations and minor anomalies in the offspring of parents with epilepsy: review of the literature. In D. Janz, M. Dam, and A. Richens (eds.), *Epilepsy, Pregnancy and the Child* (New York: Raven Press), pp. 211–222.

Jaskoll, T., Choy, H. A., Chen, H., and Melnick, M. (1996). Developmental expression and CORT-regulation of TGF-beta and EGF receptor mRNA during mouse palatal morphogenesis: correlation between CORT-induced cleft palate and TGF-beta 2 mRNA expression. *Teratology 54*, 34–44.

Jelinek, R. (1985). The rationale for models studying mechanisms of teratogenesis. *Prog. Clin. Biol. Res. 163C*, 221–226.

Johnsonbaugh, R. E., Bryan, R. N., Heirlwimmer, U. R., and Georges, L. P. (1978). Premature craniosynostosis: a common complication of juvenile thyrotoxicosis. *J. Pediatr. 93*, 188–191.

Johnston, M. C., and Bronsky, P. T. (1991). Animal models for human craniofacial malformations. *J. Craniofac. Genet. Devel. Biol. 11*, 277–391.

Johnston, M. C., and Bronsky, P. T. (1995). Prenatal craniofacial development: new insights on normal and abnormal mechanisms. *Crit. Rev. Oral Biol. Med. 6*, 25–79.

Johnston M. C., and Millicorskey G. (1985). Normal and abnormal development of the lip and palate. *Clin. Plast. Surg. 12*, 521–532.

Jones, K. L. (1986). Fetal alcohol syndrome. *Pediatr. Rev. 8*, 122–126.

Jones, K., Smith, D., and Hall, B. (1974). A pattern of craniofacial and limb defects secondary to aberrant tissue bands. *J. Pediatr. 84*, 90–95.

Jordan, R. L., Wilson, J. G., and Schumacher, H. J. (1977). Embryotoxicity of the folate antagonist methotrexate in rats and rabbits. *Teratology 15*, 73–79.

Joschko, M. A., Dreosti, I. E., and Tulsi, R. S. (1991). The teratogenic effects of nicotine in vitro in rats: a light and electron microscope study. *Neurotoxicol. Teratol. 13*, 307–316.

Juriloff, D. M., and Mah, D. G. (1995). The major locus for multifactorial nonsyndromic cleft lip maps to mouse chromosome 11. *Mammalian Genet. 6*, 63–69.

Kaartinen, V., Voncken, J. W., Shuler, C., Warburton, D., Bu, D., Heisterkamp, N., and Groffen, J. (1995). Abnormal lung development and cleft palate in mice lacking TGF-beta 3 indicates defects of epithelial-mesenchymal interaction. *Nat. Genet. 11*, 415–421.

Keeler, R. F. (1975). Teratogenic effects of cyclopamine and jervine in rats, mice and hamsters. *Exp. Biol. Med. 149*, 302–306.

Keller, H., Neuhauser, G., Durkin-Stamm, M. V., Kaveggia, E. G., Schaaff, A., and Sitzmann, F. (1978). "ADAM complex" (amniotic deformity, adhesions, mutilations)—a pattern of craniofacial and limb defects. *Am. J. Med. Genet. 2*, 81–98.

Kelly, T. E., Rein, M., and Edwards, P. (1984). Teratogenicity of anticonvulsant drugs. IV: The association of clefting and epilepsy. *Am. J. Med. Genet. 19*, 451–458.

Kelsey, J. L., Dwyer, T., Holford, T. R., and Bracken, M. B. (1978). Maternal smoking and congenital malformations: an epidemiological study. *J. Epidemiol. Comm. Health 32*, 102–107.

Kennedy, L. A., and Persand, T. V. N. (1977). Pathogenesis of developmental defects induced in the rat by amniotic sac puncture. *Acta Anat. 97*, 23–35.

Khattak, S., K-Moghtader, G., McMartin, K., Barrera, M., Kennedy, D., and Koren, G. (1999). Pregnancy outcome following gestation exposure to organic solvents; a prospective study. *J. Am. Med. Assoc. 281*, 1106–1109.

Khera, K. S., and Iverson, F. (1978). Toxicity of ethylenethiourea in pregnant cats. *Teratology 18*, 311–313.

Khoury, M. J., Gomez-Farias, M., and Mulinare, J. (1989). Does maternal cigarette smoking during pregnancy cause cleft lip and palate in offspring? *Am. J. Dis. Child 143*, 333–337.

Kim, J. C., Yun, H. I., Shin, H. C., Han, S. S., and Chung, M. K. (2000). Embryo lethality and teratogenicity of a new fluoroquinolone antibacterial DW-116 in rats. *Arch. Toxicol. 74*, 120–124.

Kochhar, D. M., and Penner, J. D. (1987). Developmental effects of isotretinoin and 4-oxo-isotretinoin: the role of metabolism in teratogenicity. *Teratology 36*, 67–75.

Kopf-Maier, P. (1985). Glucocorticoid induction of cleft palate after treatment with titanocene dichloride? *Toxicology 37*, 111–116.

Koskinen-Moffett, L., and Moffett, B. C. (1989). Sutures and intrauterine deformation. In J. A. Persing, M. T. Edgerton, and J. A. Jane (eds.), *Scientific Foundations and Surgical Treatment of Craniosynostosis* (Baltimore: Williams & Wilkins), pp. 96–106.

Kurppa, K., Holmberg, P. C., Kuosma, E., Aro, T., and Saxen, L. (1991). Anencephaly and maternal common cold. *Teratology 44*, 51–55.

Kusanagi, T. (1984). Sensitive stages and dose-response analyses of palatal slit and cleft palate in C57BL/6 mice treated with glucocorticoid. *Teratology 29*, 281–286.

Lambie, D. G., and Johnson, R. H. (1985). Drugs and folate metabolism. *Drugs 30*, 145–155.

Lammer, E. J., Chen, D. T., Hoar, R. M., Agnish, N. D., Benke, P. J., Braun, J. T., Curry, C. J., Fernhoff, P. M., Grix, A. W., Jr., and Lott, I. T. (1985). Retinoic acid embryopathy. *New Engl. J. Med. 313*, 837–841.

Lau, E. C., and Li, Z. Q. (1995). Protection of mice from teratogen-induced cleft palate by exogenous methionine. *Exp. Biol. Med. 209*, 141–145.

Laumon, B., Martin, J. L., Bertucat, I., Verney, M. P., and Robert, E. (1996). Exposure to organic solvents during pregnancy and oral clefts: a case-control study. *Reprod. Toxicol. 10*, 15–19.

Lee, M. (1985). Potentiation of chemically induced cleft palate by ethanol ingestion during gestation in the mouse. *Teratogen. Carcinogen. Mutagen. 5*, 433–440.

Leonard, C. O., Ralston, C., Carey, J. C., and Morales, L. (1987). Craniosynostosis and facial dysmorphism due to maternal Graves disease. *Clin. Res. 35*, 225A.

Levy, P. A. (1998). Amniotic bands. *Pediatr. Rev. 19*, 249.

Lidral, A. C., Romitti, P. A., Basart, A. M., Doetschman, T., Leysens, N. J., Daack-Hirsch, S., Semina, E. V., Johnson, L. R., Machida, J., Burds, A., Parnell, T. J., Rubenstein, J. L., and Murray, J. C. (1998). Association of MSX1 and TGFB3 with nonsyndromic clefting in humans. *Am. J. Hum. Genet. 63*, 557–568.

Lieff, S., Olshan, A. F., Werler, M., Strauss, R. P., Smith, J., and Mitchell, A. (1999). Maternal cigarette smoking during pregnancy and risk of oral clefts in newborns. *Am. J. Epidemiol. 150*, 683–694.

Lindhout, D., Hoppener, R. J. E. A., and Meinardi, H. (1984). Teratogenicity of antiepileptic drug combination with special emphasis on epoxidation. *Epilpesia 25*, 77–83.

Lorente, C., Cordier, S., Goujard, J., Ayme, S., Bianchi, F., Calzolari, E., De Walle, H. E., and Knill-Jones, R. (2000a). Tobacco and alcohol use during pregnancy and risk of oral clefts. Occupational Exposure and Congenital Malformation Working Group. *Am. J. Public Health 90*, 415–419.

Lorente, C., Cordier, S., Bergeret, A., DeWalle, H. E., Goujard, J., Ayme, S., Knill-Jones, R., Calzolari, E., and Bianchi, F. (2000b). Maternal occupational risk factors for oral clefts. Occupational Exposure and Congenital Malformation Working Group. *Scand J. Work Environ. Health 2000 Apr.*; *26* (2): 137–145.

Lubinsky, M. (1983). Familial amniotic bands. *J. Pediatr. 102*, 323.

Lubinsky, M. S. (1997). Classifying sex biased congenital anomalies. *Am. J. Med. Genet. 69*, 225–228.

Maestri, N. E., Beaty, T. H., Hetmanski, J., Smith, E. A., McIntosh, I., Wyszynski, D. F., Liang, K. Y., Duffy, D. L., and VanderKolk, C. (1997). Application of transmission disequilibrium tests to nonsyndromic oral clefts: including candidate genes and environmental exposures in the models. *Am. J. Med. Genet. 73*, 337–344.

Makori, N., Peterson, P. E., Blankenship, T. N., Dillard-Telm, L., Hummler, H., and Hendrickx, A. G. (1998). Effects of 13-cis-retinoic acid on hindbrain and craniofacial morphogenesis in long-tailed macaques *(Macaca fascicularis)*. *J. Med. Primatol. 27*, 210–219.

Malloy, M. H., Kleinman, J. C., Bakewell, J. M., Schramm, W. F., and Land, G. H. (1989). Maternal smoking during pregnancy: no association with congenital malformations in Missouri 1980–83. *Am. J. Public Health 79*, 1243–1246.

Marazita, M. L., Jaskoll, T., and Melnick, M. (1988). Corticosteroid-induced cleft palate in short-ear mice. *J. Craniofac. Genet. Devel. Biol. 8*, 47–51.

Marks, T. A., Ledoux, T. A., and Moore, J. A. (1982). Teratogenicity of a commercial xylene mixture in the mouse. *J. Toxicol. Environ. Health 9*, 97–105.

McCarthy, J. G., and Reid, C. A. (1980). Craniofacial synostosis in association with vitamin D resistant rickets. *Ann. Plast. Surg. 4*, 149–153.

McDonald, J. C., Lavoie, J., Coté, R., and McDonald, A. D. (1987). Chemical exposures at work in early pregnancy and congenital defect: a case-referent study. *Br. J. Industr. Med. 44*, 527–533.

McElhatton, P. R., Sullivan, F. M., and Toseland, P. A. (1977). Teratogenic activity and metabolism of primidone in the mouse. *Epilepsia 18*, 1 11.

McKee, R. H., Wong, Z. A., Schmitt, S., Beatty, P., Swanson, M., Schreiner, C. A., and Schardein, J. L. (1990). The reproductive and developmental toxicity of High Flash Aromatic Naphtha. *Toxicol. Industr. Health 6*, 441–460.

McMartin, K. I., Chu, M., Kopecky, E., Einarson, T. R., and Koren, G. (1998). Pregnancy outcome following maternal organic solvent exposure: a meta-analysis of epidemiologic studies. *Am. J. Industr. Med. 34*, 288–292.

Menking, M., Wiebel, J., Schmidt, W. T., Ebel, K. D., and Ritter, R. (1972). Premature craniosynostosis associated with hyperthyroidism in 4 children with reference to 5 further cases in the literature. *Monatsschr. Kinderheilkd. 120*, 106–110.

Miller, C. P., and Birnbaum, L. S. (1986). Teratologic evaluation of hexabrominated naphthalenes in C57BL/6N mice. *Fund. Appl. Toxicol. 7*, 398–405.

Miller, R. P., and Becker, B. A. (1975). Teratogenicity of oral diazepam and diphenylhydantoin in mice. *Toxicol. Appl. Pharmacol. 32*, 53–61.

Mills, J. L., Kirke, P. N., Molloy, A. M., Burke, H., Conley, M. R., Lee, Y. J., Mayne, P. D., Weir, D. G., and Scott, J. M. (1999). Methylenetetrahydrofolate reductase thermolabile variant and oral clefts. *Am. J. Med. Genet. 86*, 71–74.

Milunsky, A., Graef, J. W., and Gaynor, M. F., Jr. (1968). Methotrexate induced congenital malformations. *J. Pediatr. 72*, 790–795.

Mimura, J., Yamashita, K., Nakamura, K., Morita, M., Takagi, T. N., Nakao, K., Ema, M., Sogawa, K., Yasuda, M., Katsuki, M., and Fujii-Kuriyama, Y. (1997). Loss of teratogenic response to 2,3,7,8-tetrachlorodibenzo-p-dioxin (TCDD) in mice lacking the Ah (dioxin) receptor. *Genes Cells (PMID 9704006) 2*, 645–654.

Mitchell, L. E. (1997). Genetic epidemiology of birth defects: nonsyndromic cleft lip and neural tube defects. *Epidemiol. Rev. 19*, 61–68.

Mitchell, L. E., Healey, S. C., and Chenevix-Trench, G. (1995). Evidence for an association between nonsyndromic cleft lip with or without cleft palate and a gene located on the long arm of chromosome 4. *Am. J. Hum. Genet. 57*, 1130–1136.

Mitchell, L. E., Murray, J. C., O'Brien, S., and Christensen, K. (2001). Evaluation of two putative susceptibility loci for oral clefts in the Danish population. *Am. J. Epidemiol. 15*, 1007–1015.

Montenegro, M. A., Rojas, M., Dominguez, S., and Rosales, C. J. (1998). Difference in extracellular matrix components and cell density during normal and dexamethasone-treated secondary palate development in two strains of mice with different susceptibility to glucocorticoid-induced clefting. *J. Craniofac. Genet. Devel. Biol. 18*, 100–106.

Mooney, M. P., Losken, H. W., Tschakaloff, A., Siegel, M. I., Losken, A., and Lalikos, J. F. (1993). Congenital bilateral coronal suture synostosis in a rabbit and comparisons with experimental models. *Cleft Palate Craniofac. J. 30*, 121–128.

Morriss-Kay, G. M., and Sokolova, N. (1996). Embryonic development and pattern formation. *FASEB J. 10*, 961–968.

Munger, R., Isacson, P., Kramer, M., Hanson, J., Burns, T., Cherry-Holmes, K., and Hauselr, W. (1992). Birth defects and pesticide-contaminated water supplies in Iowa. *Am. J. Epidemiol. 136*, 959.

Munger, R., Lidral, A., Basart, A., Romitti, P., and Murray, J. (1995). The Msx-1 homeobox gene and risk of isolated orofacial clefts: an effect modified by maternal vitamin use? *Am. J. Epidemiol. 141*, S2.

Munger, R., Romitti, P. A., Daack-Hirsch, S., Burns, T. L., Murray, J. C., and Hanson, J. (1996). Maternal alcohol use and risk of orofacial cleft birth defects. *Teratology 54*, 27–33.

Murray, J. C. (1995). Invited editorial face facts: genes, environment, and clefts. *Am. J. Hum. Genet. 57*, 227–232.

Murray, J. C., Daack-Hirsch, S., Buetow, K. H., Munger, R., Espina, L., Paglinawan, N., Villanueva, E., Rary, J., Magee, K., and Magee, W. (1997) Clinical and epidemiological studies of cleft lip and palate in the Philippines. *Cleft Palate Craniofac. J. 34*, 7–10.

Mutchinick, O., Aizpuru, E., and Grether, P. (1992). The human teratogenic effect of cyclophosphamide. *Austr. Teratol. Abstr. 45*, 329.

Nagao, T., and Fujikawa, K. (1996). Male-mediated teratogenesis: spectrum of congenital malformations in the offspring of A/J male mice treated with ethylnitrosourea. *Teratogen. Carcinogen. Mutagen. 16*, 301–305.

Nagao, T., Golor, G., Hagenmaier, H., and Neubert, D. (1993). Teratogenic potency of 2,3,4,7,8-pentachlorodibenzofuran and of three mixtures of polychlorinated dibenzo-p-dioxins and dibenzofurans in mice. Problems with risk assessment using TCDD toxic-equivalency factors. *Arch. Toxicol. 67*, 591–597.

Nakane, K., and Kameyama, Y. (1986). Effect of maternal urethane administration on the manifestation of cleft lip and palate in CL/Fr mice. *J. Craniofac. Genet. Devel. Biol. 2* (suppl.), 109–112.

Natsume, N., Kawai, T., and Suzuki, T. (1995). Preferences for vegetables rich in β-carotene and manifestation of cleft lip and/or palate. *Plast. Reconstr. Surg. 95*, 934–935.

Natsume, N., Kawai, T., Ogi, N., and Yoshida, W. (2000). Maternal risk factors in cleft lip and palate: case control study. *Br. J. Oral Maxillofac. Surg. 38*, 23–25.

Nelson, C. J., and Holson, J. F. (1978). Statistical analysis of teratologic data: problems and advancements. *J. Environ. Pathol. Toxicol. 2*, 187–199.

Nelson, C. J., Holson, J. F., Green, H. G., and Gaylor, D. W. (1979). Retrospective study of the relationship between agricultural use of 2,4,5-T and cleft palate occurrence in Arkansas. *Teratology 19*, 377–383.

Noda, T., Morita, S., and Baba, A. (1993). Teratogenic effects of various di-n-butyltins with different anions and butyl (3-hydroxybutyl) tin dilaurate in rats. *Toxicology 85*, 149–160.

Nugent, P., and Greene, R. M. (1998). MSX-1 gene expression and regulation in embryonic palatal tissue. *In Vitro Cell. Devel. Biol. Anim. 34*, 831–835.

Numazawa, S., Shibata, M., Imaoka, S., Funae, Y., and Yoshida, T. (2000). Induction of hepatic CYP2B1/2 by a teratogenic compound cis-1-[-4-(p-menthane-8-yloxy) phenyl] piperadine (YM9429) in rats. *J. Toxicol. Sci. 25*, 57–61.

Nurminen, T., Rantala, K., Kurpaa, K., and Holmberg, P. C. (1995). Agricultural work during pregnancy and selected structural malformations in Finland. *Epidemiology 6*, 23–30.

Ohnishi, M. (1989). Craniofacial malformations induced by N-ethyl-N-nitrosourea in rat embryos in vivo and in vitro. *Kobe J. Med. Sci. 35*, 47–63.

Omnell, M. L., Sim, F. R., Keeler, R. F., Harne, L. C., and Brown, K. S. (1990). Expression of Veratrum alkaloid teratogenicity in the mouse. *Teratology 42*, 105–119.

Ortega, A., Puig, M., and Domingo, J. L. (1991). Maternal and developmental toxicity of low doses of cytosine arabinoside in mice. *Teratology 44*, 379–384.

Padmanabhan, D., and Pallot, D. J. (1995). Aspirin-alcohol interaction in the production of cleft palate and limb malformations in the TO mouse. *Teratology 51*, 404–417.

Panter, K. E., and Keeler, R. F. (1993). Quinolizidine and piperidine alkaloid teratogens from poisonous plants and their mechanism of action in animals. *Vet. Clin. North Am. Food Anim. Prac. 9*, 33–40.

Panter, K. E., Weinzweig, J., Gardner, D. R., Stegelmeier, B. L., and James, L. F. (2000). Comparison of cleft palate induction by *Nicotiana glauca* in goats and sheep. *Teratology 61*, 203–210.

Paulson, R. B., Shanfeld, J., Prause, L., Iranpour, S., and Paulson, J. O. (1991). Pre- and postconceptional tobacco effects on the CD-1 mouse fetus. *J. Craniofac. Genet. Devel. Biol. 11*, 48–58.

Persing, J. A., Babler, W. J., and Jane, J. A. (1986). Experimental unilateral coronal synostosis in rabbits. *Plast. Reconstr. Surg. 77*, 369–376.

Peterka, M., Tvrdek, M., Likovsky, Z., Peterkova, R., and Fara, M. (1994). Maternal hyperthermia and infection as one of possible causes of orofacial clefts. *Acta Chir. Plast. 36*, 114–118.

Platzek, T., and Bochert, G. (1996). Dose-response relationship of teratogenicity and prenatal-toxic risk estimation of 6-mercaptopurine riboside in mice. *Teratogen. Carcinogen. Mutagen. 16*, 169–181.

Pleet, H., Graham, J. M. J., and Smith, D. W. (1981). Central nervous system and facial defects associated with maternal hyperthermia during early gestation. *Pediatrics 67*, 785.

Porter, A., and Singh, S. M. (1988). Transplacental teratogenesis and mutagenesis in mouse fetuses treated with cyclophosphamide. *Teratogen. Carcinogen. Mutagen. 8*, 191–203.

Poswillo, D. (1968). The aetiology and surgery of cleft palate with micrognathia. *Ann. Royal Coll. Surg. Eng. 43*, 61–88.

Pratt, R. M. (1985). Receptor-dependent mechanisms of glucocorticoid and dioxin-induced cleft palate. *Environ. Health Perspec. 61*, 35–40.

Proetzel, G., Pawlowski, S. A., Wiles, M. V., Yin, M., Boivin, G. P., Howles, P. N., Ding, J., Ferguson, M. W., and Doetschman, T. (1995). Transforming growth factor-beta 3 is required for secondary palate fusion. *Nat. Genet. 11*, 409–414.

Rajini, P., Last, J. A., Pinkerton, K. E., Hendrickx, A. G., and Witschi, H. (1994). Decreased fetal weights in rats exposed to sidestream cigarette smoke. *Fund. Appl. Toxicol. 22*, 400–404.

Reddy, R. V., Mayura, K., Hayes, A. W., and Berndt, W. O. (1982). Embryocidal, teratogenic and fetotoxic effects of citrinin in rats. *Toxicology 25*, 151–160.

Reilly, B. J., Leeming, J. M., and Fraser, D. (1964). Craniosynostosis in the rachitic spectrum. *J. Pediatr. 64*, 396–405.

Riggs, W., Wilrow, R. S., and Eteldorf, J. N. (1972). Neonatal hyperthyroidism with accelerated skeletal maturation, craniosynostosis and brachydactyly. *Radiology 105*, 621–625.

Robertson, R. T., Allen, H. L., and Bokelman, D. L. (1979). Aspirin: teratogenic evaluation in the dog. *Teratology 20*, 313–320.

Rodriguez Gonzalez, M. D., and Friman Perez, M. (1985). Teratogenic effect of trifluoperazine in rats and mice. *Acta Biol. Hung. 36*, 233–237.

Rodriguez Gonzalez, M. D., Lima Perez, M. T., and Sanabria Negrin, J. G. (1983). The effect of cyproheptadine chlorhydrate on rat embryonic development. *Teratogen. Carcinogen. Mutagen. 3*, 439–446.

Rodriguez-Pinilla, E., and Martinez-Frias, M. L. (1998). Corticosteroids during pregnancy and oral clefts: a case-control study. *Teratology 58*, 2–5.

Rogers, J. M., Mole, M. L., Chernoff, N., Barbee, B. D., Turner, C. I., Logsdon, T. R., and Kavlock, R. J. (1993). The developmental toxicity of inhaled methanol in the CD-1 mouse, with quantitative dose-response modeling for estimation of benchmark doses. *Teratology 47*, 175–188.

Romitti, P. A., Lidral, A. C., Munger, R. G., Daack-Hirsch, S., Burns, T. L., and Murray, J. C. (1999). Candidate genes for nonsyndromic cleft lip and palate and maternal cigarette smoking and alcohol consumption: evaluation of genotype-environment interactions from a population-based case-control study of orofacial clefts. *Teratology 59*, 39–50.

Rosa, F. W., Wilk, A. L., and Kelsey, F. O. (1986). Teratogen update: vitamin A congeners. *Teratology 33*, 355–364.

Rowland, J. M., and Hendrickx, A. G. (1983b). Corticosteroid teratogenicity. *Adv. Vet. Sci. Comp. Med. 27*, 99–128.

Rowsell, A. R. (1988). The amniotic band disruption complex. The pathogenesis of congential limb ring-constrictions; an experimental study in the foetal rat. *Br. J. Plast. Surg. 41*, 45–51.

Roy, W. A., Iorio, R. J., and Meyer, G. A. (1981). Craniosynostosis in vitamin D resistant rickets: a mouse model. *J. Neurosurg. 55*, 265–271.

Safari, H. R., Fassett, M. J., Souter, I. C., Alsulyman, O. M., and Goodwin, T. M. (1998). The efficacy of methylprednisolone in the treatment of hyperemesis gravidarum: a randomized, double-blind, controlled study. *Am. J. Obstet. Gynecol. 179*, 921–924.

Sakanashi, T. M., Rogers, J. M., Fu, S. S., Connelly, L. E., and Keen, C. L. (1996). Influence of maternal folate status on the developmental toxicity of methanol in the CD-1 mouse. *Teratology 54*, 198–206.

Samren, E. B., van Duijn, C. M., Christiaens, G. C., Hofman, A., and Lindhout, D. (1999). Antiepileptic drug regimens and major congenital abnormalities in the offspring. *Ann. Neurol. 46*, 739–746.

Sauerbier, I. (1986). Circadian variation in teratogenic response to dexamethasone in mice. *Drug Chem. Toxicol. 9*, 25–31.

Saxen, L. (1974). Cleft lip and palate in Finland: parental histories, course of pregnancy and selected environmental factors. *Int. J. Epidemiol. 3*, 263.

Saxen, L., Holmberg, P. C., Nurminen, M., and Kuosma, E. (1982). Sauna and congenital defects. *Teratology 25*, 309–313.

Schutte, B. C., and Murray, J. C. (1999). The many faces and factors of orofacial clefts. *Hum. Molec. Genet. 8*, 1853–1859.

Schwetz, B. A., Mast, T. J., Weigel, R. J., Dill, J. A., and Morrissey, R. E. (1991). Developmental toxicity of inhaled methyl ethyl ketone in Swiss mice. *Fund. Appl. Toxicol. 16*, 742–748.

Seidman, J. D., Abbondanzo, S. L., Watkin, W. G., Ragsdale, B., and Manz, H. J. (1989). Amniotic band syndrome. Report of two cases and review of the literature. *Arch. Pathol. Lab. Med. 113*, 891–897.

Shah, R. M. (1984). Morphological, cellular, and biochemical aspects of differentiation of normal and teratogen-treated palate in hamster and chick embryos. *Curr. Top. Devel. Biol. 19*, 103–135.

Shah, R. M., Donaldson, D., and Burdett, D. (1979). Teratogenic effects of diazepam in the hamster. *Can. J. Physiol. Pharmacol. 57*, 556–561.

Shah, R. M., Schuing, R., Benkhaial, G., Young, A. V., and Burdett, D. (1991). Genesis of hadacidin-induced cleft palate in hamster: morphogenesis, electron microscopy, and determination of DNA synthesis, cAMP, and enzyme acid phosphatase. *Am. J. Anat. 192*, 55–68.

Shahinian, H. K., Jackle, R., Suh, R. H., Jarrahy, R., Aguilar, V. C., and Soojian, M. (1998). Obstetrical factors governing the etiopathogenesis of lambdoid synostosis. *Am. J. Perinatol. 15*, 281–286.

Sharma, A., and Rawat, A. K. (1986). Teratogenic effects of lithium and ethanol in the developing fetus. *Alcohol 3*, 101–106.

Shaw, D., Ray, A., Marazita, M., and Field, L. (1993). Further evidence of a relationship between the retinoic acid receptor alpha locus and nonsyndromic cleft lip with or without cleft palate (CL +/– P). *Am. J. Hum. Genet. 53*, 1156–1157.

Shaw, E. B., and Steinbach, H. L. (1968). Aminopterin-inducted fetal malformation. *Am. J. Dis. Child 115*, 477.

Shaw, G. M., and Lammer, E. J. (1999). Maternal periconceptional alcohol consumption and risk for orofacial clefts. *J. Pediatr. 134*, 298–303.

Shaw, G. M., Lammer, E. J., Wasserman, C. R., O'Malley, C. D., and Tolarova, M. M. (1995). Risks of orofacial clefts in children born to women using multivitamins containing folic acid periconceptionally. *Lancet 346*, 393–396.

Shaw, G. M., Wasserman, C. R., Lammer, E. J., O'Malley, C. D., Murray, J. C., Basart, A. M., and Tolarova, M. M. (1996). Orofacial clefts, parental cigarette smoking, and transforming growth factor–alpha gene variants. *Am. J. Hum. Genet. 58*, 551–561.

Shaw, G. M., Wasserman, C. R., Murray, J. C., and Lammer, E. J. (1998a). Infant TGF-alpha genotype, orofacial clefts, and maternal periconceptional multivitamin use. *Cleft Palate Craniofac. J. 35*, 366–370.

Shaw, G. M., Rozen, R., Finnell, R. H., Todoroff, K., and Lammer, E. J. (1998b). Infant C677T mutation in MTHFR, maternal periconceptional vitamin use, and cleft lip. *Am. J. Med. Genet. 80*, 196–198.

Shibata, M. (1993). A new potent teratogen in CD rats inducing cleft palate. *J. Toxicol. Sci. 18*, 171–178.

Shuey, D. L., Lau, C., Logsdon, T. R., Zucker, R. M., Elstein, K. H., Narotsky, M. G., Setzer, R. W., Kavlock, R. J., and Rogers, J. M. (1994). Biologically based dose-response modeling in developmental toxicology: biochemical and cellular sequelae of 5-fluorouracil exposure in the developing rat. *Toxicol. Appl. Pharmacol. 126*, 129–144.

Siegel, M. I., and Mooney, M. P. (1986). Palatal width growth rates as the genetic determinant of cleft palate induced by vitamin A. *J. Craniofac. Genet. Devel. Biol. 2* (suppl.), 187–191.

Simpson, W. J. (1957). A preliminary report on cigarette smoking and the incidence of prematurity. *Am. J. Obstet. Gynecol. 73*, 808–815.

Sivak, A. (1994). Coteratogenic effects of caffeine. *Reg. Toxicol. Pharmacol. 19*, 1–13.

Spranger, J., Benirschke, K., Hall, J. G., Lenz, W., Lowry, R. B., Opitz, J. M., Pinsky, L., Schwarzacher, H. G., and Smith, D. W. (1982). Errors of morphogenesis: concepts and terms. Recommendations of an international working group. *J. Pediatr. 100*, 160–165.

Starreveld-Zimmerman, A. A. E., van der Kolk, W. J., Elshove, J., and Meinardi, H. (1974). Teratogenicity of antiepileptic drugs. *Clin. Neurol. Neurosurg. 77*, 81–95.

Stein, J. D., Hecht, J. T., and Blanton, S. H. (1995). Exclusion of retinoic acid receptor and a cartilage matrix protein in non-syndromic CL(P) families. *Comment. J. Med. Genet. 32*, 78.

Stelnicki, E. J., Hoffman, W., and Foster, R. (1998). The in utero repair of Tessier number 7 lateral facial clefts created by amniotic band-like compression. *J. Craniofac. Surg. 9*, 557–562.

Stevenson, R. E., Hall, J. G., and Goodman, R. M. (eds.) (1993). *Human Malformations and Related Anomalies* (New York: Oxford University Press).

Strauss, O. A. (1914). Predisposing causes of cleft palate and harelip. *Trans. 6th Int. Dent. Congr. London*, pp. 470–471.

Streeter, G. L. (1930). Focal deficiencies in fetal tissue and their relation to intrauterine amputations. *Contrib. Embryol. Carnegie Inst. 22*, 1–44.

Streissguth, A. (1997). *Fetal Alcohol Syndrome* (East Peoria, IL: Paul Brookes Publishing Co).

Sulik, K. K., Johnston, M. C., Daft, P. A., Russell, W. E., and Dehart, D. B. (1986). Fetal alcohol syndrome and DiGeorge anomaly: critical ethanol exposure periods for craniofacial malformations as illustrated in an animal model. *Am. J. Med. Genet. 2* (suppl.), 97–112.

Sulik, K. K., Cook, C. S., and Webster, W. S. (1988). Teratogens and craniofacial malformations: relationships to cell death. *Development 103* (suppl.), 213–231.

Takeno, S., Nakagawa, M., and Sakai, T. (1990). Teratogenic effects of nitrazepam in rats. *Res. Commun. Chem. Pathol. Pharmacol. 69*, 59–70.

Thiersch, J. B. (1952). Therapeutic abortions with a folic acid antagonist 4 amino pteroyl glutamic, administered by the oral route. *Am. J. Obstet. Gynecol. 63*, 1298.

Thomas, D. C., Petitti, D. B., Goldhaber, M., Swan, S. H., Rappaport, E. B., and Hertz-Picciotto, I. (1992). Reproductive outcomes in relation to malathion spraying in the San Francisco Bay Area, 1981–1982. *Epidemiology 3*, 32–39.

Thomas, H. F. (1980). 2,4,5-T use and congenital malformation rates in Hungary. *Lancet 2*, 214–215.

Thomas, W. C., Seckar, J. A., Johnson, J. T., Ulrich, C. E., Klonne, D. R., Schardein, J. L., and Kirwin, C. J. (1987). Inhalation teratology studies of n-butyl mercaptan in rats and mice. *Fund. Appl. Toxicol. 8*, 170–178.

Tolarova, M., and Harris, J. (1995). Reduced recurrence of orofacial clefts after periconceptional supplementation with high-dose folic acid and multivitamins. *Teratology 51*, 71–78.

Torpin, R. (1965). Amniochorionic amesoblastic fibrous strings and amniotic bands. *Am. J. Obstet. Gynecol. 91*, 65–75.

Trasler, D. G., and Ohannessian, L. (1983). Ultrastructure of initial nasal process cell fusion in spontaneous and 6-aminonicotinamide-induced mouse embryo cleft lip. *Teratology 28*, 91–101.

Vedel-Macrander, G. C., and Hood, R. D. (1986). Teratogenic effects of nigericin, a carboxylic ionophore. *Teratology 33*, 47–51.

Vergato, L. A., Doerfler, R. J., Mooney, M. P., and Siegel, M. I. (1997). Mouse palatal width growth rates as an "at risk" factor in the development of cleft palate induced by hypervitaminosis A. *J. Craniofac. Genet. Devel. Biol. 17*, 204–210.

Walker, B. E., and Patterson, A. (1974). Induction of cleft palate in mice by tranquilizers and barbiturates. *Teratology 10*, 159–163.

Warner, R. H., and Rosett, H. L. (1975). The effects of drinking on offspring: an historical survey of the American and British literature. *J. Stud. Alcohol 36*, 1395–1420.

Weber, H., and Birnbaum, L. S. (1985). 2,3,7,8-Tetrachlorodibenzo-p-dioxin (TCDD) and 2,3,7,8-tetrachlorodibenzofuran (TCDF) in pregnant C57BL/6N mice: distribution to the embryo and excretion. *Arch. Toxicol. 57*, 159–162.

Weber, H., Harris, M. W., Haseman, J. K., and Birnbaum, L. S. (1985). Teratogenic potency of TCDD, TCDF and TCDD-TCDF combinations in C57BL/6N mice. *Toxicol. Lett. 26*, 159–167.

Webster, W. S., and Ritchie, H. E. (1991). Teratogenic effects of alcohol and isotretinoin on craniofacial development: an analysis of animal models. *J. Craniofac. Genet. Devel. Biol. 11*, 296–302.

Webster, W. S., Brown-Woodman, P. D., Snow, M. D., and Danielsson, B. R. (1996). Teratogenic potential of almokalant, dofetilide, and d-sotalol: drugs with potassium channel blocking activity. *Teratology 53*, 168–175.

Wells, P. G. (1983). Physiological and environmental determinants of phenytoin teratogenicity: relation to glutathione homeostasis, and potentiation by acetaminophen. *Prog. Clin. Biol. Res. 135*, 367–371.

Werler, M. M., Lammer, E. J., Rosenberg, L., and Mitchell, A. A. (1990). Maternal cigarette smoking during pregnancy in relation to oral clefts. *Am. J. Epidemiol. 132*, 926–932.

Werler, M. M., Lammer, E. J., Rosenberg, L., and Mitchell, A. A. (1991). Maternal alcohol use in relation to selected birth defects. *Am. J. Epidemiol. 134*, 691–698.

Werler, M. M., Hayes, C., Louik, C., Shapiro, S., and Mitchell, A. A. (1999). Multivitamin supplementation and risk of birth defects. *Am. J. Epidemiol. 150*, 675–682.

White, F. M. M., Cohen, F. G., Sherman, G., and McCurdy, R. (1988). Chemicals, birth defects and stillbirths in New Brunswick: associations with agricultural activity. *Can. Med. Assoc. J. 138*, 117–124.

Wiegand, U. W., Hartmann, S., and Hummler, H. (1998). Safety of vitamin A: recent results. *Int. J. Vitam. Nutr. Res. 68*, 411–416.

Willhite, C. C., Hill, R. M., and Irving, D. W. (1986). Isotretinoin-induced craniofacial malformations in humans and hamsters. *J. Craniofac. Genet. Devel. Biol. 2* (suppl.), 193–209.

Willis, F. R., and Beattie, T. J. (1997). Craniosynostosis in X-linked hypophosphataemic rickets. *J. Pediatr. Child Health 33*, 78–79.

Wilson, J. G. (1977). Environmental chemicals. In J. G. Wilson and F. C. Fraser (eds.), *Handbook of Teratology* (New York: Plenum), pp. 357–385.

Winn, L. M., Kim, P. M., and Wells, P. G. (1998). Investigation of the tobacco-specific carcinogen 4-(methylnitrosamino)-1-(3-pyridyl)-1-butanone for in vivo and in vitro murine embryopathy and embryonic ras mutations. *J. Pharmacol. Exp. Therapeut. 287*, 1128–1135.

Wong, W. Y., Eskes, T. K., Kuijpers-Jagtman, A. M., Spauwen, P. H., Steeger, E. A., Thomas, C. M., Hamel, B. C., Blom, H. J., and Steeger-Theunissen, R. P. (1999). Nonsyndromic orofacial clefts: association with maternal hyperhomocysteinemia. *Teratology 60*, 253–257.

Wyszynski, D. F., and Beaty, T. H. (1996). Review of the role of potential teratogens in the origin of human nonsyndromic oral clefts. *Teratology 53*, 309–317.

Wyszynski, D. F., and Diehl, S. R. (2000). Infant C677T mutation in MTHFR, maternal periconceptional vitamin use, and risk of nonsyndromic cleft lip. *Am. J. Med. Genet. 92,* 79–80.

Wyszynski, D. F., Duffy, D. L., and Beaty, T. H. (1997). Maternal cigarette smoking and oral clefts: a meta-analysis. *Cleft Palate Craniofac. J. 34,* 206–210.

Yasuda, Y., Datu, A. R., Hirata, S., and Fujimoto, T. (1985). Characteristics of growth and palatal shelf development in ICR mice after exposure to methylmercury. *Teratology 32,* 273–286.

Yoshimura, H. (1987). Embryolethal and teratogenic effects of bromofenofos in rats. *Arch. Toxicol. 60,* 319–324.

Young, D. L., Schneider, R. A., Hu, D., and Helms, J. A. (2000). Genetic and teratogenic approaches to craniofacial development. *Crit. Rev. Oral Biol. Med. 11,* 304–317.

Zackai, E., Mellman, W. J., Neidererm, B., and Hanson, J. W. (1975). The fetal trimethadione syndrome. *J. Pediatr. 87,* 280–284.

Zhang, J., and Cai, W. W. (1993). Association of the common cold in the first trimester of pregnancy with birth defects. *Pediatrics 92,* 559–563.

PART III

ANIMAL MODELING

CHAPTER 9

ANIMAL MODELS OF CRANIOSYNOSTOSIS: EXPERIMENTAL, CONGENITAL, AND TRANSGENIC MODELS

MARK P. MOONEY, Ph.D., MICHAEL I. SIEGEL, Ph.D., and
LYNNE M. OPPERMAN, Ph.D.

9.1 INTRODUCTION: THE GENERAL UTILITY OF ANIMAL MODELS FOR CRANIOFACIAL BIOLOGY

Determining an appropriate animal model for any experimental manipulation is fundamental to successful experimental design (Schmidt-Nielsen, 1961; Brown, 1963; Navia, 1977; Siegel and Mooney, 1990; Kremenak, 1990; Johnston and Bronsky, 1991). This can be a complicated procedure and a heated topic of debate (Prince et al., 1988; Rowsey, 1988; Fernandes, 1989; Loeb et al., 1989; Roach et al., 1989; English, 1989; Taylor et al., 1991; Cramer, 1998). Over a decade ago, we chaired a continuing education study session at the 42nd Annual Meeting of the American Cleft Palate–Craniofacial Association (Siegel et al., 1985). The topic for discussion was the utilization of appropriate animal models for craniofacial research. The need for this study session was based in part on the authors' experiences with the pitfalls of choosing appropriate experimental models, from discussions and interactions with a number of craniofacial biologists actively involved in experimental research, as a reference for less experienced researchers and students designing experimental studies, and as an aid to clinicians who must critically evaluate the results obtained from published animal studies for validity and potential extrapolation to the human clinical condition. As a result of this symposium (Siegel et al., 1985), we were able to propose a number of practical criteria to facilitate appropriate animal model choice for craniofacial biology studies (Siegel and Mooney, 1990).

9.1.1 Levels of Hypothesis Testing and Appropriate Animal Model Choice

Although nonhuman primates are phylogenetically closer to humans than other mammalian groups, which may make extrapolation of findings to humans theoretically "better," not all craniofacial experimental manipulations require nonhuman primate models. We suggested (Siegel and Mooney, 1990; Losken et al., 1992, 1994; Mooney and Siegel, 1993) that the choice of an appropriate animal model for cranio-

Understanding Craniofacial Anomalies: The Etiopathogenesis of Craniosynostoses and Facial Clefting, Edited by Mark P. Mooney and Michael I. Siegel, ISBN 0-471-38724-x Copyright © 2002 by Wiley-Liss, Inc.

facial biology be based, in part, on criteria initially suggested by Smith (1969), Reynolds (1969), and Goldsmith and Moor-Jankowski (1969) for determining appropriate animal models in toxicologic studies. Smith (1969), based on comparative studies of drug disposition across a large number of mammal taxa, reported great variation in drug metabolism within the various primate groups. He suggested that choosing the best animal model should be based on similar physiologic pathways between the model and the human condition, and not necessarily phyletic affinity. Smith (1969) also suggested that the level of hypothesis testing and the expected extrapolation of the results to humans be of major concern in choosing the best animal model. Based on these practical criteria, we proposed that the choice of an appropriate model for craniofacial biology also be linked to levels of hypothesis testing and/or expected extrapolation to the human clinical condition (Siegel and Mooney, 1990). Criteria important in determining animal model appropriateness should include comparative anatomic considerations and regional craniofacial growth patterns.

In our opinion, the majority of experimental manipulations in craniofacial biology research can be divided into three paradigms: manipulations of craniofacial tissue (including genes, molecules, and cells), manipulations of general craniofacial growth, and manipulations of regional cranial or facial growth. Animal model "goodness-of-fit" will vary across these paradigms as the need for extrapolation to the human condition increases.

9.1.1.1 "Generic" Mammalian Models and Basic Developmental Biology
Studies involving manipulations of craniofacial genes, molecules, cells, and tissue are viewed as the most basic. Humans, as mammals, share primitive homeobox genes

and bone and soft tissue responses with most other mammals (Young, 1989; Johnston and Bronsky, 1991, 1995; Cohen, 1993, 1997, 2000a, 2000b; Herring, 2000) (Chapters 4, 5, 6, 7, 20, 21, and 22). In general, these primitive mechanisms include similar bone and soft tissue responses to trauma (Kremenak and Searls, 1971; Navia, 1977; Carlson et al., 1982; Bardach and Kelly, 1987); basic craniofacial embryogenesis (Fraser, 1980; Johnston, 1980; Poswillo, 1980; Johnston and Bronsky, 1991; Vergato et al., 1997); and functional craniofacial skeletal adaptations (i.e., craniofacial growth and development) related, for example, to Wolff's law (Enlow and Azuma, 1975; Babler and Persing, 1982; Enlow, 1990). Thus, appropriate animal models at this level of hypothesis testing would include most "generic" laboratory animals (such as rodents and lagomorphs). The data derived from studies using the generic animal model are as valid as those derived from carnivore or primate models, given that no attempt is made to draw conclusions beyond this level of hypothesis testing.

9.1.1.2 Phylogenetically "Closer" Models and Comparative Craniofacial Growth Patterns
Studies involving manipulations of general craniofacial growth patterns present more difficult problems for choosing an appropriate animal model. Depending on the level of extrapolation to the human condition, not all animal models may be appropriate at this level of hypothesis testing. The commonly used generic animal models (such as rodents and lagomorphs) differ significantly in their basic anatomic and developmental craniofacial "bauplan" compared to phylogenetically closer models (such as some carnivores and nonhuman primates) (Gregory, 1929; De Beer, 1937; Romer and Parsons, 1986; Young, 1989) (Figs. 9.1, 9.2, and 9.3).

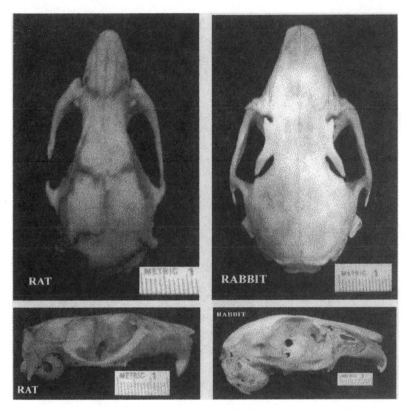

Figure 9.1 Superior and lateral views of cleaned and dried skulls from two infant generic animal models. Note the dentition, dolichocephaly, flattened cranial vaults, and lordotic cranial base angulations.

The more common generic animal models (rats and mice) are monophyodonts, exhibiting only a single set of the molar and incisor classes (Gregory 1929; De Beer, 1937; Farris and Griffith, 1949; Greene, 1968; Cooper and Schiller, 1975; Navia, 1977). These generic animal models are also dolichocephalic from limited neural cerebralization (Gregory, 1929; Kier, 1976; Romer and Parsons, 1986) (Chapter 11) and have prebasal lordosis from limited cranial base flexion (Moss et al., 1987; Smith et al., 1996) (Chapter 11 and Fig. 9.1).

In contrast, the phylogenetically closer models (carnivores and nonhuman primates) exhibit more similar craniofacial structures to humans than do the generic models (Figs. 9.2 and 9.3). They possess succedaneous dentition with varying eruption times of both the deciduous and permanent teeth (Gregory, 1929; Hartman and Strauss, 1933; De Beer, 1937; Elliot, 1963; Krogman, 1969; Schultz, 1969; Getty, 1975; Navia, 1977; Swindler and Wood, 1982; Sirianni, 1985) (Chapter 15), while nonhuman primates are brachycephalic from increased neural corticalization and gyrification (Gregory, 1929; Kier, 1976; Sirianni and Swindler, 1979; Romer and Parsons, 1986) (Chapter 11), and have prebasal kyphosis from extensive cranial base flexion (De Beer, 1937; DuBrul and Laskin, 1961; Badoux, 1966; Enlow and McNamara, 1973;

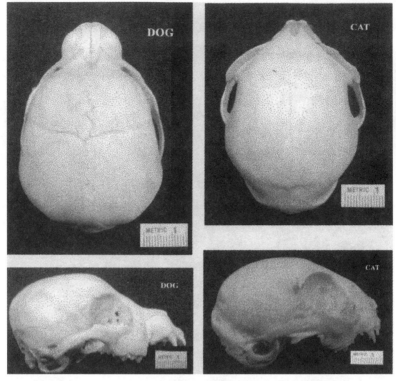

Figure 9.2 Superior and lateral views of cleaned and dried skulls from two infant phylogenetically closer animal models. Note the dentition, higher cranial vaults, and lordotic cranial base angulations.

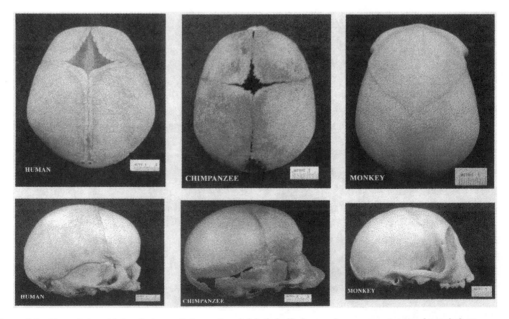

Figure 9.3 Superior and lateral views of cleaned and dried skulls from a human neonate and two infant primate animal models. Note the brachycephaly, superiorly placed cranial vaults, and kyphotic cranial base angulations.

Michejda, 1975; Enlow, 1990; Lieberman and McCarthy, 1999; Lieberman et al., 2000) (Chapter 11).

9.1.1.3 Fitting the Appropriate Animal Model to Specific Human Conditions

Studies involving the experimental manipulations of regional craniofacial growth patterns present the greatest difficulty in choosing an appropriate animal model because of direct application to clinical situations. Appropriate animal models at this level of hypothesis testing would be those that exhibit similar regional growth patterns to humans (Fig. 9.3). In general, human regional craniofacial growth patterns are characterized by a high degree of cranial base flexion (kyphosis), nasomaxillary reduction, anteroinferior displacement of the midface, and significant vertical and transverse increases in both the neurocranium and upper face (Gregory, 1929; Enlow and Azuma, 1975; Young, 1989; Enlow, 1990) (Chapters 11 and 15). The human craniofacial growth pattern differs from the primitive adult mammalian condition of a long, slender dolichocephalic craniofacial skeleton to a broad, vertically flattened brachycephalic cranium without a prominent snout (for an excellent review, see Enlow, 1990, chapter 5) (Chapters 11 and 15).

Thus, experimental studies designed to manipulate regional craniofacial growth patterns should choose an animal model based on similar regional growth vectors and patterns, and not necessarily only on phyletic affinity. Although the expectation is that phyletic affinity and similarity of regional growth patterns are reasonably related (Sirianni and Swindler, 1979; Sirianni, 1985) (Chapter 15), phyletic affinity should not be the exclusive criterion, but should be used in conjunction with similar growth patterns. Unfortunately, animal model choice at this level of hypothesis testing has been traditionally limited to nonhuman primates, based primarily on

phyletic affinity (Levy et al., 1971; Dreizen and Levy, 1977; Navia, 1977) and not homologous regional growth patterns.

9.1.2 Comparative Animal Model Growth Pattern Assessment

The lack of systematic comparative growth pattern data led us to initiate a series of studies to compare regional craniofacial growth patterns among the more commonly used animal models for the midface (Siegel and Mooney, 1990; Mooney and Siegel, 1993), mandible (Losken et al., 1992), nasal capsule (Losken et al., 1994), and cranial vault (Chapter 11). In general, it was noted that under conditions of modeling specific human craniofacial growth patterns, a single animal model may not be appropriate for all regions. For example, short-faced animals (felids and short-faced nonhuman primates) showed similar midfacial and nasal capsule growth patterns to humans (Siegel and Mooney, 1990; Losken et al., 1994), while juvenile canines showed similar mandibular growth patterns to humans (Losken et al., 1992). It was concluded from these studies that for experimental situations in which the objectives are to improve the human condition, the use of a primate model simply on the basis of phyletic affinity is not justified, and an understanding of the relationship between the growth patterns in humans and the animal under consideration is a prerequisite to appropriate animal model choice (Siegel and Mooney, 1990; Losken et al., 1992, 1994; Mooney et al., 1993).

9.2 SPECIFIC ANIMAL MODELS OF CRANIAL VAULT GROWTH AND CRANIOSYNOSTOSIS

9.2.1 Historical Perspectives

Moss (1954) may have been one of the first to challenge prevailing thought, or at least weigh in on the debates of the previous several decades. He investigated the relationship of the shape of cranial bones to the position of cranial sutures, the relationship of suture function to growth, and, more importantly, addressed the role of the suture as a site of growth (i.e., passive versus "epiphyseal center"). The major conclusions of his study on young (actively growing) Long-Evans rats were that bone shape was not predetermined by sutural location (rather that the location of sutures was determined by bones) and that the sutures were not chief sites of growth. Perhaps of more importance for the resolution of such conflicts of interpretation that arise periodically was Moss's insight of viewing the growth of the calvaria as a continuous or unified process.

Frequently, the notion of phylogenetic affinity as the basis for choice is presented as being of primary importance, and, in fact, it has been demonstrated to be true for a number of selected regions. With the advent of modern technology being applied to the field of genetics, we can now appreciate that because of "conserved" genetic material (DNA), genetic models may also compete for primacy (Wilkie and Wall, 1996; Cohen, 2000c).

In our initial attempt to improve surgical management of patients with coronal suture synostosis, we utilized a mechanical rabbit animal model with the synostosis simulated by methacrylate bonding (Losken et al., 1991a). It was during this investigation that we discovered a mutant congenital model for familial, nonsyndromic coronal suture synostosis (Mooney et al., 1993). That a phylogenetically closer (nonhuman primate) animal would seem to be better suited is supported by growth studies, but the advantages of using an animal that exhibits the same or similar alterations in the genome are multiplex. The advantages of such a model were pointed out by Cohen in 1993, and he, too, recognized that molecular techniques could lead to an understanding as well as a "cure." Yet, since around 1980, there have been only a handful of studies of an experimental nature, and immobilization by mechanical means has been the most frequent choice.

Understanding craniosynostosis does not necessitate a model of synostosis, but we are able to learn about the abnormal condition by studying the norm. Although Babler and co-workers (1982a) concluded from rabbit extirpation studies that the dura plays a role in regulating skull growth, this same conclusion had been reached some 20 years earlier by Moss (1960), who was interested in understanding the transmission of mechanical forces in the skull. Indeed, the work over four decades by Moss (and co-authors) may be viewed as both a starting point for animal models (mechanical) of craniosynostosis and fuel for debate in the quest for an understanding of the various underlying mechanisms.

At about the same time as Babler's (1982b) research, Nappen and Kokich (1983) were testing hypotheses about fetal head constraint by using cyanoacrylate to immobilize frontal and parietal bones in the skulls of rabbits. Their results rejected the hypothesis that nonsyndromic craniosynostosis was caused by constraint. A by-product of their work, however, helped clarify the role of the periosteum in the maintenance of sutures. Much of the early work would not have seemed relevant because the synostosis problem was not yet identified as such, so the model is not necessarily specific.

More recently, in 1991 (Losken et al., 1991a), experimental immobilization of the coronal suture was produced in rabbits to model bilateral coronal suture synostosis in an attempt to understand the effects of calvarial repositioning on subsequent growth. Inherent in all such studies is the drawback of arbitrary choice of event timing. It would appear, then, that a congenital model should overcome this problem.

In actuality, both models suffer the same problem: how to handle the variable of age. We can then contrast the congenital and the mechanical: In the mechanical model, you can shut down the suture and view events that follow, but you cannot do it at the age it would happen naturally. In the congenital model, you want to study the animals that are synostosed as early as you can, but you cannot identify them as early as the genome is exerting its effect. Both, then, are imperfect.

9.2.2 Levels of Appropriate Animal Models to Study Craniosynostosis

The etiology of simple, nonsyndromic craniosynostosis appears to be multifactorial and is not completely understood (Cohen, 1986a, 1989). Such etiologic heterogeneity makes the clinical study and critical assessment of the etiopathogenesis, craniofacial growth, and surgical management of craniosynostosis difficult (Marchac and Renier, 1982; Marsh and Vannier, 1985; Persing et al., 1989). Thus, the need for an etiologically homogeneous, experimental animal model exists (Persson et al., 1989; Persing et al., 1991; Babler et al., 1982b; David et al., 1982; Poswillo, 1988; Cohen, 1989; Babler, 1989; Siegel and Mooney, 1990; Mooney et al., 1993). Poswillo (1988) has suggested that "the more severe anomalies of the calvaria such as plagiocephaly, Crouzon, and Apert syndrome still defy explanation, in the absence of an appropriate animal system to study" (p. 207). Cohen (1993) has also stated that "for future studies to begin to understand the pathogenesis of craniosynostosis, two characteristics must be combined: (1) finding a genetic animal model with primary craniosynostosis and (2) applying molecular techniques to understanding the gene defect and how it causes craniosynostosis" (p. 590). The development of such appropriate animal models may be viewed as important steps in understanding the pathogenesis of this condition

(Persing, Babler, Cohen, personal communication; Poswillo, 1988; Cohen, 1993).

Levels of appropriate animal models to study craniosynostosis can be divided into (1) generic models of basic bone and suture biology, (2) phylogenetically closer models with comparable cranial vault dynamics, and (3) fitting the appropriate animal model to various craniosynostotic conditions.

9.2.2.1 Generic Mammalian Models and Basic Suture-Dura Biology: Comparative Molecular, Developmental, and Biochemical Studies

When choosing animal models to study cranial vault sutures, it is necessary to understand that contrary to the conclusions of Moss (1954, 1959), the sutures are the primary bone growth sites of the craniofacial skeleton. In humans, cranial vault sutures typically form with the interfrontal (metopic) suture between the frontal bones, the sagittal suture between the parietal bones, the paired coronal sutures between the two frontal and two parietal bones, the paired lambdoidal sutures between the supraoccipital and parietal bones, and the squamosal sutures between the parietal, temporal, and sphenoid bones. This is very similar to the arrangements seen in other species such as rabbits, mice, and rats (Fig. 9.4), which have been used as research tools to examine suture biology and pathology. The cranial vault sutures all contribute greatly to the enlarging of the cranial bones, with substantial bone growth occurring during both perinatal and postnatal periods, making these models very appropriate for examining human cranial growth.

Developing animal models to study very early development of the cranial vault must take into account the limitations of looking at very undifferentiated tissues, which are greatly under the influence of surrounding tissues (Hall, 1981; Tyler, 1983; Tyler and McCobb, 1980). These kinds of studies would involve looking at various embry-

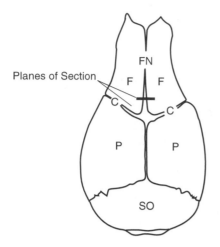

Figure 9.4 Diagram of a young adult rat skull showing position of cranial sutures and associations with accompanying cranial bones. C, coronal suture; FN, interfrontal suture; F, frontal bone; P, parietal bone; SO, supraoccipital bone. Bold lines indicate planes of section for coronal and interfrontal sutures shown in Figure 9.7. (*Source*: Reproduced from Opperman et al., 1997, with permission of the American Society for Bone and Mineral Research.)

onic stages of development, and although some egg-laying vertebrate models are available and frequently used (Xenopus, chick, zebrafish), comparisons to mammalian development may be confusing or misleading. This has been especially true with lineage studies, where use of chick and quail embryos give data that are either in conflict with what is known about mammalian development, or avian species may be different (Couly et al., 1992, 1993; Le Douarin et al., 1993; Noden, 1983, 1986a, 1986b). Mammalian models have their limitations (smaller litter size, longer gestation, inaccessible embryos) and advantages (similar placental barriers to humans, closer phylogenetic links). Each of these needs to be carefully weighed before choosing an appropriate system.

For studying basic suture-dura biology, several factors need to be taken into account. Studying suture biology in situ raises many of the issues described above.

Several models have been used (Markens and Taverne, 1978; Babler et al., 1982a, 1982b; Nappen and Kokich, 1983; Opperman et al., 1993; Roth et al., 1996; Yu et al., 1997; Mooney et al., 2001). An in vivo model was used to establish the role of the dura mater in regulating suture morphogenesis (Fig. 9.5). However, it has limitations, since it showed the dura to be essential once sutures have formed at postnatal day 1, which has subsequently been disproved (Opperman et al., 1995; Kim et al., 1998). A very basic limitation is the confounding influence of humoral factors when trying to examine both intra- and extracellular interactions between factors in regulating suture morphogenesis and growth. This is compounded by the problems of getting the right concentrations of factors applied in ways that allow them physiological relevance, but gives longterm release within the environment in which they are expected to act. This can be done using a variety of carriers, but these carriers may have distinct effects themselves.

Very useful models are those in which spontaneous mutations arise in animal species closely resembling pathologies seen in humans. The animal model most recently described is that of a spontaneously craniosynostotic rabbit, which has proved very useful in examining interactions between various tissues in regulating suture morphogenesis. The most recent experiments showed rescue of sutures from fusion in synostotic animals by transplanting dura mater from unaffected animals under the affected suture site (Fig. 9.6) (Mooney et al., 2001) or the application of transforming growth factor beta 3 (TGF-β_3) to rescue sutures that would normally undergo fusion in this model (Chong et al., 2001).

An interesting model is the normally fusing posterior interfrontal suture of rats and mice. This region of the suture fuses naturally between 15 and 30 days after birth (Fig. 9.7). Studies using both cell and

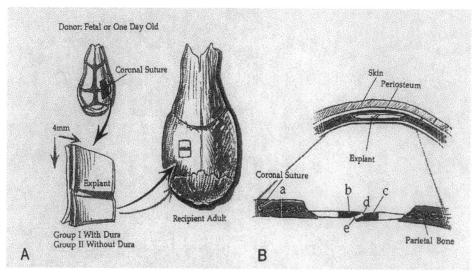

Figure 9.5 Diagram showing in vivo transplantation of fetal and neonatal coronal suture to adult host. (A) Position from which the transplant is removed from the donor site and the site of transplantation in the host animal. A 4×4 mm defect is created in the host bone without damaging the underlying dura mater. The donor tissue is placed on the host dura mater and the host periosteum is sutured closed over the donor tissue prior to closing the skin. (B) Cross-section view of transplant in the host parietal bone. Areas of measurement for histomorphometry are indicated (a–e). (*Source*: Reproduced from Opperman et al., 1993, with permission of Wiley-Liss, Inc., a subsidiary of John Wiley & Sons, Inc.)

molecular biological techniques have compared normally fusing and nonfusing sutures (Opperman et al., 1997; Roth et al., 1997). This model can also be used to compare normally fusing to abnormally fusing sutures, either induced to fuse or fusing as a result of mutation.

Several in vitro models have been developed that provide very distinct advantages over in vivo models (Opperman et al., 1995; Bradley et al., 1996b; Kim et al., 1998). In these models, fetal rat calvaria grown in culture with intact dura mater show normal coronal suture morphogenesis. When the dura mater is removed prior to culture, the sutures appear to develop normally, but then become obliterated by fusion of the bone fronts across the suture site (Fig. 9.8). In vitro culture systems provide controlled environments in which factors can be added or removed with neutralizing antibodies and responses measured, either in

morphological or biochemical and molecular biological terms. However, any removal of tissues from their in vivo environment subjects them to other factors such as wounding of tissues along dissection lines, which may influence results. An added problem when looking at basic suture-dura biology is disruption of global forces acting on the dura when the calvaria is removed from the head. This limits the efficacy of examining bone growth at the suture site, although it has been shown not to aversely affect suture morphogenesis.

A third type of model is the creation of transgenic animals, where mutated genes are either "knocked in" to a mouse strain or genes are "knocked out," making them nonfunctional. While these animals have been very useful in describing the role of genes in many developmental processes (Deng et al., 1998; Zhou et al., 2000), problems arise when the genes being altered

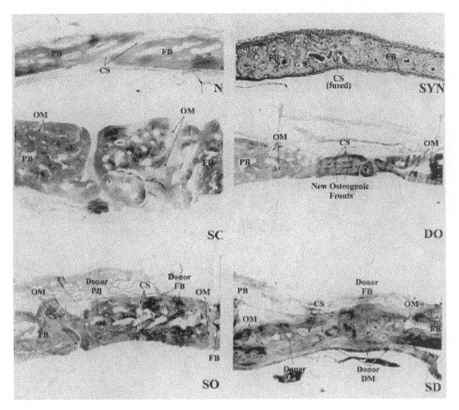

Figure 9.6   Histophotomicrographs of a normal (N—top left) and unoperated synostosed (SYN—top right) coronal suture and the suturectomy sites from a suturectomy control rabbit (SC—middle left), and rabbits with dura mater only (DO—middle right), suture only (SO—bottom left), (top left and right), and suture and dura mater (SD—bottom right) allograft transplants. In the SC specimen, note the resynostosed suturectomy site and extensive new bone (nb) formation at the frontal (FB) and parietal bones (PB) and osteotomy margins (OM). In the DO specimen, note the rudimentary, but synostosed, overlapping osteogenic fronts forming a coronal suture (CS) and the reduced new bone formation in the suturectomy site In the SO specimen, note the obliterated (synostosed) coronal suture (CS) and thickened osteotomy margins (OM). In the SD specimen, note the still patent coronal suture (CS), the thin osteogenic fronts adjacent to the osteotomy margins, and intact donor dura mater (DM) on the ectocortical side (original magnification 62.5×). (*Source*: Reproduced from Mooney et al., 2001, with permission of the American Cleft Palate–Craniofacial Association.)

code for proteins essential for early development, leading to lack of fetal development. This deficiency has been overcome in some cases by placing the gene of interest on promoters activated later in development. Creation of these animals can lead to interesting observations—for example, Liu and co-workers (1995, 1999) reported on overexpression of Msx2 transcription factor, leading to fusion of the normally nonfusing sagittal suture (Fig. 9.9). However, the information derived from these animals may be contradictory or inconclusive and needs to be interpreted with care (Liu et al., 1995, 1999; Ignelzi et al., 1995; Wilkie and Wall, 1996; Winograd et al., 1997; Maxson, Ignelzi, personal communication).

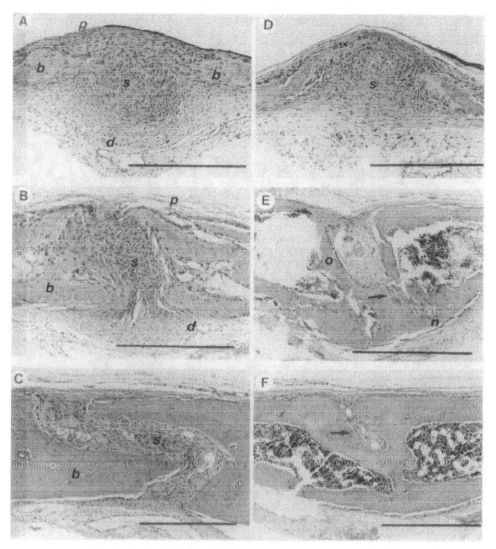

Figure 9.7 Histology of postnatal rat sagittal (A, B, C) and posterior frontal (D, E, F) sutures at different stages of development. (A and D) Patent sutures at 2 days (original magnification 525×). (B) Patent sagittal suture at 15 days (original magnification 525×). (C) Patent sagittal suture at 30 days (original magnification 430×). (D) Patent posterior frontal suture at 2 days Bar = 100 μm. (E) Fusing posterior frontal suture at 15 days. The endocranial side of the suture is obliterated by newly formed bone (n), with active fusion indicated by an arrow (original magnification 525×). (F) Fused posterior frontal suture at 30 days, with a remnant of suture connective tissue visible on the periosteal surface (arrow) (original magnification 430×). b, calvarial bone; s, suture; d, dura mater; p, periosteum; n, newly formed bone; o, active osteoblasts at osteogenic fronts. (*Source*: Reproduced from Roth et al., 1997, with permission of the American Society for Bone and Mineral Research.)

9.2.2.2 Phylogenetically Closer Animal Models and Comparative Cranial Vault Dynamics

A number of experimental animal models have been used to study the effects of premature suture immobilization or fusion on growth and development of the intracranial contents and compensatory changes of the

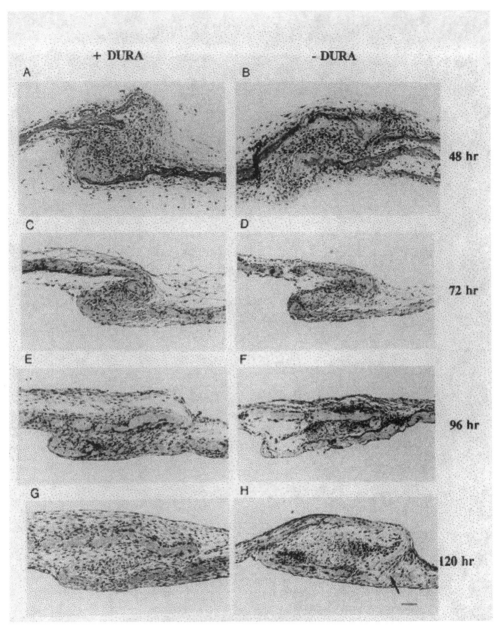

Figure 9.8   Hematoxylin and eosin histology of in vitro coronal suture morphogenesis of E19 rat calvaria cultured in the presence (+ dura) and absence (– dura) of dura mater in serum-free conditions. Calvaria were harvested after 48 (A, B), 72 (C, D), 96 (E, F), or 120 (G, H) hours in culture. Sutures remain patent up to 72 hours in culture, regardless of the presence or absence of dura mater. By 96 hours in culture, sutures cultured in the presence of dura mater remain widely patent, while sutures cultured in the absence of dura mater are markedly narrowed, with bone fronts almost touching across the suture matrix. By 120 hours in culture, sutures cultured in the presence of dura mater remain patent, while those cultured in the absence of dura mater have become breached by bone on the endocranial surface, similar to that seen during normal rat posterior frontal suture fusion in vivo (Fig. 9.6). (*Source*: Reproduced from Opperman et al., 1995, with permission of the American Society for Bone and Mineral Research.)

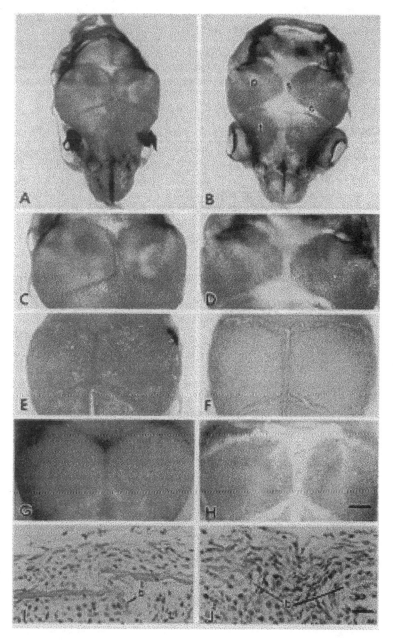

Figure 9.9 Premature closure of calvarial sutures in *Msx2* transgenic animals. Skull of TIMP1 *Msx2*Pro7His F_0 transgenic animal (A) and nontransgenic littermate (B) at 1 day postnatal. The coronal suture and sagittal suture were closed, and the lambdoidal suture was partially closed. Note that all sutures are open in the normal animal. (C) Higher magnification of the skull shown in (A). Note that the gaps (sutural spaces) among bony plates were obliterated by mineralized bone. (D) Higher magnification of the skull in (B). (E) Skull of CMV-*Msx2*Pro7His transgenic animal 12 days postnatal. All sutures except the metopic were closed. (F) Skull of a nontransgenic littermate 12 days postnatal. All sutures were open, as indicated by narrow gaps among bony plates. (G) Skull of CMV-*Msx2*WT transgenic animal. Four days after birth, when this F_0 animal died, all sutures except the metopic were closed. (H) Skull of nontransgenic littermate 4 days postnatal. All sutures were open. (I) Cross-section of the sagittal suture of the skull in (G). Note the overlap of the parietal bones, obliterating the sutural space. (J) Histological section of the sagittal suture of the skull in (H). Note that the opposing parietal bones were separated by a mesenchymal blastema (Bar of (A) and (B) = 1 mm, (C) and (D) = 1.5 mm, (E) and (F) = 1.3 mm, (G) and (H) = 0.85 mm, and (I) and (J) = 25 μm.) b, bone; c, coronal suture; f, frontal bone; i, interparietal bone; l, lambdoidal suture; p, parietal bone; s, sagittal suture. (*Source*: Reproduced From Liu et al., 1995, with permission of the author.)

adjacent craniofacial regions (reviewed by Cohen, 2000b). Animal models utilized at this level of hypothesis testing should be chosen to assess the dynamics and parameters of altered neurocapsular growth vectors on general compensatory growth mechanisms in the mammalian skull (Young, 1959; Moss, 1960, 1975; Enlow, 1990) and not to model specific craniosynostotic etiologies (Cohen, 2000b). Thus, appropriate animal models at this level of hypothesis testing would be ones that (1) exhibit a fair amount of postnatal brain growth; (2) show comparable intracranial and cerebrospinal fluid response patterns; (3) have patent cranial vault and cranial base articulations adjacent to the altered articulation, thus acting as compensatory growth sites during the neurocapsular growth phase; and (4) show similar neurocapsular and craniofacial compensatory changes following cranial vault suture immobilization.

Comparative Brain Morphology Evolutionary changes in cerebralization of the forebrain are responsible for giving definitive shape and flexion to the mammalian neuro- and chondrocranium (Young, 1959; Enlow, 1990) (Chapter 11). In rodents, lagomorphs, and carnivores, cerebralization (the development and enlargement of the cerebral hemispheres) is the dominant process resulting in dolichocephalic (long and narrow) and platycephalic (flat) head shapes (Fig. 9.1 and Chapter 11) (Enlow, 1990). In contrast, gyrification and spatial packing of the neural contents, and cranial base kyphosis, are the dominant processes (in addition to cerebralization) that are responsible for the more turri- (high-vaulted) and brachycephalic (short and wide) head shape seen in humans and most nonhuman primates (Fig. 9.3 and Chapter 11) (Enlow, 1990).

It should also be noted that there are other specific differences between gyrificated and nongyrificated models. These include (1) species-specific differences in the overall gross morphology of the fore- and hindbrain and various sensory capsules (Baer and Harris, 1969; Harel et al., 1972; Enlow and McNamara, 1973; Shek et al., 1986; Babler, 1989; Enlow, 1990) (Chapter 11); (2) variations in neurocapsular growth patterns and timing (Baer and Harris, 1969; Harel et al., 1972; Enlow and McNamara, 1973; Shek et al., 1986; Babler, 1989; Enlow, 1990); (3) the cortex-to-brain stem, cortex-to-midbrain, and cortex-to-rhinencephalon (olfactory tracts and bulbs) ratios are relatively smaller in nongyrificated models compared to humans (Baer and Harris, 1969; Harel et al., 1972; Enlow and McNamara, 1973; Shek et al., 1986; Babler, 1989; Enlow, 1990); (4) the major growth vector of the brain is in a rostrocaudal direction in nongyrificated models (Baer and Harris, 1969; Harel et al., 1972; Enlow and McNamara, 1973; Shek et al., 1986; Babler, 1989; Enlow, 1990), while in humans and other gyrificated models it is in a craniocaudal direction (Baer and Harris, 1969; Enlow and McNamara, 1973; Babler, 1989; Enlow, 1990); and (5) approximately 90% of brain growth is completed during the perinatal period in rabbits (35–42 days of age) and rodents (18–23 days of age) (Harel et al., 1972; Kier, 1976; Alberius and Selvik, 1985; Persing et al., 1986a, 1986b; Babler, 1989), as compared to about 4 to 6 years of age in humans (Baer and Harris, 1969; Enlow and McNamara, 1973; Kier, 1976; Babler, 1989; Enlow, 1990). These findings suggest that the slower and more superiorly directed brain growth in humans may be more appropriately modeled by other gyrificated nonhuman primates or perinatal nongyrificated rodent and lagomorph models.

Comparative Intracranial Pressure Changes It is thought that cranial vault growth, morphology, and/or dysmorphology are modulated by changes in intracranial pressure (ICP) within the neurocapsule (Young, 1959; Moss, 1960; Enlow, 1990; Cohen, 2000a). Thus, an appropriate animal model should exhibit similar ICP changes as those noted in affected human infants (Camfield et al., 2000).

Increased ICP in human infants with craniosynostosis is thought to be low-grade, intermittent, and chronic (Camfield et al., 2000). Camfield and Camfield (1986) reported that the frequency of increased ICP in children with craniosynostosis was highest in the first 2 years of life, which corresponds to the most rapid increase in brain volume. ICP normalized after 6 years of age, by which time adult brain volume has been reached. Renier (1989; Reneir et al., 1982) also reported that preoperative ICP was at its highest at about 6 years of age and attributed the natural decrease in ICP after 6 years of age to progressive cerebral atrophy.

Elevated ICPs were also reported in rabbits with experimentally induced hydrocephalus (Del Bigio and Bruni, 1987, 1991; Tranquart et al., 1991, 1994), rabbits with increased hypertension from increased cerebrospinal fluid (CSF) volume (Bissonnette et al., 1991; Ungersbock et al., 1995), and rabbits with uncorrected familial craniosynostosis (Mooney et al., 1998b, 1999; Fellows-Mayle et al., 2000) (Table 9.1). Longitudinal ICP in craniosynostotic rabbits (Fig. 9.10) was seen to naturally

TABLE 9.1   Selected Intracranial Pressure (ICP) Studies

Treatment	Model	Results	Comments	References
Experimentally induced hydrocephalus	Rabbit	Increased ICP	Injected viscous material into cisterna magna. Found that ICP exhibited instability and transient elevations.	Del Bigio and Bruni, 1987, 1991; Tranquart et al., 1991, 1994
Increased CSF infusion (intracranial hypertension)	Rabbit	Increased ICP	Found acute ICP increases but did not correlate with local cortical blood flow, suggesting elevated ICP may not result in neural ischemia.	Bissonnette et al., 1991; Ungersbock et al., 1995
Uncorrected familial craniosynostosis	Rabbit	Increased ICP	High postnatal ICP that decreased by 42 days of age, coincided with brain dysmorphology.	Fellows-Mayle et al., 2000
Surgically corrected familial craniosynostosis	Rabbit	Normalized ICP	Strip suturectomy performed early. ICP followed normal control pattern.	Mooney et al., 1999
Surgical and pharmacological manipulation of cranial vault and contents	Dog	Decreased ICP	Craniectomy, durotomy, and various agents all reduced ICP to varying degrees.	Bagley et al., 1996

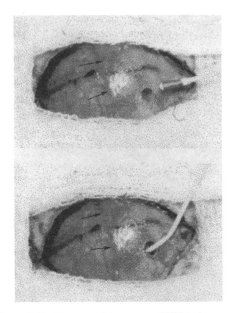

Figure 9.10   Intracranial pressure (ICP) being monitored in a rabbit calvaria with bilateral coronal suture synostosis. A burr hole is made in the parietal bone and a pressure microtransducer is positioned in the epidural space just inferior to the synostosed suture (arrows). (*Source*: Reproduced from Mooney et al., 1999, with permission of the *Journal of Craniofacial Surgery*.)

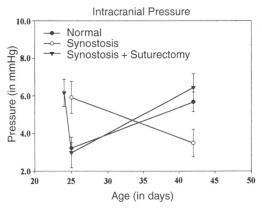

Figure 9.11   Mean (+/– S.E.) ICP changes in normal and synostotic rabbits. Note the drop in ICP in rabbits with suturectomy, the normal increasing ICP pattern from 25 to 42 days of age, and the significant drop in ICP in rabbits with uncorrected coronal suture synostosis over the same time period. (*Source*: Reproduced from Mooney et al., 1999, with permission of the *Journal of Craniofacial Surgery*.)

decrease by 42 days of age (Fig. 9.11) (Mooney et al., 1999; Fellows-Mayle et al., 2000), which coincided with secondary alterations in rabbit cranial vault morphology (Fig. 9.12) (Mooney et al., 1994b, 1998a, 2000; Burrows et al., 1995, 1999), reductions in intracranial volume (Figs. 9.13 and 9.14) (Singhal et al., 1997; Mooney et al., 1998b, 1998c, 2000), and cerebral atrophy (Fig. 9.15) (Burrows et al., 1995; Mooney et al., 1998b; Cooper et al., 1999; Fellows-Mayle et al., 2000). Reduced and asymmetric brain activity, as seen by positron emission tomography (PET) scan activity, was also noted in adult rabbits with uncorrected familial craniosynostosis (Mooney et al., 1998b). These findings demonstrate that

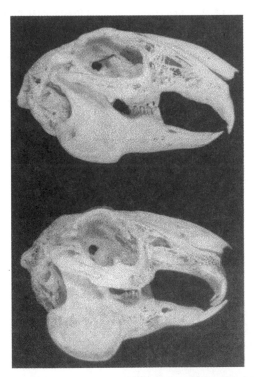

Figure 9.12   Cleaned and dried skull of a 126-day-old normal rabbit (top) and a rabbit with bilateral coronal suture synostosis (bottom). Note the malocclusion and domed cranial vault in the synostosed rabbit skull. (*Source*: Reproduced from Mooney et al., 1994a, with permission of the American Cleft Palate–Craniofacial Association.)

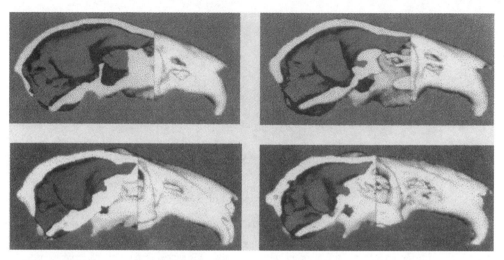

Figure 9.13   Sagittal cut-away view of three-dimensional CT scan reformations of the cranial vault and intracranial contents from 25- (top row) and 126- (bottom row) day-old normal rabbits (left column) and rabbits with coronal suture stenosis (right column). Note the superiorly displaced and inferiorly rotated cranial vaults, the thickened, hyperostotic calvariae, and the anteroposteriorly compressed intracranial reconstructions in the synostosed skulls. (*Source*: Reproduced from Singhal et al., 1997, with permission of Williams and Wilkins Publishing Co.)

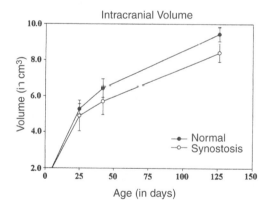

Figure 9.14   Mean (+/– S.E.) intracranial volume by group and age. Note the significant reductions in mean ICV in rabbits with synostosis by 42 days of age compared to normal controls. (*Source*: Original data from Mooney et al., 1999, 2000.)

ICP in nongyrificated models of craniosynostosis responds in a relatively similar manner to ICP in craniosynostotic infants and suggests that these models may be appropriate at this level of hypothesis testing. However, it must be stressed again that brain growth is accelerated in the nongyrificated models and should be taken into account during model choice and experimental design.

Comparative Normal Suture Fusion Timing Patterns   Another factor at this level of hypothesis testing is that an appropriate animal model should have patent cranial vault articulations adjacent to the altered articulation during the neurocapsular growth phase to allow for the study of compensatory growth mechanisms of the cranial vault. During postnatal development, most mammalian cranial vault sutures become predominantly collagenous (Ten Cate, 1989; Herring, 2000). With increasing age, fibroblasts within the sutural connective tissue decrease in number and collagen fibers become more irregularly spaced (Persson et al., 1978; Cohen, 2000a). Eventually, slender, bony spicules extend from the sutural margins, bridging the sutural gap either partially or completely, and stenose the suture

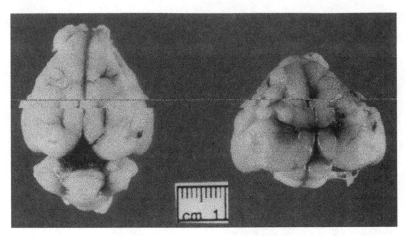

Figure 9.15 Preserved brains from a normal (left) and a synostosed (right) 42-day-old rabbit. Note the anteroposterior shortening and cerebral atrophy in the brain from the synostosed rabbit. (*Source*: Reproduced from Fellows-Mayle et al., 2000, with permission of the American Cleft Palate-Craniofacial Association.)

(Persson et al., 1978; Cohen, 1993, 2000a). Normal suture closure has been attributed to vascular, hormonal, genetic, biomechanical, and local factors (Persson et al., 1978; Persson, 1989; Cohen, 1993, 2000a; Herring, 2000). Although these factors are diverse, they produce suture fusion timing patterns that are consistent enough to be used to estimate the ages of forensic or archeological human material (Krogman, 1978), as well as animals caught in the wild (Sutton, 1972; Herring, 1974, 2000).

Normal cranial suture closure in humans typically begins in the second decade of life (with the exception of the metopic suture, which starts about the second year of life) in an endocranial to ectocranial direction (Persson et al., 1978; Kokich, 1986; Cohen, 1993, 2000a). For the most part, all the typically used animal models discussed previously have gradual and relatively late fusing cranial vault sutures relative to neurocapsular growth cessation. However, there are some taxonomic groups that are distinguished by early and complete cranial vault suture synostosis. These include the egg-laying monotremes (platypus) and minks and their relatives (but not other members of the order Carnivora) (Herring, 2000). Also, as discussed earlier, rodents (rats and mice) exhibit early posterior frontal suture synostosis, while all other cranial vault sutures remain patent (Bradley et al., 1996a, 1996b; Cohen, 2000b). Thus, rodents, lagomorphs, and carnivores may only be appropriate animal models when studying general manipulations of the cranial vault and neurocapsular matrix and not specific regions of craniosynostotic conditions.

Experimental Suture Immobilization Studies The last factor at this level of hypothesis testing is that an appropriate animal model should have similar neurocapsular and cranial vault compensatory changes following experimental suture immobilization. A number of pre- and postnatal experimental animal models of suture immobilization have been developed and utilized to study the effects of altered neurocapsular growth vectors on compensatory vault changes and are presented in Table 9.2. While the most common technique for experimental immobilization of the calvarial sutures in young animals

TABLE 9.2 **Selected Suture Immobilization Studies**

Treatment	Model	Results	Comments	References
Autogenous periosteal onlay graft	Rabbit	Bony fusion	Periosteal transplant placed over frontonasal suture.	Alhopuro et al., 1973
Bony onlay graft	Dog	Coronal stenosis	Bony disk placed over coronal suture.	Gilbin and Alley, 1944
Autologous bony onlay graft	Baboon	Synostosis of various sutures	Obliterated 12 various cranial base or vault articulations at 4–9 weeks postnatal. Founds few compensatory growth changes.	Butow, 1990
Periosteal stripping	Rat	Accelerated normal suture fusion	Caused interfrontal suture fusion and occasional coronal and sagittal suture fusion.	Moss, 1958, 1960
Demineralized bone powder	Fetal rabbit	Postnatal coronal stenosis	Put bone powder on coronal sutures at 25 days gestation. Suture fusion by 21 days postnatal.	Duncan et al., 1992
Demineralized bone powder + BMP-2	Fetal lamb	Prenatal coronal stenosis	Stenosis 20 days posttreatment and severe secondary growth changes by 70 days later.	Stelnicki et al., 1998
Methyl 2-cyanoacrylate adhesive	Rabbit, dog	Periosteal bone bridges	Adhesive results in various suture immobilizations and formation of of bony bridges but not true synostosis. Toxic properties of adhesive itself and not immobilization per se responsible for bony bridge formation.	Persson et al., 1979; Babler and Persing, 1982, 1984, 1985; Babler et al., 1982a, 1982b, 1987; Foley and Kokich, 1980; Nappen and Kokich, 1983; Babler, 1988, 1989; Persing et al., 1986a, 1986b, 1991; Losken et al., 1991a; Lalikos et al., 1995; Rosenberg et al., 1997; Christensen and Clark, 1970
Coronal suturectomy	Rabbit	Coronal stenosis	Surgical ablation of coronal suture results in regenerated but dysmorphic sutures and abnormal cranial vault growth.	Babler et al., 1982a; Mabutt and Kokich, 1975; Smith and McKeown, 1974; Antikainen et al., 1992
Bone plate fixation	Rabbit	Periosteal bone bridges	Immobilization of various sutures with predictable compensatory growth changes, but plates and screws migrate through cortical bone to dura mater. Not true synostosis. Most sutures still patent but nongrowing.	Kinney et al., 1990; Lin et al., 1991; Losken et al., 1991b; Marschall et al., 1991; Wong et al., 1991

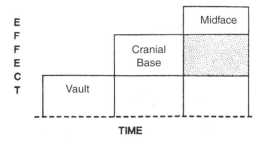

Figure 9.16. Schematic representation of the temporal sequence for appearance of secondary deformities following experimental suture immobilization (from Babler, 1989, with permission of Williams and Wilkins Publishing Co.)

involves the use of medical-grade adhesive (methyl 2-cyanoacrylate), other sutural immobilization techniques have also been reported, including autogenous periosteal and bone onlay grafts; suturectomy and demineralized bone; periosteal stripping; surgical ablation; and microplate fixation (Table 9.2).

Morphological findings from these experimental studies appear consistent with craniofacial growth changes noted in rabbits (Greene and Brown, 1932; Greene, 1933; Mooney et al., 1993, 1994a, 1994b, 1998b, 1998c, 2000; Burrows et al., 1995, 1996, 1999; Smith et al., 1996), monkeys (Cantrell et al., 1987; Schultz, 1960; Smith et al., 1977; Corner and Richtsmeier, 1992), and humans with congenital craniosynostosis (Kreiborg, 1986; Persing et al., 1989; Babler, 1989; Cohen, 2000b). Data from these experimental studies have also led to a better understanding of the temporal sequence for the appearance of secondary deformities in the neurocranium, basicranium, and midface following premature cranial suture closure (Fig. 9.16) (Babler, 1989; Losken et al., 1991a; Mooney et al., 1993).

It should also be reiterated that animal studies at this level of hypothesis testing are only modeling the effects of premature suture immobilization or fusion on growth and development of the intracranial con-

tents and compensatory changes of the adjacent craniofacial regions and not etiologic mechanisms of craniosynostosis (Cohen, 1993, 2000b). The majority of these studies deal with postnatal suture immobilization and/or suture ablation and destruction, or disruption of the surrounding tissue (see Table 9.2). In addition, many of these techniques do not produce craniosynostosis by immobilizing the suture, but instead produce bony bridging via an inflammatory wound-healing response and subsequent hyperosteogenesis (Foley and Kokich, 1980; Nappen and Kokich, 1983; Kokich, 1986; Cohen, 1993, 2000a, 2000b).

9.2.2.3 Fitting the Appropriate Animal Model to Various Primary or Secondary Craniosynostotic Conditions: Naturally Occurring, Transgenic, and Experimentally Produced Models

Naturally Occurring Models of Craniosynostosis Obviously, nonhuman primates would be the most appropriate animal to model specific etiologies of primary craniosynostosis in humans (Siegel and Mooney, 1990). Plagiocephaly has been observed in the cleaned crania of three species of wild-caught, New World monkeys (*Alouatta palliata*—the mantled howler monkey, *Ateles geoffroyi*—the black-handed spider monkey, and *Cebus capucinus*—the white-faced monkey) from museum archives (Schultz, 1960; Smith et al., 1977; Corner and Richtsmeier, 1992) (Fig. 9.17 and Table 9.3), and in one species of Old World monkey (*Cercopithecus aethiops*—the African green monkey) from a regional primate center (Cantrell et al., 1987). In the Delta Regional Primate Center, two cases of multiple suture synostosis were reported born in a breeding colony of African green monkeys, but both individuals died in the neonatal period from unrelated causes (Cantrell et al., 1987). Both individuals had the same

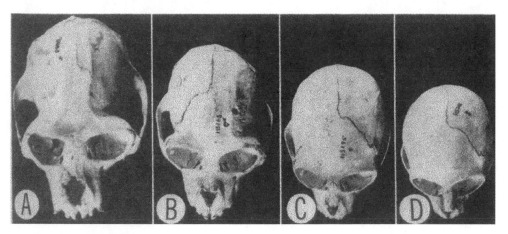

Figure 9.17 Cleaned and dried monkey skulls of *A. palliate* with unilateral coronal suture synostosis and plagiocephaly: (A) adult male, (B) adult female, (C) subadult female, (D) juvenile female. (*Source*: Reproduced from Smith et al., 1977, with permission of Karger Press.)

father, and Cohen (2000b) suggested it may be an autosomal-dominant condition or a susceptible strain. No further reports of synostosis have been seen from this colony, and this condition is presumed lost.

A strain of inbred laboratory rabbits with various degrees of congenital calvarial suture synostosis has also been documented (Greene and Brown, 1932; Greene, 1933). Phenotypic expression was variable and included oxycephaly, scaphocephaly, trigonocephaly, and plagiocephaly. Cohen (2000b) has suggested that this condition was inherited as an autosomal-dominant mutation. Unfortunately, the current location of descendants of this rabbit strain is unknown.

In 1993 (Mooney et al.), we reported on a female New Zealand rabbit born in our laboratory with congenital bilateral coronal suture synostosis. We have bred this individual and her offspring and have developed a breeding colony of rabbits with simple, familial, nonsyndromic, primary coronal suture synostosis (Mooney et al., 1994a, 1994b, 1996a, 1998a, 2000). In a series of reports, we have described the initial identification, phenotypic variability, breeding demographics, pedigree analysis,

karyotypes, fetal and postnatal craniofacial (Figs. 9.12, 9.13, and 9.18) and neurocapsule dysmorphology (Figs. 9.13, 9.15, and 9.19), coronal suture pathology (Figs. 9.20 and 9.21), prenatal synostotic progression (Figs. 9.18, 9.22, and 9.23), spatial distribution of TGF-β isoforms in the perisutural tissues, and somatic cranial vault (Figs. 9.12 and 9.24) and cranial base (Figs. 9.12 and 9.25) growth patterns of unoperated and operated animals from this colony (Mooney et al., 1993, 1994a, 1994b, 1996a, 1996b, 1998a, 1998b, 1998c, 1999, 2000, 2001; Losken et al., 1998, 1999; Burrows et al., 1995, 1996, 1999; Smith et al., 1996; Stelnicki et al., 1997; Singhal et al., 1997; Dechant et al., 1999; Poisson et al., 1999; Fellows-Mayle et al., 2000; Yang et al., 2000) (Chapter 20). The phenotypic variability in this colony ranges from partial synostosis (delayed onset), to complete unilateral or bilateral coronal suture synostosis (Fig. 9.20), to multiple suture synostosis (Fig. 9.18). Pedigree analysis suggests that this condition is inherited in an autosomal-dominant fashion with reduced penetrance and variable expression (Mooney et al., 1996a). Work is currently underway to determine the gene mutation from a pool of already

TABLE 9.3   Selected Animal Model Studies of Primary or Syndromic Craniosynostosis

Etiology	Model	Results	Comments	References
Genetic—Naturally Occurring Models				
Autosomal-dominant mutation	Rabbit	Craniostenosis	Cranial vault sutures synostosed to varying degrees and in various etiologies, combinations producing oxycephaly, scaphocephaly, trigonocephaly, and plagiocephaly.	Greene and Brown, 1932; Greene, 1933; Mooney et al., 1993, 1994a, 1996a, 1998a, 2000
Unknown (suspected genetic mutation)	Monkey (various species)	Craniostenosis and associated dental anomalies	Unicoronal stenosis, sutural bones, and plagiocephaly, possibly from closed colony inbreeding or other genetic factors.	Schultz, 1960; Smith et al., 1977; Cantrell et al., 1987; Corner and Richtsmeier, 1992
Trisomy 22	Chimp	Down syndrome	Features clinical manifestations of microcephaly, brachycephaly, and digital abnormalities.	McClure et al., 1969; Benirschike et al., 1974; Luke et al., 1995
Trisomy 16	Mouse	Down syndrome	Features clinical manifestations of microcephaly, brachycephaly, and digital abnormalities.	Richtsmeier et al., 2000; Baxter et al., 2000
Transgenic and Genetically Engineered Models				
Insertional mutation at *Fgfr/Fgf4* locus	Mouse	Cranial defects	Bulgy eye (Bey) Crouzon-like mutant.	Carlton et al., 1998
Msx1 −/−	Mouse	Cranial defects	Abnormal dental and cranial bone development, cleft palate, and perinatal lethality.	Satokata and Maas, 1994

TABLE 9.3 *Continued*

Etiology	Model	Results	Comments	References
Msx2 Pro148His engineered mutation	Mouse	Craniostenosis	Boston-type craniosynostosis with ectopic bone formation and osteogenic front hyperostosis.	Liu et al., 1995, 1999; Ignelzi et al., 1995, 1996; Winograd et al., 1997
TWIST-null mutation	Mouse	Cranial defects	Abnormalities resemble Saethre-Chotzen syndrome.	El Ghouzzi et al., 1997; Bourgeois et al., 1998
Fgfr1 Pro250ARG engineered mutation	Mouse	Cranial defects	Abnormalities resemble Pfeiffer syndrome.	Zhou et al., 2000.
Lmx 1b-null mutation	Mouse	Sutural and cranial defects	Dysmorphic frontal, coronal, and anterior lambdoidal sutures.	Chen et al., 1998
Type X collagen-null mutation	Mouse	Cranial base abnormalities	Described reduced ossification of the chondrocranium and mandible.	Chung et al., 1997
Tl-osteopetrotic mutation	Rat	Cranial base and sutural abnormalities	Described reduced ossification of the chondrocranium and nasal capsule sutures.	Marks et al., 1999a, 1999b
Growth Factor–Induced Models				
Insulin-like growth factor-1 (IGF-1)	Rat	Bony bridging	Exposed anterior frontal suture to IGF-1 for 2 weeks and found ectocortical bony bridging.	Thaller et al., 1993
Transforming growth factor-β2 (TGF-β2)	Rat	Bony bridging	Exposed posterior frontal suture to TGF-β2 and found ectocortical bony bridging.	Roth et al., 1997

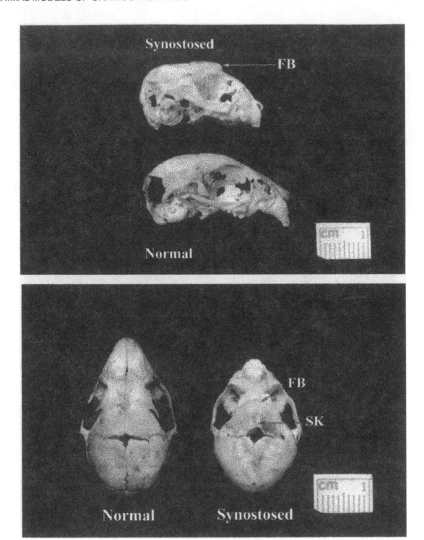

Figure 9.18   Lateral (top) and superior (bottom) views of cleaned and dried skulls of a normal 3-day-old perinatal rabbit skull and a rabbit skull with interfrontal suture synostosis. Note the trigonocephalic morphology in the affected rabbit, which is characterized by premature fusion of the interfrontal suture, excessive bone formation in the rest of the calvaria, extreme frontal bossing (FB), sagittal keeling (SK), turribrachycephaly ("turretted"-shaped skull), frontal bone and anterior cranial base constriction, mid-facial shortening, and nasal bone malformations in the rabbit skull with trigonocephaly. (*Source*: Reproduced from Mooney et al., 2000.)

identified human candidate genes, and an FGFR2 mutation has already been ruled out (Yang et al., 2000). Results from these studies demonstrate that this colony is producing craniosynostotic, but otherwise normal, healthy rabbits, and show that the pathological findings from this rabbit model are strikingly similar to clinical findings noted in human infants with primary, familial, nonsyndromic craniosynostosis (Cohen, 1993, 2000b; Mooney et al., 1994a, 1996a; Wilkie and Wall, 1996).

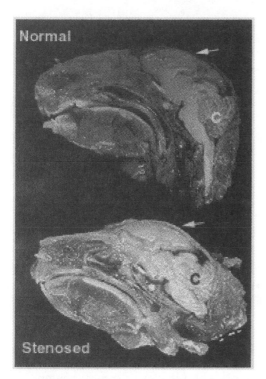

Figure 9.19 Photographs showing the right side of bisected perinatal (term) heads from a normal rabbit and a rabbit with interfrontal suture synostosis. Note the superiorly displaced cranial vault (arrow), cerebellar (C) constrictions, and brain stem reorientation in the rabbit head with synostosis. (*Source*: Reproduced from Cooper et al., 1999, with permission of the American Cleft Palate–Craniofacial Association.)

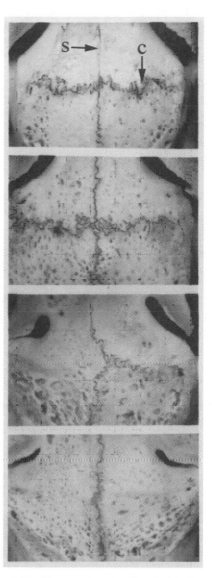

Figure 9.20 Close-up superior view (original magnification ×3) of cleaned and dried coronal (C) and sagittal (S) sutures from 126-day-old rabbits showing the range of phenotypic variability. From top to bottom: unaffected; delayed onset; unilateral; and bilateral coronal suture synostosis. (*Source*: Reproduced from Mooney et al., 1996a, with permission of Munksgaard International Publishers, Ltd. Copenhagen, Denmark.)

Transgenic and in Vivo Growth Factor Models of Craniosynostosis Recently, a number of transgenic and knockout mouse models have been developed for craniosynostotic conditions with known etiologics (Fig. 9.9 and Table 9.3). Although these models have proved very useful in understanding the etiopathogenesis of craniofacial anomalies, technical problems still exist in controlling the timing and regional specificity of the various transgenic and knockout procedures. In addition, these animals typically exhibit severely exaggerated sequelae, usually die in the perinatal period (Satokata and Maas, 1994; Ignelzi et al.,

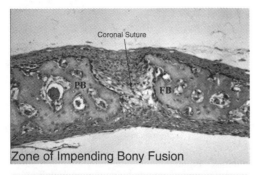

Zone of Impending Bony Fusion

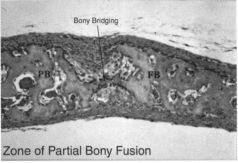

Zone of Partial Bony Fusion

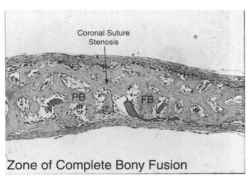

Zone of Complete Bony Fusion

Figure 9.21 Serial histophotomicrographs of the osteogenic fronts (PB, parietal bone; FB, frontal bone) of synostosed coronal sutures at the zone of impending bony union (ZIPB), the zone of partial bony fusion (ZPBF), and the zone of complete bony fusion (ZCBF). All sutures have been oriented in the same direction and serially sectioned in the sagittal plane. Note the hyperostotic osteogenic fronts and the bony bridging in the zone of partial bony fusion (original magnification 200×). (*Source:* Reproduced from Mooney et al., 1996b, with permission of the American Cleft Palate–Craniofacial Association.)

1995, 1996; Winograd et al., 1997; Cohen, 2000c), and currently are very labor intensive and expensive to produce, which may limit their utility for postnatal surgical,

cytokine, and/or biochemical manipulations (Liu et al., 1995, 1999; Ignelzi et al., 1995; Wilkie and Wall, 1996; Winograd et al., 1997; Maxson, Ignelzi, personal communication).

There have also been a few in vivo rodent studies that have manipulated normal calvarial suture exposure to high levels of various growth factors (Il-1 and TGF-β_2) in an attempt to model over-expression patterns observed in craniosynostotic infants (Thaller et al., 1993; Roth et al., 1997) (Table 9.3). Such models provide important information on in vivo suture response to these factors but suffer from a number of confounding methodological problems related to timing of growth factor exposure (postnatal versus congenital), duration of growth factor exposure (single or week exposure versus lifetime duration), and the confounding effects of surgical intervention necessary to deliver the factors with vehicles or osmotic pump.

Experimentally Produced Models of Secondary Craniosynostosis Primary craniosynostosis and craniodysostotic syndromes (i.e., affecting the cranial vault sutures or cranial base synchondroses initially) in humans are common and typically result from monogenic and chromosomal conditions (Cohen, 2000b). In secondary craniosynostosis, premature suture or synchondrosis fusion can result secondarily from known disorders such as hyperthyroidism, teratogenic insults, and various iatrogenic disorders such as fetal head constraint and shunted hydrocephalus (reviewed in Cohen, 2000b). The etiopathogenesis of craniosynostosis in these conditions varies widely and ranges from accelerated osseous maturation and premature suture fusion in certain metabolic disorders, to lack of growth stretch in microcephaly and shunted hydrocephalus, to lack of brain induction and sutural agenesis in various teratogenic disorders and fetal alcohol syndrome (Cohen, 2000b).

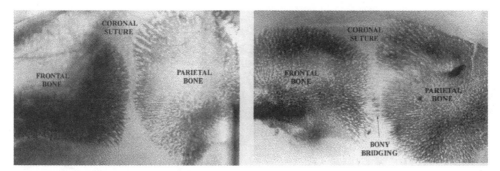

Figure 9.22   Lateral view of the presumptive coronal sutures in two 21-day-old alizarin red-stained fetal calvariae (left—normal fetus, right—synostosed fetus). Note the patent presumptive suture in the normal fetus and the bony bridging and dense bony trabeculae in the frontal and parietal bones in the synostosing coronal suture (original magnification 15×). (*Source*: Reproduced from Mooney et al., 1996b, with permission of the American Cleft Palate–Craniofacial Association.)

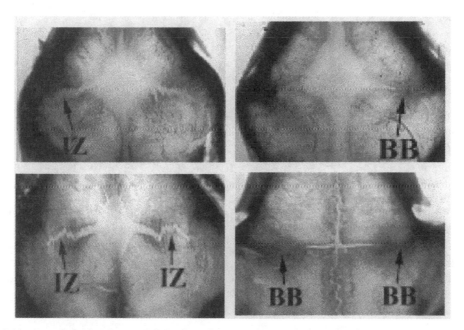

Figure 9.23   Superior view of normal (left column) and synostosed (right column) calvarial sutures at 25 and 33 days postconception (term = 31 days). Note the consistent interdigitating zone (IZ) in the middle of each patent coronal suture and the corresponding bony bridges (BB) in the synostosed coronal sutures (original magnification 15×). (*Source*: Reproduced from Mooney et al., 1996b, with permission of the American Cleft Palate–Craniofacial Association.)

There have been a number of animal models of these disorders developed that exhibit similar secondary craniosynostosis and associated skull deformities. These studies are presented in Table 9.4. Important information on the etiopathogenesis of these conditions has been derived from the models, but obviously the findings are very specific and not applicable to primary craniosynostotic conditions. Although

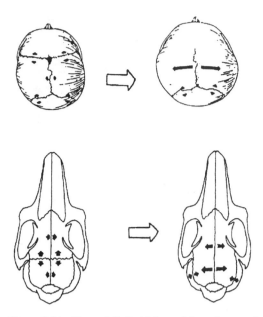

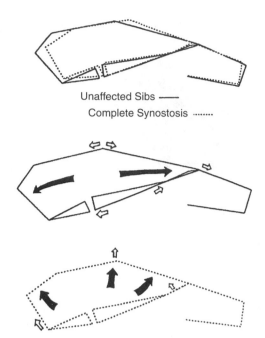

Figure 9.24   Normal (left side) cranial vault sutural growth vectors (arrows) in human and rabbit crania and similar compensatory changes following bilateral coronal suture stenosis (right side) that are responsible, in part, for producing pathological brachycephaly. (*Source*: Reproduced from Burrows et al., 1995, with permission of the American Cleft Palate–Craniofacial Association.)

Figure 9.25   Lateral head radiographic tracings from 126-day-old normal rabbits and rabbits with bilateral coronal suture synostosis. Note the altered brain growth vectors in the synostosed tracing, which have produced compensatory changes in the cranial vault and cranial base (*Source*: Reproduced from Smith et al., 1996, with permission of Munksgaard International Publishers, Ltd. Copenhagen, Denmark.)

hyperthyroidism-associated craniosynostosis in humans is well documented, the pathogenic mechanism is poorly understood (Cohen, 2000b), and experimentally induced hyperthyroidism and its effects on sutural fusion have not yet been studied in animals.

With iatrogenic disorders, it has been suggested that sporadic or isolated (nonfamilial) clinical cases of craniosynostosis may be a result of fetal head constraint from intrauterine compression (Higginbottom et al., 1980). Koskinen-Moffett and Moffett (1989) were able to experimentally produce craniosynostosis in mice by delaying parturition by 2 to 3 days, thereby causing intrauterine crowding and fetal head constraint. Unilateral craniosynostosis was noted in 88% of the 26 mouse neonates, with fusion occurring in the

coronal, squamosal, and/or squamosofrontal sutures (Koskinen-Moffett and Moffett, 1989). Plagiocephaly was most severe in the neonates located most proximally in the uterine horns, presumably where crowding was most severe.

9.3   CONCLUSION

From the preceding review it should be evident that many animal models have been developed and utilized to study normal and abnormal cranial suture and cranial vault development. It should be understood that each model has its own utility as well as its own limitations for the study of human craniofacial and sutural biology. In this chapter, we have proposed

TABLE 9.4   Selected Animal Model Studies of Secondary Craniosynostosis

Etiology	Model	Results	Comments	References
			Metabolic Disorder Models	
X-linked hypophosphatemic mutant	Mouse	Craniostenosis	Coronal suture involved with vitamin D–resistant rickets.	Roy et al., 1981
Hypocalcaemia and vitamin D–deficient diet	Rat	Suture obliteration	Internasal suture only.	Engstrom and Thilander, 1985; Engstrom et al., 1985
Defective cholesterol synthesis	Rat	Cranial defects	Model of Smith-Lemli-Opitz syndrome (SLOS)—holoprosencephaly and limb defects.	Kolf-Clauw et al., 1998; Battaile and Steiner, 2000
			Teratogenic Models	
Prenatal hypervitaminosis D	Rabbit	Craniostenosis	Features similar to Williams (supravalvular aortic stenosis) syndrome (craniosynostosis, dental, mandibular, and aortic abnormalities).	Friedman and Mills, 1969
Prenatal hypervitaminosis A	Mouse, rabbit, monkey	Craniostenosis	Coronal suture synostosis with occasional involvement of the lambdoidal or sphenotemporal sutures.	Cohlan, 1954; Kalter and Warkany, 1961; Itabashi, 1969; Wilson, 1971; Fantel et al., 1977; Yip et al., 1980
			Malformation Models	
Prenatal alcohol exposure	Monkey	Craniostenosis	Abnormalities resemble fetal alcohol syndrome (FAS) with sagittal suture synostosis and dolichocephalic head shape.	Inouye et al., 1985

TABLE 9.4 *Continued*

Etiology	Model	Results	Comments	References
Prenatal alcohol exposure	Mouse	Skull abnormalities	FAS-induced single-cavity forebrain (holoprosencephaly).	Sulik et al., 1981, 1984
Exencephaly and microcephalus	Mouse	Skull abnormalities	Inbred mouse strain with neural tube defects showed incomplete skull development and suture malformations.	Vogelweid et al., 1993
Hydranencephaly and microcephaly	Dog, monkey	Skull abnormalities	Ligated cerebral blood vessels and overlapping sutures.	Becker, 1949; Myers, 1969
Iatrogenic Disorder Models				
Fetal head constraint	Chick	Cranial base abnormalities	In vitro head restraint of shell-less chick embryos produced cranial base and squamosal bone malformations and deformations.	Jaskoll and Melnick, 1982
Delayed parturition	Mice	Sutural stenosis	Delayed birth by 2–3 days and noted coronal, squamosal, and squamosofrontal suture stenosis from intrauterine head compression.	Koskinen-Moffett and Moffett, 1989
Head binding	Rat	Cranial vault deformities	A cloth band around the crania from 5 to 30 days of age modeled human cultural deformation.	Pucciarelli, 1978
Experimental hydrocephalus	Rabbit, lamb, monkey, rat	Cranial, suture, and neural deformities	Injected viscous material into cisterna magna and observed various neural and cranial suture deformations following shunting.	Del Bigio and Bruni, 1987; 1991; Tranquart et al., 1991, 1994; Nakayama et al., 1983; Glick et al., 1984; Hoyte, 1990

a number of practical criteria for choosing and evaluating various animal models and suggested that appropriate animal model choice should also be linked to levels of hypothesis testing and/or expected extrapolation to the human clinical condition. The continued intelligent and conservative use of animal models should be based on such objective criteria and not necessarily on phyletic affinity.

ACKNOWLEDGMENTS

This work was supported in part by a Comprehensive Oral Health Research Center of Discovery (COHRCD) program/project grant from NIH/NIDCR (P60 DE13078) to the Center for Craniofacial Development and Disorders, Johns Hopkins University, Baltimore, and a publication assistance award from the Richard D. and Mary Jane Edwards Endowed Publication Fund, Faculty of Arts and Sciences, University of Pittsburgh.

REFERENCES

Alberius, P., and Selvik, G. (1985). Volumetric changes in the developing rabbit calvarium. *Anat. Rec. 213*, 207–214.

Alhopuro, S., Ranta, R., and Ritsilia, V. (1973). Growth of the rabbit snout after bony fusion of the frontonasal suture achieved by means of a free periosteal transplant. *Proc. Finn. Dent. Soc. 69*, 166–167.

Antikainen, T., Kallioinen, M., Warit, T., and Serlo, W. (1992). A new method for the creation and measurement of experimental craniosynostosis. *Child's Nerv. Sys. 8*, 457–468.

Babler, W. J. (1988). Effects of multiple suture closure on craniofacial growth in rabbits. In K. W. L. Vig and A. R. Burdi (eds.), *Craniofacial Morphogenesis and Dysmorphogenesis. Monograph 21* (Ann Arbor, MI: Center for Human Growth and Development, The University of Michigan).

Babler, W. J. (1989). Relationship of altered cranial suture growth to cranial base and midface. In J. A. Persing, M. T. Edgerton, and J. A. Jane (eds.), *Scientific Foundations and Surgical Treatment of Craniosynostosis* (Baltimore: Williams and Wilkins), pp. 87–95.

Babler, W. J., and Persing, J. A. (1982). Experimental alteration of cranial suture growth: effects on the neurocranium, basicranium, and midface. In B. G. Sarnat (ed.), *Factors and Mechanisms Influencing Bone Growth* (New York: Alan R. Liss), pp. 333–345.

Babler, W. J., and Persing, J. A. (1984). Experimental coronal synostosis: effects of surgical release on facial growth. *J. Dent. Res. 63* (special issue), 315.

Babler, W. J., and Persing, J. A. (1985). Alterations in cranial suture growth associated with premature closure of the sagittal suture in rabbits. *Anat. Rec. 211*, 14A.

Babler, W. J., Persing, J. A., Persson, K. M., Winn, H. R., Jane, J. A., and Rodheaver, G. T. (1982a). Skull growth after coronal suturectomy, periostectomy, and dural transection. *J. Neurosurg. 56*, 529–535.

Babler, W. J., Persing, J. A., Winn, H. R., Jane, J. A., and Rodheaver, G. T. (1982b). Compensatory growth following premature closure of the coronal suture in rabbits. *J. Neurosurg. 57*, 535–542.

Babler, W. J., Persing, J. A., Nagorsky, M. J., and Jane, J. A. (1987). Restricted growth of the frontonasal suture: alterations in craniofacial growth in rabbits. *Am. J. Anat. 178*, 90–98.

Badoux, D. M. (1966). Framed structures in the mammalian skull. *Acta Morphol. Neerl. Scand. 5*, 347–360.

Baer, M. J., and Harris, J. E. (1969). A commentary on the growth of the human brain and skull. *Am. J. Phys. Anthropol. 30*, 39–44.

Bagley, R. S., Harrington, M. L., Pluhar, G. E., Keegan, R. D., Greene, S. A., Moore, M. P., and Gavin, P. R. (1996). Effect of craniectomy/durotomy alone and in combination with hyperventilation, diuretics, and corticosteroids on intracranial pressure in clinically normal dogs. *Am. J. Vet. Res. 57*, 116–119.

Bardach, J., and Kelly, K. (1987). Role of animal models in experimental studies of craniofacial growth following cleft lip and palate repair. *Cleft Palate Craniofac. J. 25*, 103–113.

Battaile, K. P., and Steiner, R. D. (2000). Smith-Lemli-Opitz syndrome: the first multiple malformation syndrome associated with defective cholesterol synthesis. *Molec. Genet. Metab. 71*, 154–162.

Baxter, L. L., Moran, T. H., Richtsmeier, J. T., Troncoso, J., and Reeves, R. H. (2000). Discovery and genetic localization of Down syndrome cerebellar phenotypes using the Ts65Dn mouse. *Hum. Molec. Genet. 9*, 195–202.

Becker, H. (1949). Uber hirngefassausschaltungen. II. Intrakranielle gefa ssverschlusse. Uber experimentelle hydranencephalie (Blasenhirn). *Z. Nervenheilkd. 161*, 446–505.

Benirschike, K., Bogart, M. H., McClure, H. M., and Nelson-Rees, W. A. (1974). Fluorescence of the trisomic chimpanzee chromosomes. *J. Med. Primatol. 3*, 311–314.

Bissonnette, B., Bickler, P. E., Gregory, G. A., and Seveeringhaus, J. W. (1991). Intracranial pressure and brain redox balance in rabbits. *Can. J. Anaesthesiol. 38*, 654–659.

Bourgeois, P., Bolcato-BEllemin, A.-L., Danse, J.-M., Bloch-Zupan, A., Yoshiba, K., Stoetzel, C., and Perrin-Schmitt, F. (1998). The variable expressivity and incomplete penetrance of the *twist*-null heterozygous mouse phenotype resemble those of human Saethre-Chotzen syndrome. *Hum. Molec. Genet. 7*, 945–957.

Bradley, J. P., Levine, J. P., Roth, D. A., McCarthy, J. G., and Longaker, M. T. (1996a). Studies in cranial suture biology: IV. Temporal sequence of posterior frontal cranial suture fusion in the mouse. *Plast. Reconstr. Surg. 98*, 1039–1045.

Bradley, J. P., Levine, J. P., Blewett, C., Krummel, T., McCarthy, J. G., and Longaker, M. T. (1996b). Studies in cranial suture biology: in vitro cranial suture fusion. *Cleft Palate Craniofac. J. 33*, 150–156.

Brown, A. M. (1963). Matching the animal with the experiment. In W. Lane-Petter (ed.), *Animals for Research* (New York: Academic Press), pp. 261–280.

Burrows, A. M., Mooney, M. P., Smith, T. D., Losken, H. W., and Siegel, M. I. (1995). Growth of the cranial vault in rabbits with congenital coronal suture synostosis. *Cleft Palate Craniofac. J. 32*, 235–246.

Burrows, A. M., Mooney, M. P., Smith, T. D., Losken, H. W., and Siegel, M. I. (1996). Development of plagiocephaly in rabbits with unicoronal synostosis. *J. Neurosurg. 85*, 929–936.

Burrows, A. M., Richtsmeier, J. T., Mooney, M. P., Smith, T. D., Losken, H. W., and Siegel, M. I. (1999). Three-dimensional analysis of craniofacial form in a familial rabbit model of nonsyndromic coronal suture synostosis using Euclidean distance matrix analysis. *Cleft Palate Craniofac. J. 36*, 196–206.

Butow, K. W. (1990). Craniofacial growth disturbances after skull base and associated suture synostoses in the newborn Chacama baboon: a preliminary report. *Cleft Palate Craniofac. J. 27*, 241–251.

Camfield, P. R., and Camfield, C. S. (1986). Neurological aspects of craniosynostosis. In M. M. Cohen, Jr. (ed.), *Craniosynostosis: Diagnosis, Evaluation, and Management* (New York: Raven Press), pp. 215–225.

Camfield, P. R., Camfield, C. S., and Cohen, M. M., Jr. (2000). Neurological aspects of craniosynostosis. In M. M. Cohen Jr. and R. E. MacLean (eds.), *Craniosynostosis: Diagnosis, Evaluation, and Management* (New York: Oxford University Press), pp. 177–183.

Cantrell, C., Baskin, G., and Blanchard, J. (1987). Craniosynostosis in two African green monkeys. *Lab. Anim. Sci. 37*, 631–634.

Carlson, D. S., Ellis, E., Schneiderman, E. D., and Underleider, J. C. (1982). Experimental models of surgical intervention in the growing face: cephalometric analysis of facial growth and relapse. In J. A. McNamara, D. S. Carlson, and K. A. Ribbens (eds.), *The Effect of Surgical Intervention on Craniofacial Growth* (Ann Arbor: University of Michigan Press), pp. 11–72.

Carlton, M. B. L., Colledge, W. H., and Evans, M. J. (1998). Crouzon-like dysmorphology in the mouse is caused by an insertional mutation at the *Fgfr/Fgr4* locus. *Devel. Dynam. 21*, 242–247.

Chen, H., Ovchinikow, D., Pressman, C. L., Aulehla, A., Lun, Y., and Johnson, R. L. (1998). Multiple calvarial defects in lmx 1b mutant mice. *Devel. Genet. 22*, 314–320.

Chong, S. L., Mitchell, R., Moursi, A., Winnard, P., Losken, H. W., Ozerdem, O., Keeler, K., Opperman, L., Siegel, M. I., and Mooney, M. P. (2001). Rescue of coronal suture synostosis with TGF-β3 in craniosynostotic rabbits. *J. Dent. Res.* 80 (special issue), 89.

Christensen, F. K., and Clark, D. B. (1970). The effect of restricted suture growth on brain growth in dogs. *Surg. Forum 21*, 439–440.

Chung, K. S., Jacenko, O., Boyle, P., Olsen, B. R., and Nishimura, I. (1997). Craniofacial abnormalities in mice carrying a dominant interference mutation in type X collagen. *Devel. Dyn. 208*, 544–552.

Cohen, M. M., Jr. (1986a). History, terminology, and classification of craniosynostosis. In M. M. Cohen, Jr. (ed.), *Craniosynostosis: Diagnosis, Evaluation, and Management* (New York: Raven Press), pp. 1–20.

Cohen, M. M., Jr. (1986b). Perspectives on craniosynostosis. In M. M. Cohen (ed.), *Craniosynostosis: Diagnosis, Evaluation, and Management* (New York: Raven Press), pp. 21–58.

Cohen, M. M., Jr. (1986c). The etiology of craniosynostosis. In M. M. Cohen (ed.), *Craniosynostosis: Diagnosis, Evaluation, and Management* (New York: Raven Press), pp. 59–81.

Cohen, M. M., Jr. (1989). The etiology of craniosynostosis. In J. A. Persing, M. T. Edgerton, and J. A. Jane (eds.), *Scientific Foundations and Surgical Treatment of Craniosynostosis* (Baltimore: Williams & Wilkins), pp. 9–20.

Cohen, M. M., Jr. (1993). Sutural biology and the correlates of craniosynostosis. *Am. J. Med. Genet. 47*, 581–616.

Cohen, M. M., Jr. (1997). Transforming growth factor βs and fibroblast growth factors and their receptors: role in sutural biology and craniosynostosis. *J. Bone Miner. Res. 12*, 322–331.

Cohen, M. M., Jr. (2000a). Sutural biology. In M. M. Cohen Jr. and R. E. MacLean (eds.), *Craniosynostosis: Diagnosis, Evaluation, and Management* (New York: Oxford University Press), pp. 11–23.

Cohen, M. M., Jr. (2000b). Sutural pathology. In M. M. Cohen Jr. and R. E. MacLean (eds.), *Craniosynostosis: Diagnosis, Evaluation, and Management* (New York: Oxford University Press), pp. 95–99.

Cohen, M. M., Jr. (2000c). Twist and MSX2 mutations. In M. M. Cohen Jr. and R. E. MacLean (eds.), *Craniosynostosis: Diagnosis, Evaluation, and Management* (New York: Oxford University Press), pp. 51–68.

Cohlan, S. Q. (1954). Congenital anomalies in the rat produced by excessive intake of vitamin A during pregnancy. *Pediatrics 13*, 536–567.

Cooper, G., and Schiller, S. (1975). *Anatomy of the Guinea Pig* (Cambridge, MA: Harvard University Press).

Cooper, G. M., Mooney, M. P., Burrows, A. M., Smith, T. D., Dechant, J. J., Losken, H. W., Marsh, J. L., and Siegel, M. I. (1999). Brain growth rates in craniosynostotic rabbits. *Cleft Palate Craniofac. J. 36*, 314–321.

Corner, B. D., and Richtsmeier, J. T. (1992). Experiments of nature: premature unicoronal cranial synostosis in mantled howler monkeys *(Alouatta palliata)*. *Cleft Palate Craniofac. J. 29*, 143–151.

Couly, G. F., Coltey, P. M., and LeDouarin, N. M. (1992). The developmental fate of the cephalic mesoderm in quail-chick chimeras. *Development 114*, 1–15.

Couly, G. F., Coltey, P. M., and LeDouarin, N. M. (1993). The triple origin of skull in higher vertebrates: a study in quail-chick chimeras. *Development 117*, 409–429.

Cramer, M. (1998). Experiments using animals. *Plast. Reconstr. Surg. 102*, 926–927.

David, J. D., Poswillo, D., and Simpson, D. (1982). *The Craniosynostoses: Causes, Natural History, and Management* (Berlin: Springer-Verlag).

De Beer, G. R. (1937). *The Development of the Vertebrate Skull* (London: Oxford University Press).

Dechant, J. J., Mooney, M. P., Cooper, G. M., Smith, T. D., Losken, H. W., Mathijssen, I. M. J., and Siegel, M. I. (1999). Positional changes in the frontoparietal ossification centers in perinatal craniosynostotic rabbits. *J. Craniofac. Genet. Devel. Biol. 19*, 64–74.

Del Bigio, M. R., and Bruni, J. E. (1987). Chronic intracranial pressure monitoring in conscious hydrocephalic rabbits. *Pediatr. Neurosci. 13*, 67–71.

Del Bigio, M. R., and Bruni, J. E. (1991). Silicone oil-induced hydrocephalus in the rabbit. *Child's Nerv. Sys. 7*, 79–84.

Deng, C., Yang, J., Scott, J., Hanash, S., and Richardson, B. C. (1998). Role of the ras-MAPK signaling pathway in the DNA methyltransferase response to DNA hypomethylation. *Biol. Chem. 379*, 1113–1120.

Dreizen, S., and Levy, B. M. (1977). Monkey models in dental research. *J. Med. Primatol. 6*, 133–144.

DuBrul, E. L., and Laskin, D. M. (1961). Preadaptive potentiates of the mammalian skull: an experiment in growth and form. *Am. J. Anat. 109*, 117–132.

Duncan, B. W., Adzick, N. S., and Moelleken, B. R. (1992). An in utero model of craniosynostosis. *J. Craniofac. Surg. 3*, 70–76.

El Ghouzzi, V., Le Merrer, M., Perrin-Schmitt, F., Lajeunie, E., Benit, P., Renier, D., Bourgeois, P., Bolcato-Bellemin, A.-L., Munnich, A., and Bonaventure, J. (1997). Mutations of the TWIST gene in Saethre-Chotzen syndrome. *Nat. Genet. 15*, 42–46.

Elliot, R. (1963). *Reighard and Jennings' Anatomy of the Cat* (New York: Holt, Rinehart, and Winston).

English, D. C. (1989). Using animals for the training of physicians and surgeons. *Theor. Med. 10*, 43–52.

Engstrom, C., and Thilander, B. (1985). Premature facial synostosis: the influence of biomechanical factors in normal and hypocalcaemic young rats. *Eur. J. Orthod. 7*, 35–47.

Engstrom, C., Kiliaridis, S., and Thilander, B. (1985). Facial suture synostosis related to altered craniofacial bone remodeling induced by biomechanical forces and metabolic factors. In A. D. Dixon and B. D. Sarnat (eds.), *Normal and Abnormal Bone Growth: Basic and Clinical Research* (New York: Alan R. Liss).

Enlow, D. H. (1990). *Handbook of Facial Growth* (New York: Saunders).

Enlow, D. H., and Azuma, M. (1975). Functional growth boundaries in the human and mammalian face. *Birth Defects 11*, 217–230.

Enlow, D. H., and McNamara, J. (1973). The neurocranial basis for facial form and pattern. *Angle Orthod. 43*, 256–271.

Fantel, A. G., Shepard, T. H., Newell-Morris, L. N., and Moffett, B. C. (1977). Teratogenic effects of retinoic acid in pigtail monkeys *(Macaca nemestrina)*. *Teratology 15*, 65–72.

Farris, E. F., and Griffith, J. Q. (1949). *The Rat in Laboratory Investigation* (New York: Hafner Press).

Fellows-Mayle, W., Mooney, M. P., Dechant, J., Losken, H. W., Cooper, G. M., Burrows, A. M., Smith, T. D., Pollack, I., and Siegel, M. I. (2000). Age-related changes in intracranial pressure in rabbits with uncorrected familial coronal suture synostosis. *Cleft Palate Craniofac. J. 37*, 370–378.

Fernandes, D. (1989). Animal experimentation: necessary or not? *Cleft Palate Craniofac. J. 26*, 258–259.

Foley, W. J., and Kokich, V. G. (1980). The effects of mechanical immobilization on sutural development in the growing rabbit. *J. Neurosurg. 53*, 794–801.

Fraser, F. C. (1980). Animal models for craniofacial disorders. In M. Melnick, D. Bixler, and E. D. Shields (eds.), *Etiology of Cleft Lip and Palate*. Progress in Clinical Biological Research, Alan R. Liss, N.Y. vol. 46, pp. 1–24.

Friedman, W. F., and Mills, L. F. (1969). The relationship between vitamin D and the craniofacial and dental anomalies of the supravalvular aortic stenosis syndrome. *Pediatrics 43*, 12–18.

Getty, R. (1975). *Sisson's and Grossman's The Anatomy of the Domestic Animals* (Philadelphia: W. B. Saunders).

Gilbin, N., and Alley, A. (1944). Studies in skull growth. Coronal suture fixation. *Anat. Rec. 88*, 143.

Glick, P. L., Harrison, M. R., Halks-Miller, M., Adzick, N. S., Nakayama, D. K., Anderson, J. H., Nyland, T. G., and Edwards, M. S. (1984). Correction of congenital hydrocephalus in utero II. Efficacy of in utero shunting. *J. Pediatr. Surg. 19*, 870–881.

Goldsmith, E. I., and Moor-Jankowski, J. (eds.), (1969). Experimental medicine and surgery in primates. *Ann. N.Y. Acad. Sci. 162*, 1–704.

Greene, E. C. (1968). *Anatomy of the Rat* (New York: Hafner).

Greene, H. S. (1933). Oxycephaly and allied conditions in man and in the rabbit. *J. Exp. Med. 57*, 967–976.

Greene, H. S., and Brown, W. H. (1932). Hereditary variations in the skull of the rabbit. *Science 76*, 421–422.

Gregory, W. K. (1929). *Our Face from Fish to Man* (New York: Hafner).

Hall, B. K. (1981). Intracellular and extracellular control of the differentiation of cartilage and bone. *Histochem. J. 13*, 599–614.

Harel, S., Watanabe, K., Linke, I., and Schain, R. J. (1972). Growth and development of the rabbit brain. *Biol. Neonate 21*, 381–399.

Hartman, C. F., and Strauss, W. L. (1933). *The Anatomy of the Rhesus Monkey* (New York: Hafner).

Herring, S. W. (1974). A biometric study of suture fusion and skull growth in peccaries *Anat. Embryol. (Berlin) 146*, 167–180.

Herring, S. W. (2000). Sutures and craniosynostosis: a comparative, functional, and evolutionary perspective. In M. M. Cohen Jr. and R. E. MacLean (eds.), *Craniosynostosis: Diagnosis, Evaluation, and Management* (New York: Oxford University Press), pp. 3–10.

Higginbottom, M. C., Jones, K. L., and James, H. E. (1980). Intrauterine constraint and craniosynostosis. *Neurosurgery 6*, 39–49.

Hoyte, D. A. N. (1990). Hydrocephalus in the infant rat: a further look at the basicranial synchondroses In A. D. Dixon, B. G. Sarnat, and D. A. N. Hoyte (eds.), *Fundamental of Bone Growth: Methodology and Applications* (Boca Raton: CRC Press), pp. 489–496.

Ignelzi, M. A., Liu, Y. H., Maxson, R. E., Jr., and Snead, M. L. (1995). Genetically engineered mice: a tool to understand craniofacial development. *Crit. Rev. Oral Biol. 6*, 181–201.

Ignelzi, M. A., Liu, Y. H., Sangiorgi, F., Kundu, R., Snead, M. L., and Maxson, R. E. (1996). Exploring the relationship of fgf signaling pathways and Msx2. *J. Dent. Res. 75*, 225.

Inouye, R. N., Kokich, V. G., Clarren, S. K., and Bowden, D. M. (1985). Fetal alcohol syndrome: an examination of craniofacial dysmorphology in *Macaca nemestrina. J. Med. Primatol. 14*, 35–38.

Itabashi, M. (1969). Experimental production of congenital malformations in Japanese white rabbits by maternal treatment with hypervitaminosis A. *Congen. Anom. 9*, 143–155.

Jaskoll, T., and Melnick, M. (1982). The effects of long-term fetal constraint in vitro on the cranial base and other skeletal components. *Am. J. Med. Genet. 12*, 289–300.

Johnston, M. C. (1980). Animal models for craniofacial disorders: a critique. In M. Melnick, D. Bixler, and E. D. Shields (eds.), *Etiology of Cleft Lip and Palate*. Progress in Clinical Biological Research, Alan R. Liss N.Y. vol. 46, pp. 33–38.

Johnston, M. C., and Bronsky, P. T. (1991). Animal models for human craniofacial malformations. *J. Craniofac. Genet. Devel. Biol. 11*, 227–291.

Johnston, M. C., and Bronsky, P. T. (1995). Prenatal craniofacial development: new insights on normal and abnormal mechanisms. *Crit. Rev. Oral Biol. Med. 6*, 25–79.

Kalter, H., and Warkany, J. (1961). Experimental production of congenital malformations in strains of inbred mice by maternal treatment with hypervitaminosis A. *Am. J. Pathol. 38*, 1–21.

Kier, E. L. (1976). Phylogenetic and ontogenetic changes of the brain relevant to the evolution of the skull. In J. Bosma (ed.), *Development of the Basicranium* (Bethesda, MD: DHEW/NIH Publ. 76-989), pp. 468–499.

Kim, H. J., Rice, D. P., Kettunen, P. J., and Thesleff, I. (1998). FGF-, BMP- and Shh-mediated signaling pathways in the regulation of cranial suture morphogenesis and calvarial bone development. *Development 125*, 1241–1251.

Kinney, B. M., Resnick, J. I., and Kawamoto, H. K. (1990). The effect of miniplate fixation on cranial bone growth. In A. D. Dixon, B. G. Sarnat, and D. A. N. Hoyte (eds.), *Fundamentals of Bone Growth: Methodology and Applications* (Boca Raton: CRC Press), pp. 497–502.

Kokich, V. G. (1986). The biology of sutures. In M. M. Cohen, Jr. (ed.), *Craniosynostosis: Diagnosis, Evaluation, and Management* (New York: Raven Press), pp. 81–103.

Kolf-Clauw, M., Chevy, F., and Ponsart, C. (1998). Abnormal cholesterol biosynthesis as in Smith-Lemli-Opitz syndrome disrupts normal skeletal development in the rat. *J. Lab. Clin. Med. 131*, 222–227.

Koskinen-Moffett, L., and Moffett, B. C. (1989). Sutures and intrauterine deformation. In J. A. Persing, M. T. Edgerton, and J. A. Jane (eds.), *Scientific Foundations and Surgical Treatment of Craniosynostosis* (Baltimore: Williams & Wilkins), pp. 96–106.

Kreiborg, S. (1986). Postnatal growth and development of the craniofacial complex in premature craniosynostosis. In M. M. Cohen, Jr. (ed.), *Craniosynostosis: Diagnosis, Evaluation, and Management* (New York: Raven Press), pp. 157–190.

Kremenak, C. R. (1990). Appropriate animal models for craniofacial biology: Commentary. *Cleft Palate Craniofac. J. 27*, 25.

Kremenak, C. R., and Searls, J. C. (1971). Experimental manipulation of midfacial growth: a synthesis of five years of research at the Iowa Maxillofacial Growth Laboratory. *J. Dent. Res. 50*, 1488–1491.

Krogman, W. M. (1969). Growth changes in skull, face, jaws, and teeth of the chimpanzee. In G. H. Bourne (ed.), *Anatomy, Behavior, and Diseases of Chimpanzees*, vol. 1 (Baltimore: University Park Press), pp. 104–164.

Krogman, W. M. (1978). *The Human Skeleton in Forensic Medicine* (Springfield: Charles C. Thomas).

Lalikos, J., Tschakaloff, A., Mooney, M. P., Losken, H. W., Siegel, M. I., and Losken, A. (1995). Internal calvarial bone distraction in rabbits with experimental coronal suture immobilization: effects of overdistraction. *Plast. Reconstr. Surg. 96*, 689–698.

Le Douarin, N. M., Ziller, C., and Couly, G. F. (1993). Patterning of neural crest derivatives in the avian embryo: in vivo and in vitro studies. *Devel. Biol. 159*, 24–49.

Levy, B. M., Dreizen, S., Hampton, J. K., Taylor, A. C., and Hampton, S. H. (1971). Primates in dental research. In E. I. Goldsmith and J. Moor-Jankowski (eds.), *Proceedings of the Third Conference on Experimental Medicine and Surgery in Primates, Medical Primatology* (Basel: Karger), pp. 859–869.

Lieberman, D. E., and McCarthy, R. C. (1999). The ontogeny of cranial base angulation in humans and chimpanzees and its implications for reconstructing pharyngeal dimensions. *J. Hum. Evol. 36*, 487–517.

Lieberman, D. E., Ross, C. F., and Ravosa, M. J. (2000). The primate cranial base: ontogeny, function, and integration. *Yrbk. Phys. Anthropol. 43*, 117–169.

Lin, K. Y., Bartlett, S. P., Yarmechuk, M. J., Grossman, R. F., Udupa, J. K., and Whitaker, L. A. (1991). An experimental study on the effect of rigid fixation on the developing craniofacial skeleton. *Plast. Reconstr. Surg. 87*, 229–235.

Liu, Y-H., Kundu, R., Wu, L., Luo, W., Ignelzi, M. A., Jr., Snead, M. L., and Maxson, R. E., Jr. (1995). Premature suture closure and ectopic cranial bone in mice expressing Msx2 transgenes in the developing skull. *Proc. Natl. Acad. Sci. USA 92*, 6137–6141.

Liu, Y.-H., Tang, Z., Kundu, R., Wu, L., Luo, W., Zhu, D., Sangiorgi, F., Snead, M., and Maxson, R. (1999). Msx2 gene dosage influences the number of proliferative osteogenic cells in growth centers of the developing murine skull: a possible mechanism for MSX2-mediated craniosynostosis in humans. *Devel. Biol. 205*, 260–274.

Loeb, J. M., Hendee, W. R., Smith, S. J., and Schwarz, M. R. (1989). Human vs. animal rights: in defense of animal research. *J. Am. Med. Assoc. 262*, 2716–2720.

Losken, H. W., Tschakaloff, A., Mooney, M. P., and Losken, A. (1991a). Frontal bone advancement stability with and without the use of biodegradable microplates fixation in rabbits with experimental craniosynostosis. *Plast. Surg. Forum 14*, 206–208.

Losken, H. W., Hurwitz, D. J., Mooney, M. P., Losken, A., and Zhang, L. P. (1991b). Frontal bone advancement stability with or without microplate fixation: an experimental study in rabbits. *J. Craniofac. Surg. 2*, 22–26.

Losken, A., Mooney, M. P., and Siegel, M. I. (1992). Comparative analysis of mandibular growth patterns in seven animal models. *J. Oral Maxillofac. Surg. 50*, 490–495.

Losken, A., Mooney, M. P., and Siegel, M. I. (1994). Comparative cephalometric study of nasal cavity growth patterns in seven animal models. *Cleft Palate Craniofac. J. 31*, 17–23.

Losken, H. W., Mooney, M. P., Zoldos, J., Tschakaloff, A., Smith, T. D., Burrows, A. M., Cooper, G. M., Kapucu, M. R., and Siegel,

M. I. (1998). Internal calvarial bone distraction in rabbits with delayed onset coronal suture synostosis. *Plast. Reconstr. Surg. 102*, 1109–1119.

Losken, H. W., Mooney, M. P., Zoldos, J., Tschakaloff, A., Burrows, A. M., Smith, T. D., Cano, G., Arnott, R., Sherwood, C., Dechant, J., Cooper, G., Kapucu, R., and Siegel, M. I. (1999). Coronal suture response to distraction osteogenesis in rabbits with delayed onset coronal suture synostosis. *J. Craniofac. Surg. 10*, 29–37.

Luke, S., Gandhi, S., and Verma, R. S. (1995). Conservation of the Down syndrome critical region in humans and great apes. *Gene 161*, 283–285.

Mabbutt, L. W., and Kokich, V. G. (1975). Calvarial and sutural redevelopment following craniectomy in the neonatal rabbit. *J. Anat. 2*, 413–422.

Marchac, D., and Renier, D. (1982). *Craniofacial Surgery for Craniosynostosis* (Boston: Little Brown & Co.).

Markens, I. S., and Taverne, A. A. (1978). Development of cartilage in transplanted future coronal sutures. *Acta Anat. 100*, 428–434.

Marks, S. C., Odgren, P. R., Popoff, S. N., and Wurtz, T. (1999a). Sutures, growth plates and the craniofacial base—experimental studies in the toothless (tl-osteopetrotic) rat. *Ann. Acad. Med. Singapore 28*, 650–654.

Marks, S. C., Jundmark, C., Wurtz, T., Odgren, P. R., MacKay, C. A., Mason-Savas, A., and Popoff, S. N. (1999b). Facial development and type III collagen RNA expression: concurrent repression in the osteopetrotic (toothless, tl) rat and rescue after treatment with colony-stimulating factor-1. *Devel. Dyn. 215*, 117–125.

Marschall, M. A., Chidyllo, S. A., Figueroa, A. A., and Cohen, M. (1991). Long-term effects of rigid fixation on the growing craniomaxillofacial skeleton. *J. Craniofac. Surg. 2*, 63–71.

Marsh, J. L., and Vannier, M. W. (1985). *Comprehensive Care for Craniofacial Deformities* (St. Louis: C. V. Mosby).

McClure, H. M., Belden, K. H., Pieper, W. A., and Jacobson, C. B. (1969). Autosomal trisomy in a chimpanzee: resemblance to Down's syndrome. *Science 165*, 1010–1012.

Michejda, M. (1975). Ontogenic growth changes of the skull base in four genera of nonhuman primates. *Acta Anat. 91*, 110–117.

Mooney, M. P., and Siegel, M. I. (1993). Appropriate animal models for craniofacial biology II: their applications to clinical investigations. Paper presented at the annual meeting of the American Cleft Palate–Craniofacial Association, Pittsburgh, April.

Mooney, M. P., Losken, H. W., Tschakaloff, A., Siegel, M. I., Losken, A., and Lalikos, J. F. (1993). Congenital bilateral coronal suture synostosis in a rabbit and comparisons with experimental models. *Cleft Palate Craniofac. J. 30*, 121–128.

Mooney, M. P., Losken, H. W., Siegel, M. I., Lalikos, J. F., Losken, A., Smith, T. D., and Burrows, A. (1994a). Development of a strain of rabbits with congenital simple, nonsyndromic coronal suture synostosis. Part 1. Breeding demographics, inheritance pattern, and craniofacial anomalies. *Cleft Palate Craniofac. J. 31*, 1–7.

Mooney, M. P., Losken, H. W., Siegel, M. I., Lalikos, J. F., Losken, A., Smith, T. D., and Burrows, A. (1994b). Development of a strain of rabbits with congenital simple, nonsyndromic coronal suture synostosis. Part 2. Somatic and craniofacial growth patterns. *Cleft Palate Craniofac. J. 31*, 8–14.

Mooney, M. P., Aston, C. E., Siegel, M. I., Losken, H. W., Smith, T. D., Burrows, A. M., Wenger, S. L., Caruso, K., Siegel, B., and Ferrell, R. E. (1996a). Craniosynostosis with autosomal dominant transmission in New Zealand white rabbits. *J. Craniofac. Genet. Devel. Biol. 16*, 52–63.

Mooney, M. P., Smith, T. D., Langdon, H. L., Burrows, A. M., Stone, C. E., Losken, H. W., and Siegel, M. I. (1996b). Coronal suture pathology and synostotic progression in rabbits with congenital craniostenosis. *Cleft Palate Craniofac. J. 33*, 369–378.

Mooney, M. P., Siegel, M. I., Burrows, A. M., Smith, T. D., Losken, H. W., Dechant, J., Cooper, G., and Kapucu, M. R. (1998a). A rabbit model of human familial, nonsyndromic, unicoronal suture synostosis: part 1: synostotic onset, pathology, and sutural growth patterns. *Child's Nerv. Sys. 14*, 236–246.

Mooney, M. P., Siegel, M. I., Burrows, A. M., Smith, T. D., Losken, H. W., Dechant, J., Cooper, G., Fellow-Mayle, W., Kapucu, M. R., and Kapucu, L. O. (1998b). A rabbit model of human familial, nonsyndromic, unicoronal suture synostosis: part 2: intracranial contents, intracranial volume, and intracranial pressure. *Child's Nerv. Sys. 14*, 247–255.

Mooney, M. P., Burrows, A. M., Wigginton, W., Singhal, V. K., Losken, H. W., Smith, T. D., Dechant, J., Towbin, A., Cooper, G., Towbin, R., and Siegel, M. I. (1998c). Intracranial volume in craniosynostotic rabbits. *J. Craniofac. Surg. 9*, 234–239.

Mooney, M. P., Fellows-Mayle, W., Dechant, J., Losken, H. W., Burrows, A. M., Smith, T. D., Cooper, G. M., Pollack, I., and Siegel, M. I. (1999). Increases in intracranial pressure following coronal suturectomy in rabbits with craniosynostosis. *J. Craniofac. Surg. 10*, 104–111.

Mooney, M. P., Cooper, G. M., Burrows, A. M., Wigginton, W., Smith, T. D., Dechant, J., Mitchell, R., Losken, H. W., and Siegel, M. I. (2000). Trigonocephaly in rabbits with familial interfrontal suture synostosis: multiple effects of single-suture fusion. *Anat. Rec. 260*, 238–251.

Mooney, M. P., Burrows, A. M., Losken, H. W., Opperman, L. A., Smith, T. D., Dechant, J., Kreithen, A. M., Kapucu, R., Cooper, G. M., Ogle, R. C., and Siegel, M. I. (2001). Correction of coronal suture synostosis using suture and dura mater allografts in rabbits with familial craniosynostosis. *Cleft Palate Craniofac. J. 38*, 72–91.

Moss, M. L. (1954). Growth of the calvaria in the rat: the determination of osseous morphology. *Am. J. Anat. 94*, 333–362.

Moss, M. L. (1958). Fusion of the frontal suture in the rat. *Am. J. Anat. 102*, 141–165.

Moss, M. L. (1959). The pathogenesis of premature cranial synostosis in man. *Acta Anat. 37*, 351–370.

Moss, M. L. (1960). Inhibition and stimulation of sutural fusion in the rat calvaria. *Anat. Rec. 136*, 457–467.

Moss, M. L. (1975). Functional anatomy of craniosynostosis. *Child's Brain 1*, 22–33.

Moss, M. L., Villman, H., Moss-Salentijn, L., Sen, K., Pucciarelli, H. M., and Skalak, R. (1987). Studies on orthocephalization: growth behavior of the rat skull in the period 13–49 days as described by the finite element method. *Am. J. Phys. Anthropol. 72*, 323–342.

Myers, R. E. (1969). Brain pathology following fetal vascular occlusion: an experimental study. *Invest. Ophthalmol. 8*, 41–50.

Nakayama, D. K., Harrison, M. R., Berger, M. S., Chinn, D. H., Halks-Miller, M., and Edwards, M. S. (1983). Correction of congenital hydrocephalus in utero I. The model: intracisternal kaolin produces hydrocephalus in fetal lambs and rhesus monkeys. *J. Pediatr. Surg. 18*, 331–338.

Nappen, D., and Kokich, V. G. (1983). Experimental craniosynostosis in growing rabbits: the role of the periosteum. *J. Neurosurg. 58*, 101–109.

Navia, J. M. (1977). *Animal Models in Dental Research* (Birmingham: University of Alabama Press).

Noden, D. M. (1983). The role of the neural crest in patterning of avian cranial skeletal, connective, and muscle tissues. *Devel. Biol. 96*, 144–165.

Noden, D. M. (1986a). Origins and patterning of craniofacial mesenchymal tissues. *J. Craniofac. Genet. Devel. Biol. 2*, 15–31.

Noden, D. M. (1986b). Patterning of avian craniofacial muscles. *Devel. Biol. 116*, 347–356.

Opperman, L. A., Sweeney, T. M., Redmon, J., Persing, J. A., and Ogle, R. C. (1993). Tissue interactions with underlying dura mater inhibit osseous obliteration of developing cranial sutures. *Devel. Dyn. 198*, 312–322.

Opperman, L. A., Passarelli, R., Morgan, E. P., Reintjes, M., and Ogle, R. C. (1995). Cranial sutures require tissue interactions with dura mater to resist osseous obliteration in vitro. *J. Bone Min. Res. 10*, 1978–1987.

Opperman, L. A., Nolen, A. A., and Ogle, R. C. (1997). TGF-beta 1, TGF-beta 2, and TGF-beta 3 exhibit distinct patterns of expression during cranial suture formation and obliteration in vivo and in vitro. *J. Bone Min. Res. 12*, 301–310.

Persing, J. A., Babler, W. J., Winn, H. R., Jane, J. A., and Rodheaver, G. T. (1981). Age as a critical factor in the success of surgical correction of craniosynostosis. *J. Neurosurg. 54*, 601–606.

Persing, J. A., Babler, W. J., and Jane, J. A. (1986a). Experimental unilateral coronal synostosis in rabbits. *Plast. Reconstr. Surg. 77*, 369–376.

Persing, J. A., Babler, W. J., Nagorsky, M. J., Edgerton, M. T., and Jane, J. A. (1986b). Skull expansion in experimental craniosynostosis. *Plast. Reconstr. Surg. 78*, 594–604.

Persing, J. A., Jane, J. A., and Edgerton, M. A. (1989). Surgical treatment of craniosynostosis. In J. A. Persing, M. T. Edgerton, and J. A. Jane (eds.), *Scientific Foundations and Surgical Treatment of Craniosynostosis* (Baltimore: Williams and Wilkins), pp. 87–95.

Persing, J. A., Lettieri, J. T., Cronin, A. J., Wolcott, W. P., Singh, V., and Morgan, E. (1991). Craniofacial suture stenosis: morphologic effects. *Plast. Reconstr. Surg. 88*, 563–573.

Persson, K. M. (1989). Regulating factors in suture development, growth, and closure. In J. A. Persing, M. T. Edgerton, and J. A. Jane (eds.), *Scientific Foundations and Surgical Treatment of Craniosynostosis* (Baltimore: Williams and Wilkins), pp. 45–49.

Persson, K. M., Magnusson, B. C., and Thilander, B. (1978). Sutural closure in rabbit and man: a morphological and histochemical study. *J. Anat. 125*, 313–321.

Persson, K. M., Roy, W. A., Persing, J. A., Rodheaver, G. T., and Winn, H. R. (1979). Craniofacial growth following experimental craniosynostosis and craniectomy. *J. Neruosurg. 50*, 187–197.

Poisson, E., Mooney, M. P., Koepsel, R., Cooper, G. M., Opperman, L. A., and Sciote, J. J. (1999). TGF-β isoform expression in the perisutural tissues of craniosynostotic rabbits. *J. Dent. Res. 78* (special issue), 363.

Poswillo, D. E. (1980). Animal models for craniofacial disorders: a critique. In M. Melnick, D. Bixler, and E. D. Shields (eds.), *Etiology of Cleft Lip and Palate*. Progress in Clinical Biological Research, vol. 46, pp. 25–32.

Poswillo, D. (1988). The aetiology and pathogenesis of craniofacial deformities. In R. Thorogood and C. Tickle (eds.), *Craniofacial Development* (Cambridge: Company of Biologists Limited), pp. 207–212.

Prince, A. M., Moor-Jankowski, J., Eichberg, J. W., Schellekens, H., Mauler, R. F., Girard, M., and Goodall, J. (1988). Chimpanzees and AIDS research. *Nature 333*, 513–514.

Pucciarelli, H. M. (1978). The influence of experimental deformation on craniofacial development in rats. *Am. J. Phys. Anthropol. 48*, 455–461.

Renier, D. (1989). Intracranial pressure in craniosynostosis: pre- and postoperative recordings—correlation with functional results. In J. A. Persing, M. T. Edgerton, and J. A. Jane (eds.), *Scientific Foundations and Surgical Treatment of Craniosynostosis* (Baltimore: Williams and Wilkins), pp. 263–269.

Renier, D., Sainte-Rose, C., and Marchac, D. (1982). Intracranial pressure in craniostenosis. *J. Neurosurg. 57*, 370–375.

Reynolds, H. H. (1969). Nonhuman primates in the study of toxicological effects on the central nervous system: a review. *Ann. N.Y. Acad. Sci. 162*, 604–609.

Richtsmeier, J. T., Baxter, L. L., and Reeves, R. H. (2000). Parallels of craniofacial maldevelopment in Down syndrome and Ts65Dn mice. *Devel. Dyn. 217*, 137–145.

Roach, H. I., Shearer, J. R., and Archer, C. (1989). The choice of an experimental model: a guide for research workers. *J. Bone Joint Surg. 71-B*, 549–553.

Romer, A. S., and Parsons, T. S. (1986). *The Vertebrate Body*, 5th ed. (Philadelphia: W. B. Saunders).

Rosenberg, P., Arlis, H. R., Haworth, R. D., Heier, L., Hoffman, L., and LaTrenta, G. (1997). The role of the cranial base in facial growth: experimental craniofacial synostosis in the rabbit. *Plast. Reconstr. Surg. 99*, 1396–1407.

Roth, D. A., Bradley, J. P., Levine, J. P., McMullen, H. F., McCarthy, J. G., and Longaker, M. T. (1996). Studies in cranial suture biology: part II. Role of the dura in cranial suture fusion. *Plast. Reconstr. Surg. 97*, 693–699.

Roth, D. A., Longaker, M. T., McCarthy, J. G., Rosen, D. M., McMullen, H. F., Levine, J. P., Sung, J., and Gold, L. I. (1997). Studies in cranial suture biology: part I. Increased immunoreactivity for TGF-beta isoforms (beta 1, beta 2, and beta 3) during rat cranial suture fusion. *J. Bone Min. Res. 12*, 311–321.

Rowsey, J. J. (1988). Responsibilities in animal experimentation. *Ophthalmic. Surg. 19*, 161–162.

Roy, W. A., Iorio, R. J., and Meyer, G. A. (1981). Craniosynostosis in vitamin D–resistant rickets: a mouse model. *J. Neurosurg. 55*, 265–271.

Satokata, I., and Maas, R. (1994). Msx1 deficient mice exhibit cleft palate and abnormalities of craniofacial and tooth development. *Nat. Genet. 6*, 348–356.

Schmidt-Nielsen, B. (1961). Choice of experimental animals for research. *Fed. Proc. 20*, 902–906.

Schultz, A. H. (1960). Age changes and variability in the skulls and teeth of the Central American monkeys *Alouatta, Cebus*, and *Ateles. Proc. Zool. Soc. (London) 133*, 337–390.

Schultz, A. H. (1969). The skeleton of the chimpanzee. In G. H. Bourne (ed.), *Anatomy, Behavior, and Diseases of Chimpanzees*, vol. 1 (Baltimore: University Park Press), pp. 50–103.

Shek, J. W., Wen, G. Y., and Wisniewski, H. M. (1986). *Atlas of the Rabbit Brain and Spinal Cord* (Basel: Karger).

Siegel, M. I., and Mooney, M. P. (1990). Appropriate animal models for craniofacial biology. *Cleft Palate Craniofac. J. 27*, 18–25.

Siegel, M. I., Long, R. E., Doyle, W. J., and Mooney, M. P. (1985). Appropriate animal models for craniofacial biology. Study session presentation at the 42nd annual American Cleft Palate Association Meeting, Miami.

Singhal, V. K., Mooney, M. P., Burrows, A. M., Wigginton, W., Losken, H. W., Smith, T. D., Towbin, R., and Siegel, M. I. (1997). Age related changes in intracranial volume in craniosynostotic rabbits using 3-D CT scans. *Plast. Reconstr. Surg. 100*, 1121–1128.

Sirianni, J. (1985). Nonhuman primates as models for human craniofacial growth. In E. S. Watts (ed.), *Nonhuman Primate Models for Human Growth and Development* (New York: Liss), pp. 95–124.

Sirianni, J. E., and Swindler, D. R. (1979). A review of the postnatal craniofacial growth in old world monkeys and apes. *Yrbk. Phys. Anthropol. 22*, 80–104.

Smith, G. C. (1969). Value of nonhuman primates in predicting disposition of drugs in man. *Ann. N.Y. Acad. Sci. 162*, 600–603.

Smith, H. G., and McKeown, M. (1974). Experimental alteration of the coronal suture area: a histological and quantitative microscopic assessment. *J. Anat. 118*, 543–559.

Smith, J. D., Genoways, H. H., and Jones, J. K. (1977). Cranial and dental anomalies in three species of platyrrhine monkeys from Nicaragua. *Folia Primatol. 28*, 1–42.

Smith, T. D., Mooney, M. P., Losken, H. W., Siegel, M. I., and Burrows, A. (1996). Postnatal growth of the cranial base in rabbits with congenital coronal suture synostosis. *J. Craniofac. Genet. Devel. Biol. 16*, 107–117.

Stelnicki, E. J., Mooney, M. P., Losken, H. W., Zoldos, J., Burrows, A. M., Kapucu, R., and Siegel, M. I. (1997). Ultrasonic prenatal diagnosis of coronal suture synostosis. *J. Craniofac. Surg. 8*, 252–258.

Stelnicki, E. J., Vanderwall, K., Hoffman, W. Y., Harrison, M. R., Glowacki, J., and Longaker, M. T. (1998). A new in utero sheep model for unilateral coronal craniosynostosis. *Plast. Reconstr. Surg. 101*, 278–286.

Sulik, K. K., Johnston, M. C., and Webb, M. A. (1981). Fetal alcohol syndrome: embryogenesis in a mouse model. *Science 214*, 936–938.

Sulik, K. K., Lauder, J. M., and Dehart, D. B. (1984). Brain malformations in prenatal mice following acute maternal ethanol administration. *Int. J. Devel. Neurosci. 2*, 203–214.

Sutton, J. F. (1972). Notes on skeletal variations, tooth replacement, and cranial suture closure of the porcupine *(Erethizon dorsatum). Tulane Stud. Zool. Botany 17*, 56–62.

Swindler, D. R., and Wood, C. D. (1982). *An Atlas of Primate Gross Anatomy: Baboon, Chimpanzee, and Man* (Malabar, FL: Robert E. Krieger).

Taylor, I., Baum, M., Cooper, A., and Johnston, I. D. (1991). Dilemmas facing surgical research in the '90s. *Ann. R. Coll. Surg. Engl. 73*, 70–72.

Ten Cate, A. R. (1989). *Oral Histology*, 3rd ed. (St. Louis: C. V. Mosby).

Thaller, S. R., Hoyt, J., Tesluk, H., and Holmes, R. (1993). The effect of insulin growth factor-

1 on calvarial sutures in a Sprague-Dawley rat. *J. Craniofac. Surg. 4*, 35–39.

Tranquart, F., Berson, M., and Bodard, S. (1991). Evaluation of cerebral blood flow in rabbits with transcranial Doppler sonography: first results. *Ultrasound Med. Biol. 17*, 815–818.

Tranquart, F., DeBray, J. M., and Berson, M. (1994). Concurrent changes in intracranial pressure, cerebral blood flow velocity, and brain energy metabolism in rabbits with acute intracranial hypertension. *Child's Nerv. Sys. 10*, 285–292.

Tyler, M. S. (1983). Development of the frontal bone and cranial meninges in the embryonic chick: an experimental study of tissue interactions. *Anat. Rec. 206*, 61–70.

Tyler, M. S., and McCobb, D. P. (1980). The genesis of membrane bone in the embryonic chick maxilla: epithelial-mesenchymal tissue recombination studies. *J. Embryol. Exp. Morphol. 56*, 269–281.

Ungersbock, K., Tenckhoff, D., Heimann, A., Wagner, W., and Kempski, O. S. (1995). Transcranial Doppler and cortical microcirculation and increased intracranial pressure during the Cushing response: an experimental study on rabbits. *Neurosurgery 36*, 147–157.

Vergato, L. A., Doerfler, R. J., Mooney, M. P., and Siegel, M. I. (1997). Mouse palatal width growth rates as an "at risk" factor in the development of cleft palate induced by hypervitaminosis A. *J. Craniofac. Genet. Devel. Biol. 17*, 204–210.

Vogelweid, C. M., Vogt, D. W., Besch-Williford, C. L., and Walker, S. E. (1993). New Zealand white mice: an experimental model of exencephaly. *Lab. Anim. Sci. 43*, 58–60.

Wilkie, A. O. M., and Wall, S. A. (1996). Craniosynostosis: novel insights into pathogenesis and treatment. *Curr. Opin. Neurol. 9*, 146–152.

Wilson, J. G. (1971). Use of rhesus monkeys in teratological studies. *Fed. Proc. 30*, 104–109.

Winograd, J., Reilly, M. P., Roe, R., Lutz, J., Laughner, E., Xu, X., Hu, L., Asakura, T., Vanderkolk, C., Standberg, J. D., and Semenza, G. L. (1997). Perinatal lethality and multiple craniofacial malformations in Msx2 transgenic mice. *Hum. Molec. Genet. 6*, 369–379.

Wong, L., Dufresne, C. R., Richtsmeier, J. T., and Manson, P. N. (1991). The effect of rigid fixation on growth of the neurocranium. *Plast. Reconstr. Surg. 88*, 395–403.

Yang, Z.-W., Mooney, M. P., and Ferrell, R. E. (2000). Cloning and sequencing of the rabbit FGFR2 cDNA. *DNA Seq. 11*, 439–446.

Yip, J. E., Kokich, V. G., and Shepard, T. H. (1980). The effect of high doses of retinoic acid on prenatal craniofacial development in *Macaca nemistrina. Teratology 21*, 29–38.

Young, J. Z. (1989). *The Life of Vertebrates*, 3rd ed. (Oxford: Clarendon Press).

Young, R. W. (1959). The influence of cranial contents on postnatal growth of the skull in the rat. *Am. J. Anat. 105*, 383–414.

Yu, J. C., McClintock, J. S., Gannon, F., Gao, X. X., Mobasser, J. P., and Sharawy, M. (1997). Regional differences of dura osteoinduction: squamous dura induces osteogenesis, sutural dura induces chondrogenesis and osteogenesis. *Plast. Reconstr. Surg. 100*, 23–31.

Zhou, Y.-X., Xu, X., Chen, L., Li, C., Brodie, S. G., and Deng, C.-X. (2000). A Pro250Arg substitution in mouse Fgfr1 causes increased expression of Cbfa1 and premature fusion of calvarial sutures. *Hum. Molec. Genet. 9*, 2001–2008.

CHAPTER 10

ANIMAL MODELS OF FACIAL CLEFTING: EXPERIMENTAL, CONGENITAL, AND TRANSGENIC

VIRGINIA M. DIEWERT, D.D.S., M.Sc. and SCOTT LOZANOFF, Ph.D.

10.1 INTRODUCTION

Animal models have been and continue to be valuable for our understanding of normal mechanisms of craniofacial development, how genetic and environmental factors alter normal development, and specific features that increase or decrease susceptibility to clefting. Successful formation of the upper face, the midface, and the primary and secondary palate regions involves multiple sequences of intrinsic developmental events that must be coordinated spatially and temporally with growth and development of adjacent and supportive structures. These facial components have developmental thresholds that require their formation within a critical developmental period. Median clefts, cleft lip with or without cleft palate (CL/P), and isolated cleft palate (CP) are congenital anomalies that arise from failure of developmental processes due to many different genetic and environmental factors. During the past 50 years, animal models have been important in advancing the understanding of the complex gene–environmental interactions involved in normal and abnormal

craniofacial morphogenesis. Mouse models have been particularly important for new insights into the etiology of orofacial clefting and now provide the basis for novel future molecular and transgenic approaches. In this chapter, we focus on current in vivo models and identify features that appear similar to human development.

The craniofacial region is traditionally divided into three anatomical subcomponents based on trigeminal nerve innervation. These include the upper face, innervated by the ophthalmic division; the midface, innervated by the maxillary division; and the mandible, innervated by the mandibular division. These three components represent anatomical composites of sensory tissues derived from endoderm and ectoderm while the connective tissue surrounding these units generally arises from mesoderm and, what some consider to be the fourth germ cell constituent, neural crest. Furthermore, these neural and connective tissues must develop in synchrony to assure proper morphology and function. Failure of proper spatiotemporal interactions results not only in sensory unit dysmorphogenesis but usually in defects of

Understanding Craniofacial Anomalies: The Etiopathogenesis of Craniosynostoses and Facial Clefting, Edited by Mark P. Mooney and Michael I. Siegel, ISBN 0-471-38724-x Copyright © 2002 by Wiley-Liss, Inc.

surrounding supportive structures as well. It is becoming increasingly clear that genetic defects in the sensory and connective tissue components may share single pathways (Sarnat, 2000). Therefore, craniofacial and neural defects occur simultaneously, and defective signaling may occur very early in development with the emergence of the neural plate.

Many defects in the craniofacial supportive structures surrounding the sensory units are expressed morphologically as failures of facial prominence merging and/or fusion resulting in clefts. These defects typically are categorized into those that affect the midline (median facial clefts) and those that occur laterally (lateral facial clefts). Median facial defects occur early and probably relate closely to the initial events directing morphogenesis of the anterior midline tissue of the gastrula. Lateral facial clefts can be conceptualized as defects resulting from abnormal events usually occurring later in development once the facial primordia are in place. It is unlikely that both median and lateral facial cleft defects are simply the result of single genetic aberrations, since normal craniofacial development results from many genes inhibiting or enhancing the expression of others. Due to the wide array of genes involved in craniofacial morphogenesis, specific mechanisms driving defects have been difficult to delineate. However, the identification of specific effects exerted by the genome is becoming possible due to recent advances in recombinant DNA technology and transgenesis. The purpose of this review is to highlight some of the recent advances in the identification of mechanisms causing median and lateral facial clefts through the use of experimental mouse models.

10.2 MIDLINE CRANIOFACIAL DEVELOPMENT

10.2.1 Early Midline Craniofacial Development

Formation of the components of the face that eventually give rise to the midline tissue (maxillary and frontal regions as well as the central stem of the chondrocranium and forebrain) can be traced to the earliest stages of development in the mouse. As with other rodents, the embryonic mass of the mouse proliferates rapidly and expands into the blastocoele, forming an egg cylinder. This homologue of the embryoblast is a double-layered structure consisting of an inner layer of ectoderm and an outer layer of visceral endoderm. Subsequent formation of the primitive streak thus forms internally, and this configuration is referred to as inversion of the germ layers (Kaufman, 1992). As a result of this inversion of germ layers, the embryo assumes a U shape, with the exposed neural ectoderm directed inward. Eventually the embryo undergoes axial rotation, resulting in embryonic positioning consistent with a fetal position, which is similar with all other chordates. Mechanisms of gastrulation in the mouse provide useful insight into the process of human craniofacial and forebrain formation regardless of the initial difference in gastrula positioning.

Numerous genes are being identified in the mouse head organizers, which appear to be involved in signaling. For example, insertional mutations of *Lim1* result in all head structures anterior to rhombomere 3 to be absent while posterior structures remain normal (Shawlot and Behringer, 1995). Mice with insertional mutations in *Otx2* also fail to develop cephalic structures (Rhinn et al., 1998). *HesX1* mutations are associated with septo-optic-pituitary dysplasia structures, and these are caused by defects in the visceral endoderm (Dattani et al., 1998). *Cerebrus*-like is another visceral endodermal factor that must be expressed, or deficient rostral neural and craniofacial morphogenesis results (Belo, 1997; Piccolo et al., 1999). In addition, *Wnt3* null mutation neural ectoderm lacks proper anteroposterior patterning even though visceral endodermal markers are expressed correctly (Liu et al., 1999). These results indicate that in addition to the primitive node, the anterior visceral endoderm is pivotal in directing initial cephalization of the gastrula.

During gastrulation in the mouse, some of the mesoderm of the head process will merge with foregut endoderm to form the prechordal plate. A clear anatomical boundary between these two structures is not evident, and the term *anterior midline tissue* is applied to this composite tissue positioned anterior to the primitive node in the mouse. The anterior midline mesoderm expresses organizing genes along an anteroposterior axis with *Gsc*, *Shh*, and *Hnf3β* expressed in the presumptive prosencephalon (Chiang et al., 1996). These genes, as well as others such as *Lim1*, *Otx2*, *Bmps*, *chordin*, and *noggin*, become expressed posteriorly (Tam and Behringer, 1997; Bachiller et al., 2000). *Gsc* is of particular interest since null mutations show craniofacial defects in the vomer, palate, and sphenoid bones (Rivera-Perez et al., 1995). *Gsc-1* −/− mouse embryos with *Hnf3β* haploinsufficiency show severe ventralization of the brain along with reductions or loss in expression of *Shh* and FGF-8 (Filosa et al., 1997). Hence, *Gsc* is critical for proper development of the midline development of the murine head (Belo et al., 1998).

Of particular interest is *Shh*, which is expressed throughout the anterior midline tissue. Loss of *Shh* function has been induced in mouse embryos, resulting in the absence of ventral forebrain structures and bilateral division of the developing diencephalon and optic structures (Chiang et al., 1996). In the extreme case, holoprosen-

cephaly has been produced in the mouse as a result of *Shh* loss of function. Interestingly, Hu and Helms (1999) have shown that *Shh* gain of function results in hypertelorism. Thus, the Shh protein appears to constitute, or contributes to, the signal from the anterior midline tissue that induces ventral forebrain division and subsequent craniofacial development. Similarly, mice carrying mutations of *Gli2* show craniofacial abnormalities characteristic of mild forms of holoprosencephaly (Hardcastle et al., 1998).

Midline tissue defects can also occur slightly later in development as a result of deficient cranial neural crest migration and differentiation. Proper migration of neural crest is required to populate the facial prominences. Once seeded, facial prominences merge in the midline, requiring epithelial and mesenchymal interaction. Null mutants for *Pax 7* and *Ap2* show severe midline defects (Mansouri et al., 1996; Schorle et al., 1996), suggesting that these two factors are crucial for proper merger. Likewise, substantial evidence points to Shh as a primary molecule for establishing proper interocular distance (Hu and Helms, 1999). Interestingly, Veitch and colleagues (1999) have shown that pharyngeal arches form even in the absence of neural crest migration, suggesting that midline facial development involves both neural crest–dependent and neural crest–independent morphogenesis.

10.2.2 *Br* Mouse as a Model for Median Cleft Face

Insertional mutation analysis has done much to elucidate the actions of specific genes. However, mouse mutants that display craniofacial dysmorphologies in a heritable fashion provide useful insight into deficient developmental pathways. The *Br* mouse is a radiation-induced mutant showing inherited median midfacial clefting (Singh et al., 1998). The *Br* gene appears

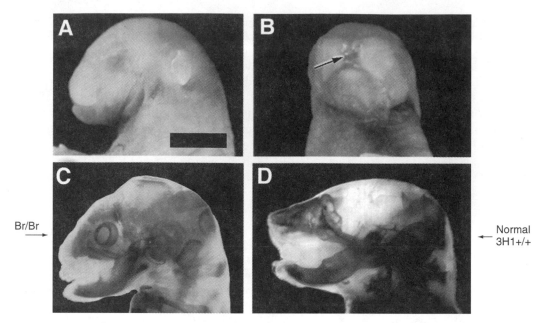

Br/Br →

← Normal
3H1+/+

Figure 10.1 Newborn 3H1 *Br/Br* mutants display midfacial retrognathia from the lateral perspective (A) and severe median facial clefting from the anterior view (B, arrow). Whole-mount staining shows that newborn 3H1 +/+ mice display a well-defined nasal capsule and midface (D), while the *Br/Br* mutants largely lack midfacial cartilages (C). Bar = 2.0 mm.

to be inherited as an autosomal-dominant mutation. The homozygote lethal condition shows severe median midfacial clefting (Lozanoff, 1993; Fig. 10.1A, B) and major reductions in the size of the midface (Fig. 10.1C, D), but animals do survive until birth. Animals show extensive midline clefting progressing posteriorly to the basisphenoid (Fig. 10.2). Although frontonasal and medial nasal prominences develop, they change little after 11 days of gestation (Fig. 10.3).

The *Br* heterozygote displays midfacial retrognathia, with growth deficiencies occurring in the midline chondrocranial derivatives, particularly the ethmoid, presphenoid, and basisphenoid (Lozanoff et al., 1994; Ma and Lozanoff, 1996). Although the heterozygotes survive, they display severe malocclusion (Lozanoff, 1993). In vitro studies suggest that the sphenoethmoidal region of the chondrocranium

shows deficient chondrocytic proliferation rather than more anterior regions encompassing the cartilaginous nasal septum. As a result, the posterior region of the anterior cranial base appears to be crucial for proper positioning of the midface relative to the mandible (Lozanoff, 1999).

Chromosomal localization of *Br* initially positioned the gene on murine chromosome 17 (Beechey et al., 1997). Subsequently, microsatellite mapping analysis placed *Br* between D17Mit122 and D17Mit190. Microsatellite mapping revealed complete penetrance of numerous microsatellite markers on distal chr17, suggesting that the mutation does not involve an extensive deletion. Distal chr17 possesses genes associated with defective craniofacial development. *Six2* and *Six3* are located at 45.5 cM and are associated with holoprosencephalic malformations in the mouse (Wallis and Muenke, 1999). The

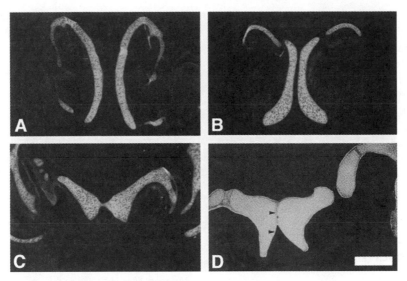

Figure 10.2 Type II collagen staining of the 3H1 *Br/Br* midface shows severe clefting of the cartilaginous anterior cranial base extending from the nasal septum anteriorly (A, B, and C) to the basisphenoid posteriorly (D). Bar = 200 µm.

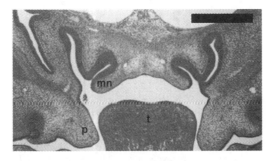

Figure 10.3 *Br/Br* mutants display medial nasal prominences with reduced embryonic growth that retain an embryonic form even in the newborn. Bar = 1.0 mm. Medial nasal prominence, mn; palatal process of the maxillary prominence, p; tongue, t.

Six1–6 genes comprise a subclass of the *Six/sine oculis* homeobox genes (Oliver et al., 1995). Interestingly, murine *Six3* maps to a homologous region of the human chromosome at 2p21, which contains a candidate gene for holoprosencephaly in the human (Wallis et al., 1999). Although heterozygous mutations in *Six3* are rare, deletions of 2p21 involving the deletion of one *Six3* allele and holoprosencephalic-associated translocation breakpoints from the 5′ end of the *Six3* coding sequence are compatible with haploinsufficiency of *Six3*-induced holoprosencephaly (Muenke and Beachy, 2000).

10.3 MIDFACIAL AND PALATAL DEVELOPMENT

Initiation of the nasal placodes and formation of the facial prominences around the stomodeum begins midfacial development. As the facial prominences enlarge around the nasal pits, they fuse and then merge to give rise to the primary palate, the tissue that forms the future upper lip and premaxillary region. Later in development, the secondary palatal shelves arise from the medial aspects of the maxillary prominences to gives rise to the hard and soft palate regions. The primary and secondary palatal regions are recognized as different embryologic entities (Trasler and Fraser, 1963; Fraser, 1970), and genetic and environmental factors that influence their closures are different. Abnormal development of the primary palate, leading to a cleft lip, may interfere secondarily with secondary palate closure to cause cleft palate. Thus, on both embryologic and genetic grounds, congenital cleft lip and cleft lip with cleft palate (CLP) appear to be etiologically related and are designated as CL(P). Isolated clefting of the secondary palate (CP) is an etiologically independent entity (Vergato et al., 1997).

10.4 PRIMARY PALATE DEVELOPMENT

Successful primary palate formation involves a sequence of local cellular events that are closely timed with spatial changes associated with craniofacial growth that must occur within a critical period in development. As facial prominences enlarge around the nasal pits to form the premaxillary region, growth of supporting craniofacial components change facial morphology and can affect the timing, location, and extent of contact between the facial prominences. Our studies of human embryos in the Carnegie and Kyoto Embryology Collections, and mouse embryos of cleft lip and noncleft strains, show that human and mouse embryos have similar phases of primary palate development and similar growth movements in the craniofacial complex. The forebrain elevates as the cranial base angle decreases, the medial nasal region narrows, and the maxilla grows forward to meet the medial and lateral nasal prominences that relocate with growth of the forebrain (Fig. 10.4).

Morphometric studies of normal human embryos revealed complex three-dimensional growth changes that occur as the primary palate develops (Diewert and Wang, 1992; Diewert and Lozanoff, 1993a, 1993b; Diewert et al., 1993c). In a study of cleft lip in human embryos in the Kyoto

collection, Diewert and Shiota (1990) showed deficient mesenchymal bridge growth and a visible deficiency of tissue in the cleft areas of embryos of stages 19 to 22, just after expected upper lip closure. When Kyoto embryos were compared to Carnegie Collection embryos, mesenchymal bridge was present at stage 17 in Carnegie embryos and at stage 18 in Kyoto embryos (Diewert, 1993). This later formation of a mesenchymal bridge indicates a potential increased liability to clefting malformation.

10.4.1 Primary Palate Development and Cleft Lip Malformation in the Mouse Model

Trasler (1968) showed that the formation of the normal lip was affected by embryonic face shape in mice. In embryos genetically predisposed to CL(P), the medial nasal prominences were more medially convergent than normal strain embryos, resulting in decreased contact with the lateral nasal prominences and a greater chance of failure of consolidation of tissues. The cleft lip strains A/J and A/WySn belong to the highly inbred A/–strain family, and the cleft lip strain CL/Fr has A/–strain ancestry (Bornstein et al., 1970). The A/ strains are derived from a cross between a Cold Spring Harbor albino and a Bagg albino more than 50 years ago (Bailey, 1978). Embryonic face shape has been shown to be a causal factor in genetic predisposition to cleft lip in mice (Juriloff and Trasler, 1976). Stock susceptible to spontaneous CL(P) had a significantly smaller distance between the nasal pits than the two stocks that were not susceptible to spontaneous CL(P). Internal developmental alterations associated with genetic liability to spontaneous CL(P) have also been reported for CL/Fr mice (Millicovsky et al., 1982) and include different orientation of medial nasal prominences similar to results reported by Trasler (1968), reduction (or

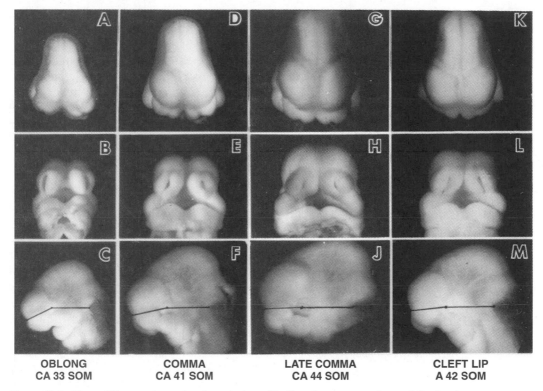

| OBLONG | COMMA | LATE COMMA | CLEFT LIP |
| CA 33 SOM | CA 41 SOM | CA 44 SOM | A 42 SOM |

Figure 10.4 Three different views of embryonic faces (A–J) at three stages of normal lip development (oblong, comma, late comma) in C57x A/WySn (CA) embryos showing narrowing of nasal pits, elevation of forebrain, and rapid enlargement of the maxillary prominences. An A/WySn cleft lip embryo showing small facial prominences (K), reduced fusion between the prominences (L), and a small maxilla (M) for a 42 somite (SOM) embryo.

absence) of epithelial activity throughout the developmental period of primary palate fusion, and hypoplasia of the lateral nasal prominences.

Although a major gene had been shown to be involved in the cause of nonsyndromic CL(P) in A/–strain mice (Juriloff, 1986; Biddle and Fraser, 1986), there was little information about the developmental effect of the gene. Recent studies of cleft lip genetics in mice (Juriloff, 1995; Juriloff and Mah, 1995; Juriloff et al., 2001) show that there is an epistatic interaction between two loci, clf1 and clf2, in the context of a genetic maternal effect. Human homologs of clf1 and clf2 are expected to be on 17q and 5q/9p (Juriloff et al., 2001).

In addition to nonsyndromic cleft lip of the A/ strain, there are other genetic anomalies that cause cleft lip in mice as a recessive trait. In Dancer mutation on chromosome 19, mouse embryos that have a liability to develop cleft lip have reduced lateral and medial nasal prominences compared to matched C57BL/J6 embryos (Trasler and Leong, 1982; Trasler and Ohannessian, 1983). In Twirler mutation on the proximal of chromosome 18, all homozygous embryos develop cleft lip (Lyon, 1958; Gong et al., 2000). The Twirler mutation on a C57BL/65 background is an interesting model, with both partial and complete cleft lip with cleft palates (Gong et al., 2000; Gong and Eulenberg, 2001).

10.4.2 Craniofacial Growth During Primary Palate Formation

The relationship of the relative position of the maxillary prominence to the cleft lip liability is still not well understood (Reed, 1933; Diewert and Shiota, 1990; Wang and Diewert, 1992; Diewert et al., 1993c). In our quantitative studies, the advancing tip of the maxillary prominence was highly correlated with enlargement of the mesenchymal bridge internally (Wang and Diewert, 1992; Diewert and Wang, 1992; Wang et al., 1995). We hypothesize that specific genes are needed to activate and regulate transcription factors, growth factors, or growth factor receptors, which are required for successful outgrowth and coalescence of prominences. The genetic defect leads to retarded forward growth of the maxillary prominence, less divergent medial nasal prominences, or deficient medial nasal prominences, and reduced areas of contact between facial prominences. Although the mechanism controlling the timing of opening of the primary choanae is unknown, opening of the oronasal membrane appears to be associated with separation of craniofacial structures as they grow (Diewert and Lozanoff, 1993b; Wang et al., 1995). The opening of the primitive choanae provides a threshold stage for definitive primary palate formation, or tearing occurs and cleft lip develops (Diewert and Wang, 1992) (Figs. 10.5 and 10.6). As the primary palate develops, the nasal pits become medially positioned, the nasal septum narrows and elongates, and the primary choanae open in both mouse and human embryos (Wang and Diewert, 1992; Diewert and Wang, 1992; Diewert and Lozanoff, 1993a, 1993b; Diewert et al., 1993a, 1993b).

10.4.3 Development of the Internal Components of the Primary Palate

Our studies of human embryos in the Carnegie and Kyoto Embryology Collec-tions and mouse embryos of cleft lip and noncleft strains showed that human and mouse embryos have similar phases of primary palate development and similar growth movements in the craniofacial complex. First, an epithelial seam forms and enlarges rapidly; then a mesenchymal bridge develops through the seam and enlarges rapidly. A robust mesenchymal bridge must form between the facial prominences before advancing midfacial growth tends to separate the facial components. The nasal pits narrow as the medial nasal region narrows and elongates vertically while the forebrain elevates and the head lifts from the thorax (Fig. 10.4). Posterior to the primary palate, the oronasal membranes rupture to form the primary choanae (posterior nares), passages between the nasal pits and the primordial oronasal cavity (Fig. 10.5). In mouse strains with cleft lip genes, the embryos have retarded maxillary growth, delayed seam formation, and delayed and deficient mesenchymal bridge formation compared with noncleft strains of mice. In the most severely affected embryos, the mesenchymal bridge fails to develop, or a small tissue bridge tears as craniofacial structures grow (Fig. 10.6).

Although different liabilities to CL(P) are known, there is only one quantitative study looking at embryogenesis of internal structures of normal noncleft and cleft-liable strains of mice (Wang et al., 1995). In this study, the sizes of the contact area, the epithelial seam, and the mesenchymal bridge were measured relative to the developmental stage marked by somite numbers. The number of somites caudal to the edge of the hind-limb bud, which lies opposite somites 23 to 28 (Butler and Juurlink, 1987), was recorded as the tail somite (TS) stage. The results show that mesenchymal bridge formation is delayed in cleft lip strains compared with normal noncleft strains (Wang et al., 1995).

Most of the embryos of noncleft lip

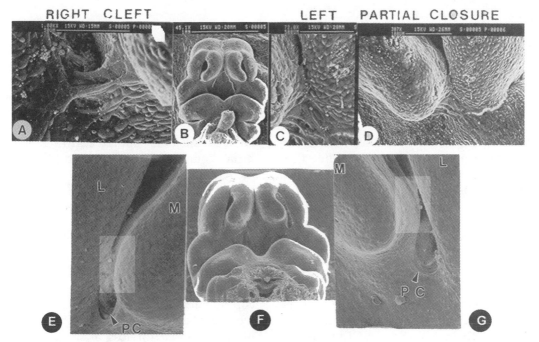

RIGHT CLEFT   **LEFT PARTIAL CLOSURE**

Figure 10.5 Scanning electron micrographs (SEMs) of a CL/Fr mouse embryo (B) showing developing cleft lip on the right (A) and partial fusion on the left (C and D). SEMs of a complete bilateral cleft lip embryo (F) showing sites of failed fusion (highlighted areas in E and F, the higher-power views) between the medial nasal (M) and lateral nasal (L) prominences immediately above the primary choanae (PC).

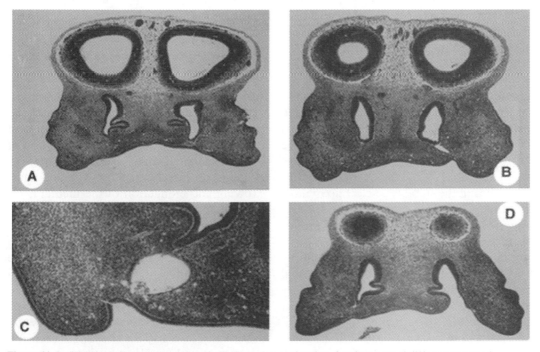

Figure 10.6 Histological sections of day 12 CL/Fr embryos showing development of different types of the primary palate defects. Partial fusion on left side that may tear (A and B). Small and abnormal mesenchymal bridge with a bleb (incomplete merging of mesenchyme) at the site of fusion (C). Complete cleft lip on the left with lateral displacement of the facial prominences (D).

strains start mesenchymal replacement at 13 TS. Cleft lip strains start later, with mesenchymal bridge formation in most A/J embryos at 14 TS, and in more than half of A/WySn and CL/Fr embryos at 16 TS. Thus, mesenchymal replacement is delayed in the cleft lip strain embryos that achieve successful primary palatal closure while their littermates have cleft lip. The replacement of mesenchyme of low (4%) clefting frequency (A/J) strain was less retarded than the high (25–30%) clefting frequency strains (A/WySn and CL/Fr) (Wang et al., 1995).

This study also provides quantitative analyses of mesenchymal–component growth from 13 to 17 TS in BALB/cByJ and C57BL/6J, from 14 to 17 TS in A/J, and from 15 to 19 TS in A/WySn and CL/Fr. Similar slopes of growth regression for the five strains suggest that growth rates of the mesenchymal components of noncleft and cleft lip strains are not different. However, significantly smaller mesenchymal area in the cleft lip strains compared to the noncleft lip strains at a given somite number indicates that mesenchymal growth in cleft lip strains is retarded relative to body development. There is a significantly smaller mesenchymal area in the high cleft lip frequency strains (CL/Fr) compared with the low cleft lip frequency strain (A/J) (Wang et al., 1995). This suggests that the formation and enlargement of the mesenchymal area is a specific indicator for determining the liability to cleft lip malformation in noncleft lip, low cleft lip frequency, and high cleft lip frequency strains of mice.

The molecular mechanisms accounting for differences in growth of the internal morphological components during primary palatogenesis in animals with and without genetic cleft lip liability remain to be clarified. The mechanisms controlling rapid expansion of the facial prominences with dramatic decreases in cell densities (Diewert, 1993; McGonnell et al., 1998),

and the mechanisms by which the mesenchyme replaces the epithelium seam and enlarges very rapidly to merge facial prominences, require more study. Early in development the facial prominences are in close proximity to each other, whereas later, as the brain and face grow rapidly, facial components move apart, but the medial nasal region narrows. The chick provides an extremely valuable model for developmental studies because it can be used for manipulations and unilateral perturbation experiments (Richman and Tickle, 1989, 1992; Richman and Delgado, 1995; McGonnell et al., 1998). In addition, there are cleft lip chick strains such as the cpp mutant (Yee and Abbott, 1978; Mac-Donald and Richman, 2001). The mechanisms proposed for loss of the epithelial seam include programed peridermal cell death and transformation of basal epithelial cells to mesenchymal cells as in the secondary palate in rodents (Fitchett and Hay, 1989) and also in the primary palate in chicks (Sun et al., 2000).

10.4.4 Primary Choanae Opening: The Critical Threshold Period

As the nasal region develops, communication between the external nares and the oronasal cavity is achieved by rupture of the oronasal membranes and formation of the primary choanae. During this period, changes in facial morphology include narrowing and elongation of the medial nasal region and forward growth of the maxillary prominences (Figs. 10.4, 10.5, and 10.6). The opening of the primary choanae at 18 to 20 TS indicates formation of a definite primary palate and established communication between the external nasal pits and oronasal cavity. Noncleft genotype embryos have a longer developmental interval (13 to 18 TS), allowing mesenchymal ingrowth and enlargement in the primary palate isthmus, compared with the 16 to 18 TS interval in CL/Fr embryos for mesenchymal bridge formation (Wang et al., 1995). These differences affect the proximity of different inbred strains to the developmental threshold and change the probability of failure leading to cleft lip formation.

In many recent studies in which time-mated inbred strains of mice are used, it is recognized that body or tail somite numbers provide more accurate comparable staging of embryos than chronological age. This is particularly important for development of structures such as the primary palate in mice with and without genetic cleft lip liability, because the failure may be a problem of temporal gene expression that leads to a growth delay, while the head continues to grow normally. The timing, location, and extent of contact between the facial prominences must all be favorable and coordinated with internal cellular and molecular changes for successful expansion, fusion, and tissue remodeling in the primary palate.

10.5 SECONDARY PALATE DEVELOPMENT

Formation of the secondary palate late in the embryonic period involves a sequence of developmental events in the palatal shelves and in the surrounding craniofacial complex. As information becomes available about normal and abnormal secondary palate formation in different mammalian embryos, critical events involved in the formation of the mammalian secondary palate are being identified. The objective of this section is to review major changes occurring in the craniofacial complex during secondary palate formation and to identify similarities and differences in developmental events in humans and in experimental animal models. By identifying specific alterations associated with palatal clefting in experimental animal models, factors that might

have relevance in human palatal clefting can be identified.

Palatal shelves of rodent embryos have the capacity to reorient to the horizontal position prior to the time they normally do in vivo when the obstructing influence of the tongue is removed (Brinkley et al., 1975, 1978; Walker and Quarles, 1976). When they were prematurely elevated in the rat fetus, they did not fuse prematurely due to tongue obstruction (Diewert, 1979a). Obliteration of cranial base integrity or segmentation of the shelves does not eliminate this capacity for horizontal positioning in vitro (Brinkley and Vickerman, 1978, 1979).

10.5.1 Craniofacial Growth During Secondary Palate Development

Morphometric studies of the craniofacial complex during progressive stages of normal secondary palate formation in mice (Diewert, 1982) and in humans (Diewert, 1983, 1986) have shown similar basic patterns of development. At stages before shelf elevation, the tongue-mandibular complex is small relative to the nasomaxillary complex, the tongue is positioned immediately ventral to the cranial base, and the head posture is flexed against the thorax (Fig. 10.7). At the time of palatal shelf elevation, the tongue and mandible extend beneath the caudal portion of the primary palate, the nasomaxillary complex lifts up and back relative to the body, and the palatal shelves elevate above the tongue to occupy the oronasal cavity space (Figs. 10.7 and 10.8). As closure of the secondary palate progresses, the prominence of mandible increases, and the tongue, attached to the anterior region of Meckel's cartilage via the genioglossus and geniohyoid muscles, also becomes positioned forward in the oral cavity (Fig. 10.7) (Diewert, 1983, 1985). During the late embryonic period, tongue contractions and swallowing-type reflex movements are

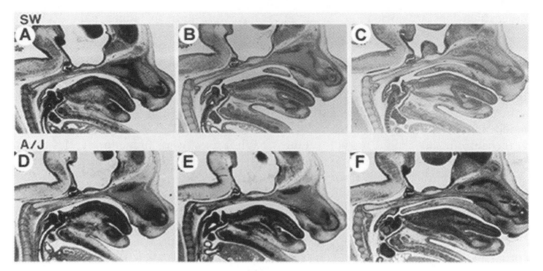

Figure 10.7 Medial sections of SWV (A, B, and C) and A/J (D, E, and F) embryonic mouse heads before (A, D), during (E, F) and after (C, F) secondary palatal shelf elevation showing extension of the tongue and tip of the mandible beneath the primary palate, flattening of the tongue, increases in oronasal vertical dimension, and elevating head positions as closure progresses.

present (Humphrey, 1971). Although embryonic facial growth in nonhuman primates has not been morphometrically studied, published photographs reveal similar patterns of increased lower facial prominence in the late embryonic and early fetal periods (Heuser and Streeter, 1941; Gasser et al., 1971; Hendrickx and Sawyer, 1975).

10.5.2 Cleft Palate Observations

During the past 50 years, experimental studies of induced and spontaneous cleft palate have enhanced our understanding of the mechanisms of clefting in experimental models (Harris, 1967; Fraser, 1968, 1976, 1980a, 1980b; Burdi et al., 1972, Diewert, 1980, 1986; Vergato et al., 1997). In most experimental animal studies, cleft palate results from failure of the palatal shelves to make contact at the time that fusion can occur, but different reasons for failure of contact have been observed. In humans and in animal models for cleft palate, wide clefts usually result when shelves remain in

the vertical position, whereas narrow clefts usually indicate elevated shelves that failed to contact and fuse, or failed to fuse even if contact was made (Diewert, 1979b, 1981; Kaartinen et al., 1995, 1997).

Major factors that have been shown to limit shelf contact are delayed shelf movement to the horizontal position and reduction in palatal shelf size (Diewert and Pratt, 1981). In different inbred strains of mice having different times of palatal closure, delayed palatal shelf movement is recognized as a factor in increased susceptibility to induced and spontaneous cleft palate (Walker and Fraser, 1956; Vekemans and Fraser, 1979). A reduced inherent capacity for shelf movement related to deficient extracellular matrix (ECM) accumulation in the shelves has been proposed (Walker and Fraser, 1956, 1957). The impact of hydration of ECM can be seen in frozen cryostat-cut sections that show the tongue being squeezed as the shelves enlarge during secondary palate closure (Diewert and Tait, 1979; Diewert and Pratt, 1981).

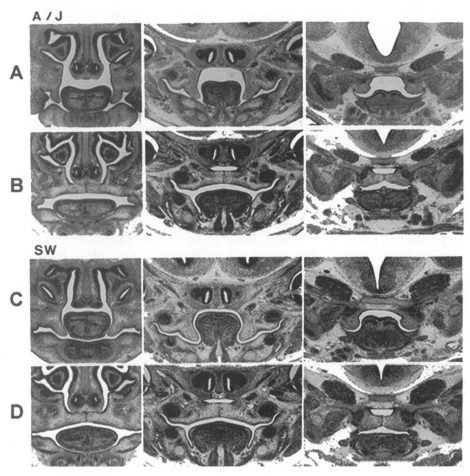

Figure 10.8 Frontal sections of A/J (A and B) and SWV (C and D) mouse heads before (A and C) and after shelf elevation (B and D). Sections through the anterior nasal, midpalatal, and soft palate regions showing contact and epithelial seam formation along the full length of the palate of mice. In A-strain embryos, shelf elevation occurs later (chronologically and developmentally) than in other strains.

A comparative study of facial growth in different inbred strains of mice showed that delayed shelf elevation was present in strains with delayed achievement of mandibular prominence, head extension, and increase in facial vertical dimension (Fig. 10.7) (Diewert, 1982). Spontaneous cleft palate is also present in a number of different mouse mutants with abnormal craniofacial morphology. Cleft palate occurs in mutants with abnormal head shape and retrognathia (Fitch, 1957, 1961), abnormal mandibular growth and chon-drodysplasia (Seegmiller and Fraser, 1977; Johnston and Nash, 1982), and abnormal first arch development (Juriloff and Harris, 1983). When cleft lip and cleft palate are present together, failure of palatal closure occurs secondarily because of altered tongue and palatomaxillary relations (Trasler and Fraser, 1963; Diewert, 1982).

In assessing the role of mandibular growth in normal palatal closure and in cleft palate, it is evident that not all teratogens cause cleft palate by interfering with mandibular growth. In addition, the absolute length of the mandible is not as important as its length compared to the oronasal cavity length to the primary palate. In a number of studies of induced cleft palate in the rat and mouse, delayed shelf movement was found to be associated with increased tongue obstruction of shelf movement secondarily because of mandibular retrognathia (Shih et al., 1974; Diewert, 1979b, 1980, 1981; Diewert and Pratt, 1979) or abnormally–flexed head posture (Diewert and Juriloff, 1983). Administration of teratogens such as the niacin antagonist 6-aminonicotinamide and the glutamine analog diazo-oxo-norleucine (DON) produced selective growth retardation in Meckel's cartilage and increased tongue obstruction to shelf movement and palatal closure (Diewert, 1979b; Diewert and Pratt, 1979). A similar pattern of increased tongue obstruction to shelf movement and contact was observed after treatment with the lathrogen, β-aminoproprionitrile, which caused retrognathia by producing deformation rather than growth retardation of Meckel's cartilage (Diewert, 1981). Abnormally flexed head posture producing a relative retrognathia and smaller vertical dimension in the oronasal cavity was observed after induction of cleft palate with methyl mercury (Diewert and Juriloff, 1983).

In the human, oligohydramnios, a deficiency in amniotic fluid volume, is associated with anomalies of the face, limbs, lungs, and skin (Gorlin et al., 1976). In experimental animals, puncture of the amniotic sac late in the embryonic period causes embryonic death and congenital anomalies such as cleft palate, clubfoot, microstomia and adactyly (Trasler et al., 1956; DeMeyer and Baird, 1969). After amniotic sac puncture, fetal head and body postures are severely constricted (Kennedy and Persaud, 1977), and a variety of types of partial clefts of the palate result (Schuepbach and Schroeder, 1984). In rodents, administration of glucocorticoids

reduces the volume of amniotic fluid and restricts normal changes in head posture (Harris, 1964; Fraser et al., 1967; Diewert and Wong, 1985). Glucocorticoids also reduce palatal shelf growth and limit the extent of shelf contact achieved (Diewert and Pratt, 1981; Pratt et al., 1980). Therefore, a combination of the effects of abnormally flexed head posture due to reduced amniotic fluid volume and reduced palatal shelf size appears to be critical in limiting capacity for shelf contact. Failure of medial epithelial breakdown has been proposed as a potential cause of cleft palate, but evidence for this mechanism in experimental models has been limited. In experimental induction of cleft palate in mice after exposure to the dioxin 2,3,7,8-tetrachlorodibenzo-p-dioxin (TCDD), palatal shelves make extensive contact, but the medial edge epithelium remains intact (Pratt et al., 1984a, 1984b).

10.6 TRANSGENIC ANIMAL MODELS FOR FACIAL CLEFTING

Development of new molecular methods to alter function of genes for transcription factors, growth factors, and receptors has stimulated new approaches to studying the mechanisms involved in normal and abnormal craniofacial development (Wilson et al., 1993; Ignelzi et al., 1995). Targeted modifications and inactivations of specific developmental genes have produced varying severities of defects in mouse embryos, ranging from early lethal effects to extremely minor developmental defects. Targeted inactivation of the transcription factor AP_2 gene causes severe cranioabdominoschesis, midline facial clefting, a contorted small axial skeleton, and severe craniofacial dismorphogensis (Schorle et al., 1996). When embryos survive to the late embryonic stages, cleft palate is the more common craniofacial defect produced, probably because successful closure has many primary and secondary contributing factors.

Misexpression of Hox genes such as Hox-2.3 and Hox-1.1 in mouse models cause craniofacial defects and clefts of the secondary palate (McLain et al., 1992; Balling et al., 1989). Other examples of targeted gene disruptions that result in palatal clefts include the gene encoding a forkhead domain-containing transcription factor, TTF-2, creating a model for thyroid dysgenesis (DeFelice et al., 1998); genes for neurotransmitter receptors such as β-3 knockout mice (GABA$_A$ β-3 subunit gene inactivated) (Condie et al., 1997; Krasowski et al., 1998); and the targeted mutation of the LIM homeobox gene Lhx8 (Zhao et al., 1999). In the transgenic mouse with col2a1 mutation, chondrodysplasia and a shortened mandible are present with cleft palate (Maddox et al., 1998). These disruptions may have primary effects on the palatal shelves, or they may affect the palate secondarily by affecting the maxilla, tongue, mandible, or head position to affect contact, fusion, seam removal, and mesenchymal expansion.

Targeted disruption of the gene for the Pax9 transcription factor affects tooth development and palatal closure (Peters et al., 1998). Pax9 disruption affects expression of BMP4 and Msx1 in the facial mesenchyme. Inactivation of MSx1 homeobox gene also affects tooth development and causes cleft palate (Satokata and Maas, 1994). Using Msx1-deficient mice as a model that has cleft palate, Chen and colleagues showed that Msx1 is required for Bmp4 and Bmp2 expression in palatal mesenchyme and Shh in the medial edge epithelium (Chen et al., 2001). Transgenic expression of human BMP4 driven by the mouse Msx1 promotor to the mutant was then used to rescue the cleft palate (Chen et al., 2001), and restore Shh and Bmp2 expression and cell proliferation.

Although the Patch deletion is quite large (Stephenson et al., 1991), the cause

of the connective tissue defects in Ph/Ph embryos (absence of PDGFa) has been confirmed in platelet-derived growth factor receptor (PDGFR)a–null mice (Soriano, 1997). The cleft face and small mandible resulting from faulty neural crest tissues were recently shown to have diminished matrix metalloproteinase–2 (MMP-2), which is regulated by PDGFA (Robbins et al., 1999). One transgenic model in which the primary defect is in the palatal shelves, is the TGFβ-3 knockout mouse model (Kaartinen et al., 1995; Proetzel et al., 1995). TGFβ-3 knockout has been shown to cause clefting by altering the ability of medial edge epithelium to fuse and be replaced by a mesenchymal bridge. Fusion of the palatal shelves was restored by adding TGFβ-3 in organ culture (Kaartinen et al., 1997; Taya et al., 1999). Using the TGFβ-3 null mutant, Blavier and colleagues (2001) showed that several matrix metalloproteinases (MMPs) were required for proteolytic degradation of the extracellular matrix as a necessary step for palatal fusion. Bulging of the medial edge epithelial cells just prior to fusion was observed using an environmental scanning electron microscope, which allows visualization of biological samples in their hydrated state (Martinez-Alvarez et al., 2000).

In our laboratory, MMP-2 was elevated in the fusion region of primary palate primordia in the mouse (Iamaroon et al., 1996a). Antibodies to growth factors TGFα and EGF, and EGF receptor were also localized in the area of epithelial seam disruption and the merging of mesenchyme (Iamaroon and Diewert, 1996; Iamaroon et al., 1996b). After numerous studies of the fate of medial edge epithelium, evidence is increasing for transformation from the epithelial- to mesenchymal-type cell (Fitchett and Hay, 1989; Shuler et al., 1992). TGFβ-3 affects the transformation of medial edge epithelial cells to transform to a mesenchymal phenotype, even in chicken palatal epithelium, which normally does not fuse (Sun et al., 1998).

The transgenic mouse models can be expected to continue to provide new methods to determine the primary and secondary causes of palatal clefting and the mechanisms for activation and regulation of essential developmental processes.

ACKNOWLEDGMENTS

The original research was supported by Medical Research Council of Canada Operating Grants to both authors. We would like to thank Dr. Ronan O'Rahilly, former director of the Carnegie Laboratories of Embryology; Dr. Kohei Shiota, director of the Kyoto Collection; and Dr. Andrew Hendrickx, director of the California Primate Research Center, for availability of the embryonic collections. We are grateful to Mrs. Ingrid Ellis for her editorial assistance in the final preparation of the manuscript.

REFERENCES

Bachiller, D., Klingensmith, J., Kemp, C., Belo, J. A., Anderson, R. M., May, S. R., McMahon, J. A., McMahon, A. P., Harland, R. M., Rossant, J., and DeRobertis, E. M. (2000). The organizer factors Chordin and Noggin are required for mouse forebrain development. *Nature 403*, 658–661.

Bailey, D. W. (1978). Sources of sub line divergence and their relative importance for sub line of six major inbred strains of mice. In H. C. Morse III (ed.), *Origins of Inbred Mice* (New York: Academic Press), pp. 197–215.

Balling, R., Mutter, G., Gruss, P., and Kessel, M. (1989). Craniofacial abnormalities induced by ectopic expression of homeobox gene Hox-1.1 in transgenic mice. *Cell 58*, 337–347.

Beechey, C., Boyd, Y., and Searle, A. G. (1997). Brahyrrhine, Br, a mouse craniofacial mutant maps to distal mouse chromosome 17 and is a candidate model for midline cleft syndrome. *Mouse Genome 95*, 692–694.

Belo, J. A. (1997). Cereberus-like is a secreted factor with neuralizing activity expressed in the anterior primitive endoderm of the mouse gatrula. *Mech. Devel. 68*, 45–57.

Belo, J. A., Leyns, L., Yamada, G., and DeRobertis, E. M. (1998). The prechordal midline of the chondrocranium is defective in Goosecoid-1 mouse mutants. *Mech. Devel. 72*, 15–25.

Biddle, F. G., and Fraser, F. C. (1986). Major gene determination of liability to spontaneous cleft lip in the mouse. *J. Craniofac. Genet. Dev. Biol. 2* (suppl.), 67–88.

Blavier, L., Lazaryev, A., Groffen, J., Heisterkamp, N., DeClerck, Y. A., and Kaartinen, V. (2001). TGF-β3-induced palatogenesis requires matrix metalloproteinases. *Molec. Biol. Cell 12*, 1457–1466.

Bornstein, S., Trasler, D. G., and Fraser, F. C. (1970). Effect of the uterine environment on the frequency of spontaneous cleft lip in CL/Fr mice. *Teratology 3*, 295–298.

Brinkley, L., and Vickerman, M. M. (1978). The mechanical role of the cranial base in palatal shelf movement: an experimental reexamination. *J. Embryol. Exp. Morphol. 48*, 93–100.

Brinkley, L. L, and Vickerman, M. M. (1979). Elevation of lesioned palatal shelves in vitro. *J. Embryol. Exp. Morphol. 54*, 229–240.

Brinkley, L., Basehoard, G., Branch, A., and Avery, J. (1975). New in vitro system for studying secondary palatal development. *J. Embryol. Exp. Morphol. 34*, 485–495.

Brinkley, L., Basehoard, G., and Avery, J. (1978). Effects of craniofacial structure on mouse palatal closure in vitro. *J. Dent. Res. 57*, 402–422.

Burdi, A., Feingold, M., Larsson, K. S, Leck, I., Zimmerman, E. F., and Fraser, F. C. (1972). Etiology and pathogenesis of congenital cleft lip and cleft palate, an NIDR state of the art report. *Teratology 6* (3), 255–270.

Butler, H., and Juurlink, B. H. J. (1987). *An Atlas for Staging Mammalian and Chick Embryos* (Boca Raton, FL: CRC Press, Inc.).

Chen, Y. P., Zhang, Z., Song, Y., Zhao, S., and Zhang, X. (2001). Transgenic Bmp 4 expression rescues cleft palate in Msx1-deficient mice. *J. Dent. Res. 80*, 230, AADR program, Abstract 1555.

Chiang, C., Litingtung, Y., Lee, E., Young, K. E., Corden, J. L., Westphal, H., and Beachy, P. A.

(1996). Cyclopia and defective axial patterning in mice lacking sonic hedgehog gene function. *Nature 383*, 407–413.

Condie, B. G., Bain, G., Gottlieb, D. I., and Capecchi, M. R. (1997). Cleft palate in mice with a targeted mutation in gamma-aminobutyric acid–producing enzyme glutamic acid decarboxylase67. *Proc. Natl. Acad. Sci. USA 94*, 11451–11455.

Dattani, M. T., Martinez-Barbera, J. P., Thomas, P. Q., Brickman, J. M., Gupta, R., Martensson, I. L., Toresson, H., Fox, M., Wales, J. K., Hindmarsh, P. C., Krauss, S., Beddington, R. S., and Robinson, I. C. (1998). Mutations in the homeobox gene HESX1/Hesx1 associated with septo-optic dysplasia in human and mouse. *Nat. Genet. 19*, 125–133.

DeFelice, M., Ovitt, C., Biffali, E., Rodriguez-Mallon, A., Arra, C., Anastassiadis, K., Macchia, P. E., Mattei, M. G., Mariano, A., Scholer, H., Macchia, V., and Di Lauro, R. (1998). A mouse model for hereditary thyroid dysgenesis and cleft palate. *Nat. Genet. 19*, 395 398.

DeMyer, W., and Baird, I. (1969). Mortality and skeletal malformations from amniocentesis and oligohydramnios in rats: cleft palate, clubfoot, microstomia, and adactyly. *Teratology 2*, 33–38.

Diewert, V. M. (1979a). Experimental induction of premature movement of rat palatal shelves in vivo. *J. Anat. 129*, 597–601.

Diewert, V. M. (1979b). Correlation between mandibular retrognathia and induction of cleft palate with 6-aminonicotinamide in the rat. *Teratology 19*, 213–228.

Diewert, V. M. (1980). The role of craniofacial growth in palatal shelf elevation. In R. M. Pratt and R. L. Christiansen (eds.), *Current Research Trends in Prenatal Craniofacial Development* (New York: Elsevier North Holland), pp. 165–186.

Diewert, V. M. (1981). Correction between alterations in Meckel's cartilage and induction of cleft palate with beta-aminoproprionitrile in the rat. *Teratology 25*, 43–52.

Diewert, V. M. (1982). A comparative study of craniofacial growth during secondary palate development in four strains of mice. *J. Craniofac. Genet. Devel. Biol. 2*, 247–263.

Diewert, V. M. (1983). A morphometric analysis of craniofacial growth showing changes in spatial relations during secondary palatal development in human embryos and fetuses. *Am. J. Anat. 167*, 495–522.

Diewert, V. M. (1985). Development of human craniofacial morphology during the late embryonic and early fetal periods. *Am. J. Orthodont. 88*, 64–76.

Diewert, V. M. (1986). Craniofacial growth during human secondary palate formation and potential relevance of experimental cleft palate observations. *J. Craniofac. Genet. Devel. Biol. 2* (suppl.), 267–276.

Diewert, V. M. (1993). Differences in primary palate formation in Carnegie and Kyoto human embryos. *Anat. Rec. 1* (suppl.), 47.

Diewert, V. M., and Juriloff, D. M. (1983). Abnormal head posture associated with induction of cleft palate with methyl mercury in C57 mice. *Teratology 28*, 437–447.

Diewert, V. M., and Lozanoff, S. (1993a). A morphometric analysis of human embryonic craniofacial growth in the medial plane during primary palate formation. *J. Craniofac. Genet. Devel. Biol. 13*, 147–161.

Diewert, V. M., and Lozanoff, S. (1993b). Growth and morphogenesis of the human embryonic midface during primary palate formation analyzed in frontal sections. *J. Craniofac. Genet. Devel. Biol. 13*, 162–183.

Diewert, V. M., and Pratt, R. M. (1979). Selective inhibition of mandibular growth and induction of cleft palate by diazo-oxo-norleucine (DON) in the rat. *Teratology 20*, 37–52.

Diewert, V. M., and Pratt, R. M. (1981). Cortisone-induced cleft palate in A/J mice: failure of palatal shelf contact. *Teratology 24*, 149–162.

Diewert, V. M., and Shiota, K. (1990). Morphological observations in normal primary palate and cleft lip embryos in the Kyoto collection. *Teratology 41*, 663–667.

Diewert, V. M., and Tait, B. (1979). Palatal process movement is demonstrated in frozen sections. *J. Anat. 128*, 609–618.

Diewert, V. M, and Wang, K.-Y. (1992). Recent advances in primary palate and midface morphogenesis research. *Crit. Rev. Oral. Biol. Med. 4*, 111–130.

Diewert, V. M., and Wong, J. (1985). Amniotic fluid reduction in glucocorticoid-treated mice. *J. Dent. Res. 64*, 271.

Diewert, V. M., Wang, K.-Y., and Tait, B. (1993a). A morphometric analysis of cell densities in facial prominences of the rhesus monkey embryo during primary palate formation. *J. Craniofac. Genet. Devel. Biol. 13*, 236–249.

Diewert, V. M., Wang, K.-Y., and Tait, B. (1993b). A new threshold model for cleftlip in mice. *Ann. New York Acad. Sci. 678*, 341–343.

Diewert, V. M., Lozanoff, S., and Choy, V. (1993c). Computer reconstructions of human embryonic craniofacial morphology showing changes in relations between the face and brain during primary palate formation. *J. Craniofac. Genet. Devel. Biol. 13*, 193–201.

Filosa, S., Rivera-Perez, J. A., Gomes, A. P., Gansmuller, A., Saskai, H., Berhringer, R. R., and Ang, S.-L. (1997). Goosecoid and HNT-3b genetically interact to regulate neural tube patterning during mouse embryogenesis. *Development 124*, 2843–2854.

Fitch, N. (1957). An embryological analysis of two mutants in the house mouse both producing cleft palate. *J. Exp. Zool. 136*, 329–361.

Fitch, N. (1961). Development of cleft palate in mice homozygous for the short head mutation. *J. Morphol. 109*, 151–157.

Fitchett, J. E., and Hay, E. D. (1989). Medial edge epithelium transforms to mesenchyme after embryonic palatal shelves fuse. *Devel. Biol. 131*, 455–474.

Fraser, F. C. (1968). Workshop on embryology of cleft lip and cleft palate. *Teratology 1*, 353–358.

Fraser, F. C. (1970). The genetics of cleft lip and cleft palate. *Am. J. Hum. Genet. 22*, 336–352.

Fraser, F. C. (1976). The multifactorial threshold concept—uses and misuses. *Teratology 14*, 267–280.

Fraser, F. C. (1980a). Animal models for craniofacial disorders. In M. Melnick, D. Bixler, and E. D. Shields (eds.), *Etiology of Cleft Lip and Cleft Palate* (New York: Alan R. Liss., Inc.), pp. 1–23.

Fraser, F. C. (1980b). The genetics of cleft lip and palate: yet another look. In R. M. Pratt and R. L. Christiansen (eds.), *Current Trends in Prenatal Craniofacial Development* (New York: North-Holland), pp. 357–366.

Fraser, F. C., Chew, D., and Verrusio, A. C. (1967). Oligohydramnios and cortisone-induced cleft palate in the mouse. *Nature 214*, 417–418.

Gasser, R. F, Hendrickx, A. G., and Bollert, J. A. (1971). Description of stages XIX, XX, XI, XXII, and XXIII. In A. G. Hendrickx (ed.), *Embryology of the Baboon* (Chicago: University of Chicago Press), pp. 127–152.

Gong, S.-G., and Eulenberg, R. (2001). Palatal development in Twirler mice. *Cleft Palate Craniofac. J. 38*, 622–628.

Gong, S.-G, White N. J., and Sakasegawa A. Y. (2000). The Twirler mouse, a model for the study of cleft lip and palate. *Arch. Oral Biol. 45*, 87–94.

Gorlin, P. J., Pindborg, J. J., and Cohen, M. M., Jr. (1976). *Syndromes of the Head and Neck*, 2nd ed. (New York: McGraw-Hill), pp. 613–617.

Hardcastle, Z., Mo, R., Hui, C.-C., and Sharpe, P. T. (1998). The Shh pathway in tooth development: defects in Gli2 and Gli3 mutants. *Development 125*, 2803–2811.

Harris, J. W. S. (1964). Oligohydramnios and cortisone-induced cleft palate. *Nature 203*, 533–534.

Harris, J. W. S. (1967). Experimental studies on closure and cleft formation in the secondary palate. *Sci. Basis Med. Annu. Rev.*, pp. 356–370.

Hendrickx, A. G., and Sawyer, R. H. (1975). Embryology of the rhesus monkey. In G. H. Bourne (ed.), *The Rhesus Monkey: Vol. II. Management, Reproduction and Pathology* (New York: Academic Press), pp. 141–169.

Heuser, C. H., and Streeter, G. L. (1941). Development of the macaque embryo. *Contrib. Embryol. Carnegie. Inst. 29*, 15–55.

Hu, D., and Helms, J. A. (1999). The role of Sonic hedgehog in normal and abnormal craniofacial development. *Development 126*, 4873–4884.

Humphrey, T. (1971). Development of oral and facial motor mechanisms in human fetuses and their relation to craniofacial growth. *J. Dent. Res. 50*, 1428–1441.

Iamaroon, A., and Diewert, V. M. (1996). Distribution of basement membrane components in the mouse primary palate. *J. Craniofac. Genet. Devel. Biol. 16*, 48–51.

Iamaroon, A., Overall, C. M., Wallon, U. M., and Diewert, V. M. (1996a). Expression of 72-kDa gelatinase (matrix metalloproteinase 2) in the developing craniofacial complex of the mouse. *Arch. Oral Biol. 41*, 1109–1119.

Iamaroon, A., Tait, B., and Diewert, V. M. (1996b). Cell proliferation and expression of EGF, TGFa and EGF receptor in developing mouse primary palate. *J. Dent. Res. 75*, 1534–1539.

Ignelzi, M. A., Liu, Y., Maxson, R. E., and Snead, M. L. (1995). Genetically engineered mice: tools to understand craniofacial development. *Crit. Rev. Oral Biol. Med. 6*, 181–201.

Johnston, L., and Nash, D. J. (1982). Sagittal growth trends in development of cleft palate in mice homozygous for the "paddle" gene. *J. Craniofac. Genet. Devel. Biol. 2*, 265–276.

Juriloff, D. M. (1986). Major genes that causes cleft lip in mice: progress in the construction of a congenic strain and in linkage mapping. *J. Craniofac. Genet. Devel. Biol. 2* (suppl.), 55–66.

Juriloff, D. M. (1995). Genetic analysis of the construction of the EDGY congenic strain indicates that nonsydromic CL(P) in the mouse is caused by two loci with epistatic interaction. *J. Craniofac. Genet. Devel. Biol. 15*, 1–12.

Juriloff, D. M., and Harris, M. J. (1983). Abnormal facial development in the mouse mutant first arch. *J. Craniofac. Genet. Devel. Biol. 3*, 317–337.

Juriloff, D. M., and Mah, D. G. (1995). The major locus for multi factorial nonsyndromic cleft lip maps to chromosome 11. *Mammalian Genome 6*, 63–69.

Juriloff, D. M., and Trasler, D. G. (1976). Test of the hypothesis that embryonic face shape is a causal factor in genetic predisposition to cleft lip in mice. *Teratology 14*, 35–42.

Juriloff, D. M., Harris, M. J., and Brown, C. J. (2001). Unraveling the complex genetics of cleft lip in the mouse model. *Mammalian Genome 12*, 426–435.

Kaartinen, V., Volchen, J. W., Shuler, C., Warburton, D., Bu, D., Heisterkamp, N., and Groffen, J. (1995). Abnormal lung develop-ment and cleft palate in mice lacking TGF-β3 indicates defects of epithelial-mesenchymal interaction. *Nat. Genet. 11*, 415–421.

Kaartinen, V., Cui, X. M., Heisterkamp, N., Groffen, J., and Shuler, C. F. (1997). Trans-forming growth factor-beta3 regulates trans-differentiation of medial edge epithelium during palatal fusion and associated degra-dation of the basement membrane. *Devel. Dyn. 209*, 225–260.

Kaufman, M. H. (1992). *The Atlas of Mouse Development* (New York: Academic Press).

Kennedy, L. A., and Persaud, T. V. N. (1977). Pathogenesis of developmental defects induced in the rat by amniotic sac puncture. *Acta Anat. 97*, 23–25.

Krasowski, M. D., Rick, C. E., Harrison, N. L., Firestone, L. L., and Homanics, G. E. (1998). A deficient of functional GABA(A) recep-tors in β3 subunit knockout mice. *Neurosci. Lett. 24*, 81–84.

Laverty, H. G., and Wilson, J. B. (1998). Murine CASK is disrupted in a sex-linked cleft palate mouse mutant. *Genomics 53* (1), 29–41.

Liu, P., Wakamiya, M., Shea, M. J., Alabrecth, U., Behringer, R. R., and Bradley, A. (1999). Requirement for Wnt3 in vertebrate axis formation. *Nat. Genet. 22*, 361–365.

Lozanoff, S. (1993). Midfacial retrusion in adult brachyrrhine mice. *Acta Anat. 147*, 125–132.

Lozanoff, S. (1999). Sphenoethmoidal growth, malgrowth and midfacial profile. In M. A. J. Chaplain, G. D. Singh, and J. C. McLachlan (eds.), *On Growth and Form: Spatiotemporal Pattern Formation in Biology* (New York: Wiley), pp. 357–372.

Lozanoff, S., Jureczek, S., Feng, T., and Padwal, R. (1994). Anterior cranial base morphology in mice with midfacial retrusion. *Cleft Palate Craniofac. J. 31*, 417–428.

Lyon, M. F. (1958). Twirler: a mutant affecting the inner ear of the house mouse. *J. Embryol. Exp. Morphol. 6*, 105–116.

Ma, W., and Lozanoff, S. (1996). Morphological deficiency in the prenatal anterior cranial base of midfacially retrognathic mice. *J. Anat. 188*, 547–555.

MacDonald, M. E., and Richman, J. M. (2001). Effect of the cleft primary palate mutation on chick craniofacial development. American Society for Developmental Biology Annual Conference, Seattle, WA, July.

Maddox, B. C., Garofalo, S., Horton, W. A., Richardson, M. D., and Trune, D. R. (1998). Craniofacial and otic capsule abnormalities in a transgenic mouse strain with a Col2a1 mutation. *J. Craniofac. Genet. Devel. Biol. 18*, 195–201.

Mansouri, A., Stoykova, A., Torves, M., and Gruss, P. (1996). Dygenesis of cephalic neural crest derivatives in Pax 7-1-mutant mice. *Development 122*, 831–838.

Martinez-Alvarez, C., Bonelli, R., Tudela, C., Gato, A., Mena, J., O'Kane, S., and Ferguson, M. W. (2000). Bulging medial edge epithelial cells and palatal fusion. *Int. J. Devel. Biol. 44*, 331–335.

McGonnell, I. M., Clarke, J. D. W., and Tickle, C. (1998). Fate map of the developing chick face: analysis of expansion of facial primordia and establishment of the primary palate. *Devel. Dyn. 212*, 102–118.

McLain, K., Schreiner, C., Yager, K. L., Stock, J. L., and Potter, S. S. (1992). Ectopic expression of Hox-2.3 induces craniofacial and skeletal malformations in transgenic mice. *Mech. Devel. 39*, 3–16.

Millicovsky, G., Ambrose, L. J. H., and Johnston, M. C. (1982). Developmental alterations associated with spontaneous cleft lip and palate in CL/FR mice. *Am. J. Anat. 164*, 29–44.

Muenke, M., and Beachy, P. A. (2000). Genetics of ventral forebrain development and holoprosencephaly. *Curr. Opin. Genet. Devel. 10*, 262–269.

Oliver, G., Mailhos, A., Wehr, R., Copeland, N. G., Jenkins, N. A., and Gruss, P. (1995). Six3, a murine homologue of the sine oculis gene, demarcates the most anterior border of the developing neural plate and is expressed during eye development. *Development 212*, 4045–4055.

Peters, H., Neubuser, A., Kratochwil, K., and Balling, R. (1998). Pax9-deficient mice lack pharyngeal pouch derivatives and teeth and exhibit craniofacial and limb abnormalities. *Genes Devel. 12*, 2735–2747.

Piccolo, S., Aguis, E., Leyns, L., Bhattacharyya, S., Grunz, H., Bouwmeester, T., and Robertis, E. M. (1999). The head inducer Cerberus is a multifunctional antagonist of Nodal, BMP and Wnt signals. *Nature 397*, 707–710.

Pratt, R. M., Salomon, D. S, Diewert, V. M., Erickson, R. P., Burns, R., and Brown, K. S. (1980). Cortisone-induced cleft palate in the brachymorphic mouse. *Teratogen. Carcinogen. Mutagen. 1*, 15–23.

Pratt, R. M., Dencker, L., and Diewert, V. M. (1984a). TCDD-induced cleft palate in the mouse: evidence for receptor mediated alterations in epithelial cell differentiation. *Teratogen. Carcinogen. Mutagen. 4*, 427–436.

Pratt, R. M., Grove, R. I., Kim, C. S., Dencker, L., and Diewert, V. M. (1984b). Mechanism of TCDD-induced cleft palate in the mouse. In *Biological Mechanisms of Dioxin Action (Banbury Report 18)* (Cold Spring Harbor, NY: Cold Spring Harbor Laboratory), pp. 61–71.

Proetzel, G., Pawlowski, S. A., Wiles, M. V., Yin, M., Boivin, G. P., Howles, P. N., Ding, J., Ferguson, M. W. J., and Doetschamn, T. (1995). Transforming growth factor-β3 is required for secondary palate fusion. *Nat. Genet. 11*, 409–414.

Reed, S. C. (1933). An embryological study of harelip in mice. *Anat. Rec. 56*, 101–110.

Rhinn, M., Dierich, A., Shawlot, W., Behringer, R. R., Le Meur, M., and Ang, S. L. (1998). Sequential roles for Otx2 in visceral endoderm and neuroectoderm for forebrain and midbrain inductions and specification. *Development 125*, 845–856.

Richman, J. M., and Delgado J. L. (1995). Locally released retinoic acid leads to facial clefts in the chick embryo but does not alter the expression of receptors for fibroblast growth factor. *J. Craniofac. Genet. Devel. Biol. 15*, 190–204.

Richman, J. M., and Tickle, C. (1989). Epithelia are interchangeable between facial primordia of chick embryos and morphogenesis is controlled by the mesenchyme. *Devel. Biol. 136*, 201–210.

Richman, J. M., and Tickle, C. (1992). Epithelial-mesenchymal interactions in the outgrowth of limb buds and facial primordia in chick embryos. *Devel. Biol.* 154, 299–308.

Rivera-Perez, J. A., Mallo, M., Gendron-Maguire, M., Gridley, T., and Behringer, R. R. (1995). Goosecoid is not an essential component of the mouse gastrula organizer but is required for craniofacial and rib development. *Development 121,* 3005–3012.

Robbins, J. R., McGuire, P. G., Wehrle-Haller, B., and Rogers, S. L. (1999). Diminished matrix metalloproteinase-2 (MM-2) in ectomesenchyme-derived tissue of Patch mutant mouse: regulation of MMP-2 by PDGF and effects on mesenchymal cell migration. *Devel. Biol. 212,* 255–263.

Sarnat, B. G. (2000). Molecular genetic classification of central nervous system malformations. *J. Child Neurol.* 15, 675–687.

Satokata, I., and Maas, R. (1994). Msx1-deficient mice exhibit cleft palate and abnormalities of craniofacial and tooth development. *Nat. Genet. 6,* 348–356.

Schorle, H., Meier, P., Buchert, M., Jaenisch, R., and Martin, P. J. (1996). Transcription factor AP-2 essential for cranial closure and craniofacial development. *Nature 381,* 235–238.

Schuepbach, P. M., and Schroeder, H. E. (1984). Prenatal repair of experimentally induced clefts in the secondary palate of the rat. *Teratology 30,* 131–142.

Seegmiller, R. E., and Fraser, F. C. (1977). Mandibular growth retardation as a cause of cleft palate in mice homozygous for the chondrodysplasia gene. *J. Embryol. Exp. Morphol. 38,* 227–238.

Shawlot, W., and Behringer, R. R. (1995). Requirement for Lim1 in head-organizer function. *Nature 30,* 425–430.

Shuler, C. F., Halpern, D. E., Guo, A. C., and Sank, A. C. (1992). Medial edge epithelium fate traced by cell lineage analysis during epithelial-mesenchymal transformation. *Devel. Biol. 154,* 318–330.

Shih, L., Trasler, D. G., and Fraser, F. C. (1974). Relation of mandible growth to palate closure in mice. *Teratology 9,* 191–202.

Singh, G. D., Johnston, J., Ma., J. W., and Lozanoff, S. (1998). Immunolocalization of laminin, fibronectin, type IV collagen and tenascin in the cleft secondary palate of embryonic *Br* mice. *Cleft Palate 35,* 219–241.

Soriano, P. (1997). The PDGFα receptor is required for neural crest development and for normal patterning of the somites. *Development 124,* 2691–2700.

Stephenson, D. A., Anderson, E., Wang, C., Mercola, M., Stiles, C., Bowen-Pope, D. F., and Chapman, V. M. (1991). The mouse mutation *Patch (Ph)* carries a deletion in the gene for platelet-derived growth factor alpha receptor. *Proc. Natl. Acad. Sci. USA 88,* 6–10.

Sun, D., Vanderburg, C. R., Odierna, G. S., and Hay, E. D. (1998). TGFβ3 promotes transformation of chicken palate medial edge epithelium to mesenchyme in vitro. *Development 125,* 95–105.

Sun, D., Baur, S., and Hay, E. D. (2000). Epithelial-mesenchymal transformation is the mechanism for fusion of the craniofacial primordial involved in morphogenesis of the chicken lip. *Devel. Biol. 228,* 337–349.

Tam, P. P. L., and Behringer, R. R. (1997). Mouse gastrulation: the formation of a mammalian body plan. *Mech. Devel. 68,* 3–25.

Taya, Y., O'Kane, S., and Ferguson, M. W. J. (1999). Pathogenesis of the cleft palate in TGF-β3 knockout mice. *Development 126,* 3869–3879.

Trasler, D. G. (1968). Pathogenesis of cleft lip and its relation to embryonic face shape in A/J and C57BL mice. *Teratology 1,* 33–50.

Trasler, D. G., and Fraser, F. C. (1963). Role of the tongue in producing cleft palate in mice with spontaneous cleft lip. *Devel. Biol. 6,* 45–60.

Trasler, D. G., and Leong, S. (1982). Mitotic index in mouse embryos with 6-aminonicotinamide-induced and inherited cleft lip. *Teratology 25,* 259–265.

Trasler, D. G., and Ohannessian, L. (1983). Ultrastructure of initial nasal process cell fusion in spontaneous and 6-aminonicotinamide-induced mouse embryo cleft lip. *Teratology 28,* 91–101.

Trasler, D. G., Walker, B. E., and Fraser, F. C. (1956). Congenital malformations produced by amniotic-sac puncture. *Science 124*, 439.

Veitch, E., Begbie, J., Schilling, T. F., Smith, M. M., and Graham, A. (1999). Pharyngeal arch patterning in the absence of neural crest. *Curr. Biol. 9*, 1481–1484.

Vekemans, M., and Fraser, F. C. (1979). Stage of palatal closure as one indication of "liability" to cleft palate. *Am. J. Med. Genet. 4*, 95–102.

Vergato, L. A., Doerfler, R. J., Mooney, M. P., and Siegel, M. I. (1997). Mouse palatal width growth rates as an "at risk" factor in the development of cleft palate induced by hypervitaminosis A. *J. Craniofac. Genet. Devel. Biol. 17*, 204–210.

Walker, B. E., and Fraser, F. C. (1956). Closure of the secondary palate in three strains of mice. *J. Embryol. Exp. Morphol. 4*, 176–189.

Walker, B. E., and Fraser, F. C. (1957). The embryology of cortisone-induced cleft palate. *J. Embryol. Exp. Morphol. 5*, 201–209.

Walker, B. E., and Quarles, J. (1976). Palate development of mouse fetuses after tongue removal. *Arch. Oral Biol. 21*, 405–412.

Wallis, D. E, and Muenke, M. (1999). Molecular mechanisms of holoprosencephaly. *Molec. Genet. Metab. 68*, 126–138.

Wallis, D. E., Roessler, E., Hehr, U., Nanni, L., Wiltshire, T., Richieri-Costa, A., Gillessen-Kaesbach, G., Zackai, E. H., Rommens, J., and Muenke, M. (1999). Mutations in the homeodomain of the human SIZ2 gene cause holoprosencephaly. *Nat. Genet. 22*, 196–198.

Wang, K.-Y., and Diewert, V. M. (1992). A morphometric analysis of craniofacial growth in cleft lip and noncleft mice. *J. Craniofac. Genet. Devel. Biol. 12* (3), 141–154.

Wang, K. Y., Juriloff, D. M., and Diewert, V. M. (1995). Deficient and delayed primary palatal fusion and mesenchymal bridge formation in cleft lip–liable strains of mice. *J. Craniofac. Genet. Devel. Biol. 15*, 99–116.

Wilson, J. B., Ferguson, M. W., Jenkins, N. A., Lock, L. F., Copeland, N. G., and Levine, A. J. (1993). Transgenic mouse model of X-linked cleft palate. *Cell Growth Differ. 4*, 67–76.

Yee, G. W., and Abbott, U. K. (1978). Facial development in normal and mutant chick embryos: scanning electron microscopy of primate palate formation. *J. Exp. Zool. 206*, 307–322.

Zhao, Y., Guo, Y., Tomac, A. C., Taylor, N. R., Grinberg, A., Lee, E. J., Huang, S., and Westphal, H. (1999). Isolated cleft palate in a targeted mutation of the LIM homeobox gene Lhx8. *Devel. Biol. 96*, 15002–15006.

PART IV

CRANIAL VAULT AND CRANIAL BASE DYSMORPHOLOGY AND GROWTH DISTURBANCES

CHAPTER 11

EVOLUTIONARY CHANGES IN THE CRANIAL VAULT AND BASE: ESTABLISHING THE PRIMATE FORM

MARK P. MOONEY, Ph.D. MICHAEL I. SIEGEL, Ph.D. TIMOTHY D. SMITH, Ph.D. and ANNIE M. BURROWS, Ph.D., D.P.T.

> Rely upon it as an indisputable fact, that nothing in the cranium exists unless designed to carry out some definite intention, or unless formed in reference to some distant and special object.
> —John Hilton, 1855

11.1 INTRODUCTION: EVOLUTIONARY CORRELATES OF HUMAN NEURO- AND CHONDROCRANIAL MORPHOLOGY

Among vertebrates, the human cranial vault and base have a unique morphology, characterized by brachycephalization (i.e., a short, wide cranial vault), deep, discernible fossae, distinctive growth patterns, and extreme flexion (kyphosis). To better understand the origins of these morphologies, this chapter reviews phylogenetic changes of the neuro- and chondrocranium in vertebrates and relates them to the phylogenetic changes in the brain and sensory capsules.

11.1.1 Bipedalism, Bracephalization, and Skull Base Kyphosis

The widely held opinion that the human neuro- and basicranium are unique among mammals is usually substantiated by claims of a relationship to upright posture (Cameron, 1925; Bolk, 1926; DuBrul, 1950; Dmoch, 1975, 1976). The argument continues that because humans walk upright their heads need to be balanced on a vertical vertebral column and thus the foramen magnum needs to be centrally located (Schultz, 1942; DuBrul, 1950; Adams and Moore, 1975). While it can be demonstrated that this correlation exists for humans, it is important to discern other correlates before concluding causality. An additional covariant of the neurobasicranial morphology has been recognized for over 50 years. Weidenreich (1941, 1945) described the role of the brain in the phylogenetic transformation of the human skull. He posited that as the frontal lobes became more superiorly directed, the temporal lobes became more everted, and the occipital lobes moved in an inferior and anterior

Understanding Craniofacial Anomalies: The Etiopathogenesis of Craniosynostoses and Facial Clefting, Edited by Mark P. Mooney and Michael I. Siegel, ISBN 0-471-38724-x Copyright © 2002 by Wiley-Liss, Inc.

direction, there was a corresponding vertical and lateral displacement of the calvaria, with occipital and frontal bossing resulting in brachycephalization and an increased kyphosis (flexion) of the cranial base. These changes are usually associated with an increase in brain size (Weidenreich, 1941, 1945) (Figs. 11.1 and 11.2). The described rolling-up or spatial packing process gains parsimony from the fact that changes toward a more spherical shape maximize volume with a minimal increase in surface area (Moss, 1958; Biegert, 1963; Enlow, 1968; Baer and Harris, 1969; Passingham, 1975; Gould, 1977; Stephan et al., 1981; Campbell, 1985; Dean, 1988; Ross and Ravosa, 1993; Spoor, 1997).

11.1.2 Neural Spatial Packing, Brachycephalization, and Skull Base Kyphosis

To see how this process relates to a unique human basicranial morphology, it is necessary to understand some of the phylogenetic changes in the vertebrate central nervous system, at least at a gross level. When examining the brain plan of fish and reptiles, there is an obvious horizontal arrangement from the olfactory lobes (and tracts) to the brain stem, which is reflected in the disposition of the anterior cranial fossa relative to the remainder of the neurocranium. While the brains of sauropsid reptiles did not fill the brain case, transitional forms (cynodonts) and early mammals had relatively larger brains, although still more reptilian than mammal-like (Ariens-Kappers et al., 1963; Hofer,

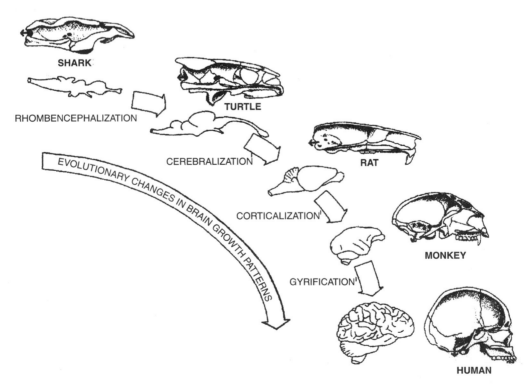

Figure 11.1 Evolutionary changes in brain growth patterns and resultant skull base morphologies from sharks through humans (small arrows denote orientation of spinal cord and the foramen magnum).

1969; Sarnat and Netsky, 1974; Kier, 1976).

Relatively larger brains present several problems for cranial reorganization. While the larger brains can simply fill larger neurocrania, there is a practical limit to unmodified size increase. Large-bodied reptiles show allometric changes in brain size compared to, say, the masticatory apparatus. With body size increases we see proportionate changes in jaws and teeth, accompanied by little change in brain size. Thus, the brain plays a smaller role in determining the internal architecture of the neurocranium. For nonhuman primates, the situation is quite different. Primates have relatively high brain-to-body size ratios, and the relatively larger brains can only be accommodated by neurocranial reorgani-

zation. In the cases of absolute large body size in primates, the masticatory apparatus shows a proportionate increase in size and is accommodated by epicranial structures (crests and ridges) rather than affecting the brain-case proper, or the cranial base. Apart from locomotor adaptation, the more flexed (kyphotic) cranial bases are associated with relative brain size increases in primates, and even the small-bodied vervets, night monkeys, and sykes monkeys have brain:body ratios (average 1:80) much higher than humans (1:47). Their foramen magnums are centrally located as well, and they are all quadrupedal (Lugoba and Wood, 1990) (Fig. 11.2).

11.1.3 Synchondrosal Fusion Timing, Brachycephalization, and Skull Base Kyphosis

How the kyphosis is established may well involve a synergistic interaction of brain expansion and growth activity at the cranial base synchondroses. The synchondroses may be viewed as growth centers (Enlow, 1990) that are similar in function to epiphyseal cartilage, and unlike bone itself, they have a higher genetic component of determination. Thus, once the cartilaginous structures are formed, the growth activity at these areas may exhibit less of an effect of the influence of growing neural tissue and be more responsive to factors regulating synchondrosal/suture maintenance or fusion. The final form of the neurocranium reflects the interaction of these two influences.

Evidence for the importance of the timing of synchondrosal fusion (or ossification) on the typical human neurobasicranial configuration comes from the clinical literature, experimental manipulation, and selective breeding studies. In all cases, increased skull base kyphosis has been evidenced in nonhuman animals with small brains and nonhuman locomotor patterns. Landauer (1927) demonstrated that when

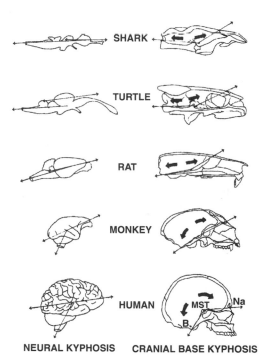

Figure 11.2 Comparative neural kyphosis (measured as the intersection between the anteroposterior cerebral axis and the brain stem axis—lower arrows; Moss et al., 1982) and its effect on skull base morphology and kyphosis (large arrows). Homologous cranial base angle landmarks: Ba, basion; MST, Midsella turcica; Na, nasion).

the spheno-occipital synchondrosis is removed in chick embryos, a more marked kyphosis of the cranial base is a consequence. This was also reconfirmed in mammals by DuBrul and Laskin (1961) with a concomitant rounding of the skull in rats. Chang (1949) and Chang and Landauer (1950) also demonstrated the same effect in other mammals when selective breeding produced early closure of the synchondrosis. Riesenfeld (1969) also discusses the consequences of pathological absence of the synchondrosis in *Cercocebus*, an Old World monkey, and describes a similar group of changes that mirror the condition described by Weidenreich (1941, 1945). Moore and Lavelle (1974) believe that phylogenetic changes in kyphosis came about as a response to increased neurocranial development, and the previous evidence argues that the typically human condition should be viewed as an interaction of timing of fusion *and* increase in relative brain size (Gould, 1977; Dean and Wood, 1984; Campbell, 1985; Ross and Ravosa, 1993; Lieberman and McCarthy, 1999; Lieberman et al., 2000). However, other factors such as facial size, facial orientation, and posture may also be important to primate basicranial morphology and variation (for an excellent review, see Lieberman et al., 2000).

To better understand the origins of the unique human neurobasicranium, this chapter describes phylogenetic changes in the vertebrate neuro- and chondrocranium and relates these modifications to evolutionary and developmental changes in the brain and sensory capsules.

11.2 PHYLOGENETIC CHANGES IN NEURO- AND CHONDROCRANIAL MORPHOLOGY

11.2.1 The Chondrocranial Template

The basis for both an embryonic and phylogenetic understanding of the cranial base begins with the chondrocranial template of fishes (Gans, 1993). A longitudinal column of fluid-filled cells, the notochord, is positioned ventral to the nerve cord during development and extends anteriorly beneath the brain to the pituitary gland. The most posterior elements of the initial chondrocranium form beneath the brain on both sides of the notochord nearly to the hypophysis or pituitary gland. These cartilaginous rods, the parachordal cartilages, increase in size, obliterating or posteriorly displacing the notochord, to form a continuous basal plate beneath the brain (Huxley, 1872; de Beer, 1937). Paired trabeculae form adjacent to both sides of the pituitary and extend beneath the forebrain (Huxley, 1872). The anterior ends unite to form the ethmoid plate in some vertebrates (Fig. 11.3). Bilaterally to the pituitary are polar cartilages that may form a bridge between the parachordal cartilages and trabeculae. The polar cartilages and trabeculae are collectively known as prechordal cartilages.

Sensory capsules form the final components of the template. The nasal capsules form and surround the developing nasal sacs and fuse to the ethmoid plate. Otic capsules form around the future inner ear (the semicircular ducts and sacs of fishes) and unite with the basal plate. The optic capsules of fishes form around the eyeballs but do not fuse with the chondrocranium. These contribute to the fibrous tunic of the eyeballs and ossify in some fishes, amphibians, reptiles, and birds. Orbital cartilages are multiple projections that unite with each other and the chondrocranium to form additional support for the eye, also surrounding associated nerves and vessels to form foramina (Fig. 11.3).

11.2.2 Derivatives of the Chondrocranial Template and Neurocranium in Vertebrates

Three basic derivations of the chondrocranial template are observed among verte-

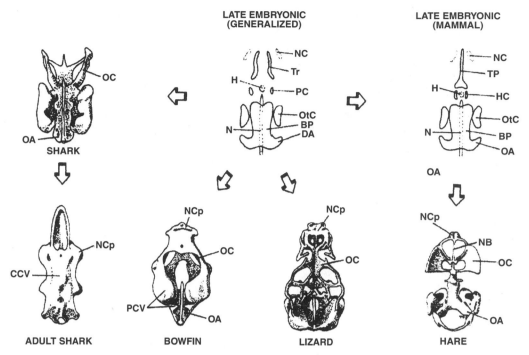

Figure 11.3   Evolutionary changes in the vertebrate embryonic chondrochordal template and fetal skull base derivatives. NC, nasal cartilages; Tr, trabeculae; PC, polar cartilages; OtC, otic capsules; H, Hypophysis; N, Notochord; BP, basal plate cartilage (fused parachordal cartilages); OA, occipital arch; HC, hypophyseal cartilages; TP, trabecular plate; NCp, nasal capsule; CCV, complete cartilaginous vault; PCV, partial cartilaginous vault; OC, orbital cartilages; NS, nasal septum. (*Source*: After de Beer, 1937.)

brates (Fig. 11.3). First, in elasmobranch fishes (sharks, rays, and skates), a cartilaginous roof (not homologous with the mammalian calvaria) fuses to the chondrocranium. This completely cartilaginous cranium never ossifies (Nelson, 1953). A second variation is seen in some Osteichthyes (bony fishes, e.g., the bowfin) (Fig. 11.3). A nearly complete chondrocranium forms, with a partial roof covering the brain dorsally. Restricted regions ossify and are reinforced by elements of the dermatocranium (Nelson, 1953; Walker and Liem, 1994). Finally, in many teleost (advanced) fishes, reptiles, and mammals, only the ventrolateral regions of the chondrocranium form, with subsequent endochondral ossification. This characterizes ancestral tetrapods (four-limbed organisms) and terrestrial vertebrates as well. Among

different taxa, there is a decreasing number of dermal (membrane) bone elements that cover the brain dorsally in ancestral tetrapods, reptiles, and mammals, respectively.

11.2.3   The Chondrocranial Template and Neurocranium of Mammals

Embryology of the basic mammalian chondrocranium initially reflects many of the more primitive morphologies seen in fishes. The parachordal cartilages of mammals obliterate the notochord in the fusion process, forming a basal plate (de Beer, 1937), whereas in primitive vertebrates the parachordal cartilages virtually form the entire ventral support for the brain, while more posterior cartilaginous elements fuse to the parachordal cartilages in more advanced vertebrates (related to vertebrate cephalization and the evolution of the head as related to brain size increase). This reflects a primitive segmentation, wherein ancestral vertebrae elements (occipital sclerotomes) were incorporated (de Beer, 1937). In placental mammals, the paired trabecular cartilages are embryologically represented by a central stem, or trabecular plate (de Beer, 1937). Trabeculae are not derived from segmented sclerotomes (Gasser, 1976). The prechordal cranial base (cranial sella turcica) appears to receive contributions from neural crest mesenchyme (Gasser, 1976; Couly et al., 1993; Kontges and Lumsden, 1996). Paired hypophyseal cartilages are additional prechordal elements of placental mammals that form bilateral to the hypophysis (Figs. 11.3 and 11.4).

The posterior part of the basal plate forms the basioccipital cartilage, and the occipital sclerotomes form the occipital arch, which surrounds the brain posteriorly and forms the boundaries of the foramen magnum (Nelson, 1953; Walker and Liem, 1994). The hypophyseal cartilages form most of the basisphenoid (including the hypophyseal fossa) in placental mammals, and the trabecular plates form the presphenoidal cartilages and the cartilaginous nasal septum. The cartilaginous lesser wings of the sphenoid are derived from orbital cartilages, and basal portions of the greater wings of the sphenoid develop from a separate center in the ala temporalis (de Beer, 1937). The otic capsule forms most of the petrous and tympanic temporal bone (de Beer, 1937; Sperber, 2001), and the nasal capsule forms the cartilaginous lateral nasal walls including conchae (de Beer, 1937). The cribriform plate forms as cartilaginous struts that run from the lateral masses to the mesethmoid. The optic capsule remains cartilaginous as the sclerotic tunic of the eye in mammals and some lower vertebrates (Nelson, 1953).

The membranous neurocranium (der-

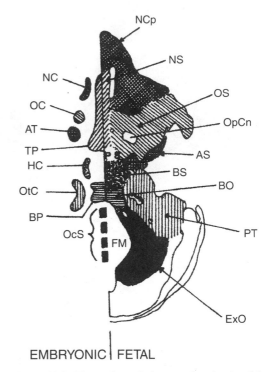

EMBRYONIC | FETAL

Figure 11.4 The embryonic human chondrochordal template and fetal skull base derivatives (similarly shaded areas). NC, nasal cartilages; OC, orbital cartilages; AT, ala temporalis; TP, trabecular plate; HC, hypophyseal cartilages; OtC, otic capsules; BP, basal plate cartilage (fused parachordal cartilages); OcS, occipital scleretomes; NCp, nasal capsule; NS, nasal septum; OS, orbitosphenoid; OpCn, optic canal; AS, alisphenoid; BS, basisphenoid; BO, basioccipital; PT, petrous temporal; ExO, exoccipital; FM, foramen magnum. (*Source*: After de Beer, 1937, and Sperber, 2001.)

matocranium) in mammals is derived from a caspsular membrane, the ectomenix, which is formed from both mesoderm and neural crest cells (Johnston and Bronsky, 1995; Sperber, 2001; see Chapters 4 and 5). Neurocranial ossification rate and location are dependent on biochemical and biomechanical interactions with the developing brain and endomenix (Sperber, 2001; Johnston and Bronsky, 1995). Developmentally, the mammalian membranous neurocranium has changed little from teleost (advanced) fishes, amphibians, and reptiles; however, in mammals there is a decreasing number of dermal (membrane) bone elements that cover the brain dorsally (de Beer, 1937, 1958; Gregory, 1963; Romer and Parsons, 1986). Mammalian neurocranial morphology is thought to have a high degree of plasticity; its resultant shape is related primarily to brain shape and secondarily to the degree of basicranial flexure, locomotor adaptations, and masticatory apparatus (Enlow, 1968, 1990; Enlow and McNamara, 1973) (Figs. 11.1 and 11.2).

Knowledge of the embryonic tissue origin of the craniofacial skeleton has become refined in recent years. The chondrocranium primarily originates from mesoderm, the dermatocranium has a dual origin from mesoderm/neural crest cells, and the viscerocranium arises from neural crest cells (Couly et al., 1993). The role of the neural crest has become especially clarified, and its progeny are apparently capable of forming cartilages that are typically derived from mesoderm (Schneider, 1999). Small populations of neural crest cells form specific muscle attachment sites of the primarily mesodermal cranial base (Kontges and Lumsden, 1996), and the prechordal cranial base in quail-chick chimeras is entirely of neural crest origin (Couly et al., 1993). The inductive signaling associated with the generation of migratory neural crest (from the anterior hindbrain and posterior midbrain) likely involves bone morphogenetic proteins (BMPs) (Kanzler et al., 2000), and the cranial destination points of migratory neural crest cells appear to reflect the rostrocaudal origin along the developing hindbrain and midbrain (Kontges and Lumsden, 1996). Thus, cellular signaling (BMPs also control aspects of Drosophila morphogensis— Kingsley, 1994) and even segmentation (e.g., migratory fate of cranial neural crest) are highly conserved aspects in the evolution of the vertebrate craniofacial complex. Furthermore, the spatial expression of HOX genes in the branchial arch and hind-

brain regions of vertebrates indicates that a highly conserved control mechanism may be at work in craniofacial development (Hunt et al., 1998a, 1998b), and primates are no exception (Vielle-Grosjean et al., 1997).

11.3 PHYLOGENETIC CHANGES IN THE NERVOUS SYSTEM AND ITS EFFECTS ON NEURO- AND CHONDROCRANIAL MORPHOLOGY

11.3.1 Cephalization

The evolutionary process of cephalization involved the polarization and developmental shift of the major sense organs to one end of the embryonic disc. This created a cephalic or cranial pole at one end of the body axis that allows higher invertebrates and all vertebrates to meet environmental stimuli head-first (Kier, 1976).

The process of cephalization resulted in a number of changes in the cranial end of the body (Kier, 1976). These included (1) the development of specialized sense organs; (2) the consequent enlargement of the brain; (3) the formation of a rigid skull base and cranium to protect the sense organs and the brain; (4) the reorganization of various branchiomeric muscle groups (e.g., extraocular, facial, and masticatory muscles); (5) the reorganization of the gill apparatus, stomodeum, and oral and nasal cavities; (6) the disappearance and reorganization of certain nerves associated with lost somites and branchial arches; (7) the shifting of the position of various organ capsules and nerve roots; and (8) the development of the forebrain and higher brain centers in mammals and primates.

The evolutionary transformation of a simple neural tube or spinal cord to a brain in vertebrates is thought to have progressed through a linear series of primary vesicular enlargements at the rostral end of the neural tube: the prosencephalon or forebrain (which will form the telencephalon/cerebral hemispheres, rhinencephalon/olfactory bulbs, and diencephalon/thalamus, hypothalamus, and epithalamus), the mesencephalon or midbrain, and the rhombencephalon or hindbrain (which will form the metencephalon/pons and cerebellum and myelencephalon/medulla oblongata). The vesicular brain undergoes dramatic evolutionary changes from fish to humans (Gregory, 1963; Sarnat and Netsky, 1974; Kier, 1976) with concomitant influences on skull base morphology, growth, and flexion at each stage (Figs. 11.1 and 11.2).

The rhombencephalic portions of the vesicular brain, or the *reptilian brain*, are most closely associated with the basioccipital cartilages of the skull base. The mesencephalon, or the *avian brain*, is most closely associated with the basisphenoidal and presphenoidal skull base cartilages in the sagittal plane. The rhinencephalon and telencephalon, or *mammalian brain*, is most closely associated with the mesethmoidal skull base cartilages. Thus, distinct evolutionary changes in the primary brain vesicles in reptiles and mammals are associated with evolutionarily and regionally distinct skull base morphologies (Fig. 11.2) (Gregory, 1963; de Beer, 1937, 1958; Romer and Parsons, 1986).

11.3.2 Cerebralization

As was previously discussed, cephalization is the process responsible for the initial development and formation of the basic vertebrate skull base necessary for support and protection of the neural contents. In contrast, cerebralization of the forebrain is responsible for giving definitive shape and flexion to the mammalian neuro- and chondrocranium (R. W. Young, 1959; J. Z. Young, 1989) (Figs. 11.1 and 11.2).

The vertebrate forebrain evolved as bilateral olfactory organs at the rostral end of the neural tube. The organs are associated with the appearance of the bilateral vesicular evaginations of the forebrain, the primordia of the cerebral hemispheres (Ariens-Kappers et al., 1963; Kier, 1976). The characteristic feature of the evolutionary changes of the mammalian brain is the functional dominance of the cerebral hemisphere, with resulting tremendous lateral, anterior, and superior expansion of the cerebral cortex (Gregory, 1963; Kier, 1976; Romer and Parsons, 1986), especially in the parietal and temporal lobes. This results in a general widening and elongation of the middle and posterior mammalian cranial fossae (Figs. 11.1 and 11.2).

11.3.3 Gyrification of the Cerebral Cortex and Spatial Packing

The most striking evolutionary feature of the primate brain is the development of large frontal and occipital lobes and the gyrification of the entire cortical surface (Weidenreich, 1941; Ziles et al., 1989) (Figs. 11.1 and 11.2). These neural changes are consistent with adaptive changes by primates to an arboreal environment requiring greater sensory control of prehensile arm-swinging limbs and a need for greater visual association and acuity (Le Gros Clark, 1965; Campbell, 1985). Neural input centers for these senses are located in the precentral gyrus of the frontal lobe and visual association centers of the occipital lobe (Gregory, 1963; Kier, 1976; Romer and Parsons, 1986). Such spatial packing of the neocortex in the neurocranium resulted in extreme kyphosis or flexion of the midsphenoidal and presphenoethmoidal synchondroses, general elongation of the anterior cranial fossa, and radial growth and elongation of the lateral side of the medial cranial fossa, giving the cranial base its characteristic hourglass configuration and leading to brachycephalization of the cranial vault (Montagu, 1943; Enlow, 1990) (Fig. 11.5).

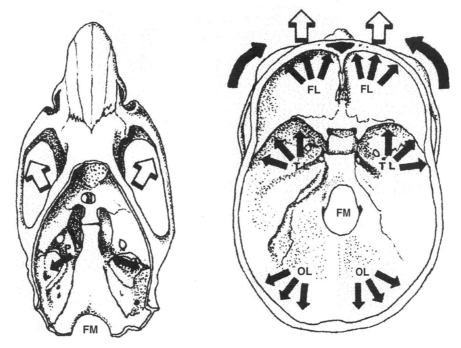

Figure 11.5 The effects of differential cerebral growth patterns on the rat (left) and human (right) skull base and the medial rotation of the orbital axes (open arrows). FL, frontal lobe; TL, temporal lobe; OL, occipital lobe; PL, parietal lobe; FM, foramen magnum. (*Source*: After Enlow, 1990.)

11.3.4 Evolutionary Changes in the Special Sense Organ Capsules

The evolution of the various special sense organs, which are housed in the cartilaginous olfactory, optic, and otic capsules, has also helped to determine skull base morphology. The limited influence of the olfactory capsule on skull base morphology in primates is consistent with the evolutionary shift in emphasis from olfaction (one of the most primitive and dominant senses in reptiles and nonprimate mammals) (Gregory, 1963; Romer and Parsons, 1986) to tactile and visual sensations, which are the dominant senses in primates (Gregory, 1963; Le Gros Clark, 1965; Enlow, 1990). The olfactory, or nasal capsules, develop as bilateral vesicular outpouchings or sacs of ciliated olfactory epithelium from the forebrain (Kier, 1976). They are located within the chondroethmoid and contribute to the for-

mation of the ectethmoid cartilages (the labyrinths), mesethmoid cartilage (the perpendicular plate), and cribriform plate of the ethmoid bone. In fish and reptiles, the linear olfactory lobes, tracts, and bulbs project anteriorly into the back of the nasal capsule, with the cribriform plate and olfactory foramina oriented perpendicularly to the oral and nasal capsules.

As the mammalian neocortex expanded anteriorly and increased in size, the olfactory cortex was displaced to the ventrolateral region of the cerebral hemisphere. With the gyrification of the neocortex in primates, the olfactory cortex was effectively buried and the olfactory tracts were significantly shortened (Baer and Nanda, 1976). These evolutionary changes in primate neural morphologies and growth patterns are consistent with, and are probably the primary causes of, (1) the horizontally rotated and shortened cribriform plate (Sirianni and Swindler, 1979; Lieberman et al., 2000); (2) the overall decreased length of the nasal capsule and reduced olfactory area (Le Gros Clark, 1965; Siegel et al., 1987; Losken et al., 1994); (3) a change in the position of the nasal capsule relative to the skull base from an anterior to an inferior position (Lozanoff et al., 1993); and (4) the downward and forward growth pattern of the midface through a redirection of the mesethmoid and cartilaginous nasal septal cartilage (Krogman, 1969, 1974; Enlow, 1990; Siegel and Mooney, 1990; Siegel et al., 1990, 1992) (Fig. 11.2).

With evolutionary changes in the eyes and supportive optic capsules from merely being light-sensitive directional organs in invertebrates and fish to becoming the dominant sense organ in arboreal arm-swinging primates, the evolution of the cartilaginous optic capsules has contributed significantly to cranial vault and skull base form in primates. The primitive optic vesicles and capsules seen in elasmobranch fishes (sharks, rays, and skates) and some Osteichthyes (bony fishes) are found as lateral projections from the linearly arranged brain vesicles (i.e., the optic sulci and optic lobes between the forebrain and the midbrain), and are surrounded completely by cartilaginous elements (Gregory, 1963; Nelson, 1953; Kier, 1976). With the evolutionary beginnings of the cerebralization and corticalization process in reptiles and mammals, the optic capsules and associated neurovascular components have been displaced anteroinferiorly and have been forced out of the protection of the chondrocranial elements of the middle cranial fossa (Fig. 11.5). In these taxa the orbital cavities are now formed by skeletal elements derived medially from the chondroethmoid, posterolaterally from the orbitosphenoid, superiorly from the dermatocranium, and inferiorly from the splanchnocranium.

The culmination of the cerebralization process, as seen in primates, resulted in extreme anteroinferior displacement and redirection of the optic capsules and orbital axes (Fig. 11.5); relocation of the optic foramina, superior orbital fissures, and the various cranial nerves (II, III, IV, V_1, VI); and formation of a true bony orbital cavity (de Beer, 1937; Romer and Parsons, 1986; Enlow, 1990; Lieberman et al., 2000; McCarthy, 2001). The anteroinferior rotation of the optic capsules, combined with a reduction in nasal capsule length, have resulted in overlapping visual fields and binocular, stereoscopic vision (Fig. 11.5), which are unique taxonomic characteristics of the order Primates (Gregory, 1963; Le Gros Clark, 1965).

The otic capsule and inner ear structures have evolved from organs that could simply "feel" sound waves in the water (the semicircular canals), as seen in sharks and bony fishes to more complex organs (the cochlea) in reptiles and mammals, which are considered to be the true organs of hearing (Gregory, 1963). With the phylogenetic addition of the cochlea, the semicircular canals now function as organs of equilibrium and velocity detection (Lieberman et al., 2000). The cartilaginous otic or auditory capsules form around the otic vesicles and the future semicircular canals of the inner ear cavity, and ossify to form the bony labyrinth and the petrous portion of the temporal bone of the chondrocranium. The evolutionary orientation of the three vestibular planes of the labyrinth is parallel to the orientation of the orbital axes, which allows "validation" of these two senses for locomotion in a horizontal plane.

With neocorticalization and extreme skull base flexion in primates, the vestibular axes of the semicircular canals and the orbital axes have rotated in tandem, and they still retain their primitive orientation in the horizontal plane. The orientation of the vestibular plane on the face is one of the few constant reference planes that can be used in comparative craniofacial studies (Delattre and Fenart, 1958, 1960; Lieberman et al., 2000) (Fig. 11.6). The anterosuperior rotation of the cerebral cortex through the attachments of the dura mater (i.e., the tentorium cerebelli) reorients the

otic capsules and the petrous portion of the temporal bone in primates, and helps to deepen the middle cranial fossa around the sella turcica (Fig. 11.5). Such reorientation of the otic capsules was also produced experimentally in bipedal rats (Moss, 1961).

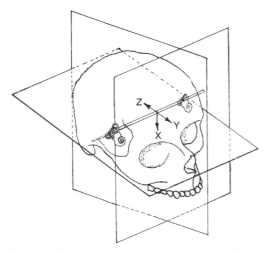

Figure 11.6 Cranial orientation using the vestibular axes in the otic capsule. (*Source*: After Stricker et al., 1990.)

11.4 SKULL BASE SYNCHONDROSES AND COMPARATIVE GROWTH

Synchondroses are the articulations of the basicranium that exist as growth centers or sites of interstitial cartilaginous expansion between ossified portions of the cranial base (Fig. 11.7). Anteroposterior growth of the skull base depends on three synchondroses: (1) the presphenoethmoidal synchondroses (between the presphenoid and cribriform plate—PSES); (2) the midsphenoidal synchondroses (between the presphenoid and basisphenoid—MSS), and (3) the spheno-occipital synchondroses (between the basisphenoid and basioccipi-

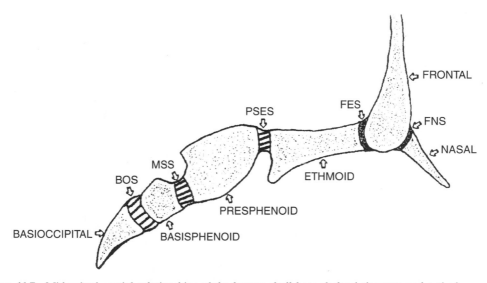

Figure 11.7 Midsagittal spatial relationships of the human skull base skeletal elements and articulations in the late fetal period. FNS, frontonasal suture; FES, frontoethmoidal suture; PSES, presphenoethmoidal synchondrosis; MSS, midsphenoidal synchondrosis; SOS, spheno-occipital synchondrosis.

tal—SOS). Anteroposterior elongation and growth at the synchondroses are thought to occur in two ways: (1) passive displacement from the enlarging neurocapsular functional matrix (i.e., brain and surrounding dura mater) (Moss and Moss-Salentijn, 1969; Moss, 1975a, 1976; Moss et al., 1982); and (2) active interstitial growth, which is under endocrine control and is mechanically similar to linear long bone growth at the epiphyseal growth plates (Baume, 1968; Hoyte, 1971a, 1971b; Van Limbourgh, 1972; Koski, 1977; Enlow, 1990).

Among mammals, the patency of the skull base articulations is variable. The timing of fusion, combined with variations in adjacent brain growth rates and vectors, is responsible for the myriad of skull base morphologies seen across the various taxa (Figs. 11.1 and 11.2) (de Beer, 1937; DuBrul and Laskin, 1961; Badoux, 1966; Michejda, 1975; Laitman et al., 1979; Enlow, 1990; Lieberman and McCarthy, 1999; Lieberman et al., 2000; McCarthy, 2001). For example, in many mammals the midsphenoidal synchondrosis exists postnatally and fuses with advanced age or not at all (Stromsten, 1937). In contrast, the human midsphenoidal synchondrosis fuses just before birth (Kvinnsland, 1971; Melsen, 1974; Sperber, 2001; Kjaer, 1990) and does not contribute to anteroposterior growth postnatally. The presphenoethmoidal synchondrosis contributes to postnatal skull base elongation until approximately 7 years of age in humans (Sperber, 2001), although fusion may occur variably between puberty and adult ages (Ford, 1958). The most prolonged anteroposterior growth in the human cranial base occurs at the spheno-occipital synchondrosis (SOS). Fusion begins on its endocranial surface between 12 and 15 years of age, and fuses ectocranially by age 20 (Powell and Brodie, 1964; Ingervall and Thilander, 1972; Sperber, 2001).

One unique series of articulations in humans, dubbed the coronal ring (Burdi et al., 1986), unites growth sites in the skull base, midface, and cranial vault, and is responsible for much of the early anteroposterior elongation of the entire cranium (Fig. 11.8). The coronal ring is a continuous set of cranial articulations beginning with the coronal suture, the lateral frontosphenoidal and the orbital frontosphenoidal articulations, and the presphenoethmoidal synchondrosis. Anteroposterior expansion of the brain enlarges the neurocranium and displaces the midface anteriorly until approximately 7 years of age. Congenital malformations of the skull base cartilages or premature fusion of the PSES have been associated with various craniofacial anomalies, most notably syndromic craniosynostosis with craniofacial dysostoses and cleft lip and palate, although controversy exists as to whether such skull base deformities are primary or secondary (Moss, 1959, 1975b; Ross, 1965; Babler and Persing, 1982; Grayson et al., 1985; Burdi et al., 1986; Marsh and Vannier, 1986; Richtsmeier et al., 1991; Cheverud et al., 1992; Harris; 1993; Lozanoff et al., 1994; Molsted et al., 1995; Skedros et al., 1995; Smith et al.,

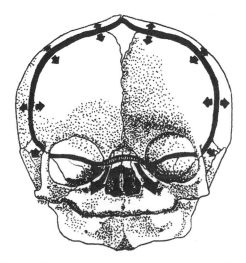

Figure 11.8 The fetal coronal ring articulations, anchored at the skull base by the presphenoethmoidal synchondrosis (slashed lines). (*Source*: After Burdi et al., 1986.)

1996). The clinical significance of this is seen in the relative plasticity of the skull base following surgical manipulation of the neurocapsular matrix, through correction of craniosynostotic sutures and congenital skull base anomalies, or following the resection of various cranial base tumors.

More than half of the anteroposterior growth of the human skull base takes place during the first 8 years of life (Bjork, 1955; Brodie, 1955; Zuckerman, 1955; Krogman and Sassouni, 1956). The basioccipit, basisphenoid, presphenoid, and mesethmoid bones all take part in a burst of anteroposterior growth during adolescence (Zuckerman, 1955; Krogman and Sassouni, 1956; Lieberman et al., 2000; McCarthy, 2001). Growth among these skeletal elements reflects (1) anteroposterior growth through displacement and drift; (2) mediolateral growth through displacement and drift; (3) superoinferior growth through drift; and (4) angulation through extension and flexion (Enlow, 1990; Lieberman et al., 2000; McCarthy, 2001). In particular, incremental growth of the human cranial base is characterized by (1) rapid growth from birth to 5 years; (2) deceleration between 5 and 12 years of age, which plateaus between 10 and 13 years of age; (3) a parapubertal acceleration; and then (4) a gradual deceleration and cessation of growth through 20 years of age (Zuckerman, 1955; Krogman and Sassouni, 1956). Although the length and the angular relationships of the individual components of the cranial base may change during a lifetime, the relative contributions of each part to the entire skull base remain virtually constant between 3 and 20 years of age (Bjork, 1955; Brodie, 1955) (Fig. 11.9). The relative contribution of the posterior component (SOS-basion) is 25%, the middle component (PSES-SOS) is 37%, and the anterior component (nasion-PSES) is 37% (Brodie, 1955).

Krogman and Sassouni (1956) reported that Welcker, in 1856, initially described the

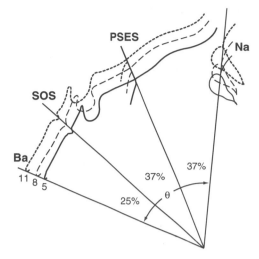

Figure 11.9 Lateral cephalograph tracings from one individual showing consistent proportional changes in the various skull base components at 5, 8, and 11 years of age. (*Source*: After Brodie, 1955.)

angle nasion (Na)–sphenoidale (a.k.a. tuberculum sellum)–basion (Ba) to measure cranial base plane inclination. It was found that Welcker's angle (Na-TS-Ba) remained constant from 3 months in utero to birth, and decreased (or flexed) throughout the growth period (Zuckerman, 1955). However, since tuberculum sellum is not a growth site, cranial base flexion is typically measured in the midsagittal plane at the midsphenoidal synchrondrosis (sella turcica) using the angle Na-MST-Ba (nasion-midsella turcica-basion) (Figs. 11.2 and 11.10) (Krogman and Sassouni, 1956; Ross, 1965). This angle is initially highly obtuse at approximately 150° in the 4-week-old (precartilage stage) embryo. At the start of ossification of the skull base, between 10 and 20 weeks, the cranial base angle decreases (flexes) to between 125° and 130°, and maintains this kyphotic angulation postnatally (Moss, 1955; Knott, 1971; Van der Linden and Enlow, 1971; Latham, 1972; Riolo et al., 1974; Broadbent et al., 1975; Schulter, 1976; Muller and O'Rahilly, 1980; Friede, 1981; Moore, 1983; Diewert, 1985; Anagnostopoulou et al.,

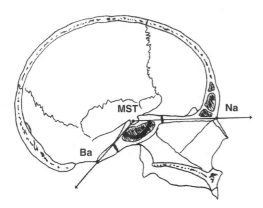

Figure 11.10   The cranial base angle that consistently measures about 130° in juveniles and adults.

1988; Sperber, 2001). The angulation changes in the cranial base are probably caused by the rapid growth and kyphosis (flexion) of the brain during the fetal and perinatal period (Moss et al., 1982; Lieberman et al., 2000), as the chondrocranium retains its embryonic shape in anencephaly (Sperber, 2001).

The main site of cranial base kyphosis or flexion in humans is located at the midsphenoidal synchondrosis in the sella turcica (Scott, 1958; Michejda, 1972; Latham, 1972; Moss et al., 1982; Sperber, 2001), which occurs very rapidly and is completed by approximately 2 years of age (Ortiz and Brodie, 1949; Bjork, 1955; Brodie, 1955; Koski, 1960; George, 1978; Lieberman and McCarthy, 1999). The early fusion of this synchondrosis is thought to be responsible for the failure of the cranial base angle to change after 2 years of age (Scott, 1958). In contrast, the midsphenoidal and sphenoethmoidal synchondroses exist postnatally as growth sites in some nonhuman primates, although for a shorter duration than the spheno-occipital (Ford, 1958; Scott, 1958; Melsen, 1971; Michejda, 1972; Dmoch, 1975; Demes, 1985). This results in an initial cranial base flexion in nonhuman primates, as also seen in human perinates, and then a postnatal cranial base extension (Angst, 1967; Cousin

et al., 1981; Lieberman and McCarthy, 1999; Lieberman et al., 2000; McCarthy, 2001). The initial cranial base flexion is probably related to an increase in brain size (deBeer, 1937; Sperber, 2001), while the later cranial base extension may be related to midfacial growth processes, since brain growth has ceased prior to cranial base extension (Dmoch, 1976; Lieberman and McCarthy, 1999; Lieberman et al., 2000; McCarthy, 2001). Thus, basicranial flexion as a growth process appears to be unique to humans among extant primates (Lieberman and McCarthy, 1999; Lieberman et al., 2000; McCarthy, 2001). In contrast, human cranial vault morphology is extremely plastic and responds to functional changes in endocranial volume (Enlow, 1968, 1990; Enlow and McNamara, 1973), cranial base growth patterns (Moss, 1958, 1959; Young, 1959; Enlow and McNamara, 1973; Enlow, 1990), and masticatory forces (Biegert, 1963; Sirianni and Swindler, 1979; Enlow, 1990) (Figs. 11.1, 11.2, and 11.5).

11.5   CONCLUSIONS

There are many phylogenetic and developmental factors that have contributed to the unique brachycephalic morphology and extreme cranial base flexion (kyphosis) of the human and nonhuman primate cranial vault and base. The major factors influencing these unique morphologies are probably the interaction of fusion timing of the various neurocranial and basicranial articulations *and* an increase in relative brain size. However, other factors such as facial size, facial orientation, and posture may also be important to primate basicranial morphology and variation.

ACKNOWLEDGMENTS

This work was supported in part by a Comprehensive Oral Health Research Center of

Discovery (COHRCD) program/project grant from NIH/NIDCR (P60 DE13078) to the Center for Craniofacial Development and Disorders, Johns Hopkins University, Baltimore, and a publication assistance award from the Richard D. and Mary Jane Edwards Endowed Publication Fund, Faculty of Arts and Sciences, University of Pittsburgh.

REFERENCES

Adams, L. M., and Moore, W. J. (1975). A biomechanical appraisal of some skeletal features associated with head balance and posture in the Hominoidea. *Acta Anat.* 92, 580–594.

Anagnostopoulou, S., Kardmiliki, D. D., and Spyropoulous, M. N. (1988). Observations on the growth and orientation of the anterior cranial base in the human foetus. *Eur. J. Orthod.* 10, 143–167.

Angst, R. (1967). Beitrag zum formwandel des craniums der ponginen. *Z. Morph. Anthrop.* 58, 109–151.

Ariens-Kappers, C. U., Huber, G. C., and Crosby, E. C. (1963). *The Comparative Anatomy of the Nervous System of Vertebrates, Including Man* (New York: Hafner).

Babler, W. J., and Persing, J. A. (1982). Experimental alteration of cranial suture growth: effects on the neurocranium, basicranium, and midface. In B. G. Sarnat (ed.), *Factors and Mechanisms Influencing Bone Growth* (New York: Alan R. Liss, Inc.), pp. 333–345.

Badoux, D. M. (1966). Framed structures in the mammalian skull. *Acta Morphol. Neerl. Scand.* 5, 347–360.

Baer, M. J., and Harris, J. E. (1969). A commentary on the growth of the human brain and skull. *Am. J. Phys. Anthrop.* 30, 39–44.

Baer, M. J., and Nanda, S. K. (1976). A commentary of the growth and form of the cranial base. In J. Bosma (ed.), *Development of the Basicranium* (Bethesda, MD: DHEW [NIH] publication 76-989), pp. 515–540.

Baume, L. J. (1968). Patterns of cephalofacial growth and development. A comparative study of the basicranial growth centers in rat and man. *Int. Dent. J. 18*, 489–513.

Biegert, J. (1963). The evaluation of the characters of the skull, hands, and feet for primate taxonomy. In S. L. Washburn (ed.), *Classification and Human Evolution* (Chicago: Aldine de Gruyter), pp. 116–145.

Bjork, A. (1955). Cranial base development. *Am. J. Orthod. 41*, 198–225.

Bolk, L. (1926). On the problem of anthropogenesis. *Kon. Akad. Wetenshc. Amsterd. Afd. Naturk. 19*, 465–475.

Broadbent, B. H., Broadbent, B. H., Jr., and Golden, W. H. (1975). *Bolton Standards of Dentofacial Developmental Growth* (St. Louis: Mosby).

Brodie, A. G. (1955). The behavior of the cranial base and its components as revealed by serial cephalometric roentgenograms. *Angle Orthod. 25*, 148–160.

Burdi, A. R., Kusnetz, A. B., Venes, J. L., and Gebarski, S. S. (1986). The natural history of the cranial coronal ring articulations: implications in understanding the pathogenesis of the Crouzon craniosynostotic defects. *Cleft Palate Craniofac. J. 23*, 28–39.

Cameron, J. (1925). The craniofacial axis of Huxley. Part II. Comparative anatomy. *Trans. Roy. Soc. Canada 19*, 129–136.

Campbell, B. (1985). *Human Evolution*, 3rd ed. (New York: Aldine).

Chang, T. K. (1949). Morphological study on the skeleton of Ancon sheep. *Growth 13*, 269–297.

Chang, T. K., and Landauer, W. (1950). Observations on the skeleton of African dwarf goats. *J. Morphol. 86*, 367–379.

Cheverud, J. M., Kohn, L. A. P., Konigsberg, L. W., and Leigh, S. R. (1992). Effects of fronto-occipital artificial vault modification on the cranial base and face. *Am. J. Phys. Anthrop. 88*, 323–345.

Couly, G. F., Coltey, P. M., and Le Douarin, N. M. (1993). The triple origin of skull in higher vertebrates: a study in quail-chick chimeras. *Development 117*, 409–429.

Cousin, R. P., Fenart, R., and Deblock, R. (1981). Variations ontogenetizues des angles basicraniens et faciaux. *Bull. Mem. Soc. Anthrop. Paris 8*, 189–212.

Dean, M. C. (1988). Growth processes in the cranial base of hominoids and their bearing on morphological similarities that exist in the cranial base of Homo and Paranthropus. In F. E. Grine (ed.), *Evolutionary History of the Robust Australopithecines* (New York: Aldine de Gruyter), pp. 107–112.

Dean, M. C., and Wood, B. A. (1984). Phylogeny, neoteny, and the growth of the cranial base in hominoids. *Folia Primatol. 43*, 157–174.

de Beer, G. R. (1937). *The Development of the Vertebrate Skull* (London: Oxford University Press).

de Beer, G. R. (1958). *Embryos and Ancestors* (New York: Clarendon Press).

Delattre, A., and Fenart, R. (1958). La methode vestibulaire. *Z. Morph. Anthrop. 49*, 90–114.

Delattre, A., and Fenart, R. (1960). *L'Hominisation Du Crane Etudiee Par La Methode Vestibulaire* (Paris: Editions CNRS).

Demes, B. (1985). Biomechanics of the primate skull base. *Adv. Anat. Embryol. Cell. Biol. 94*, 27–56.

Diewert, V. M. (1985). Development of human craniofacial morphology during the late embryonic and early fetal periods. *Am. J. Orthod. 88*, 64–75.

Dmoch, R. (1975). Beitrage zum formenwandel des primatencraniums mit bemerkungen zu den sagittalen knickungsverhaltnissen. II Dtreckenanalyse I. Neurocranium. *Geg. Morph. 121*, 521–601.

Dmoch, R. (1976). Beitrage zum formenwandel des primatencraniums mit bemerkungen zu den sagittalen knickungsverhaltnissen. V. Schlussbetrachtung, zussamenfassung, literatur und tabellenanhang. *Geg. Morph. 122*, 1–81.

DuBrul, E. L. (1950). Posture, locomotion, and the skull in Lagormorpha. *Am. J. Anat. 87*, 277–313.

DuBrul, E. L., and Laskin, D. M. (1961). Preadaptive potentiates of the mammalian skull: an experiment in growth and form. *Am. J. Anat. 109*, 117–132.

Enlow, D. H. (1968). *The Human Face* (New York: Harper and Row).

Enlow, D. H. (1990). *Handbook of Facial Growth* (New York: Saunders).

Enlow, D. H., and McNamara, J. (1973). The neurocranial basis for facial form and pattern. *Angle Orthod. 43*, 256–271.

Ford, E. H. R. (1958). Growth of the human cranial base. *Am. J. Orthod. 44*, 498–506.

Friede, H. (1981). Normal development and growth of the human neurocranium and cranial base. *Scand. J. Plast. Reconstr. Surg. 15*, 163–169.

Gans, C. (1993). Evolutionary origin of the vertebrate skull. In J. Hanken and B. K. Hall (eds.), *The Skull. Volume 2: Patterns of Structural and Systematic Diversity* (Chicago: University of Chicago Press), pp. 1–37.

Gasser, R. F. (1976). Early formation of the basicranium in man. In J. Bosma (ed.), *Development of the Basicranium* (Bethesda, MD: DHEW [NIH] Publication 76-989), pp. 29–43.

George, S. L. (1978). A longitudinal and cross-sectional analysis of the growth of the postnatal cranial base angle. *Am. J. Phys. Anthrop. 49*, 171–178.

Gould, S. J. (1977). *Ontogeny and Phylogeny* (Cambridge: Belknap Press).

Grayson, B. H., Weintraub, N., Bookstein, F. L., and McCarthey, J. G. (1985). A comparative cephalometric study of the cranial base in craniofacial anomalies: Part 1: Tensor Analysis. *Cleft Palate Craniofac. J. 22*, 75–87.

Gregory, W. K. (1963). *Our Face from Fish to Man* (New York: Hafner Publishing Co.).

Harris, E. F. (1993). Size and form of the cranial base in isolated cleft lip and palate. *Cleft Palate Craniofac. J. 30*, 170–174.

Hilton, J. (1855). *Notes on Some of the Developmental and Functional Relationships of the Cranium* (London: John Churchill), p. 4.

Hofer, H. (1969). The evolution of the brain. Its influence on the form of the skull. *Ann. NY Acad. Sci. 167*, 341–356.

Hoyte, D. A. N. (1971a). Mechanisms of growth in the cranial vault and base. *J. Dent. Res. 50*, 1447–1461.

Hoyte, D. A. N. (1971b). The modes of growth of the neurocranium: the growth of the sphenoid bone in animals. In R. E. Moyers and W. M. Krogman (eds.), *Craniofacial Growth in Man* (Oxford: Pergamon Press), pp. 77–103.

Hunt, P., Clarke, J. D. W., Buxton, P., Ferretti, P., and Thorogood, P. (1998a). Stability and plasticity of neural crest patterning and branchial arch Hox code after extensive cephalic crest rotation. *Devel. Biol. 198*, 82–104.

Hunt, P., Clarke, J. D. W., Buxton, P., Ferretti, P., and Thorogood, P. (1998b). Segmentation, crest prespecification and the control of facial form. *Eur. J. Oral Sci. 106* (suppl. 1), 12–18.

Huxley, T. H. (1872). *A Manual of the Anatomy of Vertebrate Animals* (New York: Appleton).

Ingervall, B., and Thilander, B. (1972). The human spheno-occipital synchondrosis. *Acta Odont. Scand. 30*, 349–361.

Johnston, M. C., and Bronsky, P. T. (1995). Prenatal craniofacial development: new insights on normal and abnormal mechanisms. *Crit. Rev. Oral Biol. Med. 6*, 25–79.

Kanzler, B., Foreman, R. K., and Labosky, P. A. (2000). BMP signaling is essential for development of skeletogenic and neurogenic cranial neural crest. *Development 127*, 1095–1104.

Kier, E. L. (1976). Phylogenetic and ontogenetic changes of the brain relevant to the evolution of the skull. In J. Bosma (ed.), *Development of the Basicranium* (Bethesda, MD: DHEW [NIH] Publication 76-989), pp. 468–499.

Kingsley, D. M. (1994). What do BMPs do in mammals? Clues from the mouse short ear mutation. *Trends Genet. 10*, 16–21.

Kjaer, I. (1990). Ossification of the human fetal basicranium. *J. Craniofac. Genet. Devel. Biol. 10*, 29–38.

Knott, V. B. (1971). Changes in cranial base measures of human males and females from age 6 years to early adulthood. *Growth 35*, 145–153.

Kontges, G., and Lumsden, A. (1996). Rhombencephalic neural crest segmentation is preserved throughout craniofacial ontogeny. *Development 122*, 3229–3242.

Koski, K. (1960). Some aspects of the growth of the cranial base and the upper face. *Scand. J. Odontol. 68*, 344–358.

Koski, K. (1977). The role of the craniofacial cartilages in the postnatal growth of the craniofacial skeleton. In A. A. Dahlberg and T. M. Graber (eds.), *Orofacial Growth and Development* (Hague: Mouton), pp. 45–73.

Krogman, W. M. (1969). Growth changes in skull, face, jaws, and teeth of the chimpanzee. In G. H. Bourne (ed.), *Anatomy, Behavior, and Diseases of Chimpanzees*, vol. 1 (Baltimore: University Park Press), pp. 104–164.

Krogman, W. M. (1974). Craniofacial growth and development: an appraisal. *Yrbk. Phys. Anthrop. 18*, 31–64.

Krogman, W. M., and Sassouni, V. (1956). *Syllabus in Roentgenographic Cephalometry* (Philadelphia: College Offset Press).

Kvinnsland, S. (1971). The sagittal growth of the foetal cranial base. *Acta Odont. Scand. 29*, 699–704.

Laitman, J. T., Heimbuch, R. C., and Crelin, E. S. (1979). The basicranium of fossil hominids as an indicator of their upper respiratory systems. *Am. J. Phys. Anthrop. 51*, 15–34.

Landauer, W. (1927). Utersuchungen uber Chondrodystrophie. *Roux Arch. Entw. Mech. 110*, 197–278.

Latham, R. A. (1972). The different relationship of the sella point to growth sites of the cranial base in fetal life. *J. Dent. Res. 51*, 1646–1650.

Le Gros Clark, W. E. (1965). *History of the Primates*, 5th ed. (Chicago: University of Chicago Press).

Lieberman, D. E., and McCarthy, R. C. (1999). The ontogeny of cranial base angulation in humans and chimpanzees and its implications for reconstructing pharyngeal dimensions. *J. Hum. Evol. 36*, 487–517.

Lieberman, D. E., Ross, C. F., and Ravosa, M. J. (2000). The primate cranial base: ontogeny, function, and integration. *Yrbk. Phys. Anthrop. 43*, 117–169.

Losken, A., Mooney, M. P., and Siegel, M. I. (1994). Comparative cephalometric study of nasal cavity growth patterns in seven animal models. *Cleft Palate Craniofac. J. 31*, 17–23.

Lozanoff, S., Zingeser, M. R., and Diewert, V. M. (1993). Computerized modeling of nasal capsular morphogenesis in prenatal primates. *Clin. Anat. 6*, 37–47.

Lozanoff, S., Jureczek, S., Feng, T., and Padwal, R. (1994). Anterior cranial base morphology in mice with midfacial retrusion. *Cleft Palate Craniofac. J. 31*, 417–428.

Lugoba, S. A., and Wood, B. A. (1990). Position and orientation of the foramen magnum in higher primates. *Am. J. Phys. Anthrop. 81*, 67–76.

Marsh, J. L., and Vannier, M. W. (1986). Cranial base changes following surgical treatment of craniosynostosis. *Cleft Palate Craniofac. J. 23* (suppl. 1), 9–18.

McCarthy, R. C. (2001). Anthropoid cranial base architecture and scaling relationships. *J. Hum. Evol. 40*, 41–66.

Melsen, B. (1971). The postnatal growth of the cranial base in Macaca rhesus analyzed by the implant method. *Tandlagebladet 75*, 1320–1329.

Melsen, B. (1974). The cranial base. *Acta Odont. Scand. 32*, 1–126.

Michejda, M. (1972). The role of basicranial synchondroses in flexure processes and ontogenetic development of the skull base. *Am. J. Phys. Anthrop. 37*, 143–150.

Michejda, M. (1975). Ontogenic growth changes of the skull base in four genera of nonhuman primates. *Acta Anat. 91*, 110–117.

Molsted, K., Kjaer, I., and Dahl, E. (1995). Cranial base in newborns with complete cleft lip and palate: a radiographic study. *Cleft Palate Craniofac. J. 32*, 199–205.

Montagu, M. F. A. (1943). The mesethmoid-presphenoid relationships in the primates. *Am. J. Phys. Anthrop. 41*, 129–140.

Moore, K. L. (1983). *Before We Are Born: Basic Embryology and Birth Defects*, 2nd ed. (Philadelphia: W. B. Saunders).

Moore, W. J., and Lavelle, C. L. B. (1974). *Growth of the Facial Skeleton in the Hominoidea* (London: Academic Press).

Moss, M. L. (1955). Postnatal growth of the human skull base. *Angle Orthod. 25*, 57–64.

Moss, M. L. (1958). The pathogenesis of artificial cranial deformation. *Am. J. Phys. Anthrop. 16*, 269–286.

Moss, M. L. (1959). The pathogenesis of premature cranial synostosis in man. *Acta Anat. 37*, 351–370.

Moss, M. L. (1961). Rotation of the otic capsule in bipedal rats. *Am. J. Phys. Anthrop. 19*, 301–307.

Moss, M. L. (1975a). The effect of rhombencephalic hypoplasia on posterior cranial base elongation in rodents. *Arch. Oral Biol. 20*, 489–493.

Moss, M. L. (1975b). Functional anatomy of craniosynostosis. *Child's Brain 1*, 22–33.

Moss, M. L. (1976). Experimental alteration of basi-synchondrosal cartilage growth in rat and mouse. In J. Bosma (ed.), *Development of the Basicranium* (Bethesda, MD: DHEW [NIH] Publication 76-989), pp. 541–569.

Moss, M. L., and Moss-Salentijn, L. (1969). The capsular matrix. *Am. J. Orthod. 56*, 474–490.

Moss, M. L., Moss-Salentijn, L., Vilmann, H., and Newell-Morris, L. (1982). Neuroskeletal topology of the primate basicranium: its implications for the "fetalization hypothesis." *Gegenbaurs. Morph. Jahrb. Leipzig 1*, 58–67.

Muller, F., and O'Rahilly, R. (1980). The human chondrocranium at the end of the embryonic period proper, with particular reference to the nervous system. *Am. J. Anat. 159*, 33–47.

Nelson, O. E. (1953). *Comparative Embryology of the Vertebrates* (New York: The Blakiston Co, Inc.).

Ortiz, M. H., and Brodie, A. G. (1949). On the growth of the human head from birth to the third month of life. *Anat. Rec. 103*, 319–348.

Passingham, R. E. (1975). Changes in the size and organization of the brain in man and his ancestors. *Brain Behav. Evol. 11*, 73–90.

Powell, T. V., and Brodie, A. G. (1964). Closure of the sphenooccipital synchondrosis. *Anat. Rec. 147*, 15–23.

Richtsmeier, J. T., Grausz, H. M., Morris, G. R., Marsh, J. L., and Vannier, M. W. (1991). Growth of the cranial base in craniosynostosis. *Cleft Palate Craniofac. J. 28*, 55–67.

Riesenfeld, A. (1969). The adaptive mandible: an experimental study. *Acta Anat. 72*, 246–262.

Riolo, M. L., Moyers, R. E., McNamara, J. A., and Hunter, W. S. (1974). *An Atlas of Craniofacial Growth* (Ann Arbor: Center for Human Growth and Development).

Romer, A. S., and Parsons, T. S. (1986). *The Vertebrate Body*, 5th ed. (Philadelphia: Saunders).

Ross, C. F., and Ravosa, M. J. (1993). Basicranial flexion, relative brain size, and facial kyphosis in nonhuman primates. *Am. J. Phys. Anthrop. 91*, 305–324.

Ross, R. B. (1965). Cranial base in children with lip and palatal clefts. *Cleft Palate Craniofac. J. 2*, 157–166.

Sarnat, H. B., and Netsky, M. G. (1974). *Evolution of the Nervous System* (New York: Oxford University Press).

Schneider, R. A. (1999). Neural crest can form cartilages normally derived from mesoderm during development of the avian head skeleton. *Devel. Biol. 208*, 441–455.

Schulter, F. P. (1976). Studies of the basicranial axis: a brief review. *Am. J. Phys. Anthrop. 45*, 545–552.

Schultz, A. H. (1942). Conditions for balancing the head in primates. *Am. J. Phys. Anthrop. 29*, 483–497.

Scott, J. H. (1958). The cranial base. *Am. J. Phys. Anthrop. 16*, 319–348.

Siegel, M. I., and Mooney, M. P. (1990). Appropriate animal models for craniofacial biology. *Cleft Palate Craniofac. J. 27*, 18–25.

Siegel, M. I., Mooney, M. P., Kimes, K. R., and Todhunter, J. (1987). Analysis of the size variability of the human normal and cleft palate fetal nasal capsule by means of three-dimensional computer reconstruction of histologic preparations. *Cleft Palate Craniofac. J. 24*, 190–199.

Siegel, M. I., Mooney, M. P., Eichberg, J. W., Gest, T., and Lee, D. R. (1990). Septopremaxillary ligament resection and midfacial growth in a chimpanzee animal model. *J. Craniofac. Surg. 4*, 182–186.

Siegel, M. I., Mooney, M. P., Eichberg, J. W., Gest, T. R., and Lee, D. R. (1992). Nasal capsule shape changes following septopremaxillary ligament resection in a chimpanzee animal model. *Cleft Palate Craniofac. J. 29*, 137–142.

Sirianni, J. E., and Swindler, D. R. (1979). A review of the postnatal craniofacial growth in Old World monkeys and apes. *Yrbk. Phys. Anthop. 22*, 80–104.

Skedros, D. G., Mooney, M. P., Siegel, M. I., Losken, A., Smith, T. D., and Janecka, I. P. (1995). Unilateral trigeminal nerve branch lesion (V1&2) at the skull base and nasal capsule growth in rabbits. *J. Craniofac. Surg. 6*, 40–48.

Smith, T. D., Mooney, M. P., Losken, H. W., Siegel, M. I., and Burrows, A. (1996). Post-natal growth of the cranial base in rabbits with congenital coronal suture synostosis. *J. Craniofac. Genet. Devel. Biol. 16*, 107–117.

Sperber, G. H. (2001). *Craniofacial Development* (Hamilton: B. C. Decker Inc.).

Spoor, C. F. (1997). Basicranial architecture and relative brain size of Sts 5 (*Australopithecus africanus*) and other Plio-Pleistocene hominids. *S. African J. Sci. 93*, 182–186.

Stephan, H., Frahm, H., and Baron, G. (1981). New and revised data on volumes of brain structures in insectivores and primates. *Folia Primatol. 35*, 1–29.

Stricker, M., Raphael, B., Van der Meulen, J., and Mazzola, R. (1990). *Craniofacial Development and Growth*. In M. Stricker, J. Van der Meulen, B. Raphael, and R. Mazzola (eds.), *Craniofacial Malformations* (New York: Churchhill Livingstone), pp. 61–90.

Stromsten, F. A. (1937). *Davison's Mammalian Anatomy*, 6th ed. (Philadelphia: P. Blakiston's Son & Co., Inc.).

Van der Linden, F. P. G. M., and Enlow, D. H. (1971). A study of the anterior cranial base. *Angle Orthod. 41*, 119–127.

Van Limbourgh, J. (1972). The role of genetic and local environmental factors in the control of postnatal craniofacial morphogenesis. *Acta Morphol. Neerl. Scand. 10*, 37–48.

Vielle-Grosjean, I., Hunt, P., Gulisano, M., Boncinelli, T., and Thorogood, P. (1997). Branchial HOX gene expression and human craniofacial development. *Devel. Biol. 183*, 49–60.

Walker, W. F., and Liem, K. F. (1994). *Functional Anatomy of the Vertebrates: An Evolutionary Perspective* (Philadelphia: Saunders College Publications).

Weidenreich, F. (1941). The brain and its role in the phylogenetic transformation of the human skull. *Trans. Am. Philos. Soc. 31*, 321–442.

Weidenreich, F. (1945). The brachycephalization of recent mankind. *Southwest J. Anthrop. 1*, 1–54.

Young, J. Z. (1989). *The Life of Vertebrates*, 3rd ed. (Oxford: Clarendon Press).

Young, R. W. (1959). The influence of cranial contents on postnatal growth of the skull in the rat. *Am. J. Anat. 105*, 383–390.

Ziles, K., Armstrong, E., Moser, K. H., Schleicher, A., and Stephan, H. (1989). Gyrification in the cerebral cortex of primates. *Brain Behav. Evol. 34*, 143–150.

Zuckerman, S. (1955). Age changes in the basic cranial axis of the human skull. *Am. J. Phys. Anthrop. 13*, 521–539.

CHAPTER 12

CRANIAL BASE DYSMORPHOLOGY AND GROWTH IN THE CRANIOSYNOSTOSES

ALPHONSE R. BURDI, Ph.D.

We dance around in a ring and suppose, but the secret sits in the middle and knows.

—Robert Frost, 1943

12.1 HISTORICAL PERSPECTIVE

The clinical patterns and physical appearance of craniofacial malformations associated with the spectrum of syndromic and nonsyndromic craniosynostoses have intrigued biologists and clinicians seemingly for ages. Premature closure of single or multiple calvarial sutures, especially in humans, has been linked to the pathogenesis of the craniosynostoses. Although the craniosynostoses were obviously not identified as such then, Hippocrates in 100 B.C. did observe the variability in cranial deformities and linked such deformities with disturbances in calvarial suture development. Virchow (1851) observed that skull deformities resulted from inhibition of growth at right angles to the coronal sutures and a compensatory overexpansion of the cranium at sites of the patent sutures. While the cranial base itself eluded such early attention, this focus on the premature synostoses of key cranial sutures (as primary pathogenic events) has prevailed well beyond the years of Hippocrates and Virchow.

Craniosynostosis is a feature seen in more than 100 genetic syndromes, and there is considerable variation in the kinds and number of specific calvarial sutures associated with a craniosynostotic phenotype (Simmons and Peyton, 1947; Bertelsen, 1958; Hemple et al., 1961; Anderson and Geiger, 1965; Foltz and Loeser, 1975; Cohen, 1986). In general, the degree of skull deformity in the craniosynostoses was thought to be dependent on the number and location of calvarial sutures involved and the time of onset of the sutural perturbation—that is, the earlier the synostosis, the greater the malformation. Premature or early closure of the sagittal suture may result in frontal and occipital expansion wherein the skull becomes longer and more narrow than expected for age (i.e., scaphocephaly). Premature coronal suture synostosis manifests in an anteroposterior-shortened and higher-vaulted skull (i.e., acrocephaly). Unilateral premature synostosis of the

Understanding Craniofacial Anomalies: The Etiopathogenesis of Craniosynostoses and Facial Clefting, Edited by Mark P. Mooney and Michael I. Siegel, ISBN 0-471-38724-x Copyright © 2002 by Wiley-Liss, Inc.

coronal and lambdoidal sutures is associated with the condition of plagiocephaly.

Looking specifically at Crouzon cases, about 75% of patients exhibit some degree of premature synostosis of the coronal, sagittal, and lambdoidal sutures. This condition of multiplicity of sutures is expanded upon later in this chapter. While the characteristic shape of the Crouzon calvarium can vary, its calvarium typically is foreshortened and widest in the temporal region (Cohen, 1975, 1986). With the advent of such observations, a growing awareness began to emerge bearing on the direct or indirect associations, or even causal relationships, between dysmorphogenesis among the calvarial sutures and within the cranial base. Observations similar to those reported for the human Crouzon skull were reported on the effects of experimental suture closure on the calvarium in rabbits, resulting in reduction of the length of the anterior cranial base and concomitant shortening of the posterior cranial base or clivus (Babler, 1988).

In general, the most severe cases of craniosynostoses result when more than a single suture is involved (Babler and Persing, 1984; Babler, 1988). While recognizing that the premature synostosis of calvarial sutures can manifest as abnormal calvarial shapes and size, there was an increasing awareness that the cranial base itself might be involved in the genesis of the craniosynostosis phenotype (Warkany, 1971; David et al., 1982; Jones, 1988; Gorlin et al., 1976, 2001; Smith, 1976). While a number of calvarial malformations associated with the early synostosis of specific calvarial sutures could be recognized, evidence on the direct association of cranial base abnormalities (i.e., structural and cephalometrically) with such calvarial perturbations grew even further, chiefly from a clinical or treatment perspective (Marsh and Vannier, 1986; Hoyte, 1989; Kreiborg et al., 1993). However, the cranial base sequelae of either early or late-onset calvarial synostoses are still uncertain with specific reference to the timing variable. While some evidence exists that shows early-onset or premature calvarial synostosis can be associated with cranial base abnormalities in rabbits (Babler and Persing, 1982), it has also been reported that delayed-onset calvarial synostoses in rabbits did not produce craniofacial morphology (including that of the cranial base), which was significantly different from normal at any age (Smith et al., 1996; Burrows et al., 1999). Clearly, caution needs to be exercised in directly extrapolating and transferring experimentally derived information from the laboratory animal (e.g., rabbits) to the understanding and treatment of the craniosynostoses in humans.

12.2 THE ETIOPATHOGENESIS OF CRANIAL BASE ABNORMALITIES

There continues to be, however, increasing recognition that sutural dysmorphogenesis may not be the primary etiopathogenic event in the craniosynostoses. This rethinking on the sites of etiopathogenesis in the craniosynostoses may have actually stemmed from the very early and classical observations of Bolk (1913) and Jaensch (1977), who regarded calvarial suture synostosis as a secondary event when they observed brachiocephalic and dolichocephalic skulls with patent calvarial sutures. Moss (1958) furthered this hypothesis that calvarial suture synostosis may be an extrinsically influenced event rather than a primary etiologic or causal factor. Moreover, this condition was thought to involve a lack of synchrony between the morphogenesis of the early embryonic tissue capsule surrounding the brain and the set of several discrete mesenchymal fiber tracts spreading laterally from the midline skull base as reported later by Blechschmidt (1976) and Spyropoulos and Burdi (1988).

It was later hypothesized that cranial sutures are responsive to the intrinsic tension produced by these fiber tracts, and the combination of both calvarial suture synostosis and dysmorphogenic events in the developing cranial base (i.e., the chondrocranium) may be involved as causal factors in the craniosynostoses (Moss, 1959, 1975). While such calvarial or dural stretch (tension) fields have been hypothesized (Moss and Young, 1960) as primary causal factors in the normal and abnormal morphogenesis of the human craniofacial skeleton (including the cranial base), other evidence suggests that cerebral dysmorphologies in the rabbit model are probably compensatory, secondary (postsynostotic) events and not primary causal factors in the pathogenesis of the craniosynostoses—that is, coronal suture synostosis (Cooper et al., 1999). The assignment of a primary causal role of prematurely synostosed calvarial sutures manifesting in the craniosynostosis phenotype had received even further questioning in the report of an Apert (acrocephalosyndactyly) fetus with patent calvarial—that is, coronal—sutures (Stewart et al., 1977). Actually, the basic observation of that report on the Apert fetus had some support in a study done many years earlier that speculated on the pathogenesis of Apert syndrome and theorized that the primary defect was in the mesenchymal blastema, and that it resulted in craniosynostosis and an abnormal cranial base (Park and Powers, 1920).

This continued shifting of this etiologic focus away from premature sutural synostoses toward a dysmorphology in the earlier developing cranial base was advanced further by subsequent human developmental studies. Those descriptive embryologic studies showed a spatial continuity between the coronal sutures and the midline cranial base, specifically at the sphenoethmoidal articulation. That structural linkage between the coronal sutures and the sphenoethmoidal synchondrosis

was hypothesized as the coronal ring (Venes and Burdi, 1985; Burdi et al., 1986) (see Chapter 11). In essence, this coronal ring is a closed ring involving the right and left coronal sutures that extend downward through the infratemporal fossae toward the middle cranial base at a site between the right and left optic foramina that is filled with proliferative cartilage, forming the sphenoethmoidal synchondrosis. In this coronal ring paradigm, the coronal suture segments of the ring have been associated chiefly with changes (either pre- or post-surgically) in size and shape of the calvarium, whereas the cartilaginous sphenoethmoidal synchondrosis within the midcranial base segment of the coronal ring was assigned a role much like that of an epiphyseal growth plate that contributes to increases in the length of growing bones. As such, it is the sphenoethmoidal synchondrosis that contributes chiefly to the anteroposterior and forward growth of the upper face relative to the anterior cranial base.

While the spheno-occipital synchondrosis for many years has been assigned (almost dogmatically so by orthodontists) the role of the growth plate responsible for increases in length of the entire cranial base from nasion to sella to basion, in actuality it may be the sphenoethmoidal synchondrosis rather than the spheno-occipital synchondrosis that is the growth plate for the anteroposterior of the midline cranial base up to the period of the mixed dentition (Enlow and Hans, 1996). Unlike the spheno-occipital synchondrosis, which typically fuses at 17 to 19 years of age, the sphenoethmoidal synchondrosis typically ossifies at approximately 6 to 7 years of age, with earlier closures in the Apert and Pfeiffer conditions (Taccone et al., 1989). The different growth functions hypothesized for the coronal sutures and the sphenoethmoidal synchondrosis may help explain why the characteristic Apert facies may be seen with patent coronal sutures

(Stewart et al., 1977). That is, unlike the prevailing dogma that assigns the pathogenesis of the craniosynostoses to the premature fusion of the calvarial sutures, the coronal ring paradigm points instead to varying degrees of premature synostosis of the sphenoethmoidal synchondrosis in the midline cranial base.

Fused portions of the coronal suture system may be opened through surgery in the craniosynostoses, resulting in attainment of acceptable calvarial shape and size. Yet surgical outcomes following reopening of fused calvarial sutures have not always led to the forward repositioning of the upper face. It follows, then, if the site of pathogenesis lies with the prematurely synostosed sphenoethmoidal articulation, and that region is technically beyond the field of surgery, according to the coronal ring paradigm little or no effects on upper facial growth and forward positioning might be expected from or attributed to the surgical intervention (Marchac and Renier, 1985; Kreiborg and Aduss, 1986). Moreover, a prematurely synostosed sphenoethmoidal synchondrosis that is uncorrected by craniofacial surgery would not untether the midface from its expected forward and downward repositioning from beneath the anterior cranial base. In essence, the degree of premature synostosis at, or close to, the deeply placed sphenoethmoidal synchondrosis and the severity of the characteristic facies in growing patients with Crouzon, Apert, and Pfeiffer syndromes can be associated with the form and position of the midfacial region relative to the retropositioned anterior cranial base (Posnick and Ruiz, 2000).

Recognizing pathogenesis as an outcome of expression of fundamental etiologic antecedents, and in terms of pathogenesis, the prematurely synostosed sphenoethmoidal synchondrosis has been associated with the typical craniosynostosis phenotype. However, the expanding database from craniofacial molecular biology may shed even further light on the question of whether the sphenoethmoidal synchondrosis was ever developed within the embryonic chondrocranium in the craniosynostoses. This thought of an actual failure of the sphenoethmoidal synchondrosis to be morphogenically mapped out very early in mammalian embryogenesis is not a new thought. It does rekindle a hypothesis that surfaced early in the 20th century speculating that the primary structural defect, or pathogenesis, was in the embryonic mesenchymal blastema, resulting in the fusion of the calvarial sutures and in abnormalities of the cranial base (Park and Powers, 1920). This hypothesis was subsequently supported by Cohen (1993). At the cell level, osteoprogenitor cells have be identified as the predominantly affected cell type in fused sutures (Engstrom et al., 1988).

A growing spectrum of growth factors have been suggested as determinants of both osteogenesis and suturogenesis. The bigger issue, however, may well be the genes, or combination of genes, and their expressed proteins that map out not only the cranial base synchondroses but also the various calvarial sutures. The full understanding of etiologic factors for the craniosynostoses at cell and molecular levels remains elusive. As more information emerges from studies on the molecular biology of normal and abnormal skull morphogenesis (i.e., sutures, synchondroses), increased opportunities should also emerge on a range of interventions at molecular and cellular levels in the treatment and management of the craniosynostoses.

Even though a large information chasm exists between our molecular understanding of morphogenesis and the ability to transfer that understanding to the clinical setting, a brief overview of the molecular biology of the craniosynostoses is appropriate (see also Chapters 4, 5, 6, 9, and 20). Whether the sites of pathogenesis in the craniosynostoses are along specific sutural junctions, within the cranial base, or both, recent scientific advances and resultant new discoveries focus on etiological events that are occurring at the genic and molecular levels of morphogenesis (Sperber, 1999). While the molecular and gene levels associated with skull morphogenesis are covered in more detail elsewhere in this book, it serves us well to have at least a current "snapshot" of craniofacial molecular biology pertinent to the theme of this chapter. In the world of the craniosynostoses, one of the most exciting advances in molecular biology and genetics is the discovery of the role of the fibroblast growth factors (FGFs) and fibroblast growth factor receptors (FGFRs) in furthering the understanding of skeletal dysplasias such as the craniosynostoses (Malcolm and Reardon, 1996; Gorlin, 1997). In essence, there are about nine specific genes in the FGF family and four receptors. Together they characterize or regulate basic cellular events, including proliferation, differentiation, and migration.

Important cellular signaling is mediated by cell surface receptors, which have been identified as transmembrane tyrosine kinase receptors. Each of these receptors has been linked with three extracellular domains, a transmembrane segment, and a cytoplasmic tyrosine domain. It is important to note that craniofacial FGFR1 and FGFR2 are coexpressed in prebone and precartilage regions of the developing craniofacial skeleton. This FGFR1–FGFR2 pattern is distinct from FGFR3, which is expressed in the epiphyseal cartilage plates in long bones and in the hypochondroplasia phenotype. Mutations in craniofacial FGFR receptors, which often involve only a single amino acid substitution, have been linked to specific types of craniosynostoses. Specifically, FGFR1 (on chromosome 8cen) is etiologically associated with the conditions of Pfeiffer, while the FGFR2 gene (on chromosome 10q25) is linked with Apert and Crouzon syndromes (Muenke and Schell, 1995).

The specific location of the gene that is targeted by a mutation determines how the structure of the FGFR protein is altered, which in turn influences the spectrum of defects associated with the specific kind of craniosynostosis (NIDCR Web page: http://www.nidcr.nih.gov/cranio/home.htlm). However, the genetic etiologies underlying similar phenotypes within the clinical family of craniosynostoses can differ. For example, in the craniosynostotic Saethre-Chotzen syndrome, whose features appear similar to those in Crouzon, a different gene, called TWIST, may be the causal gene having effects on the regulation of the FGFR genes. In another form of craniosynostosis, called the Boston type, the causal gene is the MSX2 gene, which is a member of the homeobox family of genes. This family of master control genes codes for proteins that bind to DNA and direct the activity of still other genes to determine the pattern in which embryos develop, including complex structures of the head. In the case of the autosomal-dominant Boston-type craniosynostosis, a single misstatement in the MSX2 homeobox gene has been associated with premature closure of one or more calvarial sutures (Jabs et al., 1993). Exactly how this mutation might be translated into premature fusion of calvarial sutures and cranial base synchondrosis closure is still not entirely known.

In terms of field patterning in very early mammalian embryogenesis, there is evidence that shows the mammalian fibroblast growth factor (FGF) family can be expressed in specific spatial and temporal manner in embryo and adult vertebrate skeletogenesis, presumably as selective determinants of joint morphogenesis (e.g., at sutures and synchondroses) in the mammalian skull (Lewanda et al., 1996; Plotnikov et al., 1999). While much of this emerging picture is on the molecular biology of calvarial sutures, such information can and should provide us with important information and approaches to our understanding of the molecular determinants of normal and the range of cranial base dysmorphogenesis observed both experimentally and clinically in the craniosynostoses (Liu et al., 1995; Wilkie, 1997).

12.3 CRANIAL BASE GROWTH PATTERNS IN CRANIOSYNOSTOSIS

Keeping with the theme of this chapter, and with this snapshot of the exciting world of craniofacial molecular biology in our vision, and to put the cranial base growth in proper perspective, it may be helpful to see the cranial base as a morphologic collage (at cell and tissue levels) in the development and growth of several contiguous bones and surrounding regions in the growing child. The overall length of the entire brain case at birth is about 63% of its expected adult total length. At 1 year, its overall length is approximately 82%; 89% at 3 years; and approximately 91% at 5 years. The anterior cranial base (basion-nasion) attains about 56% of its adult length by birth, and 70% by year 2.

With regard to cranial base width changes, at birth the width of the skull base is approximately 100 mm. At 6 months, base width is 150 mm; and 170 mm at 12 months. Thereafter, changes in cranial base width drop off to about 0.5–1.0 mm per year, from 3 to 14 years. Much of the growth in cranial base width and length in the late prenate, newborn, and child is associated with increases in brain size. In general, brain weight at birth is approximately 50% that of the adult brain. By the third year, brain weight is 80% of adult brain weight, and 90% at 5 to 8 years of age (Enlow and Hans, 1996).

The preceding descriptions of normal age-related growth in length and width of the cranial base provide a generalized human picture. Differences do occur, however, according to gender and racial variations—that is, population polymorphisms. For example, base length in young adults (18–25 years) is approximately 199 mm in blacks, 197 mm in Caucasians, and approximately 167 mm in Asians. In each of these three population groups (age-matched for group), male values are larger than female values (van der Linden, 1986; Farkas, 1994).

This picture of postnatal skull growth patterns is not without its prenatal antecedents. It has been postulated (Ford, 1958) that overall growth of the cranial base (the so-called hafting zone between the brain case and face) is influenced by growth of the cranium above and by growth of the upper facial skeleton below. The individual segments of the cranial base, however, are associated with either neural or general skeletal growth patterns. That is, the cranial base segment between sella and the foramen cecum is associated with neural growth; whereas the base segments between nasion and the foramen cecum and the segment between sella and basion follow the pattern of general skeletal growth that is associated with upper facial growth. Clinical cephalometric studies have shown that the anterior and posterior cranial base segments do not exhibit similar growth patterns, wherein the anterior base segment completes its anteroposterior growth several years before that of the posterior cranial base (Stramrud, 1959). The greater period of posterior cranial base growth (sella to basion) has been associated with increased activity at the spheno-occipital synchondrosis. Differences in anteroposterior growth of the anterior and posterior cranial base segments have also been described for second- and third-trimester human prenates (Burdi, 1965, 1969).

Let us turn now to another morphologic feature of cranial base, conventionally described as the cranial base angle. It is measured in the sagittal plane as a line connecting the cephalometric landmarks nasion to sella to basion (N-S-Ba). Changes in the values of this angle can occur over time and are initially associated with normal changes in shape (or flexures) of the developing brain, and subsequently to growth of the anterior and middle cranial fossae. For example, at about the 4th embryonic week, the human cranial base angle is about 150°, with reductions to 128° at 8 weeks, 120° at 10 weeks, and then increases to between 125° and 130° at about 15 weeks, which is sustained postnatally (Burdi, 1965; Enlow and McNamara, 1973; Riolo et al., 1974).

The character of the cranial base angle in the craniosynostoses is less clear in the craniosynostoses. Kreiborg and Bjork (1982) reported the N-S-Ba angle in a dry Crouzon skull to be quite acute (118°). It also has been observed (Kreiborg and Pruzansky, 1981; Kreiborg and Aduss, 1986) that there can be a greater variability in the cranial base angle in Crouzon, which showed a platybasia flattening or form (152°, mean at 131°), while the angle in Apert cases showed both a basilar kyphosis (range 100.5–119°) and, in some cases, a platybasia form (range 140.5–149.5°). This range of variation was attributed to increased intracranial pressure and a resultant downward displacement of the sphenoid bone. Whatever may be the shape of the cranial base angle in the various forms of craniosynostoses, it is important to keep the size of this angle in the context of surrounding regions such as the pharynx (Burdi, 1976). In a classical study, Shprintzen (1982) observed that an acute cranial base angle is associated with a forward displacement of the cranial base, which effectively shortens the anteroposterior dimension of the pharynx. Such foreshortening of the cranial base has been associated with a nearly vertical orientation of the pharynx itself. The forward displacement of the cranial base also causes

an apparent vertical lengthening of the pharyngeal airway but an anteroposterior shortening of both the nasal and oral airways.

The range of severity and variance in head (i.e., skull shape) generally seen in the craniosynostoses (either syndromic or non-syndromic types) precludes a stereotypic definition or phenotypic pattern of the conditions. This is not entirely unexpected, since the number and combination of affected calvarial sutures, and the extent to which each suture is synostosed, will differ from patient to patient despite similarity in diagnosis (Bertelsen, 1958). As noted earlier, the cranial base has been considered a developmental template upon which the face and pharyngeal regions develop, and the topographic and dimensional characteristics of the human cranial base can determine certain corresponding characteristics of the face, especially those of the midface (Enlow and Azuma, 1975; Burdi, 1976).

Experimental studies of sutural closure in growing rabbits, as noted earlier in this chapter, have also shown that premature closure of single or multiple calvarial sutures is associated with significant alterations in both the angular and linear dimensions of the cranial base (Babler and Persing, 1982). In some cases, bones of the cranial base are severely affected, further enhancing the severity of the craniosynostosis phenotype (Tessier, 1971; Pruzansky, 1977). Clearly, the outward appearance of the craniosynostotic phenotype is best understood with a summation of changing size and shape in the several contiguous regions of the calvarium, cranial base, and associated soft tissues (Cohen, 1975). Let us see how this fundamental tenet plays out in this chapter, whose focus, by design, deals with growth and morphogenesis of the cranial base in the craniosynostoses.

In Pfeiffer and Apert conditions, skulls demonstrate the typical features of premature fusion of the calvarial sutures with increased skull height, flattening of the cranial base, and a shortened or brachycephalic fronto-occipital diameter, basilar kyphosis, and shallow orbits with associated exorbitism. Also prominent are marked upward slants to the cranial base, of the lesser sphenoid wings, a depressed and widened sella turcica, with concomitant reductions in size of the anterior and posterior cranial fossae (Escobar and Bixler, 1977). Both increased angular and decreased linear changes in the cranial base appear to be consistent observations associated with experimentally produced craniosynostosis and craniosynostoses occurring naturally (Bertelsen, 1958; Moss, 1959; Murray and Swanson, 1968; Kreiborg and Pruzansky, 1981; Ousterhout and Melsen, 1982; Cohen, 1986). Any comparisons between the features seen in the human craniosynostoses and similar observations in experimental animal studies must be mindful of species-specific differences and differing patterns of craniofacial growth.

In experimental studies on the effects of suture closure on cranial base form (i.e., size and shape) in rabbits, observations similar to those observed in humans were reported, including a flattening of the cranial base, and a progressive and significant reduction in anteroposterior cranial base dimensions following premature multiple calvarial synostoses (Babler, 1988). In humans, premature suture synostosis (single or multiple calvarial sutures) has been linked, if not almost dogmatically, with the craniosynostosis phenotype. There may be premature unilateral synostosis of the coronal suture (anterior plagiocephaly) or bilateral synostoses (brachycephaly). If both coronal sutures prematurely fuse, the lateral dimensions of the skull appear widened with some concomitant bilateral displacement of the orbital rims (Williams et al., 1999). As an example, in 75% of Crouzon patients with premature synostosis, a combination of the coronal, sagittal,

and lambdoidal sutural synostoses was observed. While the actual shape of the calvarium can vary in such patients, typically the calvarium is short and widened in the temporal regions (Cohen, 1986). Similarly, in Crouzon craniosynostosis, there is a reduction in the length of the anterior cranial base and clivus (Tessier, 1971; Kreiborg and Pruzansky, 1981).

In general, the association of secondary effects with primary suture closure does not preclude a focus on the cranial base origin of deformities associated with the craniosynostosis phenotype. In several classical early studies, secondary alterations in the shape of the cranial base have been reported in studies of artificial deformation of human skulls (Moss, 1958; Bennett, 1967). However, with regard to the full understanding of the craniosynostoses, it still remains unclear whether primary and secondary effects should be assigned to naturally occurring or experimentally induced malformations of either the calvarial sutures or the cranial base in the craniosynostoses.

12.4 CONCLUSIONS

Craniofacial malformations associated with the spectrum of syndromic and nonsyndromic families of craniosynostoses have intrigued biologists and clinicians seemingly for ages. While much attention has been given to the etiology, pathogenesis, phenotypes, and treatments of the craniosynostoses, it still remains difficult, and appropriately so, to approach, understand, and treat as isolates the calvarium, cranial base, and facial regions. With the progressive emergence of newer and more precise technologies (e.g., in developmental genetics, imaging techniques), there continues to be a literal explosion in the biological understanding of the craniosynostoses.

Without doubt, the plethora of exciting advances in craniofacial molecular biology continues to expand our knowledge base on the understanding of the craniosynostoses. While there is understandable excitement about the essentiality of knowing precisely the molecular determinants of craniofacial form, especially with the craniosynostoses, we must be mindful that it is contraindicated to withhold or delay treatment until such etiologic information emerging from the world of craniofacial molecular biology becomes available. Moreover, with the advent of these newer and more precise technologies, dysmorphologists, radiologists, orthodontists, and craniofacial surgeons will be positioned even better to see, understand, and deal more effectively with the multifactorial nature of the craniosynostoses, in general, and, more specifically, with their concomitant effects on the cranial base.

REFERENCES

Anderson, F. M., and Geiger, L. (1965). Craniosynostosis: a survey of 204 cases. *J. Neurosurg. 22*, 229–270.

Babler, W. J. (1988). Effects of multiple suture closure on craniofacial growth in rabbits. In K. W. L. Vig and A. R. Burdi (eds.), *Craniofacial Morphogenesis and Dysmorphogenesis*. Monograph 21: Craniofacial Growth Series; Center for Human Growth and Development (Ann Arbor: The University of Michigan), pp. 76–89.

Babler, W. J., and Persing, J. A. (1982). Experimental alteration of cranial suture growth: effects on the neurocranium, basicranium, and midface. In A. D. Dixon and B. G. Sarnat (eds.), *Factors and Mechanisms Influencing Bone Growth* (New York: Alan R. Liss, Inc.), pp. 333–345.

Babler, W. J., and Persing, J. A. (1984). Experimental coronal synostosis: effects of surgical release on facial growth. *J. Dent. Res. 63*, 315–334.

Bennett, K. A. (1967). Craniostenosis: a review of the etiology and report of new cases. *Am. J. Phys. Anthrop. 27*, 1–10.

Bertelsen, T. I. (1958). The premature synostosis of the cranial sutures. *Acta Ophthalmol. 51* (suppl.), 1–176.

Blechschmidt, M. (1976). Biokinetics of the developing basicranium. In J. F. Bosma (ed.), *Symposium on Development of the Basicranium* (Bethesda: NIH:DHEW Publication 76-989), pp. 44–53.

Bolk, L. (1913). On the premature obliteration of sutures in the human skull. *Am. J. Anat. 17*, 495–523.

Burdi, A. R. (1965). Sagittal growth of the nasomaxillary complex during the second trimester of human prenatal development. *J. Dent. Res. 44*, 112–125.

Burdi, A. R. (1969). Cephalometric growth analysis of the human upper face during the last two semesters of gestation. *Am. J. Anat. 125*, 113–122.

Burdi, A. R. (1976). Early development of the human basicranium: its morphogenic controls, growth patterns, and relations. In J. F. Bosma (ed.), *Symposium on Development of the Basicranium* (Bethesda: NIH:DHEW Publication 76-989), pp. 81–92.

Burdi, A. R., Kusnetz, A. B., Venes, J. L., and Gebarski, S. S. (1986). The natural history and pathogenesis of the cranial coronal ring articulations: implications in understanding the pathogenesis of the Crouzon craniostenotic defects. *Cleft Palate J. 23*, 28–39.

Burrows, A. M., Richtsmeier, J. T., Mooney, M. P., Smith, T. D., Losken, H. W., and Siegel, M. I. (1999). Three-dimensional analysis of craniofacial form in a familial rabbit model of nonsyndromic coronal suture synostosis using Euclidean distance matrix analysis. *Cleft Palate Craniofac. J. 36*, 196–206.

Cohen, M. M., Jr. (1975). An etiologic and nosologic overview of the craniosynostosis syndromes. In D. Bergsma (ed.), *Birth Defects: Original Article Series. Vol. XI, No. 2* (New York: Alan R. Liss, Inc.).

Cohen, M. M., Jr. (1986). The etiology of craniosynostosis. In M. M. Cohen, Jr. (ed.), *Craniosynostosis, Diagnosis, Evaluation and Management* (New York: Raven Press), pp. 59–80.

Cohen, M. M., Jr. (1993). Sutural biology and the correlates of craniosynostosis. *Am. J. Med. Genet. 47*, 581–616.

Cooper, G. M., Mooney, M. P., Burrows, A. M., Smith, T. D., Dechant, J., Losken, H. W., Marsh, J. L., and Siegel, M. I. (1999). Brain growth rates in craniosynostotic rabbits. *Cleft Palate Craniofac. J. 36*, 196–206.

David, J. D., Poswillo, D., and Simpson, D. (1982). *The Craniosynostoses: Causes, Natural History and Management* (Berlin and New York: Springer-Verlag).

Engstrom, C., Weregeda, J. E., and Engstrom, H. (1988). Characterization of cells isolated from synostotic suture and skull bone from neonatal humans. *J. Dent. Res. 67*, 115.

Enlow, D. H. and Azuma, M. (1975). Functional growth boundaries in the human and mammalian face. In D. Bergsma (ed.), *Birth Defects: Original Article Series. Vol. XI, No. 7* (New York: Alan R. Liss, Inc.).

Enlow, D. H., and Hans, M. G. (1996). *Essentials of Facial Growth* (Philadelphia: W. B. Saunders).

Enlow, D. H., and McNamara, J. A. (1973). The neurocranial basis for facial form and pattern. *Angle Orthod. 43*, 256–262.

Escobar, V., and Bixler, D. (1977). Are the acrocephalosyndactyly syndromes variable expressions of a single gene defect? In *Birth Defects: Original Article Series. Vol. XIII, No. 3C* (New York: The National Foundation), pp. 139–154.

Farkas, L. G. (1994). *Anthropometry of the Head and Face*, 2nd ed. (New York: Raven Press).

Foltz, E. L., and Loeser, J. D. (1975). Craniosynostosis. *J. Neurosurg. 43*, 48–57.

Ford, E. H. R. (1958). Growth of the human cranial base. *Am. J. Orthod. 44*, 498–506.

Gorlin, R. J. (1997). Fibroblast growth factors, their receptors and receptor disorders. *J. Craniofac. Surg. 25*, 69–79.

Gorlin, R. J., Pindborg, J. J., and Cohen, M. M., Jr. (1976). *Syndromes of the Head and Neck*, 2nd ed. (New York: McGraw-Hill).

Gorlin, R. J., Cohen, M. M., Jr., and Hannekam, R. C. M. (2001). *Syndromes of the Head and Neck*, 3rd ed. (London: Oxford University Press).

Hemple, D. J., Harris, L. E., and Svien, H. J. (1961). Craniosynostosis involving the sagittal suture only. Guilty by association? *J. Pediatr. 58*, 342–355.

Hoyte, D. A. N. (1989). The role of the cranial base in normal and abnormal skull development. In J. A. Persing, M. T. Edgerton, and J. A. Jane (eds.), *Scientific Foundations and Surgical Treatment of Craniosynostosis* (Baltimore: Williams and Wilkins), pp. 58–76.

Jabs, E. W., Muller, U., Li, X., Luo, W., Haworth, I. S., Klisak, I., Sparkes, R., Warman, M. L., Mulliken, J. B., Snead, M. L., and Maxson, R. E. (1993). A mutation in the homeodomain of the human Msx2 gene in a family affected with autosomal dominant craniosynsotosis. *Cell 75*, 443–450.

Jaensch, P. A. (1977). Schadelsynostosch. *Zeitschrift Med. 7*, 426–432.

Jones, K. L. (1988). *Smith's Recognizable Patterns of Human Malformations* (Philadelphia: W. B. Saunders).

Kreiborg, S., and Aduss, H. (1986). Pre- and postsurgical facial growth in patients with Crouzon's and Apert's syndromes. *Cleft Palate J. 23*, 78–91.

Kreiborg, S., and Bjork, A. (1982). Description of a dry skull with Crouzon syndrome. *Scand. J. Plast. Reconstr. Surg. 16*, 245–253.

Kreiborg, S., and Pruzansky, S. (1981). Craniofacial growth in premature craniofacial synostosis. *Scand. J. Plast. Reconstr. Surg. 15*, 171–186.

Kreiborg, S., Marsh, J. L., Cohen, M. M., Jr., Liversage, M., Pedersen, M., Skovby, F., Borgesen, S. E., and Vannier, M. W. (1993). Comparative three-dimensional analysis of CT-scans of the calvaria and the cranial base in Apert and Crouzon syndromes. *J. Craniomaxillofac. Surg. 21*, 181–188.

Lewanda, A. F., Meyers, G. A., and Jabs, E. W. (1996). Craniosynostosis and skeletal dysplasias: fibroblast growth factor receptor defects. *Proc. Assoc. Am. Phys. 108*, 19–24.

Liu, Y. H., Ramendra, K., Wu, L., Luo, W., Ignelzi, M., and Snead, M. (1995). Premature suture closing and ectopic cranial bone in mice expressing MSX2 transgenes in the developing skull. *Proc. Natl. Acad. Sci. USA. 92*, 6137–6141.

Malcolm, S., and Reardon, W. (1996). Fibroblast growth factor receptor-2 mutations in craniosynostosis. *Ann. NY Acad. Sci. 785*, 164–170.

Marchac, D., and Renier, D. (1985). Craniofacial surgery for craniosynostosis improves facial growth. *Ann. Plast. Surg. 14*, 43–54.

Marsh, J. L., and Vannier, M. W. (1986). Cranial base changes following surgical treatment of craniosynostosis. *Cleft Palate J. 23*, 9–18.

Moss, M. L. (1958). The pathogenesis of artificial cranial deformation. *Am. J. Phys. Anthropol. 16*, 269–286.

Moss, M. L. (1959). The pathogenesis of premature cranial synostosis in man. *Acta Anat. 37*, 351–370.

Moss, M. L. (1975). Functional anatomy of cranial synostosis. *Child's Brain, 1*, 22–33.

Moss, M. L., and Young, R. W. (1960). A functional approach to craniology. *Am. J. Phys. Anthropol. 18*, 281–292.

Muenke, M., and Schell, U. (1995). Fibroblast growth factor receptor mutations in human skeletal disorders. *Trends Genet. 11*, 308–313.

Murray, J. E., and Swanson, L. T. (1968). Midface osteotomy and advancement for craniosynostosis. *Plast. Reconstr. Surg. 41*, 299–306.

Ousterhout, D. K., and Melsen, B. (1982). Cranial base deformity in Apert's syndrome. *Plast. Reconstr. Surg. 69*, 254–263.

Park, E. A., and Powers, G. F. (1920). Acrocephaly and scaphocephaly with symmetrically distributed malformations of the extremities. *Am. J. Dis. Child 20*, 235–315.

Plotnikov, A. N., Schlessinger, J., Hubbard, S. R., and Mohammadi, M. (1999). Structural basis for FCF receptor dimerization and activation. *Cell 98*, 641–650.

Posnick, J., and Ruiz, R. L. (2000). The craniofacial dysostosis syndromes: current surgical thinking and future directions. *Cleft Palate Craniofac. J. 37*, 433.

Pruzansky, S. (1977). Time: the fourth dimension in syndrome analysis applied to craniofacial malformations. In *Birth Defects: Original Article Series. Vol. XIII, No. 3C* (New York: The National Foundation), pp. 3–28.

Riolo, M. L., Moyers, R. E., McNamara, J. A., and Hunter, W. S. (1974). *An Atlas of Craniofacial Growth: Cephalometric Standards from the University School Growth Study, Craniofacial Growth Series, Monograph 2* (Ann Arbor: The University of Michigan Center for Human Growth and Development).

Shprintzen, R. J. (1982). Palatal and pharyngeal anomalies in craniofacial syndromes. In C. F. Salinas (ed.), *Craniofacial Anomalies: New Perspectives. Birth Defects: Original Article Series. Vol. 18, No. 1.* March of Dimes Birth Defects Foundation (New York: Alan R. Liss, Inc.), pp. 53–78.

Simmons, D. R., and Peyton, W. T. (1947). Premature closure of the cranial sutures. *J. Pediatr. 31*, 528–542.

Smith, D. W. (1976). *Recognizable Patterns of Human Malformation: Genetic, Embryologic and Clinical Aspects*, 2nd ed. (Philadelphia: W. B. Saunders Co.).

Smith, T. D., Mooney, M. P., Burrows, A. M., Losken, H. W., and Siegel, M. I. (1996). Postnatal changes in the cranial base in rabbits with congenital coronal suture synostosis. *J. Craniofac. Genet. Devel. Biol. 16*, 107–117.

Sperber, G. H. (1999). Pathogenesis and morphogenesis of craniofacial developmental anomalies. *Ann. Acad. Med. Singapore 28*, 708–713.

Spyropoulos, M., and Burdi, A. R. (1988). Morphogenic fields in craniofacial biology. In K. W. L. Vig and A. R. Burdi (eds.), *Craniofacial Morphogenesis and Dysmorphogenesis.* Monograph 21: Craniofacial Growth Series; Center for Human Growth and Development (Ann Arbor: The University of Michigan), pp. 141–148.

Stewart, R. E., Dixon, G., and Cohen, A. (1977). The pathogenesis of premature craniosynostosis in acrocephalosyndactyly (Apert's syndrome). *Plast. Reconstr. Surg. 59*, 699–707.

Stramrud, L. (1959). External and internal cranial base. A cross-sectional study of growth and association in form. *Acta Odont. Scand. 17*, 239–266.

Taccone, A., Cama, A., Ghiorizi, A., and Fondelli, P. (1989). Diagnosis and treatment of craniosynostoses: the usefulness of CT combined with 3-dimensional reconstruction. *Radiol. Med. 77*, 322–328.

Tessier, P. (1971). The definitive plastic surgical treatment of the severe facial deformities of craniofacial sysostosis: Crouzon and Apert diseases. *Plast. Reconstr. Surg. 48*, 419–442.

van der Linden, F. P. G. M. (1986). *Facial Growth and Facial Orthopedics* (Chicago: Quintessence Publ.).

Venes, J., and Burdi, A. R. (1985). Proposed role of the orbitosphenoid in craniofacial dysostosis. In R. Humphreys (ed.), *Concepts in Pediatric Neurosurgery*, vol. 5 (Basil: Karger Press), pp. 126–135.

Virchow, R. (1851). Uber den Cretinusmus, namentlich in Franken, und uber pathologische schadelformen. *Verh. Phys. Med. Gesellsch, Wurzburg 2*, 231–284.

Warkany, J. (1971). *Congenital Malformations* (Chicago: Year Book Medical Publ.).

Wilkie, A. O. (1997). Craniosynostosis: genes and mechanisms. *Hum. Molec. Genet. 6*, 1647–1656.

Williams, J., Ellenbogen, R. G., and Gruss, J. S. (1999). State of the art in craniofacial surgery: nonsyndromic synostosis. *Cleft Palate Craniofac. J. 36*, 471–485.

CHAPTER 13

CRANIAL BASE DYSMORPHOLOGY AND GROWTH IN FACIAL CLEFTING

TIMOTHY D. SMITH, Ph.D., MARK P. MOONEY, Ph.D., ANNIE M. BURROWS, Ph.D., and MICHAEL I. SIEGEL, Ph.D.

13.1 INTRODUCTION

Facial clefting can be associated with a host of regional or global craniofacial dysmorphologies. For instance, both midfacial and overall cranial reduction in length are typical in patients with cleft lip and palate (CLP) (Ross and Johnston, 1978). Therefore, investigators of the pathogenesis of facial clefting, especially CLP, have not always restricted their study to palatal or even midfacial development. Among craniofacial regions, the cranial base has presented a topic of controversy with regard to its influence on, or relationship to, the midfacial deformities (Ross and Johnston, 1978; Harris, 1993).

Several important issues revolve around the relationship between basicranial and midfacial development in cases of facial clefting (Ross and Johnston, 1978; Harris, 1993; Mølsted et al., 1995). First, depending on basicranial involvement, clefting may be regarded as an isolated defect of the midface or as part of an overall craniofacial growth disturbance. If the cranial base is indeed abnormal in some types of facial clefts, another question is how these abnor-

malities may affect the midface. Finally, the degree of alterations in development of the cranial base has profound importance on studies that assume normality of the cranial base in different types of clefting.

Conflicting results have continued, and the degree to which such varied findings depend on methodology, ontogeny, or the varied presentation of a multifactorial disorder remains unclear. This chapter reviews reports on prenatal and postnatal aspects of cranial base development in cases of cleft lip (CL), cleft palate (CP), or CLP.

13.2 PRENATAL DEVELOPMENT OF THE CRANIAL BASE IN CLEFT SAMPLES

In the process of embryonic palatogenesis, the midfacial region exhibits a merging of midline structures with more lateral portions of the developing face. Later (fetal) development of the cranial base and midface shows certain interdependencies. For instance, the nasal septum has a direct continuity with the midline cranial base via

Understanding Craniofacial Anomalies: The Etiopathogenesis of Craniosynostoses and Facial Clefting, Edited by Mark P. Mooney and Michael I. Siegel, ISBN 0-471-38724-x   Copyright © 2002 by Wiley-Liss, Inc.

the mesethmoidal cartilage (Sperber, 2001) (Chapter 4). Therefore, positioning of the cranial base has a direct bearing on midfacial form. Because of these complex changes in facial form, studies on postnatal cleft patients have addressed growth changes in sagittal, transverse, or frontal planes. Studies on prenatal cleft samples have addressed such aspects using primarily histological or radiographic data, but are sometimes limited in the latter due to the extent of ossification.

Few studies have directly addressed differences in cleft versus noncleft basicrania. The majority of studies on prenatal cleft samples have addressed midfacial growth and morphogenesis. Qualitative and quantitative reports have described prenatal midfacial structures in different ways: either as deficient or hypertrophic (Avery et al., 1957; Mooney et al., 1994; Smith et al., 1996). Whereas such descriptions appear to conflict, this may be partly attributed to differences in growth rate between various osseous or cartilaginous structures (Siegel et al., 1987; Mooney et al., 1991). One suggested consequence of heightened growth rate of certain midline structures in the midface of CLP fetuses is nasal airway constriction (Siegel et al., 1987; Mooney et al., 1994).

Volumetric examination of basicranial structures has not yet been accomplished using cleft samples, but some qualitative observations offer insight. The sphenoidal sinuses are mediolaterally compressed in fetuses with CLP compared to noncleft fetuses, even to the extent of unilateral or bilateral absence (Smith et al., 1997, 1999). Smith and colleagues (1997) attributed such differences to the hypertrophic adjacent mesethmoid cartilages of some specimens. Developmentally, the sphenoidal sinuses are a posterior constriction of the nasal fossae, rather than true pneumatizations as seen in other sinuses (Sperber, 2001). Therefore, Smith and associates (1999) considered the apparent reduction

of the sphenoidal sinuses (quantitative comparisons were impractical) to be similar to that seen in the nasal cavity (Siegel et al., 1987), possibly produced by a similar mechanism: deranged (accelerated) fetal growth of cartilage (Siegel et al., 1987; Mooney et al., 1994).

Mølsted and co-workers (1995) speculated that the chondrocranium may be deficient during palatogenesis, which is consistent with at least some prenatal observations. In a mouse model of 6-aminonicotinamide-induced CP, evidence for decreased basicranial hyperplasia was described in fetuses with cleft palate (Long et al., 1973). This may have translated positional abnormalities to the midface (see below). It is unclear whether such cellular deficiencies occur in spontaneous clefting. Cartilaginous hypoplasia has also been described in human fetuses, especially in the septal cartilages (Avery et al., 1957; Smith et al., 1996). Smith and colleagues (1996) described both hypoplastic and hypertrophic nasal septal cartilages in human fetuses between 8 and 21 weeks postmenstrual age. Such findings may be interpreted as temporal phenomena, or as cartilaginous abnormalities that show variation in a multifactorial abnormality with a variable expression. Figure 13.1 shows coronal histological sections of the anterior (mesethmoidal) region of the cranial base in embryos and fetuses with and without CLP (University of Pittsburgh sample) (Siegel et al., 1984). It is clear that the variable state (hypoplastic or hypertrophic) of cartilage also characterizes the cranial base of at least some fetuses with CLP. In general, it appears that hypoplasia exists in younger specimens of this sample, and hypertrophism in older specimens (Smith et al., 1996). This is at least consistent with the notion that cartilaginous hypoplasia may exist during palatogenesis (Mølsted et al., 1995).

Whereas there is evidence for prenatal basicranial abnormalities in CLP, it appears

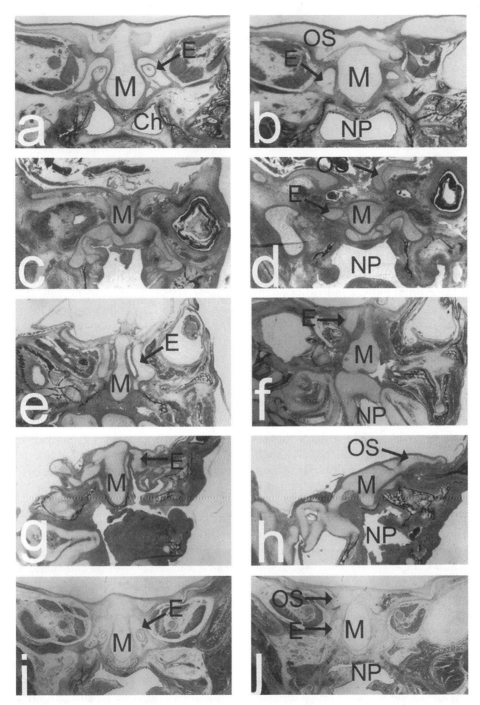

Figure 13.1 Coronal sections of embryos and fetuses with and without CLP. The left column shows the posterior end of the nasal cavity, in the region of the choanae (Ch). The right column shows the region of the nasopharynx (NP), which is slightly posterior to the level in (a). (a) and (b) show a $16\frac{1}{2}$-week (postmenstrual age) noncleft fetus (magnification ×6). The lower figures show human fetuses with clefts at 10 weeks (c, d; ×25), 12 weeks (e, f; ×25), $13\frac{1}{2}$ weeks (g, h; ×12), and $20\frac{1}{2}$ weeks (i, j; ×6). Note that the cartilages of the anterior cranial base (M, mesethmoid; E, ectethmoid; OS, orbitosphenoid) were hyoplastic, or reduced, in the 10-week fetus (c, d) compared to the noncleft fetus. In the 12- and $13\frac{1}{2}$-week fetuses, these cartilages were hypertrophic, or enlarged, compared to noncleft fetuses. In the oldest cleft fetus of this sample (i, j), these cartilages appeared relatively similar to those of noncleft fetuses.

that the cranial base of individuals with CL develops no differently than that of non-cleft populations. Sherwood and colleagues (2001) found no angular, linear, or shape differences between fetuses with CL and an age-matched sample of unaffected fetuses (Fig. 13.2).

13.3 POSTNATAL DEVELOPMENT OF THE CRANIAL BASE IN CLEFT SAMPLES

Previous studies have reported significant differences in the size and orientation (Fig. 13.3) of the cranial base of individuals with cleft lip and palate (CLP) compared to normal control individuals (Moss, 1956; Ross, 1965, 1992; Blaine, 1969; Krogman et al., 1975; Hayashi et al., 1975; Dahl et al., 1982; Grayson et al., 1987; Sandham and Cheng, 1988; Horswell and Gallup, 1992; Harris, 1993; Mølsted et al., 1993, 1995; Öztürk and Cura, 1996). However, a confusion exists regarding which aspects of the cranial base differ, and whether any such differences are biologically significant (Ross and Johnston, 1978). Since cephalometric studies of cleft samples may involve standardization with basicranial landmarks or measurements, the normality of the cranial base is a necessary assumption.

Ross and Coupe (1965) found some statistically significant differences in cranial base angles between patients with clefts and unaffected patients. These investigators also found cranial base length deficits in cleft patients, but suggested that these may best be explained as allometric differences compared to normal patients, once overall cranial length is taken into account (Ross and Johnston, 1978). Ross and Johnston (1978) disagreed with a previous report indicating angular abnormalities in children with clefts (CLP or CP) by Moss (1956), and suggested that cranial base development in CLP is "essentially normal." More recent studies have contin-

ued to conflict on the issue of cranial base normality (e.g., Mølsted et al., 1993, 1995; Krykanides et al., 2000).

Some differences in results may be related to whether the midline or more lateral regions were examined, or depending on the nature of the cleft sample (CL, CLP, CP, or pooled samples). Using lateral cephalographs, many studies have noted angular differences between cleft and noncleft groups (Ross, 1965; Blaine, 1969; Harris, 1993; Carrie et al., 2000), regardless of whether CP, CLP, or pooled samples were examined, but some conflicting reports have been made (Sandham and Cheng, 1988) (Table 13.1). Linear results are even more conflicting, at least regarding anteroposterior distances (Table 13.1). Mølsted and associates (1993) examined the spheno-occipital synchondrosis using lateral cephalographs, and found wider anteroposterior dimensions of this synchondrosis in CLP versus CL 3-month-old infants. Kyrkanides and associates (2000) found no significant difference in the degree of asymmetry in CLP and noncleft children, although it is unclear whether the authors examined cleft versus noncleft sides. The most profound differences have been described in the transverse plane. The radiographic study by Mølsted and colleagues (1995) described greater midfacial and basicranial widths, and angulation of the sphenoidal alae in CLP relative to CL children. A canine model found similar midfacial and basicranial width differences in cleft and noncleft Brittany spaniels (Richtsmeier et al., 1994).

It thus appears that some of the apparently disparate findings in studies of the cranial base are due to inherent limitations of cephalometric data sets, including those associated with registration of radiographs (Lele, 1993). It also appears that findings on corrected CLP vary more than those from uncorrected CLP (see Table 13.1). Taken together, it appears that there are angular abnormalities associated with palatal

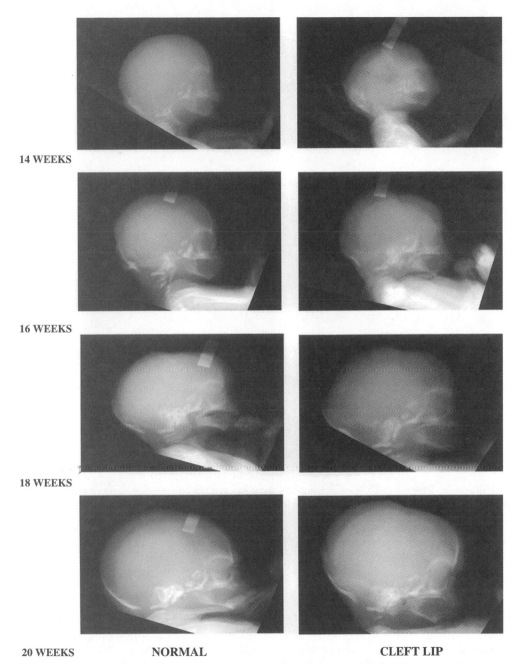

14 WEEKS

16 WEEKS

18 WEEKS

20 WEEKS　　　　　**NORMAL**　　　　　　　　　　**CLEFT LIP**

Figure 13.2　Lateral head radiographs of 14-, 16-, 18-, and 20-week normal (left) and CL fetuses. Note the similarly shaped cranial base structures between same-aged fetuses. (*Source*: Sherwood et al. (2001). Reprinted with permission from the American Cleft Palate-Craniofacial Association.)

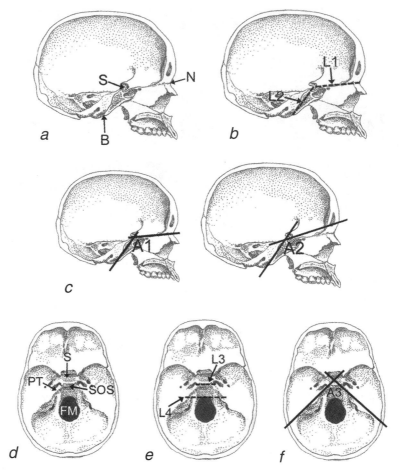

Figure 13.3 Some measurements used in studies of postnatal populations with clefts: (a) landmarks commonly used in studies on lateral cephalographs (N, nasion; S, sella; B, basion); (b) cranial base lengths (L1 = N-S; L2 = S-B); (c) cranial base angles (A1 = saddle angle; A2 = orbital angle); (d) dorsal structures (S, sella; SOS, spheno-occipital synchondrosis; PT, petrous portion of temporal bone; FM, Foramen magnum); (e) axial length measurements (L3 = SOS width; L4 = distance between pars petrosae); (f) axial angles (A3 = petrosal angle).

clefting and that length differences are more pronounced in mediolateral dimensions than in anteroposterior dimensions.

In contrast to studies using CLP or CP samples, very few differences in postnatal cranial base morphology and growth have been noted between samples with isolated cleft lip (CL) and normal control samples (Graber, 1954; Ross and Coupe, 1965; Dahl et al., 1982; Horswell and Gallup, 1992; Trotman et al., 1993; Mølsted et al., 1993;

Laatikainen et al., 1996; Hermann et al., 2000). In a comparison of intrapair cephalometric differences between twins with different types of clefts, Laatikainen's (1999) findings on CL/noncleft twin versus CLP/noncleft twin were the most discrepant. This is consistent with findings on prenatal samples (Sherwood et al., 2001) and indicates that the CL population may be reliably used as a control sample in comparison to samples exhibiting palatal cleft-

TABLE 13.1 Postnatal Comparisons of Selected Basicranial Measurements in Cleft Palate Versus Noncleft Human Samples

Author	Sample	Radiograph	Findings: Linear	Findings: Angular
Moss, 1956	103 CP or CLP[4] (age 2–45) vs. data from previous study	Lateral		"Dysostosis sphenoidalis" characterized by "malrelation between the sphenoid, ethmoid, and occipital . . ."[c]
Brader, 1957	39 CLP or CP[4] (age 4–25) vs. 60 NC children	Lateral		S-B-Pmf[b] N-S-B ns S-B-POP ns
Ross, 1965	342 children[2] (CLP, CP, CL, NC)	Lateral	N-S + SE-S + S-SOS + SOS-B ns	Orbital angle[a] Clivus-cribriform plane[b] Clivus-S-N[a] (Pooled CLP, CP)
Ross and Coupe, 1965	Six pairs of monozygotic twins (discordant for CLP or CP)[2]	Lateral and frontal	Minor differences in cranial base between twins.[c] Facial skeleton in CLP/CP is "rotated, relative to the cranial base."	
Blaine, 1969	443 CP or CLP[3] children vs. 109 NC children	Lateral	N-S and S-B appeared "proportionally similar"[c] between CP/CLP and NC.	Angles of S-B plane with N-S plane, planum plane, and orbital planes were "significantly more obtuse" in CP/CLP than NC; S-B plane-palatal. Plane was significantly more obtuse only in unoperated clefts vs. NC.
Hayashi et al., 1975	255 UCLP vs. 244 NC children (approx. 3.5–21 years)	Lateral	S-N ns at any age	N-S-B is "more flattened" in CLP than NC (some significant differences between groups at early ages).
Dahl et al., 1982	30 CL infants vs. 30 CP[1] infants	Lateral, axial	N-S[b] S-B ns Ch-Ch_a ns	Saddle angle ns N-S-CD ns

TABLE 13.1   *(Continued)*

Author	Sample	Radiograph	Findings: Linear	Findings: Angular
Grayson et al., 1987	144 CLP/CL/CP cases of various ages (1–47 years) compared to normative sample at similar ages	Lateral	Net effect of clefting is "11 percent reduction of the distance from nasion to sella."	"Cranial base angle is hardly altered" in cases of clefting.
Sandham and Cheng, 1988	30 CLP[2] children vs. 61 NC children (11–14 years)	Lateral	N-S ns S-B[b]	Saddle angle ns
Horswell and Gallup, 1992	542 CL/CP/CLP[1] children vs. 277 age-matched NC children (7–18 years)	Lateral	N-S and S-B were significantly smaller in CLP than NC from 10 to 18 years of age, but not at 7 to 8 years; no significant difference in N-S and S-B lengths between CL and NC at most ages.	Saddle angle was not significantly different in either CL or CLP compared to NC; CP had significantly more acute saddle angle at 11, 12, and 15 years.
Harris, 1993	43 CLP[2] children vs. 43 NC	Lateral	N-S ns S-B ns N-B ns	Orbital angle[b] Planum angle ns Saddle angle[b]
Mølsted et al., 1993	57 CLP[1] infants 42 CL infants	Lateral	SOS anteroposterior distance[a] SOS height ns S-SOS[b] SOS-B ns	
Trotman et al., 1993	12 pairs of monzygotic twins, discordant for CL/CLP[2] (approx. 5–11 years)	Lateral	Significantly greater infratwin differences seen in CLP compared to CL for N-S but not S-B.	No significant infratwin differences seen in CLP compared to CL for saddle angle.
Mølsted et al., 1995	52 CLP infants 48 CL infants (longitudinal study)	Axial	Maxillary width[a] Pars petrosa width[a] SOS width[a] SOS-B[a]	Maxillary angle[a] os sphenoidal[a] Pars petrosa angle[a]

Study	Sample	View	Findings	Findings
Laatikainen et al., 1996	39 pairs of twins, concordant or discordant for CL, CP, or CLP[2] (3.7–30.5 years)	Lateral		Saddle angle[a] Saddle angle: CL vs. NC ns
Öztürk and Cura, 1996	20 UCLP[2] children vs. 20 NC children	Lateral	S-N ns S-Ar ns	N-S-Ar[a]
Carrie et al., 2000	34 CP/CP[2] children vs. 26 NC children	Lateral		Sphenopalatine angle[a]
Hermann et al., 2000	53 UCL[2] children vs. 55 UCLP[2] children (22 months)	Lateral	N-B[b] N-S ns S-B[b] Width of internal or external cranial base was "within normal limits" in UCLP group.	Saddle angle ns
Kyrkanides et al., 2000	30 CLP[2] children vs. 64 NC children	Frontal	Asymmetry (r vs. l?) of vertical cranial base distance was not significantly different between groups at any age.	
Laatikainen, 1999	39 pairs of twins, concordant or discordant for CL, CP, or CLP[2] (3.7–30.5 years)	Lateral	Significantly greater intrapair differences in CLP/NC pairs vs. CL/NC pairs for N-S but not S-B.	Significantly greater intrapair differences in CLP or CP/NC vs. CL/NC for saddle angle.

Ar, articulare; B, basion; CD, condylion; Ch-Ch$_a$, distance between right and left cochlea; N, nasion; Pmf, point on posterior maxilla on palatal plane; POP, most inferior point on the occipital bone posterior to the foramen magnum; S, sella; SE, sphenoethmoidal synchondrosis; SOS, spheno-occipital synchondrosis; orbital angle, intersection of orbital and clival planes; planum angle, intersection of planum sphenoidale and clival planes; saddle angle, N-S-B; sphenopalatine angle, angle between line along endocranial surface of frontal bone and line along hard palate through the anterior nasal spine.

[a] Significantly greater in CLP compared to NC or CL.
[b] Significantly greater in NC or CL compared to CLP.
[c] No statistical comparison noted.
1 = unoperated clefts; 2 = operated clefts; 3 = combined unoperated/operated; 4 = unoperated/operated not specified.

ing. This is beneficial in settings where a normal population cannot be examined for ethical reasons (Mølsted et al., 1995).

13.4 MECHANISMS FOR BASICRANIAL INFLUENCES ON CLEFTING

The degree of basicranial involvement in populations with clefts is far from certain. However, it is clear that individuals with isolated clefts of the lip alone have little or no differences in basicranial form compared to unaffected individuals, either prenatally (Sherwood et al., 2001) or postnatally (Ross and Coupe, 1965; Dahl, 1970; Dahl et al., 1982; Horswell and Gallup; 1992; Trotman et al., 1993; Mølsted et al., 1993; Hermann et al., 2000). On the other hand, populations with palatal clefting show generally some, if variable, prenatal and postnatal basicranial dysmorphologies. The variable results can, to some extent, be attributed to the type of facial clefting. CL is associated with fewer craniofacial abnormalities than cases with palatal clefting, and CP appears to be associated with a differing manifestation of basicranial abnormalities than CLP (Grayson et al., 1987; Horswell and Gallup, 1992).

This phenotypic variation also applies to relatives of individuals with facial clefts (Ward et al., 1989; Mossey et al., 1997). This complexity typifies a multifactorial disorder—the phenotypic variation is linked to both timing of environmental factors (e.g., teratogens) and to complex genetics of facial clefting, which have revealed different epidemiology and inheritance patterns for CL +/− P versus CP (Carinci et al., 2000). It has also been suggested that phenotypic heterogeneity in craniofacial morphology is probably not one of the genetic predispositions for primary palatal clefting (Ward et al., 1989; Trotman et al., 1993; Murray, 1995; Mossey et al., 1997, 1998; Moriyama et al., 1998; Suzuki et al., 1999; Laatikainen, 1999) as compared to

secondary palatal clefting (Siegel and Mooney, 1986; Ward et al., 1989; Trotman et al., 1993; Vergato et al., 1997; Mossey et al., 1997, 1998; Moriyama et al., 1998; Suzuki et al., 1999; Sherwood et al., 2001). Therefore, the following discussion does not apply equally to all types of clefts.

The association between basicranial and midfacial abnormalities in cleft populations likely relates to the developmental and functional relationship of the two craniofacial regions. Long and colleagues (1973) discussed this relationship using a mouse model. Teratogenic effects on presphenoidal cartilage decrease cellular proliferation, and also correlate with the failure of the mouse basicranium to extend during palatogenesis. In this position, the vertical palatal shelves remain separated by the tongue and may fail to elevate. Since the junction of the neurocranium and viscerocranium occurs via the basicranium, the position of the palatal shelves is dependent on cranial base angulation. Whereas basicranial extension appears to facilitate palatogenesis in mice (Long et al., 1973), the same is not likely true for humans. The cranial base angulation (nasionsella-basion) actually becomes more acute (flexed) in humans during palatogenesis (Sperber, 2001).

The prenatal importance of the cranial base to clefting (at least for CLP) may have more basis in altered shape or size. For example, increased facial width in CLP appears to be correlated with increased basicranial width in humans (Mølsted et al., 1995) and other mammals (Richtsmeier et al., 1994). Increased facial width has been hypothesized to relate to inability of the palatal shelves to unite in the midline (Siegel and Mooney, 1986; Vergato et al., 1997), and also has been noted to characterize individuals with facial clefting (Hirschfeld and Aduss, 1974) and even their kin (Suzuki et al., 1999). In this light, it is possible that a wider cranial base could have a causal or at least correlative relationship to increased midfacial width and clefting. Investigations on the basicranium of midfacially retrusive mice (Ma and Lozanoff, 1999) make it clear that deficient chondrogenesis of the sphenoethmoidal cartilages has an indirect contribution to midfacial retrusion. Furthermore, although there is indeed a midline continuity of prenatal basicranial and midfacial cartilages, the rate of cellular proliferation is not identical throughout anteroposterior regions (Ma and Lozanoff, 1999). Until the cranial base is studied as thoroughly as midfacial regions in cases of facial clefting, our understanding of the pathogenesis of basicranial defects (e.g., altered chondrogenesis or cartilage growth) will remain incomplete.

Unfortunately, prenatal changes in basicranial width have not been compared between cleft and noncleft human fetuses (it is possible that such changes may be most significant before palatogenesis, rendering such investigations difficult in humans). There is, however, evidence for dysmorphic mesethmoid cartilages in human fetuses (Smith et al., 1999). It is also instructive to compare the findings of studies on uncorrected CLP in infants to those on older samples in which surgical correction has been done prior to the study (see Table 13.1). When interpreting the variability of results on surgically corrected children, it must be kept in mind that midfacial surgery may indirectly influence the position of anatomical landmarks used to study cranial base angles. The presence of cranial base abnormalities in infants (Mølsted et al., 1993, 1995) and fetuses (Smith et al., 1997, 1999) suggests that such malformations are congenital in nature.

13.5 CONCLUSIONS

The close anatomical link (e.g., continuity of the mesethmoid and nasal septal cartilages) between the midface and basicranium makes it a viable hypothesis that clefting is part of a "generalized growth disturbance" (Harris, 1993), and the deranged cartilage growth and development during the period of palatogenesis should be investigated further. Alterations in postnatal basicranial form may be accomplished through drift (selective resorption and apposition) or displacement (e.g., through growth across sutures or in synchondroses) (Lieberman et al., 2000). These processes, often confounded by the effects of surgical intervention, are relevant to the vastly greater number of studies on postnatal changes in the cranial base of cleft individuals. Yet questions regarding the timing (and primacy) of cranial base abnormalities might not be best answered using postnatal samples (Table 13.1). In order to elucidate the mechanism of basicranial influence on palatal clefting (if any), a renewed emphasis on prenatal morphogenesis/dysmorphogenesis is necessary.

REFERENCES

Avery, J. K., Happle, J. D., and French, W. C. (1957). Development of the nasal capsule in normal and cleft palate formation. *Cleft Palate Bull. 7*, 8–11.

Blaine, H. L. (1969). Differential analysis of cleft palate anomalies. *J. Dent. Res. 48*, 1042–1048.

Brader, A. C. (1957). A cephalometric x-ray appraisal of morphological variations in cranial base and associated pharyngeal structures: implications in cleft palate therapy. *Angle Orthodont. 27*, 179–185.

Carinci, F., Pezzetti, F., Scapoli, L., Martinelli, M., Carinci, P., and Tognon, M. (2000). Genetics of nonsyndromic cleft lip and palate: a review of international studies and data regarding the Italian population. *Cleft Palate Craniofac. J. 37*, 33–40.

Carrie, S., Sprigg, A., and Parker, A. J. (2000). Skull base factors in relation to hearing impairment in cleft palate children. *Cleft Palate Craniofac. J. 37*, 166–171.

Dahl, E. (1970). Craniofacial morphology in congenital clefts of the lip and palate. *Acta Odontol. Scand. 28* (suppl.), 57–84.

Dahl, E., Krieborg, S., Jensen, B. L., and Fogh-Andersen, P. (1982). Comparison of craniofacial morphology in infants with incomplete cleft lip and infants with isolated cleft palate. *Cleft Palate J. 19*, 258–266.

Graber, T. M. (1954). The congenital cleft palate deformity. *J. Am. Dent. Assoc. 48*, 375–395.

Grayson, B. H., Bookstein, F. L., McCarthy, J. G., and Mueeddin, T. (1987). Mean tensor cephalometric analysis of a patient population with clefts of the palate and lip. *Cleft Palate J. 24*, 267–277.

Harris, D. F. (1993). Size and form of the cranial base in isolated cleft lip and palate. *Cleft Palate Craniofac. J. 30*, 170–174.

Hayashi, I., Sakuda, M., Takimoto, K., and Miyazhi, T. (1975). Craniofacial growth in complete unilateral cleft lip and palate. A roentgencephalometric study. *Cleft Palate J. 13*, 215–235.

Hermann, N. V., Jensen, B. L., Dahl, E., Bolund, S., and Kreiborg, S. (2000). Craniofacial comparisons in 22-month-old lip-operated children with unilateral complete cleft lip and palate and unilateral incomplete cleft lip. *Cleft Palate Craniofac. J. 37*, 303–317.

Hirschfeld, W. J., and Aduss, H. (1974). Interorbital distance in cleft lip and palate: significant differences found by sign test. *J. Dent. Res. 53*, 947–952.

Horswell, B. B., and Gallup, B. V. (1992). Cranial base morphology in cleft lip and palate: a cephalometric study from 7 to 18 years of age. *J. Oral Maxillofac. Surg. 50*, 681–685.

Krogman, W. M., Mazaheri, M., Harding, R. L., Ishiguro, K., Bariara, G., Meier, J., Canter, H., and Ross, P. (1975). A longitudinal study of the craniofacial growth pattern in children with clefts as compared to normal, birth to six years. *Cleft Palate J. 12*, 59–84.

Kyrkanides, S., Klambani, M., and Subtelny, J. D. (2000). Cranial base and facial skeleton asymmetries in individuals with unilateral cleft lip and palate. *Cleft Palate Craniofac. J. 37*, 556–561.

Laatikainen, T. (1999). Etiological aspects on craniofacial morphology in twins with cleft lip and palate. *Eur. J. Oral Sci. 107*, 102–108.

Laatikainen, T., Ranta, R., and Nordströ4m, R. (1996). Craniofacial morphology in twins with cleft lip and palate. *Cleft Palate Craniofac. J. 33*, 96–103.

Lele, S. (1993). Euclidean distance matrix anlaysis (EDMA): estimation of mean form and mean form difference. *Math. Geol. 25*, 573–602.

Lieberman, D. E., Ross, C. F., and Ravosa, M. J. (2000). The primate cranial base: ontogeny, function, and integration. *Am. J. Phys. Anthrop. 43* (suppl. 31), 117–169.

Long, S. Y., Larsson, K. S., and Lohmander, S. (1973). Cell proliferation in the cranial base of A/J mice with 6-AN-induced cleft palate. *Teratology. 8*, 127–138.

Ma, W., and Lozanoff, S. (1999). Spatial and temporal distribution of cellular proliferation in the cranial base of normal and midfacially retrusive mice. *Clin. Anat. 12*, 315–325.

Mølsted, K., Kjær, I., and Dahl, E. (1993). Spheno-occipital synchondrosis in three-month-old children with clefts of the lip and palate: a radiographic study. *Cleft Palate Craniofac. J. 30*, 569–573.

Mølsted, K., Kjær, I., and Dahl, E. (1995). Cranial base in newborns with complete cleft lip and palate: radiographic study. *Cleft Palate Craniofac. J. 32*, 199–205.

Mooney, M. P., Siegel, M. I., Kimes, K. R., and Todhunter, J. (1991). Premaxillary development in normal and cleft lip and palate fetuses using three-dimensional computer reconstruction. *Cleft Palate Craniofac. J. 28*, 49–54.

Mooney, M. P., Siegel, M. I., Kimes, K., Todhunter, J., and Smith, T. D. (1994). Development of the anterior paraseptal cartilages in normal and cleft lip and palate human fetal specimens. *Cleft Palate Craniofac. J. 31*, 239–245.

Moriyama, K., Motohashi, N., Kitamura, A., and Kuroda, T. (1998). Comparison of craniofacial and dentoalveolar morphologies of three Japanese monozygotic twin pairs with cleft lip and/or palate discordancy. *Cleft Palate Craniofac. J. 35*, 173–180.

Moss, M. L. (1956). Malformations of the skull base associated with cleft palate deformity. *Plastic Reconstr. Surg. 17*, 226–234.

Mossey, P. A., McColl, J. H., and Stirrups, D. R. (1997). Differentiation between cleft lip with or without cleft palate and isolated cleft palate using parental cephalometric parameters. *Cleft Palate Craniofac. J. 34*, 27–35.

Mossey, P. A., McColl, J., and O'Hara, M. (1998). Cephalometric features in the parents of children with orofacial clefting. *Brit. J. Oral Maxillofac. Surg. 36*, 202–212.

Murray, J.C (1995). Face facts: genes, environment, and clefts. *Am. J. Hum. Genet. 57*, 227–232.

Öztürk, Y., and Cura, N. (1996). Examination of craniofacial morphology in children with unilateral cleft lip and palate. *Cleft Palate J. 33*, 32–36.

Richtsmeier, J. T., Sack, G. H., Grausz, H. M., and Cork, L. C. (1994). Cleft palate with autosomal recessive transmission in Brittany spaniels. *Cleft Palate Craniofac. J. 31*, 364–371.

Ross, B. R. (1965). Cranial base in children with lip and palate clefts. *Cleft Palate J. 2*, 157–166.

Ross, R. B. (1992). Discussion: cranial base morphology in cleft lip and palate: a cephalometric study from 7 to 18 years of age. *J. Oral Maxillofac. Surg. 50*, 686.

Ross, B. R., and Coupe, T. B. (1965). Craniofacial morphology in six pairs of monozygotic twins discordant for cleft lip and palate. *J. Can. Dent. Assoc. 31*, 149–157.

Ross, R. B., and Johnston, M. C. (1978). *Cleft Lip and Palate* (New York: Robert E. Krieger Publ. Co.).

Sandham, A., and Cheng, L. (1988). Cranial base and cleft lip and palate. *Angle Orthod. 58*, 163–168.

Sherwood, T. R., Mooney, M. P., Sciote, J. J., Smith, T. D., Cooper, G. M., and Siegel, M. I. (2001). Cranial base growth and morphology in second trimester normal and cleft lip (CL) human fetuses. *Cleft Palate Craniofac. J. 38*, 587–596.

Siegel, M. I., and Mooney, M. P. (1986). Palatal width growth rates as the genetic determinant of cleft palate. *J. Craniofac. Genet. Devel. Biol. 2* (suppl.), 187–191.

Siegel, M. I., Todhunter, J. S., Kimes, K. R., and Mooney, M. P. (1984). Fetal specimens available. *Cleft Palate J. 21*, 115–116.

Siegel, M. I., Mooney, M. P., Kimes, K. R., and Todhunter, J. (1987). Analysis of the metric variability of the normal and cleft palate human fetal nasal capsule by three-dimensional computer reconstruction of histological preparations. *Cleft Palate J. 24*, 190–200.

Smith, T. D., Siegel, M. I., Mooney, M. P., Burdi, A. R., and Todhunter, J. S. (1996). Vomeronasal organ growth and development in normal and cleft lip and palate fetuses. *Cleft Palate Craniofac. J. 33*, 385–394.

Smith, T. D., Siegel, M. I., Mooney, M. P., Burrows, A. M., and Todhunter, J. S. (1997). Formation and enlargement of the paranasal sinuses in normal and cleft lip and palate human fetuses. *Cleft Palate Craniofac. J. 34*, 483–489.

Smith, T. D., Siegel, M. I., Mooney, M. P., Burrows, A. M., and Todhunter, J. S. (1999). Development of the paranasal sinuses in cleft lip and palate human fetuses. In T. Koppe, H. Nagai, and K. W. Alt (eds.), *The Paranasal Sinuses of Higher Primates: Development, Function and Evolution* (Berlin: Quintessence Publishing Co.), pp. 65–75.

Sperber, G.H. (2001). *Craniofacial Development* (Hamilton: B. C. Decker Inc.).

Suzuki, A., Takenoshita, Y., Honda, Y., and Matsuura, C. (1999). Dentocraniofacial morphology in parents or children with cleft lip and/or palate. *Cleft Palate Craniofac. J. 36*, 131–138.

Trotman, C. A., Collett, A. R., McNamara, J. A., Jr, and Cohen, S. R. (1993). Analysis of craniofacial and dental morphology in monozygotic twins discordant for cleft lip and unilateral cleft lip and palate. *Angle Orthod. 63*, 135–139.

Vergato, L. A., Doerfler, R. J., Mooney, M. P., and Siegel, M. I. (1997). Mouse palatal width growth rates as an "at risk" factor in the development of cleft palate induced by hypervitaminosis A. *J. Craniofac. Genet. Devel. Biol. 17*, 204–210.

Ward, R. E., Bixler, D., and Raywood, E. R. (1989). A study of cephalometric features in cleft lip–cleft palate families. I: Phenotypic heterogeneity and genetic predisposition in parents of sporadic cases. *Cleft Palate J. 26*, 318–325.

CHAPTER 14

CRANIAL VAULT DYSMORPHOLOGY AND GROWTH IN CRANIOSYNOSTOSIS

JOAN T. RICHTSMEIER, Ph.D.

But the problems of variability, though they are intimately related to the general problem of growth, carry us very soon beyond our limitations.

—Sir D'Arcy Thompson (1917)

14.1 INTRODUCTION

The neurocranium or brain case consists of a complex set of bones that surround the brain. This skeletal capsule can be divided into two anatomically defined sections. The cranial vault is that part of the skull that covers the external surface of most of the cerebral hemispheres and cerebellum while the cranial base supports the most inferior portions of these neural structures as well as the pons, medulla oblongata, and brain stem (Fig. 14.1). The cranial vault differs from the cranial base in several major ways. First, the bones of the cranial vault are formed by intramembranous ossification rather than through the ossification of cartilaginous models. Second, although the anterior cranial base is known to derive from the neural crest, the origin of several cranial vault bones remains contentious (Opperman, 2000; Jiang et al., 2002)

(Chapters 4, 5, and 11). Based on information primarily from experiments that study the developing chick (Couley et al., 1993; Noden, 1986a), we must consider that the bones of the mammalian cranial vault are most likely derived from both paraxial mesoderm and neural crest while the cranial base is strictly of neural crest origin. Third, the cranial vault is phylogenetically of recent origin while the cranial base is evolutionarily ancient (Gans and Northcutt, 1983) (Chapter 11).

The neurocranium is formed by fusions between various bones belonging to the cranial base (e.g., sphenoid, petrous temporal, ethmoid, basioccipital bones) and those of the cranial vault (e.g., frontal, parietal, squamous portion of the occipital and temporal bones). Bones of the cranial base and cranial vault are structurally and functionally interrelated. Certain bones (e.g., the temporal and occipital bones) form from more than one center of ossification, certain centers forming intramembranous or dermal bone, others forming from cartilaginous models. Some classifications show the division of cranial base and cranial vault to divide single bones (see Fig. 14.1). The facial skeleton (or viscerocranium) is

Understanding Craniofacial Anomalies: The Etiopathogenesis of Craniosynostoses and Facial Clefting, Edited by Mark P. Mooney and Michael I. Siegel, ISBN 0-471-38724-x Copyright © 2002 by Wiley-Liss, Inc.

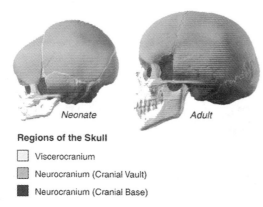

Regions of the Skull

☐ Viscerocranium

▨ Neurocranium (Cranial Vault)

■ Neurocranium (Cranial Base)

Figure 14.1 Computed tomography three-dimensional reconstruction of human infant (left) and adult (right) skulls. Anatomic regions of the skull are color coded according to the key. The assignments of several areas to either the cranial vault or cranial base are less definite than others. For example, the external surface of the greater wing of the sphenoid may be classified as part of the cranial vault or the cranial base. The squamous portion of the temporal bone and the mastoid process could also be classified as either cranial vault or cranial base according to different criteria.

tied functionally, structurally, and developmentally to the rest of the skull, but is not considered here.

The cranial vault, sometimes called the calotte or the calvaria, is made up of the paired frontal bones, the paired parietal bones, the squamous portion of the temporal bone, and the squamous portion of the occipital bone (Fig. 14.1). The greater wing of the sphenoid is considered by some to be part of the cranial vault (especially the external surface), in part because it forms in membrane and in part because of its location. Sutures that are usually devoid of ossification separate individual bones from one another, each suture fusing at a different time. Cranial vault sutures are fibrous tissues that unite the bones of the cranial vault and allow some movement between adjoining bones during parturition. They are thought of as sites of bone growth in mammals. As demonstrated by Herring (2000), the role of sutures in

mediating cranial mechanics is apparent if approached from an evolutionary perspective.

Premature closure of any of the cranial vault sutures is known as craniosynostosis. Craniosynostosis results in characteristic dysmorphology of the cranial vault, cranial base, and facial skeleton. Since Virchow (1851) formally outlined the relationship between identification of the prematurely closed suture and calvarial shape, research has focused on craniosynostosis at many levels of analysis. Due primarily to the research tools available, early investigation into craniosynostosis was largely comparative, anatomical, and biophysical in nature. Pedigree analysis of the families exhibiting syndromic craniosynostosis caused most scientists to suspect that genes were at the base of the dysmorphologies associated with craniosynostosis. However, molecular techniques and knowledge of molecular pathways had not advanced far enough to pursue this avenue of research profitably. All of this changed during the 1990s when new technology allowed the rapid accumulation of knowledge about the molecular processes involved in normal suture maintenance and fusion and identification of specific gene mutations associated with craniosynostosis.

The past decade of research into growth and development of the cranial vault has focused on the influence of genes and the proteins for which they code during formation of the cranial vault. This is in profound contrast to the classificatory approach of 19th-century scientists who could only observe the gross morphology of craniosynostosis either from museum specimens or available clinical cases. This chapter considers morphology and growth of the cranial vault and the role of growth in the production of calvarial dysmorphology in craniosynostosis. The chapter offers a historical perspective, bringing together knowledge from various fields in an attempt to determine what we know, and the important remaining questions relating to craniosynostosis.

14.2   THE RELEVANT QUESTIONS

Largely because of the clinical problems encountered in craniosynostosis, sutures are usually thought of in terms of growth, allowing the otherwise rigid cranium to enlarge. While medically appropriate, this view distorts a biological understanding of sutural function.

—Herring (2000)

The broad questions that surround craniosynostosis and resulting cranial vault dysmorphology were formulated in a clinical context. The questions remain despite significant research efforts. The craniosynostosis literature has grown enormously during the 20th century from contributions primarily by pediatricians, orthodontists, and dentists, to more recent work from these specialties as well as from craniofacial surgeons, developmental biologists, and molecular geneticists. In my view, a historical reading of the literature brings forth the following general questions relating to the cranial vault in craniosynostosis:

1. What is *the* cause of craniosynostosis?
2. Is the brain (including the meninges) or the bone primary in terms of the cause of craniosynostosis and the production of the resulting morphology?
3. How do growth patterns contribute to the establishment of craniosynostosis morphologies? Is the dysmorphology defined early in development with normal (though mechanically constrained) growth patterns contributing to the abnormal phenotype? Or, do anomalous growth patterns produce the abnormal phenotype?
4. If left untreated, do craniosynostosis morphologies worsen with time?

The search for answers to these general questions that concern the growth of the cranial vault in craniosynostosis has been executed at various levels of analysis from the molecule to the organism. The most comprehensive work about craniosynostosis is Cohen's (1986) volume. The second edition, published in 2000 (Cohen and MacLean, 2000), provides a broad summary of the progress made in the intervening 15 years. One of the figures that Cohen and MacLean (2000) provides is simple and clear, and is brought to mind on a daily basis, consciously or unconsciously, by researchers in the field (Fig. 14.2). This figure summarizes the association of specific skull morphology with the premature fusion of specific cranial vault sutures. It also depicts the sites of compensatory growth (heavy arrows oriented away from sutures) for each type of craniosynostosis. Compensatory growth is thought to occur at sutures close to the one that has undergone premature fusion. Growth at the sutures is appositional, occurring along the length of the suture; new bone is added to the edges of the bones separated by the open suture, thereby causing those bones to translate away from one another. This compensatory growth is in part responsible for the dysmorphology seen in craniosynostosis cases.

Although not necessarily intended by Cohen and MacLean (2000), the implicit meaning of this figure is that if we know which suture is closed, and we understand the principles of what occurs when a suture is closed prematurely (i.e., local appositional growth ceases and compensatory appositional growth occurs at a nearby suture), the overall shape of the cranial vault can be predicted with fairly good accuracy. Furthermore, identification of the suture that is prematurely closed can be made from gross observation of cranial vault morphology. This one-to-one correspondence of suture involvement and overall shape of the calvarium provided a

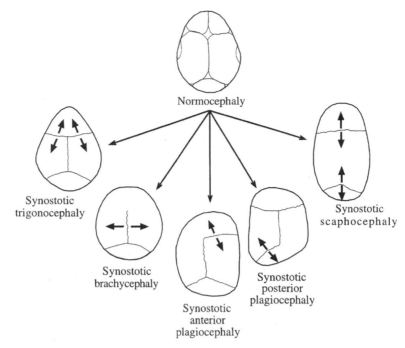

Figure 14.2   A summary of the correspondence between cranial vault shape and identity of the suture that is prematurely closed. Heavy arrows indicate the sites of compensatory growth for each type of craniosynostosis. This figure specifically uses the adjective *synostotic* to describe the cranial shape because many of these shapes are known to occur *without* premature synostosis. (*Source*: Figure and caption modified from Figure 11.5 in Cohen and MacLean, 2000. Used with kind permission from M Michael Cohen, Jr.)

false sense of knowledge to at least one researcher in the field (i.e., myself). Our ability to predict which suture was closed on the basis of skull shape suggested that we understood at least the biophysics of craniosynostosis. It seemed only a short step to identify the cause of premature closure of the suture(s). Once this was accomplished, the disease would be understood, and effective, noninvasive therapies could be designed. As with most biological phenomena, things are not that simple.

14.3   SOFT TISSUE–HARD TISSUE INTERACTIONS AS AN EXPLANATION FOR CRANIOFACIAL DYSMORPHOLOGY AND GROWTH

The cerebral capsule is formed by the tissues which surround, and are intimately

responsive to, the functional demands of the neural mass.

—Moss and Young (1960)

The early craniosynostosis literature contains both descriptive and explanatory works (Virchow, 1851; Klaauw, 1945, 1948–1952; Moss, 1957, 1958, 1959, 1960, 1975; Moss and Young, 1960; Knudson and Flaherty, 1960; Blechschmidt, 1976a, 1976b; Giuffre et al., 1978; Kreiborg and Pruzansky, 1981; David et al., 1982). Some of the earliest explanations were biophysical, mechanistic, or biomechanical in nature. Knowledge of the relationship between mechanical forces (e.g., tension, compression, shear) and the biological process of osteogenesis gained from experimental work (Glucksmann, 1942; Storey, 1955) was applied to the study of craniofacial growth and to situations in which structural and biomechanical relationships were altered and skull morphology was changed.

Researchers observed that fibrous bands of dura mater, the dural tracts, were oriented according to their attachment of the dura mater to five locations on the cranial base: the crista galli of the ethmoid bone, the paired lesser wings of the sphenoid, and the apices of the paired petrous portions of the temporal bones (Popa, 1936; Blechschmidt, 1961, 1976a, 1976b). Growth of the brain transmitted specific forces to the dura mater, resulting in the formation of cranial vault bone centers located according to their distance from folds in the dura mater, with sutures developing between these centers. Dural folds were situated directly beneath the major calvarial sutures, and the nature of biomechanical forces in these areas limited local ossification, resulting in suture patency. It was proposed that if the relative locations of sites of attachment of the dura mater to the cartilaginous cranial base were altered (e.g., through malformation of the cranial base), then the tensile forces within the oriented fiber tracts of the presumptive dura mater would be altered. This would in turn perturb the neural growth vectors, resulting in secondary, compensatory adjustments in the formation of cranial vault bones and the calvarial sutures (Moss, 1959).

These mechanisms have failed to provide a complete explanation for cause and effect in craniosynostosis. Certainly there is interdependence of the calvarium, cranial base, and central nervous system as demonstrated by observations of intentionally deformed crania from archeological sites (e.g., Antón et al., 1992; Cheverud et al., 1992; Cheverud and Midkiff, 1992; Kohn et al., 1993, 1994, 1995; Konigsberg et al., 1993; Antón and Weinstein, 1999) and from the study of postoperative response of the cranial base to cranial vault surgery (e.g., Marsh and Vannier, 1986; Richtsmeier et al., 1991; al-Qattan and Phillips, 1997; Sgouros et al., 1999; DeLeon et al., 2001). However, not only does brain growth

normally cease prior to the closure of most cranial vault sutures (altering the biomechanical forces that are proposed as the mechanism for maintaining suture patency) but the observation of Inca bones in the sutures of normal skulls is not expected under this hypothesis. It cannot be determined by the reported observations whether cranial base dysmorphology causes reorientation of the dural reflections, or premature closure of the sutures results in cranial base dysmorphology, thus reorienting the dural tracts. These counterexamples notwithstanding, the biophysical observations and their potential role in cranial vault growth and morphology should not be overlooked.

Moss's experimental work (Moss, 1957, 1958, 1960) reinforced the principle of interdependence of the cranial soft and hard tissues in growth and dysmorphogenesis of the craniofacial complex. The functional matrix (FM) hypothesis (Moss, 1957, 1958, 1960, 1971, 1975, 1981; Moss and Young, 1960; Moss and Moss-Salentijn, 1979) considers that the presence, size and shape, growth, and position in space of all skeletal tissues are secondary, compensatory, and mechanically obligatory responses to the temporally prior demands of related soft tissue functional matrices (Moss and Moss-Salentijn, 1979). Biological boundaries (e.g., the dura mater) served as transmitters of information between bone and soft tissue functional matrices.

Contemporary research has supported the role of dura mater in regulating suture patency but through processes other than biomechanical associations. Experimental systems and observations from clinical cases have clarified the role of intercellular signaling via the dura mater for maintaining cranial vault suture patency (e.g., Drake et al., 1993; Opperman et al., 1993, 1994, 1995, 1998, 2000; Bradley et al., 1996; Hobar et al., 1996; Roth et al., 1996; Levine et al., 1998; Opperman, 2000; Jiang et al., 2002). The dura mater acts to regulate the state of the suture by secreting factors that act locally to either inhibit or encourage maintenance of suture patency. It appears that the dura mater shows regional differences in its ability to regulate suture patency. Maintenance of suture patency depends on the dura mater's role in regulating a complex array of factors that may work cooperatively, antagonistically, or independently at different anatomical sites and at different points during craniofacial development. The exact nature of the signaling is not yet clear, but many of the factors involved have been identified (see Figure 4 in Opperman, 2000) (Chapter 20).

The factors involved in suture establishment and maintenance (i.e., fibroblast growth factors and their receptors, transforming growth factors, factors known to be involved in pattern formation, factors involved in epithelial-mesenchymal signaling, and TWIST and MSX2) belong to a group already known to be active in many developmental processes (Chapters 4, 5, 6, and 20). Opperman (2000) points out that much of the work on clinical pathologies involving craniosynostosis has discovered associated mutations, although the nature of the function that is affected by the mutation is not identified. Some studies try to determine the role of gene products in regulating cell function within the cranial sutures in terms of organizing and maintaining a patent suture, but ignore the fact that the suture also needs to persist as a functioning growth site (Opperman, 2000). Evolutionary studies (see below) indicate that the biomechanical or functional role of sutures needs to be considered as well (Herring, 2000).

14.4 FORMATION AND GROWTH OF CALVARIAL BONES AND THE CRANIAL VAULT

Growth is a differential developmental process. That is, different anatomical components undergo maturational changes at different times, by different regional amounts, in different regional directions and at different velocities according to different functional and architectonic conditions. The composite of separate components, however, enlarges in a closely interrelated manner. This requires precise, sensitive, responsive, adjustive give-and-take growth interactions.
—Enlow (2000)

Before molecular clues were available, early workers in craniofacial growth and craniosynostosis sought to understand how growth at the sutures was related to the totality of craniofacial growth. Cranial sutures were suspected to be growth centers—areas capable of interstitial growth and responsible for driving the growth of calvarial bone. Researchers used metallic implants and chemical markers to determine the magnitude of growth at specific suture sites and to precisely observe growth increments. Metallic implants were easily visualized by x-ray and were therefore used to quantify growth local to the area of the implant.

Animal (Gans and Sarnat, 1951; Sarnat, 1968) and human (Bjork, 1964, 1968) studies were conducted that observed the changing location of markers originally placed at suture and nonsuture locations as the skull grew. It was discovered that the distance between markers placed on either side of a cranial suture increased, and appositional growth occurred on the surface of single bones. These studies helped to change the perception of the role of the sutures in craniofacial growth and provided preliminary data on growth of cranial bones at nonsuture locations. According to Opperman and colleagues (1995), cranial sutures are best thought of as bone growth sites—secondary adaptive regions at which bone remodeling takes place without a cartilaginous intermediate. On the other hand, bone growth centers have intrinsic growth potential and tissue-separating capabilities. The most illustrative examples of bone growth centers are the cartilaginous growth plates of long bones, where both interstitial and appositional growth occurs. A growth center has intrinsic growth potential, while a growth site remains dormant until stimulated by an external signal (Opperman et al., 1995).

Enlow's exquisite histological studies and comprehensive treatment of the growing skull (especially the facial skeleton) added to data garnered from implant and marker work and proposed the idea of the skull as a differentially developing structure (Enlow, 1963, 1975). Growth of the skull requires the coordination of varying local and regional growth trajectories and growth magnitudes over different intervals involving a complex and interactive system (Enlow, 2000). Enlow provided a systematic explanation of this system that combined knowledge of the local patterns of osteoblast and osteoclast activity. He also proposed concepts summarizing the coordinated activities that enable the coadjustment of the size and shape of individual bones—for example, remodeling, displacement, relocation (Enlow, 2000; Lanyon and Rubin, 1985). Enlow's work provides a means for understanding how bones change shape while expanding in size during growth (Fig. 14.3). Importantly, his contribution provides elucidation at the cellular, tissue, and morphological levels of analysis.

Concurrent with Enlow's early work, the field of cephalometrics was widely applied in growth research and in orthodontic diagnosis and treatment evaluation. Roentgenographic cephalometry, later known as cephalometric radiography or simply cephalometrics, is a technique employing oriented radiographs for the purpose of making head measurements (Merow and Broadbent, 1990). The technique represents the combination of anthropometric principles and the ability to measure internal features of the skull in individuals over chronological time. Large

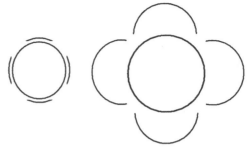

Figure 14.3   Illustration of how the calvarium adjusts to the expanding brain during growth. At left is a schematic of the neonatal brain with four cranial vault bones fitting snugly against the dura mater. At right is the same brain enlarged due to growth and the same four cranial vault bones enlarged by growth only at their periphery (i.e., at the suture). Although the addition of bone at the sutures is a necessary part of cranial vault growth, it is not sufficient. Deposition occurs on the sutural margins to maintain a tight but patent suture. Outward displacement of the calvarial bones is due to brain growth accompanied by local changes in curvature of the bones, which occurs by way of specific patterns of deposition and resorption of the ectocranial and endocranial surfaces of calvarial bones. Remodeling—the addition and subtraction of bone from surfaces—and drift—the movement of bone in space—are required for the coordinated expansion of central nervous system structures and skull components.

longitudinal data sets were amassed to study normal growth (Broadbent et al., 1975), to identify deviations from normal in patient populations, and to track abnormal growth patterns.

Many analytical tools and methods were designed for the study of cephalometric data (for a review, see Merow and Broadbent, 1990), but all rely on a plane or line of reference for superimposition of those cases being compared. It has been shown repeatedly (Moyers and Bookstein, 1979; Richtsmeier, 1985; Cole, 1997; Lele and Richtsmeier, 2001) that any comparison of forms based on superimposition cannot identify the location of the changes that occur due to growth. Instead, the patterns of differences identified vary with the superimposition scheme that is adopted.

Consequently, the results of cephalometric studies when coupled with data from histological or implant studies provide little in the way of a coherent model of cranial vault growth, normal or abnormal (Enlow, 1975).

New technologies and intense laboratory investigation have enabled an understanding of the initiation of bone formation. In a recent review, Hall and Miyake (2000) have outlined the four primary phases of skeletogenesis: (1) migration of cells to the site of future skeletogenesis; (2) epithelial-mesenchymal interactions; (3) cell condensation; and (4) differentiation of chondroblasts or osteoblasts (see also Hall, 1991; Dunlop and Hall, 1995; Hall and Miyake, 1995; Miyake et al., 1996). Hall and Miyake (2000) provide details of the molecular signaling required for initiation, modulation, growth, and termination of condensations. This work is an outstanding example of the remarkable information that is now available about the interactions and processes required to form the skeleton from an embryo that began as a single cell. The environment and the controls required for the formation of bone are fairly well understood. But more detailed work will have to be done on a complex set of developmental processes before we understand the coordinated growth of the skull. Understanding growth of the cranial vault will take more than simply knowing how individual bones grow. We will need to determine the relationship between the initiation of chondrogenesis and osteogenesis, the shape and coordinated growth of individual bones, and their integration into the functioning skull, including all functioning soft tissues and spaces.

14.5 LESSONS FROM EVOLUTION AND DEVELOPMENT

At first sight, the vertebrate head seems like an engineer's nightmare. Not only does it contain three separate skeletal systems . . . but it is animated by a veritable cat's cradle of muscles. And yet it works: through all the twists and turns of evolution, in animals as different as lampreys and chickens, the mechanical integrity of the head has been maintained.

—Ahlberg (1997)

Many of the most important findings regarding growth and development of the cranial vault come from the field of evolutionary developmental biology. The links between development and evolution have been sought for centuries, but the field of evolutionary developmental biology took on new energy and interest during the 1980s and 1990s. A second edition of Hall's textbook (Hall, 1999a) for this subdiscipline of biology provides both history and a detailed review of current knowledge.

Evolutionary developmental biology has elucidated the growth of the cranial vault by identifying the developmental processes that are required (e.g., tissue interaction, migration, pattern formation, genetic cascades) for normal development of the vertebrate head. Demonstration of the remarkable conservation of developmental programs and of the molecular controls of those programs has allowed students of human dysmorphology to focus and rely on experimental data obtained from animal models. These animal models can be manipulated during all periods of development, most of which are inaccessible in human specimens.

Relatively recent technological developments have enabled new insights into the cell populations that are important in the establishment of the skull, and especially of the cranial vault. In contrast to a filter-feeding, relatively immobile vertebrate ancestor, the modern vertebrate skull evolved primarily for the purpose of active prey capture, the jaws and various sensory placodes being the most important evolutionary developments. The evolutionary changes that established this new head (Gans and Northcutt, 1983) were made possible by the appearance of a new population of cells called the neural crest, which are the developmental source of many and varied structures (see Hall, 1999b, 2000) (Chapter 11). Synchronized specializations necessary for proper functioning of the new jaws and sense organs were made possible through the evolution of neural crest cells and their derivatives. Since neural crest cells appear early in development from the neurosomatic junction, ridges that flank the closing neural tube, developmental time frames in animal models that predate the appearance of any bony cranial vault structures will need to be considered to reveal the prime movers in cranial vault pattern formation. The neural crest cells hold answers to many questions regarding cranial vault patterning, both normal and abnormal.

It has been known for some time that bones of the facial skeleton are derived from the neural crest. However, the derivation of cranial vault bones has been more controversial. A series of experiments with quail-chick chimeras by Couly and co-workers (Couly et al., 1993) demonstrated that the bones of the cranial vault are established by migrating neural crest cells and not by paraxial mesoderm. Additional work with both bird and mouse models (Noden, 1986a, 1986b; Couly et al., 1992, 1993) suggests a neural crest origin for the cranial base and face and a mixed origin for the cranial vault.

The assumption of conserved developmental processes across vertebrates has resulted in the conveyance of experimental work determining the embryonic source of bone in bird and mouse to the human condition. A recent review of embryonic derivations of cranial vault elements lists the lateral portion of the frontal bone, the parietal bones, the temporal bones except the squama, and the occipital bone as being derived from mesoderm, while the facial bones, the mandible, the medial portion of the frontal bone, the temporal squama, and the chondrocranium are derived from neural crest (Cohen, 2000b). New experimental evidence has fine-tuned the origins of neurocranial bones and meninges and suggests that intramembranous ossification of mesodermal-derived bone requires interaction with neural crest-derived meninges, whereas ossification of neural crest-derived bone can occur independent of the meninges (Jiang et al., 2002)

Novel experimental systems have provided an even more detailed map of the specific contributions of neural crest to the cranial vault. In an elegant and highly informative analysis of rhombencephalic neural crest segmentation, Köntges and Lumsden (1996) determined the enduring fate of neural crest subpopulations that migrate from individual rhombomeres using quail-chick chimeras. Rhombomeres are locations on the developing hindbrain where neural crest is organized into seven separate segments. Because of the phylogenetic proximity of quail and chick, cells from quail can be transplanted into precise locations on the chick embryo and the developmental fate of these cells can be determined by use of an antibody labeling system. Köntges and Lumsden's (1996) transplantation experiments specified the rhombomeric source of neural crest for specific parts of the skull; demonstrated that neural crest from multiple rhombomeres sometimes combine to form single bones; and showed that when neural crest cells from multiple rhombomeres come together to form a single bone, intraskeletal boundaries are sharply defined with antibody labeling. These boundaries do not correspond to anatomical borders (e.g., they are not *between* bones), but they *are* maintained throughout development. More importantly for this review, the authors discovered a highly constrained pattern of cranial skeletomuscular connectivity. It was demonstrated that neural crest from specific rhombomeres produced musculoskeletal units. A single rhombomere sends out neural crest cells that will form the connective tissue of a muscle plus the two skeletal elements to which the ends of the muscle attach. This means that the spot of bone on the cranial vault where a particular muscle attaches (as well as the muscle itself) is made from a discrete population of neural crest cells. That spot of bone sits within a skull bone whose embryological origin is a distinct population of cells.

As pointed out by Ahlberg (1997), the functional importance of this arrangement for evolution is obvious. The findings not only clarify how the skull maintains its functional competence (muscle-bone-nerve suites) through all the morphological changes of evolution, but also elucidate how minor to moderate dysmorphogenesis might not alter functional integrity of the skull. There are certainly thresholds of shape change that would cancel these relationships (e.g., the total absence of certain tissues), but the potential for a broad class of functioning variants seems more reasonable given these findings.

Evolutionarily, sutures need to be appreciated for their ability to allow movement between bones (even if minimal) when loads are applied. Evolutionary transformations of sutures include their conversion into very flexible joints (seen in the feeding apparatuses of fishes, lizards, snakes, birds) and the solidification of the cranium (including pan-synostosis) in animals that apply large loads to their skulls (Herring, 2000). Additional evolutionary considerations may be illuminating for clinically oriented researchers working with craniosynostosis. Herring (2000) presents a synthesis of paleontological and physiological information on sutures, putting them into a broader functional and evolutionary context, and suggests that the clinical issues associated with craniosynostosis have resulted in a distortion of biologically understanding sutural function. She makes an additional, important observation: the vast array of sutural variations seen across all craniates means that evolutionary changes in sutures are easy to achieve.

Study of the evolution of the vertebrate skull demonstrates a recurring pattern of a reduction in the number of bones in the skull, apparently produced by fusions of what were separate bones into composite skull elements. The mammalian skull represents a highly modified version of the primitive synapsid plan (Kardong, 1998), its evolution characterized by a consistent pattern of loss of skull elements through evolutionary time (Gregory et al., 1935). Rice (2000) provides a model that predicts selection will, where possible, break up correlations between phenotypic characters, and he has shown this model to be consistent with the steady loss of skull bones in mammals over evolutionary time.

If it is evolutionarily important to maintain the shape of a complex structure (e.g., the cranial vault), then the size and shape of individual structures (e.g., cranial vault bones) are free to vary independently. Over time, some individual elements will get larger, some will get smaller, and some will eventually be lost by drift. To apply this model to the study of craniosynostosis in humans, we need to determine whether we can equate the loss of a cranial vault bone with the loss of an associated suture. Though the interpretation of the data is very different depending on whether we are trying to understand a modern clinical phenomenon or the transformation of the vertebrate skull through evolutionary time, the epigenetic and genetic mechanisms responsible for suture position, suture morphology, and suture patency are most likely similar, if not identical.

14.6 THE GENETIC BASIS OF CRANIOFACIAL (DYS)MORPHOLOGY

Application of the techniques of molecular genetics to human traits is beginning to show a consistent picture that makes evolutionary sense, but is not always satisfying for those wanting easy answers. Most human genes are highly polymorphic, with tens if not hundreds of alleles (DNA sequence variants). These alleles produce very many genotypes, leading to a quasi-continuous distribution of effects on traits affected by the gene. Human geneticists often cling hopefully to classical Mendelian concepts, somewhat imprisoned by nineteenth-century conceptual thinking: Mendel's observations fit the inheritance patterns of the genes themselves, but if penetrance (relationships between genotype and phenotype) is complex the distribution of traits in families may not look very Mendelian.

—Weiss and Buchanan (2000)

The past decade of craniosynostosis research can be characterized as a rapid and steady race to find the gene(s) associated with various forms of craniosynostosis, syndromic and nonsyndromic. Much of the research in human cranial dysmorphology and craniosynostosis is targeted toward understanding the molecular signaling of bone growth, primarily by studying faulty signaling in mutant or transgenic animal models (e.g., Frenkel et al., 1990; Coffin et al., 1995; Hall and Miyake, 1995; Lui et al., 1995, 1999; Kim et al., 1998; Iseki et al., 1999; Johnson et al., 2000; Lemonnier et al., 2000; Mansukhani et al., 2000). Other researchers have focused on genetic mutations present in humans with craniosynostosis (see reviews by Kingsley, 1994a, 1994b; Muenke and Schell, 1995; Opperman et al., 1995; Moloney et al., 1997; Muenke et al., 1997; Oldridge et al., 1997; Wilkie, 1997; Hehr and Muenke, 1999; Cohen and MacLean, 2000; DeLeon et al., 2000). So much terrifically valuable information has been gained about craniofacial development, growth, morphology, and dysmorphology from these studies that it is difficult to take a sober look at what we have not yet learned.

Most of the illuminating advances in the genetics of craniosynostosis have come from work with the craniosynostosis syndromes. There are several reasons for this: (1) it is not uncommon to see familial cases of these syndromes, and so traditional genetic approaches have been undertaken historically; (2) these syndromes are characterized by a constellation of abnormalities that result in a substantial burden of care to the families of affected individuals, and therefore knowledge of etiology would bring relief to many; (3) affected individuals almost always require care from major medical facilities that characteristically include research laboratories.

A recent review by Herh and Muenke (1999) illustrates quite clearly that more than one mutation is associated with any particular craniosynostosis phenotype, and that a single mutation can be associated with several different phenotypes, each occupying a different diagnostic category. For example, the single mutation Cys342Ser in exon 9 of fibroblast growth factor receptor 2 (FGFR2) can result in either Crouzon or Pfeiffer syndrome (Lui et al., 1995; Opperman et al., 1995; Hehr and Muenke, 1999). Craniosynostosis is not a special case; the human genetics literature is full of examples demonstrating that the same disease may be produced by different alleles or even different genes in different ethnic or geographic regions, the same mutation may produce (or not produce) varying diseases, and alleles of the same genes affect many different phenotypes (Terwilliger and Göring, 2000; Weiss and Buchanan, 2000). Perhaps the mutations that have been identified as being associated with craniosynostosis syndromes are more common in the general population than we currently know. Given that many of these mutations are found on genes that mediate key processes of normal growth and development (the nature of the process varying with the time of development for some), their role in the production of craniosynostosis may be incidental or associative rather than causal. Perhaps the identified mutations are simply flags that can eventually lead us to the genuine identity of the molecular basis of craniosynostosis. Perhaps these mutations produce varying dysmorphic phenotypes (or unaffected phenotypes) only in the presence of other alleles (genetic background as influenced by geographic or ethnic group) or under certain nongenetic circumstances (aspects of the pre- and post-natal environment).

The idea of environmental interaction and genetic heterogeneity is not unknown to people working in craniosynostosis research (Mulvihill, 1995; Hehr and Muenke, 1999) (Chapter 8), yet even when other contributing factors are acknowledged, the gene is held as the primary causative agent. We need to consider the possibility that the identified mutations have a less eminent role in the cause of craniosynostosis (see Terwilliger and Göring, 2000; Weiss and Buchanan, 2000; Weiss and Terwilliger, 2000, for discussions

of this view as it relates to other human diseases).

Although classic craniofacial phenotypes have been described for the craniosynostosis syndromes, similarities in the craniofacial appearance of children across the craniosynostosis syndromes have been noted. In addition, a great deal of individual variability among patients within a diagnostic category is recognized. That there is great variability in the craniofacial phenotypes within a given syndrome is not a new observation, but it is one that is gaining new interest in light of the molecular data now available. Within-diagnosis craniofacial phenotypic variability has been discussed in terms of variable expressivity, degree of penetrance, and differential impact.

If we look at the seemingly more simple case of isolated craniosynostosis, phenotypic variation is also apparent (Fig. 14.4). Since a genetic basis has not been identified for most cases of isolated craniosynostosis (but see Muenke et al., 1997), variable

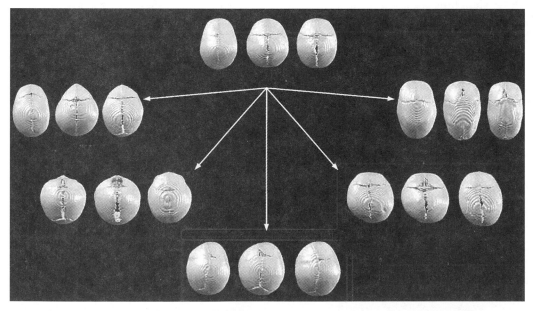

Figure 14.4. An example of the degree of morphological variability found in the cranial vault of patients with premature closure of various sutures. The cranial vault shapes are individual cases from the author's database of computed tomography scans taken for clinical purposes.

expressivity, degree of penetrance, and differential impact are not useful explanatory terms for the variation seen among most nonsyndromic craniosynostosis cases. Though we are accustomed to thinking of average shapes when contemplating the cranial vault morphologies, the variability observed among individuals with premature closure of the same suture may be as important to understanding cause and etiology as is the knowledge of any specific mutation. Objective, quantitative evaluation of the craniofacial phenotype may aid in determination of those aspects of the phenotype that are associated with premature closure of a given suture and those aspects that can be accounted for by normal population variability or background gene effects. Detailed evaluation of the craniofacial phenotype at the cellular, tissue, and morphological levels of analysis in affected individuals and that of close relatives, in association with careful genetic work, is necessary to reconcile the current discontinuity of genotype and phenotype studies. Consideration of the interaction of many genes in specific contexts is also necessary. Further understanding will come from developmental studies, but these will have to be conducted with nonhuman organisms.

14.7 CONCLUSIONS: SOME PRELIMINARY ANSWERS AND NEW QUESTIONS

The availability of computer technology has encouraged the lumping of data from many patients, presumably with the same diagnosis, for the purpose of sorting multiple variables with the expectation of gaining some clue as to cause and effect relationships. Such research designs are often illogical but the methodologies are sufficiently impressive to provide the guise of scientism. In contrast, the single case report is often scorned by the academician and dismissed with the pejorative label of being anecdotal and not neces-

sarily representative of the universe of experience.

—Pruzansky (1982)

This paper has surveyed the trends of thought and research aimed at understanding growth and development of the cranial vault, cranial vault dysmorphology associated with craniosynostosis, abnormal growth patterns, and the influence of several disciplines on craniofacial growth theory. I have briefly surveyed research pertaining to cranial vault development and morphology as a guide to what is known about craniosynostosis. The earliest students of craniosynostosis exercised then current trends of 19th-century biology by proposing and adopting a taxonomic classification of craniosynostosis. Craniofacial surgeons were probably the first investigators to document individual variation within these classes when patients responded differently to uniform surgical procedures. Recent genetic information has underscored another aspect of individual variability within diagnostic categories and has caused us to reexamine our categories and classification systems (see Cohen and MacLean, 2000, for a discussion of classification of craniosynostosis) (Chapter 2).

Partial answers to the questions that I introduced at the beginning of this chapter can be offered at this time:

1. *What is* the *cause of craniosynostosis?* The cause of craniosynostosis is more complicated than anticipated. Even though specific mutations have been identified in patients with craniosynostosis, until we can predict the phenotype with certainty from genetic information, the causative nature of the genes implicated in craniosynostosis remains unknown. Within specific types of isolated craniosynostosis and in the craniosynostosis syndromes, the true cause may lie upstream or downstream of the identified mutation(s). As suggested by Weiss and Buchanan (2000) for other human genetic

diseases, the cause may lie in the interaction of multiple genes in specific contexts.

2. *Is the brain (including the meninges) or the bone primary in terms of the cause of craniosynostosis and the production of the resulting morphology?* Development of the meninges, the brain, the membrane bones of the cranial vault, and the endochondral bone of the cranial base is directly or indirectly influenced by neural crest. These tissues develop in intimate contact and influence each other's progress using molecular signaling as well as undergoing biophysical interactions. There is clearly a relationship between central nervous system structures and the cranial vault.

One reason for surgical intervention in craniosynostosis is to relieve potentially increased intracranial pressure. In some cases of craniosynostosis, the central nervous system appears crowded with effacement of the ventricles, as though it is too big for its container. In other cases, there appears to be plenty of room for central nervous system structures even though a neighboring suture is closed prematurely. Does this mean that the central nervous system–cranial vault interaction varies according to which suture is closed? If so, does this reflect differing primary causes and developmental trajectories for the simple craniosynostoses?

3. *How do growth patterns contribute to the establishment of craniosynostosis morphologies? Is the dysmorphology defined early in development, with normal (though mechanically constrained) growth patterns contributing to the abnormal phenotype? Or, do anomalous growth patterns produce the abnormal phenotype?* The questions of growth relating to craniosynostosis have often focused on postnatal growth pattern, since that is the type of data typically available for analysis. Moreover, clinicians treat patients who are experiencing postnatal growth. It is generally thought that the severity of the dysmorphology seen in

craniosynostosis has to do with the age at which the synostosis occurs and the age of the child at diagnosis (i.e., how long the dysmorphology goes untreated). However, in my own research, we have not been able to statistically demonstrate an association between degree of dysmorphology and age in isolated sagittal synostosis (Richtsmeier et al., 1998). Extreme dysmorphology is not uncommon in newborns with isolated craniosynostosis, whereas other children are not diagnosed at birth but seem to "grow into" their dysmorphology.

Craniosynostosis can undoubtedly be established prenatally, perhaps prior to the establishment of tissues that we recognize as bone, brain, and meninges. Neural crest experiments have shown us that craniofacial growth is highly programmed and predictable but not deterministic (see Le Douarin et al., 1997). The contribution of neural crest to craniofacial development has been described as stochastic, with probabilities attached to certain developmental pathways (Hall, 2000). Craniosynostosis phenotypes are using these same pathways, but perhaps tend to access them with lower probabilities attached. If specific populations of neural crest cells are responsible for the formation of particular aspects of cranial vault structures (Hall, 2000), then we may need to focus more closely on aspects of neural crest cell migration and differentiation to further understand craniosynostosis—that is, the earliest aspects of development (i.e., stages prior to the appearance of bone) and prenatal environmental factors (e.g., maternal effects) need to be carefully considered. These considerations may be equally significant in the study of isolated and syndromic craniosynostosis.

4. *If left untreated, do craniosynostosis morphologies worsen with time?* Analyses of longitudinal data sets (including data from the adolescent years) by Kreiborg and colleagues (e.g., Kreiborg, 1981a, 1981b;

Kreiborg and Pruzansky, 1981; Kreiborg et al., 1999) have provided exciting information about the age progressivity of particular dysmorphologies. This question is becoming increasingly hard to investigate, however, since nearly all individuals diagnosed with craniosynostosis undergo some type of corrective intervention. Analyses of developing children with these conditions therefore need to consider the impact of the intervention and growth on changing craniofacial morphologies. Analytical techniques have not yet found a suitable way to discriminate between the effects of growth and the effects of intervention, since growth of the craniofacial complex remains poorly understood. Moreover, data from unaffected individuals, which are necessary to make proper comparisons, are ever more difficult to obtain.

Though some historic samples of untreated human patients exist, animal models may be our best tools for investigating the age progressivity of particular dysmorphologies. The strain of rabbits with congenital coronal synostosis (Mooney et al., 1994a, 1994b) (Chapter 9) offers a unique situation to investigate the combined effects of surgical intervention and growth in both prenatal and postnatal environments with an ideal normative sample. Many other animal models of craniosynostosis are being created and will be equally valuable in studying the interaction of these processes during ontogeny (see review of animal models in Chapter 9).

The recent history of craniofacial research has included influences from many different disciplines. The primacy of certain theories and ideas about calvarial growth during specific historic periods reflects the development and dissemination of the technological tools available to the researcher. We are currently in the molecular age where the human genome project became a primary scientific endeavor. Genes are fantastically important, but we need to recognize the identification of a mutation as a single piece in the puzzle rather than the final answer in our search to understand craniofacial growth, both normal and abnormal. We need to reformulate our approach in anticipation and acceptance of evidence from other aspects of craniofacial biology as well as other disciplines (see Weiss and Buchanan, 2000).

The complex array of interactions required to generate the skull include processes that occur very early and very late during development, those with a long evolutionary history, and those that are relatively novel due to genetic mutation and phylogenetic change. To understand morphogenesis, we need to continue to build evidence with data from all levels that span the genotype-phenotype continuum: gene, protein, cell, populations of cells, tissues, organs, and organism. This may require us to pay closer attention to the nature of the variation seen within a sample or a population (of genes, of cells, of individuals). Although labor intensive, the rewards gained from looking carefully at individual variabilities may prove worth it.

Various avenues of craniofacial biology research are producing evidence for partial answers to the questions that I listed relating to cranial vault formation, development, growth, and craniosynostosis. These answers, when accumulated, may or may not fit together nicely to form a coherent representation of the big picture of cranial vault development. Unknown and unexpected interactions and processes may be discovered, and their role in craniofacial development may only be clarified through the development of new tools. Asking the right questions, undoubtedly different from those I have listed above, will be a major step in uncovering the details of cranial vault development. Questions that I believe deserve our attention include:

- What is the role of individual (or normal) variability in the shape of the skull in craniosynostosis?

- How much of the variability seen within diagnostic categories reflects normal population variation and how much reflects aspects of the dysmorphology?
- How does the variability seen in the proposed molecular basis for certain types of craniosynostosis fit or not fit with the phenotypic variability observed?
- What aspects of the genotype, the phenotype, and the developmental program are invariant? What aspects are variable?

If we can find answers to these questions in the study of craniosynostosis in humans, we will go a long way toward answering very basic questions pertaining to evolution and development of the cranial vault.

ACKNOWLEDGMENTS

References in this paper are meant to be a guide to the potential literature and are not meant to be exhaustive. I thank Valerie Burke DeLeon, Kristina Aldridge, and Frank L. Williams for careful reading of previous versions of this manuscript. Many of the ideas about genotype-phenotype correlation in craniosynostosis were influenced through conversations with Ken Weiss, but the opinions expressed are my own. Cassio Lynm prepared Figure 14.1. Valerie Burke DeLeon prepared Figure 14.4. I thank M. Michael Cohen, Jr., for kind permission to use Figure 14.2. This paper was supported in part by PHS grant P60 DE13078, Project VI.

REFERENCES

Ahlberg, P. E. (1997). How to keep a head in order [news]. *Nature 385* (6616), 489–490.

al-Qattan, M. M., and Phillips, J. H. (1997). Clinical features of Crouzon's syndrome patients with and without a positive family history of Crouzon's syndrome. *J. Craniofac. Surg. 8*, 11–13.

Antón, S. C., and Weinstein, K. J. (1999). Artificial cranial deformation and fossil Australians revisited. *J. Hum. Evol. 36*, 195–209.

Antón, S. C., Jaslow, C. R., and Swartz, S. M. (1992). Sutural complexity in artificially deformed human (*Homo sapiens*) crania. *J. Morphol. 214*, 321–332.

Bjork, A. (1964). Sutural growth of the upper face studied by the implant method. *Trans. 40th Cong. Eur. Orthod. Soc.*, pp. 49–65.

Bjork, A. (1968). The use of metallic implants in the study of facial growth in children: method and application. *Am. J. Phys. Anthropol. 29*, 243–251.

Blechschmidt, E. (1961). *The Stages of Human Development Before Birth: An Introduction to Human Embryology* (Philadelphia: W. B. Saunders).

Blechschmidt, E. (1976a). Principles of biodynamic differentiation. In J. Bosma (ed.), *Development of the Basicranium* (Bethesda: Department of Health, Education and Welfare), DHEW Publication (NIH) 76-989, pp. 54–76.

Blechschmidt, M. (1976b). Biokinetics of the basicranium. In J. Bosma (ed.), *Development of the Basicranium* (Bethesda: Department of Health, Education and Welfare), DHEW Publication (NIH) 76-989.

Bradley, J., Levine, J., Blewett, C., Krummel, T., McCarthy, J., and Longaker, M. (1996). Studies in cranial suture biology: In vitro cranial suture fusion. *Cleft Palate Craniofac. J. 33*, 150–156.

Broadbent, B., Broadbent, B., Jr., and Golden, W. (1975). *Bolton Standards of Dentofacial Developmental Growth* (St. Louis: C. V. Mosby).

Cheverud, J. M., and Midkiff, J. E. (1992). Effects of fronto-occipital cranial reshaping on mandibular form. *Am. J. Phys. Anthropol. 87*, 167–171.

Cheverud, J. M., Kohn, L. A., Konigsberg, L. W., and Leigh, S. R. (1992). Effects of fronto-occipital artificial cranial vault modification on the cranial base and face. *Am. J. Phys. Anthropol. 88*, 323–345.

Coffin, J., Florkiewicz, R., Neumann, J., Mort-Hopkins, T., Dorn, G., Lightfoot, P., German, R., Howles, P., Kier, A., O'Toole, B., Sasse, J., Gonzalez, A., Baird, A., and Doetschman, J. (1995). Abnormal bone growth and selective translational regulation in basic fibroblast growth factor (FGF-2) transgenic mice. *Molec. Biol. Cell. 6*, 1861–1873.

Cohen, M. M., Jr. (ed.) (1986). *Craniosynostosis: Diagnosis, Evaluation, and Management* (New York: Raven Press).

Cohen, M. M., Jr. (2000a). Sutural biology. In M. M. Cohen, Jr., and R. E. MacLean (eds.), *Craniosynostosis: Diagnosis, Evaluation, and Management* (New York: Oxford University Press), pp. 11–23.

Cohen, M. M., Jr. (2000b). Merging the old skeletal biology with the new. I. Intramembranous ossification, endochondral ossification, ectopic bone, secondary cartilage, and pathologic considerations. *J. Craniofac. Genet. Devel. Biol. 20*, 84–93.

Cohen, M. M., Jr., and MacLean, R. E. (eds.) (2000). *Craniosynostosis: Diagnosis, Evaluation, and Management* (New York: Oxford University Press).

Cole, T., III (1997). Historical note: early Anthropological contributions to "geometric morphometrics." *Am. J. Phys. Anthropol. 101*, 291–296.

Couly, G., Coltey, P., and Le Douarin, N. (1992). The developmental fate of the cephalic mesoderm in quail-chick chimeras. *Development 114*, 1–15.

Couly, G., Coltey, P., and Le Douarin, N. (1993). The triple origin of skull in higher vertebrates: a study in quail-chick chimeras. *Development 117*, 409–429.

David, J. D., Poswillo, D., and Simpson, D. (1982). *The Craniosynostoses: Causes, Natural History, and Management* (Berlin: Springer-Verlag).

DeLeon, V., Jabs, E., and Richtsmeier, J. (2000). Craniofacial growth: genetic basis and morphogenetic process in craniosynostosis. In C. VanderKolk (ed.), *Surgery: Indications, Operations, Outcomes*, vol. 2 (St. Louis: Mosby-Yearbook, Inc), pp. 619–636.

DeLeon, V., Zumpano, M., and Richtsmeier, J. (2001). The effect of neurocranial surgery on basicranial morphology in isolated sagittal craniosynostosis. *Cleft Palate Craniofac. J. 38*, 134–146.

Drake, D., Persing, J. A., Berman, D., and Ogle, R. C. (1993). Calvarial deformity regeneration following subtotal craniectomy for craniosynostosis: a case report and theoretical implications. *J. Craniofac. Surg. 4*, 35–39.

Dunlop, L.-L. T., and Hall, B. (1995). Relationships between cellular condensation, preosteoblast formation and epithelial-mesenchymal interactions in initiation of osteogenesis. *Int. J. Devel. Biol. 39*, 357–371.

Enlow, D. (1963). *Principles of Bone Remodeling* (Springfield: Charles C. Thomas).

Enlow, D. (1975). *Handbook of Facial Growth* (Philadelphia: W. B. Saunders).

Enlow, D. (2000). Normal craniofacial growth. In M. M. Cohen, Jr., and R. E. MacLean (eds.), *Craniosynostosis: Diagnosis, Evaluation, and Management* (New York: Oxford University Press), pp. 454–472.

Frenkel, S., Grande, D., Collins, M., and Singh, I. (1990). Fibroblast growth factor in chick osteogenesis. *Biomaterials 11*, 38–40.

Gans, B., and Sarnat, B. (1951). Sutural facial growth of the *Macaca* rhesus monkey: a gross and serial roentgenographic study by means of metallic implants. *Am. J. Orthodont. 37*, 10–16.

Gans, C., and Northcutt, G. (1983). Neural crest and the origins of the vertebrates: a new head. *Science 220*, 268–274.

Giuffre, R., Vagnozzi, R., and Savino, S. (1978). Infantile craniosynostosis: clinical, radiological, and surgical considerations based on 100 surgically treated cases. *Acta Neurochirurgica 44*, 49–67.

Glucksmann, A. (1942). The role of mechanical stress in bone formation in vitro. *J. Anat. London 76*, 231–239.

Gregory, W., Roigneau, M., Burr, E., Evans, G., Hellman, E., Jackson, F., MacDill, M., Manter, J,. Marshak, R. (1935). "Williston's law" relating to the evolution of skull bones in the vertebrates. *American Journal of Physical Anthropology 20*, 123–152.

Hall, B. (ed.) (1991). *Bone* (Boca Raton, FL: CRC Press).

Hall, B. (1999a). *Evolutionary Developmental Biology* (Dordrecht, Holland: Kluwer Academic).

Hall, B. K. (1999b). *The Neural Crest in Development and Evolution* (New York: Springer).

Hall, B. (2000). The neural crest as a fourth germ layer and vertebrates as quadroblastic not triploblastic. *Evol. Devel. 2*, 3–5.

Hall, B. K., and Miyake, T. (1995). Divide, accumulate, differentiate: cell condensation in skeletal development revisited. *Int. J. Devel. Biol. 39*, 881–893.

Hall, B., and Miyake, T. (2000). All for one and one for all: condensations and the initiation of skeletal development. *BioEssays 22*, 138–147.

Hehr, U., and Muenke, M. (1999). Craniosynostosis syndromes: From genes to premature fusion of skull bones. *Molec. Genet. Metab. 68*, 139–151.

Herring, S. (2000). Sutures and craniosynostosis: A comparative, functional, and evolutionary perspective. In M. M. Cohen, Jr., and R. E. MacLean (eds.), *Craniosynostosis: Diagnosis, Evaluation, and Management* (New York: Oxford University Press), pp. 1–10.

Hobar, P., Masson, J., Wilson, R., and Zerwekh, J. (1996). The importance of the dura in craniofacial surgery. *Plast. Reconstr. Surg. 98*, 217–225.

Iseki, S., Wilkie, A., and Morriss-Kay, G. (1999). Fgfr1 and Fgfr2 have distinct differentiation- and proliferatiion-related roles in the developing mouse skull vault. *Development 126*, 5611–5620.

Jiang, X., Iseki, S., Maxson, R., Sucov, H., Moriss-Kay, G. (2002). Tissue origins and interactions in the mammalian skull vault. *Developmental Biology 241*, 106–116.

Johnson, D., Iseki, S., Wilkie, A., and Morriss-Kay, G. (2000). Expression patterns of *Twist* and *Fgfr1, -2* and *-3* in the developing mouse coronal suture suggest a key role for *Twist* in suture initiation and biogenesis. *Mech. Devel. 91*, 341–345.

Kardong, K. V. (1998). *Vertebrates: Comparative Anatomy, Function, Evolution*, 2nd ed. (Boston: WCB McGraw-Hill).

Kim, H., Rice, D., Kettunen, P., and Thesleff, I. (1998). FGF-, BMP-, and Shh-mediated signalling pathways in the regulation of cranial suture morphogenesis and calvarial bone development. *Development 125*, 1241–1251.

Kingsley, D. M. (1994a). The TGF-β superfamily: new members, new receptors, and new genetic tests of function in different organisms. *Genes Devel. 8*, 133–146.

Kingsley, D. M. (1994b). What do BMPs do in mammals? Clues from the mouse short-ear mutations. *Trends Genet. 10*, 16–21.

Klaauw, C. v. d. (1945). Cerebral skull and facial skull. *Arch. Neerl. Zool. 7*, 16–37.

Klaauw, C. v. d. (1948–1952). Size and position of the functional components of the skull. *Arch. Neerl. Zool. 9*, 1–559.

Knudson, H., and Flaherty, R. (1960). Craniosynostosis. *Am. J. Roentgen. 84*, 454–460.

Kohn, L. A., Leigh, S. R., Jacobs, S. C., and Cheverud, J. M. (1993). Effects of annular cranial vault modification on the cranial base and face. *Am. J. Phys. Anthropol. 90*, 147–168.

Kohn, L. A., Vannier, M. W., Marsh, J. L., and Cheverud, J. M. (1994). Effect of premature sagittal suture closure on craniofacial morphology in a prehistoric male Hopi. *Cleft Palate Craniofac. J. 31*, 385–396.

Kohn, L. A., Leigh, S. R., and Cheverud, J. M. (1995). Asymmetric vault modification in Hopi crania. *Am. J. Phys. Anthropol. 98*, 173–195.

Konigsberg, L. W., Kohn, L. A., and Cheverud, J. M. (1993). Cranial deformation and nonmetric trait variation. *Am. J. Phys. Anthropol. 90*, 35–48.

Köntges, G., and Lumsden, A. (1996). Rhombencephalic neural crest segmentation is preserved throughout craniofacial ontogeny. *Development 122*, 3229–3242.

Kreiborg, S. (1981a). Craniofacial growth in plagiocephaly and Crouzon syndrome. *Scand. J. Plast. Reconstr. Surg. 15*, 187–197.

Kreiborg, S. (1981b). Crouzon syndrome. A clinical and roentgencephalometric study. *Scand. J. Plast. Reconstr. Surg. 18* (suppl.), 1–198.

Kreiborg, S., and Pruzansky, S. (1981). Craniofacial growth in premature craniofacial synostosis. *Scand. J. Plast. Reconstr. Surg. 15*, 171–186.

Kreiborg, S., Aduss, H., and Cohen, M. M., Jr. (1999). Cephalometric study of the Apert syndrome in adolescence and adulthood. *J. Craniofac. Genet. Devel. Biol. 19*, 1–11.

Lanyon, L., and Rubin, C. (1985). Functional adaptation in skeletal structures. In H. Hildebrand (ed.), *Functional Vertebrate Morphology* (Cambridge, MA: Belknap Press of Harvard University Press).

Le Douarin, N. M., Catala, M., and Batini, C. (1997). Embryonic neural chimeras in the study of vertebrate brain and head development. *Int. Rev. Cytol. 175*, 241–309.

Lele, S., and Richtsmeier, J. (2001). *An Invariant Approach to Statistical Analysis of Shapes* (London: Chapman Hall–CRC).

Lemonnier, J., Delannoy, P., Hott, M., Lomri, A., Modrowski, D., and Marie, P. (2000). The Ser252Trp fibroblast growth factor receptor-2 (FGFR-2) mutation induces PKC-independent down regulation of FGFR-2 associated with premature calvaria osteoblast differentiation. *Exp. Cell. Res. 256*, 158–167.

Levine, J., Bradley, J., Roth, D., McCarthy, J., and Longaker, M. (1998). Studies in cranial suture biology: regional dura mater determines overlying suture biology. *Plast. Reconstr. Surg. 101*, 1441–1447.

Lui, Y., Kundu, R., Wu, L., Luo, W., Ignelzi, M., Snead, M., and Maxson, R. J. (1995). Premature suture closure and ectopic cranial bone in mice expressing Msx2 transgenes in the developing skull. *Proc. Natl. Acad. Sci. 92*, 6137–6141.

Lui, Y., Tang, Z., Kundu, R., Wu, L., Luo, W., Zhu, D., Sangiorgi, F., Snead, M., and Maxson, R. (1999). Msx2 gene dosage influences the number of proliferative osteogenic cells in growth centers of the developing murine skull: a possible mechanism for MSX2-mediated craniosynostosis in humans. *Devel. Biol. 205*, 260–274.

Mansukhani, A., Bellosta, P., Sahni, M., and Basilico, C. (2000). Signaling by fibroblast growth factors (FGF) and fibroblast growth factor receptor 2 (FGFR2)–activating mutations blocks mineralization and induces apoptosis in osteoblasts. *J. Cell Biol. 149*, 1297–1308.

Marsh, J. L., and Vannier, M. W. (1986). Cranial base changes following surgical treatment of craniosynostosis. *Cleft Palate J. 23* (suppl. 1), 9–18.

Merow, W., and Broadbent, B. (1990). Cephalometrics. In D. Enlow (Ed.), *Handbook of Facial Growth* (Philadelphia: W. B. Saunders), pp. 346–395.

Miyake, T., Cameron, A., and Hall, B. (1996). Stage-specific onset of condensation and matrix deposition for Meckel's and other first arch cartilages in inbred C57BL/6 mice. *J. Craniofac. Genet. Devel. Biol. 16*, 32–47.

Moloney, D. M., Wall, S. A., Ashworth, G. J., Oldridge, M., Glass, I. A., Francomano, C. A., Muenke, M., and Wilkie, A. O. (1997). Prevalence of Pro250Arg mutation of fibroblast growth factor receptor 3 in coronal craniosynostosis. *Lancet 349* (9058), 1059–1062.

Mooney, M. P., Losken, H. W., Siegel, M. I., Lalikos, J. F., Losken, A., Burrows, A. M., and Smith, T. D. (1994a). Development of a strain of rabbits with congenital simple nonsyndromic coronal suture synostosis. Part II: Somatic and craniofacial growth patterns. *Cleft Palate Craniofac. J. 31*, 8–16.

Mooney, M. P., Losken, H. W., Siegel, M. I., Lalikos, J. F., Losken, A., Smith, T. D., and Burrows, A. M. (1994b). Development of a strain of rabbits with congenital simple nonsyndromic coronal suture synostosis. Part I: Breeding demographics, inheritance pattern, and craniofacial anomalies. *Cleft Palate Craniofac. J. 31*, 1–7.

Moss, M. (1957). Experimental alteration of sutural area morphology. *Anat. Rec. 127*, 569–589.

Moss, M. (1958). The pathogenesis of artificial cranial deformation. *Am. J. Phys. Anthropol. 16*, 269–286.

Moss, M. (1959). The pathogenesis of premature cranial synostosis in man. *Acta Anat. 37*, 351–370.

Moss, M. (1960). Inhibition and stimulation of sutural fusion in the rat calvaria. *Anat. Rec. 136*, 457–469.

Moss, M. (1971). Functional cranial analysis and the functional matrix. In *Patterns of Orofacial Growth and Development* (Ann Arbor: American Speech and Hearing Association Reports).

Moss, M. (1975). Functional anatomy of cranial synostosis. *Child's Brain 1*, 22–33.

Moss, M. (1981). Genetics, epigenetics, and causation. *Am. J. Orthod. 80*, 366–375.

Moss, M., and Moss-Salentijn, L. (1979). The muscle-bone interface: an analysis of a morphological boundary. In D. Carlson and J. McNamara (eds.), *Muscle Adaptation in the Craniofacial Region* (Ann Arbor: Center for Human Growth and Development), pp. 39–71.

Moss, M., and Young, R. (1960). A functional approach to craniology. *Am. J. Phys. Anthropol. 18*, 281–292.

Moyers, R. E., and Bookstein, F. L. (1979). The inappropriateness of conventional cephalometrics. *Am. J. Orthodont. 75*, 599–617.

Muenke, M., and Schell, U. (1995). Fibroblast-growth-factor receptor mutations in human skeletal disorders. *Trends Genet. 11*, 308–313.

Muenke, M., Gripp, K. W., McDonald-McGinn, D. M., Gaudenz, K., Whitaker, L. A., Bartlett, S. P., Markowitz, R. I., Robin, N. H., Nwokoro, N. J., Mulvihill, J., Losken, H. W., Mulliken, J. B., Guttmacher, A. E., Wilroy, R. S., Clarke, L. A., Hollway, G., Ades, L. C., Haan, E. A., Mulley, J. C., Cohen, M. M., Jr., Bellus, G. A., Francomano, C. A., Moloney, D. M., Wall, S. A., and Wilkie, A. O. (1997). A unique point mutation in the fibroblast growth factor receptor 3 gene (FGFR3) defines a new craniosynostosis syndrome. *Am. J. Hum. Genet. 60*, 555–564.

Mulvihill, J. (1995). Craniofacial syndromes: no such thing as a single gene disease. *Nat. Genet. 9*, 101–103.

Noden, D. (1986a). Origins and patterning of craniofacial mesenchymal tissues. *Anat. Rec. 208*, 1–13.

Noden, D. (1986b). Patterning of avian craniofacial muscles. *Devel. Biol. 116*, 347–356.

Oldridge, M., Lunt, P. W., Zackai, E. H., McDonald-McGinn, D. M., Muenke, M., Moloney, D. M., Twigg, S. R., Heath, J. K., Howard, T. D., Hoganson, G., Gagnon, D. M., Jabs, E. W., and Wilkie, A. O. (1997). Genotype-phenotype correlation for nucleotide substitutions in the IgII–IgIII linker of FGFR2. *Hum. Molec. Genet. 6*, 137–143.

Opperman, L. A. (2000). Cranial sutures as intramembranous bone growth sites. *Devel. Dyn. 219*, 472–485.

Opperman, L. A., Sweeney, T. M., Redmon, J., Persing, J. A., and Ogle, R. C. (1993). Tissue interactions with underlying dura mater inhibit osseous obliteration of developing cranial sutures. *Devel. Dyn. 198*, 312–322.

Opperman, L. A., Persing, J. A., Sheen, R., and Ogle, R. C. (1994). In the absence of periosteum, transplanted fetal and neonatal rat coronal sutures resist osseous obliteration. *J. Craniofac. Surg. 5*, 327–332.

Opperman, L. A., Passarelli, R. W., Morgan, E. P., Reintjes, M., and Ogle, R. C. (1995). Cranial sutures require tissue interactions with dura mater to resist osseous obliteration in vitro. *J. Bone Miner. Res. 10*, 1978–1987.

Opperman, L. A., Chhabra, A., Nolen, A. A., Bao, Y., and Ogle, R. C. (1998). Dura mater maintains rat cranial sutures in vitro by regulating suture cell proliferation and collagen production. *J. Craniofac. Genet. Devel. Biol. 18*, 150–158.

Opperman, L. A., Adab, K., and Gakunga, P. T. (2000). Transforming growth factor-beta2 and TGF-beta3 regulate fetal rat cranial suture morphogenesis by regulating rates of cell proliferation and apoptosis. *Devel. Dyn. 219*, 237–247.

Popa, G. (1936). Mechanostruktur und Mechanofunction der Dura mater des Menschen. *Morph. Jb. 78*, 85–187.

Pruzansky, S. (1982). Craniofacial surgery: the experiment on nature's experiment. Review of three patients operated by Paul Tessier. *Eur. J. Orthod. 4*, 151–171.

Rice, S. (2000). The evolution of developmental interactions: epistasis, canalization, and integration. In J. Wolfe, E. D. Brodie, III, and M. J. Wade (eds.), *Epistasis and the Evolutionary Process* (New York: Oxford University Press).

Richtsmeier, J. (1985). A study of normal and pathological craniofacial morphology and growth using finite element methods. Ph.D. Dissertation, Northwestern University, Evanston, IL, p. 297.

Richtsmeier, J. T., Grausz, H. M., Morris, G. R., Marsh, J. L., and Vannier, M. W. (1991). Growth of the cranial base in craniosynostosis. *Cleft Palate Craniofac. J. 28*, 55–67.

Richtsmeier, J. T., Cole, T. M., III, Krovitz, G., Valeri, C. J., and Lele, S. (1998). Preoperative morphology and development in sagittal synostosis. *J. Craniofac. Genet. Devel. Biol. 18*, 64–78.

Roth, D., Bradley, J., Levine, J., McMullen, H., McCarthy, J., and Longaker, M. (1996). Studies in cranial suture biology: II. Role of the dura in cranial suture fusion. *Plast. Reconstr. Surg. 97*, 693–699.

Sarnat, B. (1968). Growth of bones as revealed by implant markers in animals. *Am. J. Phys. Anthropol. 29*, 255–264.

Sgouros, S., Natarajan, K., Hockley, A. D., Goldin, J. H., and Wake, M. (1999). Skull base growth in craniosynostosis. *Pediatr. Neurosurg. 31*, 281–293.

Smith, D., and Tondury, T. (1978). Origin of the calvaria and its sutures. *Am. J. Dis. Child 132*, 662–666.

Storey, E. (1955). Bone changes associated with tooth movement. A histological study of the effect of force in rabbit, guinea pig and rat. *Austr. J. Dent. 59*, 147–161.

Terwilliger, J. D., and Göring, H. H. (2000). Gene mapping in the 20th and 21st centuries: statistical methods, data analysis, and experimental design. *Hum. Biol. 72*, 63–132.

Thompson, D. A. W. (1917). *On Growth and Form.* (Cambridge: University Press).

Virchow, R. (1851). Uber den Cretinismus, namentlich in Franken, und uber pathologische Schadelformen. *Verh. Phys. Med. Gesellsch. Wurzburg. 2*, 230–270.

Weiss, K., and Buchanan, A. (2000). Rediscovering Darwin after a Darwinian century. *Evol. Anthrop. 9*, 187–200.

Weiss, K. M., and Terwilliger, J. D. (2000). How many diseases does it take to map a gene with SNPs? *Nat. Genet. 26*, 151–157.

Wilkie, A. O. (1997). Craniosynostosis: genesand mechanisms. *Hum. Molec. Genet. 6*, 1647–1656.

PART V

MIDFACIAL AND MANDIBULAR DYSMORPHOLOGY AND GROWTH DISTURBANCES

CHAPTER 15

EVOLUTIONARY CHANGES IN THE MIDFACE AND MANDIBLE: ESTABLISHING THE PRIMATE FORM

JOSEPH R. SIEBERT, Ph.D. and DARIS R. SWINDLER, Ph.D.

15.1 INTRODUCTION

The primates are an extraordinarily diverse group of mammals. Over 50 extant genera and 200 species are recognized (Napier and Napier, 1967). Body size is highly variable, ranging from the 80 gram (g) mouse lemur to the 160 kilogram (kg) male gorilla; maximum life spans extend from 8.8 years in the dwarf lemur to 120 years or so in humans. Ecological diversity is also well recognized: some primates inhabit very narrow niches, whereas others, most notably the human, survive in virtually every climate and habitat on earth (Bogin, 1999).

As will become apparent in this chapter, the primate craniofacial complex has played, and continues to play, a crucial role in exploiting and adapting to the environment. Given the diversity of primate morphology and natural history, it is not surprising that we lack a unified plan for craniofacial evolution. Because the ontogeny and phylogeny of the craniofacial complex are so involved, knowledge about these processes is bountiful but highly fragmented. For both extinct and extant species, individual structures and regions have been studied, and their morphologic and functional attributes elucidated. Patterns of ontogenetic and phylogenetic development and the factors that control them have been investigated. Ultimately, however, both developmental and evolutionary contributions *and their interactions* need to be understood before any final understanding can be achieved.

Unfortunately, no such synthesis has been produced. This is due to a variety of reasons, chiefly the immense morphologic and developmental complexity of the region. The craniofacial complex is formed from multiple tissues, the result of varied morphogenetic processes that are relatively independent in early life and do not necessarily correspond to adult skeletal components (Gans, 1988; Thorogood, 1988). This, and the current inability to distinguish genetic from epigenetic controls with certainty, complicates not only the study of craniofacial morphogenesis but of evolution as well.

For these reasons, we are obliged in this chapter to follow the traditional approach of examining structures, regions, and patterns more individually than we would prefer. We do so with the hope that this information will prepare the reader to understand future developments in the areas of craniofacial ontogeny and phylogeny. In this review, we will follow the formal nomenclature of paleoanthropologists. The term *hominid* derives from the family *Hominidae* and includes modern humans and all human ancestral lineages occurring from the time of the split with apes. *Pongid* (from the family *Pongidae*) refers to living apes and their ancestors. The more encompassing term *hominoid* (from the superfamily *Hominoidea*) indicates humans, apes, and their ancestors. The review is slanted toward the hominoids and, as such, gives more emphasis to the apes, australopithecines, and species of *Homo* than the more phyletically distant prosimi-

ans and monkeys. A basic timeline of primate evolution is provided in Table 15.1.

15.2 GENERAL PRINCIPLES OF MIDFACIAL ONTOGENY

15.2.1 General Principles

Much of our knowledge of craniofacial growth in the primates comes from the cross-sectional or one-time study of individuals or, in nonhuman primates, of prepared specimens. Longitudinal, or repeated, studies of primate growth have proven much more difficult but provide considerably more telling information (Gavan, 1953, 1982; Scott, 1967; Swindler et al., 1973; Watts and Gavan, 1982; Sirianni and Swindler, 1985). These and other

TABLE 15.1 Timeline of Primate Evolution

Era/Epoch	Representative Species	Boundaries of Fossil Record[a]	
		Appearance	Disappearance
Paleocene	*Plesiadapis* (rodent-like primate)	55 mya	60 mya
Eocene	*Notharctus*	58 mya	36 mya
	Adapis (early prosimians)	58 mya	36 mya
Oligocene	*Aegyptopithecus* (first true ape)	35 mya	25 mya
Miocene	*Proconsul* (dryopithecine ape)	19–20 mya	13–16 mya
Pliocene	*A. anamensis*	4.2 mya	3.9 mya
	A. afarensis	3.8 mya	2.95 mya
	A. africanus	3.05 mya	2.0–2.3 mya
	A. aethiopicus[b]	2.7 mya	2.3 mya
Pleistocene	*A. robustus*[b]	1.8–2.0 mya	1.5 mya
	P. boisei[c]	2.1–2.3 mya	1.3 mya
	H. habilis	2.0–2.4 mya	1.45 mya
	H. erectus	1.6–1.8 mya	500,000 yr
	H.s. neanderthalensis	100,000–230,000 yr	30,000 yr
	Homo sapiens	40,000 yr	—

[a] Dates are approximate and often remain under debate.
[b] South African robust form.
[c] East African robust form.
mya, million years ago; *A.*, *Australopithecus*; *P.*, *Paranthropus*; *H.*, *Homo*; *H.s.*, *Homo sapiens*.

studies have indicated that the range of variation in craniofacial features is often underestimated. Studies of wild-caught (or, in earlier times, wild-shot) animals are complicated, in that age and lineage are unknown, and specimens can be damaged or even destroyed during capture. The study of captive animals may be easier in some respects, but does not necessarily reflect primate life in the wild. Captivity can, for example, alter sexual maturation (Tanner et al., 1990) and other endocrine functions related to growth, development, and function (Bogin, 1977).

Several patterns of postnatal growth are recognized among the primates (Leigh, 1996). Some primates—for example, the marmoset—exhibit no juvenile phase of growth. Others, like the baboon, show separate infant, juvenile, and adult phases of active growth. Human growth is different yet, with a childhood period coming between infancy and the juvenile phase, and a pronounced adolescent growth spurt. Growth spurts have also been identified for particular parts of the body—for example, the snout and/or palate in the baboon and pig-tailed macaque (Swindler and Sirianni, 1973; Byrd and Swindler, 1980). Growth rates can even vary between portions of the same structure. In the baboon and macaque, the anterior palate grows at a greater rate than the posterior palate (Swindler and Sirianni, 1973). Secular trends have been observed, perhaps most notably in humans, where children often have longer, deeper, and narrower faces than their (same-sex) parents (Smith et al., 1986).

Differences in growth patterns have widespread ecological, social, and evolutionary consequences (see review by Gould, 1977). Indeed, morphologic evolution is often equated with change in developmental programs over extended periods of time. Entirely new morphological characters can be introduced, or characters that have persisted for generations may undergo changes in developmental timing—for example, stage of appearance, rate of growth, presence of a growth spurt, and so on. *Heterochrony*, the term given to such changes in timing, is widely accepted as an important mechanism of evolution. One particular type of heterochrony is *neoteny*, the retention in later generations of juvenile characters (e.g., large calvaria, flat midface, small jaws) found in earlier generations. Neoteny is also thought to be an important mechanism in human evolution, particularly that of the craniofacial complex.

15.2.2 Current Issues of Craniofacial Development

It is important in the understanding of midfacial evolution to appreciate the vast contribution that ontogenetic factors make. The contribution and interplay of genetic and epigenetic factors continues to challenge embryologists as well as dysmorphologists. Work with patterning genes has proceeded for a number of years, and is yielding much new information (see reviews by Holland, 1988; Lonai and Orr-Urtreger, 1990; Noden, 1991; Schilling, 1997; Whiting, 1997; and others). However, it remains unclear how early patterning at the cellular level, involving growth and differentiation and shown to be under genetic control, relates to later growth at the tissue level, some of which is influenced by epigenetic factors. For a further understanding of these latter issues, the reader is referred to the profuse work of Enlow (1982) and Enlow and Hans (1990) on remodeling of bone growth, Blechschmidt (1961, 1977) and Blechschmidt and Gasser (1978) on biomechanical factors, and Hall (1990) on epigenetic factors.

The discovery of homeobox genes has contributed significantly to the understanding of patterning during craniofacial morphogenesis. Homeobox genes encode transcription factors that regulate gene expression during early development, and thus specify the identity and patterning of body segments and craniofacial components (Thesleff, 1995). Combinations of Hox genes are thought to specify the patterning of particular structures. Bone morphogenetic proteins, for example, control bone, cartilage, and dentin development. Cephalic neural crest cells, which give rise to most craniofacial mesenchyme, express Hox2 homeobox genes, which probably imprint crest cells prior to migration (Hunt et al., 1991). Likewise, the branchial arches express combinations of Hox genes in highly specified ways. The ramifications of this information for evolution are, of course, extraordinary (see below).

15.3   ANATOMIC FEATURES OF THE PRIMATE MIDFACE

In this chapter, the midface is defined as the middle one-third of the cranium (measured in the cranial caudal dimension) encompassing the orbits, maxillae, nasal and oral cavities. With the inclusion of the basicranium and mandible, our discussion involves the lower two-thirds of the cranium, excluding chiefly the calvaria, brain and associated neural elements, and ears. Our discussion is limited for the most part to topographic features of the midfacial skeleton, those which are most often afforded paleontologists. For the most part, soft tissues (e.g., epithelium, nerves, skeletal muscle, and tendons) are excluded from review. They are known in extinct creatures chiefly by foramina, fossae, and irregularities left on bony surfaces, but a complete discussion is beyond the scope of this chapter.

15.3.1   Early Morphologic Studies of the Primates

The systematic understanding of primate morphology appears to have had its origins

with Aristotle (400 B.C.), followed by Galen (A.D. 200) (cited in Swindler, 1998). Indeed, Galen's studies of the Barbary ape produced the foundation for anatomic knowledge for hundreds of years. Unfortunately, his studies of this Old World monkey also formed the basis for many statements regarding human anatomy. In this way, Galen "paved the way for future confusion" (Wood et al., 1969), and not until Vesalius's work in the middle of the 16th century was human anatomy based on dissections of humans (and even then some of Vesalius's drawings were idealized). General features of the chimpanzee were probably first described by the explorer Hanno in 500 B.C.; anatomic details were put forth much later, by Tulp in 1641 (cited in Swindler, 1998). Battell wrote of the gorilla in 1625, and de Bondt described the morphology of the orangutan in 1658 (Conroy, 1990).

15.3.2   Basicranium

The cranial base, or basicranium, lies at the morphogenetic and functional interface of brain and face, and thus forms the foundation for the midface. In most mammals, the face lies anterior to the cranial vault and the foramen magnum is positioned nearer the posterior end of the cranial base, in keeping with a quadrupedal mode of locomotion. In mammals, the cranial base angle is flattened, measuring 180° or more from foramen magnum to sella turcica to anterior cranial fossae (in craniometric terms, basion-sella-nasion). Among the primates, cranial base flexure is less than 180°, although variation exists among species (Biegert, 1964, and others).

Increasing encephalization is the most commonly accepted explanation for the change in flexion of the cranial base. Increased flexion appears in the fetal period, but has been retained (neoteny) as a general characteristic of the primates. Differences in angulation among the primates are thought to be due to relative differences in growth rates and size of the cranial vault and face, although this remains under debate. The baboon (*Papio*), for instance, exhibits a small brain, large, prognathic face, and flattened cranial base. The cranial base is also flattened in the chimpanzee, presumably due to a prolonged postnatal growth trajectory (Lieberman and McCarthy, 1999). This is similar to that of other great apes, and contrasts with the human, in which a rapid postnatal growth trajectory results in comparatively marked flexion of the cranial base.

The interior aspect of the cranial base is inaccessible in most fossils, often obliterated by stony matrix depositions, and impossible to study. However, it is possible that this will change with the development or increased use of three-dimensional computed tomography (Siebert et al., 1998). The external surface of the cranial base is available for study in a number of specimens. Australopithecine crania show a more centrally positioned foramen magnum, but, like the great apes, have irregularities along the inferior aspect of the occipital bone, indicating the insertion points of large nuchal musculature (Aiello and Dean, 1990). The cranial base of gracile australopithecines shows a moderate bend at the sella (du Brul, 1977). *Paranthropus boisei* (formerly known as *Australopithecus boisei*, a robust australopithecine) has a foramen magnum that is more anterior than either *Australopithecus* or *Homo*. Angulation of the cranial base is pronounced and produces a degree of cranial shortening (Tobias, 1967).

Both early and modern forms of *Homo* are similar in having a cranial base that is shorter and wider than the apes or australopithecines. The foramen magnum is situated anteriorly, and the occipital bones sweep farther back and up, as a consequence of the expanded brain. The cranial base angle is approximately 135° to 145°.

The width of the cranial base varies in proportion to the size of the masticatory system; earlier forms of *Homo* manifest a wider base.

A more complete discussion of the evolutionary contributions of the cranial base can be found elsewhere in this volume (Chapter 11). The reader is referred to that chapter and to the classic studies of Keith (1910), Bolk (1910), Duckworth (1915), Ashton (1957), and Scott (1958).

15.3.3 Orbits

The orbital cavities serve as protective chambers for the eyes. In extant primates, orbital configuration takes two different forms (Conroy, 1990). Prosimians (lemurs, lorises) have a postorbital bar, or ridge of bone, formed by frontal and zygomatic bones, which surrounds the orbit. The marmoset, while diurnal, possesses disproportionately large orbits and eyes, with enhanced light-gathering capacity (Fig. 15.1). Anthropoids have a true orbital

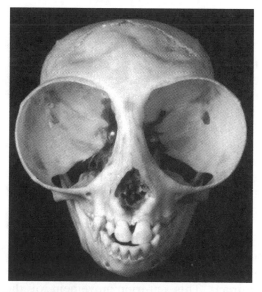

Figure 15.1 Frontal view of prepared skull of adult marmoset (*Saguinas*). Note markedly enlarged orbits; eyes in this species have highly developed light-gathering ability.

cavity, composed of a multitude of bones—frontal, zygomatic, ethmoid, sphenoid, lacrimal, palatine, and maxilla. Evolutionary explanations for these differences remain incompletely understood (see below), but seem to relate more to requirements for visual stability and acuity than to stresses of mastication as previously suggested (Ravosa et al., 2000).

The orbital cavities follow a growth pattern that correlates closely with that of the brain. This general neural pattern of growth is observed in both apes and humans. Growth of the orbits does not coincide with that of the cranium, but proceeds at a slower rate (Schultz, 1969). The growth of the chimpanzee illustrates this principle: in the newborn period, the brain case has reached 34% of its adult volume, while the orbits have attained only 21% of theirs.

Surprisingly, the growth of the orbit does not correlate precisely with that of the globe, the differences being explained by a mass effect of extraorbital muscles, periorbital adipose tissue, and other intraorbital tissues (Moore and Lavelle, 1974). In humans, the globe takes up 75% of intraorbital space during the fetal period but 24% during adulthood. This is not to say that the globe has no effect on orbital enlargement, for enucleation in infancy or experimental procedures result in a smaller-than-normal orbit (Taylor, 1939). Intraorbital adipose tissue has been assigned a considerable role by Latham and Scott (1970), who postulated that expansion of the retro-orbital fat pad contributes to the pronounced forward and downward growth of the primate snout.

In the great ape, orbital migration during the postnatal period results in positioning anterior to the neurocranium, rather than inferior, as is the case in humans. This anterior movement of the orbits leads to constriction of the neurocranium behind the orbits in apes, and is

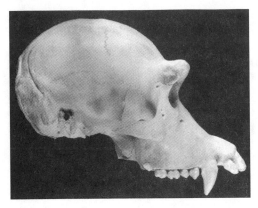

Figure 15.2 Lateral view of chimpanzee skull showing prominent supraorbital ridge.

assisted by growth of the supraorbital ridges. The ridges are especially prominent in the great apes (Fig. 15.2), and in the African great apes (chimpanzee, gorilla) they traverse the orbits, fusing in the midline to form a *supraorbital torus* (Fig. 15.3). The gorilla has the largest orbital cavity in absolute terms (although those of the chimpanzee, orangutan, and human are similar), and resembles extinct hominids in this regard. Specimens of *H. erectus* also show prominent brow ridges, with supraorbital tori.

The orbits are thought to develop at the interface of forces generated by the developing neurocranium and facial skeleton, particularly the position of the jaws (Moss and Young, 1960). In extant humans, the evolutionary reduction in the size of the dentition and superior growth of the brain case have largely eliminated this juxtaposition of forces and widespread incidence of pronounced supraorbital ridges (Moore and Lavelle, 1974). These morphological traits have been taken as evidence for an arboreal origin of the primates. Smith (1912) and Wood-Jones (1916) suggested that orbital convergence (medial migration of orbits/eyes, with increasingly anterior and parallel optical axes) occurred sec-

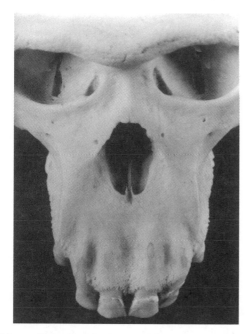

Figure 15.3 Close-up of interorbital area of chimpanzee skull demonstrating well-developed supraorbital torus.

ondary to the reduced need for olfaction and size of the snout. With more closely apposed orbits came increased stereoscopic abilities, which would confer obvious adaptive advantages for the precise locomotor needs of climbing and leaping primates. The theory has been refuted by Cartmill (1972, 1974) using the obvious argument that a number of arboreal mammals do not have close-set, anteriorly positioned orbits.

15.3.4 Zygoma

The zygoma, or zygomatic bones, lying at the inferior lateral margins of the orbits, comprise the outer limits of the primate midface. Also known as cheekbones, malar or jugal bones, they lie between the maxilla and temporal bone. The masseters originate in large part from the zygoma, so it follows that dietary demands and the responses of mastication influence the size and shape of this bone.

The gracile australopithecine, *Australopithecus africanus*, provides a case in point (Fig. 15.4a). The species manifests considerable prognathism, and is thought to have generated the bulk of its chewing forces in the premolar area. Hence, the anterolateral margin of the zygoma swings outward and forward, carrying the origin of masseter anterior as well (Aiello and Dean, 1990), providing additional chewing forces to the premolar and molar regions. The laterally flared zygoma (zygomatic prominence) gives the face of *A. africanus* its characteristic diamond shape. Zygomatic buttresses, bilateral ridges of thickened bone, extend from the zygoma to reach the maxilla at the level of the second molars, providing added strength to a face that is "not otherwise heavily structured" (du Brul, 1977).

In the robust australopithecine species, *P. boisei* (Fig. 15.4b), the zygomatic plates flare anteriorly and laterally in even more exaggerated fashion than gracile forms, producing large bony plates (du Brul, 1977). The posterior cranium is significantly narrower, producing a diagnostic appearance referred to as postorbital constriction. Interestingly, these traits are not common to all robust forms (Rak, 1983). Explanations have not been forthcoming, but it is generally agreed that the robust lineage(s) represents terminal stages that became extinct without evolving further. It is perhaps not surprising to observe highly unique features within the group.

The diamond-shaped face of gracile australopithecines contrasts with the squared appearance of the midface in great apes and modern humans. In later-appearing species of *Homo* (Figs. 15.4c,d), the zygomatic bone assumes a more vertical appearance, or in some individuals slopes backward, in parallel with the evolutionary change in diet and resultant diminution of the dentition and face (Aiello and Dean, 1990).

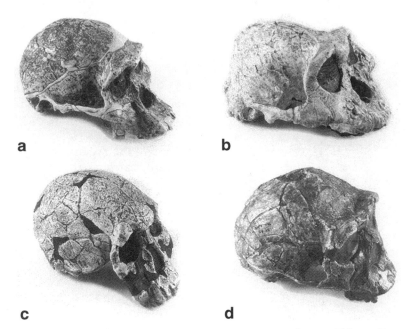

Figure 15.4 (a) Cast of fossil skull (Sts 5) belonging to *Australopithecus africanus*. This gracile species was prognathic and had an anterolaterally positioned zygoma, to which the masseter attached, enhancing the masticatory ability of premolars and molars. This specimen is probably female and is the best-preserved example of a gracile australopithecine. Popularly referred to as "Mrs. Ples," it was originally classified as belonging to the genus *Plesianthropus*. It was discovered in Sterkfontein, South Africa, in 1947, and is about 2.5 million years old. (b) Cast of fossil skull (KNM-ER-406) *Australopithecus* (or *Paranthropus*) *boisei*. This robust species had an exaggerated zygoma and more flattened midface than the gracile form. The specimen was uncovered in Kenya in 1969, and is about 1.7 million years old. (c) Cast of fossil skull (KNM-ER-1470) of *Homo habilis*. The skull is less robust than the australopithecines or *H. erectus*. It was found in Kenya in 1972 and is believed to be approximately 1.9 million years old. The taxonomic classification of this specimen remains under debate; it is referred to as *Homo rudolfensis* by some workers, or as *Australopithecus* by others. (d) Cast of fossil skull (KNM-ER-3733) of *Homo erectus*. Note the pronounced robusticity of this relatively intact skull; brow ridges and supraorbital torus are prominent.

15.3.5 Ethmoid Bone

The ethmoid bone is highly complex, being composed of numerous elements with divergent functions. The crista galli anchors the anterior extension of the falx cerebri; the frontal and cribriform plates form the floor of the anterior cranial fossae; ethmoidal air cells are part of the intracranial system of sinuses; the perpendicular plate contributes to the vertical midline support of the nasal cavity; the superior and middle turbinates are covered by mucous mem-

branes and are important to respiration. The structure is important in normal embryogenesis, forming a large portion of the cartilaginous nasal capsule; when absent, it contributes to the pathogenesis of several dysmorphogenetic conditions (Siebert, 1981; Siebert et al., 1981; Kokich et al., 1982; Souza et al., 1990).

The ethmoid bone is one of the major interorbital structures, and its configuration is closely related to biorbital breadth (Cartmill, 1971; Siebert, 1986). Ethmoidal morphology varies among the primates.

The structure of the cribriform plate relates closely to olfactory function. Monkeys show a single, narrow funnel-like structure without direct connection to the turbinates; apes and humans have more similar cribriform structures that are flattened and perforated by multiple olfactory nerves (Ishii and Takahashi, 1982). The ethmoid is larger in the chimpanzee and gorilla, giving greater separation to the orbital cavities than is found in the orangutan (Aiello and Dean, 1990).

In addition to differences in ethmoid morphology, sutural relationships vary among taxa. The medial walls of the orbit, to which the ethmoid contributes, show characteristic sutural patterns. Small animals with convergent orbits—for example, some prosimians—tend to have an ethmoid component in the medial orbital walls; the ethmoid in larger animals (i.e., gorilla) often disappears beneath the periorbital bones (Cartmill, 1971). In the orangutan, fossil hominids, and modern humans, the lacrimal bone shares a suture line with the ethmoid. However, the African apes have frontal bones that descend between these bones—a change that may be related to the anterior position of the orbits in these animals (Sonntag, 1924; Wood-Jones, 1948; Aiello and Dean, 1990).

In modern humans, the ethmoid is "characterized not only by the labyrinthine complexity, but also by asymmetry, numerous variations and deformity" (Ishii, 1982). The structure undergoes most of its ontogenetic change between ages 9 and 35 years, with deformations becoming more apparent as aging continues (Krmpotic-Nemanic et al., 1997). Continuing pneumatization is a major explanation for this change.

Much remains to be learned about the ethmoid complex in fossil hominids. The structure, by virtue of its location beneath much of the cranial surface and because it is extremely delicate, has been difficult or impossible to study.

15.3.6 External Nose, Nasal Bones, and Nasal Cavity

The external nasal apparatus in primates exhibits no single anatomic form, but instead exists as broad types that hold pivotal positions in primate taxonomy. Within the order Primates, two suborders, *Strepsirhini* and *Haplorhini*, are based on features of the nose. Extant strepsirhines (prosimians such as lemurs and lorises) have some form of tooth comb, as well as laterally split nostrils, a rhinerium (bare, moist skin between the upper lip and nose), philtrum, and frenulum. Haplorhines (tarsiers and anthropoids, including monkeys, apes, and humans) manifest unsplit nares, hairy undifferentiated skin covering the nose, absent or vestigial philtrum, and reduced or absent frenulum (Conroy, 1990). The adaptive advantages of these differences are not understood fully (see below), but it is interesting to note that strepsirhines are uniformly nocturnal, while haplorhines are diurnal. Two infraorders among the primates are based on additional features of the nose. The *Platyrrhini* (New World monkeys) manifest broad, flat noses, while the *Catarrhini* (Old World monkeys, apes, humans) have nostrils that are close together and directed downward.

A generalized shift in nasal function has occurred in primate evolution, from an emphasis on olfaction in species with longer snouts to respiration in those with shorter ones. Projection of the human nose is a unique feature and appears to have first appeared with *Homo erectus* 1.6 million years ago (Franciscus and Trinkaus, 1988). This shift in morphology remains unexplained, but the role of the external nose in thermoregulation and water conservation is appreciated widely. Warm exhaled airflow becomes turbulent as it moves through the complex web of turbinates and tends to condense in the cooler, external nose; residual moisture within the external nose and nasal cavity are then available to moisten air during inhalation. This affords active, diurnal creatures with a means of water conservation, and was an adaptation that followed the shift to full bipedalism by the earlier species *H. habilis* (Carey and Steegman, 1981; Franciscus and Trinkaus, 1988).

An overview of distinguishing anatomic

features of the primate nose is provided in Table 15.2. While soft tissues comprising the external nose cannot be recovered from fossilized remains, their presence can be inferred at times from the study of nasal elements and their relationship to other structures of the craniofacial complex of extant species. The nose of modern humans, and presumably that of *H. erectus*,

TABLE 15.2   Selected Features of the Nasal Skeleton

Species	External Nose	Nasal Bone	Relationship of Internasal and Nasomaxillary Sutures	Anterior Nasal Spine
H. sapiens	Projecting, with cartilaginous development; nasal tip bulbous; nares oriented inferiorly	Transverse convexity	Internasal suture anterior to nasomaxillary suture, producing obvious internasal angle	Positioned anteriorly; well developed
H. erectus	Probable projecting external nose, with increased volume; nares probably inferior or anteroinferior	Convex; equivocal in some specimens	Definite internasal angle	Developed; equivocal in some specimens
H. habilis	Unknown, but presumed to be flat in most	Flat or convex	Unknown in many specimens; slight internasal angle	Flattened or absent
Australopithecus (gracile and robust forms)	Unknown, but presumed to be flat (ape-like)	Similar to great apes	Sutures lie in same plane coronally, without internasal angle	Positioned posteriorly; weakly developed
Gorilla	Slight alar development; nasal tip thicker than chimp and orangutan; nares oriented anteriorly	Elongated and flat or depressed upon facial skeleton	No internasal angle	Absent
Chimpanzee, orangutan	Minimal alae and nasal tip; nares oriented anteriorly	Elongated and flat or depressed upon facial skeleton	No internasal angle	Absent

Source: After Franciscus and Trinkaus, 1988.

is distinguished from those of the apes (and Old/New World monkeys) in being more projecting. This change is reflected in the supporting nasal and adjacent skeleton (nasal bones, maxillae). The transverse convexity of nasal bones, with internasal angulation caused by an internasal suture positioned anterior to the nasomaxillary sutures, is suggestive of a protuberant nose, and is not found in those species with flat noses, such as the great apes. Nasal bones are variable among the living primates, and their taxonomic application is debated (i.e., Olson, 1985; Eckhardt, 1987).

The nasal cavity is a highly complex structure whose morphogenesis derives from diverse structures. The skeleton is formed from endochondral-derived ethmoidal components (cribriform plate, superior and middle turbinates, perpendicular plate), the inferior turbinate bones, vomer (a derivative of membranous bone), and cartilaginous nasal septum. The floor of the nasal cavity is formed by the hard and soft palates. The nasal cavity is elongated anteroposteriorly and low in mammals; it is short and high in humans and intermediate in other primates. This shape is thought to derive in large part from the changes in cranial base flexion that occur secondary to evolutionary enlargement of the brain. The enlarged nasal cavity confers a particular evolutionary advantage to humans, in that it greatly increases respiratory ability and presumably survival in the open. The great apes, by contrast, have substantially flatter faces and smaller nasal cavities, but in their arboreal, or at least forested, existence, long periods of exertion through running are not required.

The nasal septum has held an important position in primate evolution. Composed of perpendicular plate of the ethmoid, vomer, and cartilaginous nasal septum, the nasal septum plays an obvious role in providing vertical support to the nasal cavity. The septum is relatively tall and thin, and both possesses autonomous growth centers and

is influenced by the growth of adjacent structures. Anatomic deviation of the septum is more possible in the human because of its delicate shape, and can be severe enough to interfere with breathing (Takahashi, 1987).

But even more intriguing is the nasal septum's possible role in snout formation. The lemurs and baboons possess a true snout, which anatomically involves maxilla, nasal bones, vomer, and mandible, as well as the cartilaginous nasal septum and perpendicular plate of the ethmoid (Fig. 15.5). The snout appears to be more related to nasal than masticatory function, for long snouts are present in some mammalian species that have no teeth (Scott, 1963b). Other primates, apes, and humans do not possess a true snout.

Numerous authors, chief among them Scott (1953, 1963a, 1963b), have described the ontogenetic and phylogenetic roles played by nasal septal components. In fact, Scott (1953) launched the hypothesis that the cartilaginous nasal septum is the structure *primarily* responsible for the anterior and downward growth of the primate midface. Experimental support has been equivocal (e.g., Babula et al., 1970). Moss (1964), Moss and associates (1968), and

Figure 15.5   Anterolateral view of baboon skull. This species possesses a true snout, composed of maxilla, elongated nasal bones, vomer, perpendicular plate of ethmoid, cartilaginous nasal septum (not present), and mandible.

Moss and Salentijn (1969) countered Scott's hypothesis by suggesting that physiologic influences, manifesting anatomically by a number of "functional matrices," were responsible for craniofacial growth, and that septal growth occurred secondary to the demands of these matrices. Moss's own work has been examined critically (i.e., Cohen, 1993). Enlow's work (e.g., Enlow, 1966; Enlow and Azuma, 1975) also has important implications for Scott's hypothesis, and is reviewed with the discussion of maxillary growth below. The ultimate role of the nasal septum thus remains unclear, and it seems probable that final understanding will come with elucidation of genetic controls of craniofacial growth.

15.3.7 Paranasal Sinuses

Workers have identified 12 or 13 different sinuses within the craniofacial complex, involving the frontal, ethmoid, sphenoid, temporal, palatine, maxillary bones, and vomer. Of this large number, humans have 4. Sinuses are classified as primary, if they communicate directly with the nasal cavity, or secondary, if they do not (Suenaga, 1980; Takahashi, 1983). The sinuses form by outgrowths of mucous membranes lining the nasal cavity and/or middle ear, which extend into adjacent bones by pneumatization, slowly causing bone absorption. The process can continue throughout life; however, at least in humans, the sinuses generally reach maximal size during puberty.

Although paranasal sinuses are typical of mammals, they are relatively larger and anatomically more variable in the hominoids (Andrews and Martin, 1987). The ethmoid sinus is somewhat unique among humans (Takahashi, 1983). Anatomically, the ethmoid bone forms the medial walls of the orbits in humans and other primates. In nonprimate mammals, the ethmoid exists as the ethmoturbinate, and does not make contact with the orbits. Humans, gorillas,

and chimpanzees manifest enlarged ethmoidal sinuses and twin extensions, the frontal sinuses. The frontoethmoidal sinus is found among some fossil species—*Proconsul africanus* and *P. major*—which suggests an evolutionary affinity of these species with modern African apes. Physiologically, the ethmoid sinus is important to respiration, in contrast to the ethmoturbinate, a structure critical to olfaction in other mammals. Other primates exhibit small, inconsistently appearing ethmoidal air cells rather than a true frontoethmoidal sinus; examples are the lesser apes (*Hylobates*: the gibbon and siamang).

The frontal sinus is found only in modern humans, gorilla, and chimpanzee. In humans, the sinus is not present at birth, but appears first as air cells that expand into the frontal bones in the first several years of life. The frontal sinuses develop independently and are both asymmetric and highly individualized; this has afforded forensic workers a tool for identifying human remains (Harris et al., 1987; Reichs, 1993). The absence of a frontal sinus in the orangutan has been used as evidence that this species is not part of the evolutionary lineage involving African great apes and humans (Andrews and Martin, 1987). The shape and size of the frontal sinus is influenced by changes in angulation of the cranial base and shape of the cranial vault.

The sphenoid sinus lies within the sphenoid body, in the median plane; in the human, it can extend into the alisphenoid and basisphenoid and pterygoid processes. Like the frontal sinus, the space expands during the first several years of life. In most cases, it is not visible radiographically until about 4 years of age; resorption of the sphenoid body becomes evident by 6 years (Koppe et al., 1999a). The gorilla manifests a midline septum, and along with the chimpanzee, possesses a bony canal that courses through the sinus, connecting the foramen rotundum with pterygopalatine fossa (Koppe et al., 1999a). The orangutan exhibits a midline septum in the sinus as well.

The maxillary sinus has probably received the greatest attention from investigators. It is said to be a primitive character in the primate lineage (Conroy, 1990; Koppe and Nagai, 1998). The sinus is the most commonly appearing paranasal sinus (Suenaga, 1980). Both absolute and relative sinus volume is variable among primate species, and correlates weakly with measures of cranial size, suggesting that epigenetic factors are involved in determining sinus volume (Koppe et al., 1999b). The maxillary sinus is the only true paranasal sinus in Old World monkeys. It is absent in some species, small in the Japanese macaque (*Macaca fuscata*), and larger in the pigtail macaque (*Macaca nemestrina*) (Koppe et al., 1996a). The observation that the maxillary sinus is atypically small or absent in cercopithecoids suggests that the feature represents a shared, derived condition (synapomorphy) phylogenetically (Rae, 1999).

The maxillary sinus has been recognized in very early fossil reptiles. From an evolutionary perspective, enlargement of the maxillary sinus appears to represent an ancestral trait among hominoids, in that it is present in the East African Miocene fossil *Rangwapithecus*. The maxillary sinus has been identified in the fossil hominoids *Dryopithecus* and *Sivapithecus* (Rae, 1999). The space appears very early, being recognizable at 9 weeks in the human (Koppe et al., 1994). The maxillary sinus is rather large in humans, but even more enlarged in the orangutan, where it occupies the entire maxilla and at times extends into adjacent bones (Cave and Haines, 1940; Koppe and Nagai, 1998). The maxillary sinus also extends into the hard palate and premaxillary bone in the chimpanzee (Fig. 15.6) (Koppe and Nagai, 1998; Swindler, 1999).

A number of functions and selective advantages have been postulated for the paranasal sinuses (Table 15.3), yet the function of the paranasal sinuses in extant primates remains unresolved. Some feel that the sinuses have no function at al (Sperber, 1980; Aiello and Dean, 1990). Because pneumatization proceeds independent of other cranial changes in early life, others believe the sinuses play a role in development that goes beyond a mere peculiarity of ossification. Because maxillary sinus formation results from the pneumatization of several bones, some workers have

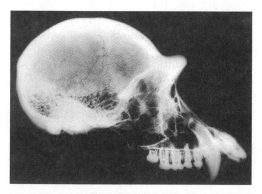

Figure 15.6 Lateral radiograph of chimpanzee skull showing prominent maxillary sinus (visible immediately above palate).

TABLE 15.3 Postulated Functions of Paranasal Sinuses During Evolution

Morphologically Based
   Reduction in weight of cranium
   Shock absorption/insulation/other protection
     for brain, other neural organs
   Allometric response to increased body and
     head size
Physiologically Based
   Thermoregulation (air conditioning)
   Air reservoir
   Ventilation
   Aid to olfaction
   Added resonance of voice
Evolutionary Based
   Residual spaces

Source: After Takahashi, 1983.

suggested that its presence may be the result of variable allometric development. Supporting this latter view are the observations by Koppe and colleagues (1994, 1996a, 1996b) that the size of the maxillary sinus is related more closely to skull or body size than to more functionally related measures such as palatal width and length. Regardless, the sinus appears to develop rather independently early in life, but then becomes more coordinated in its growth with the maxilla (Koppe et al., 1994).

Many feel that the morphology and function of early paranasal sinuses evolved as an adaptation to respiratory and olfactory needs and modifications to the cranial vault. As the need for olfaction diminished during later primate evolution and the molars enlarged, changing the shape of the dental arches and other cranial elements, the sinuses came to be more involved physiologically with respiration and thermoregulation (Takahashi, 1983). Support for this theory comes from Shea (1977), who found that maxillary sinus volume decreased in cold weather-adapted Eskimo populations.

15.3.8   Vomer

The vomer forms the posteroinferior foundation for the nasal septum. It abuts the inferior surface of the sphenoid bone and thus joins the cranial base and palate. According to Scott's (1953) hypothesis, the vomer, as part of the nasal septum, contributes to the downward growth of the midface. This has received experimental support (Latham et al., 1975), but as discussed above, has also been refuted.

The vomer attaches to the floor of the nasal cavity (superior surface of palate) and is involved in remodeling patterns of the nasal floor. This is thought to be related to the flattened cranial base in pongids. It has been suggested that the vomer, by exerting a downward force on the nasal floor during development, played a substantial role in the evolution of palatal

morphology among the australopithecines (McCollum, 1999).

Vomeral and subnasal morphology (i.e., entrance to nasal cavity, nasal sill and crest) has also been used for taxonomic purposes (e.g., Robinson, 1953, 1954; McCollum et al., 1993; McCollum, 2000). In great apes, the vomer is situated more anteriorly than in humans. Subnasal morphology is similar in gracile australopithecines, but differs in robust forms, where the insertion of the vomer takes place above the nasal surface of the premaxilla. *H. habilis* and *H. erectus* have similar subnasal patterns, although none of the fossil specimens of (African) *Homo* have a preserved vomer (McCollum, 2000).

15.3.9   Maxilla

The maxilla houses the upper dentition (except for the incisors), but also provides an important foundation of the midface. It serves as a buttress that resists mechanical forces, and it responds morphologically to those forces (Shapiro, 1977, and other workers). In nonhuman primates, the maxilla also includes the paired premaxilla. The premaxilla is, however, given separate treatment below.

The forward and downward growth of the maxilla has received much attention. The size, shape, and growth patterns of the teeth and dental arch also contribute to maxillary shape and hence to midfacial morphology. Enlow (1966) demonstrated this in comparing the growth of animals that possess snouts and those that do not. Growth of the maxillary arch is completely depository in the macaque, resulting in pronounced forward growth of the snout; in humans, the forward portion of the maxillary arch undergoes resorptive growth, which results in downward rather than forward growth, and a flattened face. The influence of the cartilaginous nasal septum upon maxillary growth (see above) has been postulated under conditions of normal development (Scott, 1953; Latham and Scott, 1970) and disease (Grymer and Bosch, 1997).

The vertically hypoplastic maxilla of the chimpanzee is said to have diverged from the general mammalian pattern of prominent midface and snout (Enlow and Azuma, 1975; Bromage, 1992). The infant chimpanzee has a flat face resembling that of modern humans. With growth, it becomes more prognathic, but is still small when compared to other mammals. These facial differences are probably related to variations in diet, in that species with higher faces (measured from midpalatal line to nasion) eat increased quantities of hard foods such as seeds or grains (Aiello and Dean, 1990).

The forces of chewing are essentially resisted by the maxilla, a process that is manifest in both extinct and extant species. In *A. africanus*, for example, blunt ridges of bone extending along the nasal apertures to the alveolar process of the maxillae are known as anterior pillars; these and the clivus form a morphological unit called the nasoalveolar triangular frame (Rak, 1983). In robust forms, a unique feature is the zygomaticomaxillary steps, which are prominent ridges that coincide with the zygomaticomaxillary sutures (Rak, 1983).

Early humans experienced a reduction in the size of the masticatory apparatus, and with it, a decrease in maxillary buttressing that persists among modern humans. *Homo habilis* (a taxon whose validity is still debated) had a maxilla that was smaller than the australopithecines, but within the range of *H. erectus* and *H. sapiens*. The Neanderthals diverged from this major evolutionary trend; with increased use of the anterior dentition, a significant midfacial prognathism evolved (Rak, 1987).

15.3.10 Premaxilla

The premaxilla, a small wedge of bone forming the anterior-most extension of the hard palate, provides the foundation for the upper central and lateral incisors. Of all the structures of the primate cranium, the bone has perhaps the most storied history, in part because of its highly variable nature—both developmental and morphologic—among the primates. As a result, the ontogenetic and phylogenetic significance—and even existence—of the premaxilla has been argued for centuries.

In primates, the anatomy of the premaxilla correlates closely with the form and function of the anterior dentition. Primates having very small incisors—for example, the lemurs—also have a small premaxilla (Swindler, 1976, 1998). Conversely, many mammalian species with large protruding incisors have a large premaxilla with prominent vertical arms (the nasomaxillary processes) that extend superiorly between the maxillae and nasal bones to the frontal bones. These processes, erroneously designated nasomaxillary bones in some studies, reach the frontal bones only rarely in primates (Swindler and Tarrant, 1973). Differential growth of the premaxilla, with persistence of the premaxillary-maxillary suture, appears to be responsible for the appearance of the diastema (expansion between lateral incisor and canine) in some primates (Schultz, 1969). This space is of considerable functional significance, enabling carnivorous animals to grip and tear meat.

The premaxilla is present in early primate fossils—for example, the North American and European plesiadapids (prosimian-like creatures from the middle Paleocene—60 million years ago)—and the East African and Eurasian dryomorphs *Proconsul*, *Rangwapithecus*, and *Dryopithecus* (apes from the Miocene period—20 million years ago). *Sivapithecus* has a premaxilla resembling that of the orangutan in curving into the nasal cavity to join the maxilla directly; both genera show a solitary communication with the incisive canal at the anterior end of the palate (Schwartz,

1983; Aiello and Dean, 1990). These features differ from the African apes and other early fossil hominids.

Extant monkeys and apes manifest a premaxilla that differs from humans in having three morphological components: a body (which supports the incisors), a palatine process, and a nasomaxillary process (Moore and Lavelle, 1974). The body is prominent on the anterior facial skeleton, forming the bony foundation of the upper lip. The premaxillary-maxillary suture has received much attention, and tends to obliterate during or shortly after infancy. This occurs in variable fashion, reaching a maximal 80% in the chimpanzee (Ashley-Montagu, 1935a, 1935b; Krogman, 1930).

The premaxilla is present in early human life, but congenitally absent (with upper incisors) in infants with median cleft lip and palate (also known as premaxillary agenesis or premaxillary dysgenesis). The premaxilla does not have the prominent anterior orientation of nonhuman primates, perhaps due, at least in part, to the reduced prognathism of humans. The premaxilla essentially disappears with obliteration of the premaxillary-maxillary suture early in life and is not identifiable as a separate bone in adult skulls. For a review of previous studies and of premaxillary embryology in humans, see the work of Kraus and Decker (1960).

15.3.11 Other Bones of the Midface

Discussion of additional bones—for example, the palatine and lacrimal—is not included in this chapter. These relatively small and delicate bones are often poorly represented in the fossil record. See the research of Cartmill (1978) for an introduction to these structures.

15.3.12 Hard Palate/Dental Arches

The hard palate is formed by the palatine processes of the maxillae and horizontal plate of the palatine bones; the premaxilla, when present, forms the anterior-most portion of the hard palate. The configuration and size of the hard palate are influenced by body, skull, and face size, genetic factors, growth and size of the dentition, tooth movement, and premaxillary topography, as well as the development, shape, and movement of the tongue, other oral and dietary habits, breathing, and the effects of disease (Koppe et al., 1996b). The formation of the secondary palate is a distinguishing feature of mammals (Koppe et al., 1996b), but comparatively little is known about the structure in fossil hominids. Discussion in this chapter is therefore limited to the hard palate.

The palate has a distinctive shape in the great apes (Fig. 15.7). The presence of a diastema (see below) gives the palate its rectangular shape; in fossil apes—for example, the dryopithecines—the palate is narrowed anteriorly, taking on a V shape (Conroy, 1990). The orientation of the premaxilla contributes to unique changes in the palates of *Sivapithecus* and the orangutan, as described above.

The palates of the australopithecines vary, depending on the evolution of the dentition. Beginning in the early Pleistocene (approximately 2 million years ago), two major changes began to occur: In robust australopithecines, the molars underwent enlargement, while the premolars became more "molarized." Gracile australopithecines underwent a progressive diminution in tooth size, with concomitant reduction in palatal size as well. Accordingly, specimens of *A. afarensis* show a palate that is quite shallow, with lateral margins that taper to the midline posteriorly; the upper dental arcades are long, narrow, and roughly parallel (Johanson and White, 1979). The oral apparatus projects in *A. africanus*; the palate is deep and hollow, and is associated with a dental arcade that is retracted posteriorly (du Brul, 1977; Rak, 1983). The palate of robust australop-

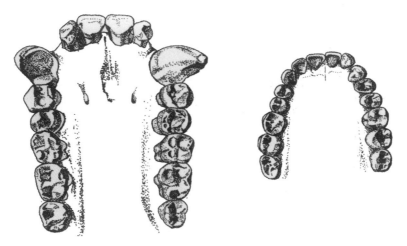

Figure 15.7 Hard palate and upper dentition of gorilla (left) and modern human (right). (Drawing by Linda Curtis—not to scale.)

ithecines is more shallow anteriorly than that of gracile forms, but deep posteriorly; the alveolar processes are straight and somewhat convergent anteriorly (Rak, 1983). The palate in robust forms resembles that of a chimpanzee with a midface "grown up" to robust size; its thickness seems to be a by-product of its vertically expanded mandibular ramus, rather than the increased mechanical demands of mastication (McCollum, 1999). In *P. boisei*, the palate is also retracted, and variably classed as shallow or deep; it is similar to that of *A. robustus* (Rak, 1983). *Homo erectus* had a massive face and proportionately large palate. The hard palate in modern *Homo* conforms to the parabolic dental arch.

The size of the maxillary (and mandibular) dental arches has been ascertained for Old World monkeys and differs according to diet (Siebert et al., 1984). Maxillary arch dimensions have also been shown to discriminate between monkey, apes, fossil hominids, and modern humans (Lavelle, 1977). Diet can have an effect on the size of the dental arcade (Siebert et al., 1984). Maxillary arch breadth, for example, decreases with chronic ingestion of soft foods (Beecher and Corruccini, 1981).

15.3.13 Dentition/Masticatory Apparatus

A great deal is known about tooth morphology in both extinct and extant primates. This is largely due to the hardness of teeth, which is sufficient to ensure their survival for millions of years. However, the ontogenetic and phylogenetic mechanisms responsible for producing these anatomic details remain elusive. Compared with many other mammalian orders, primates have a rather unspecialized dentition that has allowed a catholic diet for most species. There are exceptions, of course, as among some of the insectivorous prosimians or the leaf-eating and non-leaf-eating subfamilies of Old World monkeys. Even among these groups, the diet is never exclusively one or the other of these food types. Among hominoids, the apes are mainly vegetarians eating leaves, fruits, and seeds; among highland gorillas, bamboo shoots and roots are included. Most species supplement their diets with animal protein. Chimpanzees, for example, catch and eat young antelopes and baboons on occasion. Mills (1963, 1978) pointed out the general similarity in the chewing pattern among all extant primates, even suggesting that it may have appeared early in primate evolution. In his classical work on the evolution and development of the mammalian skull, Moore (1981, p. 165) correctly stated "that if true, this would indicate a quite remarkable retention of a relatively generalized mode of jaw usage in a large and, in many respects, diverse group of mammals."

In addition to chewing various foods, teeth also have functions as diversified as holding, carrying, ripping, tearing, fighting, cracking, and piercing. The morphology of the dentition is closely related to diet in both extinct and extant species, and fossil teeth have proven extremely useful in reconstructing the phylogenetic history of the various species of animals (in some instances, they are the only remains available for study). Some Neanderthal specimens show extreme wear on the anterior dentition, for example, suggesting that these individuals used their teeth as tools (Brace et al., 1971).

Primates, as most mammals, have a heterodont dentition—that is, the teeth are shaped differently and serve different functions from the front to the back of the oral cavity. Incisors are cutting teeth with straight incisal surfaces; the canines are conical, piercing structures with a single projecting cusp; the premolars and molars possess multiple cusps and are the crushing and grinding teeth.

The permanent dentition of hominoids has two incisors, one canine, two premolars, and three molars in each quadrant of the upper and lower jaws. Originally, early placental mammals had 44 teeth (3I-1C-4PM-3M) in each quadrant; thus, during the evolution of primates, the hominoids have lost 12 permanent teeth—that is, one incisor and the first two premolars from their respective quadrants in the upper and lower jaws. Indeed, the presence of two incisors is frequently cited as one of the most important hallmarks of the order Primates. A short discussion of the size and

morphology of the pongid dentition is presented against which the dentition of modern *Homo sapiens* is compared.

The upper central incisor of the anthropoid apes is broad when compared with the lateral incisor. In the orangutan, the lateral incisor is often quite pointed. The lower central incisor is slightly narrower than the lateral, but their morphology is similar. Incisor size relates closely to diet. Leaf-eating primates have smaller incisors than those who feed on tougher substances such as fruits (Hylander, 1975). The canines are large teeth, projecting beyond the occlusal line in both sexes; however, they are sexually dimorphic (Swindler, 1998). A wide diastema separates the canine from the lateral incisor in the upper jaw while a narrower space is usually present between the canine and lower third premolar. The upper diastema receives the lower canine, while the lower space accommodates the upper canine. Although the canines project beyond the occlusal plane, they must not overlap when the jaws are opened maximally. Therefore, canine length is closely related to jaw length and height of the temporomandibular joint above the occlusal plane (Herring, 1972; Lucas, 1981).

The upper premolars are generally bicuspid and the buccal cusp is larger than the lingual cusp. The lower premolars contrast in their size and morphology and are referred to as heterodont premolars. The lower third premolar is sectorial—that is, the crown consists of a single cusp elongated mesiodistally that occludes with the lingual surface of the upper canine, forming what is often referred to as a honing mechanism. There may be a second cusp located on the lingual side of the crown of the lower third molar. The sectorial lower third premolar has been present in pongids since the Miocene. The lower fourth premolar normally has two cusps, but there may be as many as three or four cusps.

The upper molars have four cusps of which the distolingual cusp (hypocone) is always the smallest. There is usually some development of enamel wrinkles on the occlusal surface of unworn upper and lower molars in all pongids; the most complicated patterns, however, are found in the orangutan. A connection between the mesiolingual cusp (protocone) and distobuccal cusp (metacone) is always present as the *oblique ridge* in the upper molars.

The lower molars have five cusps arranged in a Y-5 pattern that is present in *Dryopithecus*, an ape from the Miocene. The pattern consists of two buccal cusps, two lingual cusps, and one distal cusp separated by a series of sulci forming a Y-shaped configuration on the occlusal surface of the lower molars. In living pongids, the Y-5 pattern is most common on M_1, slightly less on M_2, and least common on M_3. The pattern is most consistent on all three molars in gorillas and orangutans, while showing more modifications in chimpanzees (Swindler et al., 1998).

In comparing the pongid and hominid dentitions, only the more salient differences are considered, and these are quantitative rather than qualitative. The hominid incisor is not as hypertrophied—especially the upper central incisor—as in extant pongids. Hominid canines are not as large or projecting as in pongids; rather they are smaller, more blunt, and normally project, if at all, only slightly above the occlusal plane. In contemporary human populations, there is little significant sexual dimorphism. The lower third premolar is not an elongated, sectorial tooth that hones against the lingual surface of the upper canine but is bicuspid, perhaps with as many as two additional cusps. The cusps of both upper and lower molars are more rounded and tend to be set closer together in hominids. Also, the distolingual cusp (hypocone) and the distal cusp (hypoconulid) of the upper and lower third molars are often absent, respectively. It might be noted that the molar wear pattern tends to be rather flat in hominids when

compared with the more oblique wear surface seen in pongid molars.

The hominoid dental pattern described above was present (with only slight modifications) in such pongids as *Dryopithecus*, *Lunfengpithecus*, *Otavipithecus*, *Sivapithecus*, and *Gigantopithecus* that lived in Africa, Asia, and Europe during the Middle and Late Miocene (15–5 mya). Later, two genera appeared in the Pliocene (5–1.8 mya)—*Australopithecus* and *Homo*—that represent the first members of the family Hominidae. These early hominids still possessed the basic dental morphology of their Miocene antecedents. Of course, there are some dental differences between these two families, but they are more of degree than kind; and, indeed, infraspecific differences are found within the same genus. For example, several *A. afarensis* fossils have lower canine/first premolar diastema, while only 1 of 12 *A. africanus* individuals has a diastema (Aiello and Dean, 1990). Gracile forms show large anterior teeth and moderately sized posterior ones, in contrast to robust australopithecines, which have comparatively smaller anterior teeth and very large posterior ones (du Brul, 1977).

For the australopithecines, dietary changes seem to have had a substantial influence on midfacial evolution. With a shift to more omnivorous diet, the premolars became molarized, increasing the occlusal load and demand for palatal support by the anterior (maxillary) pillars. This brought about further advancement of the face and retraction of the palate—changes that increase sequentially from gracile, omnivorous species to robust, herbivorous ones: *A. africanus*, followed by *A. robustus*, then *P. boisei* (du Brul, 1977; Rak, 1983).

Such differences do not mask the fundamental dental similarities among these early hominoid genera. Rather, this brief excursion into the dental morphology of the various members of the superfamily Hominoidea clearly supports the proposition that all members of this taxon have their roots in some group of Miocene or Pliocene apes.

15.3.14   The Masticatory Apparatus

The muscles of mastication consist of four major muscles that attach both to the cranium and the mandible. Three of these muscles are elevators that bring the mandible up until the teeth occlude; they are the masseter, temporalis, and medial pterygoid muscles. The fourth masticatory muscle is the lateral pterygoid whose function is to initially open the mouth—that is, depress the mandible. As the mouth is opened, other depressor muscles are activated that continue lowering the mandible; these are the digastric and infrahyoid muscles. These last two muscle groups are often referred to as the accessory muscles of mastication. These muscles act collectively to perform the various and complex movements of the mandible not only during mastication but any time the mandible changes position.

The muscles of mastication are similarly disposed in all hominoids, but there are differences in some of their bony attachments due to differences in skull shapes and proportions (Aiello and Dean, 1990). For example, the large, powerful temporalis muscle is attached to the sagittal crest, and occasionally the nuchal crest, in adult male gorillas and orangutans (Fig. 15.8), but rarely in the females; the sagittal crest occurs infrequently in adult male chimpanzees, never in the females, and almost never in gibbons (Ashton and Zuckermann, 1956). The development of the sagittal and nuchal crests and the supraorbital torus in anthropoid apes enhances the ability of their crania to withstand the power of their cranial and facial muscles that is far in excess of that in humans, as indicated by their greater mass, both in absolute and relative terms (Moore, 1981). Also, the neurocranial portion of the

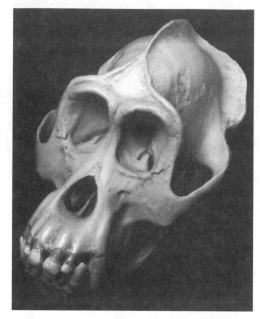

Figure 15.8   Superior oblique view of casted skull of adult male orangutan. Note the prominent sagittal and nuchal (occipital) crests. (*Source*: Swindler, 1998. Used with permission.)

modern human skull has enlarged *pari passu* with reduction of the facial skeleton to obviate the development of such superstructures. *H. erectus* deviated from this trend in having an occipital crest or torus to which large nuchal muscles attached.

15.3.15   Mandible

The mandible, while not part of the midface, interacts intimately with midfacial structures from a developmental, functional, and evolutionary standpoint. By understanding mandibular morphology and function in extant species, workers can infer a great deal about fossil specimens. The mandible has thus proven to be extremely valuable in taxonomic studies of fossil primates. Indeed, some extinct species are known, or at least were first delineated, on the basis of the discovery of mandibular fragments. However, because the form/function of the mandible and lower dentition are so closely integrated with that of the maxillae and upper dentition, the ultimate independence of mandibular development and evolution has been questioned. For this reason, some have urged a conservative approach to the interpretation of isolated cranial fragments of the mandible (Spears and Macho, 1998), or, for that matter, the entire structure (Humphrey et al., 1999).

Two primary changes occurred during the early evolution of the mandible: (1) the angular, articular, and quadrate jaw bones of reptiles evolved into the bones of the middle ear; and (2) the volume of tooth-bearing bone expanded to comprise, at least in mammals, virtually the entire mandible (Delaire, 1998). Other modifications—for example, displacement of the temporal region of the cranium, lowering of the face, and widening of the lower jaw—occurred for reasons thought to be related to changes in posture, locomotion, and mastication (Delaire, 1998).

Among the primates, the mandible shows extraordinary phenotypic variability in both extinct and extant species, from the tiny, delicate bones of the smallest species to the massive structures diagnostic of *Gigantopithecus* (Fig. 15.9) and the robust African hominids (i.e., *Australopithecus robustus*, *Paranthropus boisei*). Mandibular form, function, and articulation are related closely to diet and consequent requirements for mastication (Osborn, 1987; Varrela, 1992; Hylander et al., 1998; and others). The biomechanical attributes of mandibular function have been and continue to be studied in detail (Hylander et al., 1998; Spears and Macho, 1998). Differences in mandibular shape and condylar position and relation to the occlusal plane among the australopithecines and early *Homo* are thought to be the result of changes in diet (Osborn, 1987). This shift from a soft, chiefly frugivorous diet to a diet characterized by harder foodstuffs began in

Figure 15.9 Lateral view of cast of fossil mandibular fragment of *Gigantopithecus*. Note the extraordinary robusticity of the mandibular body and molars.

the middle Miocene, and was manifest by thickened molar enamel, enlarged incisors, and dramatically enlarged jaws (Andrews and Martin, 1987). Increasing molarization of the premolars and size of postcanine dentition during evolution has resulted in increased robusticity, with added buttressing, of the mandibular body in extant primates (Conroy, 1990). A hard diet is also thought to promote vertical growth of the mandible and anterior translocation of the maxilla, and in this way influence growth of the entire craniofacial skeleton (Varrela, 1992).

Fusion of the mandibular symphysis is an important evolutionary feature, though not limited to primates, and has been linked in some studies to increased stresses of mastication. However, among the *Adapidae* (small North American and European Eocene primates), increased levels of symphyseal fusion are associated with robust jaws and larger body size, rather than dietary effect (Ravosa, 1996).

The mandible of great apes is larger than that of humans (Fig. 15.10). The rami are taller, and because of the apes' prognathic condition, the mandibular bodies are longer (Humphrey et al., 1999). The mandible evinces parallel dental arches that are narrower than those of humans.

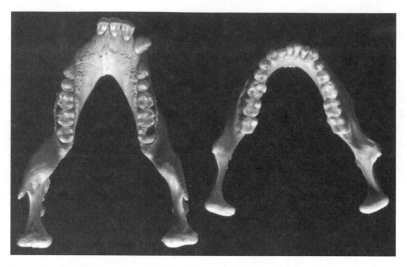

Figure 15.10 Superior view of chimpanzee (left) and human (right) mandibles. Differences in size and shape are evident. Note the characteristic parallel dental arch in the chimpanzee and the parabolic arch in the human. The left lateral incisor and left canine were extracted in the (captive) chimpanzee.

Thus, the floor of the mouth is also narrower, requiring less muscular support (Aiello and Dean, 1990). The ape mandible has no chin but rather a simian shelf (inferior transverse torus) extending along the interior surface of the mandible posterior to the symphysis to the region of the fourth premolar. In the gorilla and chimpanzee, the digastric muscles attach below this shelf (Weidenreich, 1936). The African apes differ from both the lesser apes (gibbons) and their fossil counterparts—for example, the African dryomorphs—in having a prominent shelf. A semantic curiosity is that the shelf may be absent, poorly developed, or well developed in monkeys (Scott, 1963b).

Next to the teeth, the mandible is the most commonly preserved fossil among the hominoids. It is therefore a reasonably well-understood structure among the australopithecines. The dental rows are parallel and narrowed anteriorly in *A. afarensis*, and in this way, resemble those of the great apes. The ascending ramus is broader and lower than that in the great apes, and the inferior transverse torus is much less distinct and shelf-like than in the apes. In *A. africanus*, the mandible is larger and has posterior tooth rows similar to *A. afarensis*. The robust australopithecine, *Paranthropus boisei*, has a massive mandible that is both wide and tall. The posterior tooth rows are slightly divergent rather than parallel; the molars are set back as much as 1 centimeter from the ramus, which would have allowed for transverse grinding movements during mastication (Aiello and Dean, 1990). Anteriorly, the incisors are quite small.

The mandibles of extinct and modern species of *Homo* show some variability, but in general they are narrower and shorter than those of other primates, including apes and australopithecines. The size of the mandible of *H. habilis* is intermediate between that of australopithecines and *H. erectus*. Like the apes and australopithecines, *H. erectus* has no chin but rather shows a buttress (mandibular torus) on the inner aspect of the mandible (Kennedy, 1980). The modern human mandible exhibits some variability in sexual differences, but it is not pronounced (Humphrey et al., 1999). The mandible is smaller in proportion to body size and less robust, having a rounded (parabolic) arch (Fig. 15.10) and pronounced chin without a lingual shelf (Johnson et al., 1976). In fact, the modern human is the only primate to exhibit a true chin—a structure that seems to have acquired the role of strengthening the symphyseal arch, which is a function held by the simian shelf in nonhuman primates (Scott, 1963b) or the mandibular torus in *H. erectus*. The loss of the simian shelf has been related to the acquisition of speech, but it is more likely to be related to dietary changes and reduction of sagittal and occipital crests (Scott, 1963b).

15.4 GENERAL PRINCIPLES OF EVOLUTIONARY BIOLOGY

It is possible that some readers will not be fluent in matters of evolutionary biology. The following basic tenets are therefore offered. In considering each, a healthy regard must be maintained for the difference between *observation* and *interpretation* of observations. Failure to appreciate this difference has given birth to many debates among evolutionists and, of course, other scientists as well. Like the study of medicine, the pursuit of evolutionary biology exists in part as an art form. Although now decades old, the following statement is still valid: "At the present time, evolutionary paleontology operates within a loose set of guidelines governed by tradition rather than a concisely formulated body of theory. It is a tradition defined by the belief that the key to life's history lies more in empirical observation than in inference" (Cracraft and Eldredge, 1979, p. 35).

15.4.1 Fundamental Issues

The chief goal in the study of evolution is to understand the variation—anatomic and other wise—that occurs within and between phyletic lineages over time. This pursuit is complicated by multiple factors. Change takes place in different anatomic or physiologic systems at different times and at different rates—a process termed *mosaic evolution*. An obvious complication is the large phenotypic variation apparent between and among individuals, races, species, and higher taxonomic groupings. The effects of sexual differences (dimorphism) are often understood incompletely, as are differences that arise because of geographic separation (see below for more on this issue). Specimens that represent extinct taxa may, of course, be incomplete, damaged, or undiscovered. Sample size is important to every scientific discipline, and especially to studies of evolution. As the future unfolds, sample sizes are likely to increase. This will presumably shed light on particular morphological issues, although the range of variation is likely to increase as well (McCollum et al., 1993).

15.4.2 Classification Schemes and Techniques

What constitutes a species? This is a critical issue facing taxonomists, and one that Darwin recognized in 1859, stating, "No one definition has yet satisfied all naturalists; yet every naturalist knows vaguely what he means when he speaks of a species." Among extinct creatures, differences in species are often defined in morphologic terms. The approach is different for extant species, which are defined as morphologically similar individuals that do not interbreed. The difference is a fundamental one, and sometimes difficult to reconcile.

Several different approaches to classifying species have been adopted. Unique morphologic traits, or characters, are often placed in a taxonomic hierarchy and depicted graphically in a *phylogram*. Such weighting of characters has been quantified by some workers (phenetic or numerical taxonomy). The comparison of shared and new characters (*cladistics*) can facilitate hypotheses regarding the nature of evolutionary branching patterns, which, when mapped, are called *cladograms*. Characters may be clustered and phyletic divergences from ancestral populations diagramed. Shared ancestral, or primitive, characters are termed *sympleisiomorphic*; shared, but newly arisen or derived (novel), characters are *synapomorphic*. Put differently, sympleisiomorphies are inherited from remote ancestors, whereas synapomorphies are inherited from immediate ancestors (Eldredge and Cracraft, 1980). When characters arise independently, in parallel among different groups, they are termed *convergent*. When they exist identically in ancestral and offspring groups, characters are *homologous*.

15.4.3 Speciation

Species are the basic units of evolution (Cracraft and Eldredge, 1979) and are defined phyletically as separate lineages (Simpson, 1961). Therefore, explaining the appearance of new species is a major goal of evolutionary biologists (realize that even the definition of species is still unsettled, at least in some camps, but this is a separate issue beyond the scope of the present review). The two major schools of thought involve those who believe change occurs slowly (Darwin's *descent with modification* or *phyletic gradualism*) and those who feel change occurs quickly, at least in geologic terms (*punctuated equilibrium*).

In the former model, change occurs slowly and rather uniformly, with part of a population responding to an environmental stimulus with a particular morphologic change. With time, that segment of the population undergoes behavioral, then genetic, isolation. With inbreeding, the trait becomes widespread, eventually involving all or nearly all of the population. In this model, one species evolves into another over a period of time (*sympatric speciation*). Missing stages in the fossil record are considered to be just that—missing—and, were they to be found, they would provide a more complete explanation for the gradual shifts observed in morphologic traits.

In the latter model, change occurs rapidly, say over 50,000 years or so, and gaps in the record do not represent missing data, but rather reflect comparatively sudden shifts in morphology. In this model, new species develop rapidly from splits in lineages that occur following episodes of geographic isolation from parent populations (*allopatric speciation*).

A great deal of discussion has ensued regarding these two models, but, in fact, they are not mutually exclusive, and examples of each exist. The reader should not infer that evolution necessarily has a direction—the possibilities of convergent, divergent, and parallel evolution have been discussed by Eldredge and Cracraft (1980).

15.4.4 Developmental Constraint

A basic premise of evolutionary biology is that all things are not possible, and that phylogenetic change is restricted by the finite possibilities for morphologic variation upon which natural selection can operate. As defined by Smith and co-workers (1985), developmental constraint is "a bias on the production of variant phenotypes or a limitation on phenotypic variability caused by the structure, character, composition, or dynamics of the developmental system." An example of the basic preservation of morphological relationships across taxa is demonstrated in the chimpanzee, where anatomic relationships between the orbits, jaws, and external auditory meatus remain constant during growth. This is thought to reflect an optimal positioning of structures for the purposes of sight, feeding, and auditory function (Bromage, 1992).

15.4.5 Range of Variation

Differences in morphology often form the basis for the introduction of new species. However, when this is based on a highly limited sample size, as is often the case in primate studies, the normal range of anatomic variation cannot be discerned. It then becomes problematic to proclaim the appearance of a new species, when the specimen may represent a smaller or larger version of normality. As in all branches of science, debates continue between lumper and splitter evolutionists.

15.4.6 Dating Specimens

Accurate dating is critical to the study of evolution. The appreciation of the earth's age and respect for the enormity of geologic time were major advances in scientific knowledge. At 4.6 billion years, the earth's age can barely be comprehended, nor can scientists fully understand the effect that long periods of time, and countless generations, have on evolution. It has been said that if all of earth's history were to be compressed into 1 year, primitive humans would have appeared in the evening of December 31 (Conroy, 1990).

The location of fossils, and hence their approximate age, are determined on the basis of established geologic criteria. The earth's solid materials settle out into layers over time, with older substances lying beneath younger ones, unless disrupted by some event. The study of modes of geologic and cultural deposition is called *stratigraphy*. These layers can be dated by a variety of techniques and have varying resolving powers. *Radiometric dating* is often used to date volcanic material and makes use of the

fact that radioactive elements break down (decay) at known rates to other elements— for example, potassium to argon and uranium to lead. The ratio of isotopic variants of these elements can then be used to calculate age, often to about 300,000 years. *Paleomagnetic dating* is based on the observation that the north–south polarity of the earth has changed from time to time. The duration of these shifts has been highly variable, from about 40,000 to several million years. Identifying the polarity of magnetic minerals in rock samples enables workers to determine relative sequences rather than precise geologic dates. *Carbon-14 dating*, well known to most, is only useful in dating organic substances younger than 50,000 years, and is therefore of more use in archeological than paleoanthropologic studies.

15.5 GENERAL PRINCIPLES OF CRANIOFACIAL AND MIDFACIAL EVOLUTION

A variety of models have been put forth to explain craniofacial evolution. The evolution of the cephalic neural crest and subsequent appearance of the head were revolutionary changes in vertebrates (Gans and Northcutt, 1983; Northcutt and Gans, 1983). Other major concepts are presented earlier in the chapter as they relate to specific regions or bones.

Major trends in the evolution of skull form include an increase in the size of the brain, decrease in basicranial angle (with retention of the pronounced degree of fetal flexure), change in the growth of the nasal septum (and snout) from forward to downward (Table 15.4), and reduction in jaw size. Numerous workers have debated the evolutionary significance of these changes. The increase in primate encephalization over time is accepted widely, but a complete discussion is beyond the scope of

TABLE 15.4   General Trends in Craniofacial Evolution Among the Primates

Region	Evolutionary Change	Possible Explanation(s)
Cranial (sagittal) crest	Reduction	Reduced muscles of mastication secondary to changes in diet
Calvaria	Increase in vault size	Enlargement of brain, especially neopallium
Basicranium	Increased flexion	Enlargement of brain; forward and downward rotation of face
Supraorbital (brow) ridges	Reduction	Reduced stresses on facial skeleton, with concomitant reduction in need for facial buttressing
Orbits	Forward rotation	Enhanced stereoscopic vision
Snout	Generalized reduction	Reduction in size of olfactory organs
Nasal cavity, air passages	Increased vocalization	Confers adaptive advantage
Temporomandibular joint	Elevated relative to dentition	Increasingly omnivorous diet, with generalized reduction in masticatory needs
Maxillae	Reduction	Secondary to reduction in body size
Dentition	General enlargement	Related to body size vs. specialized change

Source: After Moore and Lavelle, 1974; Swindler, 1998.

this chapter. A major school of thought holds that the failure of the cranial base angle to open has made the acquisition of upright posture possible, even necessary, with concomitant freeing up of forelimbs for purposes other than locomotion (Scott, 1963b). With the development of erect bipedalism came several changes in the craniofacial complex: cranial vaulting; cranial base angulation; retrusion with crowding of the oral complex; reshaping of the mandible, with appearance of a chin; forward shifting of foramen magnum and occipital condyles; horizontal repositioning of the nuchal plane; and reorientation of the petrous pyramids from sagittal to coronal (du Brul, 1977). The ontogenetic and phylogenetic aspects of maxillary and nasal septal growth are presented earlier in the chapter. Reductions in jaw size and changes in tooth configuration or size are thought to be related to changes in diet and increasing use of tools, especially during the Pleistocene (Brace, 1967 and others).

It should be made clear that both craniofacial development and evolution are highly complex and presumably result from the interaction of a variety of factors. For the midface, as for other skeletal structures, it is reasonable to presume that natural selection acts on a host of genetic and epigenetic mechanisms, refining and choosing among them to allow survival in changing environments (Atchley and Hall, 1991). For those species known by their fossilized remains, such models remain largely hypothetical in that they cannot be tested at the cellular or tissue level. It seems reasonable, however, to trust that evolution proceeds by biological processes that operate in the modern world, and that, as those controlling factors become better known, evolution will also be understood in more detail. All of this is not to say that evolution *must* proceed in this seemingly gradualistic way. It is also reasonable to presume that, at times, populations will become isolated, by behavioral, genetic, or geographic means, and subsequent inbreeding will result in increased gene frequencies and eventual changes in morphologic characters and even species. This would often happen in a relatively rapid manner.

15.5.1 Physiologic Contributions to Evolutionary Change

Understanding the interplay of form and function has always been a challenge to students of evolution, and nowhere is it more evident than in the primate midface. The mammalian cranium houses and protects organs essential for a variety of neuromuscular and cognitive functions, respiration, thermoregulation, moisture conservation, and mastication. Adaptations to any or all of these functions are thought to have influenced cranial evolution at the morphologic level. However, other factors must have been operational as well, for some structures (i.e., bones of mandible and middle ear) share embryologic and phyletic origins, but have remarkably dissimilar functions.

Correlations between physiologic processes and morphologic evolution have been made for many years. The role of functional matrices, for example, has been a topic of study by Moss (1968, 1971, 1972) and other investigators. The theory holds that skeletal growth occurs as a secondary response to the physiologic demands of functional spaces in the cranium, and would seem to provide some basis for the operation of natural selection upon morphologic structures. This concept continues to be examined (Cohen, 1993). Functional modules (McCollum, 1999) are an extension of Moss's work, and, like functional

matrices, would seem to be amenable to selective (i.e., adaptive) pressures. The functional stresses placed on bones have been quantified by a number of workers. The interested reader is referred to Tappen's (1969, 1970, 1971, 1976) work on split lines, Hylander and Johnson's (1992) and Hylander and co-workers' (1998) studies of strain gradients, and Herring and colleagues' (1996) investigation of bone strain.

It remains unclear how changes in morphology and function correlate with genetic change. A case in point is the evolution of snout/nasal cavity and olfactory genes. Compared to other mammals, primates have a reduced snout, which is associated with an absolute decrease in olfactory neurons and a decreased sense of smell. However, primates have also lost a large number of functional genes related to smell. Over 70% of human olfactory receptor genes, for example, have become pseudogenes that are nonfunctional (Rouquier et al., 2000). The mechanisms responsible for this change are unknown, but the role of relaxed selection (i.e., decrease in selective pressures over time) has been entertained.

15.5.2 Ecological/Behavioral Contributions to Evolutionary Change

Closely related to the physiologic attributes of midfacial evolution are those demands placed on species by their environments. Ecological factors weigh heavily on the evolution of the primates, and responses to these demands have taken both behavioral and morphological forms.

The effects of diet on craniofacial form have received a great deal of attention. The deeper and shorter configuration of the primate midface that evolved later is thought to be due to changes in diet, in the direction of increased occlusal loads. This has been observed in Old World monkeys—for example, the Japanese macaque—and in some primitive human populations (Hylander, 1977; Koppe et al., 1999a).

Additional studies have been conducted on the effects of environmental stress on skeletal and dental growth (Siegel et al., 1992). From these, it appears that perinatal exposure to a variety of stressors, including cold, heat, or noise, results in increased fluctuating asymmetry and significant reduction in a large number of tooth, long bone, and parietal measurements.

15.6 SUMMARY

A great deal is known of the ways in which craniofacial structures develop and evolve. However, a great deal also remains unknown. Evolution is indeed a vast mosaic, and thus it is unclear if a single—or true—primate form has evolved. Can a group of animals whose body mass varies by 2000 times, or whose reproductive behavior ranges from isolated females visited on occasion by free-roaming, solitary males to monogamous and altruistic lifetime pairs, be characterized by any small group of features? It would hardly seem possible, and yet for centuries taxonomists have recognized this group as the primates.

As we have demonstrated, this diverse collection of species has shown highly significant trends through millions of years of evolution and has brought us into the modern era. Species have become adept at exploiting their ecological niches and flourished, or they have failed and gone extinct. This is perhaps the chief lesson the primatologist has to offer, for the broadest niche of all, planet Earth, can now be altered irrevocably by the thrust of a bulldozer's blade or the push of a tiny button. The ability of humans to control these seemingly small acts may yet carry the greatest meaning for all the world's living creatures, including those closest to us on the evolutionary chain.

ACKNOWLEDGMENT

We thank the staff of the Department of Anthropology, University of Washington, for making casts of fossil specimens available to us for study and photography.

REFERENCES

Aiello, L., and Dean, C. (1990). *An Introduction to Human Evolutionary Anatomy* (London: Academic Press).

Andrews, P., and Martin, L. (1987). Cladistic relationship of extant and fossil hominoids. *J. Hum. Evol. 16*, 101–118.

Ashley-Montagu, M. F. (1935a). The premaxilla in the primates. *Q. Rev. Biol. 10*, 32–59.

Ashley-Montagu, M. F. (1935b). The premaxilla in the primates (concluded). *Q. Rev. Biol. 10*, 181–208.

Ashton, E. H. (1957). Age changes in the basicranial axis of the anthropoidea. *Proc. Zool. Soc. Lond. 129*, 61–74.

Ashton, E. H., and Zuckermann, S. (1956). Cranial crests in the *Anthropoidea*. *Proc. Zool. Soc. Lond. 126*, 581–634.

Atchley, W. R., and Hall, B. K. (1991). A model for development and evolution of complex morphological structures. *Biol. Rev. Cambridge Philos. Soc. 66*, 101–157.

Babula, W. J., Smiley, G. R., and Dixon, A. D. (1970). The role of the cartilaginous nasal septum in midfacial growth. *Am. J. Orthodont. 58*, 250–263.

Beecher, R. M., and Corruccini, R. S. (1981). Effects of dietary consistency on maxillary arch breadth in macaques. *J. Dent. Res. 60*, 68.

Biegert, J. (1964). In S. L. Washburn (ed.), *Classification and Human Evolution* (London: Methuen and Co. Ltd.). Chapter 6 pp. 116–145.

Blechschmidt, E. (1961). *The Stages of Human Development Before Birth* (Philadelphia: Saunders).

Blechschmidt, E. (1977). *The Beginning of Human Life* (New York: Springer-Verlag).

Blechschmidt, E., and Gasser, R. F. (1978). *Biokinetics and Biodynamics of Human*

Differentiation: Principles and Applications (Springfield: Charles C. Thomas).

Bogin, B. (1977). *Periodic Rhythm in the Rates of Growth in Height and Weight of Children and Its Relation to Season of the Year* (Ann Arbor: University Microfilms).

Bogin, B. (1999). *Patterns of Human Growth*, 2nd ed. (Cambridge: Cambridge University Press).

Bolk, L. (1910). On the slope of the foramen magnum in primates. *Verk. Acad. Wet. Amst. 12*, 525–541.

Brace, C. L. (1967). Environment, tooth form, and size in the Pleistocene. *J. Dent. Res. 46*, 809–816.

Brace, C. L., Nelson, H., Korn, N., and Brace, M. L. (1971). *Atlas of Human Evolution*, 2nd ed. (New York: Holt, Rinehart and Winston).

Bromage, T. G. (1992). The ontogeny of *Pan troglodytes* craniofacial architectural relationships and implications for early hominids. *J. Hum. Evol. 23*, 235–251.

Byrd, K. E., and Swindler, D. R. (1980). Palatal growth in *Macaca nemestrina*. *Primates 21*, 253–261.

Carey, J. W., and Steegman, A. T., Jr. (1981). Human nasal protrusion, latitude, and climate. *Am. J. Phys. Anthropol. 56*, 313–319.

Cartmill, M. (1971). Ethmoid component in the orbit of primates. *Nature 232*, 566–567.

Cartmill, M. (1972). Arboreal adaptations and the origin of the order Primates. In R. Tuttle (ed.), *Functional and Evolutionary Biology of Primates* (Chicago: Aldine).

Cartmill, M. (1974). Rethinking primate origins. *Science 184*, 436–443.

Cartmill, M. (1978). The orbital mosaic in prosimians and the use of variable traits in systematics. *Folia Primatol. 30*, 89–114.

Cave, A. J. E., and Haines, R. W. (1940). The paranasal sinuses of the anthropoid apes. *J. Anat. 72*, 493–523.

Cohen, M. M., Jr. (1993). Sutural biology and the correlates of craniosynostosis. *Am. J. Med. Genet. 47*, 581–616.

Conroy, G. C. (1990). *Primate Evolution* (New York: W. W. Norton & Company).

Cracraft, J., and Eldredge, N. (1979). *Phylogenetic Analysis and Paleontology* (New York: Columbia University Press).

Darwin, C. (1859). *On the Origin of Species* (New York: Athenum reprint 1967).

Delaire, J. (1998). The evolution of the lower jaw and the jaw joint, from reptiles to man. *Rev. Stomat. Chir. Maxillo-Fac. 99*, 3–10.

du Brul, E. L. (1977). Early hominid feeding mechanisms. *Am. J. Phys. Anthropol. 47*, 305–320.

Duckworth, W. L. H. (1915). *Morphology and Anthropology*, 2nd ed. (Cambridge: Cambridge University Press).

Eckhardt, R. B. (1987). Hominoid nasal region polymorphism and its phylogenetic significance. *Nature 328*, 333–335.

Eldredge, N., and Cracraft, J. (1980). *Phylogenetic Patterns and the Evolutionary Process* (New York: Columbia University Press).

Enlow, D. H. (1966). A comparative study of facial growth in *Homo* and *Macaca*. *Am. J. Phys. Anthropol. 24*, 293–308.

Enlow, D. H. (1982). *Handbook of Facial Growth*, 2nd ed. (Philadelphia: W. B. Saunders Company).

Enlow, D. H., and Azuma, M. (1975). Functional growth boundaries in the human and mammalian face. In J. Langman (ed.), *Morphogenesis and Malformations of the Face and Brain* (New York: The National Foundation).

Enlow, D. H., and Hans, M. G. (1990). *Facial Growth* (Philadelphia: W. B. Saunders Company).

Franciscus, R. G., and Trinkaus, E. (1988). Nasal morphology and the emergence of *Homo erectus*. *Am. J. Phys. Anthropol. 75*, 517–527.

Gans, C. (1988). Craniofacial growth, evolutionary questions. *Development 103* (suppl.), 3–15.

Gans, C., and Northcutt, R. G. (1983). Neural crest and the origin of vertebrates: a new head. *Science 220*, 268–274.

Gavan, J. A. (1953). Growth and development of the chimpanzee, a longitudinal and comparative study. *Hum. Biol. 25*, 93–143.

Gavan, J. A. (1982). Adolescent growth in nonhuman primates: an introduction. *Hum. Biol. 54*, 1–5.

Gould, S. J. (1977). *Ontogeny and Phylogeny* (Cambridge: Belknap).

Grymer, L. F., and Bosch, C. (1997). The nasal septum and the development of the midface. A longitudinal study of a pair of monozygotic twins. *Rhinology 35*, 6–10.

Hall, B. K. (1990). Evolutionary issues in craniofacial biology. *Cleft Palate J. 27*, 95–100.

Harris, A. M., Wood, R. E., Nortje, C. J., and Thomas, C. J. (1987). The frontal sinus: forensic fingerprint? A pilot study. *J. Forensic Odonto-stomatol. 5*, 9–15.

Herring, S. W. (1972). The role of canine morphology in the evolutionary divergence of pigs and peccaries. *J. Mammal. 53*, 500–512.

Herring, S. W., Teng, S., Huang, X., Mucci, R., and Freeman, J. (1996). Patterns of bone strain in the zygomatic arch. *Anat. Rec. 246*, 446–457.

Holland, P. W. (1988). Homeobox genes and the vertebrate head. *Development 103* (suppl.), 17–24.

Humphrey, L. T., Dean, M. C., and Stringer, C. B. (1999). Morphological variation in great ape and modern human mandibles. *J. Anat. 195*, 491–513.

Hunt, P., Wilkinson, D., and Krumlauf, R. (1991). Patterning the vertebrate head: murine Hox 2 genes mark distinct subpopulations of premigratory and migrating cranial neural crest. *Development 112*, 43–50.

Hylander, W. L. (1975). Incisor size and diet in anthropoids with special reference to Cercopithecidae. *Science 189*, 1095–1097.

Hylander, W. L. (1977). The adaptive significance of Eskimo craniofacial morphology. In: A. A. Dahlberg and T. M. Graber (eds.), *Orofacial Growth and Development* (Paris: Mouton).

Hylander, W. L., and Johnson, K. R. (1992). Strain gradients in the craniofacial region of primates. In Z. Davidovitch (ed.), *The Biological Mechanisms of Tooth Movement and Craniofacial Adaptation* (Columbus: Ohio University College of Dentistry).

Hylander, W. L., Ravosa, M. J., Ross, C. F., and Johnson, K. R. (1998). Mandibular corpus strain in primates: further evidence for a function link between symphyseal fusion and jaw-adductor muscle force. *Am. J. Phys. Anthropol. 107*, 257–271.

Ishii, S. (1982). Characteristics and origin of the human olfactory organ. *Auris Nasus Larynx 9*, 25–35.

Ishii, S., and Takahashi, R. (1982). Characteristics and origin of the human olfactory organ. *Rhinology 20*, 27–31.

Johanson, D. C., and White, T. D. (1979). A systematic assessment of early African hominids. *Science 203*, 321–330.

Johnson, P. A., Atkinson, P. J., and Moore, W. J. (1976). The development and structure of the chimpanzee mandible. *J. Anat. 122*, 467–477.

Keith, A. (1910). Description of a new craniometer and of certain age changes in the anthropoid skull. *J. Anat. Physiol. London 44*, 251–270.

Kennedy, G. E. (1980). *Paleoanthropology* (New York: McGraw-Hill Book Company).

Kokich, V. G., Ngim, C.-H., Siebert, J. R., Clarren, S. K., and Cohen, M. M., Jr. (1982). Cyclopia: an anatomic and histologic study of two specimens. *Teratology 26*, 105–113.

Koppe, T., Yamamoto, T., Tanaka, O., and Nagai, H. (1994). Investigations on the growth pattern of the maxillary sinus in Japanese human fetuses. *Okajimas Folia Anat. Jpn. 71*, 311–318.

Koppe, T., and Nagai, H. (1998). The maxillary sinus of extant Catarrhine primates. In K. W. Alt, F. W. Rösing, and M. Teschler-Nicola (eds.), *Dental Anthropology: Fundamentals, Limits, and Prospects* (Vienna: Springer-Verlag).

Koppe, T., Inoue, Y., Hiraki, Y., and Nagai, H. (1996a). The pneumatization of the facial skeleton in the Japanese macaque (*Macaca fuscata*)—a study based on computerized three-dimensional reconstructions. *Anthropol. Sci. 104*, 31–41.

Koppe, T., Röhrer-Ertl, O., Hahn, D., Reike, R., and Nagai, H. (1996b). The relationship between the palatal form and the maxillary sinus in orangutan. *Okajimas Folia Anat. Jpn. 72*, 297–306.

Koppe, T., Nagai, H., and Alt, K. W. (eds.) (1999a). *The Paranasal Sinuses of Higher Primates: Development, Function, and Evolution* (Chicago: Quintessence Publishing Co., Inc.).

Koppe, T., Rae, T. C., and Swindler, D. R. (1999b). Influence of craniofacial morphology on primate paranasal pneumatization. *Ann. Anat. 181*, 77–80.

Kraus, B. S., and Decker, J. D. (1960). The prenatal inter-relationships of the maxilla and premaxilla in the facial development of man. *Acta Anat. 40*, 278–294.

Krmpotic-Nemanic, J., Vinter, I., and Judas, M. (1997). Transformation of the shape of the ethmoid bone during the course of life. *Eur. Arch. Oto. Rhino. Laryngol. 254*, 347–349.

Krogman, W. M. (1930). Studies in growth changes in the skull and face of anthropoids. II. Ectocranial and endocranial suture closure in anthropoids and Old World apes. *Am. J. Anat. 46*, 315–353.

Latham, R. A., and Scott, J. H. (1970). A newly postulated factor in the early growth of the human middle face and the theory of multiple assurance. Arch. *Oral Biol. 15*, 1097–1100.

Latham, R. A., Deaton, T. G., and Calabrese, C. T. (1975). A question of the role of the vomer in the growth of the premaxillary segment. *Cleft Palate J. 12*, 351–355.

Lavelle, C. L. B. (1977). A study of the taxonomic significance of the dental arch. *Am. J. Phys. Anthropol. 46*, 415–421.

Leigh, S. R. (1996). Evolution of human growth spurts. *Am. J. Phys. Anthropol. 101*, 455–474.

Lieberman, D. E., and McCarthy, R. C. (1999). The ontogeny of cranial base angulation in humans and chimpanzees and its implications for reconstructing pharyngeal dimensions. *J. Hum. Evol. 36*, 487–517.

Lonai, P., and Orr-Urtreger, A. (1990). Homeogenes in mammalian development and the evolution of the cranium and central nervous system. *FASEB J. 4*, 1436–1443.

Lucas, P. W. (1981). An analysis of canine size and jaw shape in some Old and New World non-human primates. *J. Zool. Lond. 195*, 437–448.

McCollum, M. A. (1999). The robust australopithecine face: a morphogenetic perspective. *Science 284*, 301–305.

McCollum, M. A. (2000). Subnasal morphological variation in fossil hominids: a reassessment based on new observations and recent developmental findings. *Am. J. Phys. Anthropol. 112*, 275–283.

McCollum, M. A., Grine, F. E., Ward, S. C., and Kimbel, W. H. (1993). Subnasal morphological variation in extant hominoids and fossil hominids. *J. Hum. Evol. 24*, 87–111.

Mills, A. E. (1963). Occlusion and malocclusion in the teeth of primates. In: D. R. Brothwell (ed.), *Dental Anthropology* (New York: Macmillan).

Mills, A. E. (1978). Teeth as an indicator of age in man. In P. Butler and K. A. Joysey (eds.), *Development, Function and Evolution of Teeth* (London: Academic Press).

Moore, W. J. (1981). *The Mammalian Skull* (Cambridge: Cambridge University Press).

Moore, W. J., and Lavelle, C. L. B. (1974). *Growth of the Facial Skeleton in the Hominoidea* (London: Academic Press).

Moss, M. L. (1964). Vertical growth of the human face. *Am. J. Orthod. 50*, 359–376.

Moss, M. L. (1968). The primacy of functional matrices in orofacial growth. *Dent. Pract. 19*, 65–73.

Moss, M. L. (1971). Ontogenetic aspects of cranio-facial growth. In R. E. Moyers and W. M. Krogman (eds.), *Cranio-facial Growth in Man* (Oxford: Pergamon Press).

Moss, M. L. (1972). Twenty years of functional cranial analysis. *Am. J. Orthod. 61*, 479–485.

Moss, M. L., and Salentijn, L. (1969). The primary role of functional matrices in facial growth. *Am. J. Orthod. 55*, 566–577.

Moss, M. L., and Young, R. W. (1960). A functional approach to craniology. *Am. J. Phys. Anthropol. 18*, 281–292.

Moss, M. L., Bromberg, B. E., Song, I. C., and Eisenman, G. (1968). The passive role of nasal septal cartilage in mid-facial growth. *Plast. Reconstr. Surg. 41*, 536–542.

Napier, J. R., and Napier, P. H. (1967). *A Handbook of Living Primates* (London: Academic Press).

Noden, D. M. (1991). Vertebrate craniofacial development: the relation between ontogenetic process and morphological outcome. *Brain, Behav. Evol. 38*, 190–225.

Northcutt, R. G., and Gans, C. (1983). The genesis of neural crest and epidermal placodes: a reinterpretation of vertebrate origins. *Q. Rev. Biol. 58*, 1–28.

Olson, T. R. (1985). Cranial morphology and systematics of the Hadar formation hominids and *"Australopithecus" africanus*. In E. Delson (ed.), *Ancestors: The Hard Evidence* (New York: Liss).

Osborn, J. W. (1987). Relationship between the mandibular condyle and the occlusal plane during hominid evolution: some of its effects on jaw mechanics. *Am. J. Phys. Anthropol. 73*, 193–207.

Rae, T. C. (1999). The maxillary sinus in primate paleontology and systematics. In T. Koppe, H. Nagai, and K. W. Alt (eds.), *The Paranasal Sinuses of Higher Primates: Development, Function, and Evolution* (Chicago: Quintessence Publishing Co., Inc.).

Rak, Y. (1983). *The Australopithecine Face* (New York: Academic Press).

Rak, Y. (1987). The Neanderthal: a new look at an old face. *J. Hum. Evol. 15*, 151–164.

Ravosa, M. J. (1996). Mandibular form and function in North American and European Adapidae and Omomyidae. *J. Morphol. 229*, 171–190.

Ravosa, M. J., Noble, V. E., Hylander, W. L., Johnson, K. R., and Kowalsky, E. M. (2000). Masticatory stress, orbital orientation and the evolution of the primate postorbital bar. *J. Hum. Evol. 38*, 667–693.

Reichs, K. J. (1993). Quantified comparison of frontal sinus patterns by means of computed tomography. *Forensic Sci. Int. 61*, 141–168.

Robinson, J. T. (1953). *Telanthropus* and its phylogenetic significance. *Am. J. Phys. Anthropol. 11*, 445–501.

Robinson, J. T. (1954). The genera and species of the Australopithecinae. *Am. J. Phys. Anthropol. 12*, 181–200.

Rouquier, S., Blancher, A., and Giorgi, D. (2000). The olfactory receptor gene repertoire in primates and mouse: evidence for reduction of the functional fraction in primates. *Proc. Natl. Acad. Sci. USA 97*, 2870–2874.

Schilling, T. F. (1997). Genetic analysis of craniofacial development in the vertebrate embryo. *Bioessays 19*, 459–468.

Schultz, A. H. (1969). *The Life of the Primates* (London: Weidenfeld and Nicolson).

Schultz, A. H. (1973). Age changes, sex differences, and variability as factors in the classification of primates. In S. L. Washburn (ed.), *Classification and Human Evolution* (Chicago: Aldine).

Schwartz, J. H. (1983). Palatine fenestrae, the orangutan and hominoid evolution. *Primates 24*, 231–240.

Scott, J. H. (1953). The cartilage of the nasal septum. *Brit. Dent. J. 95*, 37–43.

Scott, J. H. (1958). The cranial base. *Am. J. Phys. Anthropol. 16*, 319–348.

Scott, J. H. (1963a). The analysis of facial growth from fetal life to adulthood. *Angle Orthodont. 33*, 110–113.

Scott, J. H. (1963b). Factors determining skull form in primates. *Symp. Zool. Soc. London 10*, 127–134.

Scott, J. H. (1967). *Dento-facial Development and Growth* (New York: Pergamon Press).

Shapiro, P. A. (1977). Responses of the nonhuman maxillary complex to mechanical forces. In J. A. McNamara, Jr. (ed.), *Factors Affecting the Growth of the Midface*. Monograph 6, Craniofacial Growth Series, Center for Human Growth and Development (Ann Arbor: The University of Michigan).

Shea, B. T. (1977). Eskimo craniofacial morphology, cold stress and the maxillary sinus. *Am. J. Phys. Anthropol. 47*, 289–300.

Siebert, J. R. (1981). The ethmoid bone: implications for normal and abnormal facial development. *J. Craniofac. Genet. Devel. Biol. 1*, 381–389.

Siebert, J. R. (1986). Prenatal growth of the median face. *Am. J. Med. Genet. 25*, 369–379.

Siebert, J. R., Kokich, V. G., Beckwith, J. B., Lemire, R. J., and Cohen, M. M., Jr. (1981). The facial features of holoprosencephaly in anencephalic human specimens. II. Craniofacial anatomy. *Teratology 23*, 305–315.

Siebert, J. R., Swindler, D. R., and Lloyd, J. D. (1984). Dental arch form in the Cercopithecidae. *Primates 25*, 507–518.

Siebert, J. R., Williams, B., Collins, D., Winkler, L. A., and Swindler, D. R. (1998). Spontaneous cleft palate in a newborn gorilla (*Gorilla gorilla gorilla*). *Cleft Palate Craniofac. J. 35*, 436–441.

Siegel, M. I., Mooney, M. P., and Taylor, A. (1992). Skeletal and dental size reduction as a consequence of environmental stress. *Acta Zool. Fennica 191*, 143–147.

Simpson, G. G. (1961). *Principles of Animal Taxonomy* (New York: Columbia University Press).

Sirianni, J. E., and Swindler, D. R. (1985). *Growth and Development of the Pigtailed Macaque* (Boca Raton, FL: CRC Press).

Smith, B. H., Garn, S. M., and Hunter, W. S. (1986). Secular trends in face size. *Angle Orthod. 56*, 196–204.

Smith, G. (1912). The origin of man. *Smithsonian Inst. Annu. Rept. 12*, 553–572.

Smith, J. M., Burian, R., Kauffman, S., Alberch, P., Campbell, J., Goodwin, B., Lande, R., Raup, D., and Wolpert, L. (1985). Developmental constraints and evolution. *Q. Rev. Biol. 60*, 265–287.

Sonntag, C. F. (1924). *The Morphology and Evolution of the Apes and Man* (London: J. Bale, Sons and Danielsson).

Souza, J. P., Siebert, J. R., and Beckwith, J. B. (1990). An anatomic comparison of cebocephaly and ethmocephaly. *Teratology 42*, 347–357.

Spears, I. R., and Macho, G. A. (1998). Biomechanical behaviour of modern human molars: implications for interpreting the fossil record. *Am. J. Phys. Anthropol. 106*, 467–482.

Sperber, G. H. (1980). Applied anatomy of the maxillary sinus. *J. Can. Dent. Assoc. 6*, 381–386.

Suenaga, Y. (1980). Comparative anatomical study on paranasal sinuses, with special references to the pneumatization mechanism. *Acta Anatom. Nippon. 55*, 551–572.

Swindler, D. R. (1976). *Dentition of Living Primates* (London: Academic Press).

Swindler, D. R. (1998). *Introduction to the Primates* (Seattle: University of Washington Press).

Swindler, D. R. (1999). Maxillary sinuses, dentition, diet, and arch form in some anthropoid primates. In T. Koppe, H. Nagai, and K. W. Alt (eds.), *The Paranasal Sinuses of Higher Primates: Development, Function, and Evolution* (Chicago: Quintessence Publishing Co., Inc.).

Swindler, D. R., and Sirianni, J. E. (1973). Palatal growth rates in *Macaca nemestrina* and *Papio cynocephalus*. *Am. J. Phys. Anthropol. 38*, 83–92.

Swindler, D. R., and Tarrant, L. H. (1973). The topography of the premaxillary-frontal region in nonhuman primates. *Folia Primatol. 19*, 18–23.

Swindler, D. R., Sirianni, J. E., and Tarrant, L. H. (1973). A longitudinal study of cephalofacial growth in *Papio cynocephalus* and *Macaca nemestrina* from three months to three years. In Symp. IVth Int. Congr. Primat., vol. 3, *Craniofacial Biology of Primates* (Basel: Karger).

Swindler, D. R., Emel, L. M., and Anemone, R. L. (1998). Dental development of the Liberian chimpanzee, *Pan troglodytes verus*. Hum. Evol. *13*, 235–249.

Takahashi, R. (1983). The formation of the human paranasal sinuses. *Acta Otolaryngol. 408* (suppl.), 1–28.

Takahashi, R. (1987). The formation of the nasal septum and the etiology of septal deformity: the concept of evolutionary paradox. *Acta Otolaryngol. 443* (suppl.), 1–160.

Tanner, J. M., Wilson, M. E., and Rudman, C. G. (1990). Pubertal growth spurt in the female rhesus monkey: relation to menarche and skeletal maturation. *Am. J. Hum. Biol. 2*, 101–106.

Tappen, N. C. (1969). The relationship of weathering cracks to split-line orientation in bone. *Am. J. Phys. Anthropol. 31*, 191–197.

Tappen, N. C. (1970). Main patterns and individual differences in baboon skull split-lines and theories of causes of split-line orientation in bone. *Am. J. Phys. Anthropol. 33*, 61–71.

Tappen, N. C. (1971). Two orientational features of compact bone as predictors of split-line patterns. *Am. J. Phys. Anthropol. 35*, 129–139.

Tappen, N. C. (1976). Advanced weathering cracks as an improvement on split-line preparations for analysis of structural orientation in compact bone. *Am. J. Phys. Anthropol. 44*, 373–377.

Taylor, W. O. G. (1939). The effect of enucleation of one eye in childhood upon the subsequent development of the face. *Trans. Ophthalmol. Soc. U.K. 59*, 361–371.

Thesleff, I. (1995). Homeobox genes and growth factors in regulation of craniofacial and tooth morphogenesis. *Acta Odont. Scand. 53*, 129–134.

Thorogood, P. (1988). The developmental specification of the vertebrate skull. *Development 103* (Suppl.), 141–153.

Tobias, P. V. (1967). *Olduvai Gorge: The Cranium and Maxillary Dentition of Australopithecus (Zinjanthropus) Boisei* (Cambridge: Cambridge University Press).

Varrela, J. (1992). Dimensional variation of craniofacial structures in relation to changing masticatory-functional demands. *Eur. J. Orthod. 14*, 31–36.

Watts, E. S., and Gavan, J. A. (1982). Postnatal growth of nonhuman primates: the problem of adolescent spurt. *Hum. Biol. 54*, 53–70.

Weidenreich, F. (1936). The mandibles of *Sinanthropus pekinensis*. *Palaeont. Sinica 7* (series D), 7, 47–60.

Whiting, J. (1997). Craniofacial abnormalities induced by the ectopic expression of homeobox genes. *Mut. Res. 396*, 97–112.

Wood, N. K., Wragg, L. E., Stuteville, O. H., and Oglesby, R. J. (1969). Osteogenesis of the human upper jaw: proof of the non-existence of a separate premaxillary centre. *Arch. Oral Biol. 14*, 1331–1339.

Wood-Jones, F. (1916). *Arboreal Man* (London: E. Arnold).

Wood-Jones, F. (1948). *Hallmarks of Mankind* (London: Baillière Tindall).

CHAPTER 16

FACIAL DYSMORPHOLOGY IN THE CRANIOSYNOSTOSES: CLINICAL IMPLICATIONS

KATHERINE W.L. VIG, B.D.S., M.S., D. Orth.

16.1 INTRODUCTION

Facial appearance and the stigma associated with recognizable patterns of dysmorphology has psychosocial and morphological implications. To be born with a birth defect and therefore to start life with an obvious deformity has different consequences than an acquired deformity. In a society that does not readily accept deviations from a normal biological variation or ideal appearance, interventions are sought early in life by the parents to attain as near normal an appearance as possible for their child.

Craniosynostosis is characterized by early fusion of one or more of the cranial vault sutures, which may also include the cranial base synostoses. This may be syndromic or nonsyndromic, but the result in the neonate and later in the infant is an early distortion in the shape, size, and position of the components of the craniofacial skeleton. If left untreated, restriction of the developing brain may occur as part of the growth and development of the neural system that is increasing in size significantly during the first year of infancy. A descrip-

tion of the clinical implications and etiopathogenesis of craniosynostosis was described by Burdi, Vig, and Reynolds in 1991 and forms the basis for this chapter. The effects of microcephaly on the growth of the brain, with consequent mental retardation, was recognized over a century ago (Virchow, 1851); however, the etiopathogenesis of premature suture closure was not known at that time. Nevertheless, recognition of the association between the rapidly enlarging brain in infants and compensatory growth of the calvarium had a profound influence on the prevailing therapeutic approach, and led logically to the concept of surgical removal of affected sutures as a method to treat craniosynostosis.

Craniectomy in North America was initially performed by Lane (1892) to release the premature closure of a child's sagittal suture at the request of the mother to "unlock her child's brain and let it grow." Although his first procedure resulted in the untimely demise of the child, this was apparently due to problems associated with the anesthetic and not with the surgery itself. Lane continued to perform craniectomies, and reported a microcephalic infant who survived the procedure with "unequivocal evidence of mental improvement." Despite Lane's self-reported success, his aggressive surgical approach was unacceptable to other clinicians. Shortly after Lane's clinical report, Jacobi (1984) soundly criticized his approach, noting that "such rash feats of indiscriminate surgery . . . are stains on your hands and sins on your soul." As Jacobi was considered the father of American pediatrics, his comments had a significant effect on the use of craniectomy as a procedure to correct craniosynostosis. It was a quarter of a century later that King (1942) reported a calvarial morcellation procedure. This was followed by a renewed interest by surgeons in preventing recurrent craniosynostosis after surgical interventions.

16.2 THEORIES OF CRANIOSYNOSTOSIS ETIOPATHOGENESIS

The etiology of craniosynostosis has been attributed to intrauterine constraint (Graham et al., 1980). Isolated and syn-dromic cases have suggested that craniosynostoses arise as malformations. Since malformations often result in embryonic death and abortion, the actual incidence of craniosynostosis is difficult to ascertain (Cohen, 1986). Three main theories have been postulated to account for the etiology of craniosynostosis. Virchow, in 1851, provided the first comprehensive description of craniosynostosis. He suggested that calvarial synostosis was the primary malformation and that secondary effects were reflected in the cranial base. According to Virchow, the characteristic distortion of the skull associated with craniosynostosis is a result of growth inhibition perpendicular to the synostosed suture. Compensatory growth at the nonsynostosed sutures in response to the enlarging neonatal brain results in a characteristic and recognizable distortion of the craniofacial complex (Fig. 16.1). An opposing theory was suggested by Park and Powers in 1920, who speculated that the primary defect associated with craniosynostosis resides in the mesenchymal blastema and results in fusion of the cranial sutures and abnormalities in the cranial base (Fig. 16.2a, b). A third theory

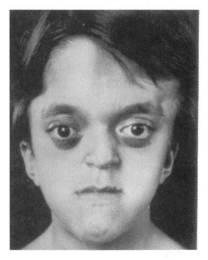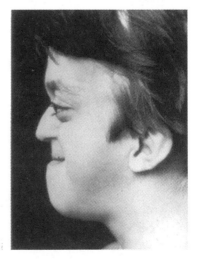

Figure 16.1   Patient with characteristic features of Crouzon syndrome. Note the exorbitism and hyperteleorism of the eyes in full face and the midface deficiency in the profile. (*Source*: With permission from the Center for Human Growth and Development, University of Michigan, 1991.)

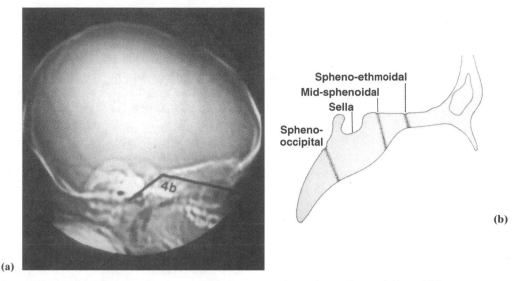

Figure 16.2 (a) Cranial base angle on scout film between anterior and posterior cranial base (4b). (b) Diagramatic representation of the cranial base with the synostoses in the anterior cranial base and the synchondrosis in the posterior cranial base.

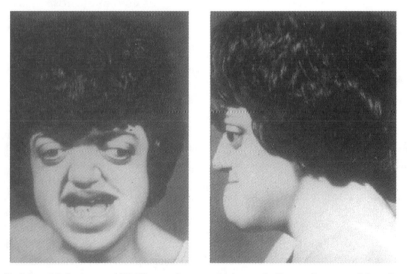

Figure 16.3 Patient with features of Pfeiffer syndrome, which are similar to Crouzon, although more severe. (*Source*: With permission from the Center for Human Growth and Development, University of Michigan, 1991.)

on the etiology of craniosynostosis was introduced by Moss in 1959. Moss and Young (1960) attributed the primary defect to the cranial base, which secondarily affects tension at the calvarial sutures via a biomechanical effect on the dura mater, resulting in premature fusion (Fig. 16.3).

These theories indicate that the cranial sutures and cranial base are the primary cause of the lack of response of the cranial vault to an increase in the size of the brain during the neonatal period. However, the mechanism by which biomechanical effects on sutures and in the cranial base are expressed has not been fully explained, although animal models have been developed to test causal relationships. Recent advances in imaging technology and reconstructive surgery have provided new information relating to mechanisms. Advances in imaging have also contributed to the planning and surgical manipulations of the disfiguring dysmorphology involving both the face and skull (Marsh and Vannier, 1985). Functional interactions associated with early chondrocranial synostoses in the cranial base and their effects on the growth of the craniofacial region continue to be of interest to craniofacial biologists, radiologists, dysmorphologists, and neurosurgeons. In evaluating normal development of the cranial base angle in skeletal class I and class II patterns between birth and 14 years, no changes were revealed in the cranial base angle. These findings did not support the previously held belief that class II skeletal pattern was associated with a more acute cranial base angle (Wilhelm et al., 2001).

The similarity or dissimilarity among the craniosynostoses has the potential to influence directly some very precise and invasive surgical approaches in the treatment of these craniofacial syndromes. With such intervention, the outcomes of treatment have important implications for the patient, family, and craniofacial team in terms of efficacy and utility of treatment (Bachmayer and Ross, 1986). However, development of new insights into the pathogenesis of craniosynostoses requires interaction with a biomedical team at the molecular biology level. Specifically, recent technological advances in molecular genetics provide information that facilitates identification of the gene locus involved in a number of conditions and ultimately lead to an explanation of the causal mechanisms

of hereditary traits (Collins, 1999; Collins and Mansoura, 2001).

Previously, treatment of patients with craniosynostoses has been based primarily on perceptions of the natural history of premature sutural fusion. With the new advances in molecular genetics and developmental biology, the treatment strategies and postsurgical growth expectations in these patients may come to be based on a more profound understanding of the causal mechanisms embracing issues of both morphogenesis and pathogenesis.

16.3 CLINICAL MANAGEMENT OF CRANIOSYNOSTOSIS

The management of patients with premature fusion of the cranial sutures developed as a consequence of the differential diagnosis of craniosynostosis. This provided a more rational approach to distinguish between a primary failure of brain growth, and its consequent effect on the growth and development of the skull in microcephalics, and the secondary effects produced by fusion of the sutures. This restraining effect on the enlarging brain may result in mental impairment and therefore requires early surgical intervention (Turvey and Gudeman, 1996).

Calvarial dysmorphogenesis has a wide spectrum of severity. Mild expression may be difficult to distinguish as minor flattening of the skull, whereas increased severity is accompanied by midface deficiency and intracranial hypertension. Early and urgent surgical intervention may be indicated in these patients (Gault et al., 1992). The etiopathogenesis is complex (Vander Kolk and Beatty, 1994), and although considerable interest in the treatment by distraction osteogenesis has provided a method for maintaining patency of the sutures, the etiology is still not fully understood (Cohen and Persing, 1998). With the resurgence of the genetic paradigm as the human genome project is nearing completion, new information is being generated at an ever-increasing rate, which may help to elucidate the syndromic from the nonsyndromic manifestations of craniosynostosis. An epigenetic or environmental influence has been suggested in those instances where intracranial pressure or extrinsic forces have resulted in calvarial deformation such as positional plagiocephaly. This type of deformation may be corrected by external forces applied with a helmet type of device to reverse and correct the initial cause of the dysmorphology (Ripley et al., 1994; Littlefield et al., 1998).

Premature fusion of cranial sutures, which may occur in utero, affects a variety of sutures. Although the primary cause of premature closure of cranial sutures remains unknown, it is generally considered that three types of craniosynostoses exist: (1) single-suture synostoses; (2) multiple-suture synostoses; and (3) craniofacial dysostoses, such as Apert, Crouzon, and Pfeiffer syndromes (Marsh and Vannier, 1985; Taccone et al., 1989).

The controversy concerning the site of the primary defect in craniosynostosis is still unresolved. The cranial sutures are considered to be reactive to the tension created by intracranial expansion of the infant's brain (Moss, 1959), and a combination of both calvarial suture synostosis and early fusion of the cranial base synchondroses may be implicated in the resulting craniofacial dysmorphology. For example, bicoronal synostosis and early fusion of the sphenoethmoidal synchondrosis have been suggested by Venes and Burdi (1985) to be causally interrelated via the coronal ring, which defines an anatomic continuity between the coronal suture system and the cranial base.

Neurosurgeons have advocated early surgical intervention, during the first 2 years of life, to prevent restriction to the enlarging neonatal brain, thus avoiding mental impairment in infants severely affected with premature suture closure (Marchac and Renier, 1982). However, reducing mental impairment has currently been a primary rationale for craniectomy; considerable attention is now also focused on the cosmetic effects of the dysmorphology and of surgery on the growth of the midface (Posnick, 1996; McCarthy, 1979). This approach was pioneered by Tessier (1971a, 1971b, 1976), who has promoted the concept of radical reconstruction of the craniofacial complex for cases of craniosynostosis associated with syndromes and also the treatment of isolated and nonsyndromic premature suture closure. The clinical manifestations of syndromic and nonsyndromic chondrocranial synostoses have been described, documented, and variously classified. The literature is replete with anecdotal reports, but because of the rarity of craniosynostosis syndromes and the difficulty in distinguishing between primary and secondary synostosis, confusion prevails in the nomenclature (Cohen, 1986).

16.4 PATTERNS OF DYSMORPHOGENESIS

The craniostenotic birth defects of Crouzon, Apert, and Pfeiffer syndromes belong to a class of conditions whose pathogenesis is characterized by premature fusion of select calvarial sutures, midfacial hypoplasia and retropositioning, and exorbitism (Kreiborg and Pruzansky, 1971; Warkany, 1971; Cohen, 1975, 1980; Ousterhout and Melsen, 1982; Vig, 1988; Gorlin et al., 1990). Surgical correction of these conditions is possible with the advent of modern surgical approaches (Tessier, 1971a, 1971b; McCarthy, 1979) and computerized tomographic imaging for preoperative planning. The Apert and Pfeiffer malformations are still thought to be more severe than Crouzon, since the craniofacial dysmorphology is less amenable to correction (Fig. 16.4).

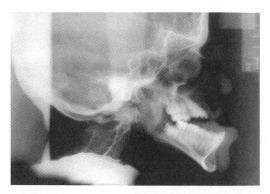

Figure 16.4 Lateral skull radiograph of a patient with Pfeiffer syndrome. Note the acute cranial base angle, midface deficiency, open bite, severe class III malocclusion, and fused cervical vertebrae.

Most surgical approaches to the treatment of Crouzon syndrome deal with one or more of the coronal ring sutures. The coronal ring (Venes and Burdi, 1985) is a continuous set of cranial articulations beginning with the coronal suture per se, the lateral frontosphenoid and the orbital frontosphenoid articulations. From a surgical perspective, this arrangement of articulations is a C-shaped sutural system with an apparent lack of structural continuity at the midline cranial base in the zone between the right and left optic foramina. In an earlier developmental description of the human skull, Blechschmidt and Gasser (1978) looked not only at the articulations between the developing calvarial bones but also at the earlier phases of mesenchymal thickenings and dural tracts in areas where cranial bones and intervening sutures were expected to develop. Based on the dural tissue tract approach, Blechschmidt and Gasser (1978) proposed that the coronal sutural system might indeed be a closed ring completely continuous at the middle cranial base so that the gap between the right and left optic foramina was spanned by a dural stretch field connecting the orbits.

Evidence to support these two alternate views, namely the C-shaped versus closed-ring arrangement of articulations, are reported in the literature. Fully formed sutures along the C-shaped system are cranial (Moss and Young, 1960; Scott, 1967; Moyers and Krogman, 1967; Crelin, 1969; DeBeer, 1971; Pierce et at., 1978; Enlow, 1982), and should affect primarily the shape and position of the cranial bones. Support for a closed-ring system of coronal articulations requires that expansion at the sutures along the C-shaped ring segments will allow expansion of the calvarium. The basilar segment of the closed ring between the right and left optic foramina may grow as a part of the cranial base and thus have an effect primarily on the position of the nasomaxillary complex.

The craniosynostoses of Crouzon, Apert, and Pfeiffer syndromes involve sites of pathogenesis either unilaterally or bilaterally along the calvarial articulations of the coronal ring. From the perspective of its natural history or pattern of morphogenesis, this coronal ring is continuous, extending on each side of the skull from the calvarial bregma downward through the infratemporal fossa into the posterior orbital regions, passing laterally across the midline sphenoethmoid articulation. The major portions of the coronal ring from the bregma to the sphenoethmoid articulation are sutural in structure; their continued growth affects the shape of the calvarium.

The shorter portion of the coronal ring that connects the inferior limbs of the C-shaped sutural components in the midline is the cartilaginous junction of the sphenoid and ethmoid bones, at the sphenoethmoid synchondrosis. The significance identified with the human sphenoethmoid synchondrosis is that this cartilaginous growth plate of the anterior cranial base affects the anteroposterior dimensions of the cranial base from sella to nasion, and affects the forward positioning and size of the midface contiguous with the anterior cranial base.

Surgical release of the premature synostoses along the cranial sutures affects mainly the calvarial form. Thus, premature synostosis of the sphenoethmoid synchondrosis, or a failure of initial differentiation into a synchondrosis, affects the anterior cranial base and indirectly the midfacial regions. If this assumption is correct, the severity of midface deficiency in the craniosynostoses, and exorbitism, is directly related to the proximity of premature synostoses to the sphenoethmoid synchondrosis.

The evidence in the literature has been tested in animal models and extrapolated to humans. If the craniosynostoses can be supported by experimental studies, then craniectomies along the sutural limbs of the coronal ring may be expected to have primary effects on the postsurgical growth of the calvarium. Craniectomies that bypass the deeper portions of the coronal ring at the sphenoethmoid synchondrosis have little or no postsurgical effects on the size and position of the midface (Marchac and Renier, 1985).

16.5 EXPERIMENTAL STUDIES

16.5.1 Animal Models

In response to the surgical distraction of cranial sutures in the human neonate, McLaughlin and colleagues (2000) studied the normal response to tensile forces in neonatal rat cranial sutures. The frontal suture differed significantly from the sagittal and coronal sutures in the rat model, which may not necessarily extrapolate to the human neonate. Previously, methylmethacrylate was used to induce calvarial synostoses and produce malformations of the calvarium in the rabbit experimental model (Babler and Persing, 1981; Persson et al., 1979). Following the coronal ring paradigm, the experimental outcomes of these studies on calvarial shape with sutures in the coronal ring, and sutures high up on the sagittal and coronal sutures of the rabbit model, resulted in similar features to those

occurring in the human phenotype. The problem of finding an appropriate animal model with the morphological characteristics of the affected human continues to be unresolved (see Chapter 9 for further discussion). Apert, Crouzon, and Pfeiffer syndromes are autosomal-dominant traits, and following the mapping of the human genome and localization to a specific gene locus on chromosome 7 may provide information on causality in the future.

16.5.2 Clinical Studies

The systematic mapping of gene loci in chromosomal disorders has located the site for many genetically transmitted diseases. The identification and isolation of gene sequences that can be linked to diseases have provided new insights into the causation of the disease at the molecular level. The international human genome initiative has already developed important information on gene linkage and association.

The degree of premature synostosis at, or close to, the deeply placed sphenoethmoid synchondrosis (see Fig. 16.2), and the severity of the characteristic facies in growing patients with Crouzon, Apert, and Pfeiffer syndromes, indicate that this synchondrosis may be implicated in both the form and position of the midfacial region contiguous with the anterior cranial base being affected.

Traditionally, descriptive studies have utilized fetal and neonatal autopsy specimens for histologic investigations (Ousterhout and Melsen, 1982). Various skulls have been discovered that portray the characteristic features associated with premature fusion of the cranial sutures. Technological advances in imaging methods are now available to follow patients' growth and development. This has been reported by Kreiborg and Aduss (1986) for Crouzon syndrome.

A retrospective radiographic study (Reynolds, 1986; Burdi et al., 1991) exam-

TABLE 16.1   Comparison of Measurements Between Apert, Crouzon, and Pfeiffer Syndromes[a]

	Apert (A) vs. Crouzon (C)	Apert (A) vs. Pfeiffer (P)	Pfeiffer (P) vs. Crouzon (C)
Greater wing of sphenoid angle	A > C**	A > P	P > C
	(A > C*)	(A > P)	(P > C*)
Optic nerve angle	A > C	A > P	P > C
	(A > C*)	(A > P)	(P > C)
Exorbitism	A > C**	A > P	P > C**
	(A > C**)	(A > P)	(P > C*)
Interorbit width	A > C*	A > P	P > C*
	(A > C*)	(A > P)	(P > C*)

* Statistically significant, $p < 0.05$.

** Statistically significant, $p < 0.01$.

[a] Comparison made for both total sample neonate to 6 years of age; top comparison and subgroup (neonate to 10 months of age given in parentheses).

ined the cranial base in three craniosynostosis syndromes characterized by early fusion of the craniofacial sutures. These patients were clinically recognizable by facial features of sagittal midface deficiency, hypertelorism, and exorbitism (Figs. 16.1 and 16.3). The sample consisted of 37 age-matched patients from the University of Michigan Hospital and the New York University Medical Center with a distribution of 14 Crouzon (8 females, 6 males), 12 Apert (5 females, 7 males), and 11 Pfeiffer (3 females, 8 males). The diagnosis was made by clinical and radiographic examination and confirmed by genetic karyotyping at both institutions. Computerized tomographic (CT) scanning films were taken at 3 millimeter slices in the neonate and increased to 5 millimeter in the older and larger infants and children. Magnification was controlled at between 1.5 and 1.8 times actual head size, with differences in magnification being resolved by using a linear scale step wedge on each magnetic scanning tape.

The findings from this study indicate differences in the cranial base that characterized the comparison of the three syndrome groups. The sample size was not adequate to allow more than identification of trends,

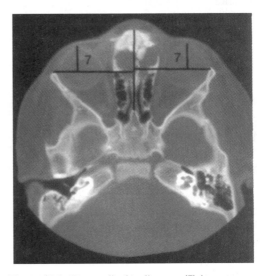

Figure 16.5   Perpendicular distance (7) is a measure of the degree of exorbitism from a line joining the most lateral aspect of the orbital wall. (*Source*: With permission from the Center for Human Growth and Development, University of Michigan, 1991.)

but these suggested that exorbitism was greater in patients diagnosed as Apert and Pfeiffer syndromes compared with the Crouzon patient sample (Table 16.1). Although the exorbitism was increased in two of the groups (Fig. 16.5), the globe size was consistent both between and among all

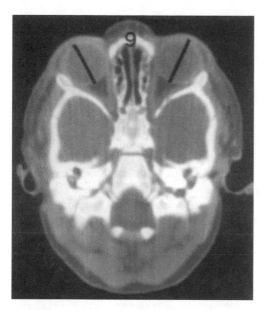

Figure 16.6 Globe size measured at the level of the optic nerve. No significant differences were found among groups or between age ranges.

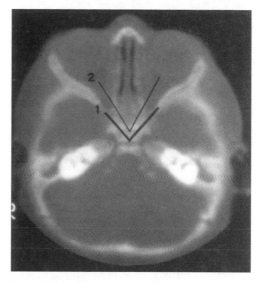

Figure 16.7 CT scan of cranial base. Angle (1) is the intersection between the greater wings of the sphenoid. Angle (2) is the intersection between the converging optic nerves. Note the cranial base synchondrosis between the sphenoid and occipital bones. (*Source*: With permission from the Center for Human Growth and Development, University of Michigan, 1991.)

three syndrome groups (Fig. 16.6). The consistency of globe size in spite of variability in the amount of exorbitism might be attributed to a characteristic difference in the volumetric size, shape, and position of the bony orbit in the three syndrome types.

The age distribution of the 37 patients was birth to 10 months of age (25 subjects), 10 months to 3 years of age (8 subjects), and 3 years to 6 years of age (4 subjects). High-resolution CT scans of the cranial base were measured to quantify the angle between the greater wings of the sphenoid and the angle between the optic nerves (Fig. 16.7). This measurement defines the lateral wall of the orbit and was found to be greater in Apert and Pfeiffer syndromes than in Crouzon (Table 16.1). The optic nerve angle was also found to be greater in the Apert group than in Crouzon, and the opening of these angles was clinically reflected in the degree of hypertelorism. Exorbitism was the linear measurement of a perpendicular distance to a line joining

the most lateral aspects of the orbital wall on the CT scan having the largest soft-tissue orbit diameter (Fig. 16.5). Both exorbitism and interorbit width were greater in Apert and Pfeiffer groups than in the Crouzon group (Table 16.1).

The birth to 10 months age group and the total sample were analyzed. Age was the covariate, and because no systematic difference was found between the subset syndrome groups, the data were pooled. The study was designed to evaluate the variability of morphological features measured in the three syndromes. Sheffe's statistical test to account for multiple comparisons was applied (Snedecor and Cochran, 1981).

An additional aspect of this study was to stage the amount of fusion of the midline cranial base synchondroses by the degree of radiopacity in the CT scans. The sphenoethmoidal synchondrosis in normal

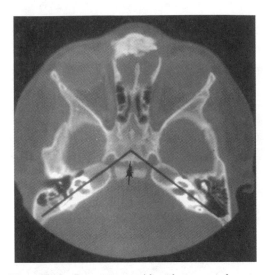

Figure 16.8 Petrous pyramid angle measured posteriorly. In the 0- to 10-month-old group, Pfeiffer was greater than Apert and Crouzon. Note the spheno-occipital synchondrosis open behind the intersection of the petrous pyramid angle.

growth and development unites between 6 and 7 years of age, and was therefore graded in the infants represented in the age group of birth to 1 year. The spheno-occipital synchondrosis normally fuses at 17 to 19 years, and was therefore scored as patent if still radiolucent (Figs. 16.7 and 16.8). No significant differences were found between the three groups, although the trend indicated that earlier closure occurred in the Apert and Pfeiffer groups, which would be consistent with the presumed etiopathogenesis.

Because of the rarity and phenotypic variance of the three syndromes compared in this study, sample sizes were small. The criteria for delineation of these syndromes are further confounded by the retrospective nature of this study, for which accurate clinical diagnosis and documentation were critical. However, as a preliminary study, trends of differences in morphology among the three syndromes were apparent. This phenotypic variance may provide a means of stratifying a larger sample to character-ize morphological features between and among the groups by a larger intercenter study design in the future.

16.6 CONCLUSIONS

Infants who survive until birth with severe craniosynostosis represent incomplete ascertainment of affected individuals. With new information in molecular genetics, autosomal-dominant syndromes have already been isolated to a specific chromosome. Gene sequencing should identify the locus responsible for the different syndromic craniosynostoses and the pathogenesis should be further elucidated in the future.

The causal mechanism for sutural synostosis remains unresolved. Previous insights into our understanding of the dysmorphic events have been provided through histopathological studies of craniosynostotic sutures (Albright and Byrd, 1981), and histomorphometry and computer imaging provide a basis from which future studies, at the molecular, genetic, and biological levels, may ultimately elucidate the causal mechanism.

ACKNOWLEDGMENTS

I am indebted to Dr. J. McCarthy and Dr. B. Grayson for generously providing access to their facility and records, and for allowing us to include their patients in the CT scan original study of a comparison between Apert, Crouzon, and Pfeiffer syndromes.

REFERENCES

Albright, A. L., and Byrd, R. P. (1981). Future pathology in craniosynostosis. *J. Neurosurg. 54*, 384–387.

Babler, W. A., and Persing, J. A. (1981). Alterations in cranial suture growth associated with premature closure of the coronal suture in rabbits. *Anat. Rec. 199*, 1–4.

Bachmayer, D. I., and Ross, R. B. (1986). Stability of Le Fort III advancement surgery in children with Crouzon's, Apert's, and Pfeiffer's syndromes. In J. L. Marsh (ed.), *Long-Term Results of Craniofacial Surgery*. Cleft Palate Journal Supplement, vol. 23, pp. 69–74.

Blechschmidt, F., and Gasser, R. F. (1978). *Biokinetics and Biodynamics of Human Differentiation* (Springfield: Charles C. Thomas).

Burdi, A. R., Vig, K. W. L., and Reynolds, R. T. (1991). The craniosynostoses: etiopathogenesis and clinical implications. In A. R. Burdi and K. W. L. Vig (eds.), *Essays in Honor of Robert E. Moyers. Craniofacial Growth Series 24* (Ann Arbor: Center for Human Growth and Development: The University of Michigan).

Cohen, M. M., Jr. (1975). An etiologic and nosologic overview of craniosynostosis syndromes. In *Birth Defects*. 9(2). (••: ••), pp. 137–189.

Cohen, M. M., Jr. (1980). Perspective on craniosynostosis (editorial review). *West. J. Med. 132*, 507–513.

Cohen, M. M., Jr. (1986). Etiology of craniosynostosis. In M. M. Cohen, Jr. (ed.), *Craniosynostosis: Diagnosis, Evaluation and Management* (New York: Raven Press), pp. 59–80.

Cohen, S. R., and Persing J. A. (1998). Intracranial pressure in single suture craniosynostosis. *Cleft Palate Craniofac. J. 35*, 194–196.

Collins, F. S. (1999). The human genome project and the future of medicine. *Ann. NY Acad. Sci. 882*, 42–55.

Collins, F. S., and Mansoura, M. K. (2001). The human genome project. *Cancer 91* (S1), 221–225.

Crelin, E. (1969). *Anatomy of the Newborn: An Atlas* (Philadelphia: Lea and Febiger).

DeBeer, G. (1971). *The Development of the Vertebrate Skull* (London: Oxford University Press).

Enlow, D. (1982). *Handbook of Facial Growth* (Philadelphia: W. B. Saunders).

Gault, D. T., Renier D., Marchac D., and Jones B. M. (1992). Intracranial pressure and intracranial volume in children with craniosynostosis. *Plast. Reconstr. Surg. 90*, 377–381.

Gorlin, R., Cohen, M. M., Jr., and Levin, L. S. (1990). *Syndromes of the Head and Neck*, 3rd ed. (New York: Oxford University Press).

Graham, J. M., Badura, R. J., and Smith, D. W. (1980). Coronal craniosynostosis: fetal head constraint as one possible cause. *Pediatrics 65*, 995–999.

Jacobi, A. (1894). "Non nocere." *Med. Rec. NY 14*, 18–23.

King, J. (1942). Oxycephaly. *Ann. Surg. 115*, 488–506.

Kreiborg, S., and Aduss, A. (1986). Pre- and postsurgical facial growth in patients with Crouzon's and Apert's syndromes. *Cleft Palate J. 23* (suppl.), 78–91.

Kreiborg, S., and Pruzansky, S. (1971). Craniofacial growth in patients with premature cranial synostosis. Presented at the 1st Symposium on the Diagnosis and Treatment of Craniofacial Malformations. New York University.

Lane, L. C. (1892). Pioneer craniectomy for relief of mental imbecility due to premature sutural closure and microcephalus. *J. Am. Med. Assoc. 18*, 49–50.

Littlefield, T. R., Beals, S. P., Manwaring, K. H., Pomatto, J. K., Joganic, E. F., Golden, K. A., and Ripley, C. E. (1998). Treatment of craniofacial asymmetry with dynamic orthototic cranioplasty. *J. Craniofac. Surg. 9*, 11–17.

Marchac, D., and Renier, D. (1982). *Craniofacial Surgery for Craniosynostosis* (Boston: Little, Brown and Co.).

Marchac, D., and Renier, D. (1985). Craniofacial surgery for craniosynostosis improves facial growth. *Ann. Plast. Surg. 14*, 43–54.

Marsh, J. L., and Vannier, M. W. (1985). *Comprehensive Care for Craniofacial Deformities* (St. Louis: C. V. Mosby Co.).

McCarthy, J. G. (1979). New concepts in the surgical treatment of the craniofacial synostosis syndromes in the infant. *Clin. Plast. Surg. 6*, 201–226.

McLaughlin, E., Zhang, Y., Pashley, D., Borke, J., and Yu, J. (2000). The load displacement characteristics of neonatal rat cranial sutures. *Cleft Palate Craniofac. J. 37*, 590–595.

Moss, J. L. (1959). The pathogenesis of premature cranial synostosis in man. *Acta Anat. 37*, 351–370.

Moss, M., and Young, R. (1960). A functional approach to craniology. *Am. J. Phys. Anthropol. 18*, 281–292.

Moyers, R. E., and Krogman, W. M. (1967). *Craniofacial Growth in Man* (London: Pergamon Press).

Ousterhout, D. K., and Melsen, B. (1982). Cranial base deformity in Apert's syndrome. *Plast. Reconstr. Surg. 69*, 254–263.

Park, E. A., and Powers, G. F. (1920). Acrocephaly and scaphocephaly with symmetrically distributed malformations of the extremities. *Am. J. Dis. Child. 20*, 235–315.

Persson, K. M., Roy, W. A., and Persing, J. A. (1979). Craniofacial growth of following experimental craniosynostosis and craniectomy in rabbits. *J. Neurosurg. 50*, 187–197.

Pierce, R., Mainen, M., and Bosma, J. (1978). *The Cranium of the Newborn Infant* (Bethesda: US-DHEW).

Posnick, J. C. (1996). Craniofacial dysostosis syndromes: a staged reconstruction approach. Chapter 26. In T. Turvey, K. W. Vig, and R. Fonseca (eds.), *Facial Clefts and Craniosynostosis—Principles and Management* (Philadelphia, W. B. Saunders), pp. 630–685.

Reynolds, R. T. (1986). Basic morphometric analyses in Crouzon, Apert and Pfeiffer defects: implications for their delineation, surgical management, and growth assessment. Master's thesis, The University of Michigan.

Ripley, C. E., Pomatto, J. K., Beals, S. P., Joganic, E. F., Manwaring, K. H., and Moss, S. D. (1994). Treatment of positional plagiocephaly with dynamic orthototic cranioplasty. *J. Craniofac. Surg. 5*, 150–159.

Scott, J. H. (1967). *Dentofacial Development and Growth* (London: Pergamon Press).

Snedecor, G. W., and Cochran, W. G. (1981). *Statistical Methods* (Ames: Iowa State University Press).

Taccone, A., Cama, A., Ghiorzi, A., and Fondelli, P. (1989). Diagnosis and treatment of craniosynostoses: the usefulness of CT combined with 3 dimensional reconstruction. *Radiologia Medica 77*, 322–328.

Tessier, P. (1971a). Relationship of craniosynostosis to craniofacial dysostosis, and to faciostenoses—a study with therapeutic implications. *Plast. Reconstr. Surg. 48*, 224–237.

Tessier, P. (1971b). The definitive plastic surgical treatment of the severe facial deformities of craniofacial dysostosis. *Plast. Reconstr. Surg. 48*, 419–442.

Tessier, P. (1976). Recent improvement in treatment of facial and cranial deformities of Crouzon's disease and Apert's syndrome. In P. Tessier (ed.), *Symposium on Plastic Surgery in the Orbital Region* (St. Louis: C. V. Mosby).

Turvey, T. A., and Gudeman, S. K. (1996). Nonsyndromic craniosynostosis. Chap. 25. In T. Turvey, K. W. Vig, and R. Fonseca (eds.), *Facial Clefts and Craniosynostosis—Principles and Management* (Philadelphia: W. B. Saunders), pp. 596–629.

Vander Kolk, C. A., and Beatty T. (1994). Etiopathogenesis of craniofacial anomalies. *Clin. Plast. Surg. 21*, 481–488.

Venes, J., and Burdi, A. R. (1985). A proposed role of the orbital sphenoid in craniofacial dysostosis. In R. Humphreys (ed.), *Concepts in Pediatric Neurology*, vol. 5, pp. 126–135 (Basel: Karger Press).

Vig, K. W. L. (1988). Orthodontic perspectives in craniofacial dysmorphology. In K. W. L. Vig and A. Burdi (eds.), *Craniofacial Morphogenesis and Dysmorphogenesis. Craniofacial Growth Series*, vol. 21 (Ann Arbor: Center for Human Growth and Development, The University of Michigan), pp. 504–536.

Virchow, R. (1851). Uber den Cretinismus, namentlich in Franken, und uber pathologische Schdelformen. *Verh. Phys. Med. Gesellsch. Wurzburg 2*, 231–284.

Warkany, J. (1971). *Congenital Malformations* (Chicago: Year Book Publishers).

Wilhelm, B. M., Beck, F. M., Lidral, A., and Vig, K. W. L. (2001). A comparison of infant cranial base growth in class I and class II skeletal patterns. *Am. J. Orthod. Dentofac. Orthop. 119*, 401–405.

CHAPTER 17

MIDFACIAL AND MANDIBULAR DYSMORPHOLOGY AND GROWTH IN FACIAL CLEFTING: CLINICAL IMPLICATIONS

BRUCE ROSS, D.D.S., Dip.Orthodontics, M.Sc., FRCD(Can)

17.1   INTRODUCTION

As noted in earlier chapters in this text, the usual location of a facial cleft is along the developmental planes where components fuse or merge with each other. The most common clefts by far are those involving the primary and secondary palates, and they are referred to as clefts of the lip and palate, or orofacial clefts. The greater part of this chapter is devoted to the many forms of cleft lip and palate, and focuses on the facial dysmorphology associated with them.

17.2   CLEFT LIP/CLEFT PALATE

Cleft lip and palate includes two distinct clefts, different in etiology, development, morphology, and growth pattern. Thus, there may be:

1. A *cleft lip* extending from the vermillion border of the lip up into the nasal sill and back into or through the anterior maxilla to the incisive foramen and floor of the nose. It may be unilateral or bilateral,

complete or incomplete, and may be submucous. If bilateral, it may be symmetrical or asymmetrical. This is a cleft of the primary palate.

2. A *cleft palate* extending from the uvula forward to the incisive foramen. The mildest cleft involves only the soft palate, and may continue into the hard palate to a variable extent. It may be submulous. When the primary palate is not clept, the palatal cleft is symmetrical in the midline. This is a cleft of the secondary palate.

3. Combinations of *cleft lip and cleft palate* combination clefts occcur with varying severity of each. A complete unilateral cleft lip and palate has a palate cleft with the nasal septum attached abnormally to the noncleft palatal shelf for some of its length, but is completely detached from the cleft-side palatal shelf, and thus is asymmetric. In the complete bilateral cleft lip and palate, the nasal septum is free of the shelves and symmetrical in the midline.

It is more convenient for clinical and research purposes to divide orofacial clefts into five somewhat separate groups:

Understanding Craniofacial Anomalies: The Etiopathogenesis of Craniosynostoses and Facial Clefting, Edited by Mark P. Mooney and Michael I. Siegel, ISBN 0-471-38724-x   Copyright © 2002 by Wiley-Liss, Inc.

1. Cleft lip (CL)
2. Unilateral cleft lip and palate (UCLP)
3. Bilateral cleft lip and palate (BCLP)
4. Cleft palate (CP), which includes the Pierre Robin sequence (PRS)
5. Submucous cleft palate, which includes the velocardiofacial syndrome (VCFS)

The physical stature of children with orofacial clefts is less than that of children without clefts (Johnson, 1960; Drillien et al., 1966; Dahl, 1970), and they may have a retarded skeletal age (Menius et al., 1966; Przezdziak, 1969). Shibasaki and Ross (1969) found evidence that the pubertal growth spurt of facial structures was delayed in patients with cleft palate. Jensen and colleagues (1983) found that males with cleft lip and palate were slightly shorter from 6 to 18 years of age. The pubertal growth maximum occurred an average of 6 months later, but continued longer, and they eventually caught up with the controls. Skeletal maturity was retarded compared to the control group.

Initial growth retardation could be explained by the preoperative feeding difficulties and the trauma associated with surgical procedures. It would be expected that these adverse environmental influences would have only a temporary effect. However, studies suggest that an intrinsic deficiency is present and that complete "catch-up" growth does not occur. Thus, at any age, smaller facial structures can be expected.

Children with orofacial clefts have a higher incidence of other associated anomalies. The major ones are readily identifiable, but there may be many less obvious problems that could affect growth and development, both physical and intellectual. It is more helpful and realistic, however, to point out that the vast majority of children with clefts of the lip and palate are, to all intents and purposes, normal except for the craniofacial structures affected by the cleft.

17.2.1 Facial Morphology

The experienced clinician is well aware that adults with clefts of the lip and palate often have unsatisfactory facial growth, so they present with esthetic and functional problems that are difficult to correct. Why are their faces different? It would be helpful in treatment planning if one could distinguish between structures that are intrinsically abnormal and those that only appear abnormal because they have become distorted or damaged by secondary environmental disturbances. The former cannot become normal; the latter may be prevented or corrected if the proper intervention can be provided.

The morphology of the adult face is determined by three main factors:

1. *Intrinsic factors*, which include the genetic inheritance, the developmental disturbance that resulted in a cleft, and the growth potential of the abnormal structures
2. *Functional activity*, which may interfere with both normal and abnormal structures, causing distortions or growth disturbances
3. *Iatrogenic factors* involving the effects of treatment

17.2.1.1 Intrinsic Deficiencies and Abnormalities

Morphology provides a framework within which growth occurs, so the starting point in an analysis of morphology and growth in cleft lip and palate is the identification of the initial dysmorphology present when a cleft develops. Before undertaking a detailed consideration of the maxillary complex in which the cleft occurs, it might be helpful to consider the other components of the craniofacial skeleton, in order to establish the extent of the anomaly.

The Orbitocranial Skeleton The cranial base is the complex of bones that forms the floor of the cranium and the roof of the face. It is a particularly valuable structure to use as a base from which to assess facial morphology. A detailed description is given in Chapter 13 of this text. Studies such as the one by Ross (1965) found that although the proportions in the cranial base of children with CLP were identical, the entire cranial base was smaller, corresponding to the smaller size of the children. A more obtuse cranial base angle occurs in CLP, but this is probably a secondary effect related to the distortion of the maxillary complex.

There is a tendency toward slightly increased interorbital width (Psaume, 1957; Graber, 1949; Ross and Coupe, 1965; Farkas and Lindsay, 1972), although this does not warrant the term *orbital hypertelorism*. Again, it is more likely a secondary characteristic induced by environmental functional forces. The zygomatic bones (Harvold, 1954; Subtelny, 1955; Coupe and Subtelny, 1960) and the pterygoid plates of the sphenoid (Van Limborg, 1964; Atkinson, 1966) are essentially normal.

Studies on cleft lip and palate have usually noted that the mandible is probably normal or slightly smaller. There are as many variations as in the noncleft population, presumably because the mandible usually is not directly involved in the production of a cleft and the normal genetic variation is expressed. The mandible is considered in more detail later in this chapter.

CONCLUSION   Other than the nasal maxillary complex, and the occasional exception of the mandible, there is no evidence that other craniofacial structures are primarily affected by the presence of a cleft of the lip, palate, or both.

The Frontonasal and Maxillary Complex
It is currrently accepted that a major factor in the development of cleft lip and cleft palate is a deficiency of mesenchyme at critical embryonic stages. This deficiency appears to persist in the child and adult as a slightly smaller maxilla. The extent of the deficiency is not precisely known, but it can be estimated, and, of course, varies from individual to individual.

A cleft is not the same as a saw cut through a normal maxilla. When clefts involve the lip and anterior maxilla (primary palate), there is, at the site of the cleft, a defect in the piriform rim of the nasal cavity, a deficiency of *basal bone*, and a deficiency of the *alveolar bone* that contains the teeth (Fig. 17.1). The inadequate alveolar bone is partly related to the absence or abnormality of tooth develop-

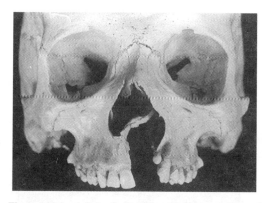

Figure 17.1 Adult skull with complete unilateral cleft of the anterior maxilla and palate. Presumably the lip was also completely cleft. There is great symmetry of the orbits, the zygomatic bones and sutures, and the basal maxilla and buccal teeth. Facial growth has been essentially normal, except for the deviation of the nasal bones, the nasal septum, and the severe deficiency of bone in the area of the cleft. The morphologic features are remarkably similar to those of the newborn condition, indicating that the initial deformity is maintained. Note that the left central incisor has been lost, but indications are that it had been sharply angled toward the cleft. The facial and dental features of this individual must have been strikingly similar to those of the boy in Figure 17.11. (*Source*: From Atkinson, 1966.)

ment adjacent to the cleft, since growth of the alveolar bone is in response to tooth development (edentulous individuals have virtually no alveolar bone). Basal bone and alveolar bone deficiencies seem to occur on both the mesial and distal margins of the cleft in cleft lip and palate.

In humans the anterior bone of the maxilla is not a separate premaxilla as in some other species (see Chapter 15), although we refer to the *premaxilla* for convenience in bilateral CLP. The premaxilla is abnormal in complete bilateral CLP because the clefts have prevented the mesenchyme of the maxillary process from migrating into the midline during embryonic development. The result is a premaxilla severely deficient in basal bone. The bulk of the premaxilla is alveolar bone that develops fairly normally in response to the developing teeth, but is somewhat restricted by the reduction in basal bone. Thus, if the teeth are removed in later life, the premaxilla is soon reduced to a tiny mass of

bone under the anterior nasal spine. The great variation in the size of the premaxilla at birth in bilateral CLP is usually an indicator of the number and size of teeth present (Fig. 17.2).

Children with only a cleft of the lip and anterior maxilla (no palate cleft) have a minimal disturbance in facial form. There are dental irregularities in the region of the cleft, and the anterior nasal spine shows some positional differences (Harvold, 1954; Coupe, 1962; Graber, 1949; Ross and Coupe, 1965; Dahl, 1970). In bilateral cleft of the lip and alveolus, particularly when the premaxilla is isolated from the remainder of the maxilla, many features of the bilateral cleft lip and palate are present, but to a lesser degree.

With clefts of the hard and soft palate (secondary palate), the relative size of the palatal bones in the fetus is difficult to ascertain, since many conflicting reports are available. It can be assumed that a deficiency in shelf mesenchyme is not clinically

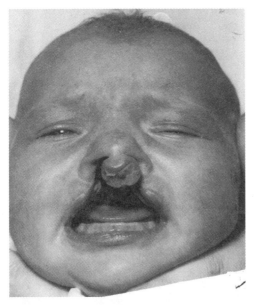
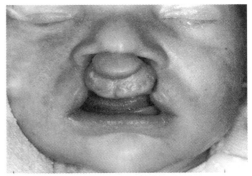

Figure 17.2   There appears to be great variation in the size of the premaxilla in complete bilateral cleft lip and palate. This is mainly related, however, to the number of teeth developing in the premaxilla and the bulk of alveolar bone that develops to support them. The basal bone is always deficient.

significant in most cases. In isolated cleft palate, the failure of palatal shelf fusion may also be related to a general midface deficiency, or small palatal shelves, or may be a mechanical interference with shelf elevation and fusion with the septum in the midline (see Chapters 4 and 5).

There are missing, malformed, and supernumerary teeth in all cleft types, as detailed in Chapter 18. With rare exceptions, the only teeth occasionally missing are the permanent lateral incisors adjacent to the cleft, and the second bicuspids and third molars in both jaws. These are the teeth often missing in the general population as well, but far more frequently in orofacial clefts.

The facial, palatal, and masticatory muscles in the infant with a cleft of the lip and palate appear to be normal. The cleft, however, causes interference with the sites of insertion of these muscles (Fig. 17.3). The anatomic arrangement of nerves, arteries, veins, skin, and mucous membranes is also disturbed (see Chapters 4 and 5).

CONCLUSION   The intrinsic abnormality is a slightly reduced overall size of the maxillary complex and local deficiencies at the

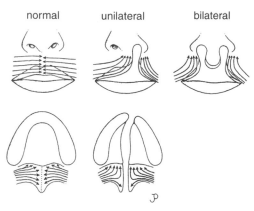

Figure 17.3   Migration paths of myoblasts entering a normal lip and palate, and a unilateral and a bilateral cleft lip and palate. The abnormal insertions and functional directions of these muscles are apparent. (*Source*: From Ross and Johnston, 1972.)

cleft margins and in the piriform rim. There are missing and hypoplastic teeth in both jaws and abnormal muscle insertions. In complete bilateral CLP, there is a gross deficiency of the basal bone of the premaxilla.

17.2.1.2 Prenatal Functional Distortions
At birth the face of the infant with cleft lip or palate has developed enormously from the early embryonic condition. The maxillary complex, with its structural continuity disrupted by the cleft, is vulnerable to distortion. Muscle activity begins very early in utero and influences the shape of the parts and their relation to each other. The different cleft types have differing responses to these forces and marked differences in morphology.

Unilateral Cleft Lip and Palate   The most striking characteristics of the infant with a complete unilateral cleft lip and palate are the distortion of the nasal cartilages and the severe deviation of the noncleft side of the maxilla away from the cleft, carrying with it the nasal structures, including the nasal septum (Figs. 17.1, 17.4, and 17.5). This deformation is the response of inadequately supported bone to distorting forces. The cheek and lip muscles on the noncleft side are abnormally inserted on the maxilla at the base of the nose, causing a rotating force on the larger segment during muscle contraction. This action is reinforced by frequent tongue protrusion into the alveolar cleft (Ross and Johnston, 1972). The tongue also causes a forward rotation of the ends of the alveolar segment and their supporting bone. Unrestrained nasal septum growth has been cited as the cause of the distortion (Latham and Burston, 1964), but this is unlikely.

The nose is pulled by muscle function toward the noncleft side, except for the alar base on the cleft side. As a result, the nostril on the cleft side is stretched and straightened. This configuration is established very

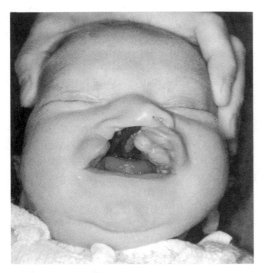

Figure 17.4 Characteristic facial morphology of a newborn with complete unilateral cleft lip and palate, showing the deflection of the anterior maxilla and the nasal structures toward the noncleft side. The cleft-side nostril has been stretched and deformed by muscle activity, which has also caused protrusion of the anterior alveolus.

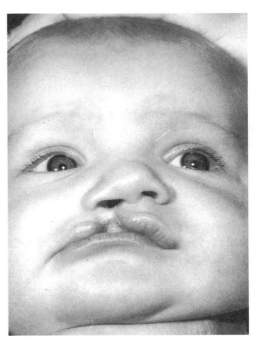

Figure 17.6 An infant with an incomplete unilateral cleft lip and palate. The distortion of the nasal structures and maxilla is considerably lessened.

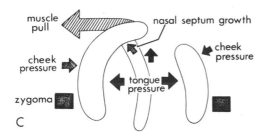

Figure 17.5 Diagram of the forces acting on the unilateral cleft lip and palate in utero. The effect is mainly on the anterior of the maxilla: the posterior tuberosities are expanded and symmetrical.

early in the embryo (Atherton, 1967; Latham, 1969), and the cleft-side alar cartilages develop and grow in this deformed matrix. By birth the deformity is permanent and the surgeon has great difficulty recontouring the cleft-side cartilages to achieve symmetrical nostrils.

Facial morphology in other types of clefts varies, but the same principles of deficiency and distortion apply (Fig. 17.6).

Infants with incomplete unilateral cleft lip and palate have all of the same tendencies but to a lesser degree. Clefts of the lip and anterior maxilla may show distortions in the alveolus and nasal structures similar to those of unilateral CLP; however, these are local disturbances and the overall morphology of the face is close to normal. The reason is simple: if there is continuity of bone in any area, the resistance to distortion is sufficient to prevent major deformity. This principle applies to incomplete unilaterals and bilaterals where part of the palate or part of the anterior maxilla is intact and less distortion is possible. Even a bridge of soft tissue across the cleft (a Simonart's band) prevents many of the expected morphological changes from occurring, particularly of the nose. The maxilla in these cases may be rotated, however, since the tongue and cheek muscles are exerting abnormal forces.

The tongue activity also expands the posterior segments, so a wider maxilla is characteristic. As a result, in the vast majority of complete CLP clefts, the maxillary gum pads are wider than the mandibular gum pads at birth. Measurements of dental models of the newborn (Peyton, 1931; Harding and Mazaheri, 1972) indicate that the arch width is greater in cleft conditions. This finding confirms work on other facial widths (Ross and Coupe, 1965) showing that the entire face is slightly wider in children with extensive clefts. It likely represents a secondary response to the expanding forces detailed above. The hypertelorism mentioned in the literature is actually a very slight increase in the interorbital distance, probably caused by the expansion of the maxilla and nasal cavity. Even in isolated cleft palate, the tuberosities are wider (Subtelny, 1955).

Complete Bilateral Cleft Lip and Palate
Although infants with complete bilateral CLP have a significantly different maxillary complex from that of infants with unilateral clefts, the differences would probably be minimal except for the activity of attached muscles and tongue. The major morphologic characteristics are the result of an altered response of the skeletal elements to muscular deformation tendencies. The premaxilla, which is relatively unsupported on the nasal septum, is not able to resist the force of the active tongue, and tilts forward (Fig. 17.7). The base (anterior nasal spine) is somewhat supported by the vomer and nasal septum and by the nose, so protrusion of the base is limited. The degree of protrusion of the premaxilla is usually not too important, since it is the dental area of the premaxilla that is rotated forward by tongue pressure. If the tongue habitually protrudes through one side of the cleft, the premaxilla is forced to the opposite side, producing an asymmetry. The premaxilla has a moderate anteroposterior mobility

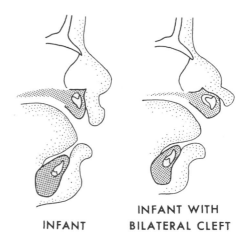

INFANT   INFANT WITH BILATERAL CLEFT

Figure 17.7 Rotation of the premaxilla in complete bilateral cleft lip and palate caused by tongue protrusion. Note that the anterior nasal spine is not advanced to any appreciable extent, but is rotated almost 90° as the lower portion of the premaxilla is pushed forward.

but great side-to-side mobility, so it can be readily displaced.

Clefts of the Hard and Soft Palates The newborn with an isolated cleft palate shows "clubbing" at the medial margin of the palatal shelves, a changing shelf angulation, and greater than normal width in the area of the posterior tuberosity (Subtelny, 1955). These are all probably caused by the tongue intruding between the shelves.

CONCLUSION The newborn with a unilateral or bilateral cleft lip and palate has an essentially normal craniofacial framework to which is attached a mildly deficient but grossly distorted nasomaxillary complex.

17.2.2 Facial Growth

17.2.2.1 Growth of the Normal Face
Before discussing facial growth in children with orofacial clefts, it would be helpful to have a review of normal facial growth. This is a complex subject that is incompletely understood at present. This section is

limited to a discussion of three important skeletal elements involved in the abnormal morphology of the face in orofacial clefting. They are the mandible, the maxilla, and the dentoalveolar component of each.

Mandibular Growth Many of the bones of the body are not single morphologic and functional units. The mandible, for example, has many components that develop relative to specific functions, all somewhat independently of each other. The major components of the mandible are (1) a central core, termed the *basal bone,* which is under genetic control for its size and shape; (2) a coronoid process that is dependent for its size and shape mainly on temporal muscle function and the position of the mandibular basal bone; (3) a gonial area dependent on and responding to the masseter and medial pterygoid muscles and mandibular core position; and (4) a dentoalveolar process rising in response to the development of the teeth and dependent on the presence of teeth (Fig. 17.8).

Growth occurs only at the posterior portion (ramus and condyle) primarily at the condyle. Increase in the length of the mandible in the clinical sense is solely due to bone increase at the condyle. Orthodontists have attempted to alter the size of the mandible by applying forces in a variety of ways but with virtually no success. As a consequence, treatment of a disproportionately large or small mandible invariably requires orthognathic surgery.

The *position* of the mandible, however, is the result of all of the muscular and soft tissue and external forces that act on it, since the bone is not firmly attached to other bones, only indirectly through the temporomandibular joints, and can swing open or closed freely to establish a posture that is in equilibrium with forces acting on it. Thus, a change in the contiguous structures or an external force can easily induce rotation to a new or unusual position. Rota-

FUNCTIONAL COMPONENTS OF MANDIBLE

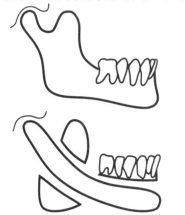

Figure 17.8 The mandible is composed of many functional areas.

tions over a long period, such as with continuous mouth breathing, are accompanied by alterations in the bony muscle attachments and the dentoalveolar component, so the mandible can be dramatically changed in its overall shape without affecting the shape or size of the central core. There are characteristic alterations in mandibular position and shape in cleft lip and palate.

Maxillary Growth The delineations between the various functional components are less obvious in the maxilla. There are orbital, nasal, palatal, zygomatic, and dentoalveolar processes, all of which are relatively independent. The basal bone (that which remains when the other processes are removed) is probably the result of a genetic blueprint, but with some critical differences from the mandible. The change from the infant to the adult maxilla requires different growth in different areas. Essentially all maxillary growth in a forward direction occurs by apposition of bone to the posterior surfaces (Enlow and Bang, 1965; Enlow, 1990; Bjork, 1966). Figures 17.9 and 17.10 indicate the areas

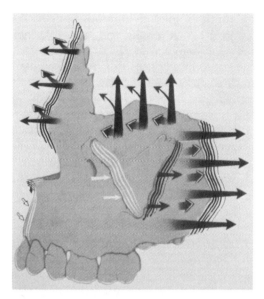

Figure 17.9 The overall growth pattern of the maxilla is indicated by the direction and size of the arrows. White arrows indicate resorption areas of bone; black arrows indicate apposition areas of bone. (*Source*: From Enlow and Bang, 1965.)

where bone is laid down to increase maxillary length. During growth, there is actually resorption of bone in the anterior basal area to permit remodeling of the surface.

Despite the apparently firm attachment of the maxilla to the skull, strong environmental forces can alter the position of the maxilla, and interference with the sutures or nasal septum can alter maxillary size. The maxilla is thus more easily inhibited in its growth than the mandible (Wieslander, 1963; Jakobsson, 1967; Droschl, 1973, Tindlund et al., 1989). Koski (1968) and others concluded that the areas of growth in the face are not primary growth centers that would force the maxilla forward. The role of the nasal septum is not completely understood, although its importance in directing postnatal maxillary growth is currently held to be of minor significance (Moss et al., 1968; Stenstrom and Thilander,

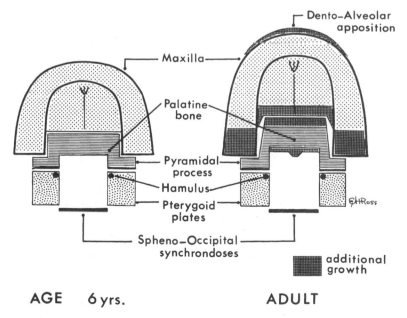

Figure 17.10 Growth in length of the maxillary basal bone is achieved by apposition to the posterior tuberosity area, with little or no contribution from the anterior maxilla, the pyramidal process of the palatine bone, or the pterygoid plates of the sphenoid.

1970). The sutures are adjustment areas that provide bone to maintain skeletal continuity. Ross (Ross and Johnston, 1972) argued that there is no major skeletal force driving the maxilla forward, that it seems rather to move forward as part of the overall genetic pattern of growth, and that as the maxilla moves forward, bone is laid down in convenient places (i.e., in the sutures and on the maxillary tuberosity).

Increase in width of the basal maxilla occurs in infancy in conjunction with activity in the midpalatine suture. The contribution of this suture is considerably reduced after the second year of life, and the width in the anterior of the maxilla remains virtually constant from this time. The apparent increase in maxillary width is due to greater posterior width accompanying apposition on the tuberosity.

Dentoalveolar Processes   The dentoalveolar processes have truly remarkable adaptive mechanisms. Alveolar bone develops in response to the presence of teeth and disappears if the teeth are absent or lost in later life. Teeth do not erupt at predetermined distances along a predetermined path into occlusion. If they did, a precise meshing of maxillary and mandibular teeth would be extremely rare. Nature has provided a mechanism to compensate for minor disharmonies between the maxilla and mandible.

A tooth erupts until it meets resistance. In the normal course, this resistance is the tooth in the opposing arch, although the tongue, thumb, or lip may interfere with full eruption. Once contact is made, an equilibrium is established.

A force exerted against a tooth is transmitted to the alveolar bone through the fibers of the periodontal ligament (as compression on one side of the tooth socket and tension on the other). The bone resorbs in response to pressure and builds in response to tension on the fibers. Thus, the tooth moves away from pressure. An erupt-

ing tooth is guided into position by the gentle pressure of the soft tissues, principally the tongue, lip, and cheek muscles. The orthodontist uses this normal biologic mechanism to correct malocclusion by applying artificial forces to move teeth.

It is essential for the attainment of a satisfactory occlusion that the dentoalveolar mechanism be free to respond to the guiding forces in the mouth. This principle is important to the understanding of the dental occlusion in operated and unoperated cleft lip and palate.

17.2.2.2 Growth Potential in Cleft Lip and Palate   As severe as facial morphology may appear at birth, it does not account for the severe facial dysmorphology that appears so frequently in adults. Perhaps the cause is an intrinsic abnormality of the facial growth pattern. Following is a discussion of that possibility.

The true growth potential of the cleft maxilla can be determined only when growth can be observed in the absence of other extrinsic influences. Such data are difficult to accumulate, since surgical repair of the cleft has an influence on growth and is routinely performed in infancy.

There have been numerous reports in the literature of children and adults with cleft lip and palate who have not received the benefits of surgical repair. Facial morphology in these individuals is the result of the interaction of an intrinsic defect, subsequent normal and abnormal environmental influences, and the expression of the growth potential of the facial skeleton. Will (2000) has recently reviewed this topic in detail. Important studies are by Mestre and colleagues (1960), Coupe (1962), Dahl (1970), Ortiz-Monasterio and associates (1974), Bishara and co-workers (1986), and Mars and Houston (1990). Almost all of the available studies have limitations, but assessing them collectively allows a good approximation of the morphology and growth potential.

Facial Width In individuals with unre-paired cleft lip and palate there is evidence that the face is wider in all areas, probably as a result of the decrease in *restraining* factors (loss of lip continuity, loss of skeletal continuity) while maintaining the *expansive* factors (tongue pressure, facial growth processes). There is also the possi-bility that an increased width is inherent and may even be an etiologic factor in the formation of a cleft (see Chapters 7 and 10). In any case, the width of the facial skeleton increases normally in the growing child with cleft lip and palate. Areas of bone apposition and suture growth are neither deficient nor inhibited if the end result is as satisfactory as is found in virtu-ally all studies on unoperated clefts of the lip and palate. Even in isolated cleft palate, the maxillary tuberosities and the tooth germs are wider (Subtelny, 1955), although not wider at higher levels—for example, the zygomatic arches or cranial base.

Facial Depth The facial growth parame-ter of most concern in cleft lip and palate is undoubtedly midface depth. Virtually all studies indicate that the basal maxilla achieves acceptable relationships with the remainder of the face in the adult with unoperated cleft lip and palate. There is strong indication that the maxilla is intrin-sically smaller than normal, but again it must be concluded that subsequent growth is not inhibited by the presence of a cleft.

Facial Height The available evidence indicates that vertical height is normal, and there is no evidence of a deficiency of midface height. Most of the increase in the normal maxilla is related to dentoalveolar growth and downward migration of the hard palate by means of an apposition-resorption mechanism (Enlow and Bang, 1965). Obviously the palate migrates in the cleft lip and palate condition, in spite of the gross disturbance of structure.

Dentoalveolar Adjustment Finally, and most significantly, the DA structures accommodate to the basal jaw relations as they do in the normal child. The maxillary posterior teeth are invariably positioned in a satisfactory relation to the mandibular teeth, but with local disturbances in the anterior area of the alveolar cleft. This observation indicates that even when a gross morphologic disturbance involves the maxilla and its alveolar ridge, teeth erupt until they encounter opposing teeth, and they can be guided into a satisfactory func-tional relationship through the influence of the adjacent soft tissues (i.e., tongue, cheek, and some lip muscles). The discrepancy of bone in the cleft area, absence of upper lip continuity, and tongue activity are suffi-cient to prevent complete dentoalveolar adaptation.

The mandible is essentially normal or slightly reduced in size. The maxilla advances with little inhibition. The teeth and alveolar bone are free to adapt to the jaw relations and the soft tissue environ-ment. The slight general midface deficiency remains, but is not really apparent, while the local dental and cleft margin deficien-cies are obvious (Fig. 17.11).

There should be no surprise when a young adult such as the boy shown in Figure 17.11 appears. Especially to be noted is the smooth alignment of the dental arches, the excess width of the maxilla, the positive overbite and overjet throughout the arch. Treated cases do not present with these characteristics, as the next section explains.

CONCLUSION   A child with an unoperated cleft lip and palate, if permitted to continue growing without interference, will have a functionally normal facial skeleton except for the presence of the cleft itself and distortions of the dentoalveolar process related to muscle activity. The growth potential is normal.

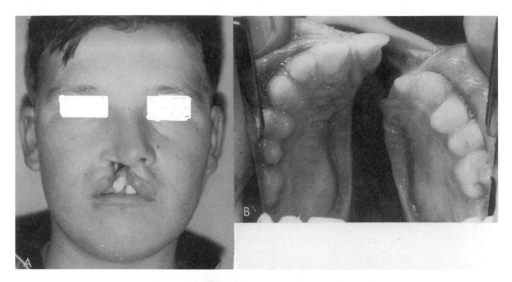

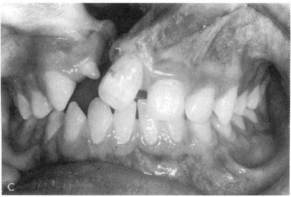

Figure 17.11 (A) A boy with complete unilateral cleft lip and palate, age 13 years, 10 months. The morphology of the maxilla and nose is similar to the usual morphology in the newborn, although muscle control has reduced the severity of the appearance somewhat. (B) The dental arch form is excellent, with regular tooth alignment. The width of the cleft appears to be as much due to expansion of the segments as to deficiency of adjacent tissue. (C) The occlusion of the teeth is, in general, excellent. The smaller cleft-side segment shows normal relations except for a slight vertical opening in the cuspid region. The larger noncleft segment is displaced away from the cleft, resulting in an excessive overlap of the upper teeth with the lower teeth. Note that the incisors tilt toward the cleft to establish more normal relations with the lower incisors.

17.2.3 Facial Growth Following Surgical Repair of Clefts: Overview

17.2.3.1 *Unilateral Cleft Lip and Palate*
The average 16-year-old male with a surgically repaired complete unilateral cleft lip and palate has a profile and skeletal pattern that deviates from the noncleft individual (Figs. 17.12 and 17.13). There is enormous variation in facial growth in children with a repaired cleft lip and palate related to the racial and familial genetic background, the type of cleft, and the nature of the surgical and orthodontic management received. The case presented in Figures 17.11 and 17.14 provides clues to the cause of the grossly abnormal morphology often noted in older children and adults. The result of

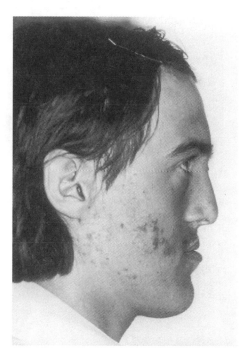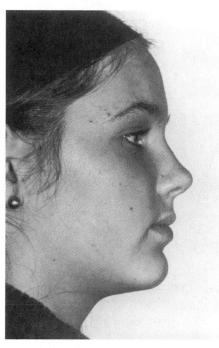

Figure 17.12 Faces of a young adult male with operated complete unilateral cleft lip and palate (left) and a noncleft female (right). Note the retruded midface in the male, expressed both in the lip and nose profile and the lack of cheek fullness. This is more severe than the average cleft patient but illustrates the characteristics encountered.

traditional surgery has changed this young man's facial pattern from excellent to poor.

The average adult with a unilateral cleft of the lip and palate who has received good surgical treatment has an anteroposterior midface deficiency of 5 to 6 millimeters. Data from virtually all cephalometric growth studies confirm this observation (Smahel and Brejcha, 1983; Ross, 1987; Rygh and Tinlund, 1996). General reviews include those by Ross and Johnston (1972), Ross (1990), Semb and Shaw (1996), and Kuijpers-Jagtman and Long (2000). Most studies note that the anterior maxilla is retruded relative to the cranial base, and this worsens with age. The midface deficiency is partly due to a decreased length of the maxilla and partly to its repositioning. Although the basal bone has an intrinsic deficiency, the deficiency observed in children or adults with operated clefts is

more than that observed in individuals with unoperated clefts. *The evidence is overwhelming that surgical repair of the lip and palate has induced a retrusion and underdevelopment of the maxilla, involving both the basal bone and the dentoalveolar process.* Transverse widths of the maxilla and dental arch are generally less than normal, so crossbites of the teeth occur in the posterior segments. Vertical development is normal or slightly less than normal in the anterior of the maxilla but deficient in the posterior.

17.2.3.2 Bilateral Cleft Lip Palate The young adult with BCLP has a reasonably normal skeletal profile, but with skeletal problems related to the isolated premaxilla and deficient posterior segments. There are many soft tissue and dental problems that are discussed in Chapters 18 and 19.

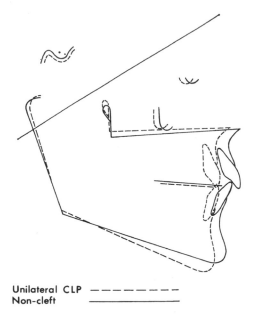

Unilateral CLP —————————
Non-cleft ⎯⎯⎯⎯⎯⎯⎯⎯⎯

Figure 17.13 Cephalometric tracings of facial morphology in 12-year-old children. Note the cleft group has maxillary vertical underdevelopment in the posterior region, maxillary underdevelopment in length, a normal mandibular length, a more open mandible with a down-and-back rotation of the chin, and acceptable jaw and tooth relations resulting from the chin retrusion.

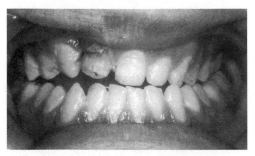

Figure 17.14 Same boy as in Figure 17.11, 3 years after surgery (age 16 years, 9 months). Deterioration of the occlusion is striking. The entire maxillary arch is in crossbite, and the vertical relations in the area of the cleft are particularly abnormal. Open bite is becoming generalized throughout. The alteration in the occlusion was partly due to maxillary contraction, partly to dental arch contraction, and partly to secondary changes in tongue position and mandibular growth changes.

17.2.3.3 Cleft Lip Only
When the cleft involves only the primary palate, facial morphology and growth are essentially normal. Irregularities of the teeth and alveolus adjacent to the cleft and of the facial soft tissues are present and vary with the extent of the clefting.

17.2.3.4 Isolated Clefts of the Hard and Soft Palate
Facial form in these clefts is characterized by a deficiency in maxillary length and a slightly retruded position of the maxilla relative to the cranial base (Graber, 1949; Ross and Coupe, 1965; Osborne, 1966; Shibasaki and Ross, 1969; Dahl, 1970). These characteristics become accentuated as growth continues to maturity (Osborne, 1966; Shibasaki and Ross, 1969). However, the mandible is slightly smaller and more open than

normal, so the chin is retruded and facial form appears to be normal. The vertical development of the maxilla is deficient in the posterior.

In the process of habilitating children with orofacial clefts, surgeons use different procedures in different areas of the midface, which have different effects on morphology and subsequent growth. The following sections detail the general effects of each.

17.2.4 Effect of Lip and Nose Reconstruction

Reconstruction of the cleft lip and nose is usually the first surgical procedure in the neonatal period. The effects on facial morphology are generally beneficial. Continuity of the orbicularis oris and the buccinator muscles is established, providing more normal functional forces on the maxilla. The repaired lip will reverse the prenatal distorting influences and encourage a growth pattern that will result in the establishment of a more normal skeletal relationship, while permitting the upper teeth to be guided into a more satisfactory relationship with the lower teeth.

17.2.4.1 Unilateral Cleft Lip and Palate

Lip repair is undertaken over a maxilla that has widely separated segments (Fig. 17.4). The tension produced by repair serves to bring the segments together, and maxillary width becomes normal (Harding and Mazaheri, 1972). The repair of the floor of the nose narrows the base of the cleft nostril to match the noncleft nostril, and the columella returns to the midline of the face. The midline of the maxilla (anterior nasal spine) is drawn back toward the midline after surgery, but invariably not completely, and it remains deviated (Harvold, 1954). The nasal septum, which had become curved toward the noncleft side, swings back after surgery and assumes the shape that is characteristic of unilateral cleft lip and palate. By the time the bony segments contact, lip tension has stabilized and is usually close to normal. The outward rotation of the premaxillary area caused by tongue activity is molded back, and a slight medial movement of the small segment occurs. These factors result in a reasonably normal maxillary arch shape that is usually well related to the mandibular dental arch as the teeth erupt (Wada and Miyazaki, 1975; Lo et al., 1999).

The repaired nasal floor and upper portion of the lip initiates pressure on the maxillary basal bone and nasal septum, while the lower portion of the repaired lip creates pressure on the teeth and alveolar bone. A well-designed lip reconstruction does not involve actively growing areas of the maxilla, nor does it directly interfere with the dentoalveolar adjustment mechanism. Thus, midfacial growth and adjustment proceed with a mild, generalized inhibition, if at all. Ross (1987) came to this conclusion in a study on partly repaired cases. Many of the "unoperated" patients reported in the literature had received a cleft lip repair. Reports indicated that lip

surgery had little adverse influence on facial growth and minimal influence on dental occlusion (Davies, 1951; Innis, 1962; Herfert, 1958; Ortiz-Monasterio et al., 1974; Bernstein, 1968; Pitanguy and Franco, 1968; BooChai, 1971; Mars and Houston, 1990). The Schweckendiek (1978) protocol (which will be discussed more completely in the section on palate repair) includes lip repair, but results in excellent jaw and tooth relations.

Poor surgical technique, however, can result in increased lip pressure on the basal maxilla, which may inhibit forward drift of the maxilla, producing a slow, progressive retrusion of the midface. A tight lower portion of the upper lip will mold the dentoalveolar structures posteriorly to an excessive degree. There have been findings from animal experiments (Bardach et al., 1979; Bardach and Eisbach, 1977) where extremely gross surgery, involving the removal of large sections of the lip and the suturing of the remaining fragments, has produced deformity. These experiments appear to have no relevance to the human cleft condition or to its treatment. An astonishing study by Capelozza Filho and co-workers (1996) concluded that lip repair was more important in growth reduction than palate repair. There appeared to be many confounding variables and arguable interpretations in their study, which engenders some skepticism.

Ross (1987) found that neither the particular reconstructive technique used nor the age at which lip surgery was performed had any appreciable effect on the skeletal morphology or growth. However Roberts-Harry and colleagues (1996) found that radical nasolabial reconstruction produced less desirable growth than did more conservative surgical reconstruction.

17.2.4.2 Bilateral Cleft Lip and Palate

The dysmorphology of the midfacial structures in complete bilateral cleft lip and palate is severe at birth, but following surgical repair of the cleft lip many favorable changes occur. The basal bone of the premaxilla, which is resistant to environmental influences, is only slightly forward at birth, and is only slowly inhibited or displaced posteriorly by the pressure of the repaired upper lip. The teeth and alveolar bone, however, are extremely sensitive to environmental forces. For this reason, thumb sucking causes severe problems in bilateral CLP, as it exacerbates the existing protrusion of the maxillary dentoalveolar process and even the basal maxilla to some extent. The same habit is a beneficial activity for infants with unilateral clefts.

Lip repair creates a soft tissue and muscle pressure on the maxilla and molds the protrusive alveolus back. If the lower portion of the repaired lip is tight and exerts excessive pressure, there is a dentoalveolar retrusion that may have almost no effect on the position of the anterior nasal spine. If the repaired lip is short and incapable of completely containing the premaxilla, severe protrusion of the alveolus may remain and the teeth will overerupt. In this situation, the lower lip falls behind the maxillary teeth, encouraging further protrusion of the teeth. Since teeth erupt until they encounter resistance from the lower teeth or the lip, and since this resistance is almost completely lacking, a great deal of excessive eruption and protrusion of the teeth may occur in rare cases. Handelman and Pruzansky (1968) noted in their patients that by age 4 an average overjet of 9.7 millimeters was present. Ross and Johnston (1972), however, found that in 38% of their cases there was incisor crossbite and rarely an overjet as great as 4 millimeters. The surgical management of the two samples was responsible for the differences in the position of the premaxilla. It would be difficult, therefore, to draw general conclusions on development from the cases of a single surgeon or technique.

There appear to be exceptional cases, however, where the morphology seems somewhat different and the nasal septum itself is very protrusive, or the septum is very thick and seems to resist retraction. Presurgical orthopedic treatment is valuable in rotating the premaxilla posteriorly and can alleviate, if not solve, the problem.

Trotman and Ross (1993) found that the premaxilla remained somewhat protrusive at 6 years of age (Trotman and Ross, 1993). Molding continued, so the premaxilla was slightly protrusive at age 12, but by age 17 it was normal or even slightly retrusive. Other studies give slightly different values, but the pattern is similar, modified by the surgeon or type of surgery. Rarely does the early protrusion persist into adulthood.

CONCLUSION Reconstruction of the cleft lip and nose, properly performed, has mainly a positive influence on the morphology and growth of the face.

17.2.5 Effect of Repair of the Cleft Alveolus

If treatment is to be satisfactory and permanent, the anterior communication between the oral and nasal cavities must be closed, and bony union of the maxillary segments must be accomplished. The latter gives the maxilla stability and permits the establishment and retention of an excellent dental occlusion. The usual method of accomplishing these goals is by soft tissue repair of the alveolus with the placement of a bone graft in the anterior maxilla.

17.2.5.1 Repair in Infancy Many surgeons repair the alveolus with soft tissue while repairing the lip. Some of the proponents of presurgical orthopedics added alveolus repair to the technique using bone grafts or periosteoplasty. They reasoned that closing the alveolar cleft and fusing the maxillary segments in a fairly normal relationship would prevent collapse of the maxilla and improve subsequent growth and dental occlusion. Different kinds of bone in various configurations have been used to partially or completely fill the alveolus, the basal bone to the pyriform fossa, and even part of the hard palate.

There is, however, an intrinsic deficiency that includes missing bone in the cleft itself. When the cleft segments are brought into contact, a cleft is closed that should probably remain open to compensate for the missing bone. Further growth or separation of segments in that area is prevented. Ross (1969) pointed out that the alveolus and area of the maxilla in which the bone graft is placed is not a site of maxillary growth in the anteroposterior or transverse dimensions (Enlow and Bang, 1965). Consequently, anteroposterior maxillary development should not be affected if grafts are confined to the alveolar area. Interference with growth might follow if the graft extended into the palate and interfered with midline growth, or added additional scar tissue to the area, inducing retraction of the anterior segment and altering the eruptive direction of the teeth. Grafting could prevent medial collapse of the bony segments, but would have no effect on the dentoalveolar medial collapse caused by residual scar tissue from the palate surgery.

Koberg (1973), Witsenburg (1985), and Kuijpers-Jagtman and Long (2000) have published excellent reviews of the use of bone grafts to fill the alveolar cleft. Results have varied. Friede and Johanson (1982) reported severe maxillary retrognathia and vertical deficiency that worsened with age. They were of the opinion that the growth inhibition was caused by interference with the maxillary vomerine suture. Dahl et al., (1981) found that some techniques of vomeroplasty did cause an increased incidence of crossbites in the primary dentition. Robertson and Jolleys (1983) had a similar experience, noting in addition that the grafts deteriorated to a small strut of bone, with no evidence of tooth migration

into it. The reports of Sameshima et al., (1996) and Smahal et al., (1996, 1998) were also discouraging. Rosenstein and associates (1982) and Nordin and associates (1983) continued to use primary bone grafts and were not dissatisfied with the maxillary growth that resulted. Trotman and co-workers (1996), however, noted maxillary retrusion and nasal depression in the Chicago (Rosenstein) sample.

Skoog (1965) developed a periosteoplasty procedure that induced bone to bridge the alveolar gap. Others (Hellquist and Ponten, 1979; Hellquist et al., 1983; Helquist and Svärdström, 1986; Rintala and Ranta, 1986; Smahel and Müllerova, 1994; Smahel et al., 1996) found less favorable facial growth using periosteoplasty. A modified technique of gingivoperiosteoplasty is currently being used in some centers, but conflicting results, mostly somewhat negative, have been reported by Millard and associates (1999), Santiago and colleagues (1998), Henkel and Gundlach (1997), Lukash et al., (1998), and Berkowitz (1996). Kuijpers-Jagtman and Long (2000) concluded that valid long-term data are required before the proponents can justify this procedure.

Repair of the alveolus in infancy with or without bone grafting or periosteoplasty caused a highly significant reduction in vertical height of the anterior maxilla (Ross, 1987). If a bone graft was used, the effect was exaggerated and anteroposterior inhibition was also noted. Even bone grafts at 4 to 10 years of age showed a marked reduction in vertical height of the maxilla at age 15.

Although there is an immediate esthetic improvement in the dental arch with infant alveolus repair, whether it is soft tissue repair, periosteoplasty, or bone grafting, the best evidence suggests that there are no long-term benefits. There are, in fact, untoward effects on midface growth. As well, later orthodontic manipulation of the segments is prevented.

17.2.5.2 Later Alveolus Repair Mixed
dentition bone grafts into an open alveolus
replace the intrinsic missing basal and
alveolar bone. Growth deficiencies were
less with bone grafts performed at 9 to 12
years; there were no anteroposterior differences from the ungrafted, although the
Oslo late grafted sample showed significantly less vertical height than the Toronto
nongrafted sample (Ross, 1995). Semb
(1988) and Daskalogianakis and Ross
(1997) had similar findings. Brattström
et al. (1991) noted that infant bone grafting resulted in attenuated growth, late
mixed bone graft results were better, and
no bone grafting at all was best. There are,
however, many potential confounding variables in all of these studies, the most important of which are the surgical techniques
being used and the skill of the surgeon.
Multiple surgical procedures seem to
increase the damage. Ross (1987) suggested that alveolar repair in unilateral
clefts should be postponed until age 9 or
later.

CONCLUSION All surgery in this area has
the potential to interfere with maxillary/
midface development. It would seem that
the best course of action would be to do no
surgery to the alveolus in infancy. Later
bone grafting into an orthodontically
expanded maxilla replaces the missing
bone in the cleft region.

17.2.6 Effect of Soft Palate Reconstruction

The Schweckendiek sample with the soft
palate (and lip) repaired, but without hard
palate repair, showed excellent maxillary
growth in length and anteroposterior position, even when compared with the totally
unoperated sample (Ross, 1987). There
was, however, a deficiency in posterior vertical height. Repair of the soft palate does
not appear to inhibit facial growth in any
clinically significant way.

17.2.7 Effect of Hard Palate Reconstruction

Repair of the hard palate has three main
effects on maxillary growth, and they are
discussed next.

17.2.7.1 Initial Contraction In the
reconstruction of a cleft palate, the surgeon
must obtain tissue to bridge the cleft. Many
techniques are employed, but they frequently involve raising mucoperiosteal
flaps from the palate and relocating them
medially and posteriorly, leaving a denuded
area of bone adjacent to the alveolar
process (Fig. 17.15). The filling in of the
denuded area produces scar tissue, which
exerts an initial contracting force on adjacent tissues (Kremenak, 1984). The initial
scar tissue contraction during healing
causes a movement of the maxillary segments until, in most cases, contact between
the segments is achieved. Following the
early stages of contraction, however, there
is only a mild reduction in arch width
(Dixon, 1966; Kremenak et al., 1970;
Harding and Mazaheri, 1972). Given the
normal growth potential noted above, if
maxillary growth could then proceed normally from this point on, the dentoalveolar
structures could easily cope with the mild

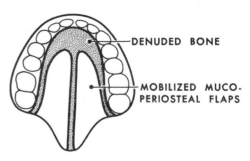

Figure 17.15 After palate repair that involves extensive flap mobilization, there is a continuum of scar
tissue created in the palate adjacent to the alveolar
process. The entire dentition is induced to develop in
a more palatal direction by the inhibiting influence of
the scarred mucosa.

discrepancy, and a fairly normal skeletal and dental pattern would result.

17.2.7.2 Long-Term Growth Interference

Data from Ross (1987) suggested that the posterior displacement of the maxilla was most likely related to the palate surgery and not to an intrinsic deficiency. One result of the posterior displacement is that the bony pharynx is shallower and the soft palate is posterior to the normal. Thus, speech should not be adversely affected.

Previously cited evidence indicates that maxillary underdevelopment is indeed progressive. Growth in length of the maxilla requires that the entire maxilla move forward freely so that bone can be deposited on the tuberosity. If during palatal repair there is undermining of tissues and hamulus fracture in the area of the pterygopalatomaxillary junction, scar tissue may form across this sensitive growth area and inhibit the forward movement of the maxilla. It is not necessary that the inhibition of growth be severe, since even a fraction of a millimeter of inhibition each year would create a severe problem by the time of facial maturity.

The transverse width of the basal maxilla probably is not appreciably reduced in most patients with repaired cleft lip and palate (Coupe and Subtelny, 1960; Dahl, 1970; Nakamura et al., 1972). The apparent narrowing is mainly in dental arch width and may also be related to the retropositioned maxilla, which places a narrower portion of the dental arch in contact with a wider portion of the mandibular dental arch. There are, of course, many patients with a true deficiency in maxillary width.

17.2.7.3 Effect on the Dentoalveolar Mechanism

The dentoalveolar component of the maxilla is affected by surgery and is probably the major factor in abnormal dental development. As was explained earlier, a free adjustment of the teeth and supporting alveolar bone is essential to the establishment of a normal dental occlusion. This mechanism is severely compromised in the child with a cleft lip and palate. The cleft maxilla is slightly smaller and more posteriorly placed than in nonclefts, and this trend increases with age. Anteroposterior jaw relations are usually found to be satisfactory in young children, but become progressively worse in older children and in patients who have had more traumatic surgical procedures (Ross, 1987).

Surgical reconstruction of the palate results in scar tissue adjacent to the alveolus (Fig. 17.15). There follows a long-term progressive constriction of the dental arch, since the suspensory mechanism of the teeth (the periodontal ligament) sends fibers into the surrounding scar tissue. Tension on the periodontal fibers during subsequent tooth eruption causes a posterior and medial deflection of the teeth. In a child in whom scar tissue has been created in the palate close to the alveolar process, the incisor teeth are deflected palatally by tension on the periodontal fibers that extend into the palatal mucosa. Free movement of the teeth to adjust to the mandibular incisors is prevented, and crossbite may develop, even in cases in which the skeletal relations are adequate (Fig. 17.16). The satisfactory relationship in the primary dentition (although there is frequently mild incisor crossbite) progressively worsens by the time of full permanent dentition. The angle ANB, a reasonably good measure of a combination of basal and dentoalveolar morphology, is well within normal range in the first decade, but becomes increasingly worse in the second decade, accounting for the deterioration of anteroposterior relations during adolescence, especially in males (Fig. 17.17). Late growth patterns can be devastating.

Mouth breathing in children with cleft lip and palate is common, and is related to the deviated nasal septum and the high incidence of upper respiratory infections that interfere with nasal breathing and cause enlarged tonsils and adenoids. Mouth breathing causes a lowered tongue posture, out of the palatal vault. In addition, the decreased size of the palatal vault (constricted dental arch and postsurgical vault

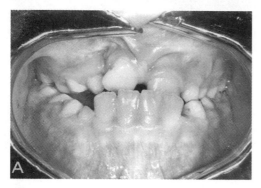

Figure 17.16 Typical mixed dentition at 8 years of age in a child with complete unilateral cleft lip and palate. The permanent incisors have erupted, but the primary molar teeth remain. The eruption of the teeth has been guided palatally by the scar tissue in the palate.

shape) inhibits proper tongue placement. Enlarged tonsils mechanically induce a forward tongue posture. Normal tongue posture in the palatal vault is particularly important to prevent maxillary constriction. If the mandible is held excessively open at rest, the mandibular incisors do not support the maxillary teeth adequately, and they are less able to resist the constricting tendency.

As the mandible rotates open, the chin point moves downward and backward and the mandibular plane becomes steeper. Altering the position of the mandible does not permanently alter muscle length; instead, the bony attachment remodels to permit the muscles to maintain their correct length for optimal function. Thus, the gonial angle (the angle formed by the mandibular body and ramus) is found to be more obtuse in cleft lip and palate, and the angle increases with age (Narula and Ross, 1970), in contrast to the normal decrease with age (Munroe, 1966). The increased vertical face height due to the more open

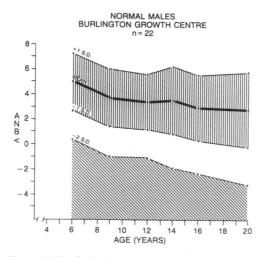

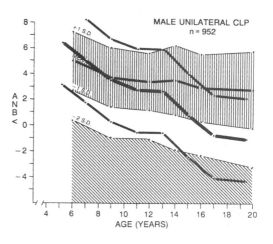

Figure 17.17   Left: Normal values for the ANB angle, which measures the skeletal relationship between the maxilla and mandible. The mean at age 6 is about 5°, and gradually the face becomes more upright in adolescence to about 3°. A value in the lower area (−2 S.D. or more) would represent a severe dysharmony with a relatively large mandible or small maxilla. Right: Superimposed values for unilateral cleft lip and palate show a normal relationship at ages 6 to 9, but a gradual severe worsening into late adolescence.

mandible is compensated for by overeruption of the teeth, particularly the maxillary posterior teeth and the mandibular anterior teeth. Vertical development of the dentoalveolar process is not inhibited in the early years. A vertical problem frequently occurs in the incisor region during the final stages of mandibular growth. Growth normally ceases in the maxilla before it does in the mandible, but compensatory dentoalveolar development maintains the occlusion. In cleft lip and palate, the maxillary incisors often are unable to make this compensation, and an open bite results.

Unilateral Cleft Lip and Palate   The retrusion of the maxillary basal bone in cleft palate and unilateral cleft lip and palate is somewhat balanced by the rotation of the mandible and consequent retrusion of the chin (Fig. 17.14). The dentoalveolar accommodation mechanism tends to fail in the permanent dentition: sometimes the maxilla only is at fault, sometimes the mandible is mainly at fault, but usually both contribute. A large sample of unilateral CLP cases were arbitrarily divided into the best, worst, and medium cases on the basis of skeletal and dental relationships. A very large number of possible indicator measurements were made to attempt to determine the most relevant ones to their classification. On average, the best cases had more protrusive maxillas and least protrusive mandibles. The worst cases had the opposite, and the medium cases were in between. These are not surprising results, except that mandibular morphology was equally important to maxillary morphology.

Bilateral Cleft Lip and Palate   Studies have shown that a posterior and vertical displacement of the posterior segments of the maxilla occurs, including the orbits and zygomatic process, plus a localized deficiency in the piriform margin. The deficiency increases slightly with age but is proportional to the increase in size of the structures measured (Narula and Ross, 1970; Trotman and Ross, 1993).

Vertical growth of the posterior maxilla may be inadequate owing to an intrinsic deficiency or, more likely, surgical interference with growth following hard and soft palate repair. In bilateral conditions the maxillary incisors are frequently inclined palatally; incisor crossbite may thus occur even when the premaxilla is protrusive. In bilateral CLP the tips of the upper incisors are retruded with respect to ANS, which is superiorly and anteriorly positioned. This anterior positioning decreases with age. The lower incisors erupt in a more upright position and maintain this position with age, while the initial retrusion of the upper incisor improves. Smahel (1984) noted that the mandibular body and ramus are shortened, and there is an obtuse gonial angle.

17.2.7.4 Different Treatment—Different Growth?   The question arises as to whether variations in surgical and orthodontic protocols and techniques can cause appreciable differences in facial morphology. Ross (1987) has shown that there are greater differences related to the kind of treatment provided than are related to having or not having a cleft. There are many variations on the surgery used to repair a cleft lip and palate. Most of the popular techniques used to accomplish palatal repair have the same effect on the growth of the facial skeleton (Ross, 1987, 1995). Leenstra et al. (1995, 1996) found a definite difference in animal experiments and human follow-up using the split-thickness supraperiosteal flaps of Perko (1974) instead of full-thickness flaps. *The most important variable to the success of cleft lip and palate surgery appears to be the surgeon performing the operation* (Ross, 1987).

To encourage less traumatic surgery, many authors have suggested repairing the soft palate early to facilitate speech, but deferring hard palate surgery (Gillies and Fry, 1921, and many others subsequently).

Late repair in theory should be less damaging than early repair, on the assumption that the more growth that has occurred, the less there remains with which to interfere. Robertson and Jolleys (1974) reported the results of a delayed hard palatal repair, and Hotz and Gnoinski (1979) made it an acceptable practice to delay hard palatal repair until the time of the mixed dentition. However, Blijdorp and Egyedi (1984) found no growth differences between children with repair of the hard palate at 3 years and 6 years of age. Noverraz et al. (1993) found no difference between repairs at 1.5 years, 4.6 years, and 9.4 years; Friede et al. (1999) found no difference between repairs at 5 and 9 years, and Rohrich et al. (1996) found no difference between 10 months and 4 years.

The sample from Schweckendieck in Marborg, where the hard palate was not operated until adolescence (Schweckendiek, 1978), provided remarkable insight into the growth-inhibiting effect of palate surgery. Olin (Bardach et al., 1984) provided data indicating that facial growth using the Schweckendiek procedure was excellent. However, Ross (1987) also reviewed these cases, noting that hard palate repair was delayed into the adolescent years when facial growth was virtually completed, or most of the records were available for analysis only at the time of surgery, and thus the sample should be considered as unoperated hard palates. The excellent facial growth in his cases should be considered a reflection of the benign effect of lip and soft palate repair, and the great impact of hard palate repair on facial growth as noted in other samples. Ross compared Schweckendieck's cases to samples from other excellent centers where both lip and palate were repaired at traditional times.

Several other samples, particularly those from Lindsay in Toronto and Manchester in Auckland, showed equally impressive growth of the maxillary *basal bone*, but with inhibition of the forward development of the *dentoalveolar process and teeth*, presumably due to the palate surgery. This suggested that palate repair causes inhibition of the dentoalveolar structures (including A point). Ross concluded that the very late hard palate repair will result in a mildly improved basal maxilla protrusion and greatly improved dentoalveolar protrusion. These data correspond almost exactly with the Sri Lankan data of Mars and Houston (1990) and are probably close to the true population parameters with "good" lip and palate surgery. Mars and Houston considered hard palate repair to be the major factor in the maxillary growth disturbances they observed in Sri Lanka. *These studies indicate that repair of the hard palate probably accounts for the major growth interference in clefts.*

A pilot study (Prahl-Anderson and Ross, 1997) noted that the Amsterdam sample (delayed palate repair at about age 8) had superior midface growth. By age 12, however, the Amsterdam sample had lost their advantage, and maxillary growth in the two samples was equal. This certainly indicates that palatal surgery, even at age 8, will inhibit growth. Ross (1987) reported that the timing of hard palatal repair did not matter within the first decade. Early repair (before 12 months of age) gave at least as satisfactory results as delayed hard palatal repair (4–7 years). *Any growth advantage from delaying hard palate closure appears to be lost after surgery.*

Meanwhile, many speech scientists have been less than enthusiastic, particularly in North America, about the results of delaying surgery past the age of early speech development (Dorf and Curtin, 1982; Witzel et al., 1984; Rohrich et al., 1996).

17.2.8 Effect of Pharyngeal Flap on Craniofacial Growth

A pharyngeal flap that in effect attaches the growing maxilla to the posterior wall of the pharynx could hypothetically have a tethering effect on the forward movement of the maxilla and reduce its overall size. The effects noted, however, are mild and unrelated to maxillary length and protrusion (Ross, 1977; Long and McNamara, 1985; Long et al., 1984; Semb and Shaw, 1990).

17.2.9 Effect of Surgical Repositioning of the Premaxilla

Some clinicians feel that surgical repositioning of the basal premaxilla in bilateral CLP is desirable in many cases. The experience of most clinicians is that these procedures may interfere with growth, so an apparently successful result in a young child may become a severe deformity in later years (Friede and Pruzansky, 1985). Modern management of difficult cases by presurgical orthopedics offers a less dangerous alternative, since the initial reduction of premaxillary protrusion can be accomplished quickly and easily, with no long-term effect on sensitive growth areas. Surgical repositioning of the premaxilla is dangerous and probably always growth inhibiting. The results are sometimes disastrous. When can the premaxilla be safely positioned surgically to a more normal position? In a small study of cases (Dijkman et al., 1998) it was found that when the premaxilla was repositioned upward and backward at age 9 to 11, there was a highly significant retrusion of the premaxilla at age 15 to 16 compared to a Toronto sample that had been identical at earlier ages. ANS was especially retruded from normal. Retrusion was known to continue in Toronto and ANS became normal at 17 to 18 years, so presumably further retrusion would occur in the operated sample with bilateral CLP. There were also more open bites and negative overjets in Dijkman's sample compared to the Toronto sample.

CONCLUSION The position of the premaxilla should not be altered surgically until the middle of the second decade.

17.2.10 Effect of Orthopedics and Orthodontics

Despite the widespread use of presurgical infant orthopedic procedures and an enormous amount of literature on their alleged benefits or lack thereof, it appears almost certain that they have no effect on facial growth and no lasting effect on the maxillary dental arch (McNeil, 1950; Hotz and Gnoinski, 1979; Ross, 1987; Kramer et al., 1994; Larson et al., 1993; Ross and McNamara, 1994; Ball et al., 1995; Joos, 1995; Winters and Hurwitz, 1995; Kuijpers-Jagtman and Prahl, 1996; Kuijpers-Jagtman and Prahl-Anderson, 1997; Kuijpers-Jagtman and Long, 2000).

This finding is also true for early orthodontic treatment during periods of primary or mixed dentition (Ross and Johnston, 1967). Facial masks for maxillary protraction (Tindlund, 1989, 1994) tend to provide a very small amount of maxillary basal advance (Sarnäs and Rune, 1987), which is only a temporary improvement (Ross, 2001). There may, however, be some permanent dental improvement. The potential exists for altering jaw growth, but the orthopedic procedures currently in use have not been shown to accomplish permanent basal bone growth increase.

17.2.11 Mandibular Growth

17.2.11.1 Orofacial Clefting The size of the basal bone of the mandible is determined almost entirely by genetic factors. It appears that only a severe environmental influence can alter the genetic blueprint, but evidence of the complete interplay of genetic and environmental forces is inadequate at present. Cephalometric studies on cleft lip and palate consistently note that the mandible is several millimeters smaller than controls in length, especially in bilateral CLP and isolated cleft palate, although the difference is not always significant. (Nakamura et al., 1972; Smahel, 1984; Ross,

1987; Dahl, 1970; Vargervik, 1983; Trotman and Ross, 1993; Venditelli and Flavelli, 1999). There are as many variations as in the noncleft population, presumably because the normal gentic variation is expressed. The one notable exception is in the Pierre Robin sequence, which is considered separately.

Mandibular morphology is affected by the maxillary skeletal and soft tissue morphology and growth. The changes have been detailed in the previous section on the results of palate and lip surgical reconstruction. There is a change in the *position* of the mandible—that is, the chin is usually swung open, so there is greater facial height in spite of normal or decreased maxillary height. The change in vertical height is mild in younger children but is especially noticeable in older children. The condyle is positioned slightly upward and forward.

17.2.11.2 Pierre Robin Sequence

The newborn with Pierre Robin sequence (PRS) has mandibular micrognathia, glossoptosis with incidents of respiratory distress, difficulty in feeding, and invariably a cleft of the secondary palate. Many authors such as Ross and Johnston (1972), Cohen (1981), and Shprintzen (1988, 1992) have cautioned against grouping all PRS patients together, since there are different syndromes that include these symptoms. There is a strong likelihood that the cleft palate associated with PRS is secondary to the micrognathic mandible and is caused by the normal-sized tongue creating mechanical blockage to palatal shelf elevation and fusion. A similar mechanical explanation has also been advanced for the high incidence of cleft palate in Klippel-Feil syndrome (Ross and Lindsay, 1965).

Walker (1959) introduced the hypothesis that compression causes mandibular retrusion and cleft palate in Pierre Robin, and there were earlier accounts of amniotic fluid loss as a possible cause. These hypotheses point to an intrinsically tiny mandible that

is not familial. Bosma (1963) theorized that the cause might also be a muscle immaturity, resulting in an inability to maintain the forward positioning of the mandible as required in the neonatal period.

In any event, the infant situation improves dramatically in the first month or two, and life-threatening respiratory incidents cease. It appears that the mandible has profuse "catch-up " growth, and it was thought that this continued until the mandible was normal in size. Muscle immaturity would explain the sudden improvement postnatally as the muscles begin to function properly. The rapid improvement could also be a mandibular growth spurt in excess of that which all babies have at this time, or a functional shift of the mandible forward related to muscle maturation and activity—probably the latter and some of the former, but there are no data on newborns to determine this. If there is a major growth spurt, we can say from clinical observation that it is almost certainly at its maximum in the first 6 weeks of life.

Rintala and Ranta (1986) felt that the mandible is extremely small in the beginning and remains small throughout life. Figueroa and associates (1991) found that the PRS group from age 3 months to 3 years exhibited a greater percent increase in mandibular length compared to cleft palate and to normal infants. It would appear from their study that any catch-up growth occurs before the first year of life. They considered it partial catch-up growth. Daskalogianakis et al. (2001) studied serial growth at 5, 10, and 15 plus years. They found no change in the mandibular growth pattern. The small mandibles remained small, and are also tipped open, which increases the retrognathic profile.

Infants born with cleft palate as part of PRS may have one of several different maxillary growth patterns, probably related to the etiology of the syndrome. In general, the maxillary complex in these children grows as it does in ordinary cleft palate: the mandible is the variable structure (Ross and Johnston, 1972).

17.3 OTHER FACIAL CLEFTS

Previous chapters in this text have classified the other facial clefts and explained their etiology and teratology. The nature and extent of the dysmorphology can usually be readily identified. The long-term growth patterns, however, are much less studied than in the more common clefts of the lip and palate.

When development goes awry, there will be one of three outcomes over time: the situation will improve, or it will worsen, or it will remain constant. There are essentially no craniofacial anomalies that improve. Infants with Robin sequence will appear to make a remarkable recovery in the first months of life, although they remain decidedly retrognathic into adulthood. On very rare occasions, a child with hemifacial microsomia will appear to improve spontaneously.

The only conditions that become worse with growth are those with direct inhibition of growth areas, such as maxillary and mandibular ankylosis. Even the dysplastic mandibular condyle in hemifacial microsomia, which causes severe asymmetry of the face in the unilateral condition, does not become appreciably worse with time. Growth of the affected side parallels that of the nonaffected, regardless of the degree of the initial deformity (Polley et al., 1997; Ross, 1999). Developmental bony ankylosis of the mandible is very rare, if it occurs at all, since neonatal or infant trauma or infection is usually the cause in young children. A decreased mobility of the mandible is caused by the soft tissue elements of the joint and does not restrict growth. Maxillary ankylosis occurs in Apert and Crouzon syndromes. Mandibulofacial dysostosis (Treacher Collins syndrome) is a condition that includes clefts in the lateral orbital margins and zygomas, as well as soft tissue defects. There is also a posterior displacement and tethering of the mandibular condyle that may be responsible for the limited mobility and unusual shape of the lower border worsening over time.

Stability is the rule in the vast majority of cases. When confronted with an infant or child with a developmental facial anomaly, a most useful rule of thumb is that *the face continues to grow in the way it has been growing*. It will invariably remain in the same configuration, with the affected areas growing proportionately to the unaffected areas. If a condition threatens the life or health of the patient, then treatment must be instituted as soon as possible. If an anomaly poses no serious problem and is stable, as it is in the vast majority of cases, we have the luxury of treating when it is most convenient for all concerned: the patient, the parents, and the health care professionals. Yet there is an urgent desire to *do something* when faced with a serious anomaly, and many clinicians micromanage their cases in an attempt to control the condition and please the parents.

Most craniofacial anomalies have functional or esthetic problems that require reconstructive surgery to the hard and soft tissues. There are usually compelling reasons to attempt treatment of these conditions as early as possible, and surgery to the growing child is often proposed. Surgery to an area generally affects the subsequent growth of that area, usually adversely. Surgery to the mandible or maxilla can attenuate or arrest subsequent growth (Huang and Ross 1982; Bachmayer et al., 1986). Some clinicians mistakenly believe that anomalies worsen, and therefore they practice early intervention procedures. In most cases, elective surgery should be delayed until facial growth is near completion unless there is reasonable certainty that surgical treatment will not inhibit or distort further growth.

One advantage of an early approach is that the child has a more attractive face throughout childhood. Surgery in the preschooler will alleviate the impact of a severe facial deformity on the child during the early school years when self-esteem is fragile and patterns of social interactions are developing. In moderate cases with some esthetic problems, the timing of treatment depends on the attitude of the patient and parents. If they are not overly concerned with the facial difference, it does not appear to be causing a psychosocial problem, and no significant functional advantage is to be gained by early correction, then it is probably preferable to wait until facial growth has neared completion and the patient enters adolescence, when social activities may cause more concern about facial differences.

It would appear that orthodontic and orthopedic treatments have very little effect on the growth of the basal bones of the face, including the mandible and maxilla. The permanent response to treatment can be measured in millimeters or fractions thereof. This is a small accomplishment in the face that is severely deficient or deformed.

17.4 SUMMARY AND CONCLUSIONS

To recapitulate the major points covered in this chapter:

- Facial morphology and growth is characteristic for each particular cleft type.
- The intrinsic defect in an individual with cleft lip and palate is mild, except in the immediate area of the cleft.
- The potential for growth of the maxillary complex is adequate to produce harmonious skeletal relationships.
- The teeth and alveolar bone have the capacity to compensate for minor deficiencies in the maxillary complex and produce a satisfactory occlusion.

- Modern lip reconstruction procedures are mainly beneficial to morphology, and only mildly affect growth.
- Early repair of the alveolus, with or without bone grafting, at best is mildly detrimental to facial growth. Some techniques are seriously detrimental.
- Palate surgery produces scar tissue that interferes with maxillary growth. It is not necessary that this be a severe restriction; any reduction in maxillary growth can be significant for children with cleft lip and palate.
- Scar tissue in the palate captures fibers from the suspensory ligaments of the teeth, causing distortion of the dental arch by deflecting the eruption of these teeth.
- The surgical procedure with the greatest inhibiting effect on growth is almost certainly the repair of the hard palate.
- The mandible is close to normal size, but tends to be displaced to a more open position.
- The timing of hard palatal repair within the first decade is not critical. There is no advantage in delaying hard palate repair until 4 to 7 years of age.
- Because of their different morphologies, bilateral cleft lip and palate tends to improve with growth, cleft palate and cleft lip both tend to remain the same, and unilateral cleft lip and palate tends to become worse.
- The most important variable in cleft palate surgery seems to be the surgeon, since traditional techniques seem to give the same results.

REFERENCES

Atherton, J. D. (1967). Morphology of facial bones in skulls with unoperated unilateral cleft palate. *Cleft Palate J. 4*, 18–30.

Atkinson, S. R. (1966). Jaws out of balance. *Am. J. Orthod. 52*, 371.

Bachmayer, D., Ross, R. B., and Munro, I. R. (1986). Maxillary growth following Le Fort III surgery in Crouzon, Apert and Pfeiffer syndromes. *Am. J. Orthod. 90*, 420–430.

Ball, J. V., Dibiase, D., and Sommerlad, B. C. (1995). Transverse maxillary arch changes with the use of preoperative orthopedics in unilateral cleft palate infants. *Cleft Palate Craniofac. J. 32*, 483–488.

Bardach, J., and Eisbach, K. J. (1977). The influence of primary unilateral cleft lip repair on facial growth. *Cleft Palate J. 14*, 88–97.

Bardach, J., Klalusner, E. C., and Eisbach, K. J. (1979). The relationship between lip pressure and facial growth after cleft lip repair. *Cleft Palate J. 16*, 137–146.

Bardach, J., Morris, H. L., and Olin, W. H. (1984). Late results of primary veloplasty: the Marburg Project. *Plast. Reconstr. Surg. 73*, 207–214.

Berkowitz, S. (1996). A comparison of treatment results in complete bilateral cleft lip and palate using a conservative approach versus Millard-Latham PSOT procedure. *Semin. Orthod. 2*, 169–184.

Bernstein, L. (1968). The effect of timing of cleft palate operations and subsequent growth of the maxilla. *Laryngoscope 78*, 1510.

Bishara, S. S. E., Jakobsen, J. R., Krause, J. C., and Sosa-Martinez, R. (1986). Cephalometric comparisons of individuals from India and Mexico with unoperated cleft lip and palate. *Cleft Palate J. 23*, 116–125.

Bjork, A. (1966). Sutural growth of the upper face studied by the implant method. *Acta Odont. Scand. 24*, 109–113.

Blijdorp, P., and Egyedi, P. (1984). The influence of age at operation for cleft on the development of the jaws. *J. Maxillofac. Surg. 12*, 193–198.

Boo-Chai, K. (1971). The unoperated adult bilateral cleft of the lip and palate. *Brit. J. Plast. Surg. 24*, 250–254.

Bosma, J. F. (1963). Maturation of function of the oral and pharyngeal region. *Am. J. Orthod. 49*, 94–104.

Brattström, V., McWilliam, J., Larson, O., and Semb, G. (1991). Craniofacial development in children with unilateral clefts of the lip, alveolus, and palate treated according to four different regimes. I. Maxillary development. *Scand. J. Plast. Reconstr. Surg. Hand. Surg. 25*, 259–267.

Capelozza Filho, F. L., Normando, A. D., and da Silva Filho, O. G. (1996). Isolated influences of lip and palate surgery on facial growth: comparison of operated and unoperated male adults with UCLP. *Cleft Palate Craniofac J. 33*, 51–56.

Cohen, M. M., Jr. (1981). A critical review of cephalometric studies of dysmorphic syndromes. *Proc. Finn. Dent. Soc. 77*, 17–25.

Coupe, T. B. (1962). A study of the morphology and growth of the nasal septum in children with clefts of the lip and palate. M.Sc. thesis, University of Toronto.

Coupe, T. B., and Subtelny, J. D. (1960). Cleft palate—deficiency or displacement of tissue. *Plast. Reconstr. Surg. 26*, 600.

Dahl, E. (1970). Craniofacial morphology in congenital clefts of the lip and palate. An X-ray cephalometric study of young adult males. *Acta Odont. Scand. 28* (suppl. 57), 1–167.

Dahl, E., Hanusardottir, B., and Bergland, O. (1981). A comparison of occlusions in two groups of children whose clefts were repaired by three different surgical procedures. *Cleft Palate J. 18*, 122–128.

Daskalogianakis, J., and Ross, R. B. (1997). Effect of alveolar bone grafting in the mixed dentition on maxillary growth in complete unilateral cleft lip and palate patients. *Cleft Palate Craniofac J. 34*, 455–458.

Daskalogiannakis, J., Ross, R. B., and Tompson, B. D. (2001). "Catch-up" mandibular growth in patients with Pierre Robin sequence: fact or myth? *Am. J. Orthod. Dentofac. Orthop. 120*, 280–285.

Davies, A. D. (1951). Unoperated bilateral complete cleft lip and palate in the adult. *Plast. Reconstr. Surg. 7*, 482–488.

Dijkman, G. E. H. M., Ross, R. B., Daskalogianakis, J. D., Kuijpers-Jagtman, A. M., and Tompson, B. D. (1998). Growth of the premaxilla in bilateral CLP following osteotomy and bone grafting at 10 years. Paper presented at American Cleft Palate–

Craniofacial Association annual meeting, Baltimore, April 23.

Dixon, D. A. (1966). Abnormalities of the teeth and supporting structures in children with clefts of lip and palate. In C. M. Drillien, T. T. S. Ingram, and E. M. Wilkinson (eds.), *The Causes and Natural History of Cleft Lip and Palate* (Edinburgh: E. & S. Livingstone), pp. 178–205.

Dorf, D. S., and Curtin, J. W. (1982). Early cleft palate repair and speech outcome. *Plast. Reconstr. Surg. 70*, 74–80.

Drillien, C. M., Ingram, T. T. S., and Wilkinson, E. M. (eds.) (1966). *The Causes and Natural History of Cleft Lip and Palate* (Edinburgh: E. & S. Livingstone).

Droschl, H. (1973). The effect of heavy orthopedic forces on the maxilla in the growing *Saimiri sciureus* (squirrel monkey). *Am. J. Orthod. 63*, 449–458.

Enlow, D. H. (1990). *Facial Growth*, 3rd ed. (Philadelphia: W. B. Saunders).

Enlow, D. H., and Bang, S. (1965). Growth and remodeling of the human maxilla. *Am. J. Orthod. 51*, 446–452.

Farkas, L. G., and Lindsay, W. K. (1972). Morphology of the orbital region in adults following the cleft lip–palate repair in childhood. *Am. J. Phys. Anthropol. 37*, 65–76.

Figueroa, A. A., Glupker, T. J., Fitz, M. G., and BeGole, E. A. (1991). Mandible, tongue and airway in Pierre Robin sequence: a longitudinal cephalometric study. *Cleft Palate J. 28*, 425–434.

Friede, H., and Johanson, B. (1982). Adolescent facial morphology of early bone grafted cleft lip and palate patients. *Scand. J. Plast. Reconstr. Surg. 16*, 41–53.

Friede, H., and Pruzansky, S. (1985). Long-term effects of premaxillary setback on facial skeletal profile in complete bilateral cleft lip and palate. *Cleft Palate J. 22*, 97–104.

Friede, H., Priede, D., Moller, M., Maulina, I., and Barkane, B. (1999). Comparisons of facial growth in patients with unilateral cleft lip and palate treated by different regimens for two-stage palatal repair. *Scand. J. Plast. Reconstr. Surg. Hand. Surg. 33*, 73–81.

Gillies, H. D., and Fry, K. W. (1921). A new principle in the surgical treatment of congenital cleft palate. *Brit. Dent. J. 42*, 293–297.

Graber, T. M. (1949). Craniofacial morphology in cleft palate and cleft lip deformities. *Surg. Gynecol. Obstet. 88*, 359–368.

Handelman, C. S., and Pruzansky, S. (1968). Occlusion and dental profile with complete bilateral cleft lip and palate. *Angle Orthod. 38*, 185–198.

Harding, R. L., and Mazaheri, M. (1972). Growth and spatial changes in the arch form in bilateral cleft lip and palate. *Plast. Reconstr. Surg. 50*, 591–599.

Harvold, E. (1954). Cleft lip and palate. Morphologic studies of the facial skeleton. *Am. J. Orthod. 40*, 493–501.

Hellquist, R., and Ponten, B. (1979). The influence of infant periosteoplasty on facial growth and dental occlusion from five to eight years of age in cases of complete unilateral cleft lip and palate. *Scand. J. Plast. Reconstr. Surg. 13*, 305–312.

Hellquist, R., and Svärdström, L. D. S. (1986). Changes in the treatment program 1964 to 1984 at the Uppsala Cleft Palate Centre. In M. Hotz, W. Gnoinski, M. Perko, H. Nussbaumer, E. Hof, and R. Haubensak (eds.), *Early Treatment of Cleft Lip and Palate* (Toronto: Hans Huber Publishers), pp. 131–134.

Hellquist, R., Svärdström, K., and Pontén, B. (1983). A longitudinal study of delayed periosteoplasty to the cleft alveolus. *Cleft Palate J. 20*, 277–288.

Henkel, K. O., and Gundlach, K. K. (1997). Analysis of primary gingivoperiosteoplasty in alveolar cleft repair. Part I: facial growth. *J. Craniomaxillofac. Surg. 25*, 266–269.

Herfert, O. (1958). Fundamental investigations into problems related to cleft palate surgery. *Brit. J. Plast. Surg. 11*, 97–101.

Hotz, M., and Gnoinski, W. (1979). Effects of early maxillary orthopedics in coordination with delayed surgery for cleft lip and palate. *J. Maxillofac. Surg. 7*, 201–210.

Huang, C. S., and Ross, R. B. (1982). Surgical advancement of the retrognathic mandible in growing children. *Am. J. Orthod. 82*, 89–103.

Innis, C. O. (1962). Some preliminary observations on unrepaired harelips and cleft palates in adult members of the Dusan tribes of North Borneo. *Brit. J. Plast. Surg. 15*, 173–182.

Jakobsson, S. O. (1967). Cephalometric evaluation of treatment effect on class II, division I malocclusions. *Am. J. Orthod. 53*, 446–452.

Jensen, B. L., Dahl, E., and Kreiborg, S. (1983). Longitudinal study of body height, radius length and skeletal maturity in Danish boys with cleft lip and palate. *Scand. J. Dent. Res. 91*, 473–481.

Johnson, R. (1960). Physical development of cleft lip and palate children. In M. A. Cox (ed.), *Five Year Report (1955–1959) of the Cleft Lip and Palate Research and Treatment Centre* (Toronto: Hospital for Sick Children), pp. 104–108.

Joos, U. (1995). Skeletal growth after muscular reconstruction for cleft lip, alveolus and palate. *Brit. J. Oral Maxillofac. Surg. 33*, 139–144.

Koberg, W. R. (1973). Present view on bone grafting in cleft palate. *J. Maxillofac. Surg. 1*, 185–191.

Koski, K. (1968). Cranial growth center: facts or fallacies. *Am. J. Orthod. 54*, 566–570.

Kramer, G. J. C., Hoeksma, J. B., and Prahl-Andersen, B. (1994). Palatal changes after lip surgery in different types of cleft lip and palate. *Cleft Palate Craniofac. J. 31*, 376–384.

Kremenak, C. R. (1984). Physiological aspects of wound healing: contraction and growth. *Otolaryngol. Clin. North Am. 17*, 437–445.

Kremenak, C. R., Huffman, W. C., and Olin, W. H. (1970). Maxillary growth inhibition by mucoperiosteal denudation of palatal shelf bone in non-cleft beagles. *Cleft Palate J. 7*, 817–825.

Kuijpers-Jagtman, A. M., and Long, R. E. (2000). The influence of surgery and orthopedic treatment on maxillofacial growth and maxillary arch development in patient treated for orofacial clefts. *Cleft Palate Craniofac. J. 37*, 527–533.

Kuijpers-Jagtman, A. M., and Prahl, C. (1996). *A Study into the Effects of Presurgical Orthopedic Treatment in Complete Unilateral Cleft Lip and Palate Patients. A Three-Center Prospective Clinical Trial in Nijmegen, Amsterdam, and Rotterdam. Interim Analysis* (Nijmegen: University Press).

Kuijpers-Jagtman, A. M., and Prahl-Andersen, B. (1997). Value of presurgical infant ortho-

pedics: an intercenter randomized clinical trial. In S. T. Lee and M. Huang (eds.), *Transactions 8th International Congress on Cleft Palate and Related Craniofacial Anomalies* (Singapore: Stamford Press), pp. 1002–1004.

Larson, M., Sallström, K. O., Larson, O., McWilliam, J., and Ideberg, M. (1993). Morphologic effect of preoperative maxillofacial orthopedics (T-traction) on the maxilla in unilateral cleft lip and palate patients. *Cleft Palate Craniofac. J. 30*, 29–34.

Latham, R. A. (1969). The pathogenesis of the skeletal deformity associated with unilateral cleft lip and palate. *Cleft Palate J. 6*, 404–411.

Latham, R. A., and Burston, W. R. (1964). The effect of unilateral cleft of the lip and palate on maxillary growth pattern. *Brit. J. Plast. Surg. 17*, 10–16.

Leenstra, T. S., Maltha, J. C., Kuijpers-Jagtman, A. M., and Spauwen, P. H. M. (1995). Wound healing in beagle dogs after palatal repair without denudation of bone. *Cleft Palate J. 32*, 363–370.

Leenstra, T. S., Kohama, G. I., Kuijpers-Jagtman, A. M., and Freihofer, H. P. M. (1996). Suprapcriosteal flap technique versus mucoperiosteal flap technique in cleft palate surgery. *Cleft Palate J. 33*, 501–506.

Lo, L. J., Huang, C. S., Chen, Y. R., and Noordhoff, M. S. (1999). Palatoalveolar outcome at 18 months following simultaneous primary cleft lip repair and posterior palatoplasty. *Ann. Plast. Surg. 42*, 581–588.

Long, R. E., Jr., and McNamara, J. A., Jr. (1985). Facial growth following pharyngeal flap surgery: skeletal assessment on serial lateral cephalometric radiograms. *Am. J. Orthod. 87*, 187–196.

Long, R. E., Jr., Gupton, S., and McGillis, M. G. (1984). Growth in width following pharyngeal flap surgery. *Angle Orthod. 54*, 55–66.

Lukash, F. N., Schwartz, M., Grauer, S., and Tuminelli, F. (1998). Dynamic cleft maxillary orthopedics and periosteoplasty: benefit or detriment? *Ann. Plast. Surg. 40*, 321–326.

Mars, M., and Houston, W. J. (1990). A preliminary study of facial growth and morphology in unoperated male unilateral cleft lip and palate subjects. *Cleft Palate Craniofac. J. 27*, 7–10.

McNeil, K. (1950). Orthodontic procedures in the treatment of congenital cleft. *Dent. Rec. 70*, 126–129.

Menius, J. A., Largent, M. D., and Vincent, C. J. (1966). Skeletal development of cleft palate children as determined by hand-wrist roentgenographs: a preliminary study. *Cleft Palate J. 3*, 67–74.

Mestre, J., DeJesus, J., and Subtelny, J. D. (1960). Unoperated oral clefts at maturation. *Angle Orthod. 30*, 78–86.

Millard, D. R., Latham, R., Huifen, X., Spiro, S., and Morovic, C. (1999). Cleft lip and palate treated by presurgical orthopedics, gingivoperiosteoplasty, and lip adhesion (POPLA) compared with previous lip adhesion method: a preliminary study of serial dental casts. *Plast. Reconstr. Surg. 103*, 1630–1644.

Moss, M. L., Bromberg, B. E., Song, I. C., and Eisenman, G. (1968). The passive role of nasal septal cartilage in midfacial growth. *Plast. Reconstr. Surg. 41*, 536–544.

Munroe, N. (1966). Radiographic cephalometric study of mandibular morphology at gonion and its relation to tongue posture in cleft palate and normal individuals (abstract). *J. Can. Dent. Assoc. 32*, 478.

Nakamura, S., Savara, B., and Thomas, D. (1972). Facial growth of children with cleft lip and or palate. *Cleft Palate J. 9*, 119–126.

Narula, J. K., and Ross, R. B. (1970). Facial growth in children with complete bilateral cleft lip and palate. *Cleft Pal. J. 7*, 239–248.

Nordin, K. E., Larson, O., Nylen, B., and Eklund, G. (1983). Early bone grafting in complete cleft lip and palate cases following maxillofacial orthopedics. I. The method and the skeletal development from seven to thirteen years of age. *Scand. J. Plast. Reconstr. Surg. 17*, 33–41.

Noverraz, A. E. M., Kuijpers-Jagtman, A. M., and Mars, M. M. (1993). Timing of hard palate closure and dental arch relationships in unilateral cleft lip and palate patients: a mixed-longitudinal study. *Cleft Palate Craniofac. J. 30*, 391–396.

Ortiz-Monasterio, F., Olmedo, A., Trigos, I., Yudovich, M., Velazquez, M., and Fuentedel-Campo, A. (1974). Final results from the delayed treatment of patients with clefts of the lip and palate. *Scand. J. Plast. Reconstr. Surg. 8*, 109–115.

Osborne, H. A. (1966). A serial cephalometric analysis of facial growth in adolescent cleft palate subjects. *Angle Orthod. 36*, 211–219.

Perko, M. A. (1974). Primary closure of the cleft palate using a palatal mucosal flap: an attempt to prevent growth inhibition. *J. Maxillofac. Surg. 2*, 40–43.

Peyton, W. T. (1931). Dimensions and growth of the palate in the normal infant and in the infant with gross maldevelopment of the upper lip and palate. *Arch. Surg. 22*, 704–714.

Pitanguy, I., and Franco, T. (1968). Facial clefts in nonoperated adults (abstract). *Plast. Reconstr. Surg. 41*, 187–196.

Polley, J. W., Figueroa, A. A., Liou, E. J., and Cohen, M. (1997). Longitudinal analysis of mandibular asymmetry in hemifacial microsomia. *Plast. Reconstr. Surg. 99*, 328–339.

Prahl-Andersen, B., and Ross, R. B. (1997). Unpublished data.

Przezdziak, B. (1969). Somatic development of children with cleft palate. *Pediatr. Pol. 44*, 1279–1286.

Psaume, J. (1957). A propos des anomalies faciales associes a des divisions palatines. *Ann. Chir. Plast. 2*, 3–11.

Rintala, A., and Ranta, R. (1986). Primary treatment of cleft lip and palate at the Finnish Red Cross Cleft Palate Center from 1966 to 1980. In M. Hotz, W. Gnoinski, M. Perko, H. Nussbaumer, E. Hof, and R. Haubensak (eds.), *Early Treatment of Cleft Lip and Palate* (Toronto: Hans Huber Publishers), pp. 140–143.

Roberts-Harry, D., Semb, G., Hathorn, I., and Killingback, N. (1996). Facial growth in patients with unilateral clefts of lip and palate: a two-center study. *Cleft Palate Craniofac. J. 33*, 489–493.

Robertson, N. R. E., and Jolleys, A. (1983). An 11-year followup of the effects of early bone grafting in infants born with complete clefts of the lip and palate. *Brit. J. Plast. Surg. 36*, 438–446.

Robertson, N. R. E., and Jolleys, A. (1974). The timing of hard palate repair. *Scand. J. Plast. Reconstr. Surg. 8*, 49–56.

Rohrich, R. J., Roswell, A. R., Johns, D. F., Drury, M. A., Grieg, G., Watson, D. J., Godfrey, A. M., and Poole, M. D. (1996). Timing of hard palate closure: a critical long-term analysis. *Plast. Reconstr. Surg. 98*, 236–246.

Rosenstein, S. W., Munroe, C. W., Kernahan, D. A., Jacobson, B. N., Griffith, B. H., and Bauer, B. S. (1982). The case for early bone grafting in cleft lip and cleft palate. *Plast. Reconstr. Surg. 70*, 297–306.

Ross, R. B. (1965). Cranial base in children with lip and palate clefts. *Cleft Palate J. 2*, 157.

Ross, R. B. (1969). The clinical implications of facial growth in cleft lip and palate. *Cleft Palate J. 7*, 37–47.

Ross, R. B. (1977). Craniofacial growth following pharyngeal flap surgery. Paper presented at the 3rd International Congress on Cleft Lip and Palate, Toronto.

Ross, R. B. (1987). Treatment variables affecting facial growth in unilateral cleft lip and palate. *Cleft Palate J. 24*, 5–71.

Ross, R. B. (1990). Craniofacial growth and development in cleft lip and palate. In J. McCarthy (ed.), *Plastic and Reconstructive Surgery* (New York: W. B. Saunders), pp. 2553–2580.

Ross, R. B. (1995). Facial growth in complete unilateral cleft lip and palate following the Malek procedure. *Cleft Palate Craniofac. J. 32*, 194–198.

Ross, R. B. (1999). Costochondral grafts replacing the mandibular condyle. *Cleft Palate Craniofac. J. 36*, 334–339.

Ross, R. B. (2001). The orthodontist and complex craniofacial anomalies. *Am. J. Orthod. Dentofac. Orthop. 119*, 92–94.

Ross, R. B., and Coupe, T. B. (1965). Craniofacial morphology in six pairs of monozygotic twins discordant for cleft lip and palate. *J. Can. Dent. Assoc. 31*, 149–156.

Ross, R. B., and Johnston, M. C. (1967). The effect of early orthodontic treatment on facial growth in cleft lip and palate. *Cleft Palate J. 4*, 157–163.

Ross, R. B., and Johnston, M. C. (1972). *Cleft Lip and Palate* (Baltimore: Williams & Wilkins).

Ross, R. B., and Lindsay, W. K. (1965). The cervical vertebrae as a factor in the etiology of cleft palate. *Cleft Palate J. 2*, 273–277.

Ross, R. B., and McNamara, C. (1994). The long term effect of presurgical infant orthopedics on facial esthetics in bilateral cleft lip and palate. *Cleft Palate Craniofac. J. 31*, 68–73.

Rygh, P., and Tindlund, R. S. (1996). Early considerations in the orthodontic management of skeletodental discrepancies. In T. A. Turvey, K. W. L. Vig, and R. J. Fonseca (eds.), *Facial Clefts and Craniosynostosis. Principles and Managment* (Philadelphia: W. B. Saunders), pp. 234–319.

Sameshima, G. T., Banh, D. S., Smahel, Z., and Melnick, M. (1996). Facial growth after primary periosteoplasty versus primary bonegrafting in unilateral cleft lip and palate. *Cleft Palate Craniofac. J. 33*, 300–305.

Santiago, P. E., Grayson, B. H., Cutting, C. B., Gianoutsos, M. P., Brecht, L. E., and Kwon, S. M. (1998). Reduced need for alveolar bonegrafting by presurgical orthopedics and primary gingivoperiosteoplasty. *Cleft Palate Craniofac. J. 35*, 77–80.

Sarnäs, K.-V., and Rune, B. (1987). Extra-oral traction to the maxilla with face mask: a follow-up of 17 consecutively treated patients with and without cleft lip and palate. *Cleft Palate J. 24*, 95–103.

Schweckendiek, W. (1978). Primary veloplasty: long-term results without maxillary deformity. A twenty-five year report. *Cleft Palate J. 15*, 268–274.

Semb, G. (1988). Effect of alveolar bone grafting on maxillary growth in unilateral cleft lip and palate patients. *Cleft Palate J. 25*, 288–295.

Semb, G., and Shaw, W. C. (1990). Pharyngeal flap and facial growth. *Cleft Palate J. 27*, 217–224.

Semb, G., and Shaw, W. C. (1996). Facial growth in orofacial clefting disorders. In T. A. Turvey, K. W. L. Vig, and R. J. Fonseca (eds.), *Facial Clefts and Craniosynostosis. Principles and Management* (Philadelphia: W. B. Saunders), pp. 28–56.

Shibasaki, Y., and Ross, R. B. (1969). Facial growth in children with isolated cleft palate. *Cleft Palate J. 6*, 290–297.

Shprintzen, R. J. (1988). Pierre Robin, micrognathia, and airway obstruction: the dependency of treatment on accurate diagnosis. *Int. Anesthesiol. Clin. 26*, 84–91.

Shprintzen R. J. (1992). The implications of the diagnosis of Robin sequence. *Cleft Palate Craniofac. J. 29*, 205–209.

Skoog, T. (1965). The use of periosteal flaps in the repair of clefts of the primary palate. *Cleft Palate J. 2*, 332–339.

Smahel, Z. (1984). Craniofacial morphology in adults with bilateral complete cleft lip and palate. *Cleft Palate J. 21*, 158–169.

Smahel, Z., and Brejcha, M. (1983). Differences in craniofacial morphology between complete and incomplete unilateral cleft lip and palate in adults. *Cleft Palate J. 20*, 113–118.

Smahel, Z., and Müllerova, Z. (1994). Facial growth and development in unilateral cleft lip and palate during the period of puberty: comparison of the development after periosteoplasty and after primary bone grafting. *Cleft Palate Craniofac. J. 31*, 106–115.

Smahel, Z., Müllerova, Z., and Horak, I. (1996). Facial development in unilateral cleft lip and palate prior to the eruption of permanent incisors after primary bone grafting and periosteal flap surgery. *Acta Chir. Plast. 38*, 30–36.

Smahel, Z., Müllerova, Z., Nejedly, A., and Horak, I. (1998). Changes in craniofacial development due to modifications of the treatment of unilateral cleft lip and palate. *Cleft Palate Craniofac. J. 35*, 240–247.

Stenstrom, S. J., and Thilander, B. L. (1970). The effects of nasal septal cartilage resections on young guinea pigs. *Plast. Reconstr. Surg. 45*, 160–168.

Subtelny, J. D. (1955). Width of the nasopharynx and related anatomic structures in normal and unoperated cleft palate children. *Am. J. Orthod. 41*, 889–908.

Tindlund, R. S. (1989). Orthopaedic protraction of the midface in the deciduous dentition. Results covering 3 years out of treatment. *J. Craniomaxillofac. Surg. 17* (suppl. 1), 17–19.

Tindlund, R. S. (1994). Skeletal response to maxillary protraction in patients with cleft lip and palate before age 10 years. *Cleft Palate Craniofac. J. 31*, 296–308.

Tindlund, R. S., Rygh, P., and Boe, O. E. (1993). Orthopedic protraction of the upper jaw in cleft lip and palate patients during the deciduous and mixed dentition periods in comparison with normal growth and development. *Cleft Palate Craniofac. J. 30*, 182–194.

Trotman, C. A., and Ross, R. B. (1993). Craniofacial growth in complete bilateral cleft lip and palate. *Cleft Palate Craniofac. J. 30*, 261–273.

Trotman, C. A., Long, R. E., Rosenstein, S. W., Murphy, C., and Johnston, L. E. (1996). Comparison of facial form in primary alveolar-bone-grafted and nongrafted unilateral cleft lip and palate patients: intercentre retrospective study. *Cleft Palate Craniofac. J. 33*, 91–95.

Van Limborg, J. (1964). Some aspects of the development of the cleft-affected face. In R. Hotz (ed.), *Early Treatment of Cleft Lip and Palate* (Toronto: Hans Huber Publishers).

Vargervik, K. (1983). Growth characteristics of the premaxilla and orthodontic treatment principles in bilateral cleft lip and palate. *Cleft Palate J. 20*, 289–302.

Venditelli, B., and Flavelli, O. (1999). A cephalometric study of mandibular morphology in different types of cleft lip and palate. Diploma thesis, University of Toronto.

Wada, T., and Miyazaki, T. (1975). Growth and changes in maxillary arch form in complete unilateral cleft lip and cleft palate children. *Cleft Palate J. 12*, 121–130.

Walker, B. E. (1959). Effects on palate development of mechanical interference with the fetal environment. *Science 130*, 981–982.

Wieslander, L. (1963). The effect of orthodontic treatment on the concurrent development of the craniofacial complex. *Am. J. Orthod. 49*, 15–21.

Will, L. A. (2000). Growth and development in patients with untreated clefts. *Cleft Palate Craniofac. J. 37*, 523–526.

Winters, J. C., and Hurwitz, D. J. (1995). Presurgical orthopedics in the surgical management of unilateral cleft lip and palate. *Plast. Reconstr. Surg. 95*, 755–764.

Witsenburg, B. (1985). The reconstruction of anterior residual bone defects in patients with cleft lip, alveolus and palate: a review. *J. Maxillofac. Surg. 13*, 197–204.

Witzel, M. A., Salyer, K., and Ross, R. B. (1984). Delayed hard palate closure: the philosophy revisited. *Cleft Palate J. 21*, 263–271.

PART VI

REGIONAL DYSMORPHOLOGY AND GROWTH DISTURBANCES

CHAPTER 18

DENTAL DEVELOPMENT AND ANOMALIES IN CRANIOSYNOSTOSES AND FACIAL CLEFTING

EDWARD F. HARRIS, Ph.D.

18.1   INTRODUCTION

Teeth, unlike other tissues such as bone and cartilage that remodel with time, provide two unique features that enhance their relevance for the study of growth and growth disturbances: (1) Once mineralized, enamel does not remodel, so each tooth provides in effect a permanent record of the individual's ability to thrive during the period of that tooth's crown formation (Goodman et al., 1988; Goodman and Rose, 1990; Hillson and Bond, 1997). (2) Teeth (20 primary, 32 permanent) form throughout most of the years from the second trimester in utero into young adulthood (Schour and Massler, 1940; Lunt and Law, 1974; Moorrees et al., 1963). Examination of a number of sequentially formed teeth affords the opportunity, then, to gather ontogenetic data on an individual's growth history. The information is not uniformly applicable, though, because groups of teeth form simultaneously and there are gaps in the record at other ages (and some tooth types are intrinsically more variable and susceptible to stressors than others) (Dahlberg, 1945; Townsend

and Brown, 1981; Kieser and Groeneveld, 1988).

Growth in a felicitous environment increases the individual's (and the formative tooth's) ability to approximate his genetic potential; growth while burdened with environmental detractions (vis-à-vis oxygen tension, nutrition, morbidity, access to nutrients, removal of waste products, genotype, psychosocial factors, and on and on) diminishes the individual's ability to thrive (and, in turn, diminishes tooth size) (Tonge and McCance, 1965, 1973; Luke et al., 1981; Ebeling et al., 1973; Suzuki, 1993; Laatikainen and Ranta, 1996a, 1996b). Striking, localized examples of this general principle are seen in children with facial hemihypertrophy and hemihypotrophy (including hemifacial microsomia), where teeth in the area with denser vascularity develop faster and are larger than on the less-vascularized side (Kogon et al., 1984; Loevy and Shore, 1985; Pollock et al., 1985; Czarnecki and Carrel, 1982; Dahlstrom and Haraldson, 1986; Fayad and Steffensen, 1994). These dramatic but unfortunate cases illustrate the power of the local envi-

Understanding Craniofacial Anomalies: The Etiopathogenesis of Craniosynostoses and Facial Clefting, Edited by Mark P. Mooney and Michael I. Siegel, ISBN 0-471-38724-x   Copyright © 2002 by Wiley-Liss, Inc.

ronment to modulate the rate and extent of crown-root growth. Several less extreme but more common effects of the environment on the dentition are discussed in this chapter.

The study of growth of children with craniofacial malformations, deformations, and disruptions (Spranger et al., 1982) is, in part, the study of how such children interact with their environment (*environment* in the broadest sense), and this interaction is reflected in their growth. There are an indefinitely large number of ways to assess growth; this chapter describes a few ways that growth influences tooth formation and how data from teeth provide means of assessing the quality of the growth that has occurred.

Cranial—specifically, pharyngeal arch—anomalies often do affect the dentition, and the effect can take on a host of interrelated manifestations (Böhn, 1963; Kraus and Jordan, 1965; Jordan et al., 1966; Ross and Johnston, 1972; Anderson and Moss, 1996). Common findings include hypodontia, reduced size of the crowns and roots (and altered crown-root ratio), aberrant root forms, simplified crown morphology, malformed teeth (e.g., atypical mamelon and cusp configurations), supernumerary teeth (notably adjacent to the cleft), delayed tooth formation and eruption.

18.2 TOOTH NUMBER

Clefting is associated with an increased risk of congenitally missing teeth. Absence of maxillary lateral incisors in the region of the cleft is to be expected (though why canines, also adjacent to the cleft, are less commonly affected is not known). There does not appear to be a sex difference in the frequency of missing incisors (Ranta, 1986). The frequency of hypodontia is greater in both dentitions and all tooth types compared to people without clefts (Böhn, 1963; Ranta, 1984), but is highest in

the cleft region of the primary dentition. Hypodontia of primary teeth outside the cleft site is comparatively infrequent, which suggests that the stress of the cleft on the child's growth is much less in utero (thus the low frequency of congenitally missing primary teeth), whereas hypodontia is common among the permanent teeth that form postnatally (where the cleft impairs the person's growth).

Among the permanent tooth types, the maxillary lateral incisors are, by far, the most likely to be congenitally absent, followed by the second premolars in both arches (Böhn, 1963; Olin, 1964; Kraus et al., 1966; Hellquist et al., 1979; Ranta, 1983, 1986; Ranta et al., 1983). While the first and second premolars undergo formation at similar times (Harris and McKee, 1990), the second premolar in each quadrant is far more likely to be congenitally absent, which suggests that the morphogenetic premolar field is very short and the second premolar is much more labile to environmental stressors (Dahlberg, 1945; Brook, 1984), but these studies do not address the cause of the observed difference. Twin studies show that hypodontia often is discordant between pairs of twins, implying that there is a weak genetic component and, instead, that environmental stressors play a significant role in modulating congenital absence (Laatikainen and Ranta, 1996c).

It has been suggested (Olin, 1964; Dixon, 1966, 1968) that the high frequency of defective maxillary incisors may be due in part to surgical trauma. Hellquist and colleagues (1979) tested this by comparing the frequencies of upper incisor defects in a sample treated by lip adhesion only and a second sample where the primary lip adhesion was combined with periosteoplasty to connect the maxillary segments, permitting new bone to repair the alveolar cleft. In fact, there was a higher frequency of normal permanent maxillary incisors (and fewer missing UI2) in the periosteoplasty group, which suggests that the union of the alveolar segments creates a better environment for the teeth in the cleft area during their development.

A popularly applied model for hypodontia involves a developmental threshold (Grüneberg, 1952; Suarez and Spence, 1974). The model suggests that local regulatory factors define and induce the sites of tooth formation in the dental lamina, but if the presumptive tooth mass is insufficient (i.e., below some critical developmental threshold), tooth formation does not progress and apoptosis removes the partially differentiated structures, leaving a congenitally absent tooth. Embryologists have provided supporting evidence for early formation followed by subsequent cell death of formative teeth (Fitzgerald, 1973; Moss-Salentijn and Moss, 1975). The supposition is that children, such as those with pharyngeal arch malformations, are stressed by the environment, thus moving them toward the threshold where the risk of hypodontia is significantly elevated. The risk of hypodontia of permanent teeth increases noticeably with severity of the cleft (CP < UCLP < BCLP) (Böhn, 1963; Olin, 1964; Hellquist et al., 1979; Ranta, 1984), which is consistent with the apparent levels of postnatal stress experienced by children with clefts (Drillien et al., 1966; Turvey et al., 1996).

In addition, where a deciduous tooth is congenitally absent, it should be anticipated that the lingual extension of the dental lamina that will form that primary tooth's permanent successor also will be absent (Avery, 1994), so there will be no successional tooth, either. Intuitively, pegged and congenitally absent maxillary lateral incisors might seem to be microforms of CLP (i.e., minor expressions that a noncleft subject is "close" to exhibiting CLP and that his risk of having relatives with CLP is elevated). The suggestion is popular because small maxillary lateral incisors are common and they are situated right at the premaxillary-maxillary suture where clefts occur (Fukuhara and Saito, 1963; Tolarova, 1969). If an abnormal UI2 is a microform of clefting, then noncleft relatives of people with cleft lip with or without cleft palate (CL ± P) ought to have an increased frequency of this anomaly. In fact, no association has been found. The incidence of UI2 anomalies was just the same (approximately 2.8%) in the parents and siblings of 142 patients with CL ± P as in 187 control families. So, too, the frequency of CL ± P was the same in 3989 relatives of 103 individuals with congenital absence of UI2 as in the general population (Woolf et al., 1965). Subsequent studies also reject the supposition that absence or diminished size of UI2 is a minor expression of clefting (Mills et al., 1968; Anderson and Moss, 1996).

Hyperdontia (supernumerary teeth) also occurs with enhanced frequency in subjects with CLP, but this is limited to the maxillary anterior segment of the arch (i.e., near the disruption of the cleft). These teeth typically are conical or incisiform (Stafne, 1932; Millhon and Stafne, 1941; Böhn, 1963).

There is a major difference in the risk of supernumerary teeth depending on whether the cleft of the lip is incomplete or complete. Nagai and co-workers (1965) found one or more supernumerary teeth in the maxillary anterior region in 51% of cases where cleft lip was incomplete. In more severe cases, with complete cleft of the lip, the frequency was much lower (0.6%). It could be that the dental lamina and/or the molecular signals that initiate tooth formation is more developed in the milder cleft forms, thereby increasing the likelihood of additional tooth germ formation.

18.3 WHAT TO MEASURE AND IMPROVING TECHNOLOGIES

Mesiodistal and buccolingual crown dimensions are the variables most commonly measured on teeth, but the rationale is centered simply on the fact that these dimensions can be defined operationally (Seipel, 1946; Moorrees, 1957; Goose, 1963), they are reasonably reproducible within and among researchers (Kieser et al., 1990), and measurements are homologous whether obtained from teeth in vivo; extracted or exfoliated teeth; teeth from forensic, archeological, or paleontological contexts; or on dental casts (Kieser, 1990; Hillson, 1996).

Still, various researchers have wondered whether other measures of a tooth might be more informative. For example, overall crown size is a conglomeration of the constituent cusp sizes; measures of the individual cusps may be more informative relative to a given biological question (Biggerstaff, 1969). Corruccini (1979; Corruccini and Potter, 1981) provides a methodology for assessing crown components on molars, and technological advances now permit digital imaging and measurements of crown details in two and three dimensions (Sekikawa et al., 1989; Peretz and Smith, 1993; Kondo et al., 1996; Peretz et al., 1996, 1999). Refinements, primarily through computer-assisted imaging and measuring systems, have provided novel information addressing how cuspal components are arranged developmentally. On the other hand, information concerning the formation of single- and multicusped teeth indicates that a fundamental process of creating more complex crown morphologies is simply to compound the molecular signaling used to create a single-cusp form (Weiss, 1990; Jernvall, 2000; Jernvall and Jung, 2000), so informational content of cusps (in a statistical sense) may be redundant in large part, though certainly not entirely (Keene, 1982). It also is generally observed that phylogenetically more recent portions of the teeth are more variable (Scott and Turner, 1997) and more susceptible to stressors in the environment (Sofaer et al., 1972; Suzuki, 1993).

Another novel approach in recent years has been to quantify the various tissue sizes that contribute to tooth size, specifically enamel thickness, dentin dimensions, and those of the pulp. This can be done by sectioning the tooth and measuring distances, areas, and volumes of the tissues in one way or another (see, e.g., reviews by Macho, 1995; Schwartz and Dean, 2000). For example, Spangler and colleagues (1998) have described the histological structure and quantity of enamel of teeth in people with tricho-dento-osseous syndrome. Nondestructive methods also have been explored using planar radiographs, such as dental periapical films (Alvesalo and Tammisalo, 1981, 1985; Stroud et al., 1994, 1998; Harris et al., 1999). Mass and co-workers (1996) studied patients with familial dysautonomia and showed that the method is reproducible and provides useful information on the differential development of enamel (ectoderm) and dentin (mesoderm). See Fitzgerald and Verveniotis (1998) for a similar study of a patient with Morquio's syndrome. It does not seem that anyone has yet studied radiographs of people with pharyngeal arch syndromes, notably CLP, to see whether the well-documented diminution in crown size (Clinch, 1963; Foster and Lavelle, 1971) is caused by thinner enamel, thinner dentin, small pulp chambers, or some combination of these.

The shortcomings of planar x-rays, particularly when using conventional clinical views, is that adequate standardization is an issue, and, in vivo, we are normally limited to views of mesial and distal crown margins. This probably will be a short-lived technical problem; increased availability of intraoral digital radiographic systems—with three-dimensional capabilities—will soon provide the clinician and researcher with the capacity for extensive data collection on vital teeth (Paurazas et al., 2000; Terkado et al., 2000).

An underexplored topic regarding craniofacial syndromes is the structure of the enamel. This can be at either of two levels: macroscopically (Goodman and Rose, 1990) or with light or electron microscopy (Hillson and Bond, 1997; Boyde, 1989; Schwartz and Dean, 2000). Developmental defects in enamel, such as enamel hypoplasia (Suckling, 1989), are formed when acute stress effectively stops an individual's growth, including amelogenesis in teeth undergoing crown formation (Goodman and Armelagos, 1985a). Subsequent enamel formation is not reparative, so permanent hypoplastic areas are left where the enamel is thin, poorly mineralized, and subject to caries. Severity of enamel defects macroscopically ranges from an opacity (hypomineralization) to a broad circumferential enamel-free band on the crown. Stressors capable of interrupting amelogenesis include febrile diseases (bacterial or viral), vitamin deficiencies, induced diabetes, acute malnutrition, parasitic infections, and other growth disturbances (Sarnat and Schour, 1941; Kreshover, 1944, 1960; Kreshover et al., 1954; Suckling et al., 1983, 1986; Goodman and Rose, 1990). The response, regardless of the kind of stressor, is the same: there appears to be no way to identify the kind of stress (Selye, 1973a, 1973b, 1976) from analysis of the enamel defect because the physiological nature of the response is generic.

The critical, useful aspect of the frequency and anatomic distribution of enamel hypoplasias is that they provide a record of when a child's growth was seriously impacted by environmental stresses (Swärdstedt, 1966). Are enamel hypoplasias more common in children with craniofacial anomalies? An increased incidence would occur because the anatomic malformations place the child in harm's way (e.g., feeding problems and ear infections in children with CLP); alternatively, it could mean that the craniofacial anomalies are overt flags that the child's overall growth and growth trajectory are less well buffered, so lessor stresses impair growth

compared to phenotypically normal children.

Because the timing of tooth formation is fairly well regimented by the genotype (Niswander, 1963; Pelsmaekers et al., 1997; Merwin and Harris, 1998), the subject's age when stressors affected enamel formation can be reconstructed in detail (Swärdstedt, 1966; Goodman and Armelagos, 1985b; Goodman and Rose, 1990). A detraction of this approach is that different tooth types differ in their susceptibility to forming enamel defects, so the record of past stresses is not recorded uniformly across the dentition. Maxillary central incisors and mandibular canines are most likely to exhibit enamel defects both in the primary and the permanent dentitions (Goodman and Armelagos, 1985a).

Enamel defects are indeed much more common in children with congenital malformations, including CLP, but more study is indicated. Kotilainen and co-workers (1995) found that enamel defects are more common in subjects with Russell-Silver syndrome (short stature with prenatal onset) than normal-growing children, and that they exhibit both pre- and postnatal onsets. Studies of children with CLP (Hellquist et al., 1979; Dahllöf et al., 1989; Vichi and Franchi, 1995; Malanczuk et al., 1999) show quite clearly that the frequency of defects is elevated and that teeth in both the primary and permanent dentitions are affected. This discloses that there is common, episodic impairment of growth, especially in infancy and early childhood.

While numerous researchers have scored the occurrence of hypoplastic teeth in subjects with CLP, none seems to have recorded the positions of defects on the tooth crowns—which would clarify whether they have a prenatal onset, or, as it seems, the stressors only occur postnatally. The marked increase in frequency of enamel hypoplasia reflects the wide-ranging health problems posed by the cleft postnatally. Documentation of prenatal

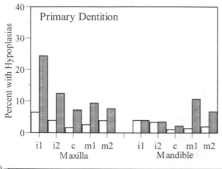

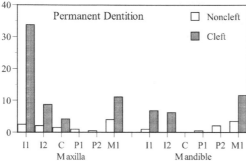

Figure 18.1 Frequencies of teeth with enamel defects in the primary and permanent dentitions. (*Source:* Data from Dixon, 1968.)

insults would be direct evidence that subjects with CLP have an intrinsically reduced growth potential (greater susceptibility, lower threshold to stress) and the problem is not just that the cleft causes health problems following birth. Enamel defects are more frequent on teeth adjacent to the cleft than on the noncleft side (Dixon, 1968; Vichi and Franchi, 1995).

One of the more thorough studies of this topic was by Dixon (1968), who examined the primary and permanent teeth of 100 children with CLP looking for readily visible gross or textbook hypoplasias (so minor expressions were ignored). Hypoplasias were more common on most of the primary tooth types in the cleft sample (Fig. 18.1), more so in the maxilla and especially on the central and lateral incisors. It would seem that growth disturbances have a prenatal onset despite maternal buffering. The early-forming permanent

teeth (I1, I2, M1) were much more frequently affected in the cleft sample, whereas the late-forming teeth (premolars) commonly were defect-free, indicating that those children with clefts who survived the environmental ravages of infancy and early childhood fared better as they grew older. Dixon's study is dated now (1968); we expect that improved health care and management of infants with clefts would substantially lower the frequency of enamel defects in contemporary patients (Dahllöf et al., 1989; Malanczuk et al., 1999).

The location, nature, and extent of enamel defects can also be assessed microscopically. Each of the three hard tissues forming a tooth, enamel, dentin, and cementum, grows in an incremental fashion, so growth disturbances will distort the width and structure of the mineralizing tissues. Enamel and dentin do not remodel, so their structure preserves a memory of the earlier episodes of their metabolism (Rose et al., 1978; Hillson, 1996). One structural enamel defect among several that may prove useful in the clinical arena is termed the Wilson band (Wilson and Shroff, 1970), which is an interval of disturbed (absent) prism structure along the striae of Retzius (Rose, 1979). Wilson bands are, like other enamel and dentin defects, nonspecific in terms of the kind of stressor that forms them, but their number and age-specific locations provide a method of reconstructing a subject's health history during the age interval of a tooth's crown formation. Rose (1977, 1979; Rose et al., 1978) and others have used data on Wilson bands to reconstruct health histories of past populations; there does not yet appear to be any clinical application of this technique.

18.4 TOOTH CROWN FORMATION

Teeth form from two sources: enamel is from ectoderm while the rest of the tooth (dentin, pulp, cementum) is from meso-derm, specifically from neural crest cells (ectomesenchyme). The first elements of the presumptive primary teeth begin forming toward the end of the fetal period, in week 7 and especially week 8, when the dental lamina (ectoderm) invades the presumptive gum epithelium horizontally from the labial lamina that, by apoptosis, later creates the labiogingival sulcus in the two arches (Fig. 18.2). There are four sites of differentiation of the dental lamina in the maxilla (Sperber, 1989): one in each maxillary process and a left-right pair of sites at the inferolateral borders of the pre-maxilla. These tissues spread and coalesce, becoming the U-shape band of lamina that follows the curve of the primitive jaw. There are, analogously, four odontogenic sites in the mandibular arch: an anterior and posterior site in each hemimandible (Sperber, 1989) that likewise coalesce into a continuous U-shape band overlying the presumptive dental arch.

Spots along the primary dental lamina are induced to condense and thicken, becoming 20 enamel organs, 5 per quadrant, that will differentiate and eventually form the 20 primary teeth. A lingual outgrowth forms early on from each of the 20 enamel organs, and, although these projections remain quiescent for a while, they eventually differentiate to form the permanent succedaneous teeth. Distal extensions of the dental lamina in each quadrant give rise to the three permanent molars (though at least one of the third molars is congenitally absent in about 10–15% of the U.S. population) (Nanda, 1954; Garn et al., 1963). Highly differentiated portions of the enamel organs produce enamel.

Development of the labiogingival (vestibular) lamina apparently is not critical for differentiation and migration of the dental lamina into the presumptive tectal ridges because in fetuses with bilateral facial clefts the medial segment of the lip generally remains attached to the dental arch (i.e., no labiogingival sulcus in the pre-maxilla), yet the dental lamina forms, giving rise to the four primary and subsequently the four permanent incisors. Ooë (1981) points out that in young embryos (about 10–15 mm CRL; approximately 6 weeks in utero) there is a gap in the dental lamina of each maxillary quadrant where the medial and lateral nasal processes normally join with the maxillary process, coinciding with the premaxillary-maxillary suture that unifies the primary and secondary palates (and runs between the lateral incisor and canine) (Jacobson, 1955). These premaxillary-maxillary gaps imply that dental lamina differentiates independently in the premaxilla long before fusion of this anterior segment that develops from the frontonasal process with the maxillary processes of the first pharyngeal arch (Sperber, 1989). In the mandible, in contrast, the dental lamina forms as a continuous horseshoe of epithelium deep to the surface of the primitive gum. Details of tooth formation, aside from the overview below, (often described in terms of the succession of bud, cap, and bell stages) are available in embryology texts (Corliss, 1976; Osborn and Ten Cate, 1976; Avery, 1994; Melfi, 1994; Ten Cate, 1994).

Tooth development is initiated by the mesenchyme's inductive influence on the overlying ectoderm. Signals from the mesenchyme determine where, how many, and what tooth type will be formed (e.g., incisor tooth, molar tooth). Induction produces an enamel organ for each presumptive tooth, which is first visible as a site of condensed ectomesenchymal cells undergoing rapid mitosis (Fig. 18.2A, B). The enamel organ projects lingual to the dental lamina and cranial to the lamina in the maxilla and caudal to it in the mandible. As the enamel organ grows, a condensation of mesenchyme develops deep to it (Fig. 18.2B, C, D). The portion of the organ next to this condensation buckles inward by enhanced peripheral and basal cell growth, thereby

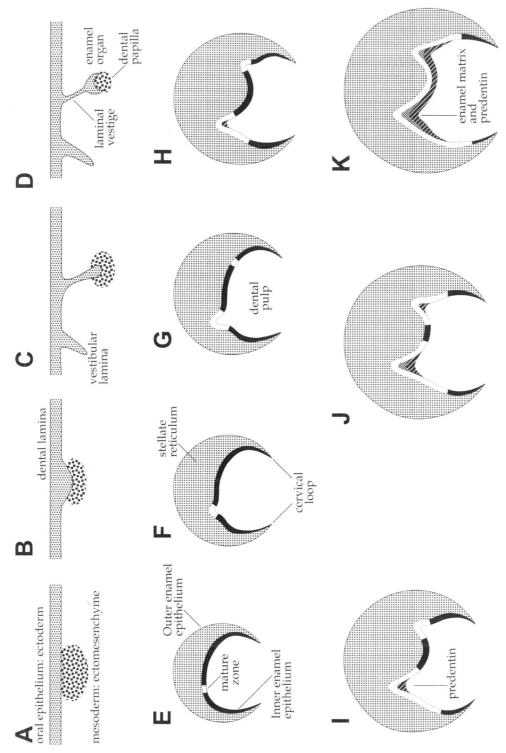

Figure 18.2 Overview of tooth crown formation. The inner enamel epithelium (IEE) is drawn greatly out of proportion to emphasize its development. Darkened areas of the IEE depict regions of active cell divisions and growth; white areas depict zones of maturity. The latter areas along the IEE cause buckling that results in cusp formation. Not shown is the basement membrane that forms just subjacent to the IEE. Odontoblasts differentiate and secrete predentin pulpward from the basement membrane, while ameloblasts differentiate and deposit enamel matrix toward the IEE. (*Sources*: Adapted from several sources, primarily Ten Cate, 1994, and Hillson, 1996.)

creating a hollow that is occupied by the condensed mesenchyme, which is the dental papilla (Fig. 18.2D). The dental papilla will differentiate into dentin and pulp. Consequently, the larger the cell mass of the enamel organ and the more quickly it grows, the larger the resulting tooth. It is supposed that reduced tooth dimensions, such as in Down syndrome (Cohen and Winer, 1965; Townsend, 1983) and Turner syndrome (Townsend et al., 1984; Midtbo and Halse, 1994a, 1994b), are the consequence of slowed growth during early tooth development.

Infolding of the organ occurs where the growth rate of the peripheral portions of the structures, particularly at the deepest portions that form the cervical loop, outstrip growth of the deepest central structures. The outer layer of the enamel organ is termed the outer enamel epithelium (Fig. 18.2E). Between this layer and the inner enamel epithelium (IEE) is the stellate reticulum (enamel pulp). Neither the outer enamel epithelium nor the enamel pulp contributes directly to tooth development, though nutrients and building materials of enamel pass through them from adjacent blood vessels (Arey, 1965). The enamel pulp also provides a protected environment for crown formation.

For present purposes, the key stage of development is in the bell stage when the enamel organ has achieved its definitive size and shape as defined by the mature internal dental epithelium (Keene, 1982). Sites in the IEE mature (i.e., cease cell division), and the IEE interacts with the ectomesenchymal cells of the dental papilla to form enamel and dentin. Dentin formation by the odontoblasts precedes enamel formation. Indeed, dentin is necessary for the induction of preameloblasts into ameloblasts to produce enamel (Sperber, 1989).

Number, position, and spatial relationships of these matured zones in the IEE evidently are controlled primarily by the genotype; they define the future number of cusps and crown conformation, which is highly regimented (Scott and Turner, 1997). The internal dental epithelium consists of a single layer of cuboidal cells (much exaggerated in size in Fig. 18.2) lining the inside of the enamel organ. This cell layer is separated from the dental papilla by a basement membrane that becomes the dentinoenamel junction. That is, the basement membrane becomes the interface between odontoblasts forming dentin, thereby displacing the dental papilla rootward, and ameloblasts building enamel prisms that develop toward the stellate reticulum. Ameloblasts develop from the cuboidal cells of the IEE that elongate into columnar cells oriented toward the stellate reticulum. Odontoblasts that have differentiated from mesenchymal cells of the dental papilla form a single layer of columnar cells at the interior margin of the basement membrane. Dentin and enamel both begin their mineralization process at the presumptive cusp tips and progress rootward, though dentin formation keeps ahead of enamel deposition.

The importance of the bell stage is that the subsequent size of the tooth is effectively defined once mineralization occurs. Mineralization starts at the cusp tips, and the sequence of cusp tip calcification probably is determined from the species' phylogenetic history (Kraus and Jordan, 1965). While the individual cusp tips are being mineralized there is still some small opportunity for growth between the cusps and at the cervical loop (i.e., the apical portion of the internal dental epithelium where the cementoenamel junction will eventually develop), but the essence of tooth size and shape has been achieved. Growth of human tooth germs has been measured by Moss and Applebaum (1964) and Butler (1967a, 1967b, 1967c, 1968), and the inflection point when definitive crown size occurs is attained when odontogenesis and amelogenesis have proceeded down the individual cusp slopes and join at the intercuspal fissures, fusing the cusps together with mineralized tissues. The only subsequent growth occurs during crown formation by the continued deposition of enamel. This proceeds in two directions: (1) outward, progressively increasing enamel thickness from the secreting end of the ameloblast to the dentinoenamel junction and (2) cervically as newly differentiated ameloblasts begin secretion, thereby increasing height of the mineralized portion of the crown (Slavkin, 1974; Hillson and Bond, 1997). Crown size is a marker of how well the individual is growing up to attainment of a tooth's bell stage.

18.5 TOOTH SIZE

The various teeth of the primary and permanent dentitions form over a span of several years (Figs. 18.3 and 18.4), and, once formed, crown size is not altered except by attrition, abrasion, or erosion. As such, much of an individual's pre- and postnatal development can be appraised from examination of teeth that mineralized at specific intervals during growth. Transient insults

will be recorded in teeth forming when the stresses occurred; chronic, systemic growth problems will affect multiple teeth. The shortcoming of this record is that different teeth are differentially susceptible to stressors (Fig. 18.5), and the relative (let alone the absolute) differences in susceptibility are poorly understood.

Earlier familial studies were viewed in light of a multifactorial threshold model (Grüneberg, 1952; Fraser, 1970, 1981; Falconer, 1989). This model, in overview, is that a trait with a dichotomous phenotypic expression, such as whether a person has a cleft lip (± cleft palate), can have an underlying continuous liability in the population based on the combinations of multiple genes. Some genotypes would be very resistant to (nonsyndromic) CL ± P, most would have average susceptibilities, and still others, at the other extreme, would be prone to developing CL ± P. A threshold exists somewhere along this genotypically defined continuum of susceptibility, beyond which a cleft occurs but below which it does not. That is, the threshold imposes a phenotypic dichotomy (normal vs. affected; noncleft vs. cleft) on the population's range of genetically modulated liability. Parents

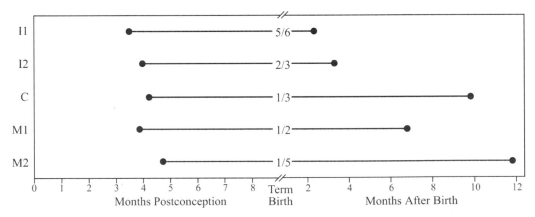

Figure 18.3 Average age spans of permanent tooth crown formation. Symbols to the left (3 to 5 months in utero) denote initial crown mineralization. Fractions at "term birth" are the proportions of the crown mineralized. Symbols to the right (2 to 12 months after birth) denote ages at crown completion. Positions of enamel defects disclose the age when the developmental insult occurred. (*Source*: Data from Lunt and Law, 1974.)

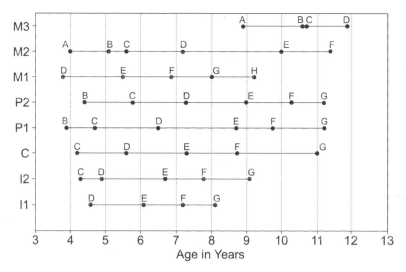

Figure 18.4 Plots of the mean ages for each permanent tooth at which mineralization stages occur. The letter codes are the eight grades described by Demirjian and colleagues (1973). These grades are illustrated in Figure 18.12. (*Source*: Data from Harris and McKee, 1990.)

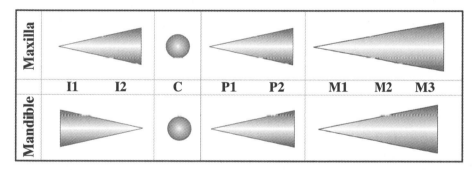

Figure 18.5 There are four morphogenetic fields in the permanent dentition: incisors, canines, premolars, and molars. There are at least two teeth within each field except for the canine. The mesial tooth is the stable or pole tooth in each field (except for the mandibular incisors) based on variability of size and shape. (*Sources*: Adapted from Dahlberg, 1945, 1951.)

with genotypes close to the threshold would have greater risk of having a child with the trait than parents far below the threshold. Parents with the trait would, *prima facie*, have genotypes above the threshold (or, at least, as modulated by the environment) and the risk of affected offspring would be higher still. Details of the quasicontinuous model have been developed by Edwards (1960), Falconer (1965),

and others. Fraser (1970, 1980, 1981, 1989) has reviewed human data on CP and CL ± P in light of a multifactorial threshold model.

The search for causation shifted in the 1990s when a number of groundbreaking studies indicated that "major" genes are likely candidates for many instances of nonsyndromic CL ± P (Carnici et al., 2000; Wong et al., 2000; Prescott et al., 2000).

Prescott and co-workers (2000) suggest that several genes (approximately 2–10) are involved in promoting the risk of CL ± P. Identification of candidate genes is currently a very active area of research, so conjectures and conclusions are bound to change in the near future. The key issue is that if major genes are responsible for a significant portion of cases with CL ± P, then first-degree relatives of probands ought to be phenotypically distinct themselves from the population at large. On the other hand, as noted by Melnick and others (1980; Fraser, 1981, 1989), expectations of a multifactorial threshold model and of a major gene with reduced penetrance and variable expressivity are in many ways coincident.

If it is assumed (as in some studies, but probably unreasonably) that individuals below the threshold in the multifactorial threshold model are phenotypically normal, then analyses of biological relatives of probands ought to help decide between the two models (though they are not mutually exclusive): relatives should be indistinguishable from the population at large

under the threshold model, whereas they should exhibit deviant means and increased phenotypic variances with the major gene model.

18.6 DIMINISHED POSTNATAL GROWTH WITH CLP

Tooth crown dimensions of the permanent teeth have been measured by several researchers to assess whether growth was affected during those intervals when the tooth crowns were forming. There is a broad consensus that tooth size is diminished in the area of the cleft, which is consistent with failed fusion and other underdeveloped, malformed structures local to the cleft.

For the permanent teeth, reduced tooth size in subjects with CLP is a systemic, dentition-wide condition (Swanson et al., 1956; Bimm et al., 1960; Clinch, 1963; Dixon, 1966; Foster and Lavelle, 1971; Werner and Harris, 1989; Haria et al., 2000). All permanent tooth types are affected (Fig. 18.6), though stunting of tooth size is

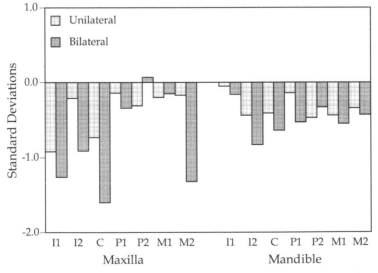

Figure 18.6 Mesiodistal crown diameters of the permanent teeth are smaller in subjects with nonsyndromic complete CLP than in noncleft controls. Stunting of tooth size is greater in cases with bilateral CLP than unilateral CLP (controlling for sexual dimorphism). (*Source*: Data from Werner and Harris, 1989.)

most apparent in the region of the cleft (I1-I2-C). Figure 18.6 underrepresents the stunting for the maxillary lateral incisor because pegged, microform (and, of course, all congenitally absent) teeth were omitted. There seems to be no sex difference in the degree of tooth size reduction in CLP (Foster and Lavelle, 1971; Werner and Harris, 1989); tooth size is diminished proportionately in males and females.

Subjects with bilateral cleft lip and palate have smaller teeth than those with unilateral CLP (Foster and Lavelle, 1971; Werner and Harris, 1989). There does not appear to be any quantitative tooth crown data for subjects with isolated cleft palate, but expectation is that they would be affected only to a minor degree.

Since permanent tooth crown dimensions throughout the dentition are smaller than noncleft controls, it might be concluded that isolated CLP is not just an anatomically localized, transient disruption in development. This is illustrated in Figure 18.7 where the early-forming first molar—which mineralizes its maximum crown diameter at about 1 year of age—is reduced just as much as the second molar that forms its maximum crown size at about 5 years of age (Moorrees et al., 1963). It would seem

that clefting is the overt manifestation of subtle dysplastic influences that affect the whole organism.

One shortcoming of this inference is that none of the permanent teeth are good measures of a subject's *prenatal* growth status. Although some permanent teeth may begin amelogenesis prior to birth (Kronfeld, 1935a, 1935b, 1935c; Kraus and Jordan, 1965; Ooë, 1981), none completes formation of those portions of the crown used to obtain maximum mesiodistal and buccolingual diameters until after birth. Consequently, the observed diminution of the permanent teeth (Foster and Lavelle, 1971; Sofaer, 1979; Werner and Harris, 1989) is confounded by serious *postnatal* health problems facing the cleft infant. Infants with clefts are much more likely to experience feeding problems, upper respiratory infections, and middle ear infections than their noncleft counterparts (Cox, 1960; Drillien et al., 1966; Cooper et al., 1979; Edwards and Watson, 1980). The slowing effect of these problems on growth are well known and are exemplified by the work of Seth and McWilliams (1988), where CLP did not affect birth weight in their study, but postnatal growth slowed significantly.

Timing of the onset of growth problems was tested (Harris, 1993) by examining tooth crown diameters of the primary teeth, which form their crowns mostly before birth (Fig. 18.3). The sample consisted of nonsyndromic children with isolated CLP. Visibly aberrant maxillary lateral incisors were not measured. The primary teeth were generally the same size in the cleft and noncleft samples (Fig. 18.8). Indeed, the arithmetic averages tended to be slightly larger in the cleft sample. Differences only achieved statistical significance in one instance (i.e., the later-forming mandibular m2, where the cleft mean was smaller).

New data on crown dimensions of primary teeth of children with CLP have become available (Narayanan et al., 1999),

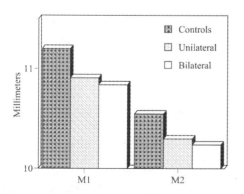

Figure 18.7 Mean mesiodistal crown diameters of the mandibular first and second permanent molars showing the consistency of size reductions in the two cleft samples between early- and late-forming teeth. (*Source*: Data from Werner and Harris, 1989.)

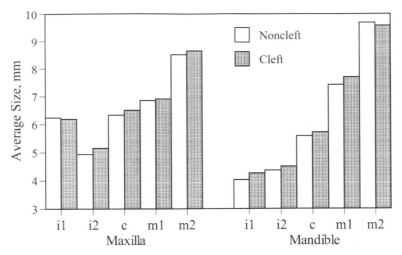

Figure 18.8 Mean mesiodistal crown sizes of deciduous cleft and noncleft children. Sample sizes were 40 cleft and 45 noncleft American white children (Harris, 1993). The clefts were complete uni- and bilateral in nature. Only mandibular m1 showed a statistically significant difference between groups, and it was larger in the cleft sample.

and they are confirmatory of the earlier pilot study (Harris, 1993). Primary crown dimensions are not stunted in size as seen in the permanent teeth that form postnatally; indeed, they tend to be slightly larger than those of noncleft controls (Narayanan et al., 1999). These data from the primary dentition imply that isolated CLP results from anatomically localized instability; aside from abnormal development at the cleft, formative capacity of the fetus seems unimpaired. This is in concert with the fact that most neonates with CLP have no other overt anomaly. The situation changes strikingly at birth as evidenced by the permanent teeth (Fig. 18.6). The newborn with a cleft experiences many feeding and airway problems (Drillien et al., 1966; Edwards and Watson, 1980; Velasco et al., 1988).

Two competing interpretations are that (1) CLP is an overt flag that there are other, albeit subtle, growth problems in children with nonsyndromic clefts or (2) the cleft, an anatomically localized and isolated malformation, creates so many growth-related problems (e.g., recurrent surgeries, increased morbidity, feeding problems, and so on)

(Drillien et al., 1966; Turvey et al., 1996) that the otherwise normal child experiences slowed growth. Unfortunately, the issues are largely confounded and untestable without data from relatives of the cleft probands. Regarding tooth size, relevant data have not been forthcoming; indeed, Crawford and Sofaer (1986) describe a cautionary tale about logistical problems in collecting adequate samples of appropriate family data. There are, however, studies using anthropometry and craniometrics of relatives of nonsyndromic subjects with CLP (Fraser and Pashayan, 1970; Coccaro et al., 1972; Kurisu et al., 1974; Nakasima and Ichnose, 1983; Ward et al., 1989; Mossey et al., 1998; Al Emran et al., 1999) including some valuable studies using animal models (Trasler, 1968; Nonaka et al., 1997). The question is not simple because of etiologic heterogeneity of samples of cleft offspring, but it seems that more attention needs to be paid to a familial, transmissible component to size, size relationships, and tempo of growth in children with CLP.

An informative study regarding the dentition was by Crawford and Sofaer (1987), who measured the left-right asymmetry of two tissue systems: finger and palm prints (dermatoglyphics) and tooth crown dimensions of the permanent teeth (odontometrics). Tooth dimensions, particularly of the later-forming premolars, were most useful in predicting group membership. Crown size fluctuating asymmetry was greatest in the cleft groups (and predictably higher in familial than sporadic cases). Odontometrics were more discriminatory than fingerprints, but buccolingual dimensions and asymmetry of early-forming teeth (incisors, canines) did not enter the model.

Perhaps the dental evidence is unsound. The inferences just put forth can be tested by examining another tissue system where, like teeth, the structures do not change after they form. Digital ridge patterns (fingerprints) are informative since they develop early in the second trimester and remain unchanged thereafter (Hale, 1952; Cummins and Midlo, 1961; Holt, 1968; Stough and Seely, 1969). Ridge patterns provide a unique view of early morphogenesis—when maternal buffering ought to minimize environmental influences.

Balgir (1986; Balgir and Mitra, 1986) scored the fingerprint patterns of subjects with isolated CLP (though he did not distinguish between CLP and CP). The cleft group had significantly simpler patterns (more loops than whorls) than controls. Recomputation of his data (sexes pooled) yields $X^2 = 8.7$ ($p = 0.01$), with a frequency of 58% loops in the CLP group and 51% among controls. The thenar and hypothenar patterns also were significantly less common in the cleft group. Subsequent studies have confirmed these simplifications in ridge patterns (Kanematsu et al., 1986; Kobyliansky et al., 1999).

In a landmark study of fluctuating asymmetry, Adams and Niswander (1967) computed the intragroup variance in the *atd* angle of palm prints (Penrose, 1954). Their CLP group showed significantly greater bilateral asymmetry than controls.

Interpretation is that decrease in developmental canalization in the CLP group stems from the embryonic period of active organogenesis.

Consequently, the picture remains unclear, probably because the critical period of the various tissue systems—when they are most susceptible to developmental stressors—occurs at different age intervals pre- and postnatally and because some structures are more resilient to perturbations of growth than others. The dermatoglyphic data make it clear that individuals with nonsyndromic CLP are not as a group wholly normal outside their cleft. Since the development of finger and palm prints is firmly established by the end of the fourth month in utero, the developmental errors recorded by them are longstanding. Tooth formation may be less sensitive to prenatal stresses than other tissues—or measuring crown diameters may yield data too gross to provide the precision needed to identify prenatal effects. At its face, though, data on tooth size indicate that nonsyndromic clefting involves comparatively little growth inhibition prior to birth, whereas the quality of growth diminishes markedly following birth.

18.7 BILATERAL ASYMMETRY

The conventional and well-tested view of polygenic traits, such as anthropometrics, craniometrics, and odontometrics, is that a person's phenotype derives from the interaction of his genotype with the environment (Waddington, 1957, 1960; Levitan and Montagu, 1977; Futuyma, 1979), where environment is the set of stressors that impact the developing phenotype to a greater or lesser extent (Selye, 1973a, 1973b, 1976). Environmental stress increases the variability of quantitative and morphological characters in a population (Reeve, 1960) because of two interrelated situations: the level of stress that an individual is exposed

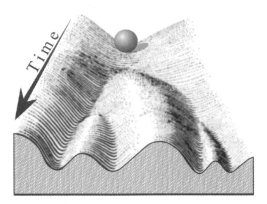

Figure 18.9 The epigenetic landscape. (*Source*: Redrawn from Waddington, 1974.)

to varies, as well as the individual's level of genetic susceptibility to stress.

Waddington (1942, 1957, 1960, 1974) provided an insightful perspective on the effects of varying environments on development. He viewed development as an adaptive landscape (Fig. 18.9), with the hills being the regulatory genetic processes and physiological feedback mechanisms that constrain and canalize an individual's development, which flows down the valleys. In individuals where these hills are larger and higher, growth is more strongly canalized, and it requires greater environmental stress (Selye, 1976) to divert growth from one channel to another, so these people tend to stay in the developmental route. Other people would have genotypes less resilient to stressors, and the analogy with the landscape is that the hills are smaller and shorter, so a stressor that would leave no mark on a well-canalized individual's growth can divert a weakly buffered subject's growth into another developmental, more aberrant channel. Greater and/or repetitive stresses would tend to push development to more peripheral channels, producing more aberrant phenotypes. If faced with sufficient stress, even a resilient genotype's developmental pathway can be altered.

Waddington noted that diversions of a person's developing phenotype should be viewed as threshold events: individuals with well-canalized growth have a steeper topography, so it requires greater stress to perturb their development. Waddington (1957) viewed the epigenetic landscape as a metaphor of developmental pathways for the histogenesis of structures; genetically founded buffering systems would permit minor stresses to be absorbed during development without consequence to the phenotype. The threshold is lower when the genotype is less capable of buffering growth, and lesser stresses can divert the subject to a more peripheral channel.

Historically, the forces of natural selection culled extreme phenotypes from the population, but progress in health care has dramatically curtailed the effects of stabilizing selection in humans (Fogh-Andersen, 1961; Fraser, 1970, 1980; Dronamraju and Bixler, 1983; Miller, 1985).

Mather (1953) elaborated on Waddington's canalization model. Like Waddington, Mather contended that the developmental stability of an organism is grounded in its genotype, and its level of canalization is reflected in the organism's ability to produce an ideal form under a given set of environmental conditions. The ideal form generally cannot be known, but one consequence of diminished canalization is left-right asymmetry—which *is* measurable—and greater asymmetry of inherently bilateral structures is roughly proportional to the combined effects of degraded genotypic canalization and the level of environmental stress experienced during development (Livshits and Kobyliansky, 1991). Tooth size and shape are fundamentally bilateral. Homologous teeth on the left and right sides of the midline normally are mirror images of one another. On finer inspection, though, the fundamental symmetry often is just approximate (Garn et al., 1966, 1967).

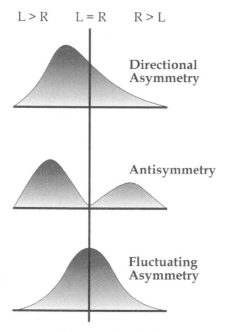

L > R   L = R   R > L

Directional
Asymmetry

Antisymmetry

Fluctuating
Asymmetry

Figure 18.10 Schematic distributions of the three kinds of bilateral asymmetry classified by Van Valen (1962). Directional asymmetry occurs when the left-right side difference summed across the data set departs significantly from zero. Antisymmetry is the same phenomenon but for traits that always occur, but have a side preference (e.g., spleen, appendix positions), or psychometric traits such as handedness. Fluctuating (random) asymmetry is symmetric about zero and is assumed to reflect local (side) differences in the growth capacity of paired structures.

18.8 TYPES OF ASYMMETRY

Van Valen (1962) classified bilateral asymmetries into three kinds: (1) directional asymmetry, (2) antisymmetry, and (3) fluctuating asymmetry (Fig. 18.10).

18.8.1 Directional Asymmetry

Directional asymmetry (DA) occurs when a structure on one side of the body is systematically larger than its antimere, or an unpaired structure characteristically is located to one side of the midline. Several of the visceral organs such as the heart and the spleen are almost invariably sinistral,

and the hemispheres of the cerebellum (Hoadley, 1929; Wada et al., 1975; Corballis and Morgan, 1978a), position of the liver, gallbladder, and appendix, lobation of the lungs, and descent of the testes (Chang et al., 1960; Mittwoch, 1976) are on the right side or are preferentially larger on the right side. These anatomic examples of DA seem to be ingrained in the developmental biology of the organism (Collins, 1977; Boklage et al., 1979; Boklage, 1984) and have a heritable basis (Dahlberg, 1943; Reeve, 1960). In contrast, other differences in laterality, such as size of muscle attachments and diaphyseal bone length and robusticity (Schultz, 1937; Jolicoeur, 1963; Chibber and Singh, 1972), may well be acquired through preferential use or disuse.

Directional asymmetries of the face and skull have received special attention for almost a century. A broad spectrum of characteristics has been examined, including persistent interest in cerebral dominance (Hochberg and LeMay, 1974; Corballis and Morgan, 1978a), facial expressions (Sackeim et al., 1978; Ekman, 1980; Monreal, 1980), sizes of the bones of the calvaria and cranial base (Woo, 1931; Pearson and Woo, 1935), cephalometric facial dimensions (Woo, 1938; Harvold, 1961), and physiologic parameters (Greene, 1931; Rogers, 1958; Pruim, 1979).

Facial clefts in humans are a prime example of directional asymmetry. All large-scale studies have found CL ± P to be twice as common on the left side as the right (Vanderas and Ranalli, 1989; Sayetta et al., 1989), which may be influenced by laterality in brain development (Corballis and Morgan, 1978a, 1978b). The common finding that males more frequently exhibit clefts than females—and that the severity of clefting tends to be more extreme in females—coincides with sex differences in susceptibility (Meskin et al., 1968; Fraser, 1989). Girls would have a higher threshold (greater genetic homeostasis) than boys, so clefts are more common in boys, and less severe developmental insults that cause clefting in boys tend to be buffered in girls, so girls tend to exhibit more severe clefts. It also follows that there should be a higher frequency of affected subjects among the relatives of females with CL ± P than among males with CL ± P; indeed, this difference has been reported by several researchers (Habib, 1978; Woolf, 1971; Crawford and Sofaer, 1987).

Few researchers have studied DA in the human dentition (Mizoguchi, 1986; Sharma et al., 1986; Harris, 1993; Townsend et al., 1999). Of note, there are systematic side preferences in tooth crown diameters, both in the primary and permanent dentitions. Left-sided dominance of crown sizes in one arch (maxilla or mandible) tends to be associated with right-sided dominance in the other. The cause of these systematic side differences remains speculative.

Since most facial clefts are unilateral, it is of interest whether crown dimensions are systematically different on the cleft side compared to the noncleft side. Sidedness does not appear to be the case for teeth developing from the buccal segments (C, P, M), but DA is evident in the premaxilla where the incisors form. Both the maxillary central and lateral incisors are significantly smaller on the cleft side (Hellquist et al., 1979; Werner and Harris, 1989). Variability of the lateral incisor probably is affected in large part by disruption at the cleft site. The central incisor on the cleft side may be diminished because of compromised vascularity (Maher, 1977, 1981).

18.8.2 Antisymmetry

Antisymmetry is exemplified by handedness (Klar, 1999). The great majority of people exhibit right- or left-hand dominance (with a 2-to-1 ratio in most populations); very few people are truly ambidextrous (Voyer, 1998). There are gradations of side dominance, so a population's distribution is bimodal: one mode for

right dominance and one for left dominance, with few people in between (Fig. 18.10). There appears to be no evidence of antisymmetry in the human dentition, though chewing-side preferences (Bouriol and Mioche, 2000) can create acquired antisymmetry in wear patterns and oral health (Rogers, 1958; Marsh et al., 1989; Vanderas and Ranalli, 1989).

18.8.3 Fluctuating Asymmetry

Fluctuating asymmetry (FA), the third type of bilateral asymmetry, is generally viewed as the consequence of diminished canalization of the growth process (Livshits et al., 1988a; Parsons, 1990; Livshits and Kobyliansky, 1991; Livshits and Smouse, 1993; Livshits et al., 1998). FA occurs when a bilateral structure on one side of the organism is larger or has a different morphology than its antimere. There is no side preference with FA; deviations in a population sample ought to sum to zero. Teeth, because they are numerous and do not change after formation, lend themselves to studies of asymmetry (Bailit et al., 1970; DiBennardo and Bailit, 1978; Doyle and Johnston, 1977). The concept is that the same genetic and environmental factors control growth of the left and right structures of the body; reduced homeostasis causes size and shape of the bilateral structures to differ. Greater asymmetry reflects less regulatory control during development. See Palmer (1996; Palmer and Strobeck, 1986) and Livshits (Livshits and Kobylianski, 1987; Livshits and Smouse, 1993) for valuable reviews of methods of quantifying FA.

FA is significantly more frequent and of greater modal expression in subjects with CLP (Fig. 18.11). In the past, this seemed an obvious consequence of the adverse growth secondary to the cleft, such as the stress of repeated surgeries, feeding problems, middle ear disorders, psychosocial problems because of the cleft and speech

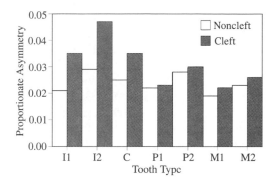

Figure 18.11 Proportionate left-right asymmetry in mesiodistal dimensions of the permanent teeth was larger in the sample of nonsyndromic subjects with complete unilateral and bilateral clefts than in noncleft controls. All cases were American whites. (*Source*: Data from Werner and Harris, 1989.)

problems, and so on. Indeed, these and other postnatal environmental stressors certainly play a role. More recent evidence also implicates a genetic basis for the generalized left-right asymmetry (i.e., reduced developmental homeostasis due to major genes in some cleft individuals) (Carnici et al., 2000; Wong et al., 2000; Prescott et al., 2000). Heterogeneity of the causes of clefting is a persistent problem in research designs; some nonsyndromic nonfamilial cases result from "major" (i.e., detectable) genes, others from a multifactorial genetic disposition, others from environmental agents, and still others from unknown combinations of these factors.

18.8.3.1 Animal Studies The ontogenetic model that bilateral differences result from impaired developmental control has been tested more extensively using the dentition than any other structure (see reviews by Sciulli et al., 1979; Parsons, 1990; Kieser, 1990; Boklage, 1992; Hillson, 1996). The research group led by Michael Siegel has documented a broad range of stressors that amplify FA in the dimensions of rodents' teeth, limb bones, and bones of the craniofacial complexes. These controlled

laboratory studies are fundamental to implicating a cause-to-effect relationship between increased levels of environmental stress and greater levels of FA. Strict comparability of a control group generally has not been obtainable for studies involving humans. The kind of physiological stressors shown to increase FA in growing animals include randomly timed bursts of high-intensity noise (Smookler et al., 1973; Siegel and Doyle, 1975a; Siegel and Smookler, 1973), extremes of hot or cold ambient air temperatures (Riesenfeld, 1973, 1976; Siegel and Doyle, 1975b; Siegel et al., 1977), malnutrition (Sciulli et al., 1979), and various drugs (Siegel et al., 1977; Slavotinek et al., 1996; Samren et al., 1999). Effects of multiple stressors on the same growing animals (with pre- and/or postnatal onsets) are synergistic (Sciulli et al., 1979), though various skeletodental variables are differentially susceptible to a given stressor (Mooney et al., 1985).

The commonality of the response of heightened FA is assumed to be tied to the generic nature of the hypothalamus-hypophysis-adrenocortical axis by the action of somatotrophic hormones, adrenocortical-stimulating hormone, or corticosteroids (Selye, 1973a, 1973b, 1976).

18.8.3.2 Applications Adams and Niswander (1967) were instrumental in applying the concept of reduced developmental homeostasis, as measured by FA, to the human situation. They assessed two tissue systems—namely dermatoglyphic patterns on the palms of the left and right hands and buccolingual crown size of the mandibular first molar in the left and right quadrants—in samples of probands with familial CL ± P, sporadic CL ± P, and a noncleft control group. Both variables are characterized by early formation without subsequent alteration. Within-subject (intrapair) variance due to asymmetry was significantly greater in the sample with familial CL ± P than controls, both for dermatoglyphics and tooth size. Variances also were greater in sporadic CL ± P cases than controls for both variables, but the ratio only achieved significance for the palmar pattern. This was the first research to document increased asymmetry in a congenital malformation. Woolf and Gianas (1976, 1977) subsequently confirmed and extended these findings, emphasizing that familial and sporadic cases of CL ± P are distinct entities: FA is much higher in familial cases as well in their first-degree relatives; FA is closer to controls in sporadic cases, and their relatives are essentially unaffected. Adams and Niswander (1967) concluded that deficiencies in canalization of development resulted in the facial cleft and the amplified asymmetry in print patterns and tooth crown–size mirrors, which decreased developmental buffering.

Homology of the physiological stress response between data on laboratory animal studies (Bader, 1965; Bader and Lehmann, 1965; Sciulli et al., 1979) and humans seems straightforward, but it is confirmatory to see that subjects known to have impaired growth and development do exhibit increased FA. For example, several studies confirm that fluctuating dental asymmetry is significantly higher in Down syndrome (trisomy 21) compared to controls with normal karyotypes (Garn et al., 1970; Barden, 1980; Townsend, 1983; Shapiro, 1983). Similarly, FA is significantly enhanced in subjects with fragile X syndrome (Peretz et al., 1988).

Illnesses of heterogeneous etiology also have been documented to show significantly greater morphological variability and heightened asymmetry, such as mental retardation (Jordan et al., 1966; Malina and Buschang, 1984; Markow and Wandler, 1986) and preterm births (Livshits et al., 1988a, 1988b).

A precept of Waddington and Lerner's models of impaired developmental stability is that there is a genetic basis for the degree of canalization. Heterozygosity enhances stability (reviewed in Livshits and Kobylianski, 1991). Human groups where ancillary evidence implies above-average inbreeding do exhibit reduced tooth size, increased population variance, and increased FA (Bailit, 1966; Ben-David et al., 1992). It is informative, then, that subjects with sporadic as well as familial CL ± P have significantly heightened FA (Adams and Niswander, 1967; Woolf and Gianas, 1976, 1977; Sofaer, 1979; Chung et al., 1980).

Amplified fluctuating asymmetry is pervasive in people with CLP, affecting all tooth types in the permanent dentition (Werner and Harris, 1989)—which is analogous to results seen for tooth size. Asymmetry is disproportionately high in and adjacent to the cleft (I1, I2, C), but there is no arcade difference outside the cleft (Fig. 18.11). No sex difference has been found for the degree of FA in CLP.

18.9 DENTAL AGE

The pace at which the formation of a tooth's crown and root proceeds is under fairly rigorous genetic control, probably more so than the regulation of bone growth (Garn et al., 1965; Tonge and McCance, 1965, 1973). Twin and family studies (Niswander, 1963; Pelsmaekers et al., 1997; Merwin and Harris, 1998) show there is much greater within-family than among-family similarity in the tempos of tooth formation. *Tempo* of growth is a term introduced by Boas (1933) to reflect the analogy with the speed at which a musical score can be performed (e.g., largo, lento, allegro, presto), and, similarly, the biological issue of how quickly an individual proceeds toward maturity. Maturity for a tooth is when the apex of the root(s) closes, which is the last event in defining its final conformation. An adverse environment can slow the tempo of development toward maturity (Eveleth and Tanner, 1990). A slower tempo of maturation normally is adaptive because an adverse environment slows mitotic growth cycles (Winick, 1970, 1971). Growth and maturation are separate but complementary issues. Growth reflects changes in size parameters and generally is linked to mitotic (or secretory) rates. Maturation involves the tempo of development. Slowing the tempo of growth—the speed at which the individual is heading toward maturity—provides a greater duration of time for growth to recover (Tanner, 1963).

The concept of physiological age involves evaluation of a person's degree of maturity assessed from biological criteria. As articulated by Tanner and co-workers (1975):

> Maturity differs in an important way from a measurement such as stature, in that the normal growth process takes every individual from one common condition of being wholly immature to another of being wholly mature. Stature lacks these common end points; a child who is "tall for his age" may be so because he is more mature than his coevals, but he may simply be a tall child of average maturity, who will eventually be a tall adult. Stature and other "size" measures can thus not be used to define maturity, except possibly in retrospect when the adult value is known.

The most common methods of assessing physiological age are to examine morphological changes of bones, notably in the hand-wrist complex (Greulich and Pyle, 1959; Pyle et al., 1971; Tanner et al., 1975) or to determine the staging of crown-root development in the dentition (Schour and Massler, 1940; Smith, 1991). Comparative data can be collected directly using dissection and histology (Kronfeld, 1935a, 1935b; Kraus and Jordan, 1965) or from radiography (Moorrees et al., 1963; Anderson et al., 1976). The two methods are complementary. Histology reveals far more of the developmental details; radiography is noninvasive. What can be seen in histological sections of developing teeth is, perforce,

different than what is imaged radiographically. This creates confusion when the nature of these different kinds of data goes unrecognized. The developing soft tissues are radiolucent, so the amount of crown-root formation seen on radiographs lags behind what is going on at the cellular fronts, and, of course, the level of detail is less refined on x-rays. On the other hand, because radiographs are noninvasive and can be taken at frequent intervals when indicated, they provide a means of dynamically monitoring a child's dental development.

18.9.1 Methods

The continuum of crown-root mineralization can be partitioned into distinguishable, ordinal-grade stages that can be compared to a person's dental status on a tooth-by-tooth basis, thus leading to evaluation of the person's dental age that can be contrasted with his chronological age (Fig. 18.12). Some of the earliest and most detailed work in this area was by Gleiser and Hunt (1955) and Fanning (1961, 1962). Various recording schemes stemming from these early works are available in the literature (Moorrees et al., 1963; Nolla, 1960; Garn et al., 1959; Demirjian et al., 1973). The procedure, in brief, is to visually compare each tooth's image as seen on a radiograph to a grading scheme and record the highest grade that the tooth achieved. Then, with reference to normative data (Moorrees et al., 1963; Anderson et al., 1976; Harris and McKee, 1990), the average chronological age at which the baseline population achieved that grade is recorded. Averaging across the subject's formative teeth provides an estimate of the child's dental age. The procedure is applicable up to 12 to 13 years of age, beyond which few teeth remain uncompleted except the third molars. Stages of third molar formation have been used to estimate the age of young adults (Garn et al., 1962; Levesque

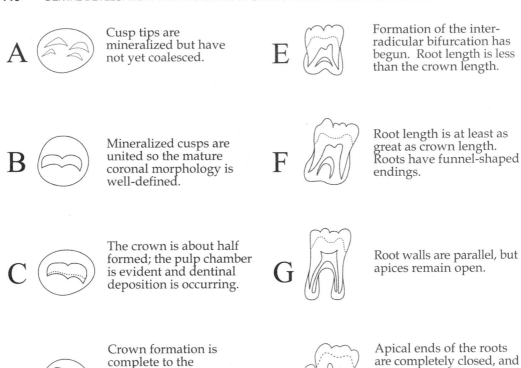

A — Cusp tips are mineralized but have not yet coalesced.

E — Formation of the inter-radicular bifurcation has begun. Root length is less than the crown length.

B — Mineralized cusps are united so the mature coronal morphology is well-defined.

F — Root length is at least as great as crown length. Roots have funnel-shaped endings.

C — The crown is about half formed; the pulp chamber is evident and dentinal deposition is occurring.

G — Root walls are parallel, but apices remain open.

D — Crown formation is complete to the dentinoenamel junction. The pulp chamber has a trapezoidal form.

H — Apical ends of the roots are completely closed, and the periodontal membrane has a uniform width around the root.

Figure 18.12 Schematic illustration and descriptions of the eight-grade scheme developed by Demirjian and co-workers (1973) to score the extent of crown-root formation of the teeth. (*Source*: Modified from Mincer et al., 1993.)

et al., 1981; Thorson and Hägg, 1991; Mincer et al., 1993), but the noteworthy variability of this tooth in humans suggests considerable caution about the precision of the estimate.

There is considerable left-right symmetry (concordance) in tooth formation (Demirjian et al., 1973), so it generally is redundant to score teeth on both sides of the mouth unless there is a reason to suspect a difference. There are also strong statistical associations among tooth types. This observation led Demirjian and co-workers (1973, 1976) to score just selected subsets of teeth in the mandible because there is no appreciable loss of information since the two arches develop in synchrony.

And, because there are fewer osseous structures superimposed on the x-ray images of the mandibular teeth, it is favored over evaluations of maxillary teeth. Other researchers also have proposed aging methods based on a reduced number of teeth (Haavikko, 1974; Bolanos et al., 2000). As an aside, dental age can be calculated from cadaveral or even skeletonized remains using essentially the same grading criteria as with radiographs (Owsley and Jantz, 1983; Liversidge et al., 1993; Liversidge and Molleson, 1999).

Some researchers contend that ordinal grading systems used to assess dental maturity are subjective and involve technical problems and that more informative data are obtainable by measuring tooth length, thereby providing ratio-scale data (Mornstad et al., 1994; Kullman, 1995; Kullman et al., 1995; Liversidge and Molleson, 1999). However, the preeminent merit of a morphological, visually assessed grading scheme is that assessments are size-indifferent. Gradation depends on shape and shape relationships (see written criteria in Fig. 18.12). This focuses on the relative extent of crown-root formation, so large teeth (and those magnified radiographically) are scored against the same criteria as small teeth (and those that are foreshortened on x-ray). There is also the assumption that completed tooth length is the same as the controls, which may be unrealistic. Still, such refinements can make the assessment of dental age more rigorous.

An important methodological concern is the relevance of the few available normative data. Commonly used standards are now decades old (Garn et al., 1956; Moorrees et al., 1963; Haavikko, 1970; Anderson et al., 1976; Hoffding et al., 1984), which raises the issue of whether secular trends (van Wieringen, 1986) that have quickened the tempos of growth in westernized countries have biased these older data (Holtgrave et al., 1997; Nadler, 1998). Probably of greater concern is that the paucity of large normative studies has led to the lulling supposition that major ethnic groups have homogeneous tempos of growth—that it is reasonable and appropriate to use one of the few normative standards as if everyone of that ethnic group grows as described by that sample. For example, it seems to hold as a broad generalization that peoples of sub-Saharan African descent form and erupt their teeth faster than Caucasians, and both have faster tempos of growth than Asians (Eveleth and Tanner, 1990). In fact, though, the few studies that have addressed this issue found considerable regional variation in the tempos of growth (Loevy and Shore, 1985; Nystrom et al., 1988; Staaf et al., 1991;

Mappes et al., 1992; Tompkins, 1996). Finding appropriate normative data in the literature is a common problem that is seldom addressed, but it can easily and substantially affect interpretation of whether a subject with a particular disorder (or a whole sample of subjects) is developing at a normal tempo.

18.9.2 Applications

Menius and colleagues (1966) and Bailit and co-workers (1968) were among the first to show that children with a cleft of the lip and palate were slow growers based on their physiological ages. An important question stemming from these early works was how the clefting related to the delayed development: Was the cleft an obvious bellwether that the child's overall growth was compromised or did the cleft impair the tempo of growth because it itself created problems (e.g., feeding difficulties, increased morbidity, including recurrent middle ear problems)? Bailit and associates were quick to point out that these causes are not mutually exclusive; adverse postnatal problems very likely exacerbate congenital problems.

Ranta (1982) studied a large series of children with CLP at 8 to 9 years of age and found a progression of delayed growth corresponding to severity of the cleft: Average delay was 0.4 years in children with CL, 0.5 years for unilateral CLP, and 0.8 years for bilateral CLP. Delayed growth in children with CP alone (0.6 years) was intermediate. Comparable rankings have been reported by others (Bailit et al., 1968; Pöyry et al., 1989). Ranta (1982, 1984, 1986), as in other studies, did not find a difference in the amount of delay between boys and girls, nor did he find a significant difference between those with and without a family history of clefting. Ranta did point out a substantive difference in the extent of delayed growth based on the ancillary feature of hypodontia (i.e., congenitally missing teeth excluding the cleft region and third molars). Children with a cleft who were also positive for hypodontia were half-again as delayed as those not missing teeth (about an 8% delay compared to about 5% in children without hypodontia). One suggestion is that these three factors—the cleft, delayed growth, and hypodontia—are facets of the same problem of diminished canalization, and those with greater expressions of each are even less well buffered (Bailit et al., 1968; Ranta, 1983).

Delayed dental age is a systemic effect, with all of the formative teeth being delayed in their tempos of maturation (Ranta, 1984, 1986; Pöyry et al., 1989; Harris and Hullings, 1990). It also is frequently remarked that older children with clefts are absolutely more delayed than younger children. Actually this has been observed with various other growth disturbances (Harris et al., 1993; Kaloust et al., 1997) and is attributable to the observation that delay is proportional to age (Garn et al., 1958), so a delay of X% corresponds to a small amount in infancy by much more in adolescence.

Harris and Hullings (1990) calculated dental ages in a sample of nonsyndromic children with complete CLP. The early-forming incisors were not studied. Dental age was delayed in 89% (48/54) of the children. There was no discernible sex difference or difference between unilateral and bilateral cases. The average delay was 0.9 years, which was a 10% retardation compared to the sample's mean chronological age (Fig. 18.13). Tooth-specific assessments showed considerable concordance in the degree of developmental delay between maxillary and mandibular teeth. Of interest, there appeared to be a temporal gradient in the degree of involvement (Fig. 18.14); teeth forming during the early prenatal period (C, M1) were most affected, later-forming elements were less delayed (P1, P2), and the teeth forming latest (M2,

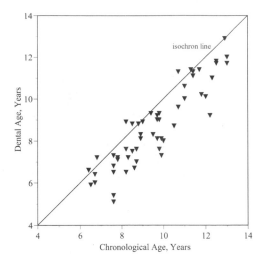

Figure 18.13 Scatterplot showing the characteristic delayed dental age in children with complete clefts (n = 54 children). (*Source*: Data from Harris and Hullings, 1990.)

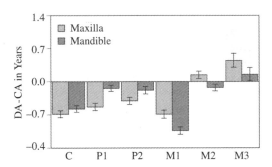

Figure 18.14 Mean differences (±1 SEM) between chronological and dental ages, by tooth, in a series of children with complete uni- or bilateral clefts. (*Source*: Data from Harris and Hullings, 1990.)

M3) were least affected. It is likely that the delay in maxillary canine formation was partly attributable to its location adjacent to the cleft, so it shared in the localized growth distortions of the clefting anomaly, but this would not explain the similar delay of the mandibular canine.

Tooth formation normally occurs synchronously between sides (Demirjian, 1978), but asynchrony is amplified in chil-

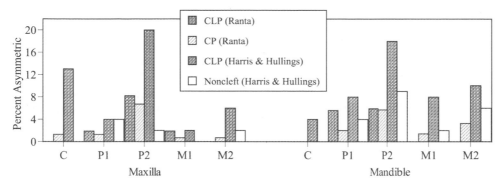

Figure 18.15 Percentages of samples showing asymmetric tooth formation in children with clefts. (*Source*: Data combined from Ranta, 1973, and Harris and Hullings, 1990.)

dren with clefts (Ranta, 1973). This breakdown in development occurs among all tooth types in both jaws, but it is particularly obvious for the second premolar (Fig. 18.15). Asymmetric tooth formation (where homologous teeth differ in formation by a stage or more) is equally common in both jaws once the incisors are omitted (Ranta, 1973; Harris and Hullings, 1990). Frequencies of asymmetric development mirror the severity of the clefting: CLP > CP > noncleft (Fig. 18.15). Ranta and Rintala (1982) suggest that asymmetric formation is linked etiologically with hypodontia. They found that hypodontia of a given tooth is associated with overt delay of formation of the antimeric tooth.

18.10 SYNDROMIC CRANIOSYNOSTOSES

Probably because of the infrequency of syndromes where craniosynostosis is a characteristic feature (e.g., Crouzon, Apert, Pfeiffer, Saethe-Chozten, and others) (Jones, 1997), little is known about the associations of premature suture closure and development of the dentition. Most of the comments in the following sections are based on case reports.

Of note, molecular research is beginning to disclose why people with syndromic craniosynostosis often also exhibit abnormalities of tooth development. Fibroblast growth factors (FGFs) which are known mitogens, concentrate in the enamel knot and underlying mesenchyme during tooth morphogenesis (Jernvall et al., 1994; Jernvall and Jung, 2000). The enamel knot is a localized thickening of the inner enamel epithelium where mitotic activity has ceased (Fig. 18.2E); enamel knots appear to be instrumental in defining the number, locations, and spatial relationships of tooth cusps. FGFs also are expressed in the developing sheets of condensing mesenchyme that form the cranial and facial bones. So, the association of malformed bones and teeth is due to pleiotropic effects of FGFs during embryogenesis (Chan and Thorogood, 1999).

18.10.1 Pfeiffer Syndrome

The obvious signs of Pfeiffer syndrome are brachycephaly brought on by coronal craniosynostosis and partial (generally cutaneous) syndactyly. Cohen (1993) recognized three subtypes with increasing severity of anomalies from types 1 to 3, which aides in prognosis but may not parallel the genetic basis of the condition.

There is a tendency for dental crowding, precocious tooth emergence (natal teeth), and supernumerary teeth. As in all of the syndromic craniosynostoses (Jones, 1997), the midface is hypoplastic, often accentuated by frontal bossing (Pfeiffer, 1964; Asnes and Morehead, 1969; Naveh and Friedman, 1976). This underdevelopment of the maxillae typically leads to an anterior open bite because the palatal plane is canted up anteriorly. Anterior open bites are obvious in the deciduous dentition and become progressively larger with growth unless corrected orthodontically. "Irregularly spaced teeth" is the most common dental finding reported in the literature (Pfeiffer, 1964; Saldino et al., 1972; Naveh and Friedman, 1976), but this is pretty vague.

18.10.2 Apert Syndrome

Apert syndrome (acrocephalosyndactyly) is, as the formal name implies, characterized by a short, high skull shape due to fusion of parts of several calvarial sutures and syndactyly (generally osseous as well as cutaneous) of the hands and feet (Kreiborg and Cohen, 1998; Kreiborg et al., 1999). Underdevelopment of the midface, especially anteroposteriorly, characteristically leads to anterior open bite. Retained deciduous teeth, delayed emergence of permanent teeth, ectopic eruption, impacted teeth, and supernumerary teeth in combination with congenital absence of other teeth are all common findings (Ferraro, 1991; Kreiborg and Cohen, 1992). Kaloust and co-workers (1997) documented delayed tooth formation in Apert syndrome. Because delayed growth generally progresses as a proportion of chronological mature (Garn et al., 1958), the absolute difference increases as patients age. Kreiborg and Cohen (1992) suggest a different growth pattern in Apert syndrome, namely that the tempo of growth becomes progressively delayed during childhood, but then deceleration becomes even more pronounced during adolescence (typically more than 2 years).

Posterior lingual crossbites are common in Apert syndrome—as in all of the syndromic craniosynostoses—because of inadequate mediolateral growth of the maxilla, which also involves a narrow vaulted palate (so-called Byzantine form of the palate in frontal view) with diminished volume of the nasopharynx that predisposes for chronic mouth breathing (Kreiborg and Pruzansky, 1981; Peterson-Falzone et al., 1981).

18.10.3 Crouzon Syndrome

As with other syndromic craniosynostoses, midface hypoplasia is an anticipated finding (Turvey et al., 1979), along with the likelihood of anterior crossbite and pseudomandibular prognathism. Thus, a deficient mandible coupled with a normal-sized mandible gives the impression of a strong longer jaw, but the maxilla is the problem, producing a posterior lingual crossbite and a Class III sagittal molar relationship (Atkinson, 1937).

Dental crowding resulting in ectopic tooth eruption, particularly in the later-emerging teeth, also is common. Tooth rotations and displacements occur more frequently in the upper arch, probably due to its anteroposterior growth deficiency causing greater arch size–tooth size discrepancies (O'Donnell, 1985). Tooth size and morphology of the extant teeth seem to be within normal limits, but no large sample has been analyzed to confirm this. On the other hand, hypodontia and delayed emergence (and nonemergence) of teeth in both the deciduous and permanent dentitions is common (Kelln et al., 1960; Stein and Wahl, 1969).

18.10.4 Saethre-Chozten Syndrome

This autosomal-dominant condition is occasionally termed acrocephalosyndactyly type III. There is little in the literature regarding the dental sequelae of this syndrome. Cranial features are brachycephaly, midface hypoplasia, and syndactyly. Midface hypoplasia is characteristic, along with acrocephaly (Friedman et al., 1977), so the same occlusal problems (anterior open bite, posterior lingual crossbite, pseudomandibular prognathism) seen in other familial craniosynostoses would be anticipated. Wide mesiodistal crown dimensions with narrow, thin roots have been reported, involving all of the permanent teeth (Goho, 1998).

18.11 DENTAL RESPONSE TO STRESS

Tooth structure develops in a well-regimented fashion. This is reflected in the observation that heritability estimates for tooth size as well as the tempo of tooth formation are high (Townsend and Brown, 1978; Pelsmaekers et al., 1997; Dempsey et al., 1995, 1999). Teeth appear to develop in a well-buffered environment, probably more so than bone (Garn et al., 1965; Tonge and McCance, 1965, 1973).

Obviously, though, tooth formation can be perturbed, either by intrinsic or environmental factors. At the cellular level, growth of tooth tissues (enamel, dentin, pulp) responds to the local environment just as other organ systems and structures (Winick, 1970, 1971): a favorable environment (in terms of a sound genotype, proper nutrition, oxygen supply, waste removal, and the like) enhances growth. Stressors, by definition, compromise development. The growth response of a tooth's tissues is, however, generic; there are few conditions that can be identified from their distinctive impact on tooth formation (Itchell and Fahmy, 1968; Hillson et al., 1998; Kumar et al., 2000; Rwenyonyi et al., 2000). Most of the sequelae in Table 18.1 have been documented to occur in subjects with facial clefts and syndromic craniosynostosis (though we know much less about the latter).

A common theme among the majority of dental responses to stress listed in Table 18.1 is diminished growth. The etiology of

clefting in a given sample of subjects with CLP typically is heterogeneous. From case to case, the etiology can be genetic (major gene effects or multifactorial), teratogenic, or poor environment brought on by the condition itself. Regardless, tooth formation responds generically, characteristically by slowing down and losing developmental canalization (Jordan et al., 1966).

18.12 CLP: A DEVELOPMENTAL MODEL

A developmental model suggested by analyses of dental data is shown in Figure 18.16. This paradigm was developed by Sofaer (1979) and can be elaborated on in light of subsequent studies. If CL ± P were an anatomically localized and isolated error of development, tooth size and bilateral tooth crown asymmetry outside of the cleft itself should be the same in those with and without a cleft. Tooth size, for example, would be disrupted at the cleft in the maxilla, perhaps involving the adjacent teeth (i1 and c) as well, but teeth elsewhere in the arches would be indistinguishable from controls. If, instead, clefting is an obvious consequence of a generalized loss of developmental canalization, then the response would be a more-or-less systemic, dentition-wide influence on the teeth—increased hypodontia and asymmetry, diminished tooth size and morphology, and slowed tooth formation (Table 18.1).

The data distinguishing between these alternatives have been reviewed in this chapter. In brief, the data are consistent with the latter scenario, but the effects are dramatically different between the deciduous and permanent dentitions because these two sets of teeth approximate the shift in timing of crown development from prenatal to postnatal ages. The cleft per se does not impact fetal growth as evidenced by normal crown dimensions of the teeth forming in utero (Narayanan et al., 1999).

TABLE 18.1 Common Responses of Teeth to Stressors

Increased incidence of hypodontia (congenital absence of one or more teeth)

Reduced crown and/or root dimensions

Altered crown-root ratios, often because of diminished root length

Increased size variability among tooth types within individuals

Increased size variability among individuals for the same tooth type

Greater size gradients within a morphogenetic field; disproportionate reduction of posterior teeth in a field

Simplified crown morphology; fewer accessory cusps and cingular features

Altered root number (fused roots, accessory roots)

Increased morphological variability

Altered size and shape of pulp chamber

Amplified bilateral asymmetry of homologous teeth

Decreased similarities among siblings and between parents and offspring because of major genes or reduced canalization

Delayed crown and root formation (delayed dental age)

Delayed tooth eruption, including impactions, noneruption, and ectopic eruption

Diminished sexual dimorphism

Increased frequency of enamel defects

Increased frequency of microscopic enamel and dentin disturbances

Prenatal growth is, however, not completely normal, as shown by amplified asymmetry of the deciduous teeth (and other prenatal tissue systems—e.g., dermatoglyphics).

The nonsyndromic cleft child's developmental stability as viewed from the dentition can be schematized as in Figure 18.16 for the permanent dentition. For completeness, there is a narrow band of variability that corresponds to the lack of perfect bilateral symmetry common to everyone (Garn et al., 1966, 1967; Livshits and Smouse, 1993). There is a second narrow band representing (1) the observable but modest increase in FA in the deciduous teeth (Sofaer, 1979), (2) amplified asymmetry local to the cleft, and (3) elevated asymmetry of later-forming teeth, notably m2, which undergo most of their crown formation after birth (Fig. 18.3).

The cleft itself is a major debilitating force after birth (Drillien et al., 1966; Turvey et al., 1996). The permanent teeth, which predominantly form postnatally (Fig. 18.4), reflect the consequences of environmental (plus any genetic) stresses on developmental stability. A host of dental features disclose the obviously harsher postnatal environment, where the cleft itself degrades the child's growth potential:

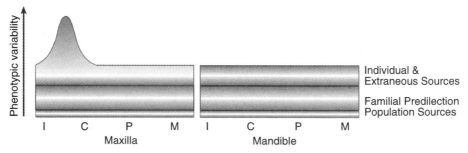

Figure 18.16 Patterns of permanent tooth variability in subjects with CL ± P. The three sources of variability (shaded bars) are cummulative. All individuals show some bilateral asymmetry and other markers of developmental stress, but the magnitude generally is low (bottom bar). Based either on a multifactorial threshold model or one due to major genes, relatives of CL ± P probands show elevated variability, and this greater variability is also expressed by the cleft subjects (middle bar). Individuals with a cleft show amplified dental variability throughout the dentition, though the cleft area is most dysmorphic. (*Source*: Adapted from Sofaer, 1979.)

diminished size, structural abnormalities, delayed and asymmetric growth, amplified bilateral asymmetry. Localized dysplasia at the cleft site continues to be obvious as evaluated from the permanent dentition that completes its crown formation at 6 to 7 years of age (Fig. 18.4).

18.13 PERSPECTIVE

Study of the dentition can play an important role in evaluating a patient's health history because tooth size, shape, and structure preserve a permanent record of the child's ability to thrive during intervals of tooth formation. Enamel and dentin are sensitive recorders of the timing and extent of stress on the child's growth. The twofold intent in this chapter is (1) to describe some of the common methods of extracting data from the dentition as it relates to health history and (2) to review the literature of how dental data have assisted understanding of the timing, tempo, and severity of stressors (intrinsic and acquired) on the cleft child's growth and development. Little is known about tooth development in craniofacial syndromes aside from CLP.

The postnatal environment greatly affects children with clefts as evidenced by reduced tooth size, frequent enamel defects and hypodontia, amplified bilateral asymmetry, and slowed tempo of tooth development. Stressors seem to create most damage during infancy and early childhood; late-forming teeth tend to be less affected, at least according to some criteria such as enamel defects. Other measures (size, tempo of growth) show no evidence of improvement. What is not yet known, though, is whether formative teeth, once stressed, have the capacity for subsequent catch-up growth (Harris et al., 1993).

The prenatal picture—when the cleft per se seems to be inconsequential to the fetus's ability to thrive—remains more problematic. Sizes of the deciduous tooth crowns are at least as large as controls, which parallel the normal body weight and length of most neonates with isolated CL ± P (Becker et al., 1998). Still, family analyses strongly favor the inference that the probands' growth is abnormal *because* the noncleft relatives' growth is abnormal and the risk of more peripheral developmental pathways has been transmitted across generations. More work needs to be done to understand the prognostic risks of subtle growth aberrations such as bilateral asymmetry of skeletodental structures and slowed tempos of maturation.

REFERENCES

Adams, M. S., and Niswander, J. D. (1967). Developmental "noise" and a congenital malformation. *Genet. Res. 10*, 313–317.

AlEmran, S. E., Fatani, E., and Hassanain, J. E. (1999). Craniofacial variability in parents of children with cleft lip and cleft palate. *J. Clin. Pediatr. Dent. 23*, 337–341.

Alvesalo, L., and Tammisalo, E. (1981). Enamel thickness in 45, X females' permanent teeth. *Am. J. Hum. Genet. 33*, 464–469.

Alvesalo, L., and Tammisalo, E. (1985). Enamel thickness in 47, XYY males' permanent teeth. *Ann. Hum. Biol. 12*, 421–427.

Anderson, D. L., Thompson, G. W., and Popovich, F. (1976). Age of attainment of mineralization stages of the permanent dentition. *J. Forensic Sci. 21*, 191–200.

Anderson, P. J., and Moss, A. L. H. (1996). Dental findings in parents of children with cleft lip and palate. *Cleft Palate Craniofac. J. 33*, 436–439.

Arey, L. B. (1965). *Developmental Anatomy: A Textbook and Laboratory Manual of Embryology* (Philadelphia: W. B. Saunders).

Asnes, R. S., and Morehead, C. D. (1969). Pfeiffer syndrome. *Birth Defects Orig. Art. Ser. 5*, 198–203.

Atkinson, F. R. B. (1937). Hereditary craniofacial dysostosis or Crouzon's disease. *Presse Med. 195*, 118–121.

Avery, J. K. (ed.) (1994). *Oral Development and Histology*, 2nd ed. (New York: Thieme Medical Publishers, Inc.).

Bader, R. S. (1965). Fluctuating asymmetry in the dentition of the house mouse. *Growth 29*, 291–300.

Bader, R. S., and Lehmann, W. H. (1965). Phenotypic and genotypic variation in odontometric traits of the house mouse. *Am. Midland Naturalist 74*, 28–38.

Bailit, H. L. (1966). Tooth size variability, inbreeding, and evolution. *Ann. NY Acad. Sci. 134*, 616–623.

Bailit, H. L., Doykos, J. D., and Swanson, L. T. (1968). Dental development in children with cleft palates. *J. Dent. Res. 47*, 664.

Bailit, H. L., Workman, P. L., Niswander, J. D., and MacLean, C. J. (1970). Dental asymmetry as an indicator of genetic and environmental conditions in human populations. *Hum. Biol. 42*, 626–638.

Balgir, R. S. (1986). Dermatoglyphic features in congenital cleft lip and cleft palate anomalies. *Indian Med. Assoc. 84*, 369–372.

Balgir, R. S., and Mitra, S. (1986). Congenital cleft lip and cleft palate anomalies: a dermatoglyphic study. *J. Postgrad. Med. 32*, 18–23.

Barden, H. S. (1980). Fluctuating dental asymmetry: a measure of developmental instability in Down syndrome. *Am. J. Phys. Anthropol. 52*, 169–173.

Becker, M., Svensson, H., and Kallen, B. (1998). Birth weight, body length, and cranial circumference in newborns with cleft lip and palate. *Cleft Palate Craniofac. J. 35*, 255–261.

Ben-David, Y., Hershkovitz, I., Rupin, D., Moscona, D., and Ring, B. (1992). Inbreeding effects on tooth size, eruption age, and dental directional and fluctuating asymmetry among South Sinai Bedouins. In P. Smith and E. Tchernov (eds.), *Structure, Function and Evolution of Teeth* (London: Freund Publishing House, Ltd.), pp. 361–389.

Biggerstaff, R. H. (1969). The basal area of posterior tooth crown components: the assessment of within tooth variations of premolars and molars. *Am. J. Phys. Anthropol. 31*, 163–170.

Bimm, J. A., Eisner, D. A., and Ibanez, C. D. (1960). Cleft palate: morphology of the human mandible. *Am. J. Orthod. 46*, 791–792.

Boas, F. (1933). Studies in growth II. *Hum. Biol. 5*, 429–444.

Böhn, A. (1963). Dental anomalies in harelip and cleft palate. *Acta Odont. Scand. 38* (suppl.), 1–109.

Boklage, C. E. (1984). Differences in protocols of craniofacial development related to twinship and zygosity. *J. Craniofac. Genet. Devel. Biol. 4*, 151–169.

Boklage, C. E. (1992). Method and meaning in the analysis of developmental asymmetries. In J. R. Lukacs (ed.), *Culture, Ecology and Dental Anthropology* (Delhi: Kamla-Raj Enterprises), pp. 147–156.

Boklage, C. E., Elston, R. C., and Potter, R. H. (1979). Cellular origins of functional asymmetries: evidence for schizophrenia, handedness, fetal membranes, and teeth in twins. In J. Gruzelier and P. Flor-Henry (eds.), *Hemisphere Asymmetries of Function in Psychopathology* (Amsterdam: Elsevier/North Holland Biomedical Press), pp. 79–104.

Bolanos, M. V., Manrique, M. C., Bolanos, M. J., and Briones, M. T. (2000). Approaches to chronological age assessment based on dental calcification. *Forensic Sci. Int. 110*, 97–106.

Bouriol, P., and Mioche, L. (2000). Correlations between functional and occlusal tooth-surface areas and food texture during natural chewing sequences in humans. *Arch. Oral Biol. 45*, 691–699.

Boyde, A. (1989). Enamel. In *Handbook of Microscopic Anatomy, Vol. V6, Teeth* (Berlin: Springer-Verlag), pp. 309–473.

Brook, A. H. (1984). A unifying aetiological explanation for anomalies of human tooth number and size. *Arch. Oral Biol. 29*, 373–378.

Butler, P. M. (1967a). The prenatal development of the human upper permanent molar. *Arch. Oral Biol. 12*, 551–563.

Butler, P. M. (1967b). Relative growth within the human first upper permanent molar during the prenatal period. *Arch. Oral Biol. 12*, 983–992.

Butler, P. M. (1967c). Comparison of the development of the second deciduous molar and first permanent molar in man. *Arch. Oral Biol. 12*, 1245–1260.

Butler, P. M. (1968). Growth of the human second lower deciduous molar. *Arch. Oral Biol. 13*, 671–682.

Carnici, F., Pezzetti, F., Scapoli, L., Matinelli, M., Carinci, P., and Tognon, M. (2000). Genetics of nonsyndromic cleft lip and palate: a review of international studies and data regarding the Italian population. *Cleft Palate Craniofac. J. 37*, 33–40.

Chan, C. T. J., and Thorogood, P. (1999). Pleiotropic features of syndromic craniosynostosis correlate with differential expression of fibroblast growth factor receptors 1 and 2 during human craniofacial development. *Pediatr. Res. 45*, 46–53.

Chang, K. S. F., Hsu, F. K., Chan, S. T., and Chan, Y. B. (1960). Scrotal asymmetry and handedness. *J. Anat. 94*, 543–548.

Chibber, S. R., and Singh, I. (1972). Asymmetry in muscle weight in the human upper limb. *Acta Anat. 81*, 462–465.

Chung, C. S., Rao, D. C., and Ching, G. H. (1980). Population and family studies of cleft lip and palate. *Prog. Clin. Biol. Res. 46*, 325–352.

Clinch, L. M. (1963). A longitudinal study of the mesiodistal crown diameter of the deciduous teeth and their permanent successors. *Trans. Eur. Orthod. Soc.* 202–215.

Coccaro, P. J., D'Amico, R., and Chavoor, A. (1972). Craniofacial morphology of parents with and without cleft lip children. *Cleft Palate J. 9*, 28–42.

Cohen, M. M., Jr. (1993). Pfeiffer syndrome update, clinical subtypes, and guidelines for differential diagnosis. *Am. J. Med. Genet. 45*, 300–307.

Cohen, M. M., Jr., and Winer, R. A. (1965). Dental and facial characteristics in Down's syndrome (mongolism). *J. Dent. Res. 44*, 197–208.

Collins, R. L. (1977). Origins of the sense of asymmetry: Mendelian and non-Mendelian models of inheritance. *Ann. NY Acad. Sci. 299*, 283–305.

Cooper, H. K., Hardking, R. L., and Krogman, W. M. (eds.) (1979). *Cleft Palate and Cleft Lip: A Team Approach to Clinical Management and Rehabilitation of the Patient* (Philadelphia: W. B. Saunders).

Corballis, M. C., and Morgan, M. J. (1978a). On the biological basis of human laterality: I. evidence for a maturational left-right gradient. *Behav. Brain. Sci. 2*, 261–269.

Corballis, M. C., and Morgan, M. J. (1978b). On the biological basis of human laterality: II. the mechanisms of inheritance. *Behav. Brain. Sci. 2*, 270–277.

Corliss, C. E. (1976). *Patton's Human Embryology: Elements of Clinical Development* (New York: McGraw-Hill Book Company).

Corruccini, R. S. (1979). Molar cusp-size variability in relation to odontogenesis in hominoid primates. *Arch. Oral Biol. 24*, 633–634.

Corruccini, R. S., and Potter, R. H. (1981). Developmental correlates of crown component asymmetry and occlusal discrepancy. *Am. J. Phys. Anthropol. 55*, 21–31.

Cox, M. A. (ed.) (1960). *Toronto Research Institute, the Cleft Lip and Palate Treatment Centre: A Five Year Report, 1955–1959* (Toronto: Hospital for Sick Children).

Crawford, F. C., and Sofaer, J. A. (1986). Difficulties in obtaining a sample of young adult cleft lip and palate subjects and their relatives. *Commun. Dent. Oral Epidemiol. 14*, 195–197.

Crawford, F. C., and Sofaer, J. A. (1987). Cleft lip with or without cleft palate: identification of sporadic cases with a high level of genetic predisposition. *J. Med. Genet. 24*, 163–169.

Cummins, J., and Midlo, C. (1961). *Finger Prints, Palms and Soles: An Introduction to Dermatoglyphics* (New York: Dover Publications, Inc.).

Czarnecki, E. S., and Carrel, R. (1982). Hemifacial hypertrophy. *J. Pedod. 6*, 270–280.

Dahlberg, A. A. (1945). The changing dentition of man. *J. Am. Dent. Assoc. 32*, 676–690.

Dahlberg, A. A. (1951). The dentition of the American Indian. In W. S. Laughlin (ed.), *Papers on the Physical Anthropology of the American Indian* (New York: Viking Fund, Inc.), pp. 138–176.

Dahlberg, G. (1943). Genotypic asymmetries. *Proc. Roy. Soc. Edinburgh 63*, 20–31.

Dahllöf, G., Ussisoo-Joandi, R., Ideberg, M., and Modeer, T. (1989). Caries, gingivitis, and dental abnormalities in preschool children with cleft lip and/or palate. *Cleft Palate Craniofac. J. 26*, 233–237.

Dahlstrom, L., and Haraldson, T. (1986). Facial asymmetry and unstable occlusion due to poliomyelitis: report of a case. *Swed. Dent. J. 10*, 171–174.

Demirjian, A. (1978). The dentition. In F. Falkner and J. M. Tanner (eds.), *Human Growth, Vol 2. Postnatal Growth* (New York: Plenum Publishers), pp. 413–444.

Demirjian, A., and Goldstein, H. (1976). New systems for dental maturity based on seven and four teeth. *Ann. Hum. Biol. 3*, 411–421.

Demirjian, A., Goldstein, H., and Tanner, J. M. (1973). A new system of dental age assessment. *Hum. Biol. 45*, 211–227.

Dempsey, P. J., Townsend, G. C., Martin, N.G., and Neale, M. C. (1995). Genetic covariance structure of incisor crown size in twins. *J. Dent. Res. 74*, 1389–1398.

Dempsey, P. J., Townsend, G. C., and Martin, N. G. (1999). Insights into the genetic basis of human dental variation from statistical modelling analyses. *Perspect. Hum. Biol. 4*, 9–17.

DiBennardo, R., and Bailit, H. L. (1978). Stress and dental asymmetry in a population of Japanese children. *Am. J. Phys. Anthropol. 48*, 89–94.

Dixon, D. A. (1966). Anomalies of the teeth and supporting structures in children with clefts of lip and palate. In C. M. Drillien, T. X. Ingram, and E. M. Wilkinson (eds.), *The Causes and Natural History of Cleft Lip and Palate* (Edinburgh: E&S Livingstone, Ltd.), pp. 178–205.

Dixon, D. A. (1968). Defects of structure and formation of the teeth in persons with cleft palate and the effect of reparative surgery on the dental tissues. *Oral Surg. Oral Med. Oral Pathol. 25*, 435–446.

Doyle, W. J., and Johnston, O. (1977). On the meaning of increased fluctuating dental asymmetry: a cross populational study. *Am. J. Phys. Anthropol. 46*, 127–134.

Drillien, C. M., Ingram, T. X., and Wilkinson, E. M. (eds.) (1966). *The Causes and Natural History of Cleft Lip and Palate* (Edinburgh: E&S Livingstone, Ltd.).

Dronamraju, K. R., and Bixler, D. (1983). Fetal mortality in oral cleft families: data from Indiana and Montreal. *Clin. Genet. 23*, 350–353.

Ebeling, C. F., Ingervall, B., Hedegard, B., and Lewin, T. (1973). Secular changes in tooth size in Swedish men. *Acta Odont. Scand. 31*, 141–147.

Edwards, J. H. (1960). The simulation of Mendelism. *Acta Genet. Statist. Med. 10*, 63–70.

Edwards, M., and Watson, A. C. H. (1980). *Advances in the Management of Cleft Palate* (New York: Churchill Livingstone).

Ekman, P. (1980). Asymmetry in facial expression. *Science 209*, 833–834.

Eveleth, P. B., and Tanner, J. M. (1990). *Worldwide Variation in Human Growth*, 2nd ed. (Cambridge: Cambridge University Press).

Falconer, D. S. (1965). The inheritance of liability to certain diseases, estimated from the incidence among relatives. *Ann. Hum. Genet. 29*, 51–76.

Falconer, D. S. (1989). *Introduction to Quantitative Genetics*, 3rd ed. (New York: Wiley), pp. 300–312.

Fanning, E. A. (1961). A longitudinal study of tooth formation and root resorption. *N.Z. Dent. J. 57*, 202–217.

Fanning, E. A. (1962). Effect of extraction of deciduous molars on the formation and eruption of their successor. *Angle Orthod. 32*, 44–53.

Fayad, S., and Steffensen, B. (1994). Root resorptions in a patient with hemifacial atrophy. *J. Endodont 20*, 299–303.

Ferraro, N. F. (1991). Dental, orthodontic, and oral/maxillofacial evaluation and treatment of Apert syndrome. *Clin. Plast. Surg. 18*, 291–307.

Fitzgerald, J., and Verveniotis, S. J. (1998). Morquio's syndrome: a case report and review of clinical findings. *NY State Dent. J. 64*, 48–50.

Fitzgerald, L. R. (1973). Deciduous incisor teeth of the mouse *(Mus musculus). Arch. Oral Biol. 18*, 381–389.

Fogh-Andersen, P. (1961). Incidence of cleft lip and palate: constant or increasing? *Acta Chir. Scand. 122*, 106–111.

Foster, T. D., and Lavelle, C. L. B. (1971). The size of the dentition in complete cleft lip and palate. *Cleft Palate J. 8*, 177–184.

Fraser, F. C. (1970). The genetics of cleft lip and cleft palate. *Am. J. Hum. Genet. 22*, 336–352.

Fraser, F. C. (1980). The William Allan memorial awards address: evolution of a palatable multifactorial threshold model. *Am. J. Hum. Genet. 32*, 796–813.

Fraser, F. C. (1981). Genetics Society of Canada award of excellence lecture. The genetics of common familial disorders—major genes or multifactorial? *Can. J. Genet. Cytol. 23*, 1–8.

Fraser, F. C. (1989). Research revisited. *Cleft Palate J. 26*, 255–257.

Fraser, F. C., and Pashayan, H. (1970). Relation of face shape to susceptibility to congenital cleft lip. *J. Med. Genet. 7*, 112–117.

Friedman, J. M., Hanson, J. W., Graham, C. B., and Smith, D. W. (1977). Saethre-Chotzen syndrome: a broad and variable pattern of skeletal malformations. *J. Pediatr. 91*, 929–923.

Fukuhara, T., and Saito, S. (1963). Possible carrier status of hereditary cleft palate with cleft lip: report of cases. *Bull. Tokyo Med. Dent. Univ. 10*, 333–345.

Futuyma, D. J. (1979). *Evolutionary Biology* (Sunderland, MA: Sinauer Associates).

Garn, S. M., Lewis, A. B., and Shoemaker, D. W. (1956). The sequence of mineralization of molar and premolar teeth. *J. Dent. Res. 35*, 555.

Garn, S. M., Lewis, A. B., Koski, K., and Polacheck, D. L. (1958). The sex difference in tooth calcification. *J. Dent. Res. 38*, 561–567.

Garn, S. M., Lewis, A. B., and Polacheck, D. L. (1959). Variability of tooth formation. *J. Dent. Res. 38*, 135–148.

Garn, S. M., Lewis, A. B., and Bonné, B. (1962). Third molar formation and its developmental course. *Angle Orthod. 32*, 270–279.

Garn, S. M., Lewis, A. B., and Vicinius, J. H. (1963). Third molar polymorphism and its significance to dental genetics. *J. Dent. Res. 42*, 1344–1363.

Garn, S. M., Lewis, A. B., and Blizzard, R. M. (1965). Endocrine factors in dental development. *J. Dent. Res. 44*, 243–258.

Garn, S. M., Lewis, A. B., and Kerewsky, R. S. (1966). The meaning of bilateral asymmetry in the permanent dentition. *Angle Orthod. 36*, 55–62.

Garn, S. M., Lewis, A. B., and Kerewsky, R. S. (1967). Buccolingual size asymmetry and its developmental meaning. *Angle Orthod. 37*, 186–193.

Garn, S. M., Cohen, M. M., Jr., and Geciauskas, M. A. (1970). Increased crown-size asymmetry in trisomy G. *J. Dent. Res. 49*, 465.

Gleiser, I., and Hunt, E. E., Jr. (1955). The permanent mandibular first molar: its calcification, eruption and decay. *Am. J. Phys. Anthropol. 13*, 253–284.

Goho, C. (1998). Dental findings in Saethre-Chotzen syndrome (acrocephalosyndactyly type III): a case report. *ASDC J. Dent. Child 65*, 136–137.

Goodman, A. H., and Armelagos, G. J. (1985a). Factors affecting the distribution of enamel hypoplasias within the human permanent dentition. *Am. J. Phys. Anthropol. 68*, 479–493.

Goodman, A. H., and Armelagos, G. J. (1985b). The chronological distribution of enamel hypoplasia in human permanent incisor and canine teeth. *Arch. Oral Biol. 30*, 503–507.

Goodman, A. H., and Rose, J. C. (1990). Assessment of systemic physiological perturbations from dental enamel hypoplasias and associated histological findings. *Yrbk. Phys. Anthropol. 33*, 59–110.

Goodman, A. H., Armelagos, G. J., and Rose, J. C. (1980). Enamel hypoplasias as indicators of stress in three prehistoric populations from Illinois. *Hum. Biol. 52*, 515–528.

Goodman, A. H., Thomas, R. B., Swedlund, A. C., and Armelagos, G. J. (1988). Bicultural perspectives of stress in prehistoric, historical, and contemporary population research. *Yrbk. Phys. Anthropol. 31*, 169–202.

Goose, D. H. (1963). Dental measurement: an assessment of its value in anthropological studies. In D. R. Brothwell (ed.), *Dental Anthropology* (New York: Pergamon Press), pp. 125–148.

Greene, D. (1931). Asymmetry of the head and face in infants and in children. *Am. J. Dis. Child 41*, 1317–1326.

Greulich, W. W., and Pyle, S. I. (1959). *Radiographic Atlas of Skeletal Development of the Hand-Wrist*, 2nd ed. (Stanford: Stanford University Press).

Grüneberg, H. (1952). Genetical studies of the skeleton of the mouse. IV. quasicontinuous variations. *J. Genet. 51*, 95–114.

Haavikko, K. (1970). The formation and the alveolar and clinical eruption of the permanent teeth: an orthopantomographic study. *Proc. Finn. Dent. Soc. 66*, 103–70.

Haavikko, K. (1974). Tooth formation age estimated on a few selected teeth: a simple method for clinical use. *Proc. Finn. Dent. Soc. 70*, 15–19.

Habib, Z. (1978). Factors determining occurrence of cleft palate and cleft palate. *Surg. Gynecol. Obstet. 146*, 105–110.

Hale, A. R. (1952). Morphogenesis of volar skin in the human fetus. *Am. J. Anat. 91*, 147–153.

Haria, S., Noar, J. H., and Sanders, R. (2000). An investigation of the dentition of parents of children with cleft lip and palate. *Cleft Palate Craniofac. J. 37*, 395–405.

Harris, E. F. (1993). Timing of cleft insults as viewed from development of the dentition. *Am. J. Phys. Anthropol. 16* (suppl.), 104.

Harris, E. F., and Hullings, J. G. (1990). Delayed dental development in children with isolated cleft lip and palate. *Arch. Oral Biol. 35*, 469–473.

Harris, E. F., and McKee, J. H. (1990). Tooth mineralization standards for blacks and whites from the middle southern United States. *J. Forensic Sci. 35*, 859–872.

Harris, E. F., Barcroft, B. D., Haydar, S., and Haydar, B. (1993). Delayed tooth formation in low birthweight American black children. *Pediatr. Dent. 15*, 30–35.

Harris, E. F., Hicks, J. D., and Barcroft, B. D. (1999). Absence of sexual dimorphism in enamel thickness of human deciduous molars. In J. T. Mayhall and T. Heikkinen (eds.), *Dental Morphology* (Oulu, Finland: University of Oulu Press), pp. 338–349.

Harvold, E. P. (1961). Asymmetries of the upper facial skeleton and their morphological significance. *Trans. Eur. Orthod. Soc. 25*, 63–78.

Hellquist, R., Linder-Aronson, S., Norling, M., Ponten, B., and Stenberg, T. (1979). Dental abnormalities in patients with alveolar clefts, operated upon with or without primary periosteoplasty. *Eur. J. Orthod. 1*, 169–180.

Hillson, S. (1996). *Dental Anthropology* (Cambridge: Cambridge University Press).

Hillson, S., and Bond, S. (1997). Relationship of enamel hypoplasia to the pattern of tooth crown growth: a discussion. *Am. J. Phys. Anthropol. 104*, 89–103.

Hillson, S., Grigson, C., and Bond, S. (1998). Dental defects of congenital syphilis. *Am. J. Phys. Anthropol. 107*, 25–40.

Hoadley, M. F. (1929). Measurement of internal diameters of the skull in relation to the "pre-eminence" of the left hemisphere. *Biometrika 21*, 94–123.

Hochberg, F. H., and LeMay, M. (1974). Arteriographic correlates of handedness. *Neurology 25*, 218–222.

Hoffding, J., Maeda, M., Yamaguchi, K., Tsuji, H., Kuwabara, S., Nohara, Y., and Yoshida, S. (1984). Emergence of permanent teeth and onset of dental stages in Japanese children. *Commun. Dent. Oral Epidemiol. 12*, 55–58.

Holt, S. B. (1968). *The Genetics of Dermal Ridges* (Springfield: Charles C. Thomas).

Holtgrave, E. A., Kretschmer, R., and Muller, R. (1997). Acceleration in dental development: fact or fiction. *Eur. J. Orthod. 19*, 703–710.

Itchell, D. F., and Fahmy, H. (1968). A test of seven agents for vital dye and intrinsic dental staining activity. *J. Oral Ther. Pharmacol. 4*, 378–381.

Jacobson, A. (1955). Embryological evidence for the non-existence of the premaxilla in man. *J. Dent. Assoc. South Africa 10*, 189–210.

Jernvall, J. (2000). Linking development with generation of novelty in mammalian teeth. *Proc. Natl. Acad. Sci. USA 97*, 2641–2645.

Jernvall, J., and Jung, H. S. (2000). Genotype, phenotype, and developmental biology of molar tooth characters. *Yrbk. Phys. Anthropol. 43*, 171–190.

Jernvall, J., Kettunen, P., Karavanova, I., Martin, L. B., and Thesleff, I. (1994). Evidence for the role of the enamel knot as a control center in mammalian tooth cusp formation: nondividing cells express growth stimulating Fgf-4 gene. *Int. J. Devel. Biol. 38*, 463–469.

Jolicoeur, P. (1963). Bilateral symmetry and asymmetry in limb bones of *Martes americana* and man. *Can. Rev. Biol. 22*, 409–432.

Jones, K. L. (1997). *Smith's Recognizable Patterns of Human Malformation*, 5th ed. (Philadelphia: W. B. Saunders).

Jordan, R. E., Kraus, B. S., and Neptune, C. M. (1966). Dental abnormalities associated with cleft lip and/or palate. *Cleft Palate J. 3*, 22–55.

Kaloust, S., Kazuhiro, I., and Vargervik, K. (1997). Dental development in Apert syndrome. *Cleft Palate Craniofac. J. 34*, 117–121.

Kanematsu, N., Yoshida, Y., Kishi, N., Kawata, K., Kaku, M., Maeda, K., Taoka, M., and Tsutsui, H. (1986). Study on abnormalities in the appearance of finger and palm prints in children with cleft lip, alveolus, and palate. *J. Maxillofac. Surg. 14*, 74–82.

Keene, H. J. (1982). The morphogenetic triangle: a new conceptual tool for application to problems in dental morphogenesis. *Am. J. Phys. Anthropol. 59*, 281–287.

Kelln, E. C., Chaudhry, A. P., and Gorlin, F. J. (1960). Oral manifestations of Crouzon's disease. *Oral Surg. Oral Med. Oral Pathol. 13*, 1245–1248.

Kieser, J. A. (1990). *Human Adult Odontometrics: The Study of Variation in Adult Tooth Size* (Cambridge: Cambridge University Press).

Kieser, J. A., and Groeneveld, H. T. (1988). Patterns of variability in South African Negro dentition. *J. Dent. Assoc. South Africa 43*, 105–110.

Kieser, J. A., Groeneveld, H. T., McKee, J., and Cameron, N. (1990). Measurement error in human dental mensuration. *Ann. Hum. Biol. 17*, 523–528.

Klar, A. J. (1999). Genetic models for handedness, brain lateralization, schizophrenia, and manic-depression. *Schizophr. Res. 39*, 207–218.

Kobyliansky, E., Bejerano, M., Yakovenko, K., and Katznelson, M. B. (1999). Relationship between genetic anomalies of different levels and deviations in dermatoglyphic traits. Part 6: Dermatoglyphic peculiarities of males and females with cleft lip (with or without cleft palate) and cleft palate—family study. *Coll. Anthropol. 23*, 1–51.

Kogon, S. L., Jarvis, A. M., Daley, T. D., and Kane, M. F. (1984). Hemifacial hypertrophy affecting the maxillary dentition: report of a case. *Oral Surg. Oral Med. Oral Pathol. 58*, 549–553.

Kondo, S., Wakatsuki, E., Sun-Te, H., Sheng-Yen, C., Shibazaki, Y., and Arai, M. (1996). Comparison of the crown dimensions between the maxillary second deciduous molar and the permanent first molar. *Okajimas Folia Anat. Jpn. 73*, 179–184.

Kotilainen, J., Holtta, P., Mikkonen, T., Arte, S., Sipila, I, and Pirinen, S. (1995). Craniofacial and dental characteristics of Silver-Russell syndrome. *Am. J. Med. Genet. 56*, 229–236.

Kraus, B. S., and Jordan, R. E. (1965). *The Human Dentition Before Birth* (Philadelphia: Lea and Febiger).

Kraus, B. S., Jordan, R. E., and Pruzansky, S. (1966). Dental abnormalities in the deciduous and permanent dentition of individuals with cleft lip and palate. *J. Dent. Res. 45*, 1736–1746.

Kreiborg, S., and Cohen, M. M., Jr. (1992). The oral manifestations of Apert syndrome. *J. Craniofac. Genet. Devel. Biol. 12*, 41–48.

Kreiborg, S., and Cohen, M. M., Jr. (1998). Is craniofacial morphology in Apert and Crouzon syndromes the same? *Acta Odont. Scand. 56*, 339–341.

Kreiborg, S., and Pruzansky, S. (1981). Craniofacial growth in patients with premature craniosynostosis. *Scand. J. Plast. Reconstr. Surg. 15*, 171–186.

Kreiborg, S., Aduss, H., and Cohen, M. M., Jr. (1999). Cephalometric study of the Apert syndrome in adolescence and adulthood. *J. Craniofac. Genet. Devel. Biol. 19*, 1–11.

Kreshover, S. J. (1944). The pathogenesis of enamel hypoplasia: an experimental study. *J. Dent. Res. 23*, 231–238.

Kreshover, S. J. (1960). Metabolic disturbances in tooth formation. *Ann. NY Acad. Sci. 86*, 161–167.

Kreshover, S. J., Clough, O. W., and Hancock, J. A. (1954). Vaccinia infection in pregnant rabbits and its effect on maternal and fetal dental tissues. *J. Am. Dent. Assoc. 49*, 549–562.

Kronfeld, R. (1935a). Development and calcification of the human deciduous dentition. *The Bur. 15*, 18–25.

Kronfeld, R. (1935b). First permanent molar: its condition at birth and its postnatal development. *J. Am. Dent. Assoc. 22*, 1131–1155.

Kronfeld, R. (1935c). Postnatal development and calcification of the anterior permanent teeth. *J. Am. Dent. Assoc. 22*, 1521–1536.

Kullman, L. (1995). Accuracy of two dental and one skeletal age estimation method in Swedish adolescents. *J. Forensic Sci. Int. 30*, 225–236.

Kullman, L., Martinsson, T., Zimmerman, M., and Welander, U. (1995). Computerized measurements of the lower third molar related to chronologic age in young adults. *Acta Odont. Scand. 53*, 211–216.

Kumar, J., Swango, P., Haley, V., and Green, E. (2000). Intra-oral distribution of dental fluorosis in Newburgh and Kingston, New York. *J. Dent. Res. 79*, 1508–1513.

Kurisu, K., Niswander, J. D., Johnston, M. C., and Mazaheri, M. (1974). Facial morphology as an indicator of genetic predisposition to cleft lip and palate. *J. Hum. Genet. 26*, 702–714.

Laatikainen, T., and Ranta, R. (1996a). Occurrence of the Carabelli trait in twins discordant or concordant for cleft lip and/or palate. *Acta Odont. Scand. 54*, 365–368.

Laatikainen, T., and Ranta, R. (1996b). Taurodontism in twins with cleft lip and/or palate. *Eur. J. Oral Sci. 104*, 82–86.

Laatikainen, T., and Ranta R. (1996c). Hypodontia in twins discordant or concordant for cleft lip and/or palate. *Scand. J. Dent. Res. 102*, 88–91.

Levesque, G. Y., Demirjian, A., and Tanguay. R. (1981). Sexual dimorphism in the development, emergence, and agenesis of the mandibular third molar. *J. Dent. Res. 60*, 1735–1741.

Levitan, M., and Montagu, A. (1977). *Textbook of Human Genetics*, 2nd ed. (New York: Oxford University Press).

Liversidge, H. M., and Molleson, T. I. (1999). Developing permanent tooth length as an estimate of age. *J. Forensic Sci. 44*, 917–920.

Liversidge, H. M., Dean, M. C., and Molleson, T. I. (1993). Increasing human tooth length between birth and 5.4 years. *Am. J. Phys. Anthropol. 90*, 307–313.

Livshits, G., and Kobyliansky, E. (1987). Dermatoglyphic traits as possible markers of developmental processes in humans. *Am. J. Med. Genet. 26*, 111–122.

Livshits, G., and Kobyliansky, E. (1991). Fluctuating asymmetry as a possible measure of developmental homeostasis in humans: a review. *Hum. Biol. 63*, 441–466.

Livshits, G., and Smouse, P. E. (1993). Multivariate fluctuating asymmetry in Israeli adults. *Hum. Biol. 65*, 547–578.

Livshits, G., Davidi, L., Kobyliansky, E., Ben-Amitai, D., Levi, Y., and Merlob, P. (1988a). Decreased developmental stability as assessed by fluctuating asymmetry of morphometric traits in preterm infants. *Am. J. Med. Genet. 29*, 793–805.

Livshits, G., Davidi, L., Kobyliansky, E., Ben-Amitai, D., Levi, Y., and Merlob, P. (1988b). Some biological and social factors of risk associated with birth of preterm infants. *Genet. Epidemiol. 5*, 137–149.

Livshits, G., Yakovenko, K., Kletselman, L., Karasik, D., and Kobyliansky, E. (1998). Fluctuating asymmetry and morphometric variation of hand bones. *Am. J. Phys. Anthropol. 107*, 125–136.

Loevy, H. T., and Shore, S. W. (1985). Dental maturation in hemifacial microsomia. *J. Craniofac. Genet. Devel. Biol. 1* (suppl.), 267–272.

Luke, D. A., Tonge, C. H., and Reid, D. J. (1981). Effects of rehabilitation on the jaws and teeth of protein-deficient and calorie-deficient pigs. *Acta Anat. 110*, 299–305.

Lunt, R. C., and Law, D. B. (1974). A review of the chronology of calcification of deciduous teeth. *J. Am. Dent. Assoc. 89*, 599–606.

Macho, G. A. (1995). The significance of hominid enamel thickness for phylogenetic and life-history reconstruction. In J. Moggi-Cecchi (ed.), *Aspects of Dental Biology, Palaeontology, Anthropology and Evolution* (Florence: Angelo Pontecorbi), pp. 51–68.

Maher, W. P. (1977). Distribution of palatal and other arteries in cleft and non-cleft human palates. *Cleft Palate J. 14*, 1–12.

Maher, W. P. (1981). Artery distribution in the prenatal human maxilla. *Cleft Palate J. 18*, 51–58.

Malanczuk, T., Opitz, C., and Retzlaff, R. (1999). Structural changes of dental enamel in both dentitions of cleft lip and palate patients. *Orofac. Orthop. 60*, 259–268.

Malina, R. M., and Buschang, P. H. (1984). Anthropometric asymmetry in normal and mentally retarded males. *Ann. Hum. Biol. 11*, 515–531.

Mappes, M. S., Harris, E. F., and Behrents, R. G. (1992). An example of regional variation in the tempos of tooth mineralization and hand-wrist ossification. *Am. J. Orthod. Dentofac. Orthop. 101*, 145–151.

Markow, T. A., and Wandler, K. (1986). Fluctuating dermatoglyphic asymmetry and the genetics of liability to schizophrenia. *Psychiatry Res. 19*, 323–328.

Marsh, J. L., Baca, D., and Vannier, M. W. (1989). Facial musculoskeletal asymmetry in hemifacial microsomia. *Cleft Palate J. 26*, 292–302.

Mass, E., Zilberman, U., and Gadoth, N. (1996). Abnormal enamel and pulp dimensions in familial dysautonomia. *J. Dent. Res. 75*, 1747–1752.

Mather, K. (1953). Genetical control of stability in development. *Heredity 7*, 297–336.

Melfi, R. C. (1994). *Permar's Oral Embryology and Microscopic Anatomy*, 9th ed. (Philadelphia: Lea & Febiger).

Melnick, M., Shields, E. D., and Bixler, D. (1980). Studies of cleft lip and cleft palate in the population of Denmark. In M. Melnick, D. Bixler, and E. D. Shields (eds.), *Etiology of Cleft Lip and Cleft Palate* (New York: Alan R. Liss), pp. 225–248.

Menius, J. A., Largent, M. D., and Vincent, C. J. (1966). Skeletal development of cleft palate children as determined by hand-wrist roentgenographs: a preliminary study. *Cleft Palate J. 3*, 67–75.

Merwin, D. R., and Harris, E. F. (1998). Sibling similarities in the tempo of tooth mineralization. *Arch. Oral Biol. 43*, 205–210.

Meskin, L. H., Pruzansky, S., and Gullen, W. H. (1968). An epidemiologic investigation of factors related to the extent of facial clefts: 1. sex of patient. *Cleft Palate J. 5*, 23–29.

Midtbo, M., and Halse, A. (1994a). Tooth crown size and morphology in Turner syndrome. *Acta Odont. Scand. 52*, 7–19.

Midtbo, M., and Halse, A. (1994b). Root length, crown height, and root morphology in Turner syndrome. *Acta Odont. Scand. 52*, 303–314.

Miller, A. (1985). Infant mortality in the U.S. *Sci. Am. 253*, 21–27.

Millhon, J., and Stafne, E. (1941). Incidence of supernumerary and congenitally missing lateral incisor teeth in eighty-one cases of harelip and cleft palate. *Am. J. Orthod. 27*, 599–604.

Mills, L. F., Niswander, J. D., Mazaheri, M., and Brunelle, J. A. (1968). Minor oral and facial defects in relatives of oral cleft patients. *Angle Orthod. 38*, 199–204.

Mincer, H. H., Harris, E. F., and Berryman, H. E. (1993). The A.B.F.O. study of third molar development and its use as an estimator of chronological age. *J. Forensic Sci. 38*, 379–390.

Mittwoch, U. (1976). Lateral asymmetry and the function of the mammalian Y chromosome. In K. Jones and P. E. Brandham (eds.), *Current Chromosome Research* (Amsterdam: North-Holland Publishing Company), pp. 195–201.

Mizoguchi, Y. (1986). Correlated asymmetries detected in the tooth crown diameters of human permanent teeth. *Bull. Natl. Sci. Mus. (Tokyo) 12*, 24–45.

Monreal, F. J. (1980). Asymmetric crying facies: an alternative explanation. *Pediatrics 65*, 146–149.

Mooney, M. P., Siegel, M. I., and Gest, T. R. (1985). Prenatal stress and increased fluctuating asymmetry in the parietal bones of neonatal rats. *Am. J. Phys. Anthropol. 68*, 131–134.

Moorrees, C. F. A. (1957). *The Aleut Dentition: A Correlative Study of Dental Characteristics in an Eskimoid People* (Cambridge: Harvard University Press).

Moorrees, C. F. A., Fanning, E. A., and Hunt, E. E., Jr. (1963). Age variation of formation stages for ten permanent teeth. *J. Dent. Res. 42*, 1490–1502.

Mornstad, H., Staaf, V., and Welander, U. (1994). Age estimation with the aid of tooth development: a new method based on objective measurements. *Scand. J. Dent. Res. 102*, 137–143.

Moss, M. L., and Applebaum, E. (1964). Differential growth analysis of vertebrate teeth. *Am. J. Orthod. 48*, 504–529.

Mossey, P. A., McColl, J., and O'Hara, M. (1998). Cephalometric features in the parents of children with orofacial clefting. *Br. J. Oral Maxillofac. Surg. 36*, 202–212.

Moss-Salentijn, L., and Moss, M. L. (1975). Studies on dentin. 2. Transient vasodentin in the incisor teeth of a rodent *(Perognathus longimembris). Acta Anat. 91*, 386–404.

Nadler, G. L. (1998). Earlier dental maturation: fact or fiction. *Angle Orthod. 68*, 535–538.

Nagai, I., Fujiki, Y., Fuchihata, H., and Yohimoto, T. (1965). Supernumerary tooth associated with cleft lip and palate. *J. Am. Dent. Assoc. 70*, 642–647.

Nakasima, A., and Ichnose, M. (1983). Characteristics of craniofacial structures in parents of children with cleft lip and/or palate. *Am. J. Orthod. 84*, 140–146.

Nanda, R. S. (1954). Agenesis of the third molar in man. *Am. J. Orthod. 40*, 698–706.

Narayanan, A., Smith, S., and Townsend, G. (1999). Dental crown size in individuals with cleft lip and palate. *Perspectives Hum. Biol. 4*, 61–70.

Naveh, Y., and Friedman, A. (1976). Pfeiffer syndrome: report of a family and review of the literature. *J. Med. Genet. 13*, 277–280.

Niswander, J. D. (1963). Effects of heredity and environment on the development of dentition. *J. Dent. Res. 42*, 1288–1296.

Nolla, C. M. (1960). Development of the permanent teeth. *ASDC J. Dent. Child. 27*, 254–266.

Nonaka, K., Sasaki, Y., Watanabe, Y., Yanagita, K., and Nakata, M. (1997). Effects of fetus weight, dam strain, dam weight, and litter size on the craniofacial morphogenesis of LC/Fr mouse fetuses affected with cleft lip and palate. *Cleft Palate Craniofac. J. 34*, 325–330.

Nystrom, M., Ranta, R., and Silvola, H. (1988). Comparisons of dental maturity between the rural community of Kuhmo in northeastern Finland and the city of Helsinki. *Commun. Dent. Oral Epidemiol. 16*, 215–217.

O'Donnell, D. (1985). Dental management problems related to self-image in Crouzon's syndrome. *Aust. Dent. J. 30*, 355–357.

Olin, W. H. (1964). Dental anomalies in cleft lip and palate patients. *Angle Orthod. 34*, 119–123.

Ooë, T. (1981). *Human Tooth and Dental Arch Development* (Tokyo: Ishiyaku Publishers, Inc.).

Osborn, J. W., and Ten Cate, A. R. (1976). *Advanced Dental Histology*, 3rd ed. (Bristol: John Wright and Sons, Ltd.).

Owsley, D. W., and Jantz, R. L. (1983). Formation of the permanent dentition in Arikara Indians: timing differences that affect dental age assessments. *Am. J Phys. Anthropol. 61*, 467–471.

Palmer, A. R. (1996). From symmetry to asymmetry: phylogenetic patterns of asymmetry variation in animals and their evolutionary significance. *Proc. Natl. Acad. Sci. USA 93*, 4279–4286.

Palmer, A. R., and Strobeck, C. (1986). Fluctuating asymmetry: measurement, analysis, patterns. *Ann. Rev. Ecol. Syst. 17*, 391–421.

Parsons, P. A. (1990). Fluctuating asymmetry: an epigenetic measure of stress. *Biol. Rev. 65*, 131–145.

Paurazas, S. B., Geist, J. R., Pink, F. E., Hoen, M. M., and Steiman, H. R. (2000). Comparison of diagnostic accuracy of digital imaging by using CCD and CMOS-PAS sensors with E-speed film in the detection of periapical bony lesions. *Oral Surg. Oral Med. Oral Pathol. Oral Radiol. Endod. 89*, 356–362.

Pearson, K., and Woo, T. L. (1935). Further investigation of the morphometric characters of the individual bones of the human skull. *Biometrika 27*, 424–465.

Pelsmaekers, B., Loos, R., Carels, C., Derom, C., and Vlietinck, R. (1997). The genetic contribution to dental maturation. *J. Dent. Res. 76*, 1337–1340.

Penrose, L. S. (1954). The distal triradius *t* on the hands of parents and sibs of mongol imbeciles. *Ann. Hum. Genet. 19*, 10–38.

Peretz, B., and Smith, P. (1993). Morphometric variables of developing primary mandibular second molars. *Arch. Oral Biol. 38*, 745–749.

Peretz, B., Ever-Hadani, P., Casamassimo, P., Eidelman, E., Shellhart, C., and Hagerman, R. (1988). Crown size and asymmetry in males with FRA(X) or Martin-Bell syndrome. *Am. J. Med. Genet. 30*, 185–190.

Peretz, B., Shapira, J., Farbstein, H., Arieli, E., and Smith, P. (1996). Modification of tooth size and shape in Down's syndrome. *J. Anat. 188*, 167–172.

Peretz, B., Katzenel, V., and Shapira, J. (1999). Morphometric variables of the primary second molar in children with Down syndrome. *J. Clin. Pediatr. Dent. 23*, 333–336.

Peterson-Falzone, S. J., Pruzansky, S., Parris, P. J., and Laffer, J. L. (1981). Nasopharyngeal dysmorphology in the syndromes of Apert and Crouzon. *Cleft Palate J. 18*, 237–250.

Pfeiffer, R. A. (1964). Dominant erbliche akrocephalosyndakylie. *Zeitschr. Kinderheitkunde 90*, 301–320.

Pollock, R. A., Newman, M. H., Burdi, A. R., and Condit, D. P. (1985). Congenital hemifacial hyperplasia: an embryologic hypothesis and case report. *Cleft Palate J. 22*, 173–184.

Pöyry, M., Nystrom, M., and Ranta, R. (1989). Tooth development in children with cleft lip and palate: a longitudinal study from birth to adolescence. *Eur. J. Orthod. 11*, 125–130.

Prescott, N. J., Lees, M. M., Winter, R. M., and Malcom, S. (2000). Identification of susceptibility loci for nonsyndromic cleft lip with or without cleft palate in a two stage genome scan of affected sib-pairs. *Hum. Genet. 106*, 345–350.

Pruim, G. J. (1979). Asymmetries of bilateral static bite forces in different locations on the human mandible. *J. Dent. Res. 58*, 1685–1687.

Pyle, S. I., Waterhouse, A. M., and Greulich, W. W. (1971). *A Radiographic Standard of Reference for the Growth of the Hand and Wrist* (Chicago: Year Book Medical Publishers, Inc.).

Ranta, R. (1973). Development of asymmetric tooth pairs in the permanent dentition of cleft-affected children. *Proc. Finn. Dent. Soc. 69*, 71–75.

Ranta, R. (1982). Comparison of tooth formation in noncleft and cleft-affected children with and without hypodontia. *ASDC J. Dent. Child. 49*, 197–199.

Ranta, R. (1983). Hypodontia and delayed development of the second premolars in cleft palate children. *Eur. J. Orthod. 5*, 145–148.

Ranta, R. (1984). Associations of some variables to tooth formation in children with isolated cleft palate. *Scand. J. Dent. Res. 92*, 496–502.

Ranta, R. (1986). A review of tooth formation in children with cleft lip/palate. *Am. J. Orthod. Dentofac. Orthop. 90*, 11–18.

Ranta, R., and Rintala, A. (1982). Tooth anomalies associated with congenital sinuses of the lower lip and cleft lip/palate. *Angle Orthod. 52*, 212–221.

Ranta, R., Stegars, T., and Rintala, A. E. (1983). Correlations of hypodontia in children with isolated cleft palate. *Cleft Palate J. 20*, 163–165.

Reeve, E. C. R. (1960). Some genetic tests on asymmetry of sternopleural chaeta number in Drosophila. *Genet. Res. 1*, 151–172.

Riesenfeld, A. (1973). The effects of extreme temperature and starvation on the body proportions of the rat. *Am. J. Phys. Anthropol. 39*, 426–460.

Riesenfeld, A. (1976). Compact bone changes in cold exposed rats. *Am. J. Phys. Anthropol. 44*, 111–112.

Rogers, W. M. (1958). The influence of asymmetry of the muscles of mastication upon the bones of the face. *Anat. Rec. 131*, 617–632.

Rose, J. C. (1977). Defective enamel histology of prehistoric teeth from Illinois. *Am. J. Phys. Anthropol. 46*, 439–446.

Rose, J. C. (1979). Morphological variations of enamel prisms within abnormal striae of Retzius. *Hum. Biol. 51*, 139–151.

Rose, J. C., Armelagos, G. J., and Lallo, J. W. (1978). Histological enamel indicator of childhood stress in prehistoric skeletal samples. *Am. J. Phys. Anthropol. 49*, 511–516.

Ross, R. B., and Johnston, M. C. (1972). *Cleft Lip and Palate* (Baltimore: Williams & Wilkins).

Rwenyonyi, C. M., Birkeland, J. M., and Haugejorden, O. (2000). Age as a determinant of severity of dental fluorosis in children residing in areas with 0.5 and 2.5 mg fluoride per liter in drinking water. *Clin. Oral Invest. 25*, 157–161.

Sackeim, H. A., Gur, G. C., and Savey, M. C. (1978). Emotions are expressed more intensely on the left side of the face. *Science 202*, 434–436.

Saldino, R. M., Steinbach, H. L., and Epstein, C. J. (1972). Familial acrocephalosyndactyly (Pfeiffer syndrome). *Am. J. Roentgenol. Radium Ther. Nucl. Med. 116*, 609–622.

Samren, E. B., van Duijn, C. M., Christiaens, G. C., Hofman, A., and Lindhout, D. (1999). Antiepileptic drug regimens and major congenital abnormalities in the offspring. *Ann. Neurol. 46*, 739–746.

Sarnat, B. G., and Schour, I. (1941). Enamel hypoplasias (chronic enamel aplasia) in relationship to systemic diseases: a chronological, morphological and etiological classification. *J. Am. Dent. Assoc. 28*, 1989–2000.

Sayetta, R. B., Weinrich, M. C., and Coston, G. N. (1989). Incidence and prevalence of cleft lip and palate: what we think we know. *Cleft Palate J. 26*, 242–248.

Schour, I., and Massler, M. (1940). Studies in tooth development: the growth pattern of human teeth. *J. Am. Dent. Assoc. 27*, 1918–1931.

Schultz, A. H. (1937). Proportions, variability and asymmetries of the long bones of the limbs and clavicles in man and apes. *Hum. Biol. 9*, 281–328.

Schwartz, G. T., and Dean, C. (2000). Interpreting the hominid dentition: ontogenetic and phylogenetic aspects. In P. O'Higgins and M. J. Cohn (eds.), *Development, Growth and Evolution: Implications for the Study of the Hominid Skeleton* (London: Academic Press), pp. 207–233.

Sciulli, P. W., Doyle, W. J., Kelley, C., Siegel, P., and Siegel, M. I. (1979). The interaction of stressors in the induction of increased levels of fluctuating asymmetry in the laboratory rat. *Am. J. Phys. Anthropol. 50*, 279–284.

Scott, G. R., and Turner, C. G. (1997). *The Anthropology of Modern Human Teeth: Dental Morphology and Its Variation in Recent Human Populations* (Cambridge: Cambridge University Press).

Seipel, C. (1946). Variation in tooth position: a metric study of variation and adaptation in the deciduous and permanent dentitions. *Svensk. Tandl. Tidskr. 39*, 1–176.

Sekikawa, M., Namura, T., Kanazawa, E., Ozaki, T., Richards, L. C., Townsend, G. C., and

Brown, T. (1989). Three-dimensional measurement of the maxillary first molar in Australian whites. *Nichidai. Koko. Kagaku. 15*, 457–464.

Selye, H. (1973a). The evolution of the stress concept. *Am. Sci. 61*, 692–699.

Selye, H. (1973b). Homeostasis and heterostasis. *Perspect. Biol. Med. 16*, 441–445.

Selye, H. (1976). *Stress in Health and Disease* (Boston: Butterworth).

Seth, A. K., and McWilliams, B. J. (1988). Weight gain in children with cleft palate from birth to two years. *Cleft Palate J. 25*, 146–150.

Shapiro, B. (1983). Down syndrome: a disruption of homeostasis. *Am. J Med. Genet. 14*, 241–269.

Sharma, K., Corruccini, R. S., and Potter, R. H. Y. (1986). Genetic and environmental influences of bilateral dental asymmetry in Northwest Indian twins. *Int. J. Anthropol. 4*, 349–360.

Siegel, M. I., and Doyle, W. J. (1975a). The differential effects of prenatal and postnatal audiogenic stress on fluctuating dental asymmetry. *J. Exp. Zool. 191*, 211–214.

Siegel, M. I., and Doyle, W. J. (1975b). The effects of cold stress on fluctuating asymmetry in the dentition of the mouse. *J. Exp. Zool. 193*, 385–389.

Siegel, M. I., and Smookler, H. H. (1973). Fluctuating dental asymmetry and audiogenic stress. *Growth 37*, 35–39.

Siegel, P., Siegel, M. I., Krimmer, E. C., Doyle, W. J., and Barry, H. (1977). Fluctuating dental asymmetry as an indicator of the stressful prenatal effects of Delta-9 THC in the laboratory rat. *Toxicol. Appl. Pharmacol. 42*, 339–344.

Slavkin, H. C. (1974). Embryonic tooth formation: a tool for developmental biology. *Oral Sci. Rev. 4*, 7–136.

Slavotinek, A., Hellen, E., Gould, S., Coghill, S. B., Huson, S. M., and Hurst, J. A. (1996). Three infants of diabetic mothers with malformations of left-right asymmetry—further evidence for the aetiological role of diabetes in this malformation spectrum. *Clin. Dysmorphol. 18*, 241–247.

Smith, B. H. (1991). Standards of human tooth formation and dental age assessment. In

M. A. Kelley and S. P. Larsen (eds.), *Advances in Dental Anthropology* (New York: Wiley-Liss), pp. 143–168.

Smookler, H. H., Goebel, K. J., Siegel, M. I., and Clarke, D. E. (1973). Hypertensive effects of prolonged auditory, visual and motion stimulation. *Fed. Proc. 32*, 2105–2110.

Sofaer, J. A. (1979). Human tooth-size asymmetry in cleft lip with or without cleft palate. *Arch. Oral Biol. 24*, 141–146.

Sofaer, J. A., MacLean, C. J., and Bailit, H. L. (1972). Heredity and morphological variation in early and late developing human teeth of the same morphological class. *Arch. Oral Biol. 17*, 811–816.

Sokal, R. R., and Rohlf, F. J. (1995). *Biometry: The Principles and Practice of Statistics in Biological Research*, 3rd ed. (San Francisco: W. H. Freeman and Company).

Spangler, G. S., Hall, K. I., Kula, K., Hart, T. C., and Wright, J. T. (1998). Enamel structure and composition in the tricho-dento-osseous syndrome. *Connect. Tissue Res. 39*, 165–175.

Sperber, G. H. (1989). *Craniofacial Embryology*, 4th ed. (Boston: Wright).

Spranger, J., Benirschke, K., Hall, J. G., Lenz, W., Lowry, R. B., Opitz, J. M., Pinsky, L., Schwarzacher, H. G., and Smith, D. W. (1982). Errors of morphogenesis: concepts and terms. *J. Pediatr. 100*, 160–165.

Staaf, V., Mornstad, H., and Welander, U. (1991). Age estimation based on tooth development: a test of reliability and validity. *Scand. J. Dent. Res. 99*, 281–286.

Stafne, E. C. (1932). Supernumerary teeth. *Dent. Cosmos 74*, 653–659.

Stein, G. M., and Wahl, H. (1969). Partial anodontia of both the deciduous and permanent dentitions in a case of Crouzon's disease. *Oral Surg. Oral Med. Oral Pathol. 28*, 808–812.

Stough, T. R., and Seely, J. R. (1969). Dermatoglyphics in medicine. *Clin. Pediatr. 8*, 32–41.

Stroud, J. L., Buschang, P. H., and Goaz, P. W. (1994). Sexual dimorphism in mesiodistal dentine and enamel thickness. *Dentomaxillofac. Radiol. 23*, 169–171.

Stroud, J. L., English, J., and Buschang, P. H. (1998). Enamel thickness of the posterior dentition: its implications for nonextraction treatment. *Angle Orthod. 68*, 141–146.

Suarez, B. K., and Spence, M. A. (1974). The genetics of hypodontia. *J. Dent. Res. 53*, 781–785.

Suckling, G. (1989). Developmental defects of enamel—historical and present-day perspectives of their pathogenesis. *Adv. Dent. Res. 3*, 87–94.

Suckling, G., Elliott, D. C., and Thurley, D. C. (1983). The production of developmental defects of enamel in the incisor teeth of penned sheep resulting from induced parasitism. *Arch. Oral Biol. 28*, 393–399.

Suckling, G., Elliott, D. C., and Thurley, D. C. (1986). The macroscopic appearance and associated histological changes in the enamel organ of hypoplastic lesions of sheep incisor teeth resulting from induced parasitism. *Arch. Oral Biol. 31*, 427–439.

Suzuki, N. (1993). Generational differences in size and morphology of tooth crowns in the young modern Japanese. *Anthropol. Sci. 101*, 405–429.

Swanson, L. T., McCollum, D. W., and Richardson, S. D. (1956). Evaluation of the dental problems in the cleft palate patient. *Am. J. Orthod. 42*, 749–765.

Swärdstedt, T. (1966) *Odontological Aspects of a Medievel Population from the Province of Jamtland/Mid-Sweden* (Stockholm: Tiden Barnangen).

Tanner, J. M. (1963). The regulation of human growth. *Child Devel. 34*, 817–847.

Tanner, J. M., Whitehouse, R. H., Marshall, W. A., Healy, M. J., and Goldstein, H. (1975). *Assessment of Skeletal Maturity and Prediction of Adult Height (TW2 Method)* (London: Academic Press).

Ten Cate, A. R. (1994). *Oral Histology: Development, Structure, and Function*, 4th ed. (St Louis: Mosby).

Terakado, M., Hashimoto, K., Arai, Y., Hondo, M., Sekiwa, T., and Sato, H. (2000). Diagnostic imaging with newly developed ortho cubic super-high resolution computed tomography (ortho-CT). *Oral Surg. Oral Med. Oral Pathol. Oral Radiol. Endod. 89*, 509–518.

Thorson, J., and Hägg, U. (1991). The accuracy and precision of the third molar as an indicator of chronological age. *Scand. Dent. J. 15*, 15–22.

Tolarova, M. (1969). Microforms of cleft lip and/or cleft palate. *Acta Chir. Plast. 11*, 96–107.

Tompkins, R. L. (1996). Human population variability in relative dental development. *Am. J. Phys. Anthropol. 99*, 79–102.

Tonge, C. H., and McCance, R. A. (1965). Severe undernutrition in growing and adult animals. 15. The mouth, jaws and teeth of pigs. *Brit. J. Nutr. 19*, 361–372.

Tonge, C. H., and McCance, R. A. (1973). Normal development of the jaws and teeth in pigs, and the delay and malocclusion produced by calorie deficiencies. *J. Anat. 115*, 1–22.

Townsend, G. C. (1983). Fluctuating dental asymmetry in Down syndrome. *Aust. Dent. J. 28*, 39–44.

Townsend, G. C., and Brown, T. (1978). Inheritance of tooth size in Australian aboriginals. *Am. J. Phys. Anthropol. 48*, 305–314.

Townsend, G. C., and Brown, T. (1981). Morphogenetic fields within the dentition. *Aust. Orthod. J. 7*, 3–12.

Townsend, G. C., Jensen, B. L., and Alvesalo, L. (1984). Reduced tooth size in 45,X (Turner syndrome) females. *Am. J. Phys. Anthropol. 65*, 367–371.

Townsend, G., Dempsey, P., and Richards, L. (1999). Asymmetry in the deciduous dentition: fluctuating and directional components. *Perspect. Hum. Biol. 4*, 45–52.

Trasler, D. G. (1968). Pathogenesis of cleft lip and its relation to embryonic face shape in A/J and C57BL mice. *Teratology 1*, 33–50.

Turvey, T. A., Long, R. E., and Hall, D. J. (1979). Multidisciplinary management of Crouzon syndrome. *J. Am. Dent. Assoc. 99*, 205–209.

Turvey, T. A., Vig, K. W. L., and Fonseca, R. J. (1996). *Facial Clefts and Craniosynostosis: Principles and Management* (Philadelphia: W. B. Saunders).

Vanderas, A. P., and Ranalli, D. N. (1989). Evaluation of craniomandibular dysfunction in children 6 to 10 years of age with unilateral cleft lip or cleft lip and palate: a clinical diagnostic adjunct. *Cleft Palate J. 26*, 332–337.

Van Valen, L. (1962). A study of fluctuating asymmetry. *Evolution 16*, 125–142.

van Wieringen, J. C. (1986). Secular growth changes. In F. Falkner and J. M. Tanner (eds.), *Human Growth: A Comprehensive Treatise*, 2nd ed., vol. 3 (New York: Plenum Press), pp. 307–331.

Velasco, M. G., Ysynza, A., Hernandez, X., and Marquez, C. (1988). Diagnosis and treatment of submucous cleft palate: a review of 108 cases. *Cleft Palate J. 25*, 171–173.

Vichi, M., and Franchi, L. (1995). Abnormalities of the maxillary incisors in children with cleft lip and palate. *ASDC J. Dent. Child 62*, 412–417.

Voyer, D. (1998). On the reliability and validity of noninvasive laterality measures. *Brain Cogn. 36*, 209–236.

Wada, J. A., Clarke, R., and Hamm, A. (1975). Cerebral hemispheric asymmetry in humans. *Arch. Neurol. 32*, 239–246.

Waddington, C. H. (1942). Canalisation of development and the inheritance of acquired characters. *Nature 150*, 563–565.

Waddington, C. H. (1957). *The Strategy of the Genes* (London: Allen and Unwin).

Waddington, C. H. (1960). Experiments on canalizing selection. *Genet. Res. 1*, 140–150.

Waddington, C. H. (1974). A catastrophe theory of evolution. *Ann. NY Acad. Sci. 231*, 32–42.

Ward, R. E., Bixler, D., and Raywood, E. R. (1989). A study of cephalometric features in cleft lip–cleft palate families: I. phenotypic heterogeneity and genetic predisposition in parents of sporadic cases. *Cleft Palate J. 26*, 318–326.

Weiss, K. M. (1990). Duplication with variation: metameric logic in evolution from genes to morphology. *Yrbk. Phys. Anthropol. 33*, 1–24.

Werner, S. P., and Harris, E. F. (1989). Odontometrics of the permanent teeth in cleft lip and palate: systemic size reduction and amplified fluctuating asymmetry. *Cleft Palate J. 26*, 36–41.

Wilson, D. F., and Shroff, F. R. (1970). The nature of the striae of Retzius as seen with the optical microscope. *Aust. Dent. J. 15*, 162–171.

Winick, M. (1970). Cellular growth in intrauterine malnutrition. *Pediatr. Clin. North Am. 17*, 69–78.

Winick, M. (1971). Cellular changes during placental and fetal growth. *Am. J. Obstet. Gynecol. 109*, 166–176.

Wong, F. K., Hagberg, C., Karsten, A., Larson, O., Gustavsson, M., Juggare, J., Larsson, C., The, B. T., and Linder-Aronson, S. (2000). Linkage analysis of candidate regions in Swedish nonsyndromic cleft lip with or without cleft palate families. *Cleft Palate Craniofac. J. 37*, 357–362.

Woo, T. L. (1931). On the asymmetry of the human skull. *Biometrika 22*, 324–352.

Woo, T. L. (1938). A biometrical study of the human malar bone. *Biometrika 29*, 113–123.

Woolf, C. M. (1971). Congenital cleft lip: a genetic study of 496 propositi. *J. Med. Genet. 8*, 65–82.

Woolf, C. M., and Gianas, A. D. (1976). Congenital cleft lip and fluctuating dermatoglyphic asymmetry. *Am. J. Hum. Genet. 28*, 400–403.

Woolf, C. M., and Gianas, A. D. (1977). A study of fluctuating dermatoglyphic asymmetry in the sibs and parents of cleft lip propositi. *Am. J. Hum. Genet. 29*, 503–507.

Woolf, C. M., Woolf, R. M., and Broadbent, T. R. (1965). Lateral incisor anomalies: microforms of cleft lip and palate? *Plast. Reconstr. Surg. 35*, 543–547.

CHAPTER 19

VOCAL TRACT ANATOMY AND FUNCTIONAL CONSEQUENCES IN CLEFT LIP/PALATE AND SYNDROMES OF CRANIOSYNOSTOSIS

SALLY J. PETERSON-FALZONE, Ph.D. and PATRICIA K. MONOSON, Ph.D.

19.1 INTRODUCTION

The basic scientist or the clinical practitioner who expects a predictable relationship between cause and effect is often challenged by human variability in structure and function. In addition, anatomy, physiology, and behavior all change as the child grows, whether that child's anatomy is normal or abnormal. Grasping the potential relationship(s) between structure and function is a difficult task even in the absence of congenital anomalies. Structural abnormalities of the supralaryngeal vocal tract can challenge anatomists and clinicians alike to rethink their science.

In this chapter, we first review normal development of the human vocal tract in the early years of childhood. We then take that perspective of early normal development and "perturb" it in the ways it is altered by cleft lip and palate, by three multianomaly disorders involving clefts, and by three syndromes of premature craniofacial synostosis.

19.2 NORMAL ANATOMY OF THE VOCAL TRACT

19.2.1 Introduction

Anatomical differences in the vocal tract of infants as compared to adult morphology include (1) the shape of the tract or tube and thus the course and insertion of muscles of the tube, and (2) the position of the larynx in the neck. The tube or vocal tract is comprised of 15 muscles attached to 6 bones (mandibular, maxillary, sphenoidal, occipital, temporal, and hyoid) and 4 laryngeal cartilages (thyroid, arytenoid, corniculate, and cricoid). The infant's vocal tract is about one-half the total length of the adult tract, and has an obtuse angle at the junction of the oral and pharyngeal portions of the tube. Characteristics of the infant's structure change gradually, and the change in the angle of the oral and pharyngeal portions of the vocal tract becomes a key factor in the ability to produce speech.

The head of the infant is about one-fourth his total body (crown-heel) length, whereas in the adult the head is about one-eighth the body length (Bosma, 1986). The

---

Understanding Craniofacial Anomalies: The Etiopathogenesis of Craniosynostoses and Facial Clefting, Edited by Mark P. Mooney and Michael I. Siegel, ISBN 0-471-38724-x Copyright © 2002 by Wiley-Liss, Inc.

muscles of the mouth, pharynx, and larynx are well developed at birth, but the mandible is proportionately smaller in comparison to the adult model. The upper face in the infant is almost two-thirds the size of what it will be when the infant has become an adult, but the midfacial and lower facial height are only one-third the size of what they will become; the length of the mandible is also one-third the adult size.

Simultaneous with the changes in the bony skeleton in childhood, the muscles of the vocal tract change course and grow in volume. The muscles of the head and trunk constitute about 40% of the body weight, in comparison to 20% to 30% of the body weight in an adult (Bosma, 1986). The skeletal muscle fibers are all present at birth, so the growth is in volume and length of the fibers, not in numbers (Crelin, 1973).

19.2.2   The Vocal Tract at Birth

The newborn's vocal tract consists of the oral cavity (4 centimeters long) and the oro- and laryngopharynx (less than 3 centimeters long). The tongue fills the oral cavity, and the laryngeal vestibule is at the level of the third cervical vertebra, which means the larynx is riding high in the neck and there is little angulation between the oral and pharyngeal portions of the vocal tract. By contrast, the adult vocal tract has virtually a 90° angle between the oral cavity and the pharynx.

The changes in length and shape of the vocal tract are due to three structural reorganizations. First, the oral cavity increases in volume and comes forward as a result of the growth of the lower and midface. Second, the cranial base grows with the developing brain and contributes to forward growth of the viscerocranium. Third, the larynx descends in the neck from birth to 10 years of age (Fig. 19.1) and the vertebrae grow vertically. The overall pattern of growth is forward and down-

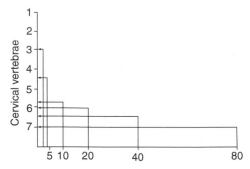

Figure 19.1   Vertical descent of the larynx during life. (Based on Wind, J. *On the Phylogeny and Ontogeny of the Human Larynx.* Groningen: Wolters-Noordhoof Publishing, 1979.) (*Source*: From Zemlin, 1998, p. 179.)

ward, but some bony landmarks maintain a relatively stable position.

In general, bony growth occurs by deposition and resorption of bone, both processes reshaping bone in the growth process and creating the adult morphology. However, because it is bone deposition that accounts for forward movement of the face and increase in the length of the vocal tract, this discussion is limited to the sites of deposition.

19.2.3   The Pharynx

At birth the pharynx is only about 6 centimeters long. The pharynx in an adult is a 12 centimeter tubular structure that is broadest at its upper end and narrowest at the laryngeal vestibule. It is suspended from the cranium and from the mandible by multiple muscles and ligaments. As the baby grows, a bend develops at the juncture of the oral cavity and the vertical pharyngeal portion of the vocal tract. This developmental differentiation is primary in both its ontogenetic and phylogenetic importance.[1]

[1] There have been some historically entertaining arguments between linguists who believed that nothing in the development of man's ancestors could lead to the ability to produce speech-like sounds and

In the infant, the lower part of the pharynx is attached to a flexible bony skeleton (hyoid bone, larynx, and vertebrae). Thus, the muscles have to be well developed in order to maintain an adequate airway, especially when the baby is in the supine position. In the adult, the uppermost segment of the pharynx is attached both posteriorly and anteriorly to bones, the anterior attachments being to the sphenoid alae and the palatine bones. The forward displacement of the palate with growth and the changes in the cranial base account for the increase in anterior-posterior (AP) depth that occurs in infants. Much of that increase is accomplished by $1\frac{1}{2}$ years of age (Muller, 1963). In later developmental stages, the pharynx continues to shift forward, but without significant depth change because its posterior bony structures (primarily the anterior tubercle of the atlas) are also growing forward.

After age 2, even though there is continued forward and downward growth of the face, the anterior-posterior depth of the oropharynx remains relatively stable. This is because the lower segment of the superior constrictor muscle is attached to the posterior surface of the pterygoid laminae, a growth site where new bone is being deposited, thus keeping the bony anchor or point of attachment in a relatively stable position. At the same time, the cervical vertebrae are growing in the anterior-to-posterior dimension as well as vertically, contributing to the forward shift of the posterior pharyngeal wall. Overall, there is less than a 4 millimeter increase in AP depth of the oropharynx from infancy to age 16 (King, 1952).

AP depth of the lower part of the pharynx (laryngopharynx) also shows little change in dimension during childhood. The lower pharyngeal constrictor muscle fibers

then speech itself, opposed to those who believed the development of speech was as inevitable as all the other developmental changes that produced modern man.

attach anteriorly to the hyoid bone and the laryngeal cartilages. Measurement of the anterior-posterior changes in the hyoid bone is made difficult because of the influence of head position on hyoid bone position. Even if the hyoid showed anterior displacement during maturation, the posterior growth of the greater cornu would nullify that change.

By far the greatest growth of the pharynx is in the vertical dimension. The changes are due to vertical growth of the vertebrae and the descent of the hard palate, pterygoid processes, mandible, and hyoid bone (Fig. 19.2). Of all these structures, the hyoid bone shows the most vertical change, descending from an initial position higher than the mandibular symphysis to a level below the mandible (King, 1952).

19.2.4 The Hyoid Bone

The hyoid bone is primarily a cartilaginous structure at birth and remains flexible throughout the first 2 years of life. The lesser cornu is attached via ligaments to the styloid process of the temporal bone. The course of the ligaments and muscles in a newborn is much less vertical than that of the adult. The muscles of the pharynx that attach to the hyoid show marked change as the bone descends in the neck. Specifically, the stylohyoideus, posterior belly of the digastricus, and stylopharyngeal muscles course less horizontally as the child grows and assume a more oblique orientation.

19.2.5 The Larynx

The larynx, positioned higher in the neck in the baby than in the adult (Manson, 1968), is initially funnel-shaped, with the vertical axis of the lower part of the laryngeal cavity extending more posteriorly than in the adult larynx. In the baby, the hyoid bone and the superior surface of the thyroid cartilage nearly approximate. This approxima-

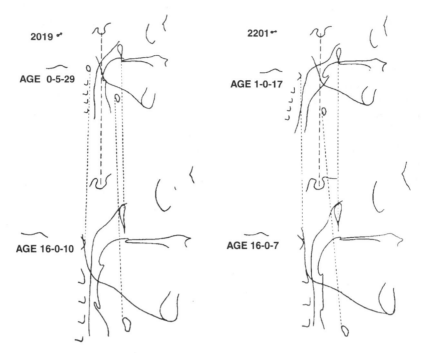

Figure 19.2 Tracings of the same individuals at different ages showing the degree of change between selected landmarks with age. (*Source*: From King, 1952, p. 34.)

tion accounts for the relatively short laryngopharynx of the newborn in comparison to the adult. The position and angle of the larynx in the neck contribute to the obtuse angle of the infant vocal tract. Ossification of the larynx occurs well into adulthood and is completed during the fifth decade of life.

19.2.6 The Vertebral Column

The vertebral column is flexible at birth and only develops its characteristic curve at the time an infant is able to lift its head. The first vertebra (atlas) is fixed to the occipital bone, and the second (axis) is closely coupled to the first via the odontoid process and its ligaments. The remaining cervical vertebrae have three ossification centers joined by hyaline cartilage, and become fully ossified between the third and sixth postnatal year. The atlas and axis become ossified by the ninth year of life.

The bodies of the vertebrae double their transverse and sagittal dimensions between birth and adulthood (Manson, 1968).

19.2.7 Mandibular Growth

Postnatal growth of the mandible occurs on the posterior portion of the ramus and condyle and at the midline symphysis. In addition, bony growth of the mandible occurs as the alveolar bone forms around the developing teeth (Enlow, 1968; Manson, 1968).[2] Figure 19.3 illustrates the growth of the mandible from infancy to

[2] Teeth and alveolar bone have an interdependent relationship: teeth must have alveolar bone to erupt into, and alveolar bone bulk is maintained only where teeth are present. Think about the facial appearance of an elderly person whose teeth might have been lost to decay or extraction: that drawn-mouth appearance is due to the lack of normal alveolar bone bulk (from reabsorption) in the maxilla and/or mandible, which is due in turn to loss of teeth.

adulthood. The growth sites account for the forward and downward growth of the mandible and the increase in vertical height of the lower face. The angle of the mandible (the angle between the ramus and body) is about 170° at birth and about 140° at age 4. In adulthood, the angle is between 110° and 120° (Enlow, 1968). The forward and downward growth of the mandible extends the length of the vocal tract and similarly affects the positions of the attached connective tissues, muscles, tendons, and ligaments. The gradual forward positioning of

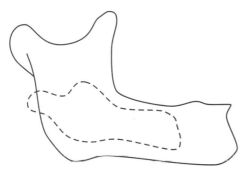

Figure 19.3 Comparison of the newborn mandible (dotted line figure) with the adult mandible (solid line figure).

the mandible accounts for some of the differences in the bent tubular vocal tract of the adult.

19.2.8 Cranial Base Changes

From birth to age 10, the cranial base angle (the relationship between the anterior cranial base as defined by a line from nasion to the midpoint of sella turcica) and the posterior cranial base (as defined by a line from the midpoint of sella to basion) (Fig. 19.4) changes to accommodate the expanding brain (see Chapters 11, 12, and 13). While undergoing this change, the cranial base must maintain the position of the foraminae and fossae for passage of blood, nerves, and the spinal column. The anterior portion of the developing brain expands to a greater extent than the posterior portion. This differential expansion accounts for the flexure of the cranial base.

The expansion of the basicranium is complex. The cranial base angle remains relatively constant from age 3 through puberty, while the whole cranial base moves downward and forward due to brain

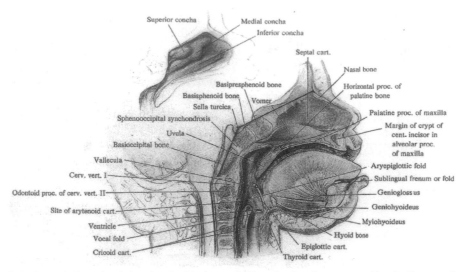

Figure 19.4 Lateral view of an infant head, with the cranial base landmarks connected by the lines used to measure the anterior and posterior base. (*Source*: Adapted from Bosma, 1986, p. 44.)

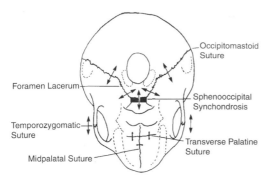

Figure 19.5 Direction of growth of the cranial base at suture sites. (*Source*: From Sperber and Tobias, 1989, p. 115.)

growth (May and Sheffer, 1999; Meredith, 1959; Riolo et al., 1974). Although the pooled data suggest little change in the angle during childhood, many individuals show a marked increase or decrease in angulation as they grow (Brodie, 1961). George (1978) presented data demonstrating some angular reduction in the first $1\frac{1}{2}$ years of life. That pattern had also been noted earlier by Brodie (1955).[3]

With flexure of the base and anterior expansion of the cranium, two changes take place in the vocal tract: the oral portion of the tract increases in length, and the pharyngeal portion expands its length in the nasopharynx. The basicranium growth that affects the forward growth of the face is found in the posterior and middle cranial fossae, specifically the spheno-occipital synchondrosis (Fig. 19.5).

19.2.9 Facial Growth

The postnatal growth sites in the midface of the viscerocranium are found at the suture lines of the facial bones,[4] the posterior maxillary tuberosity, and the alveolar

margins (Sperber and Tobias, 1989) (Chapters 15, 16, and 17). The *height* of the midface is attributed to the deposition of new bone creating the alveolar processes, and the eruption of the teeth in the maxilla and mandible. The lower face becomes continuously more adult-like in its morphology as the teeth erupt. Of particular significance in this process is the eruption of the 6- and 12-year molars.

Midface postnatal growth is complex, with changes in the *width* of the face occurring at the zygomaticomaxillary and intermaxillary sutures. Increasing *height* of the midface is a result of growth of the nasal septum along with the deposition of bone at the frontomaxillary, frontozygomatic, frontonasal, ethmoidmaxillary, and frontoethmoidal sutures. These are the growth sites accounting for the depth and height changes in the maturing midface. Some investigators have hypothesized that the deposition of new bone along suture lines is preceded by separation of the suture resulting from external forces.[5] For example, Scott (1953) and Muller (1963) postulated that the nasal septum with its cartilaginous interstitial growth patterns is the forward displacement mechanism in the midface. While this theory does provide some explanation for the displacement and thus the sutural deposition of bone, the factors contributing to the normal forward displacement of the midface are probably more complex.

The mean magnitude of change in facial *depth* with growth is about 9% (Meredith, 1959). Changes in facial depth average 3% less for girls than for boys from the ages of 5 to 11, with considerable within-gender variability. Overall, there is relatively little correlation among the three dimensions of facial growth during childhood.

---

[3] What is always questioned in these studies is the longitudinal stability of the measurement points, especially nasion and sella turcica.

[4] That is, the facial bones grow by the mechanism of new bone being laid down along either side of suture lines.

[5] It is this principle that underlies the currently popular treatment technique of distraction osteogenesis used to restructure the face in acquired or congenital craniofacial defects.

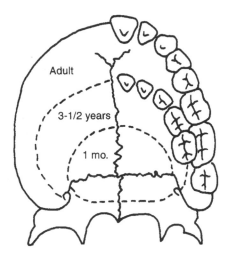

Figure 19.6 Palatal growth from 1 month of age to adulthood. (*Source*: From Zemlin, 1998, p. 290.)

19.2.10 Maxillary Changes

The width of the maxilla increases by 6 millimeters from birth to age 2. The hard palate of the newborn is approximately 3 centimeters in length. It is composed of the palatine processes of the maxilla and the horizontal plates of the palatine bone. The palatine bone anteriorly is approximately 1 centimeter in the vertical dimension as it blends with the alveolar arch. Palatal growth occurs appositionally at the medial palatal sutures, increasing the anterior-to-posterior dimension. By the end of the sixth year, the palate has obtained most of its growth (Zemlin, 1998). Figure 19.6 is an image of the transverse and horizontal growth of the palate. The palatal plane does not change its angular relationship to the viscerocranium throughout life.

While palatal growth accounts for depth changes of the roof of the vocal tract, change in depth of the midface is also accomplished by deposition of bone at the pterygoid laminae. Because of the growth of the maxilla and deposition of bone in the posterior aspects of the midface, the pharynx does not deepen to the same extent as the forward growth would predict. This lack of 1:1 relationship exemplifies the complex nature of the growth of the head.

19.2.11 Muscular Changes

The relationship of the muscles within the vocal tract depends on bony growth, muscle function, and neural maturity. As an example, the tongue at birth apposes (touches) the lips, buccal wall, and soft palate (Moyers, 1970). As the lower and midface grow forward and downward, the tongue no longer apposes the lips and soft palate, thus contributing to the independence of the tongue from the jaw, which is necessary for speech. Further, the primary functions of the oral and pharyngeal cavities are for respiration and mastication—two primitive reflexive patterns that must be overridden for volitional speech movement.

Feeding and breathing require (1) stabilization of craniocervical posture of the tract to maintain the airway and (2) specific movement patterns to meet progressive, evolving needs for feeding from liquids to solid foods. In the newborn, the uvula and epiglottis approximate, allowing protective closure of the oral cavity from the oropharynx so that the infant can simultaneously suckle and breathe. According to Enlow (1990), the lower pharynx has more mobility in the infant than in the adult. The anterior attachments of the lower and middle pharyngeal muscles are to the cartilaginous larynx and hyoid, while the superior constrictor attaches to the less flexible bony maxilla. In the lowest extent of the pharynx, the paired cricopharyngeus muscle lies in approximation to the rest of the pharyngeal muscles, while in an adult this muscle shows some separation from the rest of the pharyngeal muscle bundles. For speech, the implications of this developmental change are unclear.

One frequently overlooked factor that may play a role in speech production is the positioning of the muscles of the vocal tract. While substantial attention has been focused on the muscles of the soft palate, especially in relation to aberrations of insertions in persons with facial anomalies, less attention has been focused on the rest of the muscles of the vocal tract. An example of the directional change is the three styloid muscles that attach into the pharyngeal musculature and onto the hyoid bone.

The pharyngeal constrictors in an infant are thick and fairly undifferentiated, unlike those of the adult. In addition, a number of the hyoid sling muscles commingle with the pharyngeal constrictors. There is evidence of this interdigitation in the infant larynx as well as in the adult (Zemlin, 1998). Interestingly, the criocothyroid muscle does not blend with the pharyngeal muscles, and its innervation is also different from that of the other laryngeal muscles. Thus, it is the only airway muscle that remains distinct from the other musculature throughout the growth period.

At birth, the facial muscles and the oral musculature are very well developed in contrast to the pharyngeal and laryngeal muscles (Cole, 1970; Cox, 1970). The reciprocal influence of the oral musculature on the anterior facial structures and of the bony structures on the musculature is not well understood. It is thus risky to assume that an abnormality in one (bone or muscle) is directly responsible for an abnormality in the other.

19.2.12 Lessons from Comparative Anatomy

Comparative anatomy studies reveal that infants and higher-level primates, specifically chimpanzees, have similarly shaped upper airways. According to Lieberman (1975, p. 109), the similarities are as follows:

the tongue in its resting state is completely contained within the oral cavity; the soft palate can be approximated to the epiglottis; there is almost no supralaryngeal portion of the pharynx; and the vocal cords are at the level of the third and fourth cervical vertebrae. These similarities between the newborn human and chimpanzee, in contrast to the adult human, may explain why there is a limited repertoire of speech sounds in babies. The bent tube in human adults allows tongue movement to change the configuration of the oral cavity and pharynx simultaneously. In the newborn and chimpanzee, the tongue cannot constrict the pharyngeal space because it does not make up the anterior pharyngeal wall.

Brodie (1961) presented other evidence that the newborn and ape have similar structure when he traced the changes in the airway in quadrupeds and bipeds. He stated that since gravity aids the quadruped in maintaining a patent airway, the need for muscles is less important. The bipedal human must use muscles to keep the airway open. Without the airway changes seen in the adult human, the tongue would be forced backward in the oral cavity and obstruct the pharynx. He further stated that while the adult human anatomy of the temporomandibular joint is not very different from that of the ape, the manner of movement of the mandible is. Specifically, the angled musculature of the airway must operate the mandible using a different type of rotary movement than found in apes. This difference is necessary because of the broadening of the mandible to accommodate the tongue. In addition, the hyoid sling muscles must assist in stabilizing the hyoid bone and larynx for maximum jaw opening. If the newborn airway is shaped like the quadrupedal ape airway, then the muscles needed to stabilize the airway are not in the optimal position to do so. This lack of stability may indeed affect speech function that depends on jaw mobility and tongue movement. If an airway remains in the more ape-like or newborn shape without a bent tube, then perhaps some aspects of speech disorders in children and adults with anatomical abnormalities can be attributed to the airway shape and muscle positioning as described by Brodie (1961).

19.3 CONGENITAL ANOMALIES OF THE HUMAN SUPRALARYNGEAL VOCAL TRACT

19.3.1 Cleft Lip and Palate

The most common congenital anomaly of the craniofacial complex in humans is cleft lip and palate.[6] The frequency of different types of clefts varies with sex, ethnic group, and etiology, but overall 25% of clefts are of the lip (with or without the alveolus) only, 25% are clefts of the secondary palate only, and 50% are either unilateral or bilateral clefts of the lip and palate (Peterson-Falzone et al., 2001). Abnormalities of the supralaryngeal vocal tract have not been documented in clefts of the lip with or without the alveolus only.[7] There have been at least five decades of clinical reports and research into the effects of cleft palate on the development of communication ability (Peterson-Falzone et al., 2001). The present discussion is focused on possible structural differences in the vocal tract and how they may affect speech output.

It is important to realize at the outset that cleft lip with or without cleft palate is not an isolated tissue deficiency. Consider the following: (1) more than 40 years ago, Fletcher (1960) pointed out that the physical causes of "hypernasality" should be

considered evidence of a "regional birth defect"; (2) clinical and radiographic studies have delineated abnormalities in the nasopharynx and other oral and pharyngeal structures even in nonsyndromic clefts (Moss, 1956; Osborne et al., 1971; Subtelny, 1955);[8] (3) in about half of the total aggregate of individuals with clefts, careful clinical examination will reveal at least one minor or one major associated anomaly (Shprintzen et al., 1985a); and (4) clefts are often part of multianomaly disorders (Peterson-Falzone et al., 2001; Shprintzen et al., 1985b).

All of the above makes the attribution of specific speech problems to specific anatomic and/or functional problems in individuals with clefts risky and perhaps an invitation to scientific folly. In the literature, the traditional concerns have been (1) the inability to close off the nasal airway from the oral airway, (2) the likely effects of this problem on the child's phonetic, phonologic, and language development, and, similarly, (3) the pluripotential effects of all of the above on the child's ability to relate to his world (i.e., talk, listen, understand, respond). We know that speech development in children whose clefts are closed early enough to facilitate normal development of phonology and language will be essentially what we expect in children without clefts (Peterson-Falzone et al., 2001). But we also know that the child for whom palatal closure is delayed for some reason may (1) develop maladaptive compensatory articulations because he is ready to talk and his physical system is not capable of what he needs it to do (Trost, 1981; Trost-Cardamone, 1990; Trost-Cardamone and Bernthal, 1993); (2)

---

[6] Clefts of the lip and palate may occur separately or together.

[7] Some individuals with repaired cleft lip do have deficient length and/or mobility of the upper lip, making bilabial closure difficult and perhaps contributing to difficulty in rounding the lips in speech, but to date no one has produced scientific data on the effect of such factors on articulation.

[8] Subtelny (1955) documented abnormally increased width of the cranial base in children with clefts and conjectured how this might be related to the occurrence of clefts. Osborne et al. (1971) found that abnormalities of the upper cervical spine contributed to increased pharyngeal depth in speakers with velopharyngeal inadequacy for speech.

exhibit early delays in phonologic and language development because of the phonetic limitations laid down by physical inadequacy of the velopharyngeal mechanism (O'Gara and Logemann, 1988; O'Gara et al., 1994); and thus (3) potentially mislead the scientist who is trying to examine the relationship(s) between clefts and the function of the supralaryngeal vocal tract. If you are studying a youngster whose cleft was repaired after his developing phonology and language needed an intact oropharyngeal mechanism, what you are hearing may have nothing to do with that child's *current* oropharyngeal mechanism.

In the last two decades, the bulk of research on the effects of unrepaired, badly repaired, and well-repaired clefts of the palate on speech has focused on articulation and phonologic development. Prior to that time, some acoustic studies focused on the spectrographic correlates of nasality (Curtis, 1968; Dickson, 1962; Fant, 1960; Hattori et al., 1958). These studies showed the major characteristics of nasalization found on an output spectrum to be (1) a reduction in intensity of the first formant; (2) the appearance of antiresonances; (3) the appearance of extra resonances, most notably between the first and second formants; and (4) a shift in the center frequencies of the formants. Phillips and Kent (1984) reported an extra nasal formant below the first formant in nasalized vowels, a weakening of intensity, and a slight upward shift in frequency of the second formant; an overall weakening of acoustic energy across the spectrum; and an increase in formant band widths. These early spectrographic studies delineated some of the factors that may account for difficulty in understanding the speech of someone with an inadequately repaired cleft: vowels carry a great deal of the informational content of speech, and the abnormalities in the acoustic information from nasalized vowels may make it difficult for

the listener to differentiate one vowel from another. Thus, not all of the difficulties the listener has in understanding the speech of a person with an inadequately repaired cleft or other velopharyngeal problem may be due to structurally based (phonetic) problems in consonant production, or to compensatory articulation patterns. Fortunately, surgical success in the repair of clefts is now much more prevalent than surgical failure, so we now expect to see very few children and adults with repaired clefts suffer from abnormal speech development (Peterson-Falzone et al., 2001).

While the focus of this chapter is on *supralaryngeal* structural differences in speakers with congenital craniofacial anomalies, we cannot ignore the effect of laryngeal output (excitation source). For example, in speakers with clefts, there is a high incidence of abnormal laryngeal vocal quality, which has been repeatedly documented in the clinical literature (McWilliams et al., 1969, 1973). The increased vulnerability of children with inadequate velopharyngeal closure to vocal cord abnormalities is thought to be due to overdrive of the laryngeal mechanism in the unconscious effort to make up for the inefficient escape of some of the vocal airflow through an incompetent velopharyngeal port. It is difficult for clinicians to establish that the occurrence of breathiness or hoarseness in this population is more than could be expected in any population of boisterous, school-aged children, but the findings of McWilliams and colleagues (1969, 1973) still alert clinicians that the potential for vocal abuse among these children may be higher than in noncleft children.

Overall, the current expectation for children with clefts in the United States is for essentially normal development of speech and language skills, with allowance for some problems with precision of articulation related to dental and occlusal problems. If gross misapproximations to pressure consonants pervade a child's speech, physical management of the cleft has been too late, inadequate, or sabotaged by other physical or developmental factors. Abnormalities in resonance are attributed to inadequate velopharyngeal closure (hypernasality), overcorrection of VP inadequacy with a too-large pharyngeal flap or prosthetic speech bulb (hyponasality), and sometimes the combination of inadequacy of velopharyngeal closure in the presence of nasal obstruction (cul-de-sac resonance). Other than the possible role of cervical spine anomalies contributing to an abnormally deep pharynx (Osborne et al., 1971), the differences in cephalometric findings in individuals with cleft (Krogman, 1961; Moss, 1956; Subtelny, 1955) have not been specifically tied to differences in resonance characteristics or articulation.

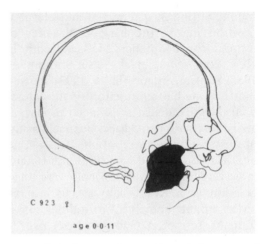

Figure 19.7 Severe micrognathia and the retracted and elevated tongue position in an infant with Robin sequence (*Source*: From Peterson-Falzone et al., 2001, p. 33.)

19.3.2 Multianomaly Disorders of the Craniofacial Complex

Within the confines of this chapter, we cannot discuss all the congenital multi-anomalies of the head and face that affect both the structure and function of the human supralaryngeal vocal tract. We will present information on Robin sequence, which is not a syndrome but is a multi-anomaly condition that is frequently a part of syndromes in addition to being a significant clinical entity on its own; on three syndromes of which Robin sequence is a part but which entail other significant anomalies; and on three of the syndromes of craniofacial synostosis. Each of these clinical groups entails anomalies of the supralaryngeal vocal tract, although the extent to which these anomalies have been documented to cause abnormal speech or abnormal transfer function (Fant, 1960) of the acoustic spectrum is variable.

19.3.2.1 *Pierre Robin Sequence*
Pierre Robin was a French stomatologist who described a combination of micro-

gnathia, glossoptosis, and respiratory distress in a series of neonates. His original report did not include cleft palate (1923), but a later report did (1934). There is widespread disagreement among dysmorphologists as to whether a palatal cleft must be present for the Robin label to be applied. The physical distortions of the vocal tract in infants are somewhat variable, but typically include the small and retruded mandible and the tongue sitting in a retracted position with the dorsum elevated (Fig. 19.7). If the baby is nonsyndromic, there is a good possibility that the mandible will grow forward sufficiently in the first few years of life so that retrognathism or micrognathism is much less severe or no longer applies. However, this is not necessarily the case when the Robin sequence is part of a multianomaly syndrome.[9]

Prior to the 1980s, most investigators reported that the cause of upper airway obstruction in Robin infants was the

---

[9] Robin sequence occurs in more than 30 syndromes or associations of congenital anomalies of the head and face (Gorlin et al., 1990).

tongue sitting in a retracted and elevated position in the oropharynx. A series of reports from Montefiore Medical Center in New York (Sher et al., 1986; Shprintzen, 1988; Shprintzen and Singer, 1992) showed that the medial constriction of the lateral pharyngeal walls or a sphincteric pharyngeal constriction could also be mechanisms of obstruction in these infants. To date, no one has determined how often the tongue may remain in an abnormally retracted position as a Robin baby grows into an older infant and then a toddler. If it remains retracted (and elevated) over a sufficiently long time, does this habitual posture affect vowel quality or ability to consistently reach appropriate articulatory targets for consonants? Timing of closure of the palatal cleft, when present, is another variable that may affect tongue posture. In some treatment centers, teams are reluctant to surgically close the cleft within the first year of life if there have been significant respiratory problems in the neonatal period. Other teams have found that surgical closure of the cleft may help to normalize tongue posture (because it can no longer go up into the cleft itself) and may further alleviate crowding of the pharynx by getting the uvular halves out from behind the tongue.

If the tongue in a Robin child has a tendency to remain in a more retracted position in the early years of life, the theoretical conclusion is that there may be an effect on vowel quality (location of formants) and precision of articulation. However, to date none of the studies on either resonance or articulation in children with clefts has carefully segregated children with cleft palate (with or without cleft lip) from children who may have had Robin sequence.

19.3.2.2 Some Syndromes Involving Robin Sequence
The findings of micrognathia, glossoptosis, and early respiratory problems with or without a cleft of the secondary palate occur in many multi-anomaly syndromes, including Stickler syndrome, velocardiofacial syndrome, and mandibulofacial dysostosis. Each of these syndromes is fraught with abnormalities of the vocal tract, some more fully documented than others.

Stickler Syndrome   Stickler syndrome, also known as hereditary arthro-ophthalmopathy (Stickler et al., 1965; Stickler and Pugh, 1967), is an autosomal-dominant disorder with the major findings being high myopia in early childhood, retinal detachment and cataracts, progressive sensorineural hearing loss, cleft palate, and progressive orthopedic problems involving the long bones and epiphyses. The facial features vary in severity from patient to patient, but typically include a flat midface secondary to maxillary hypoplasia, prominent eyes, low nasal bridge, long philtrum, and mandibular hypoplasia.

Shprintzen (1992) reported that 34 of 100 cases of Robin sequence had Stickler syndrome. Although every child with Robin sequence should be evaluated for possible Stickler syndrome, and there is a high likelihood that a child with Stickler syndrome has the features of the Robin sequence, each can occur without the other. The palatal anomalies include overt clefts, submucous clefts (either obvious or "occult"), and inadequate velopharyngeal closure in the absence of either overt or submucous clefts, but some Stickler patients do not have palatal anomalies. Estimates of frequency of palatal problems vary across reports, perhaps as a function of how the patient was first ascertained—for example, through an ophthalmology clinic versus through a cleft palate clinic. Lucarini and co-workers (1987) found that 8 of 14 patients who were admitted to a hospital because of retinal detachment had either submucous or overt clefts.

When Stickler syndrome is present, the mandible may not exhibit the "catch-up" growth in early childhood that is more typical of Robin sequence in isolation. The airway may remain impaired for a longer period of time, meaning that tracheotomies are more likely to be needed (Fig. 19.8).

As of this date, there have been two radiographic studies to delineate the craniofacial measurements in patients with Stickler syndrome (Glander and Cisneros, 1992; Saksena et al., 1983). The earlier of these two studies examined a large number of linear and angular measurements from lateral cephalometric films. The authors reported "markedly shortened cranial base

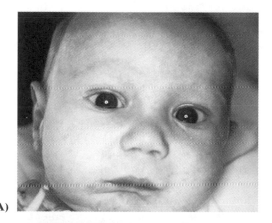

(A)

(B)

Figure 19.8 The typical facial features of a child with Stickler syndrome. Note the flat midface, retruded mandible, depressed nasal bridge, and prominent eyes. Also note his tracheotomy. His mother (holding him in B) also has Stickler; note her thick glasses, required due to the progressive visual problems in this syndrome.

length, midfacial depth and height, maxillary depth, and mandibular depth, but significantly larger total and lower facial height measurements" (p. 19). In theory, any of these abnormalities could have affected the function of the supralaryngeal vocal tract, but the study did not include information on speech.

The study by Glander and Cisneros (1992) was a four-way comparison between patients with Stickler with and without Robin sequence, as compared to patients with velocardiofacial syndrome with and without Robin sequence. This structure made the results difficult to interpret, which was exacerbated by the fact that the age of patients across whom data were pooled encompassed early childhood (3–4 years) to adulthood (42–52 years). Although the amount of numeric cephalometric measurements was painstaking, the results are nearly impossible to interpret. The authors concluded (1) "the Robin features in VCF may be caused by hypotonia rather than any craniofacial or physical obstruction of the airway"; (2) "Stickler and VCF are similar in craniofacial morphology but show marked differences in pharyngeal and airway morphology"; and (3) "should not be the sole prognosticator of the Robin sequence and its association with Sticker and VCF." However, the functional importance of the wealth of measurements they reported was lost in their approach to data analysis. Even prior to the publication of these two studies, we knew that cephalometric data in children with either of these syndromes were apt to show abnormalities in the cranial base and velopharyngeal measurements. Thus, these morphometric studies showed nothing of specific importance to the supralaryngeal vocal tract other than what was already known.

Similarly, studies on the speech of Stickler children have not derived any syndrome-specific findings, but we know these speakers are vulnerable to multiple problems stemming from (1) prolonged tracheotomy, (2) abnormal tongue posture due to lack of normal growth of the mandible, and (3) velopharyngeal problems. An additional factor is the progressive sensorineural hearing loss, although the loss may not be severe enough in early childhood to affect speech and language acquisition.

Velocardiofacial Syndrome This syndrome was probably first described by Sedlackova (1955), who reported a series of patients with dysmorphic facial features, reduced facial expression, velopharyngeal problems, minor ear malformations, and digital anomalies. There were several subsequent reports from other clinicians during the 1960s and 1970s (see Peterson-Falzone et al., 2001, for a bibliography). Shprintzen and associates (1978) documented a much longer list of physical findings, including cardiac defects, and coined the term *velocardiofacial syndrome* for what would come to be known as the most frequently occurring syndrome involving palatal anomalies. In subsequent years, the list of physical and functional problems[10] in this syndrome has grown to include nearly every major organ system of the body. This is another autosomal-dominant disorder,[11] and estimates of frequency of occurrence have accelerated as clinicians have come to recognize the wide range of severity of findings (thus recognizing milder cases). With specific regard to the craniofacial complex, the features include (Fig. 19.9) a vertically long face with a broad nasal

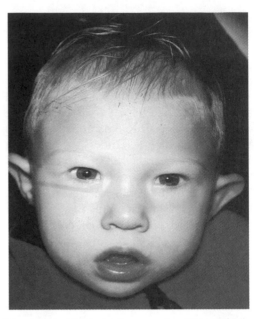

Figure 19.9 The facial features of a child with velocardiofacial syndrome.

root, narrow alar base, flattened malar eminences, narrow and downward-slanted palpebral fissures, abundant scalp hair, retruded mandible, and palatal anomalies in over 90% of cases including overt clefts, submucous clefts, occult submucous clefts, and deficient velopharyngeal movement in the absence of clefts. Additional craniofacial findings are pharyngeal hypotonia, lymphoid tissue hypoplasia, obtuse cranial base, minor ear anomalies, microcephaly (40%), medial displacement of the internal carotids (25%), and variable eye findings (Goldberg et al., 1993).[12] There have also been a few reports of cleft lip. A variety of laryngeal abnormalities have also been reported (see Solot et al., 2000, for a comprehensive list of reports on this point).

The vocal tract in velocardiofacial syndrome is subject to a fairly long list of

[10] The functional problems in this syndrome include nearly universal occurrence of learning disabilities, a high probability of mental retardation, and a very high likelihood of late-onset psychosis, with genetic overlap with schizophrenia (Chow et al., 1994; Goldberg et al., 1993; Karayiorgou et al., 1995; Moss et al., 1999; Pulver et al., 1994; Shprintzen et al., 1992).

[11] About 75% of cases show a microdeletion on the 22nd chromosome, at 22q11.2, and the syndrome is now frequently given this label in the literature (e.g., Solot et al., 2000).

[12] Please see Peterson-Falzone and co-workers (2001) for a more detailed listing of the physical findings reported in the aggregate of studies on this syndrome and a more complete list of published studies.

possible abnormalities: laryngeal abnormalities, inadequate velopharyngeal closure, pharyngeal hypoplasia, adenoid hypoplasia, small mandible, and possibly increased pharyngeal depth[13]—together a high likelihood of hypotonia. In many patients, especially in the early childhood years (Solot et al., 2000), reasonable assessment of the effect of physical differences of the vocal tract on speech production capabilities is largely negated by the effects of abnormal intellectual development. Virtually every speech pathologist who has become familiar with this syndrome has learned the multiplicity of physical and functional problems (Solot et al., 2000), and thus the risks of attributing specific speech characteristics to specific physical findings.

Mandibulofacial Dysostosis This syndrome, also known as Treacher Collins syndrome, involves anomalies of the ears, eyes, maxilla, and mandible (Fig. 19.10). Like Stickler syndrome and velocardiofacial syndrome, it is autosomal dominant, meaning that every child of every affected individual has a 50% chance of having the disorder. Also like Robin sequence, Stickler syndrome, and velocardiofacial syndrome, the range of severity from patient to patient is so great that mildly affected individuals may not be correctly diagnosed.[14] The facial anomalies are described

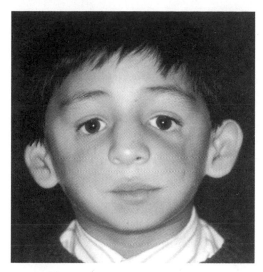

Figure 19.10 The facial features of a youngster with mandibulofacial dysostosis.

in detail by Peterson-Falzone and colleagues. (2001). The ear anomalies involve the outer and middle ears; the hearing loss is conductive and varies from mild to maximum.

There are several bony and soft tissue abnormalities in this syndrome that may affect speech. The mandible usually has a characteristic curvature that becomes worse over time, contributing to an open bite (Fig. 19.11) that may affect articulation. There is about a 30 to 35% chance of overt cleft palate, and an additional 30 to 40% chance of other velopharyngeal problems (submucous clefts, inadequate velopharyngeal closure without a cleft). In addition, the pharynx is small in size (Peterson-Falzone and Pruzansky, 1976; Shprintzen et al., 1979), partially because of posterior displacement of the tongue secondary to the mandibular deformity. Another factor contributing to the pharyngeal crowding may be a progressive bending of the cranial base, moving the posterior pharyngeal wall forward (Peterson-Falzone and Figueroa, 1986) (Fig. 19.12). All these factors create distortions in the supralaryngeal vocal tract.

[13] Goldberg and colleagues (1993) reported that an obtuse cranial base was present in 75% of their 120 patients, but in an earlier report from the same group (Arvystas and Shprintzen, 1982) the cranial base was misidentified, leading one to question the conclusion drawn in the later report.

[14] This is an important point with regard to many disorders, but with regard to mandibulofacial dysostosis there is an additional concern: The facial features are so similar to Nager syndrome that all but the most experienced diagnosticians may easily confuse the two. In Nager syndrome, there are digital anomalies that are not present in MFD, but these may be so minor that they are missed. For the speech scientist or speech pathologist, the more important factor is the high likelihood of complete agenesis of the velum in Nager individuals (Peterson-Falzone et al., 2001).

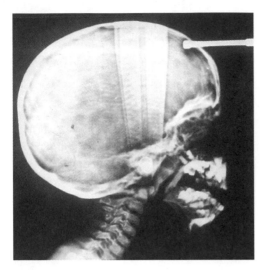

Figure 19.11 Lateral x-ray of an individual with mandibulofacial dysostosis, demonstrating the curvature of the lower border of the mandible, which becomes worse with time.

Figure 19.12 The progressive bending of the cranial base in mandibulofacial dysostosis. (*Source*: From Peterson-Falzone et al., 2001, p. 45.)

Perceptual descriptions of speech in MFD have used the adjectives "muffled" and "hot potato" trying to describe the listener's impression of speech produced with a very crowded vocal tract. One acoustic study of 12 patients (Forner et al., 1984) showed a general damping of formants, with an overall reduction in the acoustic energy represented in the spectrum, and some shifts in the central frequencies of the second and third formants (the formants that carry most of the information on vowel identity to the listener). In general, the spectral information seemed to be reduced in specific acoustic cues to vowel identity and in overall energy. The acoustic information found in these spectral analyses appeared to reflect the consequences of supralaryngeal vocal tracts that were not able to accomplish the full range of movement found in normal vocal tracts.[15]

It is difficult for researchers to find MFD cases in whom the effects of the aberrant shape of the vocal tract can be studied in isolation without the contaminating factor of inadequate velopharyngeal closure, especially because the extremely small airway can hinder or even prevent successful surgical or prosthetic treatment of VPI.

19.3.3 Three Syndromes of Craniofacial Synostosis

There is a family of severe congenital anomalies of craniofacial growth and development known under the general rubric of craniofacial synostosis. The family is a large one, including more disorders than can be described in this chapter. Each of the disorders discussed here is rare in occurrence, but worthy of mention for two reasons: (1) speech pathologists have a greater chance of encountering these children in their caseloads now than was the

---

[15] Interestingly, other etiological sources of vocal tract abnormalties can also produce the perceptual judgments and aberrant spectrographic measurements found in mandibulofacial dysostosis. In a series of

studies on patients with obstructive sleep apnea, Monoson and colleagues (Fox et al., 1989; Monoson and Fox, 1987; Monoson et al., 1991) reported that spectral analysis of speakers with sleep apnea showed abnormally broadened bandwidths, reduced formant amplitudes, and overall flatttening of the spectrum. Perhaps a crowded, boggy airway produces similar abnormalities in the speech output spectrum even when the causes of the abnormal airway dimensions are disparate.

case many years ago because better early intervention has placed the children in the mainstream of the public school system, and (2) the distortions of the supralaryngeal vocal tract in these disorders challenge both clinicians and basic scientists who try to relate structure to function.

The syndromes of Apert, Crouzon, and Pfeiffer share similar dysmorphic features both in terms of facial appearance and the likely distortions of the supralaryngeal vocal tract. Each is characterized by premature synostosis of one or more of the cranial sutures, plus sutures of the facial skeleton. The shape of the skull is determined by the particular sutures that are synostosed. In each of these three syndromes, there is a lack of appropriate forward growth of the midface due to the premature closure of the sutures of the cranial base and the spheno-occipital synchondrosis. The common facial features include ocular proptosis, beak-shaped nose with depressed nasal bridge, and a retruded midface that contributes to both the proptosis (due to the deficiency of infraorbital bony support for the eyes and also to flattened orbits) and class III malocclusion.

Apert syndrome (Figs. 19.13 and 19.14), also known as acrocephalosyndactyly type I, is characterized by bilateral coronal suture synostosis, with variable involvement of other cranial sutures and synostosis of the skull base. There is comparatively little patient-to-patient variation in the severity of the craniofacial dysmorphology. The maxilla is markedly hypoplastic and compressed into a V shape in both the coronal and transverse planes. Contributing to this distorted shape is the progressive accumulation of soft tissue along the palatal shelves, producing a configuration that may be mistaken for a palatal cleft (Fig. 19.15) (Peterson and Pruzansky, 1974). True clefts, either overt or submucous, occur in an estimated 30% of patients. In addition to the craniofacial abnormalities, Apert patients have severe anomalies

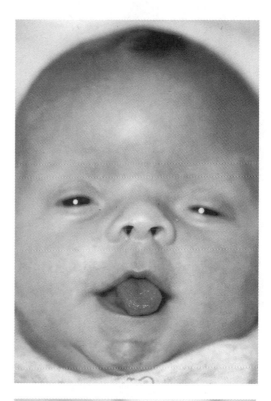

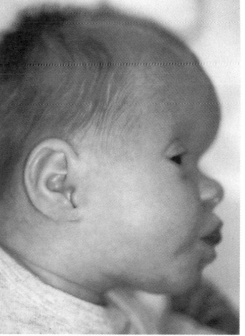

Figure 19.13 Frontal and lateral views of a baby with Apert syndrome. Note the protrusion of the tongue, necessitated by inadequate intraoral space, and the beak-shaped nose with depressed nasal bridge.

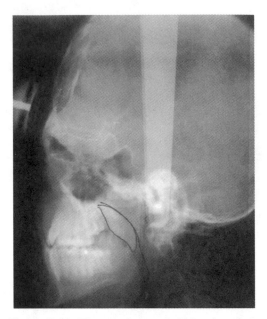

Figure 19.14 The very long and thick velum (outlined in black) in a teenager with Apert syndrome. Note the towering skull and the retruded position of the midface.

of the hands and feet (syndactyly with progressive calcification) and progressive synostosis of the elbows, shoulders, hips, and knees.

There are multiple aberrations of the shape and function of the vocal tract in Apert syndrome.[16] The midface hypoplasia results in a lack of normal nasopharyngeal and oropharyngeal depth; added to this is the fact that the velum tends to be overly long and thick (Fig. 19.14), further crowding the already diminished pharyngeal space (Peterson and Pruzansky, 1974; Peterson-Falzone et al., 1981). Some children require tracheotomy because even open-mouth breathing cannot provide them with enough air exchange. The nasal airway is reduced not only as a result of

[16] Even the subglottic airway is at risk in the syndromes of Apert, Crouzon, and Pfeiffer, due to tracheal malformations (see Peterson-Falzone et al., 2001).

the nasopharyngeal crowding but by the frequently associated finding of choanal atresia. Many patients are markedly hyponasal. The accumulation of soft tissue along the palatal shelves literally obliterates the palatal vault. The tongue may not have adequate room to reach clearly differentiated articulatory targets (e.g., linguo-alveoluars versus velars) or to retroflex for /r/. Articulation is also imperiled by the class III malocclusion and open bite. Additional features that can affect speech include recurrent ear disease and conductive hearing loss due to nasopharyngeal crowding, as well as mental retardation, which is a frequent finding in this population.

Peterson-Falzone and Landahl (1981) reported the cumulative results of a series of studies on vowel formants in speakers with the syndromes of Apert and Crouzon. They found abnormal formant frequency distributions with intervowel relationships that resembled the norm but intravowel relationships (degree of convergence and divergence of formants) that were significantly different from the norm. They also found (1) constriction of the vowel space that seemed to reflect the crowded oral and pharyngeal cavities; (2) marked confusion or overlap of F_2 versus F_1 data points in comparison to the classic vowel ellipses of Peterson and Barney (1952); (3) a tendency toward elevated fundamental frequencies that made spectrographic measurement of F_1 difficult; and (4) appearance of unexplained spectrographic features such as antireonances and extra resonances usually associated with hypernasality despite the fact that none of these speakers had an inadequate velopharyngeal closure. In general, the F_2 versus F_1 data points clustered around a neutral or *schwa* position, again probably reflecting the reduced ability of the oral structures to reach the clearly distinct positions necessary to produce normal vowels. In the 20 years since these studies were done, no percep-

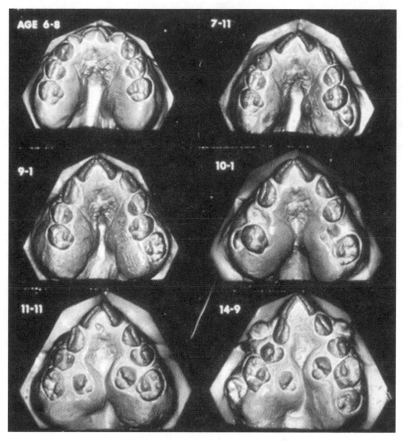

Figure 19.15 Serial dental casts from a child with Apert syndrome, showing the progressive accumulation of soft tissue along the lateral palatine shelves. Note also how the V shape of the maxilla becomes more remarkable over time. (*Source*: From Peterson-Falzone et al., 2001, p. 54.)

tual studies of the intelligibility of individuals with syndromes of premature craniofacial synostosis have focused on the vowels. On the other hand, it would probably be inappropriate for the speech pathologist trying to help a Crouzon or Apert speaker to focus on acoustic vowel space. The ability to change that space is dependent on physical changes of the oral and pharyngeal portions of the vocal tract.

Crouzon disease, also called craniofacial dysostosis, is much more variable in severity from patient to patient. This syndrome entails synostosis of the coronal suture, with a high probability of synostosis of the sagittal and lamdoidal sutures. The facial dysmorphology ranges from mild to severe (Figs. 19.16 and 19.17). Mental retardation is not a common feature of the syndrome, and many patients lead normal lives in society. However, the threats to development in early childhood can be just as plentiful and complex as in Apert syndrome. The same nasopharyngeal dysmorphology and oropharyngeal crowding lead to respiratory problems that may require tracheotomy. The nasopharyngeal crowding makes the child vulnerable to recurrent ear disease.

The same midface hypoplasia and malocclusion occur as in Apert syndrome. Many patients show accumulation of soft

Figure 19.16 Facial features in a youngster with Crouzon disease. (*Source*: From Peterson-Falzone et al., 2001, p. 57.)

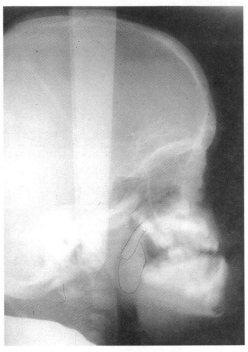

Figure 19.17 Lateral cephalometric film of an individual with Crouzon disease, demonstrating the midface retrusion and the very crowded airway. In this patient, the enlarged tonsil exacerbates the airway problem. (*Source*: From Peterson-Falzone et al., 2001, p. 57.)

tissue along the lateral palatine shelves, but this does not occur as often as in Apert, and, when it does occur, it may not be quite as severe. The velum, again, tends to be long and thick. Thus, the features that might affect speech include the maxillary hypoplasia, nasopharyngeal crowding, possible prolonged tracheotomy, malocclusion, and hearing loss. To date, there have been no acoustic studies of speakers with Crouzon.

Pfeiffer syndrome, in its classic form,[17] entails synostosis of the coronal suture with very similar facial features to those of Apert syndrome (Fig. 19.18). Clinically, these patients are recognizable by the presence of broad thumbs and great toes, and partial soft tissue syndactyly of the hands. Pfeiffer patients are subject to the same nasopharyngeal crowding and respiratory

[17] Cohen (1993) segregated Pfeiffer patients into three categories, his type 1 being the classic form described here. Only type 1 is considered compatible with life; patients in types 2 and 3 have such severe anomalies, including central nervous system anomalies, that the expectation is for early death.

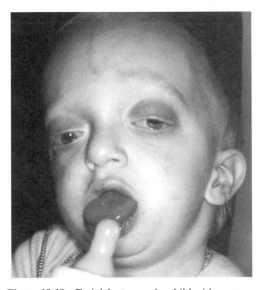

Figure 19.18 Facial features of a child with a severe expression of Pfeiffer syndrome. (*Source*: From Peterson-Falzone et al., 2001, p. 59.)

problems as seen in Apert patients and many Crouzon patients, and also to tracheal anomalies. As with Apert and Crouzon patients, the size of the airway is not always effectively increased with midface advancement (Peterson-Falzone et al., 2001). The nasopharyngeal crowding contributes to ear disease, but there are also cases of stenosis or atresia of the auditory canal, middle ear hypoplasia, and ossicular hypoplasia (Vallino-Napoli, 1996). Again, to date there have been no acoustic studies of the speech of Pfeiffer patients.

There have been a few articles presenting perceptual descriptions of the speech of patients with the craniofacial synostosis syndromes described here (Elfenbein et al., 1981; Peterson, 1973) and a limited number of descriptions of changes in speech pursuant to midface advancement (McCarthy et al., 1977; Guyette et al., 1996). The latter studies indicated that articulation may be improved when the jaw relationship is improved via surgical advancement or advancement by distraction osteogenesis, and that velopharyngeal closure for speech is not apt to be imperiled in patients whose anomalies do not include cleft palate. This body of literature is still growing, particularly as the result of the current popularity of distraction osteogenesis.

19.4 SUMMARY

The distortions of the supralaryngeal vocal tract in the congenital anomalies discussed here present us with the opportunity to compare the effects on speech of abnormal anatomy and physiology to the speech output of the physically normal system. The challenge for both speech scientists and speech pathologists is to decide what structural parameters are necessary, or must at least be approximated, if a patient is to have normal speech. We have a substantial amount of physical data (cephalometric, videofluoroscopic, computerized tomographic, endoscopic, aerodynamic, acoustic) to tell us what is "normal" in both structure and function. Are there inviolable physical parameters for speech? If so, why can some speakers who do not have all the normal physical equipment maintain intelligibility? What information can we give to surgeons (and other specialists who can change the physical speech mechanism) to help bring an individual with any of these anomalies up to the norm for communicative function in our society?

One relatively recent advance in this endeavor is the advent of magnetic resonance imaging (MRI), which is taking a more visible role in the diagnosis of probems of the velopharyngeal system as well as the rest of the supralaryngeal vocal tract (Baer et al., 1987, 1990; Johns and Rohrich, 1995; Moon, 1993; Moore, 1992; Sulter et al., 1992). Currently, such studies are extemely expensive, but (1) they carry no known risk to the speaker (as opposed to radiographic imaging), (2) the technology now includes real-time imaging, and (3) for many children, this imaging technique should be far less frightening than nasopharyngoscopy. The new information derived with MRI, and with whatever advanced imaging technique follows it, should provide speech scientists and speech pathologists with a wealth of information that can be used to bring speakers with congenital craniofacial anomalies increasingly closer to the norm in communicative function.

REFERENCES

Arvystas, M., and Shprintzen, R. J. (1982). Craniofacial morphology in Treacher Collins syndrome. *J. Craniofac. Genet. Devel. Biol.* 4, 39–45.

Baer, T., Gore, J. C., Boyce, S., and Nye, P. W. (1987). Application of MRI to the analysis of speech production. *Magn. Reson. Imag. 5,* 1–7.

Baer, T., Gore, J. C., Gracco, L. C., and Nye, P. W. (1990). Analysis of vocal tract shape and dimensions using magnetic resonance imaging: vowels. *J. Acoust. Soc. Am. 90*, 799–828.

Bosma, J. F. (1986). *Anatomy of the Infant Head* (Baltimore: Johns Hopkins Press).

Brodie, A. G. (1955). The behavior of the cranial base and its components as revealed by serial cephalometric roentgenograms. *Angle Orthod. 25*, 148–160.

Brodie, A. G. (1961). Comparative anatomy of the pharynx. In S. Pruzansky (ed.), *Congenital Anomalies of the Face and Associated Structures* (Springfield: Charles C. Thomas), pp. 251–268.

Chow, E. W. C., Bassett, A. S., and Weksberg, R. (1994). Velo-cardio-facial syndrome and psychotic disorders: implications for psychiatric genetics. *Am. J. Med. Genet. 54*, 107–112.

Cohen, M. M., Jr. (1993). Pfeiffer syndrome update: clinical subtypes and guidelines for differential diagnosis. *Am. J. Med. Genet. 45*, 300–307.

Cole, R. M. (1970). Speech. In *ASHA Report #6: Patterns of Orofacial Growth and Development* (Washington, DC: American Speech and Hearing Association), pp. 79–95.

Cox, R. L. (1970). Muscular development and maturation of the dentofacial complex: normal and abnormal. In *ASHA Report #5: Speech and the Dentofacial Complex* (Washington, DC: American Speech and Hearing Association), pp. 20–32.

Crelin, E. S. (1973). *Functional Anatomy of the Newborn* (New Haven, CT: Yale University Press).

Curtis, J. F. (1968). Acoustics of speech production and nasalization. In D. C. Spriestersbach and D. Sherman (eds.), *Cleft Palate and Communication* (New York: Academic Press), pp. 27–60.

Dickson, D. R. (1962). An acoustic study of nasality. *J. Speech Hear. Res. 5*, 103–111.

Elfenbein, J., Waziri, M., and Morris, H. L. (1981). Verbal communication skills of six children with craniofacial anomalies. *Cleft Palate J. 18*, 59–64.

Enlow, D. H. (1968). *The Human Face: An Account of the Postnatal Growth and Development of the Craniofacial Skeleton* (Evanston, IL: Harper and Row–Hoeber Medical Division).

Enlow, D. H. (1990). *Facial Growth*, 3rd ed. (Philadelphia: W. B. Saunders).

Fant, G. (1960). *Acoustic Theory of Speech Production* (The Hague: Mouton).

Fletcher, S. G. (1960). Hypernasal voice as an indication of regional growth and developmental disturbances. *J. Speech Hear. Res. 3*, 3–12.

Forner, L. L., Peterson-Falzone, S. J., and Kretschmer, L. W. (1984). Acoustic characteristics of speech in mandibulofacial dysostosis. Presented before the Annual Convention of the American Speech-Language-Hearing Association, San Francisco.

Fox, A. W., Monoson, P. K., and Morgan, C. D. (1989). Speech dysfunction of obstructive sleep apnea. *Chest 96*, 589–595.

George, S. I. (1978). A longitudinal and cross-sectional analysis of the growth of the postnatal cranial base angle. *Am. J. Phys. Anthropol. 49*, 171–178.

Glander, K., and Cisneros, G. J. (1992). Comparison of the craniofacial characteristics of two syndromes associated with the Pierre Robin sequence. *Cleft Palate Craniofac. J. 29*, 210–219.

Goldberg, R. B., Moszkin, B., Marion, R., Scambler, P. J., and Shprintzen, R. J. (1993). Velo-cardio-facial syndrome: a review of 120 patients. *Am. J. Med. Genet. 45*, 313–319.

Gorlin, R. J., Cohen, M. M., Jr., and Levin, L. S. (1990). *Syndromes of the Head and Neck* (New York: Oxford University Press).

Guyette, T. W., Polley, J. W., Figueroa, A. A., and Cohen, M. N. (1996). Mandibular distraction osteogenesis: effects on articulation and velopharyngeal function. *J. Craniofac. Surg. 7*, 186–191.

Hattori, S., Yamamoto, K., and Fujimura, O. (1958). Nasalization of vowels in relation to nasals. *J. Acoust. Soc. Am. 29*, 267–274.

Johns, D. F., and Rohrich, R. J. (1995). Functional magnetic resonance imaging of the velopharynx: technique and future application.

Presented before the American Cleft Palate–Craniofacial Association, Tampa.

Karayiorgou, M., Morris, M. A., Morrow, B., Shprintzen, R. J., Goldberg, R. B., Borrow, J., Gos, A., Nestadt, G., Wolyniec, P. S., Lasseter, V. K., Eisen, H., Childs, B., Kazazian, H. H., Kucherlapati, R., Antonarakis, S. E., Pulver, A. D., and Housman, D. E. (1995). Schizophrenia susceptibility associated with interstitial deletions of chromosome 22q11. *Proc. Natl. Acad. Sci. USA 92*, 7612–7616.

King, E. W. (1952). A roentgenographic study of pharyngeal growth. *Angle Orthod. 22*, 23–27.

Krogman, W. M. (1961). The growth of the head and face studied craniometrically and cephalometrically, in normal and cleft palate children. In S. Pruzansky (ed.), *Congenital Anomalies of the Face and Associated Structures* (Springfield: Charles C. Thomas), pp. 208–250.

Lieberman, P. (1975). *On the Origins of Language* (New York: Macmillan).

Lucarini, J. W., Liberfarb, R. M., and Eavey, R. D. (1987). Otolaryngological manifestations of the Stickler syndrome. *Int. J. Pediatr. Otorhinolaryngol. 14*, 215–222.

Manson, J. D. (1968). *A Comparative Study of the Postnatal Growth of the Mandible* (London: Henry Kimpton).

May, R., and Sheffer, D. B. (1999). Growth changes in internal and craniofacial flexion measurements. *Am. J. Phys. Anthropol. 110*, 47–56.

McCarthy, J. G., Coccaro, J. P., and Schwartz, M. D. (1977). Velopharyngeal function following maxilary advancement. *Plast. Reconstr. Surg. 64*, 180–189.

McWilliams, B. J., Bluestone, C. D., and Musgrave, R. H. (1969). Diagnostic implications of vocal cord nodules in children with cleft palate. *Laryngoscope 79*, 2072–2080.

McWilliams, B. J., Lavorato, A. S., and Bluestone, C. D. (1973). Vocal cord abnormalities in children with velopoharyngeal valving problems. *Laryngoscope 83*, 1745–1753.

Meredith, H. V. (1959). A longitudinal study of growth in face depth during childhood. *Am. J. Phys. Anthropol. 17*, 125–135.

Monoson, P. K., and Fox, A. W. (1987). Preliminary observation of speech disorder in obstructive and mixed sleep apnea. *Chest 92*, 670–675.

Monoson, P. K., Prosek, F. A., and Fox, A. W. (1991). Selected acoustic features of the speech of talkers with speech apnea. Presented before the American Speech-Language-Hearing Association, Atlanta.

Moon, J. B. (1993). Evaluation of velopharyngeal function. In K. T. Moller and C. D. Starr (eds.), *Cleft Palate: Interdisciplinary Issues and Treatment* (Austin, TX: Pro-Ed), pp. 251–306.

Moore, C. A. (1992). The correspondence of vocal tract resonance with volumes obtained from magnetic resonance images. *J. Speech Hear. Res. 35*, 1009–1023.

Moss, E. M., Batshaw, M. L., Solot, C. B., Gerdes, M., McDonald-McGinn, D. M., Driscoll, D. A., Emanuel, B. S., Zackai, E. H., and Wang, P. P. (1999). Psychoeducational profile of the 22q11.2 microdeletion: a complex pattern. *J. Pediatr. 134*, 193–198.

Moss. M. (1956). Malformation of the skull base associated with cleft palate deformity. *Plast. Reconstr. Surg. 17*, 226–234.

Moyers, R. E. (1970). Postnatal development of the orofacial musculature. In *ASHA Report #6: Patterns of Orofacial Growth and Development* (Washington, DC: American Speech and Hearing Association), pp. 38–47.

Muller, G. (1963). Growth and development of the middle face. *J. Dent. Res. 42*, 385–399.

O'Gara, M. M., and Logemann, J. A. (1988). Phonetic analysis of the speech development of babies with cleft palate. *Cleft Palate J. 25*, 122–134.

O'Gara, M. M., Logemann, J. A., and Rademaker, A. W. (1994). Phonetic features by babies with unilateral cleft lip and palate. *Cleft Palate Craniofac. J. 31*, 446–451.

Osborne, G., Pruzansky, S., and Koepp-Baker, H. (1971). Upper cervical spine anomalies and osseous nasopharyngeal depth. *J. Speech Hear. Res. 14*, 14–22.

Peterson, G., and Barney, H. (1952). Control methods used in a study of the vowels. *J. Acoust. Soc. Am. 24*, 175–184.

Peterson, S. J. (1973). Speech pathology in craniofacial malformations other than cleft lip and palate. In *ASHA Report 8: Orofacial Anomalies: Clinical and Research Implications* (Washington, DC: American Speech and Hearing Association), pp. 111–131.

Peterson, S. J., and Pruzansky, S. (1974). Palatal anomalies in the syndromes of Apert and Crouzon. *Cleft Palate J. 11*, 394–403.

Peterson-Falzone, S. J., and Figueroa, A. A. (1986). Longitudinal changes in cranial base angulation in mandibulofacial dysostosis. *Cleft Palate J. 26*, 114–122.

Peterson-Falzone, S. J., and Landahl, K. L. (1981). Effect of aberrant supralaryngeal vocal tracts on transfer function. *Speech Lang.: Adv. Bas. Res. Prac. 6*, 265–303.

Peterson-Falzone, S. J., and Pruzansky, S. (1976). Cleft palate and congenital palatopharyngeal incompetency in mandibulofacial dysostosis: frequency and problems in treatment. *Cleft Palate J. 23*, 354–360.

Peterson-Falzone, S. J., Pruzansky, S., Parris, P. J., and Laffer, J. L. (1981). Nasopharyngeal dysmorphology in the syndromes of Apert and Crouzon. *Cleft Palate J. 19*, 237–250.

Peterson-Falzone, S. J., Hardin-Jones, M. A., and Karnell, M. P. (2001). *Cleft Palate Speech*, 3rd ed. (St. Louis: C. V. Mosby).

Phillips, B. J., and Kent, R. D. (1984). Acoustic-phonetic descriptions of speech production in speakers with cleft palate and other velopharyngeal disorders. *Speech Lang.: Adv. Bas. Res. Prac. 11*, 113–168.

Pulver, A. E., Nestadt, G., Goldberg, R., Shprintzen, R. J., Lamacz, J., Wolyniec, P. S., Morrow, B., Karayiorgou, M., Antonarakis, S. E., Houslman, D., and Kucherlapati, R. (1994). Psychotic willness in patients diagnosed with velo-cardio-facial syndrome and their relatives. *J. Nerv. Ment. Dis. 182*, 476–478.

Riolo, M. L., Moyers, R. E., McNamara, J. A., and Hunter, W. S. (1974). *An Atlas of Craniofacial Growth: Cephalometric Standards from the University School Growth Study*. Monograph 2: Cranial Facial Growth Series: 41 (Ann Arbor: Center for Human Growth and Development, University of Michigan).

Robin, P. (1923). La chute de la base de la langue consideree comme und veile cause do gene dans la respiration naso-pharyngienne. *Bull. Acad. Nat. Med.* (Paris) *89*, 37–41.

Robin, P. (1934). Glossoptosis due to atresia and hypotrophy of the mandible. *Arch. Pediatr. Adol. Med. 48*, 541–547.

Saksena, S. S., Bixler, D., and Yu, P. (1983). Stickler syndrome: a cephalometric study of the face. *J. Craniofac. Genet. Devel. Biol. 3*, 19–28.

Scott, J. H. (1953). The cartilage of the nasal septum (a contribution to the study of facial growth). *Brit. Dent. J. 95*, 37–43.

Sedlackova, E. (1955). Insuficience patrohitanoveho zaveru; jako vyvojova porucha (Insufficiency of the palatolaryngeal passage disorder). *Casopis Lekaru Ceskvch (Prague) 94*, 1304–1307.

Sher, A., Shprintzen, R. J., and Thorpy, M. J. (1986). Endoscopic observations of obstructive sleep apnea in children with anomalous upper airways: predictive and therapeutic value. *Int. J. Pediatr. Otorhinolaryngol. 11*, 135–146.

Shprintzen, R. J. (1988). Pierre Robin, micrognathia, and airway obstruction: the dependency of treatment on accurate diagnosis. *Int. Anesth. Clin. 26*, 64–71.

Shprintzen, R. J. (1992). The implications of the diagnosis of Robin sequence. *Cleft Palate Craniofac. J. 29*, 205–209.

Shprintzen, R. J., and Singer, L. (1992). Upper airway obstruction and the Robin sequence. *Int. Anesth. Clin. 30*, 109–114.

Shrpintzen, R. J., Goldberg, R. B., Lewin, M., Sidoti, E., Berkman, M., Argamaso, R., and Young, D. (1978). A new syndrome involving cleft palate, cardiac anomalies, typical facies, and learning disabilities: velo-cardio-facial syndrome. *Cleft Palate J. 15*, 56–62.

Shprintzen, R. J., Croft, C., Berkman, M. D., and Rakoff, S. (1979). Pharyngeal hypoplasia in Treacher Collins syndrome. *Arch. Otolaryngol. 105*, 127–131.

Shprintzen, R. J., Siegel-Sadewitz, V. L., Amato, J., and Goldberg, R. B. (1985a). Anomalies associated with cleft lip, cleft palate, or both. *Am. J. Med. Genet. 20*, 585–595.

Shprintzen, R. J., Siegel-Sadewitz, V. L., Amato, J., and Goldberg, R. B. (1985b). Retrospective diagnoses of previously missed syndromic disorders among 1000 patients with cleft lip, cleft palate, or both. *Birth Defects: Original Article Series 21*, 85–92.

Shprintzen, R. J., Goldberg, R. B., Golding-Kushner, K. J., and Marion, R. (1992). Late-onset psychosis in velo-cardio-facial syndrome [letter]. *Am. J. Med. Genet. 42*, 141–142.

Solot, C. B., Knightly, C., Handler, S. D., Gerdes, M., McDonald-McGinn, M., Moss, E., Wang, P., Cohen, M., Randall, P., LaRossa, D., Driscoll, D. A., Emanuel, B. S., and Zackai, E. H. (2000). Communication disorders in the 22q11.2 microdeletion syndrome. *J. Commun. Dis. 33*, 187–204.

Sperber, G. H., and Tobias, P. H. (1989). *Craniofacial Embryology* (London: Wright).

Stickler, G., and Pugh, D. (1967). Hereditary progressive arthro-ophthalmolopathy II. Additional observations on vertebral abnormalities, a hearing defect, and a report of a similar case. *Mayo Clin. Proc. 42*, 495–500.

Stickler, G., Bealu, P., Farrell, F., Jones, J., Pugh, D., Steinberg, A., and Ward, L. (1965). Hereditary progressive arthroophthalmopathy. *Mayo Clin. Proc. 40*, 433–455.

Subtelny, J. D. (1955). Width of the nasopharynx and related anatomic structures in normal and unoperated cleft palate children. *Am. J. Orthod. 41*, 889–909.

Sulter, A. M., Miller, D. G., Wolf, R. F., Schutte, H. K., Wit, H. P., and Mooyaart, E. L. (1992). On the relation between the dimensions and resonance characteristics of the vocal tract: a study with MRI. *Magn. Reson. Imag. 10*, 365–373.

Trost, J. E. (1981). Articulatory additions to the classical descriptions of the speech of persons with cleft palate. *Cleft Palate J. 18*, 193–203.

Trost-Cardamone, J. E. (1990). The development of speech: assessing cleft palate misarticulations. In D. A. Kernahan and S. W. Rosenstein (eds.), *Cleft Lip and Palate: A System of Management* (Baltimore: Williams and Wilkins).

Trost-Cardamone, J. E., and Bernthal, E. (1993). Articulation assessment and procedures and treatment decisions. In K. T. Moller and C. D. Starr (eds.), *Cleft Palate: Interdisciplinary Issues and Treatment* (Austin: Pro-Ed), pp. 307–336.

Vallino-Napoli, L. D. (1996). Audiology and otologic characteristics of Pfeiffer syndrome. *Cleft Palate Craniofac. J. 33*, 524–529.

Zemlin, W. R. (1998). *Speech and Hearing Science: Anatomy and Physiology* (Boston: Allyn and Bacon).

PART VII

MOLECULAR BIOLOGY STUDIES AND FUTURE DIRECTIONS

CHAPTER 20

MOLECULAR STUDIES OF CRANIOSYNOSTOSIS: FACTORS AFFECTING CRANIAL SUTURE MORPHOGENESIS AND PATENCY

LYNNE A. OPPERMAN, Ph.D. and ROY C. OGLE, Ph.D.

20.1 INTRODUCTION: REGULATION OF SUTURE MORPHOGENESIS AND PATENCY

Regulation of cranial suture morphogenesis and patency is a prolonged process initiated during early embryogenesis and completed upon fusion of adjacent bones during adulthood. Suture morphogenesis begins when the expansion and mineralization of mesenchymal cell blastemas result in the expanding bone fronts approximating one another and either overlapping or butting up against one another. Sutures in the cranial midline (interfrontal, sagittal) generally form as butt sutures, while those on either side of the midline (coronal, lambdoid, frontonasal) form overlapping sutures (see Chapter 9, especially Fig. 9.4).

For the rat coronal suture, the approximating bone fronts are initially separated by presumptive suture mesenchyme (Fig. 20.1, E19). Suture formation is initiated over the next 72 hours by the appearance of a highly cellular suture blastema be-

tween the overlapping bone fronts (Fig. 20.1, P1). As the suture becomes stabilized, it is remodeled to a narrower, more fibrous tissue layer separating the frontal and parietal bones (Fig. 20.1, P5 and P21). Bone growth occurs through intramembranous bone formation at the leading edges of the bone fronts in response to extrinsic signals, such as the expanding neurocranium, which drive the bone fronts apart. These signals appear to be transmitted via the dura mater.

For the suture to continue to contribute to the expansion of the cranial bones, it must remain unossified. However, sufficient new osteoblasts need to be generated at the bone fronts to contribute to the production and mineralization of bony matrix.

20.1.1 General Principles

Abnormal suture biology can arise from several possible causes. Failure of the bone fronts to approximate one another will result in absence of suture formation and wide-open fontanels, such as seen in clei-

Understanding Craniofacial Anomalies: The Etiopathogenesis of Craniosynostoses and Facial Clefting, Edited by Mark P. Mooney and Michael I. Siegel, ISBN 0-471-38724-x Copyright © 2002 by Wiley-Liss, Inc.

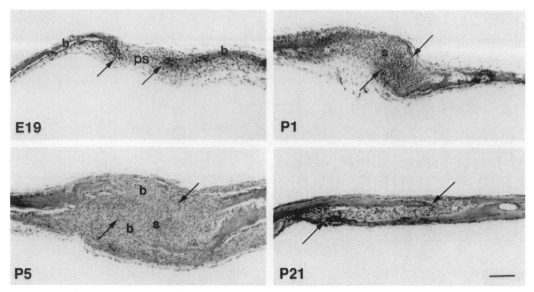

Figure 20.1 Photomicrographs of hematoxylin and eosin stained histologic sections through normal rat coronal sutures, showing various stages of morphogenesis. Tissues are oriented with periosteal surface at the top and dura mater underneath. Arrows indicate the leading edges of the bone fronts (b). E19, embryonic day 19 suture showing frontal and parietal bones (b) widely separated by presumptive suture (ps) mesenchyme. Note the highly cellular nature of the sutural edges of the bone fronts (arrows). P1, postnatal day 1 suture showing overlapping frontal and parietal bones separated by a highly cellular suture blastema (s). P5, postnatal day 5 suture showing gradual thickening of the bones on either side of the suture and increased overlap of the bone fronts compared to P1. P21, postnatal day 21 sutures showing dramatic remodeling of the suture matrix to a narrow strip of fibrous, cellular material separating the overlapping frontal and parietal bones. Bar = 100 μm. (*Source*: Reprinted from Opperman, 2000.)

docranial dysplasia. Inability of bone fronts to appropriately overlap or butt up against one another during early suture formation can result in bony obliteration of the suture site. Lastly, inability to sustain a suture once it is formed will result in premature osseous obliteration of the suture site. While failure of bone fronts to approximate one another is likely due to inhibition or delay of bone formation, bony obliteration of the suture site is caused by premature or accelerated bone formation within the fibrous suture matrix.

20.1.2 Role of Dura Mater in Regulating Suture Patency

As the bone fronts overlap one another during initiation of coronal suture forma-

tion, the presumptive mesenchyme will become divided, with the mesenchyme between the overlapping bone fronts becoming the suture blastema, the outer mesenchyme becoming the ectoperiosteal layer, and the inner mesenchyme becoming the dura mater (Kokich, 1986). Early experiments examined the contribution of either the outer periosteum or the inner dura mater to suture formation and maintenance. These experiments used rat embryonic day 19 (E19) and postnatal day 1 (P1) coronal sutures with surrounding frontal and parietal bones transplanted into a bony defect created in the middle of adult rat parietal bones (see Chapter 9, especially Fig. 9.5), either with dura mater or periosteum present or removed prior to transplant. It was found that neither dura mater nor ectoperiosteum was required for initial suture formation, as the bone fronts over-lapped each other even in the absence of these tissues (Fig. 20.2). However, the presence of the dura mater was required to maintain the suture long term (Fig. 20.3). Removal of the periosteum did not affect long-term maintenance of the sutures. From these experiments, it was hypothesized that the dura mater provided a stabilizing signal to the newly formed suture (Fig. 20.4B), and once the suture became stabilized, it in turn induced an osteoinhibitory signal within the dura underneath the suture (Fig. 20.4C). Failure of suture stabilization or failure to produce an osteoinhibitory signal by the suture would then result in suture oblitera-tion (Fig. 20.4D).

When these experiments were repeated in vitro, it was confirmed that the presence of the dura was required for initial stabi-lization of the newly formed suture using E19 rat fetuses (Opperman et al., 1995) (see Chapter 9, especially Fig. 9.5). How-ever, it was found that P1 rat coronal sutures no longer required the presence of dura mater to maintain the sutures in their patent state. Data were supported by similar findings in coronal sutures of mice (Kim et al., 1998).

20.1.3   Role of Growth and Transcription Factors

Several clinical craniofacial pathologies with abnormal cranial sutures (either pre-maturely obliterated, present but not func-tional as bone growth sites, or as wide-open midline defects) have now been found to be associated with mutations in a variety of factors (Table 20.1). Mutations are associ-ated with various regions of the fibroblast growth factor (*FGF*) receptor genes that encode a family of tyrosine kinases that bind FGFs. Interestingly, although the mutations resulted in a variety of receptor malfunc-tions, such as enhanced binding of

ligand or constitutive activation, all known mutations resulted in increased receptor activity. In all cases, the resulting pathology included premature obliteration of cranial suture sites, most commonly the coronal sutures.

The other known mutations occur in the genes of a variety of transcription factors, namely *MSX2*, *TWIST*, and *RUNX2* (pre-viously called *CBFA1*). Mutations result-ing in enhanced DNA binding of MSX2 (the equivalent of increased presence of transcription factor) result in premature obliteration of several cranial sutures. Conversely, mutations resulting in haploin-sufficiency of MSX2 result in a midline cranial defect in which sutures remain widely patent. Similarly, mutations result-ing in haploinsufficiency of RUNX2 result in cleidocranial dysplasia, which presents with widely patent sutures. All reported mutations in *TWIST* also result in haploin-sufficiency, and these patients present with craniosynostosis.

While mutations do occur in other regions of *FGFR* as well as in other tran-scription factors, these mostly appear to occur with low frequency and are more adequately described in recent reviews and books (Cohen, 2000; Elmslie and Reardon, 1998; Hehr and Muenke, 1999; Jabs, 1998; Nuckolls et al., 1999; Sperber, 1999; Webster and Donoghue, 1997; Wilkie, 1997).

20.2   MOLECULAR REGULATION OF SUTURE MORPHOGENESIS AND PATENCY

20.2.1   Growth Factors

Growth factors are small peptides manu-factured by cells in response to extracel-lular signals received by the cells. These factors are secreted from the cell, usually in the form of an inactive pro-peptide and often associated with latency binding

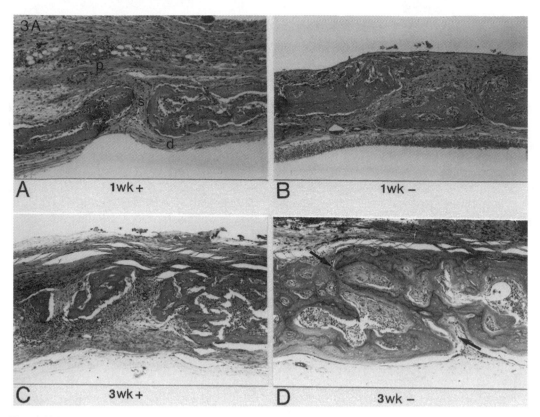

Figure 20.2A Histology of E19 in vivo transplants. (A, B) The osteoinductive fronts have overlapped one another to form a suture after 1 week in the host regardless of presence (1 wk +) or absence (1 wk −) of the fetal dura mater. (C, D) The suture is still present after 3 weeks in the host when the fetal dura mater is left intact (3 wk +), but osseous union has obliterated the fibrous suture in the absence of the fetal dura mater (arrows, 3 wk −). d, dura mater, p, periosteum, s, suture.

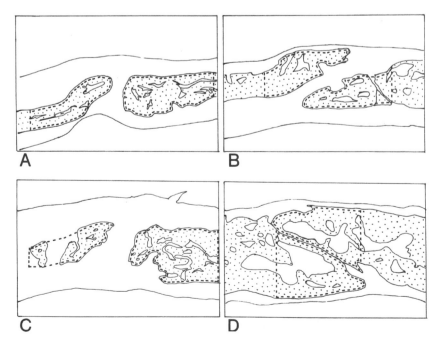

Figure 20.2B Line drawing of sections shown in Figure 20.2A, indicating the cross-sectional area of bone measured (⋯⋯). Position of the suture prior to obliteration in the absence of the fetal dura mater is indicated between arrows in Figure 20.2A (D). (*Source*: Reproduced from Opperman et al., 1993.)

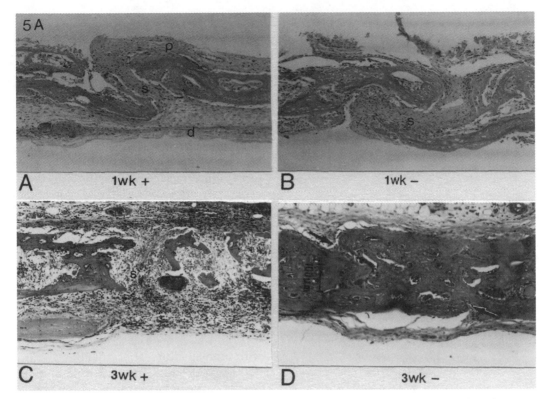

Figure 20.3A Histology of P1 in vivo transplants. (A, B) A fully formed suture is present after 1 week in the host both in the presence (1 wk +) and absence (1 wk −) of neonatal dura mater. (C, D) A suture remains after 3 weeks in the host animal in explants containing neonatal dura mater (3 wk +); removal of the neonatal dura mater prior to transplantation results in fusion (3 wk −) similar to E19 transplants (Figure 20.2). d, dura mater, p, periosteum, s, suture.

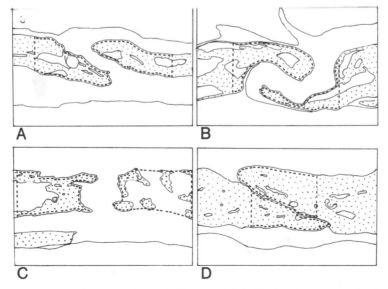

Figure 20.3B Line drawings of sections shown in Figure 20.3A indicating the cross-sectional area of bone measured (⋯⋯). Position of the suture prior to obliteration in the absence of the fetal dura mater is indicated between arrows in Figure 20.3A (D). (*Source*: Reproduced from Opperman et al., 1993.)

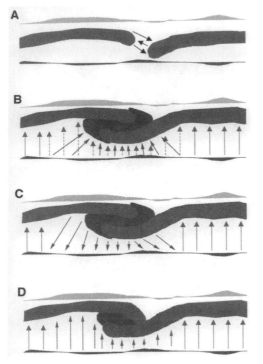

Figure 20.4 Diagramatic representation of various stages of suture morphogenesis (A, B, and C) and suture fusion (D). (A) Inductive signals (arrows) arising from the approaching bone fronts allow the bone fronts to deflect away from each other or butt up against each other without obliterating the suture. These signals are independent of signals from the dura mater or periosteum. (B) Once the bone fronts have overlapped one another, a signal (arrows) arising from the dura mater maintains the presence of the newly formed suture. Osteogenic signals (arrows) from the dura mater cause the bones to become thickened by depositing and mineralizing new osteoid on the periosteal surface. These osteogenic signals may be continuous along the dura mater prior to formation of the suture (dotted arrows). (C) Once the suture is stabilized, it signals (arrows) the local underlying dura mater not to produce osteogenic signals. (D) In the absence of osteoinhibitory signals from the suture, the underlying dura mater remains continuously osteogenic (arrows), overriding signals within the suture and resulting in osseous obliteration of the suture. Periosteum is adjacent. (*Source*: Reprinted from Opperman, 2000.)

proteins. The growth factors remain in the local environment and bind to cell surface receptors upon activation, causing them to co-localize with other specific cell surface receptors, at which time receptor phosphorylation is induced, triggering a cascade of intracellular signaling events and resulting in increased transcriptional activity.

Several growth factors have been identified to be actively involved in the process of suture morphogenesis and in the maintenance of the suture as a functional bone growth site. Each growth factor is discussed separately, after which a unifying hypothesis on how growth and transcription factors interact with one another to regulate suture growth and patency is discussed.

20.2.1.1 Transforming Growth Factors Beta (Tgf-βs) and Their Receptors

While no mutations in *TGF*-βs or their receptors have been described that present with cranial suture abnormalities, there is good evidence that Tgf-β1, Tgf-β2, and Tgf-β3 play crucial roles in regulating suture patency once the sutures have formed. All three Tgf-βs are present in the dura and in the osteoblasts lining the dural and periosteal surfaces of the cranial bones throughout suture morphogenesis and after the suture is fully formed (Opperman et al., 1997; Roth et al., 1997b). However, they have different distributions within the suture matrix and bone fronts during initial suture formation, when the suture is fully formed, and during suture obliteration (Fig. 20.5). While Tgf-βs are absent from or low in suture matrix during initial suture formation and in the patent sutures, suture matrix of fusing sutures contains high levels of Tgf-β1 and Tgf-β2 (Most et al., 1998; Opperman et al., 1997; Roth et al., 1997a, 1997b). The osteogenic bone fronts on either side of the suture contain Tgf-β1 and Tgf-β3 during initial suture formation, and all three Tgf-βs are present in the bone fronts of fully formed sutures. However, Tgf-β3 is absent from the bone fronts of fusing sutures (Opperman et al., 1997; Roth et al., 1997b).

The presence of growth factors in tissues

TABLE 20.1 Common Mutations Associated with Abnormal Cranial Suture Biology

Gene	Protein	Mutation Type	Mechanism	Phenotype	Reference
MSX2	Transcription factor	Missense	Enhanced DNA binding	Boston-type craniosynostosis[a]	Jabs et al., 1993; Ma et al., 1996
MSX2	Transcription factor	Gene deletion	Haploinsufficiency	Wide-open sutures[b]	Wilkie et al., 2000
TWIST	Transcription factor	Various	Haploinsufficiency	Saethre-Chotzen[a]	el Ghouzzi et al., 1997, 1999, 2000; Howard et al., 1997; Paznekas et al., 1998
RUNX2 (CBFA1)	Transcription factor	Missense	Haploinsufficiency	Cleidocranial dysplasia[b]	Mundlos et al., 1997; Quack et al., 1999; Zhang et al., 2000a
FGFR1	Tyrosine kinase receptor	Missense	Enhanced ligand receptor binding	Pfeiffer[a]	Muenke et al., 1994
FGFR2	Tyrosine kinase receptor	Missense	Enhanced ligand receptor binding	Apert[a,b]	Wilkie et al., 1995
FGFR2	Tyrosine kinase receptor	Various	Constitutive activation	Crouzon[a]	Reardon et al., 1994
FGFR2	Tyrosine kinase receptor	Various	Constitutive activation	Jackson-Weiss[a]	Jabs et al., 1994; Park et al., 1995; Tartaglia et al., 1997
FGFR3	Tyrosine kinase receptor	Missense	Enhanced receptor binding (predicted)	Muenke craniosynostosis[a]	Muenke et al., 1997

[a] Fused sutures.
[b] Midline skull defect.

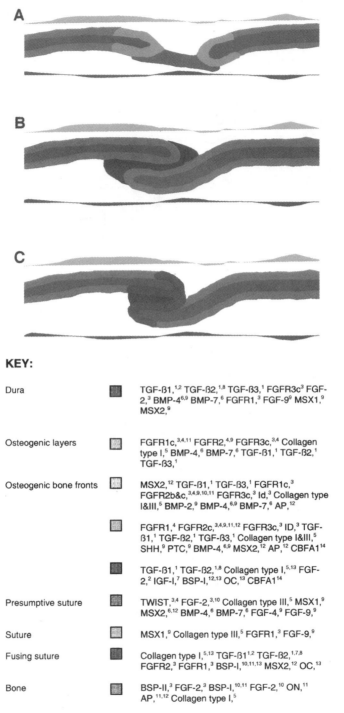

KEY:

Dura		TGF-β1,[1,2] TGF-β2,[1,8] TGF-β3,[1] FGFR3c[3] FGF-2,[3] BMP-4[6,9] BMP-7,[6] FGFR1,[3] FGF-9[9] MSX1,[9] MSX2,[9]
Osteogenic layers		FGFR1c,[3,4,11] FGFR2,[4,9] FGFR3c,[3,4] Collagen type I,[5] BMP-4,[6] BMP-7,[6] TGF-β1,[1] TGF-β2,[1] TGF-β3,[1]
Osteogenic bone fronts		MSX2,[12] TGF-β1,[1] TGF-β3,[1] FGFR1c,[3] FGFR2b&c,[3,4,9,10,11] FGFR3c,[3] Id,[3] Collagen type I&III,[5] BMP-2,[9] BMP-4,[6,9] BMP-7,[6] AP,[12]
		FGFR1,[4] FGFR2c,[3,4,9,11,12] FGFR3c,[3] ID,[3] TGF-β1,[1] TGF-β2,[1] TGF-β3,[1] Collagen type I&III,[5] SHH,[9] PTC,[9] BMP-4,[6,9] MSX2,[12] AP,[12] CBFA1[14]
		TGF-β1,[1] TGF-β2,[1,8] Collagen type I,[5,13] FGF-2,[2] IGF-I,[7] BSP-I,[12,13] OC,[13] CBFA1[14]
Presumptive suture		TWIST,[3,4] FGF-2,[3,10] Collagen type III,[5] MSX1,[9] MSX2,[6,12] BMP-4,[6] BMP-7,[6] FGF-4,[9] FGF-9,[9]
Suture		MSX1,[9] Collagen type III,[5] FGFR1,[3] FGF-9,[9]
Fusing suture		Collagen type I,[5,13] TGF-β1[1,2] TGF-β2,[1,7,8] FGFR2,[3] FGFR1,[3] BSP-I,[10,11,13] MSX2,[12] OC,[13]
Bone		BSP-II,[3] FGF-2,[3] BSP-I,[10,11] FGF-2,[10] ON,[11] AP,[11,12] Collagen type I,[5]

Figure 20.5 Diagramatic representation of a presumptive suture (A), a fully formed suture (B) and a fusing suture (C). The gray scale–coded regions of the sutures, osteogenic bone fronts, and bone are given in the accompanying key. The key lists the growth and transcription factors, receptors, and extracellular matrix components known to be present in each of the stages of suture morphogenesis. ([1]Opperman et al., 1997; [2]Most et al., 1998; [3]Rice et al., 2000; [4]Johnson et al., 2000; [5]Marks et al., 1999; [6]Rice et al., 1999; [7]Roth et al., 1997a; [8]Roth et al., 1997b; [9]Kim et al., 1998; [10]Iseki et al., 1997; [11]Iseki et al., 1999; [12]Liu et al., 1999; [13]Lemmonier et al., 2000; [14]Zhou et al., 2000.) (*Source*: Reprinted from Opperman, 2000.)

gives very little information about what role, if any, they play in regulating suture morphogenesis or patency. Experiments in which neutralizing antibodies to Tgf-βs were placed into cultures of E19 rat calvaria demonstrated that removal of Tgf-β2 rescued fusing sutures from obliteration, while removal of Tgf-β3 induced obliteration in sutures normally remaining patent (Opperman et al., 1999). In the reverse experiments, in which growth factors were added to sutures, addition of Tgf-β2 induced suture fusion, both in vitro (Opperman et al., 2000) and in vivo (Roth et al., 1997a), while addition of Tgf-β3 to normally fusing sutures in vitro rescued them from obliteration (Opperman et al., 2000). Addition of Tgf-β3 in vivo to posterior frontal sutures of rats, which normally fuse between 15 and 21 days after birth, also delays obliteration until at least 24 days after birth (Cooper et al., 2001). In similar experiments done in a congenitally craniosynostotic rabbit model, Tgf-β3 rescued partially fused sutures from obliteration (Chong et al., 2001).

The mechanisms by which Tgf-βs regulate suture patency are beginning to be elucidated. Suture obliteration, either due to removal of dura mater or Tgf-β3 or addition of Tgf-β2, is accompanied by increased cell proliferation (Opperman et al., 1998, 2000). Conversely, rescue of sutures from obliteration by addition of Tgf-β3 is accompanied by decreased cell proliferation (Opperman et al., 2000). Furthermore, suture obliteration is accompanied by decreased levels of apoptosis within the suture matrix, while suture patency is associated with high levels of apoptosis within the suture matrix (Opperman et al., 2000). It appears that Tgf-βs regulate suture patency by controlling the number of cells in the suture and surrounding bone fronts by regulating cell proliferation and apoptosis. Any abnormal alteration in the balance of these growth factors can change the balance of cell numbers present,

either resulting in premature osseous obliteration of the sutures, or delaying suture formation by reducing the numbers of osteoblasts present, resulting in wide-open fontanels.

Since both these factors are present in sutures, it is likely that the interplay between them and other factors present in sutures regulates suture patency. It has also been shown that receptors for Tgf-βs (Tβr-I and Tβr-II) are elevated in dura and osteoblasts lining the bone fronts of fusing sutures (Mehrara et al., 1999). Conversely, addition of an increasing concentration of Tgf-β3 to coronal sutures in vitro reduces expression of Tβr-I (Opperman et al., 2002). Interestingly, no mutations in *TβR-I* and *TβR-II* have yet been identified in association with any craniofacial anomalies. Current understanding of the complex interplay between growth factors and their receptors in regulating suture patency is summarized in Section 20.3.

20.2.1.2 Fibroblast Growth Factors and Their Receptors

Several craniofacial syndromes present with craniosynostosis as a pathological phenotype. In recent years, many of these have been associated with mutations in fibroblast growth factor (*FGF*) receptor (*FGFR*) genes. There have been four FGFRs identified to date, and mutations associated with craniosynostosis have been identified in *FGFR1*, *FGFR2*, and *FGFR3*. No mutations associated with craniofacial anomalies have been identified in *FGFR4*. All mutations described to date can be referred to as activating mutations, as they mimic the effect of increased stimulation of receptor activity. However, the nature of the activation is markedly different for different mutations.

The FGFRs are similar in sequence and structure, all containing a series of three immunoglobulin (IgG)-like loops interspersed by linker regions in the extracellular domain (Fig. 20.6). These are linked to a transmembrane domain and an intracel-

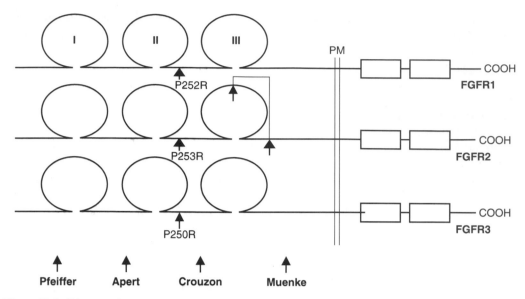

Figure 20.6 Diagramatic representation of FGFR1, FGFR2, and FGFR3 showing the locations of the most commonly occurring mutations (arrows) found in association with Pfeiffer, Apert, and Crouzon syndromes and with Muenke craniosynostosis. PM, plasma membrane; COOH represents the intracellular carboxy terminal of the receptor proteins, I, II, III, extracellular immunoglobulin-like domains of the receptor proteins.

lular signaling domain. Mutations in the linker region between IgG loops II and III, associated with Apert craniosynostotic phenotype, are a result of a selective decrease in FGF2 dissociation kinetics from mutant FGFR (Anderson et al., 1998). It is assumed that the same mechanism is operating in Pfeiffer[1] and Muenke syndromes. With these mutations, the receptors are still responsive to the presence of ligand, but signaling is prolonged in the presence of ligand. Mutations found within IgG domain III of FGFR2 are associated with constitutive activation of the receptor, resulting in continuous signaling by the receptor. However, in these mutations, the signaling no longer requires the presence of ligand and is termed ligand-independent receptor signaling.

[1] The Pfeiffer mutation found on FGFR1 is in the linker region. Over 30 mutations of Pfeiffer syndrome are also found within the third IgG loops of FGFR2.

Most mutations show coronal suture synostosis and result in similar facial dysmorphology, with midface hypoplasia being the defining feature of these syndromes. Interestingly, in a mutant mouse model, an insertional mutation in the region between the *Fgf3/Fgf4* locus produces a mouse with facial shortening, increased hypertelorism, and premature craniosynostosis (Carlton et al., 1998). In this mutant mouse, expression of both Fgf3 and Fgf4 is upregulated, confirming that locally elevated levels of growth factor produce similar phenotypes to those seen with mutant receptors. Coffin and co-workers (1995) demonstrated that a mouse strain overexpressing Fgf2 produced a similar phenotype, with "calvaria enlarged over the occipital bones." However, these mice exhibited marked long bone growth plate defects; hence it is likely they had prematurely fused cranial bases, which may be solely responsible for the facial pathology. These authors did not report looking at the sutures or the cranial base in their study.

Interestingly, several mutations in *FGFR3* have been associated with achondroplasia. These mutations produce ligand-independent stimulation of kinase activity, resulting in abnormally shortened long bones and cranial bases (Webster and Donoghue, 1996), similar to that seen in mice with an activated *Fgfr3* transgene (Naski et al., 1998). Confirmation of the role for Fgfr3 in cartilage growth plate development comes from studies done by Deng and associates (1996). These authors disrupted expression of *Fgfr3*, which resulted in prolonged endochondral bone growth, with resultant elongated long bones. However, these authors did not find any abnormalities in intramembranous bones of the craniofacial skeleton.

During suture morphogenesis, *Fgfr1*, *Fgfr2*, and *Fgfr3* (Iseki et al., 1997, 1999; Johnson et al., 2000; Kim et al., 1998; Rice et al., 2000) are expressed in the osteogenic cell layers of newly formed bone and in the advancing and eventually overlapping bone fronts on either side of the suture (Fig. 20.5). During approximation of the bone fronts, no *Fgfr*s are found in cells of the presumptive suture matrix; however, once the suture is formed, *Fgfr1* is present in cells in this region. In prematurely fusing sutures, both *Fgfr1* and *Fgfr2* can be localized to cells within the suture matrix (Rice et al., 2000).

During suture morphogenesis, Fgf4 and Fgf9 are present in presumptive suture tissues (Kim et al., 1998), but only Fgf9 is present in the fully formed suture (Kim et al., 1998). While Fgf2 is localized to the dura and mature bone throughout suture morphogenesis (Rice et al., 2000), it becomes highly expressed in the osteogenic bone fronts of prematurely fusing sutures (Most et al., 1998).

To examine the mechanism by which *FGFR* mutations affect cell function, several groups have grown osteoblasts derived from human tissue with mutant receptors in culture. In these studies, most of which have been conducted on cells derived from Apert patients, cells appear to have a normal response to FGF2; however, they show accelerated differentiation, such as upregulated expression of markers for osteoblastic phenotype (Fragale et al., 1999; Lemonnier et al., 2000; Lomri et al., 1998). Similar findings have been reported in osteoblasts from patients with nonsyndromic craniosynostosis (De Pollak et al., 1996). Interestingly, the *FGFR1* mutation associated with the Pfeiffer phenotype results in transiently increased cell proliferation, followed by accelerated expression of markers for osteoblast differentiation such as Runx2 (Zhou et al., 2000).

In contrast to this last finding, increased presence of Fgf2, associated with both normal and prematurely induced suture closure, appears to be connected with a decline in cell proliferation and a change in receptor expression by osteoblasts from *Fgfr2* to *Fgfr1* and upregulation of expression of Bsp-1 (Iseki et al., 1999). These authors also demonstrated that *Fgfr2* expression co-localized with cells showing BrdU uptake, a marker for cell proliferation, while *Fgfr1* co-localized in cells expressing markers for osteoblast differentiation. It is likely, then, that mutations in the *Fgfr*s cause cells to transition more rapidly through their division cycles and differentiate prematurely.

20.2.1.3 Bone Morphogenetic Proteins
Bone morphogenetic proteins (BMPs) have critical functions both during initial embryonic development and during bone formation. During embryonic development, BMPs are essential for establishing appropriate pattern formation in many organ systems, such as the central nervous system (Lewandoski et al., 1997; Martinez et al., 1999; Meyers and Martin, 1999; Neubuser et al., 1997; Trumpp et al., 1999; Wassarman et al., 1997), teeth (Aberg et al., 1997;

Jernvall et al., 1998; Keranen et al., 1998, 1999; Thesleff and Aberg, 1999; Thesleff and Sharpe, 1997), and limbs (Ganan et al., 1996; Macias et al., 1997, 1999; Martin, 1998; Niswander and Martin, 1993; Niswander et al., 1993; Sun et al., 2000; Thomas et al., 1997; Vogel et al., 1995). Recently, the group led by Thesleff has begun looking at the role of BMPs in cranial suture morphogenesis (Kim et al., 1998; Rice et al., 1999). They have demonstrated the presence of Bmp2, Bmp4, and Bmp7 in the osteogenic bone fronts of developing sutures and that the presence of these growth factors within these tissues declines postbirth, once the sutures have been formed (Fig. 20.5). Bmp4 and Bmp7 were also shown to be present in the presumptive suture mesenchyme and the underlying dura, but it is unclear whether their presence is continuous throughout suture development.

In experiments designed to establish a role for Bmp4 during suture development, beads soaked in Bmp4 were placed either on the osteogenic fronts or on the midsutural mesenchyme of developing mouse sutures (Kim et al., 1998). When beads were placed on the osteogenic bone fronts, accelerated suture closure was observed. However, placing beads on the midsutural mesenchyme resulted in increased tissue volume, with no effect on suture closure. Interestingly, Bmp4 induces expression of *Msx1* and *Msx2* in sutural tissue. Since all of these factors are involved in initial epithelial-mesenchymal tissue interactions and in pattern formation, it can be predicted that disruption of expression of Bmps would lead to disruption of initial events in suture formation. Recently, it was found that Bmp4 and Bmp7 are located within midsutural mesenchyme, at the same location as cells labeled positive for apoptosis during normal suture development. It was suggested that since Bmps act in conservative signaling pathways leading to apoptosis, expression of Bmps and their regulation of apoptosis is integral to nor-

mal suture development, and therefore disrupting apoptosis by disrupting the Bmp signaling pathways could lead to either premature or delayed suture closure (Rice et al., 1999).

20.2.2 Transcription Factors

Transcription factors are a more recently described group of factors that reside within cells, either within the cytoplasm or within the nucleus. Upon cellular stimulation, either by growth factors, extracellular matrix signaling, or hormones, those transcription factors in the cytoplasm can translocate to the nucleus, either alone or after co-localizing with other transcription factors. Alternately, inhibitory transcription factors can bind to activating factors, preventing their migration to the nucleus. Once in the nucleus, further co-localization of factors can occur, after which the transcription factors can bind to specific consensus sequences on DNA, where they can either up- or downregulate transcriptional activity.

Mutations in several transcription factors have been described which dramatically influence suture morphogenesis and patency. Some factors, such as Msx2 are known to act early in development, while others such as Runx2 (formerly Cbfa1) act later. Interactions between these transcription factors and growth factors are beginning to be described and where such interactions are known, these will be described in section 20.3 below, which discusses the interactions between growth and transcription factors in regulating suture morphogenesis and patency.

20.2.2.1 Msx2 The gene for *Msx2* belongs to the family of genes known as homeobox genes. These genes contain a coding region that encodes the homeodomain region of the protein. This region of the protein enables it to bind to its DNA binding site on genes, influencing their rate of transcription. *Msx* genes get their name from their homology to the *Drosophila Msh* (muscle segment homeobox) gene, and two such genes are identified in humans, namely *MSX1* and *MSX2*. While some functional redundancy of the *Msx* genes can occur, as may be the case in cranial suture development (Satokata et al., 2000), they play sequential roles in tooth development (Chen et al., 1996; Satokata and Maas 1994).

Both *Msx1* and *Msx2* have been described in the underlying dura (Kim et al., 1998) and presumptive suture mesenchyme (Liu et al., 1999; Rice et al., 1999) of developing mouse sutures (Fig. 20.5). Msx2 is also present in fusing sutures (Liu et al., 1999). Mutations in *MSX2* resulting in prolonged binding of MSX2 to its DNA binding site are responsible for a syndrome known as Boston-type craniosynostosis, in which several cranial vault sutures are fused prematurely (Ma et al., 1996). Prolonged binding of MSX2 to the promoter regions of type I collagen and osteocalcin leads to inhibition of their transcription (Dodig et al., 1996; Towler et al., 1994). While it appears contradictory that MSX2 inhibits osteoblast differentiation and yet results in premature osseous obliteration of sutures, this can be explained by overexpression studies. In mice overexpressing *Msx2*, premature suture obliteration is accompanied by elevated levels of cell proliferation (Liu et al., 1995, 1999). It may be that overexpression or mutated Msx2 allows cells to go through several more rounds of cell proliferation prior to differentiating and that this results in too many cells within the suture environment, resulting in suture obliteration.

A human syndrome has also been described with functional haploinsufficiency of MSX2. Not unexpectedly, this mutation results in delayed suture formation and wide-open fontanels (Wilkie et al., 2000). It has been shown in mice with Msx2 haploinsufficiency that the cranial base is extremely shortened (Satokata et al., 2000).

Therefore, it is possible that the wide-open fontanels represent compensatory growth of the cranium from the suture sites to accommodate the expanding brain (Opperman, 2000).

20.2.2.2 Twist

The human *TWIST* gene is the homologue of *Drosophila Twist* gene and belongs to a group of genes encoding helix-loop-helix proteins. These are regulatory proteins, generally activating transcription of several other genes upon binding to DNA. Twist protein is also shown to regulate expression of Fgfr in *Drosophila* (Shishido et al., 1993) and in mice (Johnson et al., 2000; Rice et al., 2000); hence it is thought to be in the same signaling pathway as Fgfr. Several mutations have been ascribed to the *TWIST* gene, all of which result in protein haploinsufficiency (Table 20.1).

Twist has a very specific tissue location in the presumptive suture mesenchyme during initial suture formation (Johnson et al., 2000; Rice et al., 2000). Its expression appears transient, with downregulation of *Twist* expression as the bone fronts overlap one another to form the initial suture (Fig. 20.5). Twist downregulates osteoblast differentiation, similar to Msx2. It also upregulates Fgfr2 expression, which is associated with proliferating cells. Since Twist appears early in suture morphogenesis, it may well be that Twist haploinsufficiency prematurely switches off cell proliferation and initiates cell differentiation, resulting in bony obliteration of the suture.

20.2.2.3 Runx2/Cbfa1

Runx2 belongs to the runt domain or core binding family of factors. These factors have homology to *Drosophila melanogaster* pair rule gene *runt*, which plays a role in segmental pattern formation. While several mammalian *Runx* genes have been described, *Runx2* appears to be an essential transcription factor for osteoblast differentiation. The creation of *Runx2* knockout mice produced a phenotype in which offspring totally lack bone, both intramembranous and endochondral. The intramembranous portions of the craniofacial skeleton remains as fibrous tissue, while the cartilaginous portions remain unossified. The homozygous offspring die at birth due to difficulties with breathing and suckling (Komori et al., 1997).

During cranial suture development, *Runx2* is found in the osteoblasts lining the osteogenic bone fronts at the edges of the suture and is responsible for inducing the osteoblastic phenotype (Komori et al., 1997; Otto et al., 1997; Zhou et al., 2000). Several mutations in *RUNX2* have been identified that result in haploinsufficiency or inability of the protein to bind to DNA (Mundlos et al., 1997; Otto et al., 1997; Zhou et al., 1999). These mutations all give rise to a human condition known as cleidocranial dysplasia (CCD). Patients present with wide-open fontanels among other anomalies, likely due to alterations in intramembranous bone formation. However, several groups have reported defects in endochondral cartilage hypertrophy (Mundlos et al., 1997; Zhou et al., 1999), so a contribution from abnormal growth of the cranial base to CCD cranial pathology cannot be excluded.

Recently, a mutation in Fgfr1 created in mice, orthologous to the Pfeiffer mutations in humans, resulted in accelerated osteoblast differentiation. This was accompanied by dramatically increased expression of Runx2, suggesting that Runx2 may be a downstream target of Fgfr1 signals (Zhou et al., 2000).

20.3 INTERACTIONS BETWEEN GROWTH AND TRANSCRIPTION FACTORS IN REGULATING SUTURE MORPHOGENESIS AND PATENCY

Figure 20.7 represents an attempt to synthesize what is currently known about the interplay between the various growth factors and transcription factors in regulating suture morphogenesis and growth. It is known that Bmp2 upregulates *Id*, which in turn downregulates *Twist* (Rice et al., 2000). Furthermore, Bmp4 upregulates the transcription factor *Msx2* (Kim et al., 1998). Since Msx2 is known to downregulate Fgf2 expression, and since decreased Fgf2 expression results in decreased presence of Twist, these factors have been put into a common pathway. However, it is not known whether Id affects Msx2 expression or whether this represents an alternate pathway. Since continued expression of Twist inhibits *Fgfr* expression (Rice et al., 2000), decreasing Twist should elevate Fgfr expression. Elevated Fgfr1 has been associated with increased Runx2 expression (Zhou et al., 2000), resulting in suture obliteration. Runx2 expression is also upregulated by Bmp7 (Ducy et al., 1997), increasing osteopontin and osteocalcin promoter activity.

Interestingly, increased presence of Fgf2 is associated with suture obliteration (Iseki et al., 1997; Most et al., 1998). Prolonged elevation of Fgf2 leads to increased Tgf-β2 expression, which is known to result in premature suture fusion (Opperman et al., 1999, 2000; Roth et al., 1997b) and is accompanied by increased osteocalcin production and matrix mineralization (Debiais et al., 1998). While no direct link between Tgf-β2 and Runx2 activity has been demonstrated in suture obliteration, it is likely this link exists, since Runx2 interacts with Smads, enhancing Runx2 transactivating ability, a pathway that is disrupted in cases of cleidocranial dysplasia, where suture formation is inhibited (Zhang et al., 2000a, 2000b).

It is now apparent that two distinct but likely interlinked pathological processes can result in premature suture obliteration. The first process is linked to elevated FGF expression, mutated FGFRs (of which all current mutations appear to be activating

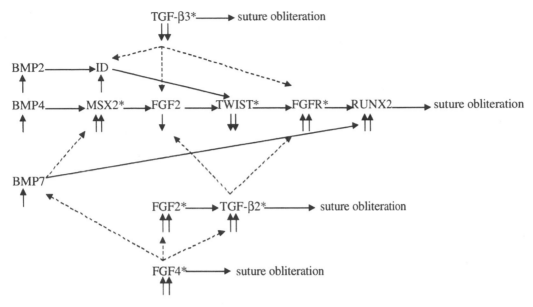

Figure 20.7 Schematic representation of the known (solid long arrows) and some potentially interesting unknown (dashed arrows) associations between factors linked with craniosynostosis. Numbers of short arrows show degree of upregulation (↑↑) or downregulation (↓↓) of each factor known to result in craniosynostosis. Asterisks show factors that are known to result in craniosynostosis when perturbed independently of upstream regulators. Factors shown more than once indicate alternate pathways or pathways where possible links are not yet known. (*Source*: Reprinted in modified form from Opperman, 2000.)

mutations), or overexpressed Runx2. In this pathway, suture obliteration is a consequence of accelerated differentiation of cells into osteoblasts (Debiais et al., 1998; Iseki et al., 1997, 1999; Lemonnier et al., 2000; Marks et al., 1999). This may be accompanied by transient increases in cell proliferation, as seen with Fgfr2 and Twist mutations (Iseki et al., 1999). The other process is linked to Msx2 and Tgf-β expression, where suture obliteration is accompanied by elevated levels of cell proliferation (Liu et al., 1999; Opperman et al., 2000) and decreased levels of apoptosis within the suture (Opperman et al., 2000). Premature obliteration of sutures will occur when any of these processes is disrupted. Normal suture morphogenesis and maintenance of intact sutures therefore requires a balancing act between having the appropriate number of cells present to permit bone growth from the suture site and not allowing too many cells too accumulate, which would trigger osteoblastic differentiation.

20.4 CONCLUSIONS, FUTURE STUDIES, AND DIRECTIONS

It is important to understand the links between different possible pathways, since using these alternate pathways provides the possibility of molecular interventions. These pathways currently provide contradictory information about the regulatory processes involved. For example, data indicate that both increased and decreased Fgf2 expression results in premature suture obliteration. The complicating factor in these studies is establishing the normal levels against which these factors are measured, since some studies use normal animal models, whereas others rely on the availability of tissues carrying genetic mutations, either from human subjects or transgenically created animal models.

Intramembranous bone growth at the suture site is still understudied when compared to the study of endochondral growth plates. The regulation of intramembranous bone growth from sutures is likely as complicated as that occurring at the endochondral growth plate, providing interesting new challenges and new research opportunities.

REFERENCES

Aberg, T., Wozney, J., and Thesleff, I. (1997). Expression patterns of bone morphogenetic proteins (Bmps) in the developing mouse tooth suggest roles in morphogenesis and cell differentiation. *Devel. Dyn. 210*, 383–396.

Anderson, J., Burns, H. D., Enriquez-Harris, P., Wilkie, A. O. M., and Heath, J. K. (1998). Apert syndrome mutations in fibroblast growth factor receptor 2 exhibit increased affinity for FGF ligand. *Hum. Molec. Genet. 7*, 1475–1483.

Carlton, M. B., Colledge, W. H., and Evans, M. J. (1998). Crouzon-like craniofacial dysmorphology in the mouse is caused by an insertional mutation at the Fgf3/Fgf4 locus. *Devel. Dyn. 212*, 242–249.

Chen, Y., Bei, M., Woo, I., Satokata, I., and Maas, R. (1996). Msx1 controls inductive signaling in mammalian tooth morphogenesis. *Development 122*, 3035–3044.

Chong, S. L., Mitchell, R., Moursi, A., Winnard, P., Losken, H. W., Ozerdem, O., Keeler, K., Opperman, L. A., Siegel, M. I., and Mooney, M. P. (2001). Rescue of coronal suture synostosis with TGF-β3 in craniosynostotic rabbits. *J. Dent. Res. 80*, 89.

Coffin, J. D., Florkiewicz, R. Z., Neumann, J., Mort-Hopkins, T., Dorn, G. W., II, Lightfoot, P., German, R., Howles, P. N., Kier, A., and O'Toole, B. A. (1995). Abnormal bone growth and selective translational regulation in basic fibroblast growth factor (FGF-2) transgenic mice. *Molec. Biol. Cell. 6*, 1861–1873.

Cohen, M. M., Jr. (2000). *Craniosynostosis, Diagnosis, Evaluation and Management*, 2nd ed. (New York.: Oxford University Press), p. 454.

Cooper, G. M., Sayne, J. R., Williams, A., Moursi, A., and Opperman, L. A. (2001). TGF-β3 in collagen gels rescues rat posterior frontal sutures from fusion in vivo. *J. Dent. Res. 80*, 229.

Debiais, F., Hott, M., Graulet, A. M., and Marie, P. J. (1998). The effects of fibroblast growth factor-2 on human neonatal calvaria osteoblastic cells are differentiation stage specific. *J. Bone Min. Res. 13*, 645–654.

Deng, C., Wynshaw-Boris, A., Zhou, F., Kuo, A., and Leder, P. (1996). Fibroblast growth factor receptor 3 is a negative regulator of bone growth. *Cell 84*, 911–921.

De Pollak, C., Renier, D., Hott, M., and Marie, P. J. (1996). Increased bone formation and osteoblastic cell phenotype in premature cranial suture ossification (craniosynostosis). *J. Bone Min. Res. 11*, 401–407.

Dodig, M., Kronenberg, M. S., Bedalov, A., Kream, B. E., Gronowicz, G., Clark, S. H., Mack, K., Liu, Y. H., Maxon, R., Pan, Z. Z., and Upholt, W. B. (1996). Identification of a TAAT-containing motif required for high level expression of the COL1A1 promoter in differentiated osteoblasts of transgenic mice. *J. Biol. Chem. 271*, 16,422–16,429.

Ducy, P., Zhang, R., Geoffroy, V., Ridall, A. L., and Karsenty, G. (1997). Osf2/Cbfa1: a transcriptional activator of osteoblast differentiation. *Cell 89*, 747–754.

el Ghouzzi, V., Le Merrer, M., Perrin-Schmitt, F., Lajeunie, E., Benit, P., Renier, D., Bourgeois, P., Bolcato-Bellemin, A. L., Munnich, A., and Bonaventure, J. (1997). Mutations of the TWIST gene in the Saethre-Chotzen syndrome. *Nat. Genet. 15*, 42–46.

el Ghouzzi, V., Lajeunie, E., Le Merrer, M., Cormier-Daire, V., Renier, D., Munnich, A., and Bonaventure, J. (1999). Mutations within or upstream of the basic helix-loop-helix domain of the TWIST gene are specific to Saethre-Chotzen syndrome. *Eur. J. Hum. Genet. 7*, 27–33.

el Ghouzzi, V., Legeai-Mallet, L., Aresta, S., Benoist, C., Munnich, A., de Gunzburg, J., and Bonaventure, J. (2000). Saethre-Chotzen mutations cause TWIST protein degradation or impaired nuclear location. *Hum. Molec. Genet. 9*, 813–819.

Elmslie, F. V., and Reardon, W. (1998). Craniofacial developmental abnormalities. *Curr. Opin. Neurol. 11*, 103–108.

Fragale, A., Tartaglia, M., Bernardini, S., Di Stasi, A. M., Di Rocco, C., Velardi, F., Teti, A., Battaglia, P. A., and Migliaccio, S. (1999). Decreased proliferation and altered differentiation in osteoblasts from genetically and clinically distinct craniosynostotic disorders. *Am. J. Pathol. 154*, 1465–1477.

Ganan, Y., Macias, D., Duterque-Coquillaud, M., Ros, M. A., and Hurle, J. M. (1996). Role of TGF betas and BMPs as signals controlling the position of the digits and the areas of interdigital cell death in the developing chick limb autopod. *Development 122*, 2349–2357.

Hehr, U., and Muenke, M. (1999). Craniosynostosis syndromes: from genes to premature fusion of skull bones. *Molec. Genet. Metab. 68*, 139–151.

Howard, T. D., Paznekas, W. A., Green, E. D., Chiang, L. C., Ma, N., Ortiz de Luna, R. I., Garcia-Delgado, C., Gonzalez-Ramos, M., Kline, A. D., and Jabs, E. W. (1997). Mutations in TWIST, a basic helix-loop-helix transcription factor in Saethre-Chotzen syndrome. *Nat. Genet. 15*, 36–41.

Iseki, S., Wilkie, A. O., Heath, J. K., Ishimaru, T., Eto, K., and Morriss-Kay, G. M. (1997). Fgfr2 and osteopontin domains in the developing skull vault are mutually exclusive and can be altered by locally applied FGF2. *Development 124*, 3375–3384.

Iseki, S., Wilkie, A. O., and Morriss-Kay, G. M. (1999). Fgfr1 and Fgfr2 have distinct differentiation- and proliferation-related roles in the developing mouse skull vault. *Development 126*, 5611–5620.

Jabs, E. W. (1998). Toward understanding the pathogenesis of craniosynostosis through clinical and molecular correlates. *Clin. Genet. 53*, 79–86.

Jabs, E. W., Muller, U., Li, X., Ma, L., Luo, W., Haworth, I. S., Klisak, I., Sparkes, R., Warman, M. L., and Mulliken, J. B. (1993). A mutation in the homeodomain of the human MSX2 gene in a family affected with autosomal dominant craniosynostosis. *Cell 75*, 443–450.

Jabs, E. W., Li, X., Scott, A. F., Meyers, G., Chen, W., Eccles, M., Mao, J. I., Charnas, L. R., Jackson, C. E., and Jaye, M. (1994). Jackson-Weiss and Crouzon syndromes are allelic with mutations in fibroblast growth factor receptor 2. *Nat. Genet. 8*, 275–279.

Jernvall, J., Aberg, T., Kettunen, P., Keranen, S., and Thesleff, I. (1998). The life history of an embryonic signaling center: BMP-4 induces p21 and is associated with apoptosis in the mouse tooth enamel knot. *Development 125*, 161–169.

Johnson, D., Iseki, S., Wilkie, A. O. M., and Morriss-Kay, G. M. (2000). Expression patterns of *Twist* and *Fgfr1, -2* and *-3* in the developing mouse coronal suture suggest a key role for *Twist* in suture initiation and biogenesis. *Mech. Devel. 91*, 341–345.

Keranen, S. V., Aberg, T., Kettunen, P., Thesleff, I., and Jernvall, J. (1998). Association of developmental regulatory genes with the development of different molar tooth shapes in two species of rodents. *Devel. Genes Evol. 208*, 477–486.

Keranen, S. V., Kettunen, P., Aberg, T., Thesleff, I., and Jernvall, J. (1999). Gene expression patterns associated with suppression of odontogenesis in mouse and vole diastema regions. *Devel. Genes Evol. 209*, 495–506.

Kim, H. J., Rice, D. P., Kettunen, P. J., and Thesleff, I. (1998). FGF-, BMP- and Shh-mediated signalling pathways in the regulation of cranial suture morphogenesis and calvarial bone development. *Development 125*, 1241–1251.

Kokich, V. G. (1986). The biology of sutures. In M. M. Cohen, Jr. (ed.), *Craniosynostosis, Diagnosis, Evaluation and Management* (New York: Raven Press), pp. 81–103.

Komori, T., Yagi, H., Nomura, S., Yamaguchi, A., Sasaki, K., Deguchi, K., Shimizu, Y., Bronson, R. T., Gao, Y. H., Inada, M., and Sato, M. (1997). Targeted disruption of Cbfa1 results in a complete lack of bone formation owing to maturational arrest of osteoblasts. *Cell 89*, 755–764.

Lemonnier, J., Delannoy, P., Hott, M., Lomri, A., Modrowski, D., and Marie, P. J. (2000). The

Ser252Trp fibroblast growth factor receptor-2 (FGFR-2) mutation induces PKC-independent downregulation of FGFR-2 associated with premature calvaria osteoblast differentiation. *Exp. Cell Res. 256*, 158–167.

Lewandoski, M., Meyers, E. N., and Martin, G. R. (1997). Analysis of Fgf8 gene function in vertebrate development. *Cold Spring Harbor Symposia on Quantitative Biology 62*, 159–168.

Liu, Y. H., Kundu, R., Wu, L., Luo, W., Ignelzi, M. A., Jr., Snead, M. L., and Maxson, R. E., Jr. (1995). Premature suture closure and ectopic cranial bone in mice expressing Msx2 transgenes in the developing skull. *Proc. Natl. Acad. Sci. USA 92*, 6137–6141.

Liu, Y. H., Tang, Z., Kundu, R. K., Wu, L., Luo, W., Zhu, D., Sangiorgi, F., Snead, M. L., and Maxson, R. E. (1999). Msx2 gene dosage influences the number of proliferative osteogenic cells in growth centers of the developing murine skull: a possible mechanism for MSX2-mediated craniosynostosis in humans. *Devel. Biol. 205*, 260–274.

Lomri, A., Lemonnier, J., Hott, M., de Parseval, N., Lajeunie, E., Munnich, A., Renier, D., and Marie, P. J. (1998). Increased calvaria cell differentiation and bone matrix formation induced by fibroblast growth factor receptor 2 mutations in Apert syndrome. *J. Clin. Invest. 101*, 1310–1317.

Ma, L., Golden, S., Wu, L., and Maxson, R. (1996). The molecular basis of Boston-type craniosynostosis: the Pro148 → His mutation in the N-terminal arm of the MSX2 homeodomain stabilizes DNA binding without altering nucleotide sequence preferences. *Hum. Molec. Genet. 5*, 1915–1920.

Macias, D., Ganan, Y., Sampath, T. K., Piedra, M. E., Ros, M. A., and Hurle, J. M. (1997). Role of BMP-2 and OP-1 (BMP-7) in programmed cell death and skeletogenesis during chick limb development. *Development 124*, 1109–1117.

Macias, D., Ganan, Y., Rodriguez-Leon, J., Merino, R., and Hurle, J. M. (1999). Regulation by members of the transforming growth factor beta superfamily of the digital and interdigital fates of the autopodial limb mesoderm. *Cell Tissue Res. 296*, 95–102.

Marks, S. C., Jr., Lundmark, C., Wurtz, T., Odgren, P. R., MacKay, C. A., Mason-Savas, A., and Popoff, S. N. (1999). Facial development and type III collagen RNA expression: concurrent repression in the osteopetrotic (toothless, tl) rat and rescue after treatment with colony-stimulating factor-1. *Devel. Dyn. 215*, 117–125.

Martin, G. R. (1998). The roles of FGFs in the early development of vertebrate limbs. *Genes Devel. 12*, 1571–1586.

Martinez, S., Crossley, P. H., Cobos, I., Rubenstein, J. L., and Martin, G. R. (1999). FGF8 induces formation of an ectopic isthmic organizer and isthmocerebellar development via a repressive effect on Otx2 expression. *Development 126*, 1189–1200.

Mehrara, B. J., Steinbrech, D. S., Saadeh, P. B., Gittes, G. K., and Longaker, M. T. (1999). Expression of high-affinity receptors for TGF-beta during rat cranial suture fusion. *Ann. Plast. Surg. 42*, 502–508.

Meyers, E. N., and Martin, G. R. (1999). Differences in left-right axis pathways in mouse and chick: functions of FGF8 and SHH. *Science 285*, 403–406.

Most, D., Levine, J. P., Chang, J., Sung, J., McCarthy, J. G., Schendel, S. A., and Longaker, M. T. (1998). Studies in cranial suture biology: up-regulation of transforming growth factor-beta1 and basic fibroblast growth factor mRNA correlates with posterior frontal cranial suture fusion in the rat. *Plast. Reconstr. Surg. 101*, 1431–1440.

Muenke, M., Schell, U., Hehr, A., Robin, N. H., Losken, H. W., Schinzel, A., Pulleyn, L. J., Rutland, P., Reardon, W., and Malcolm, S. (1994). A common mutation in the fibroblast growth factor receptor 1 gene in Pfeiffer syndrome. *Nat. Genet. 8*, 269–274.

Muenke, M., Gripp, K. W., McDonald-McGinn, D. M., Gaudenz, K., Whitaker, L. A., Bartlett, S. P., Markowitz, R. I., Robin, N. H., Nwokoro, N., Mulvihill, J. J., and Losken, H. W. (1997). A unique point mutation in the fibroblast growth factor receptor 3 gene (FGFR3) defines a new craniosynostosis syndrome. *Am. J. Hum. Genet. 60*, 555–564.

Mundlos, S., Otto, F., Mundlos, C., Mulliken, J. B., Aylsworth, A. S., Albright, S., Lindhout, D., Cole, W. G., Henn, W., Knoll, J. H., and

Owen, M. J. (1997). Mutations involving the transcription factor CBFA1 cause cleidocranial dysplasia. *Cell 89*, 773–779.

Naski, M. C., Colvin, J. S., Coffin, J. D., and Ornitz, D. M. (1998). Repression of hedgehog signaling and BMP4 expression in growth plate cartilage by fibroblast growth factor receptor 3. *Development 125*, 4977–4988.

Neubuser, A., Peters, H., Balling, R., and Martin, G. R. (1997). Antagonistic interactions between FGF and BMP signaling pathways: a mechanism for positioning the sites of tooth formation. *Cell 90*, 247–255.

Niswander, L., and Martin, G. R. (1993). Mixed signals from the AER: FGF-4 and Bmp-2 have opposite effects on limb growth. *Prog. Clin. Biol. Res. 383B*, 625–633.

Niswander, L., Tickle, C., Vogel, A., Booth, I., and Martin, G. R. (1993). FGF-4 replaces the apical ectodermal ridge and directs outgrowth and patterning of the limb. *Cell 75*, 579–587.

Nuckolls, G. H., Shum, L., and Slavkin, H. C. (1999). Progress toward understanding craniofacial malformations. *Cleft Palate Craniofac. J. 36*, 12–26.

Opperman, L. A. (2000). Sutures as intramembranous bone growth sites. *Devel. Dyn. 219*, 472–485.

Opperman, L. A., Sweeney, T. M., Redmon, J., Persing, J. A., Ogle, R. C. (1993). Tissue interactions with underlying dura mater inhibit osseous obliteration of developing cranial sutures. *Devel. Dyn. 198*, 312–322.

Opperman, L. A., Passarelli, R. W., Morgan, E. P., Reintjes, M., and Ogle, R. C. (1995). Cranial sutures require tissue interactions with dura mater to resist osseous obliteration in vitro. *J. Bone Min. Res. 10*, 1978–1987.

Opperman, L. A., Nolen, A. A., and Ogle, R. C. (1997). TGF-beta 1, TGF-beta 2, and TGF-beta 3 exhibit distinct patterns of expression during cranial suture formation and obliteration in vivo and in vitro. *J. Bone Min. Res. 12*, 301–310.

Opperman, L. A., Chhabra, A., Nolen, A. A., Bao, Y., and Ogle, R. C. (1998). Dura mater maintains rat cranial sutures in vitro by regulating suture cell proliferation and collagen production. *J. Craniofac. Genet. Devel. Biol. 18*, 150–158.

Opperman, L. A., Chhabra, A., Cho, R. W., and Ogle, R. C. (1999). Cranial suture obliteration is induced by removal of transforming growth factor (TGF)-beta 3 activity and prevented by removal of TGF-beta 2 activity from fetal rat calvaria in vitro. *J. Craniofac. Genet. Devel. Biol. 19*, 164–173.

Opperman, L. A., Adab, K., and Gakunga, P. T. (2000). TGF-β2 and TGF-β3 regulate fetal rat cranial suture morphogenesis by regulating rates of cell proliferation and apoptosis. *Devel. Dyn. 219*, 237–247.

Opperman, L. A., Galanis, V., Williams, A. R., Adab, K. (2002). Transforming growth factor-beta 3 (Tgf-β3) down-regulates Tgf-β receptor type 1 (Tβr-1) on cells during rescue of cronial sutures from osseous obliteration. *Ortho. Craniofac. Res. 5*, 5–16.

Otto, F., Thornell, A. P., Crompton, T., Denzel, A., Gilmour, K. C., Rosewell, I. R., Stamp, G. W., Beddington, R. S., Mundlos, S., Olsen, B. R., and Selby, P. B. (1997). Cbfa1, a candidate gene for cleidocranial dysplasia syndrome, is essential for osteoblast differentiation and bone development. *Cell 89*, 765–771.

Park, W. J., Meyers, G. A., Li, X., Theda, C., Day, D., Orlow, S. J., Jones, M. C., and Jabs, E. W. (1995). Novel FGFR2 mutations in Crouzon and Jackson-Weiss syndromes show allelic heterogeneity and phenotypic variability. *Hum. Molec. Genet. 4*, 1229–1233.

Paznekas, W. A., Cunningham, M. L., Howard, T. D., Korf, B. R., Lipson, M. H., Grix, A. W., Feingold, M., Goldberg, R., Borochowitz, Z., Aleck, K., and Mulliken, J. (1998). Genetic heterogeneity of Saethre-Chotzen syndrome, due to TWIST and FGFR mutations. *Am. J. Hum. Genet. 62*, 1370–1380.

Quack, I., Vonderstrass, B., Stock, M., Aylsworth, A. S., Becker, A., Brueton, L., Lee, P. J., Majewski, F., Mulliken, J. B., Suri, M., and Zenker, M. (1999). Mutation analysis of core binding factor A1 in patients with cleidocranial dysplasia. *Am. J. Hum. Genet. 65*, 1268–1278.

Reardon, W., Winter, R. M., Rutland, P., Pulleyn, L. J., Jones, B. M., and Malcolm, S. (1994). Mutations in the fibroblast growth factor receptor 2 gene cause Crouzon syndrome. *Nat. Genet. 8*, 98–103.

Rice, D. P., Kim, H. J., and Thesleff, I. (1999). Apoptosis in murine calvarial bone and suture development. *Eur. J. Oral Sci. 107*, 265–275.

Rice, D. P., Aberg, T., Chan, Y., Tang, Z., Kettunen, P. J., Pakarinen, L., Maxson, R. E., and Thesleff, I. (2000). Integration of FGF and TWIST in calvarial bone and suture development. *Development 127*, 1845–1855.

Roth, D. A., Gold, L. I., Han, V. K., McCarthy, J. G., Sung, J. J., Wisoff, J. H., and Longaker, M. T. (1997a). Immunolocalization of transforming growth factor beta 1, beta 2, and beta 3 and insulin-like growth factor I in premature cranial suture fusion. *Plast. Reconstr. Surg. 99*, 300–309; discussion 310–316.

Roth, D. A., Longaker, M. T., McCarthy, J. G., Rosen, D. M., McMullen, H. F., Levine, J. P., Sung, J., and Gold, L. I. (1997b). Studies in cranial suture biology: Part I. Increased immunoreactivity for TGF-beta isoforms (beta 1, beta 2, and beta 3) during rat cranial suture fusion. *J. Bone Min. Res. 12*, 311–321.

Satokata, I., and Maas, R. (1994). Msx1 deficient mice exhibit cleft palate and abnormalities of craniofacial and tooth development. *Nat. Genet. 6*, 348–356.

Satokata, I., Ma, L., Ohshima, H., Bei, M., Woo, I., Nishizawa, K., Maeda, T., Takano, Y., Uchiyama, M., Heaney, S., and Peters, H. (2000). Msx2 deficiency in mice causes pleiotropic defects in bone growth and ectodermal organ formation. *Nat. Genet. 24*, 391–395.

Shishido, E., Higashijima, S., Emori, Y., and Saigo, K. (1993). Two FGF-receptor homologues of Drosophila: one is expressed in mesodermal primordium in early embryos. *Development 117*, 751–761.

Sperber, G. H. (1999). Pathogenesis and morphogenesis of craniofacial developmental anomalies. *Ann. Acad. Med. Singapore 28*, 708–713.

Sun, X., Lewandoski, M., Meyers, E. N., Liu, Y. H., Maxson, R. E., Jr., and Martin, G. R. (2000). Conditional inactivation of Fgf4 reveals complexity of signalling during limb bud development. *Nat. Genet. 25*, 83–86.

Tartaglia, M., Di Rocco, C., Lajeunie, E., Valeri, S., Velardi, F., and Battaglia, P. A. (1997). Jackson-Weiss syndrome: identification of two novel FGFR2 missense mutations shared with Crouzon and Pfeiffer craniosynostotic disorders. *Hum. Genet. 101*, 47–50.

Thesleff, I., and Aberg, T. (1999). Molecular regulation of tooth development. *Bone 25*, 123–125.

Thesleff, I., and Sharpe, P. (1997). Signalling networks regulating dental development. *Mech. Devel. 67*, 111–123.

Thomas, J. T., Kilpatrick, M. W., Lin, K., Erlacher, L., Lembessis, P., Costa, T., Tsipouras, P., and Luyten, F. P. (1997). Disruption of human limb morphogenesis by a dominant negative mutation in CDMP1. *Nat. Genet. 17*, 58–64.

Towler, D. A., Rutledge, S. J., and Rodan, G. A. (1994). Msx-2/Hox 8.1: a transcriptional regulator of the rat osteocalcin promoter. *Molec. Endocrinol. 8*, 1484–1493.

Trumpp, A., Depew, M. J., Rubenstein, J. L., Bishop, J. M., and Martin, G. R. (1999). Cre-mediated gene inactivation demonstrates that FGF8 is required for cell survival and patterning of the first branchial arch. *Genes Devel. 13*, 3136–3148.

Vogel, A., Roberts-Clarke, D., and Niswander, L. (1995). Effect of FGF on gene expression in chick limb bud cells in vivo and in vitro. *Devel. Biol. 171*, 507–520.

Wassarman, K. M., Lewandoski, M., Campbell, K., Joyner, A. L., Rubenstein, J. L., Martinez, S., and Martin, G. R. (1997). Specification of the anterior hindbrain and establishment of a normal mid/hindbrain organizer is dependent on Gbx2 gene function. *Development 124*, 2923–2934.

Webster, M. K., and Donoghue, D. J. (1996). Constitutive activation of fibroblast growth factor receptor 3 by the transmembrane domain point mutation found in achondroplasia. *EMBO J. 15*, 520–527.

Webster, M. K., and Donoghue, D. J. (1997). FGFR activation in skeletal disorders: too much of a good thing. *Trends Genet. 13*, 178–182.

Wilkie, A. O. (1997). Craniosynostosis: genes and mechanisms. *Hum. Molec. Genet. 6*, 1647–1656.

Wilkie, A. O., Slaney, S. F., Oldridge, M., Poole, M. D., Ashworth, G. J., Hockley, A. D., Hayward, R. D., David, D. J., Pulleyn, L. J., and Rutland, P. (1995). Apert syndrome results from localized mutations of FGFR2 and is allelic with Crouzon syndrome. *Nat. Genet. 9*, 165–172.

Wilkie, A. O., Tang, Z., Elanko, N., Walsh, S., Twigg, S. R., Hurst, J. A., Wall, S. A., Chrzanowska, K. H., and Maxson, R. E., Jr. (2000). Functional haploinsufficiency of the human homeobox gene MSX2 causes defects in skull ossification. *Nat. Genet. 24*, 387–390.

Zhang, Y. W., Yasui, N., Ito, K., Huang, G., Fujii, M., Hanai, J., Nogami, H., Ochi, T., Miyazono, K., and Ito, Y. (2000a). A RUNX2/PEBP2alpha A/CBFA1 mutation displaying impaired transactivation and Smad interaction in cleidocranial dysplasia. *Proc. Natl. Acad. Sci. USA 97*, 10,549–10,554.

Zhang, Y. W., Yasui, N., Kakazu, N., Abe, T., Takada, K., Imai, S., Sato, M., Nomura, S., Ochi, T., Okuzumi, S., and Nogami, H. (2000b). PEBP2alphaA/CBFA1 mutations in Japanese cleidocranial dysplasia patients. *Gene 244*, 21–28.

Zhou, G., Chen, Y., Zhou, L., Thirunavukkarasu, K., Hecht, J., Chitayat, D., Gelb, B. D., Pirinen, S., Berry, S. A., Greenberg, C. R., and Karsenty, G. (1999). CBFA1 mutation analysis and functional correlation with phenotypic variability in cleidocranial dysplasia. *Hum. Molec. Genet. 8*, 2311–2316.

Zhou, Y.-X., Xu, X., Chen, L., Li, C., Brodie, S. G., and Deng, C.-X. (2000). A Pro250Arg substitution in mouse Fgfr1 causes increased expression of Cbfa1 and premature fusion of calvarial sutures. *Hum. Molec. Genet. 9*, 2001–2008.

CHAPTER 21

MOLECULAR STUDIES OF FACIAL CLEFTING: FROM MOUSE TO MAN

MICHAEL MELNICK D.D.S. Ph.D., and TINA JASKOLL, Ph.D.

21.1 INTRODUCTION

The human lip and palate form as a result of the cell proliferation (growth), apposition, and fusion of embryonic facial processes between the 5th and 12th weeks of gestation. This requires that the processes appear in the correct place, achieve the correct shape and size, and have no obstruction to fusion. Given the complex nature of this oral development, we can readily imagine a long list of potential mishaps (Trasler and Fraser, 1977). Indeed, oral clefts are a major public health problem worldwide (Marazita et al., 1986; Melnick, 1992; Wyszynski et al., 1996) (Chapters 3 and 7).

Cleft lip with or without cleft palate (CL ± P) has an incidence at birth of about 1 in 500–1000 that varies by population; people of Asian descent often are at higher risk than those of Caucasian or African descent (Marazita et al., 1986; Melnick, 1992; Murray, 1995; Wyszynski et al., 1996). In all populations there are significantly more males born with CL ± P than females. The incidence at birth for cleft palate alone (CP) is relatively uniform across populations at about 1 in 2000; signifi-cantly more females are born with CP than males (Shields et al., 1981; Wyszynski et al., 1996). It has clearly been established that CL ± P and CP are etiologically distinct (Melnick and Shields, 1982). People with CL ± P very rarely have relatives with CP and vice versa. What CL ± P and CP do share is that despite 50 years of intense study, the etiologies of both are largely an enigma.

21.2 ETIOLOGY OF NONSYNDROMIC ORAL CLEFTS

In 1875, Charles Darwin wrote: "Although many congenital monstrosities are inherited, of which examples have already been given, and to which may be added the lately recorded case of the transmission during a century of hare-lip with a cleft-palate in the writer's own family [Sproule, 1863], yet other malformations are rarely or never inherited. Of these latter cases, many are probably due to injuries in the womb or egg, and would come under the head of non-inherited injuries or mutilations." So the matter stood until the "discovery" of Mendel 25 years hence.

Understanding Craniofacial Anomalies: The Etiopathogenesis of Craniosynostoses and Facial Clefting, Edited by Mark P. Mooney and Michael I. Siegel, ISBN 0-471-38724-x   Copyright © 2002 by Wiley-Liss, Inc.

At the outset of the 20th century in England, there arose a venomous dispute between Mendelian geneticists, such as Bateson at Cambridge, and anti-Mendelian biometricians, such as Pearson at the Galton Laboratory in London, over the genetic etiology of such "physical deformities" as cleft lip and palate (Melnick, 1997). To Pearson and colleagues, such traits were an expression of physical and racial degeneracy that could be traced to polygenically poor protoplasm. To Bateson and his associates, such traits were Mendelian unit characters whose segregation could be seen in carefully constructed family pedigrees. Bateson dismissed the work of the anti-Mendelians as "unsound in construction" and predicted such thinking would inevitably lead to "brutal" control of those who the larger society deemed unfit. History proved Bateson astutely prescient.

By 1925, there was a growing response to both sides of this argument. This is well represented by the writings of H. S. Jennings, Henry Walters professor of zoology and director of the Zoological Laboratory, Johns Hopkins University. In *Prometheus* (Jennings, 1925), he dismisses Pearsonism and takes on the shortcomings of the Bateson camp:

> These facts—the [so-called "unit characters" of Mendelism]—gave rise to a general doctrine, a philosophy of heredity and development, a doctrine which has had and still has a very great influence on general views of life. It is to this doctrine that the prevailing ideas as to the relation of heredity and environment, as to the relative powerlessness of environment, are due. But it has turned out to be a completely mistaken one. . . . The doctrine is dead—though as yet, like the decapitated turtle, it is not sensible of it. . . . What recent investigation has shown is this: the [genes] interact, in complex ways, for long periods; and every later characteristic is a long-deferred and indirect product of this interaction. Into the production of any

characteristic has gone the activity of hundreds of the genes . . . ; and many intermediate products occur before the final one is reached. . . . The genes then are simply chemicals that enter into a great number of complex reactions, the final upshot of which is to produce the completed body. . . . In, producing the structures [nerve, muscle, bone, gland, and other tissues], the genes interact, not only with each other, with the cytoplasm, with the oxygen from the surrounding medium, and with the food substances in the cytoplasm; but also, what is most striking and important, with products from the chemical processes in neighboring cells. . . . What any given cell shall produce, what any part of the body shall become; what the body as a whole shall become—depends not alone on what it contains—its "heredity"—but also on its relation to many other conditions; on its environment.

This early explication of the epigenotype, a series of interrelated developmental pathways through which the adult form is realized (King and Stansfield, 1990), informed Jennings's understanding of the etiology of complex human diseases. He wrote:

> If a characteristic is observed in a given case to be inherited as a sex-linked character, we cannot be certain that it will be sex-linked in other cases. If it is recessive in some stocks, it may be dominant in others. . . . [H]undreds of genes are required to make a mind—even a feeble mind. . . . Doubtless feeble-mindedness is produced in hundreds of different ways—some sorts heritable according to one set of rules, others according to other sets of rules. . . . It is a commonly received dogma that if two parents are defective in the same hereditary characteristic, all the offspring will have this defect. But this need not occur. It will be true only if the defective characteristic is due to a peculiarity of the same gene in the two parents. Where it is due to defects in different genes in the two parents [genetic hetero-

geneity], then the latter supplement each other, and none of the offspring have the defective feature. . . . Heredity is not the simple, hard-and-fast thing that old-fashioned Mendelism represented it. Further, more attentive observation has revealed that any single one of the genes affects, not one characteristic only, but many [pleiotropy]. . . . The idea of representative hereditary units, each standing for a single later characteristic, is exploded; it should be cleared completely out of mind.

Hyperbole aside, the kernal of Jennings's argument is that complex human traits (organ formation, mentation, etc.) are the result of tissue-specific epigenotypes, and these are related to multigenic inheritance and gene–environment interactions. Lip and palate formation are but two examples of this.

In 1942, Poul Fogh-Andersen published his seminal study of hundreds of CL ± P and CP families. He concluded that oral clefts are Mendelian autosomal-dominant disorders with greatly reduced penetrance. In other words, they were Mendelian/Batesonian "unit characters." A few years later, Curt Stern (1949) reached a different conclusion from the published data: "*Harelip* and *cleft-palate* are developmental abnormalities which have a genetic basis. In many pedigrees, they depend on the cooperation of specific alleles at several autosomal loci, and, in addition, require the presence of mostly uncontrolled environmental factors." Stern's conclusion was consistent with the epigenetic model put forth by Jennings (1925). Two decades later, the multifactorial/threshold model (MF/T) was introduced to the problem (Carter, 1969). The MF/T model provided a clever and innovative first approximation to a solution of what was clearly a difficult problem: it recognized that the genetic component was likely non-Mendelian and that environment played an important role as well. The MF/T model logically gave rise to a series of testable predictions of population and family data. Unfortunately, these predictions were rarely, if ever, satisfied when subjected to statistical analysis (Marazita et al., 1992; Nemana et al., 1992; Melnick, 1992; Wyszynski et al., 1996).

Since the mid-1970s, other etiologic models have been proposed (see review by Melnick, 1992): (1) single-gene inheritance of environmental susceptibility, (2) stochastic single-gene inheritance, (3) pure chance, (4) allelic restriction, and (5) emergenic inheritance. In essence, all of these are but elaborations (mathematical and otherwise) on Stern's insight from more than 50 years ago. From the weight of the current evidence, it is clear that there are important major gene effects (Melnick, 1992; Murray, 1995; Wyszynski et al., 1996); these tentatively appear to involve genes related to growth or fusion of facial processes (Lidral et al., 1998; Machida et al., 1999; Tanabe et al., 2000). Nevertheless, the inheritance patterns of CL ± P and CP are not classically Mendelian, exhibiting phenocopies, incomplete penetrance, genetic heterogeneity within and between populations, and the influence of modifier genes and diverse environmental factors such as folic acid deficiency (Loffredo et al., 2001; Martinelli et al., 2001; Piedrahita et al., 1999) and corticosteroid exposure (Park-Wyllie et al., 2000). This is well illustrated by the Fraser-Juriloff paradigm (Fraser, 1980) of differences in susceptibility to an environmental teratogen resulting from a genetically determined difference in *normal* oral development (Figure 21.1).

21.3 INVESTIGATING THE FRASER-JURILOFF PARADIGM

The elucidation of genetic factors for complex etiologies, such as those for oral clefts, is proving to be frustrating. There have been many reports of genes or loci that might be linked or associated with clefting, but none have been unequivocal

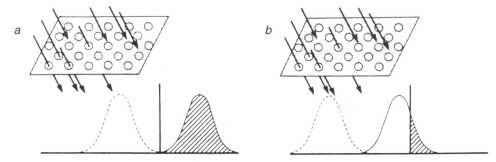

Figure 21.1 Fraser-Juriloff model of CP susceptibility. The roof with holes in it represents the maternal barrier between teratogen (arrows) and embryo. The x-axis represents the phenotypic distribution, normal to the left of the vertical threshold and abnormal to the right; the threshold separates palate closure from palate nonclosure. (a) Palate closure is normally late (*slow growth*), so the phenotypic distribution for this genotype (dashed curve) is near the threshold, and the delaying effect of the teratogen causes all embryos (solid curve) of this genotype to fall beyond the threshold and be affected (hatched area). (b) In an early closing (*faster growth*) genotype, the same delay causes a minority of embryos to be affected. Of course, these two cases are the outer boundaries of the model, and there will be many genotypes (dashed curves) at varying distances to the left of the threshold (Fraser, 1980).

(Ardinger et al., 1989; Hecht et al., 1991; Lidral et al., 1998; Machida et al., 1999; Tanabe et al., 2000). The modest nature of the identified gene effects for oral clefts likely explains the contradictory and inconclusive claims about their identification (Melnick, 1992; Murray, 1995). Despite the small effects of such genes, the magnitude of their attributable risk (the proportion of people affected due to them) may be large because they are quite frequent in the population (Melnick, 1992; Risch and Merikangas, 1996). Knowing this does not lessen the frustrations so far encountered in the search for etiologic solutions to human clefting.

Molecular geneticists looking for ways to understand human disease have increasingly turned to the mouse as biomedicine's model mammal (Malakoff, 2000). However, even the search for mouse models of nonsyndromic oral clefting has proved frustrating, at least as an attempt to confirm or inform the human data. Transforming growth factor-α (*TGFA*) was purported to be associated with nonsyndromic clefting (Ardinger et al., 1989). Not only has this not been confirmed in humans (Hecht et al., 1991; Lidral et al., 1998; Machida et al., 1999) but *Tgfa* knockout mice do not exhibit a cleft phenotype (Luetteke et al., 1993). Knockout of the receptor for Tgfa (Egfr) results in syndromic clefting, including micrognathia and other facial abnormalities (Miettinen et al., 1999). Transforming growth factor-β3 (*TGFB3*) and the transcription factor *MSX1* were also purported to be associated with CL ± P, and *MSX1* with CP, in humans (Lidral et al., 1998). Again, this has not been confirmed (Tanabe et al., 2000). Further, while *Msx1* and *Tgfb3* knockout mice exhibit clefting, the clefting is *syndromic* (Kaartinen et al., 1995; Proetzel et al., 1995). Similarly, while human nonsyndromic CL ± P has been significantly associated with *TGFB2* (Tanabe et al., 2000), *Tgfb2* knockout mice exhibit *syndromic* clefting, including cardiac, lung, limb, spinal, eye, urogenital, and other craniofacial defects (Sanford et al., 1997).

Human linkage/association studies and mouse studies of transgenic models have provided limited insight into the etiology of oral clefting. Considering Stern's (1949) modeling of the human data and the

Fraser-Juriloff paradigm (Fraser, 1980), it is little wonder that there is uncertainty. Clearly we have far more work to do in mouse studies, searching for more embryologically and genetically appropriate analogies and homologies before we return to human studies. The central problem is determining what is signal and what is noise by understanding what detail at the level of individual units is essential to understanding more macroscopic regularities (Levin et al., 1997). In this regard, it is instructive to recount the emerging story of CP susceptibility in H-2 congenic mice, if only as a "proof" of a 25-year-old paradigm (Melnick and Shields, 1976).

21.3.1 H-2 Haplotype: Maternal and Embryonic Effects

The exposure of embryonic mice to corticosteroids (CORT) has long been known to result in CP (Fraser and Fainstat, 1951). Studies in our laboratory, and others, have shown consistently that CORT-induced CP is related to genetic variation at or near the H-2 complex on mouse chromosome 17 (Melnick et al., 1981a; Goldman, 1984; Gasser et al., 1991). H-2 congenic mice were originally developed by George Snell in the late 1940s. They share identical genetic backgrounds, with the exception of a 3–18 cM region of chromosome 17 (i.e., the congenic region), which encompasses the H-2 complex and defines each H-2 haplotype (Fig. 21.2) (Vincek et al., 1990).

These mice are an important tool for investigating the contribution of specific congenic genes to develoment, including palate morphogenesis.

By using the H-2 congenic mice, we have shown that B10.A ($H\text{-}2^a$) mice are ninefold more susceptible to CORT-induced CP than B10 ($H\text{-}2^b$) mice (Table 21.1); reciprocal hybrid studies have demonstrated a significant maternal effect (Table 21.1) (Melnick et al., 1981a). In reciprocal crosses between two inbred strains, the two types of F_1 females are genetically identical. If two inbred strains differ in response to a teratogen, one being susceptible and the other resistant, we can test for cytoplasmic effects by backcrossing the two types of reciprocal F_1 females to the susceptible strain males (Fig. 21.3) (Melnick et al., 1983). If there are statistically significant differences in the frequency of developmental abnormality in the two types of treated backcross offspring (in the direction of the line of the mother's mother), a genetic difference in maternal physiology is ruled out as a reasonable explanation and a cytoplasmic factor (mitochondrial genes) is suggested (Biddle and Fraser, 1977). In fact, there is a significant increase in developmental anomaly frequency (DAF) when the mother's mother is of the susceptible B10.A strain (Fig. 21.3) (Melnick et al., 1983).

One explanation for this interesting finding is the recently demonstrated interaction between nuclear and mitochondrial

TABLE 21.1 CORT-Induced Cleft Palate in H-2 Congenic Mice

Cross Dam × sire	CORT mg/kg	Embryo (haplotype)	Cleft Palate Frequency (%)
B10.A × B10.A	0	B10.A ($H\text{-}2^a/H\text{-}2^a$)	0
B10.A × B10.A	2	B10.A ($H\text{-}2^a/H\text{-}2^a$)	45
B10 × B10	0	B10 ($H\text{-}2^b/H\text{-}2^b$)	0
B10 × B10	2	B10 ($H\text{-}2^b/H\text{-}2^b$)	5
B10.A × B10	2	B10.A.B10 ($H\text{-}2^a/H\text{-}2^b$)	31
B10 × B10.A	2	B10.B10.A ($H\text{-}2^b/H\text{-}2^a$)	21

Significance tests: (1) B10.A (2 mg/kg) vs. B10 (2 mg/kg): $p < 0.001$; (2) B10.A.B10 vs. B10.B10.A: $p < 0.05$.

Chromosome Seventeen

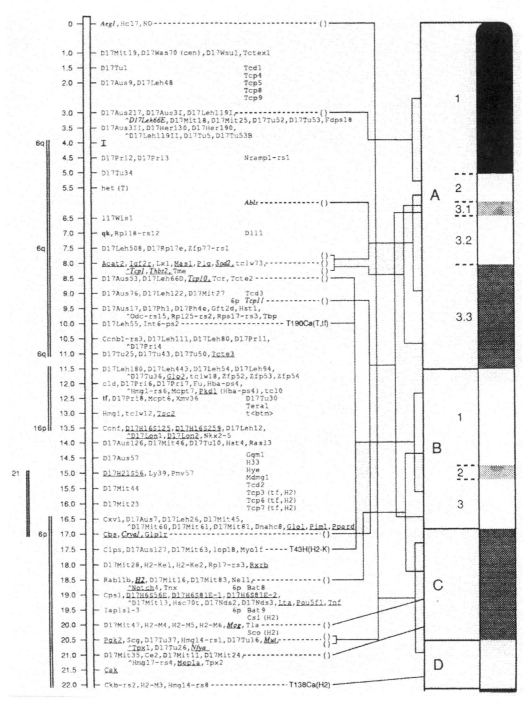

Figure 21.2 Mouse gene map, chromosome 17 (Mouse Gene Informatics—Jackson Laboratory: www.informatics.jax.org). Note the positions of *Igf2r* and *Plg* at 8.0 cM, as well as *H2* at 18.5 cM (all of which have corresponding genes on human chromosome 6).

genomes (Johnson et al., 2001). In 1982, Ferris and co-workers examined mouse mitochondrial DNA from various locales in the Northern Hemisphere using restriction enzymes that cut the molecule at an average of 150 sites and found a high level of restriction-site polymorphism in wild mice but no variation among the "old" inbred strains commonly found in the laboratory, including the C57BL strain, which is the background of the H-2 congenic mice described above. Ferris and colleagues (1982) suggest that all the "old" inbred strains of laboratory mice are descendants of one single female who lived sometime between 1200 B.C. and A.D. 1920. Perhaps what is seen in the experiments outlined in Figure 21.3 is the phenotypic effects of "gene-gene interaction" between variant congenic genotypes and an invariant mitochondrial genotype. The key, then, is finding the relevant gene(s) in the congenic genotypes.

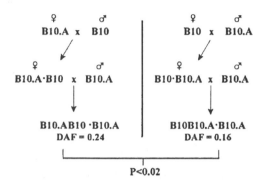

Figure 21.3 Testing for a cytoplasmic factor (mitochondrial genes). Reciprocal F_1 females are backcrossed to males of the susceptible strain (B10.A). There is a significant difference in the frequency of developmental abnormality (DAF) in the direction of the line of the mother's mother, and, thus, evidence of a cytoplasmic (mitochorial) effect.

21.3.2 Corticosteroids and Gene Regulation

Corticosteroid hormone signaling is somewhat unique among signal transduction mechanisms (Fig. 21.4). CORT is lipophilic and crosses the cell membrane, where it

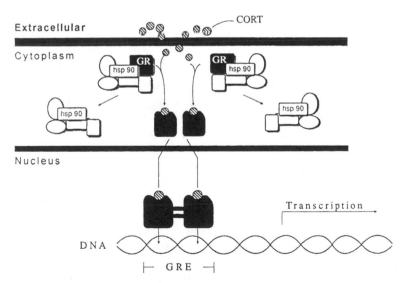

Figure 21.4 Corticosteroid signal transduction pathway. Lipophilic glucocorticoid (CORT) translocates across the plasma membrane to the cytoplasm and binds to the glucocorticoid receptor (GR). Activation of GR releases the hsp90-dominated complex to which it is bound. CORT/GR translocates to the nucleus, homodimerizes, and binds to a DNA glucocorticoid response element (GRE) in the regulatory region of target genes to up- or downregulate transcription.

binds to its cognate cytoplasmic receptor (GR). Activated ligand-bound receptor is translocated to the nucleus, dimerizes with other ligand-bound receptors, and binds to response elements (GREs) in the regulatory region of target genes. In essence, the CORT/GR complex serves as a transcription factor, up- and downregulating gene expression.

The TGF-β family of proteins is involved in regulating cell proliferation, differentiation, and extracellular matrix formation and degradation (Dunker and Krieglstein, 2000). The three mammalian TGF-β isoforms are TGF-β1, TGF-β2, and TGF-β3, each encoded by different genes on different chromosomes. There is a significant increase in TGF-β1 and TGF-β3 transcript levels and a significant decline in TGF-β2 transcript levels with progressive palatal development (Jaskoll et al., 1996). All TGF-β isoforms signal via the same cell membrane–bound heteromeric receptor complex: TGF-β receptor type I and ligand binding TGF-β receptor type II (Fig. 21.5). Signal transduction from the receptor to the nucleus is mediated by intracellular effector molecules termed SMADs (Fig. 21.5). TGF-β2, the only isoform primarily localized in the palatal mesenchyme (Jaskoll et al., 1996), inhibits palatal mesenchymal cell proliferation and thus palatal shelf growth (Ferguson, 1988; Jaskoll et al., 1996). Downregulation of Cdk4-mediated cell division results from TGF-β2/SMAD-induced upregulation of the transcription factor p27 (Fig. 21.5). For palates to grow, then, TGF-β2 must be downregulated. It has been shown that CORT-induced delay in the normal downregulation of TGF-β2 transcription is a key event in the pathogenesis of CORT-induced CP in B10.A embryos (Jaskoll et al., 1996).

21.3.3 The TGF-β/IGF Connection

H-2 haplotype-specific differences in the rate of embryonic development in B10.A and B10 congenic mice have been studied extensively. Significant strain differences in the number of embryonic day 12 (E12) embryos that reach the appropriate Theiler developmental stage (Theiler, 1989) are seen routinely (Fig. 21.6). In addition, B10.A mice produce smaller embryos, with delayed palatal development, lung maturation, H-2 antigen expression, and skeletal development compared with B10 mice at identical Theiler stages (Good et al., 1991; Hu et al., 1990; Jaskoll et al., 1991; Melnick and Jaskoll, 1992; Melnick et al., 1981b, 1982). Thus, if CORT inhibits palatogenesis to the same degree in both strains via TGF-β2 regulation (Jaskoll et al., 1996), then it is not suprising that the slower-developing B10.A embryo is more vulnerable to abnormal palatogenesis than the faster-developing B10 embryo, as predicted by the Fraser-Juriloff paradigm (Fig. 21.1). The question is what is the link between genes in the 3–18 cM congenic region of chromosome 17 and TGF-β2, the gene for which is on chromosome 1 and thus identical in B10.A and B10 mice.

Viewing the map of chromosome 17 in the potential congenic region (Fig. 21.2), we quickly see that the best candidate is the gene for insulin-like growth factor receptor type 2 (*Igf2r*). *Igf2r* maps to approximately 8 cM from the centromere and 10 cM from the more telomeric H-2. IGF-IIR is a large, membrane-bound glycoprotein (~300 kDa) that contains distinct binding sites for two ligands: insulin-like growth factor type 2 (IGF-II) and mannose-6-phosphate (M6P)–bearing molecules such as lysosomal enzymes and latent TGF-β (Jones and Clemmons, 1995; Vignon and Rochefort, 1992) (Fig. 21.5). IGF-IIR does not appear to transduce mitogenic signals (Moats-Staats et al., 1995); instead, it sequesters IGF-II from type I IGF receptors, which mediate IGF-II growth signal transduction (Ballard et al., 1986; Lau et al., 1994; Wang et al., 1994; for reviews, see Barlow, 1995, and Haig and Graham, 1991). This seques-

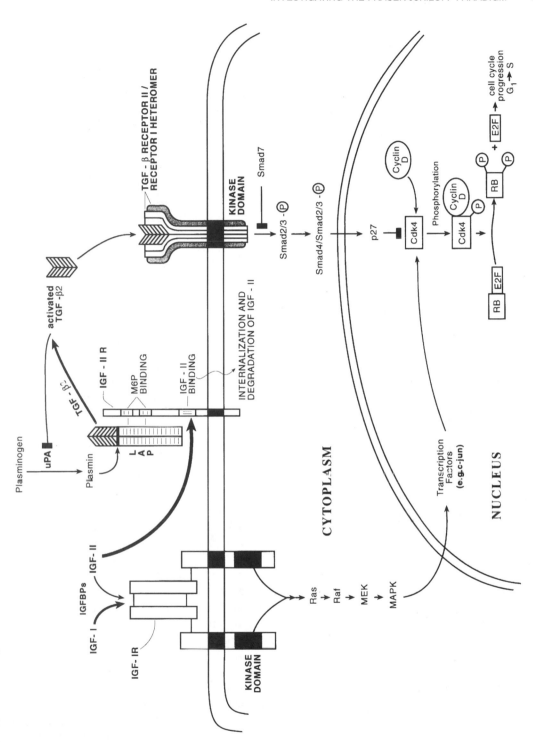

Figure 21.5 Epigenetic network of information processing relative to cell proliferation. IGF, insulin-like growth factor; R, receptor; TGF-β, transforming growth factor-beta; LAP, latency-associated protein; M6P, mannose-6-phosphate; uPA, urokinase-type plasminogen activator; Ras, a G protein; Raf, Ras-activated factor; MEK, a protein kinase; MAPK, mitogen-activated protein kinase; Smad, a family of transcription factors; p27, a transcription factor.

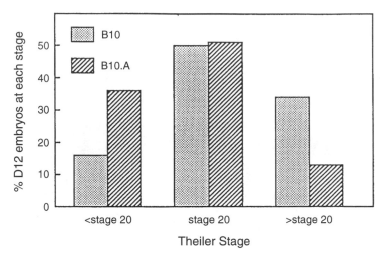

Figure 21.6   Theiler staging of B10 and B10.A embryos on day 12 of gestation, when 50% of both B10 and B10.A embryos are at stage 20. However, there is a significantly greater percentage of B10.A embryos at less than stage 20 compared with B10 embryos, and a significantly smaller percentage of B10.A embryos are at greater than stage 20 compared with B10 embryos.

tration of IGF-II by IGF-IIR regulates the levels of IGF-II ligand available to IGF-IR for use in promoting growth (Barlow, 1995; Ellis et al., 1996; Filson et al., 1993) (Fig. 21.5).

As discussed in detail below, it is critical to note that *Igf2r* is genomically imprinted. Primarily, the maternal copy is expressed in postimplantation embryos, giving rise to the widespread belief that imprinting serves to control embryonic growth in utero (Barlow, 1995; Barlow et al., 1991). Mouse embryos that inherit a nonfunctional maternal *Igf2r* gene confirm that the IGF-IIR is crucial for regulating normal embryonic growth and also for regulating the levels of free IGF-II ligand (Lau et al., 1994). IGF-IIR is also key to regulating the levels of activated TGF-β2 in developing palates (Melnick et al., 1998) (Fig. 21.5).

21.3.4   IGF and Palate Morphogenesis

The presence of IGF-II, IGF-IR, and IGF-IIR in all cellular types of the embryonic palate in all developmental stages (E12–E15) indicates that the IGF-II signal transduction pathway (IGF-II + IGF-IR → cell division) and the IGF-IIR negative regulation of the IGF-II pathway (Fig. 21.5) are involved in regulating palatal growth (Melnick et al., 1998). On embryonic day 14 (E14), which is a critical day for palatal growth, the slower-growing B10.A embryonic palates contain 82% more IGF-IIR transcript than faster-growing B10 palates (Fig. 21.7). This significant elevation of IGF-IIR levels in B10.A embryonic palates reduces the concentration of IGF-II ligand available to growth-promoting IGF-IR, resulting in a decreased growth rate of B10.A palates (Fig. 21.8). In terms of the Fraser-Juriloff paradigm (Fig. 21.1), this would place the B10.A genotype closer to the threshold than the B10 genotype; thus, exposure to even equivalent CORT-induced downregulation of palatal growth results in far greater adverse phenotypic outcomes for B10.A than for B10 embryonic palates, as previously noted (Table 21.1) (Melnick et al., 1981a).

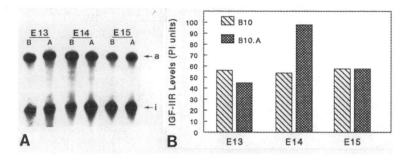

Figure 21.7 IGF-IIR transcript levels in developing B10 and B10.A palates. (A) An Rnase protection assay to compare the steady-state levels of IGF-IIR mRNA in E13, E14, and E15 B10(B) and B10.A(A) palates. (B) Bars represent mean phosphor imaging (PI) units of IGF-IIR mRNA; E14 B10.A levels are 82% greater than E14 B10 ($p < 0.05$), while all other mean levels are equivalent (Melnick et al., 1998).

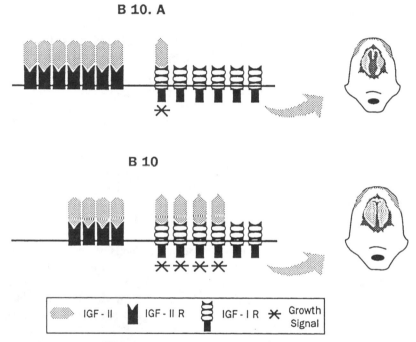

Figure 21.8 Schematic of the IGF-IIR role in palate development. IGF-II binds to IGF-IR and IGF-IIR, having a higher affinity for IGF-IIR than for IGF-IR. The lower level of IGF-IIR in the E14 B10 palate results in increased availability of IGF-II, thereby enabling more of the ligand to bind the IGF-IR in B10 palates compared with B10.A palates. Because only IGFII/IGF-IR binding transduces a mitogenic signal (asterisks), the result is an accelerated rate of morphogenesis in the B10 palate compared with the B10.A palate.

21.3.5 IGF-IIR and TGF-β2 Activation

All the mammalian TGF-βs are 25 kD homodimers in their biologically active form; they show a high level of sequence conservation (Lawrence, 1996). The larger, latent TGF-β is comprised of one mature TGF-β molecule noncovalently bound to the proregion dimer, the latency-associated peptide (LAP) (Fig. 21.5); TGF-β1 and

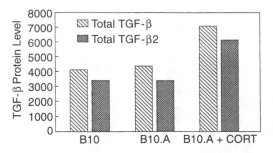

Figure 21.9 Mean total TGF-β and total TGF-β2 protein levels in E14 palates. There is no substantial difference between strains in total TGF-β levels; CORT treatment induces a 62% increase in total TGF-β levels and a 56% increase in total TGF-β2 levels (Melnick et al., 1998).

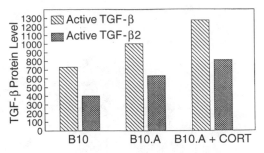

Figure 21.10 Mean total active TGF-β and total active TGF-β2 protein levels in E14 palates. B10.A palates exhibit a 37% greater level of active TGF-β than B10 and a 57% greater level of active TGF-β2; elevated CORT levels significantly increase the levels of activated TGF-β, 64% of which is TGF-β2 (Melnick et al., 1998).

TGF-β2 LAPs contain M6P residues (Brunner et al., 1992; Dennis and Rifkin, 1991; Gleizes et al., 1997). Cellular activation of latent TGF-β appears to require binding to the M6P binding site of the IGF-IIR and is plasmin and plasminogen activator dependent (Dennis and Rifkin, 1991; Gleizes et al., 1997). Plasminogen and plasminogen activators are all found in the embryonic palate (Melnick et al., 1998). It is probably more than coincidence that the polymorphic plasminogen gene, *Plg*, is closely linked to *Igf2r* (Fig. 21.2) (Barlow et al., 1991; Friezner et al., 1990).

The relationship between TGF-β2 activation and IGF-IIR is key. Because embryonic day 14 (E14) is a critical stage of palatal growth that is dependent on the downregulation of TGF-β2 (Jaskoll et al., 1996), and known strain differences in growth rate (Melnick and Jaskoll, 1992) are associated with TGF-β2 (Jaskoll et al., 1996) and IGF-IIR (Melnick et al., 1998) expression, it is not surprising that greater availability of the IGF-IIR receptor in B10.A embryonic palates (Fig. 21.7) would result in a higher level of activated TGF-β2 (Melnick et al., 1998). Approximately one-fifth of the total TGF-β is activated, and approximately two-thirds of the total activated TGF-β is TGF-β2 (Figs. 21.9 and 21.10). It is particularly noteworthy that E14 B10.A embryonic palates have a 57%

greater level of active TGF-β2 than B10 embryonic palates (Fig. 21.10), even though their total TGF-β2 levels are nearly identical (Fig. 21.9). Thus, the more IGF-IIR (receptor), the more active TGF-β2.

As noted above, TGF-β2-induced inhibition of palatal mesenchymal cell proliferation is related to arrest of the $G_1 \rightarrow S$ transition of the cell cycle through SMAD/p27-mediated downregulation of Cdk4 (Fig. 21.5) and, perhaps, other G_1 factors such as cyclins D and E and Cdk2 (Derynck, 1994). Thus, an inverse relationship is seen between levels of active TGF-β2 and Cdk4 (Melnick et al., 1998). E14 B10 palates with a lower level of active TGF-β2 (Fig. 21.10) have a 52% greater level of Cdk4 transcript (Fig. 21.11) than B10.A with a higher level of active TGF-β2. Thus, the variation of TGF-β2/IGF-IIR-mediated growth inhibition in the late G_1 phase of the cell cycle (Figs. 21.5 and 21.7–21.11) (Melnick et al., 1998) appears to account for the slower growth and development of B10.A embryonic palates relative to its B10 congenic partner (Melnick and Jaskoll, 1992).

21.3.6 CORT-Induced Palate Pathogenesis

It is well established that CORT induces CP many-fold in B10.A embryos relative to

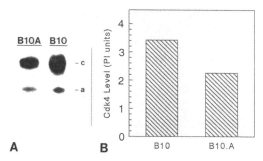

Figure 21.11 Cdk4 mRNA levels. (A) An Rnase protection assay was used to compare the steady-state levels of Cdk4 mRNA in E14 B10 and B10.A palates. (B) Bars represent E14 palatal mean phosphor imaging (PI) units of Cdk4 mRNA; B10 palates exhibit a significant 52% increase in Cdk4 mRNA levels compared with B10.A palates (Melnick et al., 1998).

its B10 congenic partner when exposed on E12 (Melnick et al., 1981a). CORT exposure has been shown to inhibit palatal mesenchyme cell proliferation, resulting in smaller palatal processes and CP (Potchinsky et al., 1996; Salomon and Pratt, 1979). CORT also delays by 1 day the downregulation of palatal TGF-β2 transcription normally seen on E14 (Jaskoll et al., 1996). Further, in E14 B10.A palates, elevated CORT exposure significantly increases TGF-β protein levels, 87% of which is TGF-β2, as well as the levels of active TGF-β, 64% of which is TGF-β2 (Figs. 21.10 and 21.11) (Melnick et al., 1998). This enhances the TGF-β2/IGF-IIR-mediated growth inhibition via downregulation of Cdk4 in late G_1 of the cell cycle.

Thus, we have an outline of the pathogenetic mechanism in B10.A embryos: Slower-growing B10.A embryos have an upregulation of IGF-IIR that serves to sequester IGF-II from the growth-promoting IGF-IR and to bind more CORT-upregulated, latent TGF-β2 for subsequent plasmin-dependent activation. Higher levels of TGF-β2 signaling lead to palatal growth inhibition at a critical stage of palatogenesis and, thus, subsequent CP. In terms of the Fraser-Juriloff paradigm

(Fig. 21.1), B10.A embryos that are already close to the threshold of abnormality are pushed beyond that threshold with the CORT-induced upregulation of activated TGF-β2.

21.4 IGF-IIR/TGF-β2 EPIGENETIC NETWORK

Figure 21.5 is a model of information processing as it relates to cell proliferation of mesenchyme in embryonic palates. Ligand-receptor binding is the first step in pathways of signal processing that effect specific gene expression and phenotypic change. Typically signaling pathways are studied as though information processing were linear. However, it is becoming increasingly apparent that pathways interact with one another, and the final biologic response is shaped by this interaction (Bhalla and Iyengar, 1999; Strohman, 1997). This results in signaling *networks* of great complexity and nonlinearity. Such networks are *epigenetic* networks in that they include feedback to the genome and changing patterns of gene expression (Strohman, 1997).

Although this representation of the molecular control of cell proliferation (Fig. 21.5) is not strictly reducible to its parts (labeled boxes, ovals, arrows, etc.), some of the known factors that make this complex epigenetic network nonlinear and adaptive include (1) IGF-IIR binds IGF-II with a very significantly greater affinity than IGF-IR (Jones and Clemmons, 1995); (2) although ligand binding of the IGF-IR is not the *sine qua non* for cell cycle progression, it is probably required for the cell cycle to be maintained at a normal rate (LeRoith et al., 1995); (3) since IGF-II and MP6-bearing molecules (e.g., latent TGF-β2) competitively bind to their cognate IGF-IIR sites because of steric hinderance or conformational change, any imbalance in ligand(s) and receptor concentration is likely to alter associated biological functions, such as IGF-II degradation, IGF-II/IGF-IR binding, and

TGF-β2 binding and activation (Vignon and Rochefort, 1992); (4) TGF-β decreases the mRNA expression of both uPA and tPA plasminogen activators and may stimulate PA inhibitor production (Agrawal and Brauer, 1996; Keski-Oja et al., 1988); (5) plasmin-dependent activation of TGF-β is modulated by surface localization of uPA by its recptor (Odekon et al., 1994). Functionally, then, the IGF-IIR/TGF-β2 epigenetic network is a dynamical network that uses a continuous logic to learn its rules from changing conditions. As such, it can be modeled mathematically as an artificial neural network.

21.4.1 Genes and Development

Nearly four decades ago, Maruyama (1963) clearly outlined the developmental biologist's nightmare:

> [I]t is not necessary for the genes to carry all the information regarding the adult structure, but it suffices for the genes to carry a set of rules to generate the information.

> The amount of information to describe the resulting pattern is much more than the amount of information to describe the generating rules and the positions of the initial tissues. The pattern is generated by the rules and by the interaction between the tissues. In this sense, the information to describe the adult individual was not contained in the initial tissues at the beginning but was generated by their interactions.

> [I]t is in most cases impossible to discover the simple generating rules after the pattern has been completed, except by trying all possible sets of rules. When the rules are unknown, the amount of information needed to discover the rules is much greater than the amount of information needed to describe the rules. This means that there is much more waste, in terms of the amount of information, in tracing the process backwards than in tracing it forward.

Maruyama's dilemma, and ours, is that we cannot reduce emergent developmental phenomena to nucleotide sequences. It has been mathematically demonstrated that merely 40 genes *could* produce entirely specific cell lineages for about one million differentiated states (Gierer, 1973). Certainly this is not a reality. Nevertheless, an emerging theme in developmental biology is that defined sets of epigenetic circuits are used in multiple places, at multiple times, for similar and sometimes different purposes during organogenesis (Melnick and Jaskoll, 2000). In the context of palate development and the IGF/TGF-β epigenetic network, we might also benefit by looking briefly at several other emerging themes in developmental biology: cellular automata, differential methylation, and submolecular information processing.

21.4.2 Game of Life

Important to the Fraser-Juriloff paradigm (Fig. 21.1) is the demonstration of significant measurable differences between inbred strains (or human individuals) for growth-related anatomic variables relevant to palate development. In support of this, Diewert (1982) finds significant quantitative differences between A/J and C57BL/6 inbred mice for several growth variables related to palatal shelf elevation, contact, and closure. These include the length of Meckel's cartilage relative to that of the bronchonasal cavity, the height and width of the oral cavity, and the width of the maxilla. Similar differences are seen in other inbred strain comparisons (Ciriani and Diewert, 1986), as well as in B10.A and B10 congenic mice (Melnick and Jaskoll, 1992). In sum, certain strains (A/J, B10.A) are slower growing, and ultimately somewhat smaller, than other strains (C57BL/6, B10).

We have seen above how this heterochronic palatal development is related to the IGF/TGF-β epigenetic network. To put

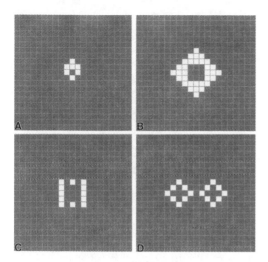

Figure 21.12 Game of Life. This is a cellular automaton that was invented by Cambridge mathematician John Conway. It consists of a collection of cells which, based on a set of simple rules, can live, die, or multiply; depending on the initial conditions, the cells form specific patterns from one generation to the next. (A) Initial condition "small exploder." (B) "Small exploder" seven generations later. (C) Initial condition "exploder." (D) "Exploder" seven generations later. (See www.bitstorm.org/game of life.)

"Developmental Generations"

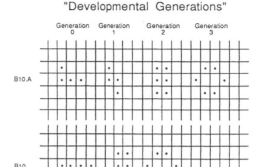

Figure 21.13 Conway's Game of Life and palatal development. Different initial conditions (patterns) can result in identical final stages, albeit over different time periods.

this relationship in a more amenable conceptual framework, it is convenient to call on John Conway's "Game of Life" (www.bitstorm.org/gameoflife/), a popular example of a two-dimensional cellular automaton. Using the same set of rules to determine the fate of future generations, it can be seen that any given initial condition (or starting pattern of occupied cells) will inevitably lead to a specific sequel pattern over any specified number of generations (Fig. 21.12). What is important to the present developmental problem is the property that a given configuration can have several preceding sequels due to several different initial conditions, but only one future.

Viewing Figure 21.13, it can be seen that different initial conditions (patterns) can lead to identical final states, albeit with different chronologies. While viewing a particular final state on a computer screen

does not allow us to deduce its predecessors, the tools of developmental biology allow us to make a first approximation regarding organogenesis. Nevertheless, we have to always remember that the rules of the game are set by the epigenotype, a cellular informational system that integrates genetic and environmental information into a dynamical process able to generate responses that are functionally adaptive (Strohman, 1997). This is why the "Game of Life" (Fig. 21.12) is such an apt computational metaphor, and so informative.

Suppose strains B10.A and B10 are as represented in Figure 21.13; suppose the rules of the "game" are identical in each strain, and these are set by a well-conserved epigenotype; suppose the initial conditions (patterns) are influenced by IGF-IIR transcript and protein levels. Then we may look at Figure 21.13 in the following way:

1. Different initial conditions (phenotypes, "game patterns") are tightly associated with epigenetically determined differences in IGF-IIR levels in each strain (see Fig. 21.7 and methylation discussion below).

2. Different initial conditions (phenotypes, "game patterns") are associated with different chronologies; namely, B10.A requires three "developmental generations" to reach the final state, while B10 requires only two (Melnick and Jaskoll, 1992).

3. Identical final states (phenotypes, "game patterns") in both sequences (strains) have similar, if not identical, predecessor states (phenotypes, "game patterns"), though initial conditions (phenotypes, "game patterns") are quite different (Melnick, 1992).

From this list we can clearly visualize the larger organismal picture, namely that strain differences are largely chronologic differences, and these are determined by epigenetically mediated initial state (phenotype) differences that ultimately play their hand with a set rule book. It is tempting to think of this as knowledge, but game theory is a metaphor. Like most metaphors, it captures some aspects of the truth but leads us astray if we take it as anything but a first approximation.

21.4.3 IGF-IIR Imprinting and Methylation

Mammals exhibit the unique (and non-Mendelian) process of genomic imprinting. In this epigenetic process, a gene on one chromosome is silenced (imprinted), while its homologous allele on the other homologous chromosome is expressed. There are several known mechanisms of gene silencing, the best characterized being DNA methylation (see reviews by Pfeifer, 2000; Reik and Walter, 2001).

DNA methylation refers to the addition of methyl groups to cytosine residues of cytosine-guanine dinucleotide (CpG) repeats in the DNA sequence of specific genes. DNA methylation is mediated by several well-characterized methyl trans-

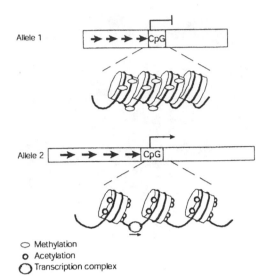

○ Methylation
○ Acetylation
◯ Transcription complex

Figure 21.14 Imprinted genes. This is a diagramatic representation of a homologous pair of imprinted alleles (say the alleles for the gene *Igf2r*). Characteristics of imprinted genes such as CpG islands and repeats (arrows) are noted. The enlargement below each cognate chromosomal location illustrates the allele-specific epigenetic changes, such as nucleosomal condensation (via deacetylation) and methylation (allele 1) and increased nucleosomal spacing (via acetylation), demethylation, and the binding of a transcriptional complex for gene expression (allele 2). (*Source*: Adapted from Reik and Walter, 2001.)

ferases. Gene silencing by methylation is associated with a characteristic change in chromatin structure (Fig. 21.14) (Razin, 1998; Wolffe, 1998). Methylation-associated allelic repression is usually quite stable, but it can be reversed (Ramechandani et al., 1999). DNA demethylase also shows CpG specificity, but it catalyzes the cleavage of a methyl residue from 5-methyl cytosine and its release as methanol. Thus, demethylase performs the reverse reaction to DNA methyltransferase and would seem to be a natural partner in shaping the site-specific methylation pattern of genomes, a *sine qua non* of normal mammalian embryogenesis (Li et al., 1992; Ramchandani et al., 1999).

Excluding X-chromosome inactivation (lyonization), the relatively small number

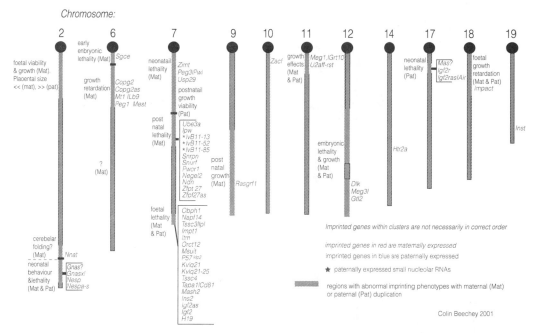

Figure 21.15 Mouse imprinted genes, regions, and phenotypes. (*Source*: From Beechey et al., 2001.)

of imprinted genes exhibit allelic expression differences that depend on parental origin (Fig. 21.15). Imprinted genes are found in clusters (Fig. 21.15), and these clusters are conserved between mouse and human (Fig. 21.16), leading to the proposition that the clustering of imprinted genes is essential to their regulation (Thorvaldsen and Bartolomei, 2000). Perusing the mouse genomic imprinting map (Fig. 21.15), it is easy to see how aberrant imprinting disturbs embryonic development. Such is certainly the case regarding *Igf2r*.

The significant increase in B10.A IGF-IIR is transient and specific to E14 (Fig. 21.7), a day that is critical to mouse palatal growth. The most likely mechanism is a switch from mostly monoallelic expression of *Igf2r* to more biallelic expression—a switch that results from more than relaxation of methylation (Lerchner and Barlow, 1997). Lerchner and Barlow show that the paternal *Igf2r* allele is repressed in mice from E6.5 onward; however, a low

level of paternal expression remains in tissues that highly express the maternal allele from E7.5 onward. Functional polymorphism (monoallelic/biallelic) exists with the parental imprinting of the human *Igf2r* gene as well (Xu et al., 1993). Lerchner and Barlow propose that mere DNA methylation is not sufficient to cause monoallelic expression and must occur by a multifactorial process. This is supported by Smrzka and colleagues (1995), who found that the human *IGF2R* gene has the classic imprinting characteristics of monoparental methylation and replication asynchrony but does not show unequivocal monoallelic expression. Finally, additional genetic or epigenetic control of allelic expression is also supported by congenic mouse matroclinus reciprocal hybrid cross data (Melnick et al., 1998). B10.A/B10.A embryos had 60% greater IGF-IIR levels (P < 0.01) than B10.A/B10 embryos, which in turn were equivalent to B10/B10.A and B10/B10 embryos (Fig. 21.17). This can only

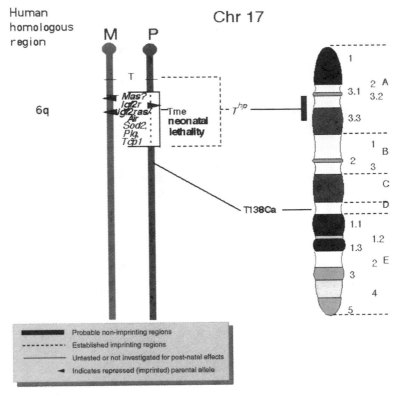

Figure 21.16 Mouse imprinting map of chromosome 17. M, maternal chromosome; P, paternal chromosome. (*Source*: From Beechey et al., 2001.)

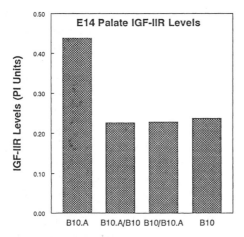

Figure 21.17 IGR-IIR mRNA levels (mean PI units) in B10 and B10.A incrosses, as well as matroclinus reciprocal hybrid crosses.

occur if the control of monoallelic expression is both biparental and B10 dominant—but how?

One possible explanation for the data in Figure 21.17 is that B10 and B10.A *Igf2r* genes have important sequence differences in the promoter regions. However, sequencing of a 900 bp region that includes the *Igf2r* promoter reveals identity in the B10–B10.A congenic pair (Fig. 21.18). Another possibility is that the critical strain differences on E14 are correlated with monoallelic *Igf2r* expression in B10 embryos and biallelic *Igf2r* expression in B10.A embryos. Thus, relaxation of methylation in B10 palates should result in IGF-IIR mRNA levels equivalent to normal

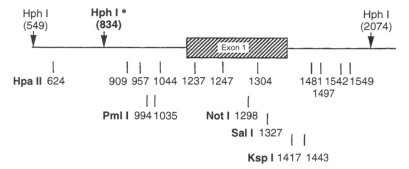

Figure 21.18   Restriction enzyme map of the *Igf2r* promotor region (not to scale). Genomic DNA and cDNA, prepared from E14 B10 and B10.A palates, were amplified by PCR and sequenced for the presence of polymorphic sites in the promotor region of the *Igf2r* genes. There is 100% identity between these congenic strains (Hoffman et al., unpublished data).

B10.A palates. Relaxation of methylation is achieved by administering the demethylating 5-Aza-2′-deoxycytidine (d-AZA) to pregnant B10 females and comparing E14 palatal IGF-IIR mRNA levels to palatal levels in untreated B10 and B10.A embryos (Melnick and Jaskoll, unpublished data). d-AZA-treated B10 E14 palates exhibit a 35% greater level of IGF-IIR transcript than untreated B10 palates and are nearly identical to untreated B10.A palate levels (Fig. 21.19). The increase in IGF-IIR transcripts in d-AZA-treated B10 and untreated B10.A is correlated with a 20+% decline in Cdk4 transcripts, and thus diminished palatal growth (Fig. 21.20).

These results (Figs. 21.19 and 21.20) suggest an important genetic and/or epigenetic regulation of *Igf2r* imprinting during a critical stage of palatogenesis—one that is B10 dominant. A number of *cis*-acting sequences are being defined that are also important for the control of imprinting. A *cis*-acting locus is a specific region of nucleotide sequence that affects the activity of gene(s) on that same DNA molecule; *cis*-acting loci generally do not encode proteins. Methylation differences are found in two regions of the *Igf2r* gene (Reik and Constancia, 1997). Region 1 is in the pro-

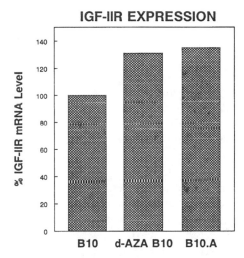

Figure 21.19   IGF-IIR gene expression in B10 and B10.A incrosses, as compared with d-AZA-induced relaxation of methylation in B10 incrosses.

moter, and it is methylated when the paternal allele is not expressed. Region 2 is downstream in the second intron, and it is methylated in the expressed maternal allele, suggesting that this region contains an "imprinting box" with silencer sequences that can be suppressed by DNA methylation. Indeed, it has been demonstrated that intron 2 produces an antisense

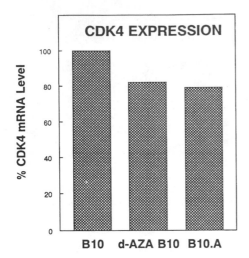

Figure 21.20 Cdk4 gene expression in B10 and B10.A incrosses, as compared with d-AZA-induced relaxation of methylation in B10 incrosses.

that is expressed from the paternal chromosome (*Igfras* or *Air*, for antisense *Igf2r* RNA; see Fig. 21.16) (Wutz et al., 1997); region 2 is the promoter for this antisense RNA (Lyle et al., 2000). When this region 2 promoter is methylated in the maternal allele, the repressor antisense is not available and the *Igf2r* gene proper is transcribed because its promoter is not methylated. The opposite obtains in the paternal allele. Antisense RNAs may regulate expression of their cognate sense mRNAs via transcriptional interference or expression competition; it is not certain if *Air* expression is a cause of *Igf2r* repression or a consequence of the repression mechanism (Lyle et al., 2000). Should it be the former, we should expect sequence differences between B10.A and B10 in the region 2 promoter for *Air*. Since a low level of paternal *Igf2r* allele expression is found in tissues that highly express the maternal allele (Lerchner and Barlow, 1997), it is reasonable to surmise that the matroclinus reciprocal hybrid cross data (Fig. 21.17) (Melnick et al., 1998) can be explained by strain differences in *Air* expression.

21.4.4 Submolecular Biology and Quantum Computing

In 1960, between the discovery of the DNA double helix and the discovery of the genetic code, two-time Nobel Laureate, Albert Szent-Györgyi, wrote a monograph titled *Introduction to a Submolecular Biology*. In this work he makes the case that biochemistry needs to follow "its parent science, chemistry, allying itself with physics and mathematics, [making] a dive into a new dimension, that of the submolecular or subatomic dimension of electrons, a dimension the happenings of which can no longer be described in the terms of classic chemistry, the rules of which are dominated by quantum, or wave, mechanics. . . . What admits no doubt in my mind is that the Creator must have known a great deal of quantum mechanics and solid state physics, and must have applied them. Certainly, he did not limit himself to the molecular level when shaping life just to make it simpler for the biochemist." Using a variety of examples, he convincingly argues that "distinguishing between structure and function, classic chemical reactions and quantum mechanics, or the sub- and supramolecular, only shows the limited nature of our approach and understanding." Were Szent-Györgyi alive today, he would still be ahead of his time. He would be heartened, though, to see that an increasing number of his colleagues view biology as a manifestation of information and computation at the cellular, molecular, and submolecular levels, and that the organism is essentially a huge, if mysterious, supercomputer parallel processing millions of bits of information (see review by Siegfried, 2000).

There is information and there is information processing. This chapter contains information coded in letters at one level and, at a higher level, in agreed-upon meanings of words, syntax, and grammar. It will stay with me unless I send these codes

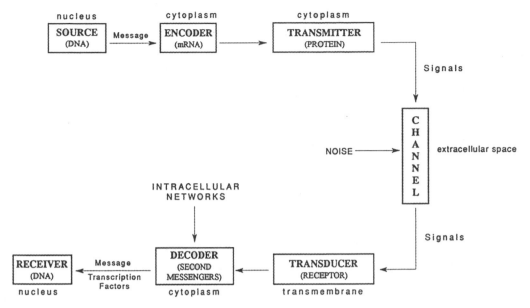

Figure 21.21 Biocybernetics: adapting information theory.

to you by some means, electronic or otherwise. Thus, there is an important difference between information coded in DNA and transfer of that information to sites elsewhere, between storing Chopin's music in your head and transferring its beauty by piano to my auditory system with some fidelity. Regarding the IGF/TGF-β epigenetic network (Fig. 21.5), we can acquire a sense of the transfer of information if we place it in the context of an adapted information theory (Fig. 21.21).

Fifty years ago, Shannon and Weaver (1949) presented a mathematical theory of how information in the form of a message or signal from a SOURCE transmitted to a RECEIVER is influenced by the CHANNEL through which that signal must pass. They concluded that in a closed system in which nothing comes in from the outside, there can never be more information presented to the RECEIVER than was initially signaled by the SOURCE. In fact, given that there is almost always "noise" in the CHANNEL, there is likely to be less information presented to the

RECEIVER unless compensated for by repeated signals of the same type.

We can adapt this general scheme (Fig. 21.21) to accommodate our current knowledge (Fig. 21.5). Several features of Figure 21.21 are worth noting. First, and most important, the system is necessarily *open*. It is made so by intracellular networking, and so the RECEIVER may be presented with *more* information than was initially signaled by the SOURCE. Second, noise in this context is comprised of the vagaries of extracellular space. As such, the efficiency of the signal transmission is reduced and a general redundancy of signal is a given. Finally, the output of the RECEIVER is responsible for a number of distinct phenotypic effects (pleiotropy). Thus, proteins in epigenetic networks such as Figure 21.5 have as their primary function the transfer and processing of information; these circuits perform a variety of simple computational tasks including amplification, integration, and information storage (Bray, 1995).

Protein molecules are in priniciple able to perform a variety of logical or computational operations. As circuits they have been modeled as artificial neural networks that use binary bits (0, 1) for data storage and processing and Boolean logic gates (AND, OR, NOT) for program execution. In fact, the mathematical formalism of artificial neural networks is a more accurate approximation for networks of protein molecules than for networks of real neurons (Bray, 1995).

The network of Figure 21.5 is but a small part of a much larger connections map (Fig. 21.22) that functions to regulate cell proliferation, cell quiescence, and cell death. As argued by Bray (1995) and others, the most important defining characteristic of protein-based neural networks is that they are governed by diffusive processes. Signals pass by means of physical contact between molecules in crowded conditions, and their dispersion through the cytoplasm is limited by the random thermal motion of molecules. Even though proteins remain the fundamental units of computation in an artificial neural net, cellular computing is more accurately viewed as quantum, not Boolean (Siegfried, 2000). This is because information storage and processing is submolecular (atoms, etc.), and as such obeys quantum rather than classical laws. This has interesting implications for the functioning of the IGF/TGF-β epigenetic network (Fig. 21.5).

Since quantum mechanics appears to be an accurate formulation of nature, it governs all physical systems, including submolecular information processing. While a classical bit must be either 0 or 1, a single quantum mechanical bit, a *qubit*, simultaneously contains a 0 component and a 1 component, and n quantum mechanical bits can simultaneously represent 2^n bits at once. This "quantum parallelism" would process larger amounts of information faster than classical bit processing.

In classical Boolean logic, one simple device is a Probabilistic NOT Gate, say a gate that models a fair coin flip. In this case, we use a logic gate that completely ran-

domizes its input (0 = tails, 1 = heads), producing a 0 or 1 output with equal probability. This function is represented by the transformation matrix:

$$
\begin{array}{cc}
 & \text{output} \\
 & \begin{array}{cc} 0 & \quad 1 \end{array} \\
\text{input} \begin{array}{c} 0 \\ 1 \end{array} & \left| \begin{array}{cc} \frac{1}{2} & \frac{1}{2} \\ \frac{1}{2} & \frac{1}{2} \end{array} \right.
\end{array}
$$

Thus, a "tail" on the first flip has no influence on the outcome of the second flip of the same coin or the flip of a different fair coin.

A quantum mechanical gate is quite different because qubits simultaneously contain a 0 state and a 1 state (for review, see Hayes, 1995). Quantum states and their superpositions are represented by a notational device called a ket ($|\rangle$). A quantum coin flip would be represented by the following transformation matrix:

$$
\begin{array}{cc}
 & \text{output} \\
 & \begin{array}{cc} |0\rangle & \quad |1\rangle \end{array} \\
\text{input} \begin{array}{c} |0|\rangle \\ |1\rangle \end{array} & \left| \begin{array}{cc} 1/\sqrt{2} & -1/\sqrt{2} \\ 1/\sqrt{2} & 1/\sqrt{2} \end{array} \right.
\end{array}
$$

The probability of each transition is the square of the corresponding value; thus, all values in the probability matrtix of this quantum coin flip are $\frac{1}{2}$ (Brassard, 1994). It would appear, then, that there is no practical difference between the classic coin-flip gate and the quantum coin-flip gate. But this is deceiving.

Probability calculations for a series of classic coin-flip gates are simply made by using the multiplication rule (e.g., probability of three heads = $\frac{1}{2} \times \frac{1}{2} \times \frac{1}{2} = \frac{1}{8}$). However, quantum coin-flip gates in series work quite differently, and the implications for information processing in epigenetic networks (e.g., Fig. 21.5) is quite interesting. Using what we reviewed above, we can derive the following sequence:

$$\uparrow \text{TGF-}\beta 2 \rightarrow \uparrow \text{p27} \rightarrow \downarrow \text{cyclin/Cdk}$$

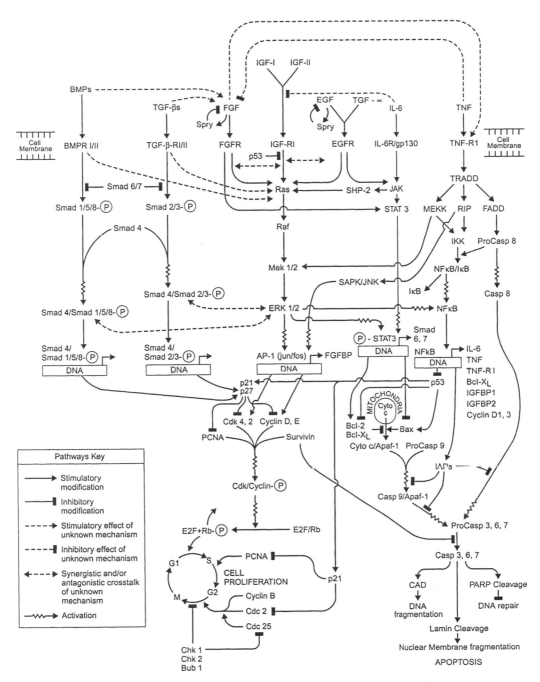

Figure 21.22   Connections map: signal transduction, cell proliferation, and apoptosis.

Thus, an increase in TGF-β2, via an increase in p27, results in a downregulation of cyclin/Cdk, and hence declining palatal mesenchymal cell proliferation. If this information were processed by a classic coin-flip probabilistic NOT gate, in response to the surmized Poissonian fluctuation of small numbers of macromole-

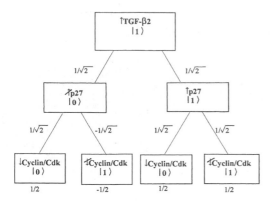

Figure 21.23 TGF-β2 upregulation, quantum NOT gates, and information processing. Note that two of the paths through a pair of quantum gates have amplitudes that interfere destructively, making downregulation of Cdk4 the certain outcome.

cules (Spudich and Koshland, 1976), then the outcomes for cyclin/Cdk (downregulation and no change) would be equally probable. However, using a quantum NOT gate (i.e., √NOT), the outcome (Fig. 21.23), downregulation, is *certain*! Thus, quantum gates in series, or the way submolecular information is processed in networks, is quite deterministic.

All this is quite intriguing and would be worth a considerable effort in computer modeling. However, since any interaction between a qubit and the macroscopic world results in decoherence, a collapsing of the quantum superposition to a classical state, such modeling has proven difficult (Taubes, 1997). In 1982, Feynman conjectured that a quantum computer would be better suited to the task and that such a quantum computer could be programed to simulate any local quantum system. Feynman's conjecture was recently shown to be correct (Lloyd, 1996). The march toward building a functioning quantum computer is steady but, unfortunately, slow (Bouwmeester et al., 2000). Nonetheless, the explosion of work in quantum information processing (e.g., Orlov et al., 2001) promises to provide valuable insight into such biologic processes

as epigenetic networks (Figs. 21.5, 21.22, and 21.23).

21.5 CONCLUSION

The great Edinburgh physician, John William Ballantyne (1861–1923), capped a lifetime of investigation into abnormal human development by publishing the *Manual of Antenatal Pathology and Hygiene, The Embryo* (1904). He made the following observations about the pathogenesis and etiology of oral clefts:

> Nearly every one is prepared to admit that these fissures [clefts] are the result of arrested development, even if all are by no means at one as to the precise mechanism by which the arrest is initiated. . . . Apparently simple explanations of matters embryological and teratological have, however, on more occasions than one turned out to be fallacious, and it was so with hare-lip. . . . The facts of embryology must first be thoroughly investigated (teratological developments being utilised as hints to direct research), and then the general principles of teratogenesis must be applied to the scrutiny of the results; if this be done, I feel sure that the actual model of production of hare-lip [pathogenesis] and all other malformations will be made plain. Of course this does not mean that the cause [the etiology] which leads to the arrested developments will be discovered, although we may be in an infinitely better position to make surmises regarding its nature.

What is remarkable about Ballantyne's observations is that nearly 100 years later we cannot say a great deal more about the etiology and pathogenesis of oral clefts in humans. We can add a few cell and molecular biology terms and concepts from our work in mice, but still all we can say is that clefting results from "arrested development," and all of us who have investigated this matter for the last 25 years "are by no means at one as to the precise mechanism" or the etiology.

This essay and its authors make no claim to knowing the etiology and pathogenesis of human cleft lip and cleft palate. The foregoing is not presented necessarily for the truth of the matter but as a humble "proof of paradigm"—a paradigm proffered by Melnick and Shields (1976) 25 years ago after thinking about the elegant work of Holliday (Holliday and Pugh, 1975), among others, on the control of gene expression during development.

REFERENCES

Agrawal, M., and Brauer, P. R. (1996). Urokinase-type plasminogen activator regulates cranial neural crest cell migration in vitro. *Devel. Dyn. 207*, 281–290.

Ardinger, H. H., Buetow, K. H., Bell, G. I., Bardach, J., Van Demark, D. R., and Murray, J. C. (1989). Association of genetic variation of the transforming growth factor-alpha gene with cleft lip and palate. *Am. J. Hum. Genet. 45*, 348–353.

Ballantyne, J. W. (1904). *Manual of Antenatal Pathology and Hygiene. The Embryo* (Edinburgh: William Green & Sons), pp. 384–385.

Ballard, F. J., Read, L. C., Francis, G. L., Bagley, C. J., and Wallace, J. C. (1986). Binding properties and biological potencies of insulin-like growth factors in L6 myoblasts. *Biochem. J. 233*, 223–230.

Barlow, D. P. (1995). Gametic imprinting in mammals. *Science 270*, 1610–1613.

Barlow, D. P., Stoger, R., Herrman, B. G., Saito, K., and Schweifer, N. (1991). The mouse insulin-like growth factor type 2 receptor is imprinted and closely linked to the *Tme* locus. *Nature 349*, 84–87.

Beechey, C. V., Cattanach, B. M., and Selley, R. L. (2001). MRC Mammalian Genetics Unit, Harwell, Oxfordshire, World Wide Web Site—Genetic and Physical Imprinting Map of the Mouse (URL: http://www.mgu.har.mrc.ac.uk/).

Bhalla, U. S., and Iyengar, R. (1999). Emergent properties of networks of biological signaling pathways. *Science 283*, 381–387.

Biddle, F. G., and Fraser, F. C. (1977). Maternal and cytoplasmic effects in experimental teratology. In J. G. Wilson and F. C. Fraser (eds.), *Handbook of Teratology*, vol. 3 (New York: Plenum Press), pp. 3–33.

Bouwmeester, D., Ekert, A., and Zeilinger, A. (2000). *The Physics of Quantum Information: Quantum Cryptography, Quantum Teleportation, Quantum Computation* (New York: Springer-Verlag).

Brassard, G. (1994). Cryptology column—Quantum computing: The end of classical cryptography? *SIGACT News 25*, 15–21.

Bray, D. (1995). Protein molecules as computational elements in living cells. *Nature 376*, 307–312.

Brunner, A. M., Lioubin, M. N., Marquardt, H., Malacko, A. R., Wang, W. C., Shapiro, R. A., Newbauer, M., Cook, J., Madisen, L., and Purchio, A. F. (1992). Site-directed mutagenesis of glycosylation sites in the transforming growth factor-β1 (TGFβ1) and TGFβ2 (414) precursors and of cysteine residues with mature TGFβ1: effects on secretion and bioactivity. *Molec. Endocrinol. 6*, 1691–1700.

Carter, C. O. (1969). Genetics of common disorders. *Brit. Med. Bull. 25*, 52–57.

Ciriani, D., and Diewert, V. M. (1986). A comparative study of development during primary palate formation in A/WYSN, C57BL/6 and their F_1 crosses. *J. Craniofac. Genet. Devel. Biol. 6*, 369–377.

Darwin, C. (1875). *The Variation of Animals and Plants Under Domestication*, vol. 1 (New York: Appleton), p. 466.

Dennis, P. A., and Rifkin, D. B. (1991). Cellular activation of latent transforming growth factor β requires binding to the cation-independent mannose-6-phosphate/insulin-like growth factor type II receptor. *Proc. Natl. Acad. Sci. USA 88*, 580–584.

Derynck, R. (1994). TGF-β-receptor mediated signaling. *Trends Biochem. Sci. 19*, 548–553.

Diewert, V. M. (1982). A comparative study of craniofacial growth during secondary palate development in four strains of mice. *J. Craniofac. Genet. Devel. Biol. 2*, 247–263.

Dunker, N., and Krieglstein, K. (2000). Targeted mutations of transforming growth factor-β genes reveal important roles in mouse development and adult homeostasis. *Eur. J. Biochem. 267*, 6982–6988.

Ellis, M. J. C., Lev, B. A., Yang, Z., Rasmussen, A., Pearce, A., Zweibel, J. A., Lippman, M. E., and Cullen, K. J. (1996). Affinity for the insulin-like growth factor-II (IGF-II) receptor inhibits autocrine IGF-II activity in MCF-7 breast cancer cells. *Molec. Endocrinol. 10*, 286–297.

Ferguson, M. W. J. (1988). Palate development. *Development 103*, 41–60.

Ferris, S. D., Sage, R. D., and Wilson, A. C. (1982). Evidence for mtDNA sequences that common laboratory strains of inbred mice are descended from a single female. *Nature 295*, 163–165.

Feynman, R. P. (1982). Simulating physics with computers. *Int. J. Theor. Phys. 21*, 467–488.

Filson, A. J., Louvi, A., Efstratiadis, A., and Robertson, E. J. (1993). Rescue of the T-associated maternal effect in mice carrying null mutations of *Igf2* and *Igf2r*, two reciprocally imprinted genes. *Development 118*, 731–736.

Fogh-Andersen, P. (1942). *Inheritance of Hare Lip and Cleft Palate* (Copenhagen: Nyt Nordisk Forlag Arnold Busck).

Fraser, F. C. (1980). Animal models for craniofacial disorders. In M. Melnick, D. Bixler, and E. D. Shields (eds.), *Etiology of Cleft Lip and Cleft Palate* (New York: Alan R. Liss, Inc.), pp. 1–23.

Fraser, F. C., and Fainstat, T. D. (1951). Production of congenital defects in the offspring of pregnant mice treated with cortisone. *Pediatrics 8*, 527–533.

Friezner, R., Degen, S. J., Bell, S. M., Schaefer, L. A., and Elliott, R. W. (1990). Characterization of the cDNA coding for mouse plasminogen and localization of the gene to mouse chromosome 17. *Genomics 8*, 49–61.

Gasser, D. L., Goldner-Sauve, A., Katsumata, M., and Goldman, A. S. (1991). Restriction fragment length polymorphisms, glucocorticoid receptors and phenytoin-induced cleft palate in congenic strains with steroid susceptibility. *J. Craniofac. Genet. Devel. Biol. 11*, 366–371.

Gierer, A. (1973). Molecular models and combinatorial principles in cell differentiation and morphogenesis. *Cold Spring Harbor Symp. Quant. Biol. 38*, 951–961.

Gleizes, P. E., Munger, J. S., Nunes, I., Harpel, J. G., Mazzieri, R., Noguera, I., and Rifkin, D. B. (1997). TGF-β latency: biological significance and mechanisms of activation. *Stem Cells 15*, 190–197.

Goldman, A. S. (1984). Biochemical mechanism of glucocorticoid- and phenytoin-induced cleft palate. In E. F. Zimmerman (ed.), *Current Topics in Developmental Biology* (New York: Academic Press), pp. 217–239.

Good, L., Jaskoll, T., Melnick, M., and Minkin, C. (1991). The major histocompatibility complex and murine fetal development. *J. Dent. Res. 70*, 579.

Haig, D., and Graham, C. (1991). Genomic imprinting and the strange case of the insulin-like growth factor II receptor. *Cell 64*, 1045–1046.

Hayes, B. (1995). The square root of NOT. *Am. Scientist. 83*, 304–308.

Hecht, J. T., Wang, Y., Blanton, S. H., Michels, V. V., and Daiger, S. P. (1991). Cleft lip and palate: no evidence of linkage to transforming growth factor alpha. *Am. J. Hum. Genet. 49*, 682–686.

Holliday, R., and Pugh, J. E. (1975). DNA modification mechanisms and gene activity during development. *Science 187*, 226–232.

Hu, C. C., Jaskoll, T., Minkin, C., and Melnick, M. (1990). The mouse major histocompatibility complex (H-2) and fetal lung development: implications for human pulmonary maturation. *Am. J. Med. Genet. 35*, 126–131.

Jaskoll, T., Hu, C. C., and Melnick, M. (1991). Mouse major histocompatibility complex and lung development: haplotype variation, H-2 immunolocalization and progressive maturation. *Am. J. Med. Genet. 39*, 422–436.

Jaskoll, T., Choy, H. A., Chen, H., and Melnick, M. (1996). Developmental expression and CORT-regulation of TGF-β and EGF receptor mRNA during mouse palatal morphogenesis: correlation between CORT-induced cleft palate and TGF-β2 mRNA expression. *Teratology 54*, 34–44.

Jennings, H. S. (1925). *Prometheus or the Advancement of Man* (New York: Dutton).

Johnson, K. R., Zheng, Q. Y., Bykhovskaya, Y., Spirina, O., and Fischel-Ghodsian, N. (2001). A nuclear-mitochondrial DNA interaction affecting hearing impairment in mice. *Nat. Genet. 27*, 191–194.

Jones, J. I., and Clemmons, D. R. (1995). Insulin-like growth factors and their binding proteins: biological actions. *Endocrinol. Rev. 16*, 3–34.

Kaartinen, V., Voncken, J. W., Shuler, C., Warburton, D., Bu, D., Heisterkamp, N., and Groffen, J. (1995). Abnormal lung development and cleft palate in mice lacking TGF-β3 indicates defects of epithelial-mesenchymal interaction. *Nat. Genet. 11*, 415–421.

Keski-Oja, J., Blasi, F., Leof, E. B., and Moses, H. L. (1988). Regulation of the synthesis and activity of urokinase plasminogen activator in A549 human lung carcinoma cells by transforming growth factor-β. *J. Cell Biol. 106*, 451–459.

King, R. C., and Stansfield, W. D. (1990). *A Dictionary of Genetics* (New York: Oxford University Press), p. 106.

Lau, M., Stewart, C. E. H., Liu, Z., Bhatt, H., Rotwein, P., and Stewart, C. L. (1994). Loss of imprinted IGF2/cation-independent mannose-6-phosphate receptor results in fetal overgrowth and perinatal lethality. *Genes. Devel. 8*, 2953–2963.

Lawrence, D. A. (1996). Transforming growth factor-β: a general review. *Eur. Cytokine Netw. 7*, 363–374.

Lerchner, W., and Barlow, D. P. (1997). Paternal repression of the imprinted mouse *Igf2r* locus occurs during implantation and is stable in all tissues of the postimplantation mouse embryo. *Mech. Devel. 61*, 141–149.

LeRoith, D., Werner, H., Beitner-Johnson, D., and Roberts, C. T. (1995). Molecular and cellular aspects of insulin-like growth factor I receptor. *Endocrinol. Rev. 16*, 143–163.

Levin, S. A., Grenfell, B., Hastings, A., and Perelson, A. S. (1997). Mathematical and computational challenges in population biology and ecosystems science. *Science 275*, 334–343.

Li, E., Bestor, T. H., and Jaenisch, R. (1992). Targeted mutation of the DNA methyltransferase gene results in embryonic lethality. *Cell 69*, 915–926.

Lidral, A. C., Romitti, P. A., Basart, A. M., Doetschman, T., Leysens, N. J., Daack-Hirsch, S., Semina, E. V., Johnson, L. R., Machida, J., Burds, A., Parnell, T. J., Rubenstein, J. L. R., and Murray, J. C. (1998). Association of MSX1 and TGFB3 with nonsyndromic clefting in humans. *Am. J. Hum. Genet. 63*, 557–568.

Lloyd, S. (1996). Universal quantum simulators. *Science 273*, 1073–1078.

Loffredo, L. C. M., Souza, J. M. P., Freitas, J. A. S., and Mossey, P. A. (2001). Oral clefts and vitamin supplementation. *Cleft Palate Craniofac. J. 38*, 76–83.

Luetteke, N. C., Qiu, T. H., Peiffer, R. L., Oliver, P., Smithies, O., and Lee, D. C. (1993). TGF alpha deficiency results in hair follicle and eye abnormalities in targeted and waved-1 mice. *Cell 73*, 263–278.

Lyle, R., Watanabe, D., te Vruchte, D., Lerchner, W., Smrzka, O. W., Wutz, A., Schageman, J., Hahner, L., Davies, C., and Barlow, D. P. (2000). The imprinted antisense RNA at the *Igf2r* locus overlaps but does not imprint *Mas1. Nat. Genet. 25*, 19–21.

Machida, J., Yoshiura, K., Funkhauser, C. D., Natsume, N., Kawai, T., and Murray, J. C. (1999). Transforming growth factor-α (TGFA): genomic structure, boundary sequences, and mutation analysis in nonsyndromic cleft lip/palate and cleft palate only. *Genomics 61*, 1–6.

Malakoff, D. (2000). The rise of the mouse, biomedicine's model mammal. *Science 288*, 248–253.

Marazita, M. L., Spence, M. A., and Melnick, M. (1986). Major gene determination of liability to cleft lip with or without cleft palate: a multiracial view. *J. Craniof. Genet. Devel. Biol. 2* (suppl.), 89–97.

Marazita, M. L., Hu, D. N., Spence, M. A., Liu, Y. E., and Melnick, M. (1992). Cleft lip with or without cleft palate in Shanghai, China: evidence for an autosomal major locus. *Am. J. Hum. Genet. 51*, 648–653.

Martinelli, M., Scapoli, L., Pezzeti, F., Carinci, F., Carinci, P., Stabellini, G., Bisceglia, L., Gombos, F., and Tognon, M. (2001). C677T variant form at the MTHFR gene and CL/P: a risk factor for mothers? *Am. J. Med. Genet. 98*, 357–360.

Maruyama, M. (1963). The second cybernetics: deviation-amplifying mututal causal processes. *Am. Scientist 51*, 164–179.

Melnick, M. (1992). Cleft lip (± cleft palate) etiology: A search for solutions. *Am. J. Med. Genet. 42*, 10–14.

Melnick, M. (1997). Cleft lip and palate etiology and its meaning in early 20th century England: Galton/Pearson vs. Bateson; polygenically poor protoplasm vs. Mendelism. *J. Craniofac. Genet. Devel. Biol. 17*, 65–79.

Melnick, M., and Jaskoll, T. (1992). H-2 haplotype and heterochronic orofacial morphogenesis in congenic mice: consideration as a possible explanation for differential susceptibility to teratogenesis. *J. Craniofac. Genet. Devel. Biol. 12*, 190–195.

Melnick, M., and Jaskoll, T. (2000). Mouse submandibular gland morphogenesis: a paradigm for embryonic signal processing. *Crit. Rev. Oral Biol. Med. 11*, 199–215.

Melnick, M., and Shields, E. D. (1976). Allelic restriction: a biologic alternative to multifactorial threshold inheritance. *Lancet 1*, 176–179.

Melnick, M., and Shields, E. D. (1982). Cleft lip and cleft palate. In M. Melnick, E. D. Shields, and N. J. Burzynski (eds.), *Clinical Dysmorphology of Oral-Facial Structures* (Boston: John Wright-PSG), pp. 360–372.

Melnick, M., Jaskoll, T., and Slavkin, H. C. (1981a). Corticosteroid-induced cleft palate in mice and H-2 haplotype: maternal and embryonic effects. *Immunogenetics 13*, 443–450.

Melnick, M., Jaskoll, T., and Slavkin, H. (1981b). The association of H-2 haplotype with implantation, survival, and growth of murine embryos. *Immunogenetics 13*, 303–308.

Melnick, M., Jaskoll, T., and Marazita, M. (1982). Localization of H-2K^k in developing mouse palates using monoclonal antibody. *J. Embryol. Exp. Morphol. 70*, 45–60.

Melnick, M., Marazita, M., and Jaskoll, T. (1983). Corticosteroid-induced abnormality in fetal mice and H-2 haplotype: evidence of a cytoplasmic effect. *Immunogenetics 17*, 141–146.

Melnick, M., Chen, H., Buckley, S., Warburton, D., and Jaskoll, T. (1998). Insulin-like growth factor II receptor, transforming growth factor-β, and Cdk4 expression and the developmental epigenetics of mouse palate morphogenesis and dysmorphogenesis. *Devel. Dyn. 211*, 11–25.

Miettinen, P. J., Chin, J. R., Shum, L., Slavkin, H. C., Shuler, C. F., Derynck, R., and Werb, Z. (1999). Epidermal growth factor receptor function is necessary for normal craniofacial development and palate closure. *Nat. Genet. 22*, 69–73.

Moats-Staats, B., Price, W. A., Xu, L., Jarvis, H. W., and Stiles, A. D. (1995). Regulation of the insulin-like growth factor system during rat lung development. *Am. J. Res. Cell. Mol. Biol. 12*, 7035–7056.

Murray, J. C. (1995). Face facts: genes, environment and clefts. *Am. J. Hum. Genet. 46*, 486–491.

Nemana, L. J., Marazita, M. L., and Melnick, M. (1992). Genetic analysis of cleft lip with or without cleft palate in Madras, India. *Am. J. Med. Genet. 42*, 5–9.

Odekon, L. E., Blasi, F., and Rifkin, D. B. (1994). Requirement for receptor-bound urokinase in plasmin-dependent cellular conversion of latent TGF-β to TGF-β. *J. Cell Physiol. 158*, 398–407.

Orlov, A. O., Kummamuru, R. K., Ramasubramaniam, R., Toth, G., Lent, C. S., Bernstein, G. H., and Snider, G. L. (2001). Experimental demonstration of a latch in clocked quantum-dot cellular automata. *Appl. Phys. Lett. 78*, 1625–1627.

Park-Wyllie, L., Mazotta, P., Pastuszak, A., Moretti, M. E., Beique, L., Hunnisett, L., Friesen, M. H., Jacobson, S., Kasapinovic, S., Chang, D., Diav-Citrin, O., Chitayat, D., Nulman, I., Einarson, T. R., and Koren, G. (2000). Birth defects after maternal exposure to corticosteroids: prospective cohort study and meta-analysis of epidemiologic studies. *Teratology 62*, 385–392.

Pfeifer, K. (2000). Mechanisms of genomic imprinting. *Am. J. Hum. Genet. 67*, 777–787.

Piedrahita, J. A., Oetama, B., Bennett, G. D., van Waes, J., Kamen, B. A., Richardson, J., Lacey, S. W., Anderson, R. G. W., and Finnell, R. H.

(1999). Mice lacking the folic acid–binding protein Folbp1 are defective in early embryonic development. *Nat. Genet. 23*, 228–232.

Potchinsky, M., Nugent, P., Lafferty, C., and Greene, R. M. (1996). Effects of dexamethasone on the expression of transforming growth-factor-beta in mouse embryonic palatal mesenchymal cells. *J. Cell Physiol. 166*, 380–386.

Proetzel, G., Pawlowski, S. A., Wiles, M. V., Yin, M., Boivin, G. P., Howles, P. N., Ding, J., Ferguson, M. W., and Doetschman, T. (1995). Transforming growth factor-beta 3 is required for secondary palate fusion. *Nat. Genet. 11*, 409–414.

Ramchandani, S., Bhattacharya, S. K., Cervoni, N., and Szyf, M. (1999). DNA methylation is a reversible biological signal. *Proc. Natl. Acad. Sci. USA 96*, 6107–6112.

Razin, A. (1998). CpG methylation, chromatin structure and gene silencing—a three-way connection. *EMBO J. 17*, 4905–4908.

Reik, W., and Constancia, M. (1997). Making sense or antisense? *Nature 389*, 669–671.

Reik, W., and Walter, J. (2001). Genomic imprinting: parental influence on the genome. *Nat. Rev. Genet. 2*, 21–32.

Risch, N., and Merikangas, K. (1996). The future of genetic studies of complex human diseases. *Science 273*, 1516–1517.

Salomon, D. S., and Pratt, R. M. (1979). Involvement of glucocorticoids in the development of the secondary palate. *Differentiation 13*, 141–154.

Sanford, L. P., Ormsby, I., Gittenberger-de Groot, A. C., Sariola, H., Friedman, R., Boivin, G. P., Cardell, E. L., and Doetschman, T. (1997). TGFβ2 knockout mice have multiple developmental defects that are non-overlapping with other TGFbeta knockout phenotypes. *Development 124*, 2659–2670.

Shannon, C., and Weaver, W. (1949). *The Mathematical Theory of Communication* (Urbana: University of Illinois Press).

Shields, E. D., Bixler, D., and Fogh-Andersen, P. (1981). Cleft palate: a genetic and epidemiologic investigation. *Clin. Genet. 20*, 13–24.

Siegfried, T. (2000). *The Bit and the Pendulum* (New York: Wiley).

Smrzka, O. W., Faé, I., Stöger, R., Kurzbauer, R., Fischer, G. F., Henn, T., Weith, A., and Barlow, D. P. (1995). Conservation of a maternal-specific methylation signal at the human *IGF2R* locus. *Hum. Molec. Genet. 4*, 1945–1952.

Sproule, J. (1863). Hereditary nature of hare-lip. *Brit. Med. J. 1*, 412.

Spudich, J. L., and Koshland, D. E. (1976). Non-genetic individuality: chance in the single cell. *Nature 262*, 467–471.

Stern, C. (1949). *Principles of Human Genetics* (San Francisco: W. H. Freeman), pp. 292–293.

Strohman, R. C. (1997). The coming Kuhnian revolution in biology. *Nat. Biotech. 15*, 194–200.

Szent-Györgyi, A. (1960). *Introduction to a Submolecular Biology* (New York: Academic Press).

Tanabe, A., Taketani, S., Endo-Ichikawa, Y., Tokunaga, R., Ogawa, Y., and Hiramoto, M. (2000). Analysis of the candidate genes responsible for non-syndromic cleft lip and palate in Japanese people. *Clin. Sci. 99*, 105–111.

Taubes, G. (1997). Putting a quantum computer to work in a cup of coffee. *Science 275*, 307–309.

Theiler, K. (1989). *The House Mouse. Atlas of Embryonic Development* (New York: Springer-Verlag).

Thorvaldsen, J. L., and Bartolomei, M. S. (2000). Mothers setting boundaries. *Science 288*, 2145–2146.

Trasler, D. G., and Fraser, F. C. (1977). Time-position relationships. In J. G. Wilson and F. C. Fraser (eds.), *Handbook of Teratology*, vol 2 (New York: Plenum Press), pp. 271–292.

Vignon, F., and Rochefort, H. (1992). Interactions of pro-cathepsin D and IGF-II on the mannose-6-phosphate/IGF-II receptor. *Breast Cancer Res. Treat. 22*, 47–57.

Vincek, V., Sertie, J., Zaleska-Rutezynska, Z., Figuero, F., and Klein, J. (1990). Characterization of H-2 congenic strains using DNA markers. *Immunogenetics 31*, 45–51.

Wang, Z. Q., Fund, M. R., Barlow, D. P., and Wagner, E. F. (1994). Regulation of embryonic growth and lysosomal targeting by the imprinted *Igf/Mpr* gene. *Biochem. Biophys. Res. Commun. 197*, 747–754.

Wolffe, A. P. (1998). Packaging principle: how DNA methylation and histone acetylation control the transcriptional activity of chromatin. *J. Exp. Zool. 282*, 239–244.

Wutz, A., Smrzka, O. W., Schweifer, N., Schellanders, K., Wagner, E. F., and Barlow, D. P. (1997). Imprinted expression of the *Igf2r* gene depends on an intronic CpG island. *Nature 389*, 745–749.

Wyszynski, D. F., Beaty, T. H., and Maestri, N. E. (1996). Genetics of nonsyndromic oral clefts revisited. *Cleft Palate Craniofac. J. 33*, 406–417.

Xu, Y., Goodyear, C. G., Deal, C., and Polychronakos, C. (1993). Functional polymorphism in the parental imprinting of the human *IGF2R* gene. Biochem. *Biophys. Res. Commun. 197*, 747–754.

CHAPTER 22

MOLECULES AND FACES: WHAT IS ON THE HORIZON?

HAROLD C. SLAVKIN, D.D.S., Ph.D.

22.1 INTRODUCTION

We are living in a time of incredible achievements by the international biomedical research community. In fact, the incredibility itself is the noteworthy characteristic of our shared time in human history. The progress in reading and understanding the human genetic lexicon, the advances in bioinformatics, and the utilization of stem cells to regenerate biological tissues—each of these accomplishments was once thought to be beyond human possibilities, and they have become the new benchmark for scientific progress. This is particularly evident when considering the scientific progress in two significant problem areas.

In this chapter we explore the scientific principles and emerging scientific evidence in these two areas: (1) the molecular basis of craniofacial morphogenesis, and (2) biomimetics for repair and regeneration of craniofacial and skeletal tissues. The emerging intellectual synthesis between craniofacial morphogenesis and biomimetics provides a number of potentially important perspectives to consider what is on the horizon as we anticipate the next decade of the 21st century.

The formation of concepts, in a scientific sense, about the nature of craniofacial morphogenesis started with Hippocrates in the fifth century B.C. Soon thereafter, Aristotle defined the fundamental question that would fuel this intellectual adventure for almost 2000 years to follow (see Aristotle et al., 1526). Aristotle asked if all parts of embryos come into existence at the same time, or if they appear in some sequence of events. He asked if everything was preformed and predetermined from the beginning, or if growth and development (morphogenesis) is the result of many sequential processes that are integrated (somehow) into the forming organism. Aristotle rejected preformation and favored epigenesis!

Aristotle's questions were answered in the last 20 years of the 20th century. The scientific concepts and evidence supporting epigenesis have finally coalesced—all embryos develop and grow through epigenesis. As we shall consider, DNA encodes the master blueprint within discrete genes, which in turn are orchestrated to control morphogenesis. Aristotle's proposed hypothesis of epigenesis has been validated through experimental scientific inquiry.

Understanding Craniofacial Anomalies: The Etiopathogenesis of Craniosynostoses and Facial Clefting, Edited by Mark P. Mooney and Michael I. Siegel, ISBN 0-471-38724-x Copyright © 2002 by Wiley-Liss, Inc.

The fundamental connection between multiple genes and morphogenesis has been established—all genes specify proteins, and genes specify proteins that determine and coordinate cellular fates, features, and patterns. Changes or mutations in gene(s) activities vis-à-vis misspellings in their protein products, the subsequent protein-protein and protein-gene interactions, and gene(s)-environment interactions are reflected in phenotype—the structure and function of organelles, cells, tissues, organs, systems, and behaviors in normal or abnormal craniofacial morphogenesis (see Genebank: http://www.ncbi.nlm.nih.gov/Web/Search/index.html; OMIM: http://www3.ncbi.nlm.nih.gov/Omim/searchomim.html; Genome Project: http://gdbwww.gdb.org/) (Chapters 4–7).

Morphogenesis begins prior to fertilization with the "seeding" of significant molecular information (instructions) in the cytoplasm of the unfertilized egg—a number of "masked" maternal RNAs encode instructions for early signal transduction processes that are a prerequisite for morphogenesis. Development continues at fertilization within a single cell, and thereafter results in the formation of increasingly smaller and smaller cells until a steady state in cell size is reached during gastrulation. Subsequent development continues during neurulation and organogenesis. Each of the early embryonic stages is characterized by a biological process whereby one cell generates many types of cells that result in numerous migrations and rearrangements, boundary conditions between communities of cells, overt boundaries as observed in histogenesis or tissue formations, and cytodifferentiation. The timing and positional controls for these sequential biological processes are found in *pattern formations* and *morphogenesis*.

Programatic information required for craniofacial morphogenesis is regulated by thousands of genes that experience exquis-

ite interactions with one another (epistasis) as well as with the environment. The genome has an exceedingly complex architecture (both mitochondrial and nuclear genomes), and unique combinations of genes and their products, as well as biomechanical processes, are required to produce morphogenesis.

As of June 26, 2000, all three billion bases (A, adenosine; T, thymidine; G, guanosine; and C, cytosine) that comprise the human genome were identified (see also Chapter 4). These billions of bits of information are assembled into 30,000 genes, and these structural and regulatory genes are packaged within 23 pairs of chromosomes (Genome Project: http://gdbwww.gdb.org/). Ironically, the 30,000 encoded genes represent only 5% of the total human genome. Determining functions for the noncoding regions of the human genome is a major scientific opportunity. (see Venter et al., 2001; and The International Human Genome Sequencing Consortium, 2001).

In December 1999, the entire map of chromosome 21 was completed. In February 2000, the fruit fly genome found within 4 chromosome pairs was completed (Flybase: http://flybase.bio.indiana.edu:82/). By the end of the year 2000, all genes had been identified and mapped to their precise positions in 10 of the 23 pairs of chromosomes. All of the genes will be identified and mapped to all of the human chromosomes by the year 2002. In tandem, public/private partnerships are also identifying, mapping, and completing the genomes for many microbial, plant, and animal genomes, resulting in knowledge bases that can be mined for significant evolutionary significance. In addition, these public/private consortia are also producing discrete molecular tools that will be invaluable for the identification of discrete changes or mutations in the various genomes related to either normal variance or inherited and acquired mutations that result in single- or multiple-gene-related diseases or disorders. These new molecular tools are termed *single nucleotide polymorphisms* (SNPs) and will soon become essential to studies of the molecular epidemiology of human variance or polymorphisms, and craniofacial and skeletal dysmorphogenesis, diseases, and disorders (see http://www.ncbi.nlm.nih.gov/SNP/).

On another front of inquiry, several additional scientific concepts are required for understanding craniofacial morphogenesis; concepts that integrate genetics, developmental biology, evolutionary biology, mathematics, physics, and clinical medicine and dentistry (Fig. 22.1). For example, consider that each gene or allele is present in two copies (a state of diploidy) in the genome of each human somatic cell, in addition to those genes that constitute the maternally derived mitochondrial genome (prefertilization). The phenotype expressed does not always reveal the dual genetic character encoded in the genome. Certain genes or alleles can be dominant, recessive, or can combine to form a new trait. The activity of one gene can mask or obscure the function of a different gene. Using a variety of experimental techniques, such as transgenic mouse models presenting null or dominant negative mutations and no apparent clinical phenotype, perplexing evidence emerged to support the concept of developmental redundancy: multiple or redundant molecular controls conserved through evolution that regulate particular processes related to craniofacial morphogenesis.

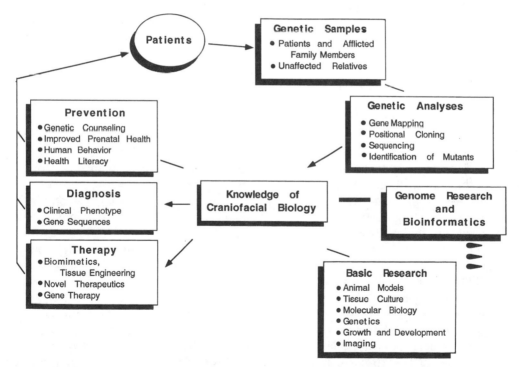

Figure 22.1 New Approaches for craniofacial dysmorphogenesis. Knowledge is achieved through multifaceted research combining information from the communities, clinics, and laboratories. This knowledge is transformed into benefits for patients and families. The rate of this cycle from scientific knowledge to health care and health promotion has greatly increased by progress in the study of the human genome, database management strategies for the analyses and dissemination of bioinformatics, and an increased appreciation for evidence-based decisions in the management of health care.

Another perplexing issue for future scientific inquiry is to explain how slightly different mutations in one gene can produce different craniofacial syndromes, such as those associated with FGFR2 (fibroblast growth factor receptor 2) (see recent review by Nuckolls et al., 1999). Finally, traits that are particularly interesting to those of us who think about craniofacial dysmorphogenesis are not necessarily determined by one gene at one gene locus, but rather by intricate and synergistic actions between many genes located in many different gene loci—often described as polygenic, quantitative inheritance, or complex human diseases and disorders (see critical discussions in Cohen, 1997a, 1997b; Hollaway et al., 1997; Jabs, 1998; Mulvihill, 1995; Neilson and Friesel, 1995).

The fruits of scientific inquiry during the last decade produced significant advances in our understanding of many complex biological processes associated with craniofacial morphogenesis (see review by Gorlin and Slavkin, 1997). A reasonable translation of this fundamental scientific understanding is the capacity to mimic biological processes in order to design and fabricate biomaterials. This concept is now represented in the term *biomimetics*, and is often associated with efforts to produce new cells, tissues, and organs through nanotechnology, bioengineering, and tissue engineering strategies. Biomimetics is an emerging interdisciplinary field at the interface between biology, chemistry, physics, mathematics, and engineering, as well as clinical dentistry and medicine (Sarikaya and Aksay, 1995). Biomimetics is concerned with the architecture and fabrication of molecules, organelles, membranes, extracellular matrices, cell, tissues, and organs for functional restoration or for lost or impaired function. The emerging field of nanotechnology in part emphasizes biotechnology at the *nano* level of biological organization—the scale of atomic and molecular interactions (Amato, 1997).

Meanwhile, an important lesson has been learned that biological repair and regeneration processes recapitulate morphogenesis. Inductive or instructive epigenetic signals and their transcriptional controls required for normal growth and development provide an intellectual scaffold for candidate molecules and circuits to be used in biomimicry (Slavkin, 1995, 1998a). Biomimetic products have already emerged, for example: (1) the creation of artificial skin, cornea, and mucosa; (2) bone and cartilage morphogenetic proteins used for bone and cartilage formation; (3) cementum, dentin, and enamel morphogenetic proteins used for dental tissue repair and regeneration; (4) antiangiogenesis molecules to reduce tumor formation; and (5) a number of other applications to clinical craniofacial, oral, and dental diseases, dysmorphology, and trauma (Urist, 1965; Reddi and Binderman, 1996; Slavkin, 1996; Slavkin, 1998b; Slavkin et al., 1999).

22.2 MOLECULAR ADVANCES IN CRANIOFACIAL MORPHOGENESIS

Craniofacial morphogenesis through embryonic and fetal development, infancy, childhood, and adolescence is topographically and geometrically complex. What is now emerging through a remarkable coalition of international scientific investigations is a molecular understanding of craniofacial developmental and evolutionary biology, genetics, and clinical medicine and dentistry (Slavkin, 1979, 1998b; Winter, 1995; Wilkie, 1997) (Chapters 4, 5, 21, and 22). At the molecular level, morphoregulatory and structural genes and their controls are being rapidly identified, sequenced, and mapped to precise chromosomal locations and analyzed for their translation products or protein structure and function during craniofacial morphogenesis (see discussions in Nuckolls et al., 1998). Therefore, it is becoming increasingly reasonable to ascertain the precise structure and function of genes and their gene products required for morphogenesis. This scientific odyssey also enables the molecular understanding of those mechanisms required for the specification and molding of forms within changing microenvironments—morphogenesis. The intent of this chapter's discussion is to provide the scientific ideas that will build paradigms for craniofacial morphogenesis and dysmorphogenesis in the future.

A number of scientific discoveries have already changed the ways we consider craniofacial morphogenesis. First, the genotype or genome describes the genetic lexicon encoded within the 30,000 genes that are found in the nucleus of every somatic cell in the human body (Collins et al., 1998; Venter et al., 2001; and the International Human Genome Sequencing Consortium, 2001; Genome Database: OMIM:http://www3.ncbi.nlm.nih.gov/Omim/searchomim.html). These human genes are often remarkably conserved during evolution, and near identical genes are known to be homologous to fruit flies, fish, amphibians, reptiles, birds, and scholars (Gorlin and Slavkin, 1997; Nuckolls et al., 1998, 1999). The remarkable homology further supports the value of using many different animal models for investigations into the mechanisms of craniofacial morphogenesis.

Second, the phenotype describes the expression of one or more genes associated with one of many biological processes during the life span—from differential gene expression of insulin in the pancreas to the multiple genes associated with a particular human behavior. The sequence and rate of gene expression is controlled by an exquisite series of gene-gene and gene-environment interactions. Epigenesis describes a biological process in which the adult develops from an undifferentiated condition. The genotype within the fertilized egg contains the genetic program for producing a human being, and the process by which genotype generates phenotype is epigenesis, with philosophical roots that extend to Aristotle. The challenge is to explain how the changing microenvironments during gastrulation, neurulation, organogenesis, and all of the following developmental stages regulate cell fate, cell determination, cell patterns, and differential gene expression characteristic for a particular phenotype.

What are the epigenetic signals (steroid-like and polypeptide-like molecules)? How are these signals received (plasma membrane-, cytoplasmic-, and nuclear-specific receptors)? What are the signal pathways that lead to and from the genome? How are these dynamic bioactivities integrated into circuits or feedback loops? How do various levels of biological structure and function—organelle, membrane, cell, tissue, organ, system, family, community, and population—translate into craniofacial morphogenesis?

What are the epigenetic signals? One emerging scientific discovery is that the unfertilized egg contains maternally derived molecular information in the form of specific transcription factors, growth factors (BMPs, TGF-betas, IGFs), vitamins (retinoic acid, vitamin D), trace elements (selenium, zinc) or ions (calcium, magnesium), and other regulatory factors that signal early gene activities. During early embryogenesis, the fertilized egg is sequentially divided into increasingly smaller cells and an increased number of cells. Following implantation, clusters of cells during gastrulation emerge with different rates of cell division, positional information, motility, and gene expression—and boundaries between populations of cells, the so-called germ layers (ectoderm, mesoderm, and endoderm).

Less obvious, however, in this embryonic landscape are the molecular gradients of master genes or transcription factor genes, growth factors and their cognate receptor genes, vitamins such as retinoids, and a complex group of signal pathway genes that collectively define the epigenetic landscape and the numerous circuits that control gastrulation, neurulation, organogenesis, and what follows. Curiously, the emerging scientific evidence considers molecular rules that control morphogenesis and further suggests that combinations of transcriptional factors, ligands or signals, receptors, and the many intracellular molecules required for communication between the cell surface and the nucleus provide the exquisite sensitivity and specificity that enables development.

The remarkable progress being made in the human genome, coupled with exquisite functional genomic investigations in various animal models such as the fruit fly, zebra fish, and mouse, has resulted in the discovery of morphoregulatory genes that determine craniofacial morphogenesis: (1) genes that determine anterior-posterior or cephalic-caudal axis; (2) genes that determine segmentation in early-forming neurulation, branchial arches, and paired somites for the occipital and cervical processes (e.g., pair rule and segment polarity such as *Sonic*, *Indian*, and *desert hedgehog* genes); (3) genes that determine bilateral symmetry and right and left axes; (4) genes that determine rates of cell division; (5) genes that determine dorsal-ventral and medial-distal axes; and (6) genes that determine cell lineage and cell fate (see Table 22.1).

These morphoregulatory genes are now assumed to be controlled or regulated through complex combinatorial processes that often utilize homo- and heterodimer as well as homo- and heterotetramer assemblies. These combinations reflect protein-DNA and protein-protein interactions. For example, a transcriptional control may utilize 10 to 20 different proteins. Presumably these combinations between homologous and/or heterologous proteins assemble and function within milliseconds to seconds. Rather than considering one gene, one receptor, and one message or ligand in a linear process, we now observe

that combinations of molecules within interdependent circuits gain or release energy and propagate signals from the microenvironment of the cell to and from the nucleus as well as between a community of cells. Biological controls and specificity are becoming problems that can be addressed using physical chemistry, computational biology, molecular modeling, and novel imaging technology.

Consequences of the molecular combinations within and between circuits appear to represent the epigenetic landscape of morphogenesis. The molecular controls that regulate specification during development are being defined for the pattern formations of bilateral symmetry for determining the cranial-caudal (anteroposterior) and the dorsoventral axes, the unique segmentation patterns of forebrain, midbrain, and hindbrain with rhombomeres, the sequential somite pairs, the branchial arches, the specifications for limbs, dentition, and various other organs, and the subsequent features of growth and development. Our challenge is to explain and understand these dynamic and interrelated biological processes in a hierarchy of levels. To understand human growth, development, form, and dysmorphogenesis requires a step-by-step reductionism or hierarchical reductionism within the grand context of evolutionary biology.

How are these signals received? Biological information can be physical and/or chemical. Individual cells, tissues, and organs possess receptor-mediated processes designed to identify and respond to epigenetic signals. Signal transduction describes ligand binding to receptors and thereby initiating a complex series of biochemical reactions that mediate the extracellular signals into transcription controls at the gene level of organization. Such signal transduction mediates growth factor, cytokine, and chemokine instructions during growth and development. At the gene level, the response may be

TABLE 22.1   Craniofacial Disorders Caused by Mutations in Evolutionarily Conserved Genes[a]

Fruit Fly Gene	Function	Human Gene	Human Genetic Disorder	OMIM* Number
Breathless (btl)	Fibroblast growth factor receptor	FGFR1	Pfeiffer syndrome	101600
		FGFR2	Apert syndrome	101200
			Beare-Stevenson syndrome	123790
			Crouzon syndrome	123500
			Jackson-Weiss syndrome	123150
			Pfeiffer syndrome	101600
		FGFR3	Achondroplasia	100800
			Craniosynostosis, nonsvnd romic	134934
			Crouzon-syndrome with acanthosis nigricans	134934
			Hypochondroplasia	146000
			Thanatophoric dysplasia	187600
Cubitus interruptus (ci)	Zinc finger transcription factor	GLI-3	Greig cephalopolysyndactyly syndrome	175700
			Pallister-Hall syndrome	146510
			Postaxial polydactyly, type A	174200
Decapentaplegic (dpp)	Extracellular signaling molecule	BMP4 (Bone Morphogenetic Protein)	Fibrodysplasia ossificans progressiva	135100
Distal-less (dll)	Homeodomain transcription factor	DLX 3	Trichodento-osseous syndrome	190320
Eyeless (ey)	Paired box transcription factor	PAX6	Aniridia, type 2	106210
Eyes absent (eya)	Transcription factor	EYA1	Branchio-otorenal dysplasia	113650
			Branchiotic syndrome	601653
Hedgehog (hh)	Extracellular signaling molecule	SHH (Sonic hedgehog)	Holoprosencephaly, type 3	142945
Muscle segmentation homeobox (msh)	Homeodomain transcription factor	MSX1	Wolf-Hirschhorn syndrome	142983
			Familial tooth agenesis	106600
		MSX2	Craniosynostosis, Boston type	123101
Twist (twi)	Basic helix-loop-helix transcription factor	TWIST	Saethre-Chotzen syndrome	101400

[a] Techniques of genetic manipulation in the fruit fly (*Drosophila melanogaster*) have led to identification and characterization of networks of genes that regulate growth and development. Positional cloning and mutation analysis revealed the genetic basis of numerous human craniofacial and skeletal disorders. Similarities between the sequence, structure, and function of the human and *Drosophila* gene products allow for the application of information from *Drosophila* genetic studies toward an understanding of the etiology of human diseases. (See OMIM: http://www3.ncbi.nlm.nih.gov/Omim/searchomim.html.)

positive or inhibitory, and it results in cell division, differentiation, or programed cell death. These responses are common themes in growth and development (see Gorlin and Slavkin, 1997; Graham et al., 1994; Nuckolls et al., 1998; Sperber, 1992; Slavkin, 1998a, for extensive discussions of these processes during embryogenesis).

Subtle misspellings in one or more of the genes associated with a particular process may result in craniofacial dysmorphogenesis. For example, misspellings in several transcription factors (MSX1 and TWIST), growth factor receptors (FGFR1, FGFR2, and FGR3), and structural proteins (FBN1 or fibrillin) produce a common clinical phenotype of craniosynostosis (see reviews by Cohen, 1997a, 1997b; Mulvihill, 1995; Nuckolls et al., 1998, 1999) (Chapters 6, 9, 12, and 21). Mutations in TWIST are associated with deletions or mutations in the short arm of chromosome 7 and Saethre-Chotzen syndrome (OMIM, 2000). Mutations in FGFR2 are associated with several types of Apert, Crouzon, and Pfeiffer syndromes. Fibroblast growth factor (FGF) binds to the FGF receptor and mediates instructions related to cranial suture formation. A minor misspelling or point mutation in one or more of the several different FGF receptors can produce craniosynostosis associated with a number of craniofacial syndromes (Table 22.1). In this example, the fibroblast growth factor ligand is abundant, but the conformation of the cognate receptor molecule is sufficiently abnormal that a clinical defect is expressed. Interactions between FGFR2 and adjacent neural cell adhesion molecules (CAMs) may also be implicated. Gain or loss of function mutations are additional possibilities to explain these clinical phenotypes (diagnosis) and may also serve to build the necessary platforms for biomimetic approaches (treatments and therapeutics) to repair or regenerate the tissues associated with various types of craniosynostosis. The

precise physical-chemical mechanism for this aberrant signal transduction model is currently under active investigation (see Cohen, 1997a, 1997b, for recent reviews).

Candidate genes identified in the fruit fly, *Drosophila melanogaster*, have rapidly been tested for their homology to mammalian genes including human (see Winter, 1995). The fruit fly genome was completed in February 2000 (Flybase: http://flybase.bio.indiana.edu:82/). Since the mid-1980s, literally thousands of genes have been identified in the fruit fly or earth worm that have striking homology and significance toward understanding the molecular basis of craniofacial morphogenesis (see excellent analysis by Lawerence, 1992). For example, based on homology to specific motifs in the fruit fly genome, several different gene families have been identified in mouse and human craniofacial processes that encode specific DNA binding domains such as the homeodomain, the basic helix-loop-helix domain, the zinc finger domain, the fork head domain, and the POU domain. Each of these families of genes encodes transcription factors that regulate the expression of downstream target genes as mediated through sequence-specific DNA recognition of regulatory sequences during embryonic craniofacial morphogenesis. This has been especially significant toward understanding transcriptional factors that function as either positive regulators or antagonists for gene expression. Table 22.1 provides a few examples of these remarkable discoveries. For more comprehensive detail, the reader is encouraged to use the Internet and investigate the various genetic databases now readily available (e.g., Flybase, Genebank, OMIM [Web addresses earlier in this chapter], and Zebrafish Genome Project: http://zfish. uoregon.edu/zf_info/zfmap.html).

Over the next decade, essentially all of the 30,000 genes that comprise the human genome, as well as those characteristic of most of the currently available animal models (e.g., fruit fly, sea urchin, zebra fish, frog, chick, mouse, rat, and monkey), will be identified, mapped, and functionally elucidated as to their roles in craniofacial morphogenesis. Moreover, scientific activities are already progressing rapidly in the "postgenomic era," a phase of science that will emphasize functional genomics—the structural biology and function of combinations of genes and their gene products required to generate a specific phenotype.

22.3   BIOMIMETICS FOR REPAIR AND REGENERATION OF CRANIOFACIAL TISSUES

Biomimetics is a hybrid science that develops strategies and methods for the design and fabrication of new tissue or organ replacements, and the repair or regeneration of dysfunctional cells, tissues, organs, or systems. By definition, biomimetics requires integration between recombinant DNA technology, immunology, physiology, engineering, physics, mathematics, informatics, biomechanics, bioengineering, chemistry, and the principles of cellular, molecular, and developmental biology.

A number of advances in science have resulted in the capacity to develop ideas and to implement these ideas with nanotechnology—designs, processes, and products created at the level of one-billionth of a millimeter. Of course, nanotechnology is the level of activity that best characterizes most cellular processes. Atomic force tunnel microscopy, along with recent progress in high-resolution nuclear magnetic resonance imaging (MRI), enables visualization of individual as well as aggregate tertiary and quaternary molecular structures under physiological conditions. These technologies and many others, coupled with changing habits of mind, are serving to accelerate significant research and development in biomimetics.

22.4   POSSIBILITIES FOR CRANIOFACIAL AND SKELETAL GROWTH MODIFICATIONS

The possibilities for craniofacial and skeletal growth modifications are profound in the new millennium (Slavkin et al., 1999). The scientific ideas, the paradigms of thought, and the necessary technologies are converging around biological solutions to biological problems with the advent of biomimetics. For example, bone-regenerating options for craniofacial skeletal regeneration presently utilize bone grafts, alloplastics, electrical stimulation, distraction osteogenesis, guided bone regeneration, and combinations of growth factors such as bone morphogenetic factors (BMPs), platelet-derived growth factor (PDGF), insulin-like growth factor (IGF), transforming growth factor-beta (TGF-beta), and fibroblast-derived growth factors (FGFs). One recent innovation is to utilize chondrogenic cells that can be readily grown outside the body in glass or plastic dishes as monolayers and produce sheets of cartilage. Further, histocompatible chondrogenic cells can be grown in extracellular matrix scaffolds or molds and can produce three-dimensional structures such as the cartilaginous form of an ear or condyle. Histocompatible mesenchymal stromal stem cells can be used to induce intramembranous and endochondral bone formations under comparable conditions. Xenografts can be used to assist these clinical biomimetic strategies. Bone morphogenetic proteins BMPs in various combinations and delivery systems can induce new bone, dentin, or cementum formation. So-called knockout strategies have been used to produce striking mutations for craniofacial dysmorphogenesis in animal models, and gene-mediated corrective strategies have been accomplished that eliminate the dysmorphogenesis in these transgenic animal models. These and many other emerging habits of thought and technologies, such as the biomechanics of rigid fixation and bone repair and regeneration, are serving to catalyze biomimetics and clinical advances.

22.5 PROSPECTUS FOR THE 21ST CENTURY

Consider craniofacial dysmorphogenesis challenges such as those presented in hemifacial microsomia, micrognathia, macrognathia, craniosynostosis, and anodontia. We appreciate that each and every example is a consequence of abrogation in genetic controls, whether due to the Mendelian inheritance of one or more genetic misspellings or to acquired and multiple gene misspellings resulting from environmentally induced mutations. During the last decade, more than 1000 different genes associated with craniofacial morphogenesis have been identified. The vast majority of these craniofacial genes discovered in fruit fly or zebra fish have been found to be highly conserved and are also found in mouse and human craniofacial morphogenesis. Therefore, access to the bioinformation encoded within various animal craniofacial genomes becomes an essential requirement for mining significant knowledge bases for applications to gene-based diagnostics, treatments, and therapeutics. And the rate of discovery is increasing with the recent emergence of several technologies including the DNA microchip and the microarray strategies used for improved and faster DNA sequencing in human and animal genomics, and with applications to functional genomics for screening thousands of genes from the same stage of development, from the same developing process, and at the same time within the same experiment (Hieter and Boguski, 1997).

As we advance into the 21st century, consider the innovations, boldness, and advances in the understanding of the genetics of life, molecular and developmental biology, the advent of biotechnology, and the future of clinical medicine and dentistry. We are on the verge of understanding craniofacial morphogenesis—the capacity to decipher how cells read and edit the chemical texts found within each cell in the human body. We are learning the rules for making a human face, and we are learning the rules for making a dentition (Thompson, 1952; Thorogood, 1997; Slavkin et al., 1999; Winter, 1995). In tandem, the remarkable progress toward understanding biological and biomechanical mechanisms is fostering biomimetic approaches that can produce smarter, real-time functional MRI, diagnostics, treatments, therapeutics, and biomaterials. These advances directly contribute to craniofacial dysmorphology, head and neck trauma, and head and neck cancer teams for the management of many diseases and disorders or conditions.

What is on the horizon? Succinctly, rapid advances and rapid change! Major science and technology advances will continue to serve as foundations for molecular approaches to craniofacial-oral-dental morphogenesis. Scientific and technological seeds planted in the last century are now blossoming. The completion of microbial, animal, plant, and human genomes; the rapidly expanding studies of functional genomics; bioinformatics and the capacity to access and mine complex knowledge bases; globalization and the era of biotechnology; and numerous innovations in diagnosis, treatments, and therapeutics for craniofacial diseases, disorders, and conditions are all becoming a reality. Significant advances are being made and will continue to be made in gene-, cell-, and tissue-mediated therapeutics, and the design and fabrication of biomaterials. Innovative imaging and molecular tools will also complement craniofacial plastic and reconstructive surgery, oncology, otolaryngology, oral and maxillofacial surgery, neurology, genetics, speech pathology and therapy, and orthodontics, dental orthopedics, and prosthodontics in the next decade.

ACKNOWLEDGMENTS

I want to take this opportunity to thank my laboratory colleagues and students with whom so much has been learned over more than three decades of research in craniofacial developmental biology and genetics in Los Angeles as well as in Bethesda. I am particularly grateful to my many mentors, research colleagues, and students for their numerous intellectual contributions to my understandings of craniofacial morphogenesis and biomimetics, and to the National Institute of Dental and Craniofacial Research (NIDCR) at the National Institutes of Health (NIH) for their support over these many years.

REFERENCES

Amato, I. (1997). Candid cameras for the nanoworld. *Science 276*, 1982–1985.

Aristotle Gazes, T., Joannes, P., and Petreius, N. (1526). *Quinque de generatione animalium libri* (Venetiis, Per Joannem Antium & Stephanum ac fraters de Sabio).

Cohen, M. M., Jr. (1997a). Transforming growth factor betas and fibroblast growth factors and their receptors: role in sutural biology and craniosynostosis. *J. Bone Miner. Res. 12*, 322–331.

Cohen, M. M., Jr. (1997b). *The Child with Multiple Birth Defects*, 2nd ed. (New York: Oxford University Press).

Collins, F. S., Patrinos, A., Jordan, E., Chakravarti, A., Gesteland, R., and Watters, L. (1998). New goals for the US human genome project. *Science 282*, 682–689.

Gorlin, R., and Slavkin, H. C. (1997). *Congenital Anomalies of the Ear, Nose, and Throat* (New York: Oxford University Press).

Graham, A., Francis-West, P., Brickell, P., and Lumsden, A. (1994). The signaling molecule BMP4 mediates apoptosis in the rhombencephalic neural crest. *Nature 372*, 684–686.

Hieter, P., and Boguski, M. (1997). Functional genomics: it's all how you read it. *Science. 278*, 601–602.

Hollway, G. E., Suthers, G. K., Haan, E. A., Thompson, E., David, D. J., Gecz, J., and Mulley, J. C. (1997). Mutation detection in FGFR2 craniosynostosis syndromes. *Hum. Genet. 99*, 251–255.

International Human Genome Sequencing Consortium (2001). Initial sequencing and analysis of the human genome. *Nature 409*, 860–865.

Jabs, E. W. (1998). Toward understanding the pathogenesis of craniosynostosis through clinical and molecular correlates. *Clin. Genet. 53*, 79–86.

Lawrence, P. A. (1992). *The Making of a Fly* (London: Blackwell Scientific Publications).

Mulvihill, J. J. (1995). Craniofacial syndromes: no such thing as a single gene disease. *Nat. Genet. 9*, 101–103.

Neilson, K. M., and Friesel, R. E. (1995). Constitutive activation of fibroblast growth factor receptor-2 by a point mutation associated with Crouzon syndrome. *J. Biol. Chem. 270*, 26037–26040.

Nuckolls, G. H., Shum, L., and Slavkin, H. C. (1998). *Ectodermal Dysplasia: A Synthesis Between Evolutionary, Developmental, and Molecular Biology and Human Clinical Genetics. Molecular Basis of Epithelial Appendage Morphogenesis* (Georgetown: R. G. Landes Company).

Nuckolls, G. H., Shum, L., and Slavkin, H. C. (1999). Progress towards understanding craniofacial malformations. *Cleft Palate Craniofac. J. 36*, 12–26.

Reddi, H., and Binderman, I. (1996). Bone morphogenetic proteins in context. In M. M. Cohen, Jr. and B. J. Baum (eds.), *Studies in Stomatology and Craniofacial Biology* (Amsterdam: IOS Press).

Sarikaya, M., and Aksay, I. A. (1995). *Biomimetics: Design and Processing of Materials* (Woodbury: American Institute of Physics).

Slavkin, H. C. (1979). *Developmental Craniofacial Biology* (Philadelphia: Lea and Febiger).

Slavkin, H. C. (1995). Recombinant DNA technology in the diagnosis and therapeutics of oral medicine. In D. Chambers (ed.), *DNA, the Double Helix: Perspective and Prospective at Forty Years* (New York: New York Academy of Sciences).

Slavkin, H. C. (1996). Biomimetics: replacing body parts is no longer science fiction. *J. Am. Dent. Assoc. 127*, 1254–1257.

Slavkin, H. C. (1998a). Regulation of embryogenesis. In R. Polin and W. Fox (eds.), *Fetal and Neonatal Physiology*, 2nd ed. (Philadelphia: W. B. Saunders).

Slavkin, H. C. (1998b). Toward molecularly based diagnostics for the oral cavity. *J. Am. Dent. Assoc. 129*, 1138–1143.

Slavkin, H. C., Panagis, J. S., and Kousvelari, E. (1999). Future opportunities for bioengineering research at the National Institutes of Health. *Clin. Orthop. 367*, S17–S30.

Sperber, G. H. (1992). Current concepts in embryonic craniofacial development. *Crit. Rev. Oral Biol. Med. 4*, 67–72.

Thompson, D. W. (1952). *On Growth and Form* (London: Cambridge University Press).

Thorogood, P. (1997). *Embryos, Genes and Birth Defects* (New York: Wiley).

Urist, M. R. (1965). Bone: formation by autoinduction. *Science 150*, 893–899.

Venter, J. C., Adams, M. D., Myers, E. W. (2001). The sequence of the human genome. *Science 291*, 1304–1310.

Wilkie, A. O. (1997). Craniosynostosis: genes and mechanisms. *Hum. Molec. Genet. 6*, 1979–1984.

Winter, R. M. (1995). What's in a face? *Nat. Genet. 12*, 124–129.

INDEX